Magnetic Ferrites and Related Nanocomposites

Magnetic Ferrites and Related Nanocomposites

Ali Ghasemi

ELSEVIER

Elsevier
Radarweg 29, PO Box 211, 1000 AE Amsterdam, Netherlands
The Boulevard, Langford Lane, Kidlington, Oxford OX5 1GB, United Kingdom
50 Hampshire Street, 5th Floor, Cambridge, MA 02139, United States

Notices

Knowledge and best practice in this field are constantly changing. As new research and experience broaden our understanding, changes in research methods, professional practices, or medical treatment may become necessary.

Practitioners and researchers must always rely on their own experience and knowledge in evaluating and using any information, methods, compounds, or experiments described herein. In using such information or methods they should be mindful of their own safety and the safety of others, including parties for whom they have a professional responsibility.

To the fullest extent of the law, neither the Publisher nor the authors, contributors, or editors, assume any liability for any injury and/or damage to persons or property as a matter of products liability, negligence or otherwise, or from any use or operation of any methods, products, instructions, or ideas contained in the material herein.

ISBN: 978-0-12-824014-4

Publisher: Matthew Deans
Acquisitions Editor: Edward Payne
Editorial Project Manager: John Leonard
Production Project Manager: Nirmala Arumugam
Cover Designer: Greg Harris

Typeset by TNQ Technologies

To my wife Azadeh
for her continued love, patience, and support
AND
my children
Reza, Ava, and Nava,
who are the reason for the mercy of God

Contents

Preface

When I decided to write this book, the world was reacting to the unwanted COVID-19 pandemic. Now that the final pages are finished, the fourth peak of this pandemic is spreading all over the world and affecting many people. It has been a cause of great sorrow witnessing innocent people perishing unexpectedly because of this virus. I wish those who have lost their lives a beatifying and blessed soul. I sincerely believe that, with the aid of science, humankind will completely eliminate this pandemic from the face of the Earth.

In modern societies today, considerable attention has been paid to the development and analysis of magnetic materials. Magnetic ferrites are among the most important materials and play important roles in many applications, including microwave-absorbing media, high-density recording magnets, bioscience, telecommunication devices, magnetoelectric sensors, electromagnetic noise suppressors, exchange-spring magnets, permanent magnets, photocatalysts, and many components of microwave devices.

Despite there being many published books on this subject, no detailed book has been introduced that deals exclusively with the magnetic properties of ferrite nanocomposites. Those who become interested in the science of magnetism usually have quite different professional backgrounds. They may be metallurgists, physicists, electrical engineers, chemists, geologists, or ceramists. Consequently, although each new individual to the field has a different view of such fundamentals as atomic theory, crystallography, electric circuits, and crystal chemistry, this book covers all of these topics. Furthermore, recent developments in ferrite nanocomposites are reviewed. The main highlights include new experimental approaches and findings related to the mechanisms of magnetic features within ferrite nanocomposites.

The book comprises eight chapters. Their sequence is chosen to give readers an insight into the behavior of magnetic ferrite nanocomposites. Chapter 1 provides a quick review of magnetism and related phenomena. The origin of magnetism, the definition of magnetic parameters, and important equations are introduced. The importance of different types of ferrites such as hexagonal ferrites, spinel ferrites, garnets, and orthoferrites with industrially related applications is emphasized. Chapter 2 discusses the subject of techniques for characterizing ferrites. The chapter provides insights into phase identification and structural and thermal analysis of ferrites. Measurement approaches for static and dynamic magnetic features and high-frequency permeability, permittivity, and reflection loss are also reviewed. Chapter 3 develops general information about substituted ferrites. Structural and magnetic features of different types of hexagonal ferrites, including M, W, Y, U, X, and Z types, spinel ferrite, iron oxides, and garnets are studied. The role of substituting cations on the magnetic properties of ferrite nanoparticles, bulks, and thin films is evaluated. Chapter 4 covers the magnetoelectric (ME) components of ferrite–ferroelectric nanocomposites used in fabricating spintronic devices, nanoscale electronics, sensors, information storage, and magnetic MRAMs. Also, different ferroelectric and magnetic ferrites with various heterostructural configurations that are mostly employed to prepare ME components are introduced. Chapter 5 deals with exchange-spring ferrite nanocomposites. Nanostructure ferrites designed with specific ratios of hard-to-soft magnetic phase are introduced for enhancing maximum-energy products. The role of effective processing parameters, the nanocomposite configuration of soft and hard phases, and $(BH)_{max}$ variation are also studied. Chapter 6 is devoted to ferrite–carbon nanocomposites with different compositional and morphological characteristics used in microwave absorption media. The role of carbon material additions in balancing permeability and permittivity is investigated in a wide frequency range to reach the highest reflection loss values. Chapter 7 starts with the basic definition of photocatalysts. The photocatalytic activation of typical ferrites is investigated, and the effect of adsorbent dose on the photodegradation performance of dyes using ferrite is evaluated. The roles of shape factors and effective parameters in the catalytic activity of ferrites are also evaluated. Chapter 8 presents various techniques for fabricating magnetic ferrites, such as particles, bulks, coatings, and fibers.

The references employed comprise relevant books, classical papers, review articles, and conference proceedings and are provided at the end of each chapter.

This book is anticipated to be useful for materials scientists, physicists, electronic engineers, and chemists involved in developing and researching high-quality magnetic ferrites. As the author of this book, I am privileged to acquaint readers with the scientific aspects of the subject while also introducing the outcome of my many experimental and research endeavors presented in this book. I sincerely hope the contribution will assist readers in proliferating their knowledge of ferrites and related nanocomposites.

Acknowledgment is made of the support of Prof. Q. Taghizadeh, president of the MUT, for providing laboratory facilities and financial support. I would like to express my gratitude to Eng. Mohammad Reza Nasr Isfahani and Eng. Behnoush Alirezaei for spending considerable time procuring the required permissions and revising the photographs and diagrams. Their thoughtful efforts significantly reduced the burden of preparing the final version of this book. I have benefited enormously from the assistance of Dr. Tahmineh Sodaee in providing useful information about the photocatalytic performance and synthesizing techniques of ferrites. Dr. Samira Samanifar provided literature about FORC, and Dr. Ali Rostamnejadi helped to interpret magnetic susceptibility. I also would like to thank Dr. Amir Hossein Montazer for his critical comments and constructive suggestions on parts of the book. Further, I owe my colleagues a deep debt of gratitude; in particular, Dr. Ebrahim Paimozd and Dr. Gholamreza Gordani for providing a friendly laboratory environment and assistance in purchasing the raw materials needed over the course of the experiment. I thank Dr. Arkom Kaewrawang, Prof. Vladimir Sepelak, and Prof. Andrea Paesano Jr. for their contributions in original research works that became the bases of my experimental setups. I acknowledge Simon Holt, John Leonard, and Nirmala Arumugam for their patience during the course of writing this book. I am very grateful to the late Prof. Ardeshir Hosseinpour, who initially encouraged me to pursue my interest in the field of magnetic ferrites. It is a pleasure to express my deep appreciation to all individuals who helped me promote my knowledge of magnetic ferrites, especially Prof. Akimitsu Morisako and Prof. Xiaoxi Liu. I am also obliged to the many publishers and individuals who have given me permission to reproduce their figures and tables. Last but by no means least, I would like to thank Elsevier for their encouragement and steady collaboration.

Ali Ghasemi
Shahin Shahr, Isfahan
May 2021

CHAPTER

Fundamentals of ferrites

1

1.1 Introduction

This chapter explains the principles of magnetic phenomena and related theories to provide a background for those with limited familiarity with magnetism and related effects. Magnetic materials of various natures, including metallic, ceramic, and composite, encompass many categories and applications. Metallic specimens such as SmCo, NdFeB, FePt, and TbFeCo have an important role in developing hard magnetic features. Soft metallic magnets such as iron cobalt, mu-metal, permalloy, and superalloy have applications in shielding environments and recording heads. The most important magnetic ceramics are the ferrites used in communication and electronic engineering fields. Ferrites are usually known as ferrimagnetic oxides. Ferrites are electrically insulating and exhibit low loss within high-frequency fields, making them appropriate candidates for communication and microwave devices. The crystal structure of ferrite contains an interlocking network of positively charged cations and negatively charged divalent oxygen anions. The site occupancy of ions and the crystal structure of ferrite play an important role in determining magnetic features. Large varieties of ferrites in different configurations are fabricated to be employed in many new technologies with various requirements. The frequency applications of ferrites at which any electronic device can function range from DC to microwave.

1.2 Brief history of magnets

Lodestone rocks rich in Fe_3O_4 ferrites as permanent magnets were known by priests and people in "Sumer, China, pre-Columbian America, and ancient Greece." The Chinese used magnetic compasses sometime before 2500 BCE. The first device based on a lodestone was developed by the Chinese and named a "South pointer" (Coey, 2010). In the 6th century BCE, The Greek philosopher Thales carried out the earliest observation of magnetism for which a lodestone or magnetite attracted iron. The first manufactured magnets were fabricated by rubbing iron needles on magnetite to form the essential parts of a compass (Goldman, 2006). In 1064, Zheng Gongliang carried out thermoremanent magnetization through special heat

Magnetic Ferrites and Related Nanocomposites. https://doi.org/10.1016/B978-0-12-824014-4.00009-3

treatment of iron. Shen Kua invented the navigation compass in 1088. The perpetuum mobile was described by Petrus Peregrinus in the first European text on magnetism (Coey, 2010). William Gilbert found that melted and forged steel bars cooled under the magnetic field of the earth contain magnetic features. He published the first systematic experimental study on magnetism in 1600. He found that the Earth is the greatest magnet in the world and identified it as the source of the magnetic field that affects the compass needle and determines the direction of navigation systems (Özgür et al., 2009a).

The horseshoe magnet was invented by Daniel Bernoulli in 1743. In 1819, Oersted found that a wire carrying current could deflect a compass needle, representing a kind of correlation between magnetism and electricity. Then, in Paris, Andre-Marie Ampere and Dominique-Francois Arago wound wire into a coil and demonstrated that a magnetic field could be generated by applying current to the coil (Coey, 2010).

William Sturgeon introduced the iron-core electromagnet in 1824, which was more effective than a weak permanent magnet in electric motors and generators. Michael Faraday discovered electromagnetic induction in 1821 and the magneto-optic effect in 1845. The previous experimental investigation inspired James Clerk Maxwell to formulate the theory of electricity and magnetism in 1864 (Coey, 2010).

The first hysteresis loop was plotted by Warburg for an iron magnet in 1880.

Curie found the transition temperature from ferromagnetism to paramagnetism (Curie law) in 1895. The theory of diamagnetism and paramagnetism was discussed by Langevin in 1905 (Özgür et al., 2009a).

The theory of ferromagnetism was proposed by Pierre-Ernest Weiss in 1907 to describe the transition temperature (Curie temperature) (Özgür et al., 2009a).

Regarding magnetic materials, it is well known that carbon steel was the first important permanent magnet during 1820–1900. The 1920s was the beginning of quantum mechanics and the physics of magnetism, which were used for a deeper explanation of magnetic phenomena. At the beginning of the 20th century, the development of permanent magnets rapidly grew by adding other magnetic or nonmagnetic elements such as Mn, Co, W, Al, and Ni into the composition while controlling processing conditions, especially the quenching steps of the heat-treatment process. During 1900–1935, the Alnico was developed. Ferrite permanent magnets were introduced in the 1930s and used in loudspeakers and later for motors in portable appliances. Substituting various cations in ferrites and making new nanocomposites are still rapidly expanding areas. The first rare earth cobalt permanent magnet was fabricated in the 1960s, and the NdFeB hard magnet was developed in the 1980s (Goldman, 2006; Özgür et al., 2009a). Novel achievements in permanent magnets, magnetic recording, and high-frequency materials have made progress in telecommunications, computer science, and microwave systems. In recent decades, the impact of nanomagnetism and spin electronics in modern electronics has been considerable.

1.3 Basic science of magnetism

In this section, some important magnetic parameters and well-known laws describing magnetic phenomena are explained.

1.3.1 Origin of magnetism

According to the basic science of atomic configuration, atomic magnetic moment originates from the orbital motion of the electron around the nucleus and the spin motion of the electron around its own axis. Consider a simple model of an atom in which the electron moves in a circular loop with radius r and angular velocity of ω. The total magnetic moment produced by circular and spin motions is

$$\mu = -g\frac{\mu_0 e}{2m}P \tag{1.1}$$

where m and e are the mass and electric charge of a single electron, respectively ($e/m = 1.76 \times 10^{11}$ C/kg), μ_0 is the magnetic permeability of free space ($\mu_0 = 4\pi \times 10^{-7}$ H/m), and P denotes the angular momentum of the circular motion. The factor of g is called "gyromagnetic ratio" or simply g factor, which is 1 and 2 for orbital and spin motions, respectively. The unit of orbital magnetic moment and spin magnetic moment is the Bohr magneton (Chikazumi & Graham, 2009).

Note that in magnetic ceramics such as ferrites, the crystalline field (electric field) resulting from the surrounding ions would cause the quenching of orbital moments, in which case the orbital moment of electrons is far smaller than spin moments (Smit, 1959, pp. 278–280).

1.3.2 Coulomb and Lorentz forces

Consider two magnetic poles with distance r in meter, as well as strengths of m_1 (weber) and m_2 (weber). The force exerted in newtons from one pole to the other, known as the Coulomb force, is given by

$$F = \frac{m_1 m_2}{4\pi\mu_0 r^2} \tag{1.2}$$

The force on the magnetic pole can also be imposed by employing an electric current. For example, by considering an infinitely long coil or solenoid carrying a current of I, the uniform magnetic field is generated and defined by

$$H = NI \tag{1.3}$$

where N denotes the number of turns per unit length (along the axis) of the solenoid (Chikazumi & Graham, 2009; Morrish, 2001).

On the other hand, by considering a particle with charge q moving with a velocity of v and subjected to magnetic and electric fields of B and E, the Lorentz force of $f = q(E + v \times B)$ can be defined. Cyclotrons and other particle accelerators, cathode ray tube televisions, and generators are based on the Lorentz force equation.

1.3.3 Definition of fundamental magnetic parameters

When a magnetic field (H) is applied to a material, the response of materials is called magnetic induction (B), which heavily depends on the nature of the specimen. The relation of B and H defines the properties and statements of magnetic material types. In free space and some limited materials, the relation is linear, while in many cases, it is very complicated and not even single-valued in some magnets. An important parameter that can be extracted from the relation of magnetic induction with a magnetic field is permeability, which can be represented as follows:

$$\mu = B/H \tag{1.4}$$

Permeability indicates how permeable the material is to the magnetic field. The relation of magnetic induction with magnetic field is expressed in the following:

$$B = \mu_0(H + M)(\text{SI system}) \tag{1.5}$$

$$B = H + 4\pi M(\text{cgs system}) \tag{1.6}$$

where μ_0 is the B/H ratio measured in a vacuum, and M is the magnetization of the material. In cgs units, the permeability of free space is unity and so does not appear in the equation (Cullity & Graham, 2011; Spaldin, 2010).

Magnetization is a very important parameter of magnetic materials. Magnetization strongly depends on both the individual magnetic moments of the constituent ions, atoms, or molecules and the type and strength of interaction of magnetic moments with each other in a medium. In general, magnetization is defined as the magnetic moment (μ) per volume unit (V):

$$M = \mu/V \tag{1.7}$$

Magnetic polarization (J) indicates the intensity of magnetization and is given by

$$J = \mu_0 M \tag{1.8}$$

The pole strength for the area unit perpendicular to M is found by the following:

$$\sigma = M \cdot n \tag{1.9}$$

where n is the unit vector normal to the surface and σ is the pole strength per unit of area.

The magnetization-to-magnetic field ratio is called magnetic susceptibility and reflects how responsive a material is to an applied magnetic field. The susceptibility equation in both units is expressed by

$$\chi = M/H \tag{1.10}$$

The relations between permeability and susceptibility in SI and cgs systems are given by (Cullity & Graham, 2011; Morrish, 2001)

$$\mu = \mu_0(1 + \chi)(\text{SI system}) \tag{1.11}$$

$$\mu = 1 + 4\pi\chi(\text{cgs system}) \tag{1.12}$$

The magnetic flux Φ is defined as the flux of magnetic induction through a surface area A; that is,

$$\Phi = \int_A B \cdot ndS \tag{1.13}$$

where n is the unit normal. A large amount of flux density in materials reflects the high permeability value of the medium (Morrish, 2001).

The units of magnetic induction, magnetic field, magnetization, magnetic polarization, permeability, susceptibility, and magnetic flux, along with the conversion factors, are summarized in Table 1.1.

1.3.4 Ampère's law

Andre-Marie Ampère discovered an important law in 1826 which relates the magnetic field along a closed loop in which current I carried through the loop. The law is as follows:

$$\oint_C B \cdot dl = \mu_0 \iint_s j \cdot dS = \mu_0 I \tag{1.14}$$

where j represents the current density. Ampère's law can be employed for calculating the magnetic field of current distributions with a high degree of symmetry. In general, using this law, the magnetic field generated by the current can be calculated for different geometries, such as straight lines and coaxial cable. Note that the current should be steady and not change with time, and only currents carrying across the path section are considered for finding the magnetic field (Morrish, 2001; O'handley, 2000).

Table 1.1 The relationship between some magnetic parameters in cgs and SI units.

Quantity	cgs units	SI units	Conversion factor (cgs to SI)
Magnetic induction (B)	G	T	10^{-4}
Applied field (H)	Oe	A/m	$10^3/4\pi$
Magnetization (M)	emu/cm^3	A/m	10^3
Magnetization ($4\pi M$)	G	A/m	$10^3/4\pi$
Magnetic polarization (J)	emu/cm^3	T	$4\pi \times 10^{-4}$
Permeability (μ)	Dimensionless	H/m	$4\pi \times 10^{-7}$
Susceptibility (χ)	emu/cm^3	Dimensionless	4π
Magnetic flux (Φ)	maxwell (Mx), G.cm^2	weber (Wb), volt. second (V.s)	10^{-8}

where, A, *Ampere;* emu, *electromagnetic unit;* G, *Gauss;* H, *Henry;* J, *Joule;* Oe, *Oersted;* T, *Tesla;* Wb, *Weber.*

1.3.5 Faraday's law

In 1831, Michael Faraday discovered that time-varying magnetic fields could generate an electric field. The following equation was obtained experimentally:

$$\oint_{\text{Circuit}} E \cdot ds = -\frac{d}{dt}\int_{A} B \cdot n dS \tag{1.15}$$

where A is a surface which has the circuit as its boundary, n denotes the normal vector, and E represents the electrostatic field. The phenomenon is known as electromagnetic induction. He demonstrated that the electromagnetic force (emf) ε induced in a coil is proportional to the negative rate of changes in the magnetic flux:

$$\varepsilon = -d\Phi/dt \tag{1.16}$$

For a coil consisting of N loops, the emf is given by (Morrish, 2001; O'handley, 2000)

$$\varepsilon = -Nd\Phi/dt \tag{1.17}$$

1.3.6 Lenz's law

Lenz's law determines the direction of an induced current. The induced current generates magnetic fields that tend to oppose variations in magnetic flux with the fields imposing such currents. Consider a conducting loop placed in a uniform magnetic field, with the positive direction of the normal vector defined. Then, by finding the rate of flux change $d\Phi/dt$ through differentiation, three possibilities can be determined. If $d\Phi/dt$ is positive, then $\varepsilon < 0$; if it is negative, the induced emf is positive; finally, considering $d\Phi/dt = 0$, $\varepsilon = 0$. The right-hand rule is used to determine the direction of the induced current. By pointing the thumb in the direction of the area vector and curling the fingers around the closed loop, the induced current flows in the same direction as the direction of fingers curl by considering positive induced emf, while it is the opposite direction if $\varepsilon < 0$ (Cullity & Graham, 2011; Morrish, 2001).

1.3.7 Maxwell's equations

In response to the experiments carried out by Ampere and Faraday, Maxwell developed the following equations:

$$\nabla \cdot D = \rho \tag{1.18}$$

$$\nabla \cdot B = 0 \tag{1.19}$$

$$\nabla \times E = -\frac{\partial B}{\partial t} \tag{1.20}$$

$$\nabla \times H = \left(j + \frac{\partial D}{\partial t}\right) \tag{1.21}$$

where $\nabla \cdot$ and $\nabla \times$ are vector operators of divergence and curl, respectively. E is the electrical field, D shows the displacement vector, j denotes the current density, ρ represents the electric charge volume density, and t is time. The displacement vector is related to the electrical field by the following equation:

$$D = \varepsilon_0 E + P = \varepsilon E \tag{1.22}$$

where ε_0 is the permittivity of free space ($\varepsilon_0 = 8.85 \times 10^{-12}$ F m^{-1}), P denotes the polarization of electric dipole per unit volume, and ε is the dielectric constant (Morrish, 2001).

1.4 Classes of magnetic materials

All materials can be classified in terms of their magnetic susceptibility into five major groups (O'handley, 2000):

1. diamagnetism
2. paramagnetism
3. ferromagnetism
4. antiferromagnetism
5. ferrimagnetism

1.4.1 Diamagnetism

Michael Faraday discovered the diamagnetic nature in September 1845. Diamagnetism is the weakest form of magnetism. This magnetism is related to the orbital rotation of electrons around the nuclei induced electromagnetically within an external magnetic field. In diamagnetic substances, applying an external magnetic field induces a magnetic moment that opposes the external magnetic field causing it; thus, a negative magnetization is produced where the susceptibility of a diamagnetic material is negative and very weak (Chikazumi & Graham, 2009). The susceptibility in diamagnetic materials is temperature independent. Superconductors are an ideal type of diamagnets that have great applications in modern industries. By applying an external magnetic field, depending on field strength and working temperature, superconductors expel the field lines from their interiors. All noble gases, bismuth, and pyrolytic graphite are diamagnets.

1.4.2 Paramagnetism

In many cases, paramagnetic materials contain magnetic atoms or ions with a magnetic moment due to unpaired electrons in partially filled orbitals. The spins are isolated from their surrounding environment and can somehow freely align in the direction of the applied field. The interaction between magnetic moments is very weak, and the order of magnitude of magnetic susceptibility is 10^{-3} to 10^{-5}.

In paramagnetic materials at finite temperatures, the spins are agitated, and thermal energy causes misalignment of spins. With an elevation of temperature, the thermal agitation will increase, causing vibration in atomic magnetic moments that reduces susceptibility. This behavior is described by the Curie law, in which susceptibility has an opposite trend to temperature. The Curie law is given by (Cullity & Graham, 2011)

$$\chi = C/T \tag{1.23}$$

where T is the absolute temperature and C is a material constant called the Curie's constant. Later experiments demonstrated that the susceptibility of some materials is fitted by Curie—Weiss law:

$$\chi = C/(T - \vartheta) \tag{1.24}$$

where ϑ is a constant. In this equation, ϑ can be positive, negative, or zero. Clearly, when $\vartheta = 0$, then the Curie—Weiss law equates to the Curie law. When ϑ is nonzero, then there is an interaction between magnetic moments of neighboring atoms, and the materials are only paramagnetic above a certain transition temperature. If ϑ is positive, then the material is ferromagnetic below the transition temperature where the value of ϑ corresponds to the transition temperature (Curie temperature, T_C). If ϑ is negative, then the material is antiferromagnetic below the transition temperature (Néel temperature, T_N), though the value of ϑ does not relate to T_N. Note that this equation is only valid when the material is in a paramagnetic state. It is not valid for many metals as the electrons contributing to the magnetic moment are not localized. However, the law does apply to some metals, e.g., rare earth, where the 4f electrons that create the magnetic moment are closely bound (Chikazumi & Graham, 2009).

Depending on the strength and direction of the external magnetic field, partial alignment of the atomic magnetic moments occurs, resulting in a net positive magnetization and positive susceptibility. After applying a magnetic field to a paramagnetic system with no interaction between atomic magnetic moments, the potential energy U_H is given by

$$U_H = -\mu H \cos\theta \tag{1.25}$$

where θ is the angle between the direction of applied field H and atomic magnetic moment μ. By considering $\cos\theta = 1$ and magnetic field strength of 10^6 A/m, the magnitude of energy is on the order of 10^{-23} J. The order of magnitude of thermal energy k_BT, where k_B is Boltzmann constant at room temperature, is 4.1×10^{-21} J. When the two energies are compared, the thermal energy is larger by factors of 10^2 to 10^3 compared with the potential energy of the magnetic field (Chikazumi & Graham, 2009). This indicates that at room temperature for a noninteracting paramagnetic system, the thermal energy causes random fluctuation of magnetic moments and provides very weak magnetization. For reaching saturation in practice, the temperature must be very low for freezing the spins, while the intensity of the magnetic field should be very high.

Paramagnetic materials obey from important theories, including Langevin and Pauli models. The Langevin model, which is true for materials with noninteracting localized electrons, states that every atom has a magnetic moment that is randomly oriented as a result of thermal agitation. By considering N atomic moments per unit volume, the magnetization can be calculated based on the Langevin theory. In this case, let $n(\theta)d\theta$ denote the number of atomic moments in the unit volume providing an angle between θ and $\theta + d\theta$ with the direction of the applied field, which is proportional to the solid angle $2\pi\sin\theta d\theta$ and Boltzmann factor $exp((\mu H\cos\theta)k_BT)$ according to the following:

$$n(\theta)d\theta = 2\pi n_0 e^{(\mu H\cos\theta)/k_BT}\sin\theta d\theta \tag{1.26}$$

where n_0 is a proportionality factor determined by considering that the total density of atomic moment is N:

$$\int_0^\pi n(\theta)d\theta = 2\pi n_0\int_0^\pi e^{(\mu H\cos\theta)/k_BT}\sin\theta d\theta = N \tag{1.27}$$

The intensity of magnetization is expressed by

$$M = \int_0^\pi \mu\cos\theta n(\theta)d\theta \tag{1.28}$$

By combining the above equations, one can reach

$$M = N\mu\frac{\int_{-1}^{1} e^{\alpha x}xdx}{\int_{-1}^{1} e^{\alpha x}dx} = N\mu\left(\frac{e^\alpha + e^{-\alpha}}{e^\alpha - e^{-\alpha}} - \frac{1}{\alpha}\right) = N\mu\left(coth\alpha - \frac{1}{\alpha}\right) \tag{1.29}$$

By setting $\mu H/k_BT = \alpha$ and $\cos\theta = x$, the following expression is obtained:

$$M = N\mu\frac{\int_{-1}^{1} e^{\alpha x}xdx}{\int_{-1}^{1} e^{\alpha x}dx} = N\mu\left(\frac{e^\alpha + e^{-\alpha}}{e^\alpha - e^{-\alpha} - \frac{1}{\alpha}}\right) = N\mu\left(\coth\alpha - \frac{1}{\alpha}\right) \tag{1.30}$$

The function in the parentheses is the Langevin function, $L(\alpha)$. For $\alpha << 1$, the Langevin function can be expanded as

$$L(\alpha) = \alpha/3 - \alpha^3/45 + \cdots. \tag{1.31}$$

Considering the first term in the above equation, the intensity of magnetization can be simplified to

$$M = \frac{N\mu^2}{3k_BT}H \tag{1.32}$$

The magnetic susceptibility can also be determined by the following:

$$\chi = \frac{M}{H} = \frac{N\mu^2}{3k_BT} \tag{1.33}$$

The above relation demonstrates that magnetic susceptibility has an inverse relationship with temperature, which is the well-known Curie law of a paramagnetic substance. Note that in such calculations, it was assumed that each individual atomic moment is considered in an arbitrary direction. On the other hand, for actual cases where each atom contains gJ spins, where J is the total angular momentum number including the spin and orbital angular momenta and g is the g factor, the total magnetization of substances becomes

$$M = N\mu \frac{\sum_{J_z} \frac{J_z}{J} e^{J_z \mu H / J k_B T}}{\sum_{J_z} e^{J_z \mu H / J k_B T}} = N\mu \left(\frac{2J+1}{2J} \coth \frac{2J+1}{2J} \alpha - \frac{1}{2J} \coth \frac{\alpha}{2J} \right) \tag{1.34}$$

The function in parentheses is called Brillouin function $B_J(\alpha)$, which is reduced to the Langevin function at very large J (Chikazumi & Graham, 2009; Morrish, 2001):

$$\frac{M}{N\mu_B} = gJB_J(\alpha) \tag{1.35}$$

The Pauli model of paramagnetism also works for materials in which the electrons are free and interact to form a conduction band valid for most paramagnetic metals. It is well described by the band structure theory of collective electrons.

1.4.3 Ferromagnetism

In a paramagnetic state, by considering $\alpha = 7$, the required field for magnetizing an assembly of free atomic moments is on the order of 10^7 Oe, which is a very large magnetic field. In ferromagnets, such as permalloys, the magnetic field required to reach the saturation state is on the order of 0.013 Oe, and for Alnico, the typical permanent magnet is about 630 Oe (Chikazumi & Graham, 2009). Clearly, the external magnetic field for magnetizing a ferromagnet is far smaller than that for a paramagnet substance. This is because the atomic magnetic moments in ferromagnetic materials tend to align themselves parallel to each other, which is caused by the inner field. This existence of a strong inner magnetic field was explained in classical theory, which was first postulated by Weiss in 1907, based on the presence of a molecular field within the ferromagnetic materials (Chikazumi & Graham, 2009). The molecular field H_m can be expressed as

$$H_m = wM \tag{1.36}$$

where w is the proportionality factor.

Heisenberg described the nature of molecular fields in 1928 (Chikazumi & Graham, 2009). Based on his theory, the force inducing parallel alignment of magnetic moments is an exchange force. The exchange force is a quantum mechanical phenomenon that results in the parallel alignment of moments, causing large net magnetization even in the absence of a magnetic field. The potential energy (w_{ij}) between two atoms with spins S_i and S_j is expressed by

$$w_{ij} = -2JS_i \cdot S_j \tag{1.37}$$

where J is the exchange integral. In the case of positive J, the energy is minimum when the S_i is parallel to S_j; by considering a negative value for J, the spins are arranged in antiparallel. Exchange forces are very large, equivalent to a field on the order of 1000 T, or approximately 100 million times the strength of the Earth's magnetic field. Although electronic exchange forces in ferromagnets are sufficiently large, thermal energy at Curie temperature eventually overcomes the exchange and imposes random arrangement of spins. Below the Curie temperature, the ferromagnet is ordered, while above it, the spin orientation is disordered.

By applying an external magnetic field in the direction of the magnetization vector of the system, the potential energy of an individual atomic moment is given by

$$U_H = -\mu(H + wM)\cos\theta \tag{1.38}$$

The probability of atomic moment to have this energy is proportional to the Boltzmann factor, $e^{(-U_H/k_BT)}$, with the average magnetization expressed by

$$M = N\mu \frac{\int_0^{\pi} e^{(\mu(H+wM)\cos\theta)/k_BT}\cos\theta\sin\theta d\theta}{\int_0^{\pi} e^{(\mu(H+wM)\cos\theta)/k_BT}\sin\theta d\theta} = N\mu L\left(\mu\frac{H+wM}{k_BT}\right) \tag{1.39}$$

The analytical solution for finding magnetization versus magnetic field and temperature is very difficult, where the best solution is to seek a graphical solution (Chikazumi & Graham, 2009; Morrish, 2001).

1.4.4 Antiferromagnetism

Antiferromagnetic materials, similar to paramagnetic materials, have weak magnetism with small positive susceptibility. The curve of susceptibility versus temperature is characterized by the occurrence of a kink called Néel temperature. Fig. 1.1 reflects the Néel temperature. It should be noted that susceptibility for polycrystalline material is calculated by averaging parallel and perpendicular susceptibility.

Below the Néel temperature, the spin arrangement is antiparallel, and there are two sublattices, namely A and B, where there is a strong negative interaction causing the plus and minus spins to completely cancel each other. Above the Néel temperature, the random spin arrangement causes diminished susceptibility versus temperature and follows the Curie—Weiss law.

In the periodic table, the only element exhibiting antiferromagnetism at room temperature is chromium ($T_N = 37°C$). MnO is also representative of antiferromagnetic materials. Van Vleck and later Néel and Anderson proposed the theoretical treatment of antiferromagnetism phenomena (Chikazumi & Graham, 2009). Consider an A site occupied by positive spins and a B location arranged by negative spins. According to the Weiss theory of ferromagnetism, the role of exchange can be expressed by molecular field theory. The molecular field which acts at the A sublattice H_{mA} is expressed by

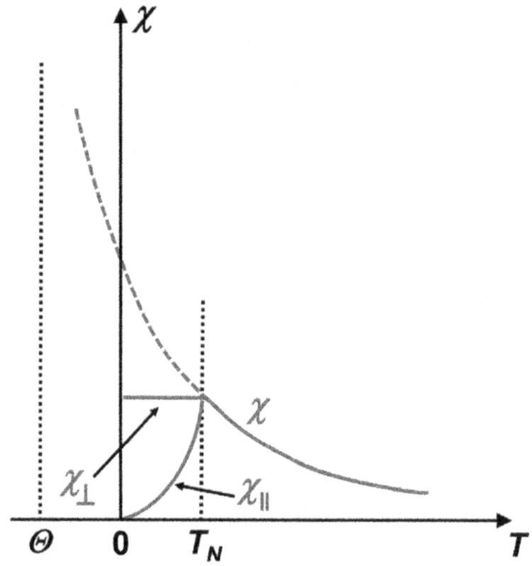

FIGURE 1.1

Variations of susceptibility versus temperature in antiferromagnetic substances.

From Böer, K. W., Pohl, U. W., Böer, K. W., & Pohl, U. (2014). Magnetic semiconductors *(pp. 1–24). Springer International Publishing. https://doi.org/10.1007/978-3-319-06540-3_9-1.*

$$H_{mA} = w_{AA}M_A + w_{AB}M_B \tag{1.40}$$

where M_A and M_B are the magnetization of A and B sublattices, respectively. The coefficient w_{AB} is always negative. The molecular field for B site H_{mB} is given by

$$H_{mB} = w_{BA}M_A + w_{BB}M_B \tag{1.41}$$

Since A and B sites are symmetrical in antiferromagnetism, it can be written for simplicity as

$$w_{AA} = w_{BB} = w_1, w_{AB} = w_{BA} = w_2 \tag{1.42}$$

On the other hand, in the absence of an external magnetic field, the magnetizations of sublattices A and B are equal:

$$M_A = -M_B \tag{1.43}$$

Through simplification, one can reach the following:

$$H_{mA} = (w_1 - w_2)M_A \tag{1.44}$$

$$H_{mB} = (w_1 - w_2)M_B \tag{1.45}$$

The thermal equilibrium values of magnetization in the A and B sites are expressed by

$$M_A = \frac{N\mu}{2}L\left(\frac{\mu(w_1 - w_2)M_A}{k_B T}\right) \tag{1.46}$$

$$M_B = \frac{N\mu}{2} L\left(\frac{\mu(w_1 - w_2)M_B}{k_B T}\right) \tag{1.47}$$

where N is the number of magnetic atoms per unit volume and L is the Langevin function. According to the equation, magnetizations in A and B sublattices decrease with an increase in temperature, whereas at the Néel temperature, T_N suddenly disappears:

$$T_N = \frac{N\mu^2(w_1 - w_2)}{6k_B} \tag{1.48}$$

By applying an external magnetic field, M_A and M_B are not symmetrical, and the assumption of $M_A = -M_B$ is not valid. In that case, by applying a magnetic field in the plus direction, the magnetizations of both sites are given by (Chikazumi & Graham, 2009; Morrish, 2001)

$$M_A = \frac{N\mu}{2} L\left(\frac{\mu(H + w_1 M_A + w_2 M_B)}{k_B T}\right) \tag{1.49}$$

$$M_B = \frac{N\mu}{2} L\left(\frac{\mu(H + w_2 M_A + w_1 M_B)}{k_B T}\right) \tag{1.50}$$

1.4.5 Ferrimagnetism

As mentioned, the magnetization vectors of both sublattices in antiferromagnetism cancel each other, and in the absence of an applied field, the net magnetization is zero. Consider that the intensity of magnetization in one sublattice is larger than in the other. In that case, although the spins in both sublattices are antiparallel, the net magnetization is relatively large. Materials with such spin configurations are called ferrimagnetic materials and have large susceptibility. In general, strong magnetism is seen in ferromagnetic and ferrimagnetic substances. Ferrimagnetism is normally observed in compounds with more complex crystal structures than pure elements. The ferrimagnetic term was proposed by Néel to explain the magnetic nature of ferrites (Chikazumi & Graham, 2009). Various ferrites, including spinels, hexagonal, and garnets, have a ferrimagnetic nature. Normally, in ferrite, the sublattice A contains the plus direction of spins, while the B sites are occupied by down spins. The number of magnetic cations along with the magnitude of spins of individual cations is different on A and B sites; thus, spontaneous magnetization is produced without applying an external magnetic field. At Curie temperature, because of thermal agitation, the spins become completely random, and the spontaneous magnetization vanishes. Above the Curie temperature, the materials are transformed into paramagnetic substances, and the susceptibility diminishes with an increase in temperature (Coey, 2010). The variation in inverse susceptibility versus temperature is plotted in Fig. 1.2, where the linear extrapolation of inverse susceptibility at high temperatures would intersect the temperature axis on its negative side, which is called the asymptotic Curie point (Chikazumi & Graham, 2009).

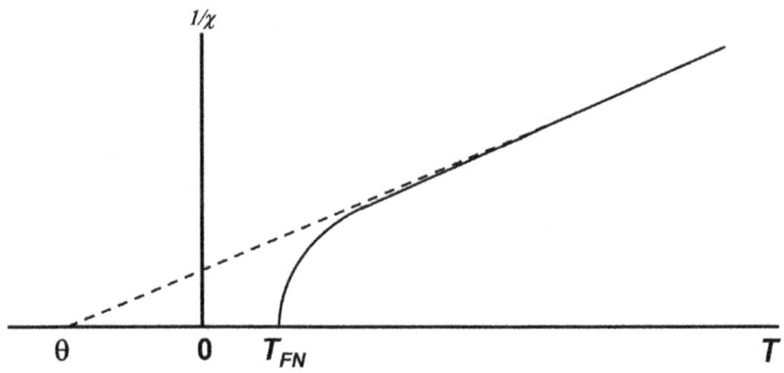

FIGURE 1.2

Variations of inverse magnetic susceptibility versus temperature in ferrimagnetic substances.

From Morrish, A. H. (2001). The physical principles of magnetism.

The molecular fields for ferrimagnetic substances are similar to those of antiferromagnetic materials and can be given by

$$H_{mA} = -w_{AA}M_A - w_{AB}M_B \tag{1.51}$$

$$H_{mB} = -w_{BA}M_A - w_{BB}M_B \tag{1.52}$$

In this case, at equilibrium the $w_{AB} = w_{BA}$ as earlier; however, since the sublattices are crystallographically inequivalent, then $w_{AA} \neq w_{BB}$, and the magnetization in both sublattices is not equal. By applying an external magnetic field of H, the magnetic fields at sites A and B are expressed by

$$H_A = H - w_{AA}M_A - w_{AB}M_B \tag{1.53}$$

$$H_B = H - w_{AB}M_A - w_{BB}M_B \tag{1.54}$$

Since the interaction between two crystallography sites is antiparallel, w_{AB} has a positive value. In principle, w_{AA} and w_{BB} could be negative or positive, though for many ferrimagnetic materials, they are positive and usually smaller than w_{AB}:

$$w_{AA} = \alpha w_{AB} \text{ and } w_{BB} = \beta w_{AB} \tag{1.55}$$

where α and β are the proportional constants. The magnetization of sublattice A in thermal equilibrium is expressed by

$$M_A = \sum_i N_i g_i \mu_B S_i B_{S_i}(x_A) \tag{1.56}$$

where

$$x_A = \frac{S_i g_i \mu_B}{k_B T} H_A \tag{1.57}$$

$$B_{S_i}(x_A) = \frac{2S_i + 1}{2S_i} \cot h \frac{2S_i + 1}{2S_i} x_A - \frac{1}{2S_i} \cot h \frac{x_A}{2S_i} \tag{1.58}$$

The magnetization of site B is also written as

$$M_B = \sum_j N_j g \mu_B S_i B_{S_i}(x_B) \tag{1.59}$$

where N_i is the number of atoms per unit volume with a spin quantum number of S_i, with the expression of x and $B_{S_i}(x)$ being similar for both sites (Morrish, 2001).

1.5 Importance of ferrites

Professor Takeshi Takei, with the collaboration of Professor Yogoro Kato, synthesized the first ferrite compounds with promising properties in 1930 in Japan. In 1933, Kato and Takei produced cobalt ferrite by cooling it in a magnetic field. Thereafter, in 1945, J. L. Snoek of the Phillips Research Laboratories in the Netherlands provided a "soft" ferrite for practical applications.

Ferrite with a permanent feature, also called ferroxdure, was developed by Went and marketed by Philips in 1952. Then E. Albers Schonberg reported microwave ferrites in the United States. About 60% of soft ferrites were produced in Japan in the 1970s, while in the 1980s, Western Europe and Southeast Asia entered the supplier market. The large-scale productions of ferrites were used for TV tube deflection yokes and high-voltage flyback transformers (Goldman, 2006). The global ferrite market confirmed the growing global demand for ferrites. The iron element claims about 5% by weight of the crust and is the most abundant element overall in the crust (Coey, 2010). Thus, the cheapest metal claims a considerable percentage of ferrites, making ferrite a suitable and available candidate for many applications. This is in contrast to rare earth magnets, which normally contain cobalt and rare earth elements 100 times more expensive and 1000 times scarcer than iron.

International conferences on ferrites (ICFs) have provided a forum for contributing to the basic knowledge and the latest experimental information regarding the nature and magnetic features of ferrites. Many scientists and professors, who are experts in the field of physics, material science, chemistry, electronics and devices, biology, and medical sciences, participate in and exchange information on the latest knowledge and technological developments of ferrites. The earliest ICFs were held in Kyoto (1970, 1980, 2000), Paris (1976), San Francisco (1984, 2004), Mumbai (1989), Tokyo (1992), Bordeaux (1996), Chengdu (2008), Okinawa (2013), and Boston (2019).

1.6 Fundamentals of ferrite crystal structures

Ferrites crystallize into four crystal types: spinel, magnetoplumbite, garnet, and orthoferrites. The classification of various ferrites is shown in Table 1.2.

Table 1.2 Classification of various ferrites.

Ferrites	Crystal structure	Composition
Spinel	Cubic	M^{2+} Fe_2^{3+} O_4 $M^{2+} = Ni^{2+}$, Mn^{2+}, Zn^{2+}, Mg^{2+}, Co^{2+}, Cu^{2+}, Cd^{2+}, Fe^{2+} ions or a mixture of these ions
Garnet	Cubic	R^{3+} Fe_5^{3+} O_{12} $R = Y^{3+}$, Gd^{3+}, Sm^{3+}, Eu^{3+}, Tb^{3+}, Dy^{3+}, Ho^{3+}, Tm^{3+}, Yb^{3+}, Lu^{3+}
Hexaferrite	Hexagonal magnetoplumbite	$AFe_{12}O_{19}$ M-type, $AM_2Fe_{16}O_{27}$ W-type, $AM_2Fe_{23}O_{46}$ X-type, $A_2M_2Fe_{16}O_{27}$Y-type, $A_3M_2Fe_{24}O_{41}$ Z-type, $A_4M_2Fe_{36}O_{60}$ U-type, $M^{2+} = Fe^{2+}$, Ni^{2+}, Mn^{2+}, Zn^{2+}, Mg^{2+} ions or a mixture of these ions
Orthoferrite	Orthorhombic perovskite	$R^{3+}Fe^{3+}O_3$ $R^{3+} = Gd^{3+}$, Y^{3+}, Nd^{3+}, Sm^{3+}, Lu^{3+}

From Kotnala, R. K., Shah, J., & Buschow, K. H. J. (2015). Chapter 4 – Ferrite materials: nano to spintronics regime *(Vol. 23, pp. 291–379). Elsevier. https://doi.org/10.1016/B978-0-444-63528-0.00004-8.*

1.6.1 Spinel ferrites

The chemical formula of spinel ferrite is MFe_2O_4. The M contains elements such as Fe, Ni, Mn, Zn, Mg, Co, Cu, and Cd. The ions are placed at tetrahedral (*A*-site) or octahedral sites (*B*-site) of the spinel structures. Fig. 1.3 displays the unit cell of a spinel structure and the sites of the different ions. Ferrites with M^{2+} in tetrahedral sites and Fe^{3+} in octahedral sites are known as normal spinel structures. Ferrites in which the M^{2+} are on octahedral sites, and Fe^{3+} are equally divided between tetrahedral and octahedral sites are referred to as inverse spinel structures, as shown in Fig. 1.4.

The cation distributions on tetrahedral and octahedral sites are demonstrated by

$$\left[M_\delta^{2+}Fe_{1-\delta}^{3+}\right]_A\left[M_{1-\delta}^{2+}Fe_{1+\delta}^{3+}\right]_B O_4^{2-} \tag{1.60}$$

where $\delta = 0$ for inverse spinel ferrites, $\delta = 1$ for normal spinel ferrites, and $0 < \delta < 1$ for mixed spinel ferrites. The ionic radii of the specific ions, size of the interstices, as well as processing and sintering parameters define the filling of these sites, which in turn influence the magnetic properties. The site preferences of ions in various kinds of spinel ferrites are shown in Table 1.3.

1.6.2 Hexagonal ferrites

Ferrites with hexagonal structures are usually known as hexaferrites. The hexaferrites are classified into six various types according to chemical composition and crystal structure, including M, Y, W, Z, X, and U. The chemical formula and stacking order of various hexaferrites are reported in Table 1.4 (Pullar, 2012). The

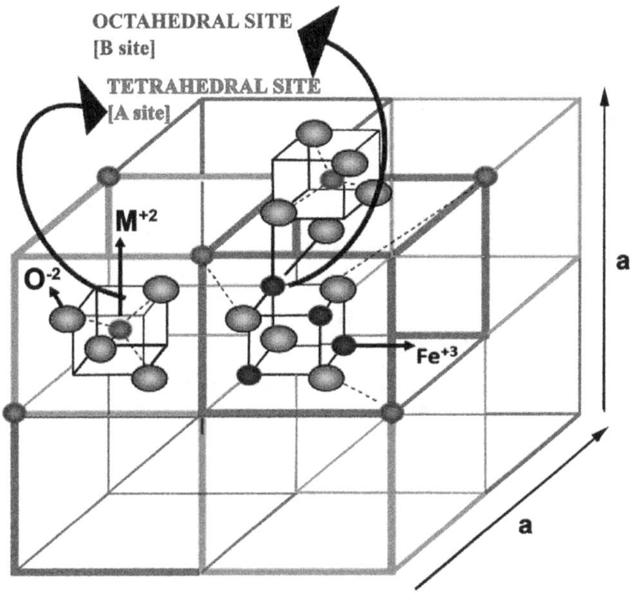

FIGURE 1.3

The unit cell of a spinel structure and the sites of the different ions.

From Kotnala, R. K., Shah, J., & Buschow, K. H. J. (2015). Chapter 4 — Ferrite materials: nano to spintronics regime *(Vol. 23, pp. 291–379). Elsevier. https://doi.org/10.1016/B978-0-444-63528-0.00004-8.*

building blocks of hexaferrites can be described as stacking sequences of three blocks—S [$M_2Fe_4O_8$], R [AFe_6O_{11}], and T [$A_2Fe_8O_{14}$]—as shown in Fig. 1.5.

The S block contains two spinel units with a unit formula of $M_2Fe_4O_8$, where M is a divalent metal cation. In a single S block, there are two layers of four oxygen anions. The cations are in four octahedral sites surrounded by six nearest neighbors of oxygen anion and in two tetrahedral sites where four oxygen anions surround the cation (Pullar, 2012). To define the structure of R and T blocks, typical barium ferrite is considered. The R block with a unit formula of $BaFe_6O_{11}$ is developed by three hexagonally packed layers. Each layer contains four oxygen anions. The oxygen anions in the middle layer are substituted by a barium cation of almost the same size to form the unit formula. The block contains five octahedral sites and one trigonal bipyramidal site. The T block consists of four oxygen layers, and for developing the unit formula of $Ba_2Fe_8O_{14}$, the oxygen anions are substituted by barium cations in two middle layers. This block contains six octahedral and two tetrahedral sites.

The unit cells of hexagonal ferrites normally contain multiple molecular units. The M, W, and Z-type ferrites contain two molecular units, while the X and Y-type ferrites have three molecular units, and U-type ferrite involves one molecular unit.

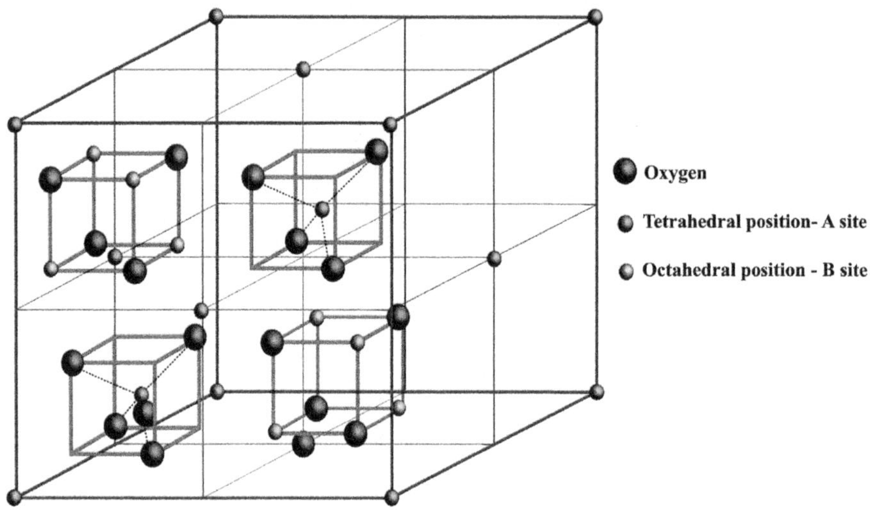

FIGURE 1.4

Schematic representation of crystal strcture inverse spinel.

From Deepapriya, S., Devi, S. L., Vinosha, P. A., Rodney, J. D., Raj, C. J., Jose, J. E., & Das, S. J. (2019). Estimating the ionicity of an inverse spinel ferrite and the cation distribution of La-doped NiFe two O four nanocrystals for gas sensing properties. Applied Physics A, 125*(10), 683. https://doi.org/10.1007/s00339-019-2978-x.*

1.6.2.1 The M structure

$SrFe_{12}O_{19}$, $BaFe_{12}O_{19}$, and $PbFe_{12}O_{19}$ ferrites contain magnetoplumbite (M-type) structure. In the hexaferrites with an M structure, the Fe ions fill five crystallographic sites. Three of them are octahedral sites (12k, $4f_2$, 2a), one is a tetrahedral site ($4f_1$), and one is a trigonal site (2b). The 12k sites are apportioned to the R- and S-units. The $4f_2$ sites are found in block R, the 2a sites are in block S, and the $4f_1$ sites are also found in block S. Fig. 1.6 depicts the typical unit cell of the M-type hexaferrite. The arrows exhibited on the right in Fig. 1.6 demonstrate the direction of spin polarization of the magnetic structure of hexaferrites (Liu et al., 2019). The lattice parameters of barium hexaferrite are $c = 23.17$ Å and $a = 5.89$ Å, while those for strontium ferrites with a similar crystal structure and unit cell ordering, yet with a smaller size of Sr^{2+} cations are $c = 23.03$ Å and $a = 5.86$ Å (Braun et al., 1957; Pullar, 2012). Since the size of a Pb^{2+} ion is also smaller than that of Ba^{2+}, the lattice parameters of lead ferrites $c = 23.02$ Å, $a = 5.88$ Å are also smaller than those of barium ferrite (Adelskold, 1938).

1.6.2.2 The W structure

The crystal structure of W-type ferrite is shown in Fig. 1.7 (Pullar, 2012). It consists of two S blocks and one R block. The two S blocks are placed above and under the R block, and there is a mirror plane in the R block. The unit cell consists of

Table 1.3 Site preference of ions in spinel structure and magnetic moment of some ferrites.

Ferrites	Ionic distribution		Magnetic moment of tetrahedral ions in (μ_B)	Magnetic moment of octahedral ions in (μ_B)	Magnetic moment per molecule in (μ_B)	
	Tetrahedral site	Octahedral site			Theoretical	Experimental
Fe_3O_4	Fe^{3+}	$Fe^{2+} + Fe^{3+}$	5	$4 + 5$	4	4.1
$NiFe_2O_4$	Fe^{3+}	$Ni^{2+} + Fe^{3+}$	5	$5 + 2$	2	2.3
$Li_{0.5}Fe_{2.5}O_4$	Fe^{3+}	$Fe^{3+}_{1.5} + Li^{+}_{0.5}$	5	$7.5 + 0$	2.5	2.6
$MgFe_2O_4$	Fe^{3+}	$Mg^{2+} + Fe^{3+}$	5	$0 + 5$	0	1.1
$CoFe_2O_4$	Fe^{3+}	$Co^{2+} + Fe^{3+}$	5	$3 + 5$	3	3.7
$MnFe_2O_4$	Fe^{3+}	$Mn^{2+} + Fe^{3+}$	5	$5 + 5$	5	4.6
$CuFe_2O_4$	Fe^{3+}	$Cu^{2+} + Fe^{3+}$	5	$1 + 5$	1	1.3
$CdFe_2O_4$	Cd^{2+}	$Fe^{3+} + Fe^{3+}$	0	$5 - 5$	0	1
$ZnFe_2O_4$	Zn^{2+}	$Fe^{3+} + Fe^{3+}$	0	$5 - 5$	0	1

From Kotnala, R. K., Shah, J., & Buschow, K. H. J. (2015). Chapter 4 – Ferrite materials: nano to spintronics regime *(Vol. 23, pp. 291–379). Elsevier. https://doi.org/10.1016/B978-0-444-63528-0.00004-8.*

Table 1.4 The chemical formula and stacking order of various hexaferrite.

Hexaferrite type	Chemical formula	Stacking order	c (Å)	Space group
M	$AFe_{12}O_{19}$	RSR*S*	~23	*P63/mmc*
Y	$A_2M_2Fe_{12}O_{22}$	TSTSTS	~43	*R—3m*
W	$AM_2Fe_{16}O_{27}$	RSSR*S*S*	~33	*P63/mmc*
Z	$A_3M_2Fe_{24}O_{41}$	RSTSR*S*T*S*	~52	*P63/mmc*
X	$A_2M_2Fe_{28}O_{46}$	RSR*S*S*	~84	*R—3m*
U	$A_4M_2Fe_{36}O_{60}$	RSR*S*T*S*	~113	*R—3m*

**Indicates a rotation of the corresponding block by 180 degrees about the c-axis.*

SSRS*S*R* ordering. The three-dimensional model of the W-type ferrite unite cell is shown in Fig. 1.7.

1.6.2.3 The X structure

The structure of typical $Ba_2M_2Fe_{28}O_{46}$ X-type ferrite, which is very similar to the W-type, is shown in Fig. 1.8. Since the unit cell involves three molecular units of M and W, the illustration is not a complete unit cell. The crystal structure contains

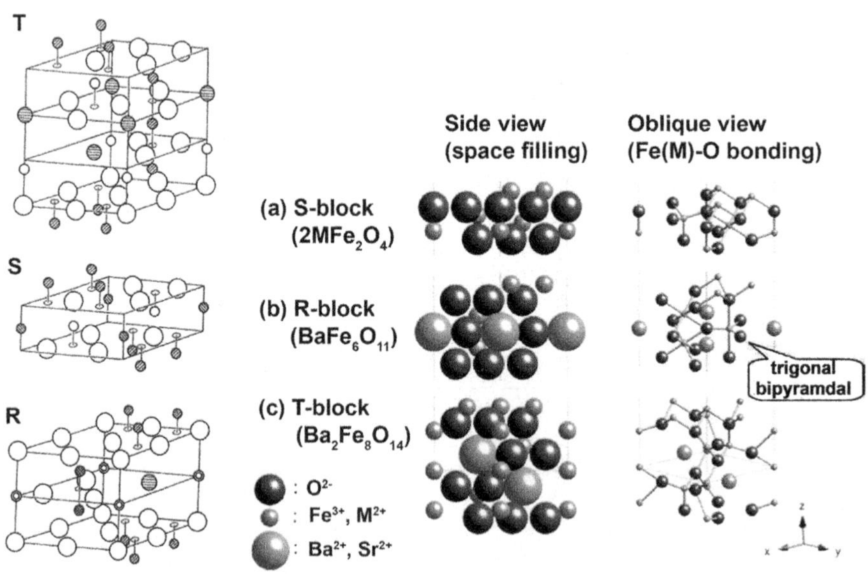

FIGURE 1.5

The cation and anion occupancies in S, R, and T blocks.

From Pullar, R. C. (2012). Hexagonal ferrites: a review of the synthesis, properties and applications of hexaferrite ceramics. Progress in Materials Science, 57(7), 1191–1334. https://doi.org/10.1016/j.pmatsci.2012.04.001.

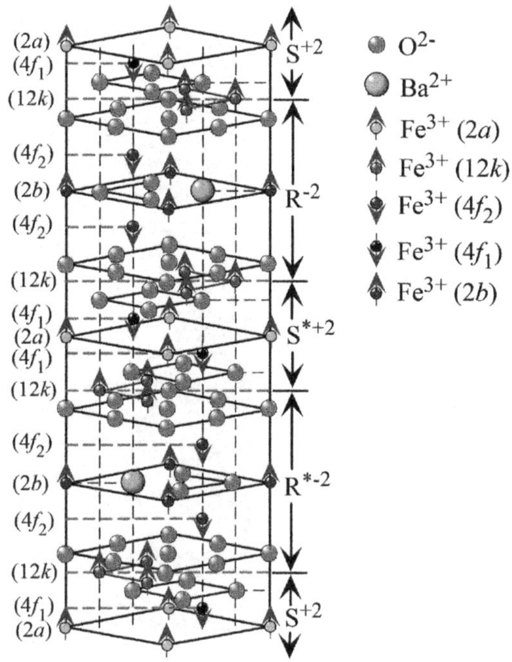

FIGURE 1.6

The unit cell of the M-type hexaferrite.

From Özgür, Ü., Alivov, Y., & Morkoç, H. (2009). Microwave ferrites, part 1: fundamental properties. Journal of Materials Science: Materials in *Electronics, 20(9), 789–834. https://doi.org/10.1007/s10854-009-9923-2.*

the sequence ordering of SRS*S*R*. In this crystal structure, the W block rotates 180° in relation to the M block (Braun et al., 1957; Pullar, 2012).

1.6.2.4 The Y structure

The crystal structure of typical Y ferrite ($Ba_2M_2Fe_{12}O_{22}$) is shown in Fig. 1.9. The crystal structure is developed by one S and one T unit. It consists of six layers. The unit cell is formed by three of these units, and the unit cell formula is 3(ST) (Braun et al., 1957; Pullar, 2012). Note that the T block has no mirror plane, and a series of three T blocks are required to fulfill the overlapping of hexagonal and cubic packed layers.

1.6.2.5 The Z structure

The crystal structure of Z-type ferrite of $Ba_3M_2Fe_{24}O_{41}$ is shown in Fig. 1.10. The crystal structure consists of Y + M and is built by ordering ST + SR with the mirror plane placed in the R block and a repeated distance of 11 oxygen layers. A single unit cell of Z ferrite is developed by two molecular units in which one unit rotates 180 degrees around the c-axis of another unit. The two molecular unit formula is $STSRS^*T^*S^*R^*$.

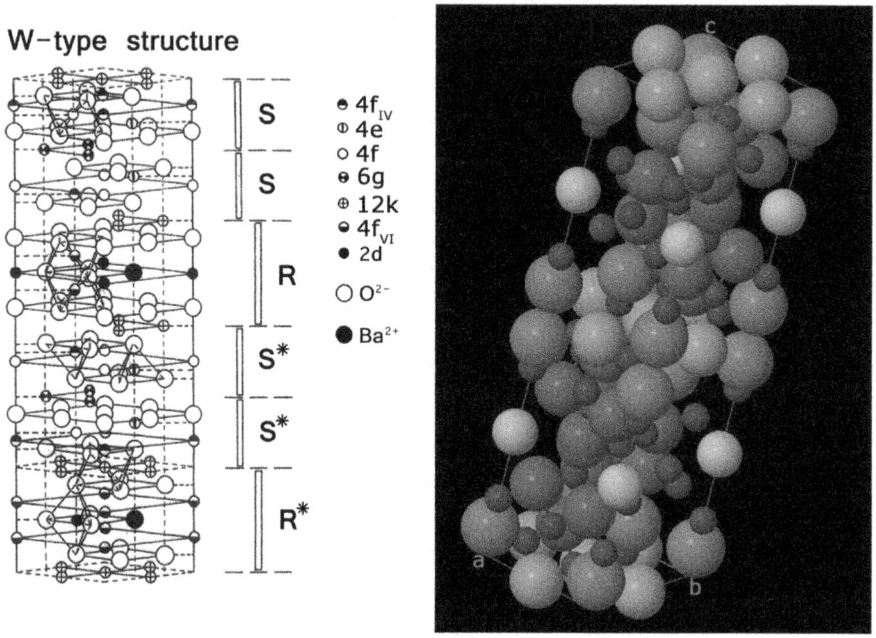

FIGURE 1.7

Perspective view of typical $BaM_2Fe_{16}O_{27}$ with W structure and three-dimensional model of a W ferrite unit cell.

From Pullar, R. C. (2012). Hexagonal ferrites: a review of the synthesis, properties and applications of hexaferrite ceramics. Progress in Materials Science, *57(7), 1191–1334. https://doi.org/10.1016/j.pmatsci.2012.04.001.*

1.6.2.6 The U structure

The crystal structure of U-type ferrite ($Ba_4M_2Fe_{36}O_{60}$) is developed by Z + M, or M + Y + M, as shown in Fig. 1.11. As can be seen, it contains the block ordering of $RSR^*S^*T^*S^*$. The unit cell forms from one molecular unit in which the blocks of R and S rotate 180 degrees relative to the M block (Kohn et al., 1971; Pullar, 2012).

1.6.3 Garnet structure

Garnet has the chemical formula $3M_2O_3 \cdot 5Fe_2O_3$. Four units of $3M_2O_3 \cdot 5Fe_2O_3$ form the entire crystal structure. In the formula, M represents trivalent rare earth such as yttrium (nonmagnetic cation) or magnetic rare earth ranging from lanthanum through ytterbium. Garnet has a very complicated crystal structure, and compared with the spinel structure involving two metallic sites of octahedral and tetrahedral, it contains octahedral (*a* site), tetrahedral (*d* site), and dodecahedral (*c* site in which trivalent cations are surrounded by eight oxygen anions). The 96 oxygen anions in a unit cell exist in the *h* site, where the corners of a tetrahedron, octahedron, and two large polyhedrals meet. Fig. 1.12 displays only an octant of the crystal structure of

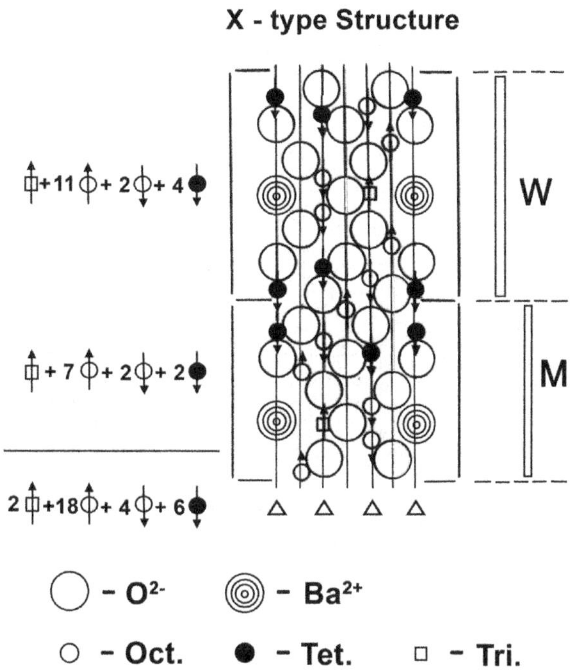

FIGURE 1.8

The X-type structure of $Ba_2M_2Fe_{28}O_{46}$ ferrite in which the arrows reflect the orientation of magnetic moments.

From Pullar, R. C. (2012). Hexagonal ferrites: a review of the synthesis, properties and applications of hexaferrite ceramics. Progress in Materials *Science, 57(7), 1191–1334. https://doi.org/10.1016/j.pmatsci.2012.04.001.*

garnet. The magnetic interaction between tetrahedral and octahedral sites is antiparallel. The net magnetic moment is antiparallel to the magnetic moment of rare earth cations, and the final magnetic ordering is ferrimagnet (Özgür et al., 2009b).

1.6.4 Orthoferrite structure

The orthoferrite or perovskite structure is another type of ferrite with the formula of $MFeO_3$, where M represents a rare earth cation such as yttrium ion. The typical crystal structure of gadolinium orthoferrite is shown in Fig. 1.13. In orthoferrites, the net magnetization is weak due to canting alignment of antiferromagnetically coupled ions (Goldman, 2006).

1.7 Superexchange interaction

Magnetic cations in ferrimagnetic substances, including ferrites, are separated by oxygen anions, which are shown schematically in Fig. 1.14. In such materials,

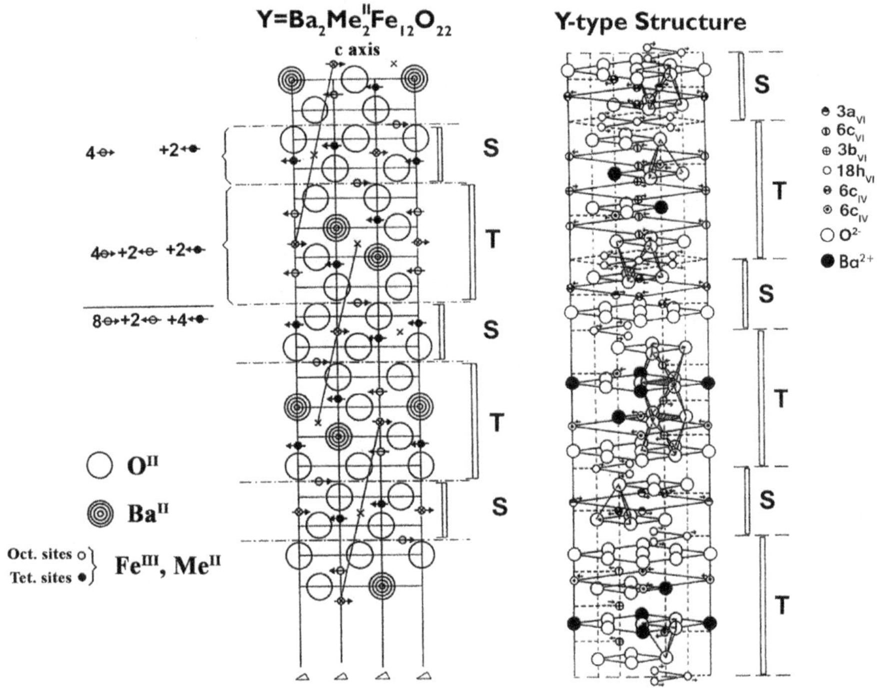

FIGURE 1.9

The structure of typical Y ferrite ($Ba_2M_2Fe_{12}O_{22}$) and the related perspective view. The arrows show the direction of the magnetic moments of the cations relative to the c-axis.
From Pullar, R. C. (2012). Hexagonal ferrites: a review of the synthesis, properties and applications of hexaferrite ceramics. Progress in Materials Science, 57*(7), 1191–1334. https://doi.org/10.1016/j.pmatsci.2012.04.001.*

superexchange interaction is predominant, referring to the interaction of spin moments of metal cations on the opposite site of oxygen anion through the *p* orbit of oxygen.

Consider that a system consists of two metal cations, M^{n+} and M^{m+}, separated by the oxygen anions O^{2-}. There is the possibility that one electron of oxygen anion is excited and transferred to the neighboring cation of M^{n+}, and at the same time, the unpaired left electron in the *p* orbit of anion interacts with M^{m+}. According to Pauli's principle, the two electrons in the *p* orbits must have opposite spins, and both magnetic cations should have an antiparallel configuration to satisfy Hund's role. The interactions between *A* and *B* sublattice depend on the distances between two metal ions and the oxygen ion and on the angle (Φ) between the two metal ions and the oxygen ion. The superexchange interaction is expected to be strongest when $M^{n+} - O^{2-}M^{m+}$ lie along a straight bond. In this configuration, the *p* orbit of the oxygen anion can overlap with the orbits of M^{n+} and M^{m+}. On the contrary, when the bonding angle is 90°, the strength of superexchange interaction is weakest.

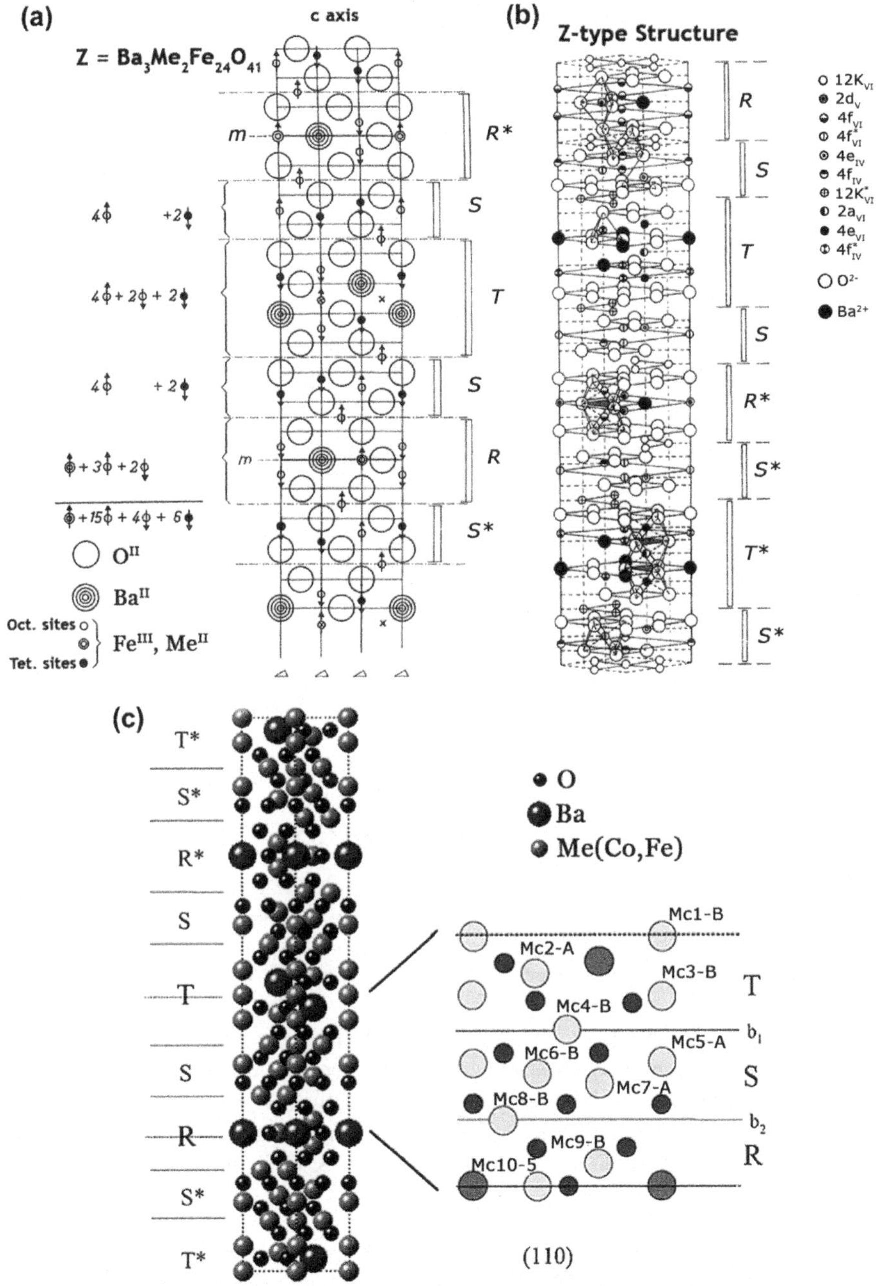

FIGURE 1.10

(A) The crystal structure of Z-type ferrite of $Ba_3M_2Fe_{24}O_{41}$ along with the orientation of magnetic moments and related perspective view. (B) and (C) Perspective views of the Z-type structure.

From Pullar, R. C. (2012). Hexagonal ferrites: a review of the synthesis, properties and applications of hexaferrite ceramics. Progress in Materials Science, 57*(7), 1191–1334. https://doi.org/10.1016/j.pmatsci.2012.04.001.*

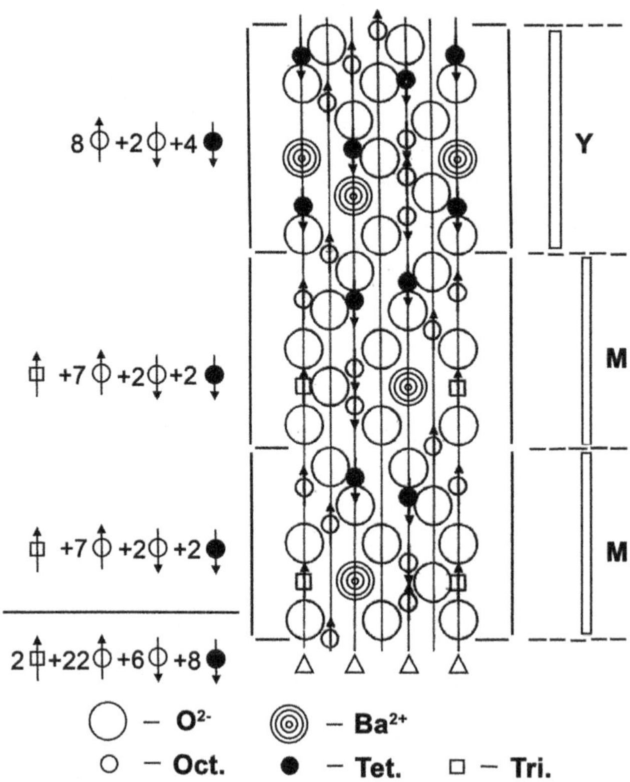

FIGURE 1.11

Structure of the U ferrite ($Ba_4M_2Fe_{36}O_{60}$) and the orientation of magnetic moment.

From Pullar, R. C. (2012). Hexagonal ferrites: a review of the synthesis, properties and applications of hexaferrite ceramics. Progress in Materials Science, 57*(7), 1191–1334. https://doi.org/10.1016/j.pmatsci.2012.04.001.*

When the d orbital of the metal ions is half or more than half-filled, these superexchange interaction energies are negative and therefore enforce an antiferromagnetic orientation. It has been empirically found that when the d orbital of the metal ions is less than half-filled, these interaction energies are positive, thereby inducing a ferrimagnetism orientation (Chikazumi & Graham, 2009).

1.8 Applications of ferrites

Ferrites have many applications in modern technology, such as automotive industry and hybrid cars, radar-absorbing materials (RAMs), photocatalysts, energy harvesting, satellite communication, power computer component and digital devices, sensors, anechoic chamber tiles, ferrofluid, hyperthermia, delay lines, high-density recording media, bubble memory, telecommunication component, microwave

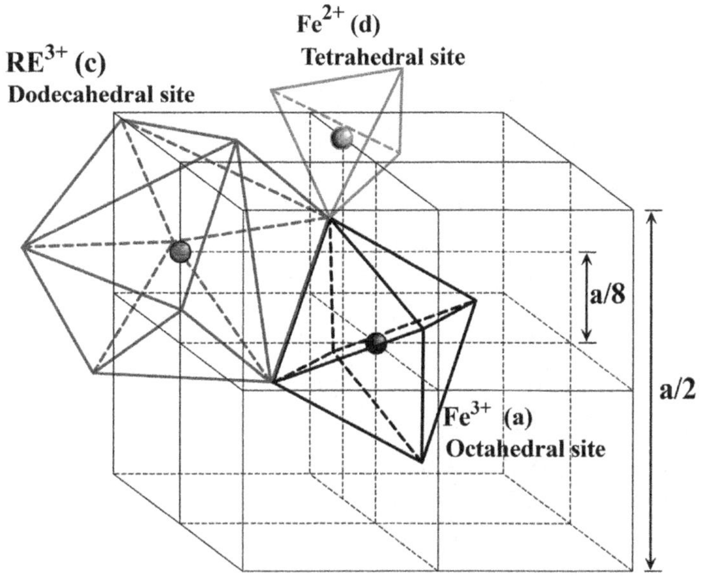

FIGURE 1.12

An "octant" of a garnet crystal structure.

From Özgür, Ü., Alivov, Y., & Morkoç, H. (2009). Microwave ferrites, part 1: fundamental properties. Journal of Materials Science: Materials in Electronics, 20*(9), 789–834. https://doi.org/10.1007/s10854-009-9923-2.*

systems, electromagnetic interference shielding, as well as TV and radio markets (Valenzuela, 2012). Some important ferrite applications are briefly explained in the following.

1.8.1 Applications of ferrites in microwave-absorbing media

Different approaches have been used to camouflage military vehicles including airplanes and submarines. The most important method is to reduce the radiation cross-section (RCS) of a vehicle by designing the shape of the object and eliminating the unnecessary edges and sharp points (Knott et al., 2004). Another effective method is utilizing electromagnetic wave absorbing materials, especially in the microwave region. As shown in Fig. 1.15, the combination of RAMs and designing RCS techniques could be very effective for application in stealth technology. The first research in this regard was carried out in 1936 by employing carbon black as an absorber and titanium oxide with a large admittance coefficient for decreasing the thickness of the absorber. Germans used carbonyl iron with a thickness of 3 inches to enhance the absorption coefficient of submarine camouflage at 3 GHz in the second world war. The important features for obtaining suitable absorbing media include the optimized composition of the absorber, low thickness and density, wide bandwidth, large reflection loss, suitable adhesion and corrosion resistance,

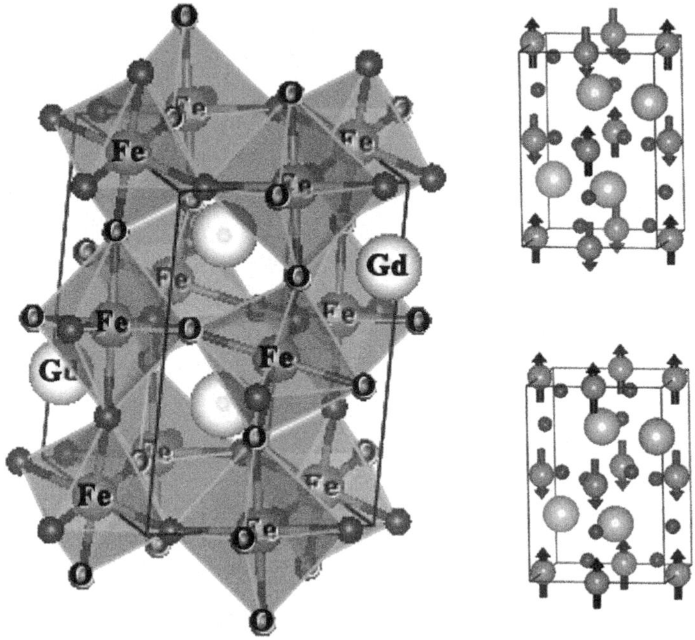

FIGURE 1.13

Crystal structure of gadolinium orthoferrite ($GdFeO_3$).

From Mekam, D., Kacimi, S., Djermouni, M., Azzouz, M., & Zaoui, A. (2012). Ab initio calculations on RE–TM–O_3 perovskites: A comparative study of cation effect. Results in Physics, 2, *156–163. https://doi.org/10.1016/j.rinp.2012.09.008.*

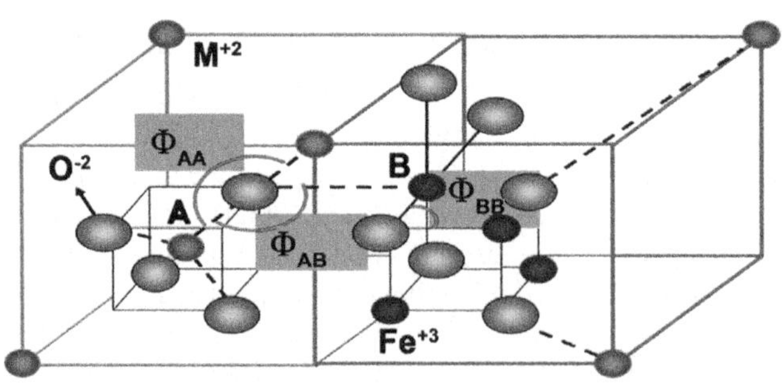

FIGURE 1.14

The schematic presentation of superexchange interactions of A and B sublattice via oxygen anions.

From Kotnala, R. K., Shah, J., & Buschow, K. H. J. (2015). Chapter 4 – Ferrite materials: nano to spintronics regime *(Vol. 23, pp. 291–379). Elsevier. https://doi.org/10.1016/B978-0-444-63528-0.00004-8.*

FIGURE 1.15

Combination utilization of designing radiation cross-section and radar-absorbing material for fabricating a stealth unmanned combat aerial vehicle.

From Sepulveda, E., & Smith, H. (2017). Technology challenges of stealth unmanned combat aerial vehicles. The Aeronautical Journal, 121*(1243), 1261–1295. https://doi.org/10.1017/aer.2017.53.*

low environmental contamination, and large tensile as well as compression strength. Ferrite-carbon materials nanocomposites with large permeability and permittivity could provide high reflection loss over a wide range of microwave frequencies.

In a separate application, ferrite tile absorbers with high absorption characteristics, immune to humidity, corrosion, and chemical agents are used for the anechoic chamber, as shown in Fig. 1.16.

FIGURE 1.16

Ferrites used for foam absorber in anechoic chamber.

1.8.2 Applications of ferrites in hard disk drives

A very simple hard disk drive (HDD) shown in Fig. 1.17 is a storage device in which digital information is stored on spinning magnetic hard disks, called magnetic media. Normally, HDD consists of recording media, read/write heads, a spindle, and an actuator. The spindle motor attached to the recording media will spin the disk. Read and write heads are attached to a swinging arm for reading and writing the information (Hoagland, 2003).

The random access method of accounting and control (RAMAC), the first HDD involving longitudinal recording technology, was invented by IBM in 1956 (Hoagland, 2003). This drive (Fig. 1.18) contained 50 of 24-inch platters with a total capacity of 5 MB and a recording density of 2 kb/in^2. The areal density of HDDs continues to increase, with the recording density reaching approximately 10 Gb/in^2 in the year 2000, i.e., an increase by a factor of five million. With the continuing development of new magnetic materials, HDDs with a typical capacity of 400 Gb/in^2 areal density were fabricated in 2009. Recently, the approach to reaching higher capacity (more than 6 Tb/in^2) has continued to be a significant challenge.

Many research efforts have been made in improving the storing capacity and enhancing the signal-to-noise ratio of the reading and writing process. Magnetic recording media were developed using hard magnetic materials such as hard ferrite (strontium ferrite and barium ferrite), SmCo, NdFeB, TbFeCo, and FePt. However, the trend will be limited by the thermal instability of magnetic materials on the nanoscale, which is called superparamagnetic. To fulfill storage capacity enhancement and obtain high-density recording media, patterned media, in which data are stored in a dot array, are widely discussed (Chou, 1997; Dobin & Richter, 2006; Victora & Shen, 2005; White et al., 1997). In a patterned media system, transition noise is

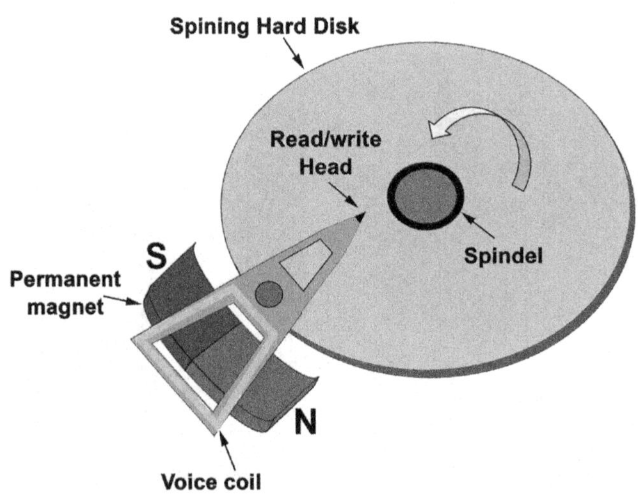

FIGURE 1.17

Schematic representation of the anatomy of a hard disk drive.

FIGURE 1.18

First HDD involving longitudinal recording technology.

From Oh, K. W., Lee, P., & Choi, Y. W. (2015). Enhanced unlatch operation of disk drive for low temperature environment. TRIZ and Knowledge-Based Innovation in Science and Industry, 131, *906–913. https://doi.org/10.1016/j.proeng.2015.12.402.*

eliminated, and the storage density is far higher than in a conventional hard drive. Ferrite dot array with epitaxial growth could be grown on a substrate by choosing an appropriate underlayer. The schematic of the ferrite dot array is shown in Fig. 1.19. Each element stores one bit, depending on its magnetization state; for instance, magnetization up could represent 1, and down could denote 0.

Recording devices based on the Faraday or Kerr effects are also being developed for enhancing recording density and eliminating some mechanical parts of HDD, such as spindle motors. The techniques require thin films with a large magneto-optical Kerr effect and unidirectional alignment of magnetization. Garnets such as $Gd_3Fe_5O_{12}$ are promising candidates for magneto-optical memories; however, since the absorption coefficient is small, higher energy is required for writing. Bismuth iron garnet with a large Faraday rotation coefficient is a suitable candidate for this kind of recording mode (Gomi et al., 1991).

In addition to the recording media, ferrites can be used in the magnetic recording head. The perpendicular recording mode is illustrated in Fig. 1.20. For this purpose, ferrites must have high saturation magnetization, low coercivity, and remanence, as well as high permeability to induce a magnetic field with sufficient strength in the head gap. Low remanence is required to avoid any writing process when the signal of the coil is zero. Ferrites such as Ni–Zn ferrites with high saturation magnetization and permeability as well as superior corrosion and wear resistance were used in the

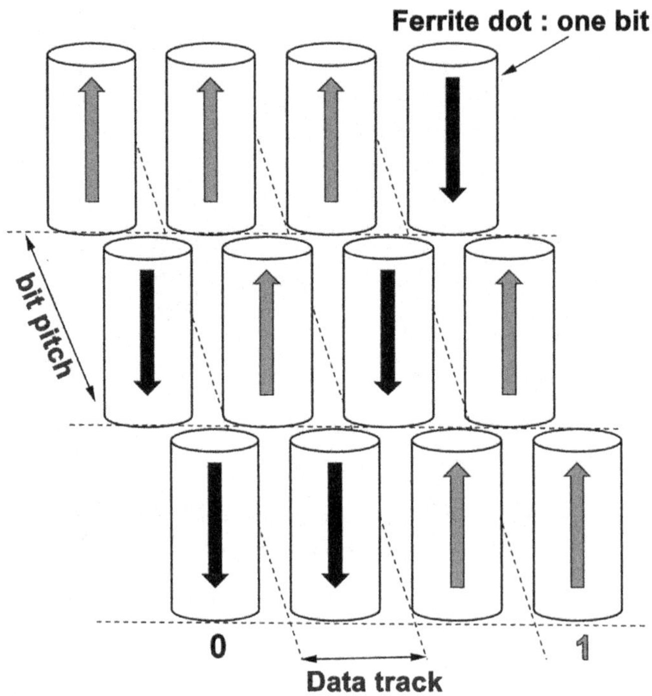

FIGURE 1.19

Schematic representation of bit pattern media.

audio and VTR recording heads (Kugimiya, 1990; Kugimiya et al., 1974). Metallic magnets with high saturation magnetization such as iron cobalt and permalloy with very large permeability are well suitable for the recording head; however, low corrosion and wear resistance cause some technical difficulties.

1.8.3 Applications of ferrites in biosciences

Cancer is one of the most widespread causes of death whose incidence is progressively increasing. Theranostics is a treatment involving both terms therapeutics and diagnostics by employing one radioactive drug to characterize (diagnose) and a second radioactive drug to deliver therapy to treat the original tumor and any metastatic tumors. Indeed, this field integrates therapeutics and diagnostics into one platform.

Magnetic nanoparticles (MNPs) are very promising candidates for effective diagnosis and drug delivery for several kinds of human diseases, including cancer. The application of MNPs in bioscience is shown in Fig. 1.21. In general, MNPs would be able to detect and diagnose the tumor, conduct the drug delivery to the tumor, and allow the monitoring of the treatment through magnetic resonance imaging (MRI).

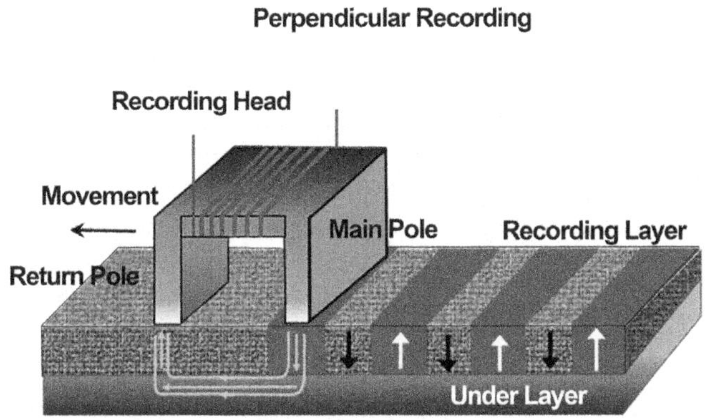

FIGURE 1.20

Schematics of perpendicular recording mode.

From Tudu, B., & Tiwari, A. (2017). Recent developments in perpendicular magnetic anisotropy thin films for data storage applications. Vacuum, 146, *329–341. https://doi.org/10.1016/j.vacuum.2017.01.031.*

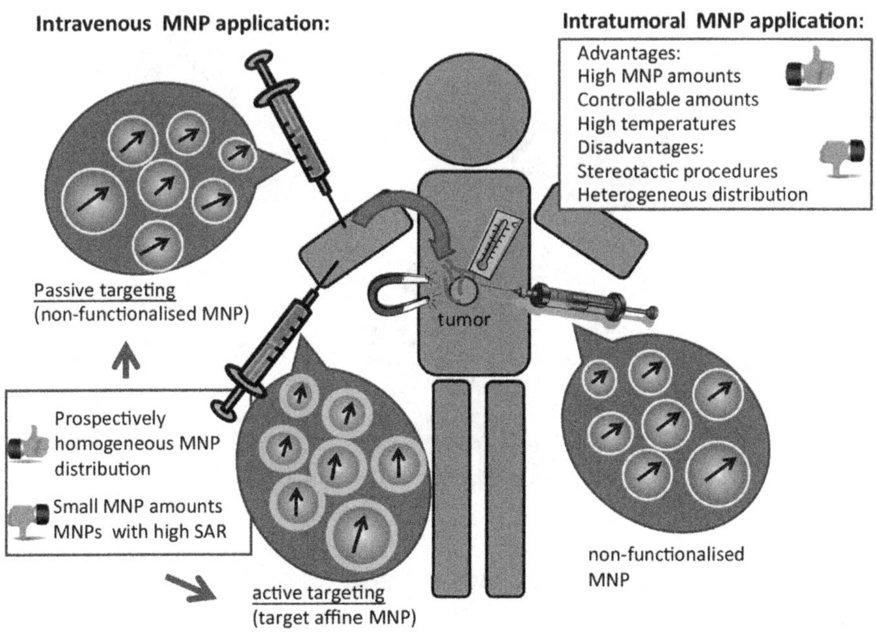

FIGURE 1.21

The application of magnetic nanoparticles in bioscience.

From Hilger, I. (2013). In vivo applications of magnetic nanoparticle hyperthermia. International Journal of Hyperthermia, 29*(8), 828–834. https://doi.org/10.3109/02656736.2013.832815.*

One of the effective methods to curb the growth rate of superficial cancer is employing a hyperthermia heat source directly to the site of the tumor. MNPs of ferrite such as Fe_3O_4 or γ-Fe_2O_3 with large biocompatibility can be concentrated at the tumor site where cancer cells could be exclusively heated by applying an alternating magnetic field to reach hyperthermic temperatures of 42–45°C (Khot et al., 2015). Iron oxide, with inherent toxicity, ease of synthesis, narrow size distribution, specific surface chemistry, as well as unique physical, thermal, and magnetic features, is widely used in many medical applications such as biomedicine, magnetic resonance technology, cancer therapy, drug delivery, and high gradient magnetic separation.

Iron oxide ferrofluids offer practical applications in medical science since their iron cations also exist in the blood hemoglobin. The size and feature of coated MNPs in ferrofluids are shown in Fig. 1.22. The heating ability is measured by specific absorption rate (SAR) and the biocompatibility of ferrite ferrofluids, which are key factors that must be considered. There are several research papers on SAR of

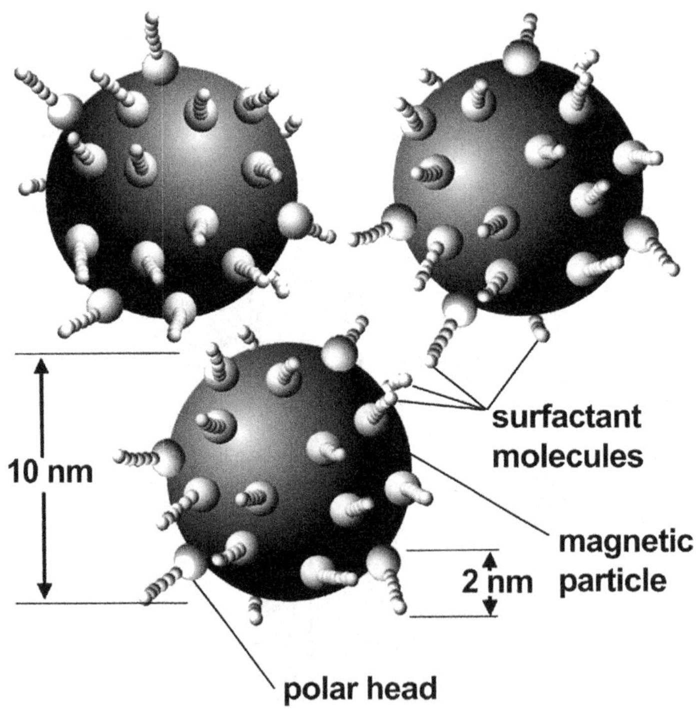

FIGURE 1.22

Magnetic nanoparticle with attached surfactant molecules used in ferrofluids.

From Sharifi, I., Shokrollahi, H., & Amiri, S. (2012). Ferrite-based magnetic nanofluids used in hyperthermia applications. Journal of Magnetism and Magnetic Materials, 324*(6), 903–915. https://doi.org/10.1016/j.jmmm.2011.10.017.*

nanoparticles of Fe_3O_4 or γ-Fe_2O_3 particles, though there are no sufficient data available for other types of ferrites.

Iron oxide superparamagnetic particles are also used in MRI as they have the ability to change the rate of proton decay from the excited level to the ground state. These nanoparticles are selectively related to the healthy parts of tissue and could provide a different dark contrast from the other region of the tissue (Kobayashi, 2011).

1.8.4 Applications of ferrites in the environment

Microorganisms, anions, dyes, phenols, toxic metal particles such as Cd (II) and Pb (II), pesticides, and detergents normally contaminate water and induce different diseases. Many practical efforts such as electrochemical medication, ion exchange, precipitation, biosorption process, film filtration, advanced oxidation, electrolysis, invert osmosis, buoyancy, and particle trade are often made to treat the wastewater and produce a good source of potable water (Salman et al., 2017). In the mentioned processes, the disposal requires a large area, and normally the cost-effectiveness is low, and absorbent recovery is also difficult. Spinel ferrite nanoparticles and related nanocomposites are one of the most effective candidates for wastewater and groundwater treatment through the easy removal of metallic pollutants such as arsenic, chromium, cobalt, iron, and uranium. Using ferrites and with a simple magnetic process, catalysts from the reaction medium can also be removed (Cunha et al., 2016). Ferrites possess many advantages compared with nonmagnetic treatment options. The approaches for synthesizing ferrites are almost easy and cost-effective. Fast absorption with high sensitivity and large absorption capacity are the effective parameters that are well satisfied through ferrite composites. Meanwhile, ferrites contain an excellent redox pair based on Fe^{2+}/Fe^{3+} and have the ability to be doped with different cations such as Cu^{2+}, Co^{2+}, Mn^{2+}, and Ni^{2+}, allowing to control of electrical, magnetic, and catalytic/redox activity. The required amounts of ferrites for treatment are small, and they can also be reused with recovered absorbents for several cycles. For example, the composite of nickel ferrite with reduced graphene oxide is used for the removal of heavy metals of Pb (II) and Cr (III) from water. The composites have the capability to be reused for four cycles with more than 99% adsorption efficiency. The role of Fe_3O_4 in the removal of Cu (II) ions from the copper-ammonia system is illustrated in Fig. 1.23.

1.8.5 Applications of ferrites as permanent magnets

Permanent barium and strontium ferrites have many applications. For example, ring-shaped ferrites with axial magnetization are used in loudspeakers for generating magnetic flux, as shown in Fig. 1.24. There is a coil fixed at the end of the speaker cone with the signal which flows through the coil being converted to sound. The interaction between magnetic flux and the current imposes a force on the speaker

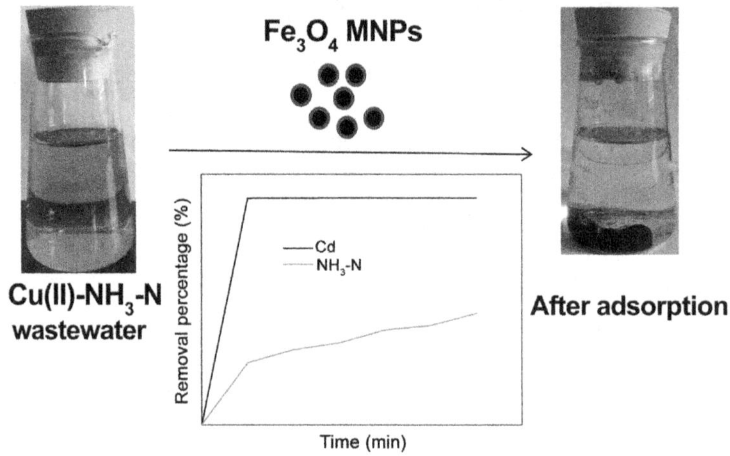

FIGURE 1.23

Schematic illustration of the removal of Cu (II) ions from copper-ammonia system.

From Liu, F., Zhou, K., Chen, Q., Wang, A., & Chen, W. (2019). Application of magnetic ferrite nanoparticles for removal of Cu(II) from copper-ammonia wastewater. Journal of Alloys and Compounds, 773, *140–149. https://doi.org/10.1016/j.jallcom.2018.09.240.*

cone, which can generate vibration according to the electric signal. Loudspeaker systems made with strontium ferrite are smaller, lighter, simpler, and cheaper than the equivalent system using Alnico 5 magnet (Valenzuela, 2012).

It is well known that magnetic fields can be generated by flowing current through a coil or using a permanent magnet. Magnetic fields can apply torque on the rotor and induce movement in electric motors. Ferrites were used in stepper motors of floppy disk drivers, printers, and other computer components. In small direct current (DC) motors, for instance, motors of windshields, fans, and windows of expensive cars, the use of a permanent magnet such as strontium ferrite is more convenient and cost-effective. The use of ferrites in DC motors is schematically shown in Fig. 1.25. The ferrite element is a curved segment with perpendicular magnetization to its plane. Normally, soft metallic magnets such as iron are used for providing mechanical structure and for completing the magnetic circuit.

In another application, refrigerator gaskets (Fig. 1.26) are made of ferrite. These ferrites are produced through c-axis oriented magnetization of ferrite particles (anisotropic ferrite) dispersed in a rubber matrix and eventually produced by the rolling or extrusion process (Valenzuela, 2005).

1.8.6 Applications of ferrites in modern electronics

Ferrites have many applications in modern electronics. In this section, some of the most important utilizations are briefly described.

FIGURE 1.24

Configuration of ferrite in the loudspeaker.

From https://en.wikipedia.org. (n.d.). https://en.wikipedia.org/wiki/Loudspeaker#/media/File:Loudspeaker-bass.png.

1.8.6.1 Ferrite yokes

The first application of ferrite elements in modern electronics was in TV deflection yokes to provide a strong magnetic field and focus electrons through the interaction with the field and create an image on the screen (Valenzuela, 2005). Typical ferrite yokes are shown in Fig. 1.27. Kobayashi and colleagues proposed Cu–Mg–Ti substituted nickel-zinc ferrite for high-frequency deflection yoke (Kobayashi et al., 1992).

1.8.6.2 Ferrites in switched-mode power supplies

Power supplies are designed for operation in steady-state conditions; however, transient conditions, including startup, shutdown, and often load transients, induce catastrophic stress on the components. In modern electronics, using modern transistor switching circuits with a power transformer, it is possible to design and fabricate smaller, more efficient, and lighter devices. Ferrites are frequently used in switched-mode power supplies (SMPSs) of computers and small instruments. Basically, in SMPS, the signal of main power is rectified, smoothed, and switched as rectangular pulses into a ferrite transformer to reach the desired voltage (Valenzuela,

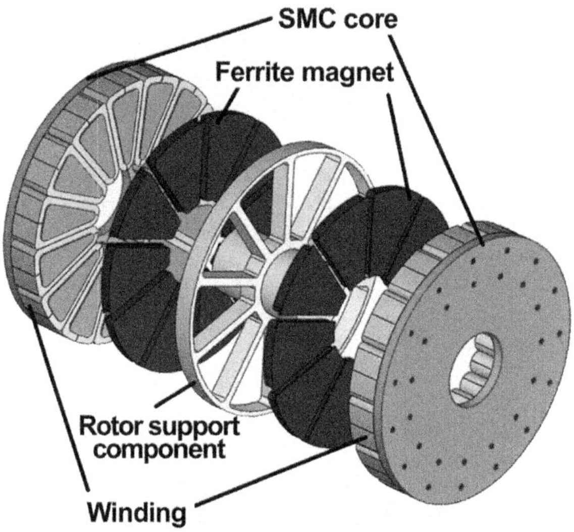

FIGURE 1.25

Ferrite magnets employed in a direct current motor.

From Chiba, K., Chino, S., Takemoto, M., & Ogasawara, S. (2014). Fundamental analysis of a ferrite permanent magnet axial gap motor with coreless rotor structure. IEEJ Journal of Industry Applications, 3*(1), 47–54. https://doi.org/10.1541/ieejjia.3.47.*

2005). The employed ferrites should have low coercivity, large saturation of magnetization, low magnetostriction coefficient, large Curie temperature, and high electrical resistivity. Normally, Ti^{4+}– Co^{2+} substituted Mn–Zn ferrites are a suitable candidate for this application (Hanke & Neusser, 1984).

FIGURE 1.26

Ferrite rubber used in refrigerator gaskets.

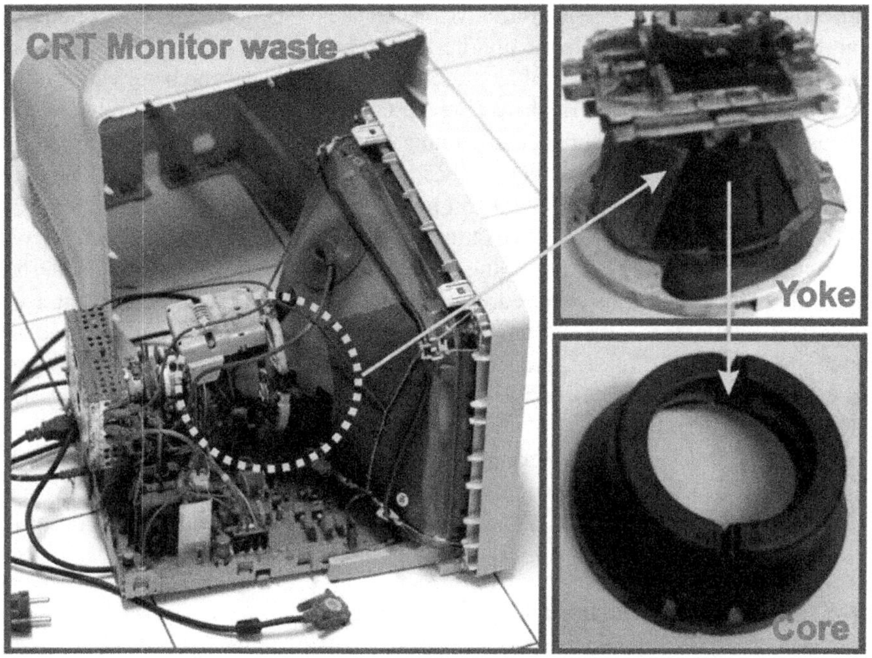

FIGURE 1.27

Schematics of TV deflection yoke.

From Ubaidillah, U., Mazlan, S. A., Sutrisno, J., Yahya, I., & Imaduddin, F. (2016). Physicochemical properties and stress-strain compression behaviors of a waste based magnetorheological elastomers. Scientia Iranica, *23(3), 1144–1159. https://doi.org/10.24200/sci.2016.3885.*

1.8.6.3 Ferrites in electromagnetic interference suppressors

Frequent use of electronic devices such as high-speed digital interfaces in notebooks and computers, scanners, digital cameras, laptops, and several types of miniaturized electronic equipment and wireless communication in small areas could cause serious difficulties in the performance of electronic systems by electromagnetic disturbance originating from electric and magnetic fields (Valenzuela, 2012). The frequency of generated noise is normally higher than the working frequency of the circuit. The interference should be reduced by employing electromagnetic interference suppressors.

For reducing electromagnetic interference and enhancing the electromagnetic compatibility within the electronic circuit, several approaches, including inductors, chokes, capacitors, integrated passive devices, and ferrite beads, are frequently utilized. Inductive components of many electronic circuits such as filters, low-noise amplifiers, impedance matching networks, and voltage-controlled oscillators employ ferrites. Through the miniaturization approach and increasing the integration density, and using higher clock frequency, small-size filters with higher efficiency are required. Ferrite multilayers such as Ni–Zn ferrite were developed to fulfill this requirement within the middle-frequency range of 20–200 MHz. With an increase in frequency, the type of ferrites and values of losses and bandwidth should be tuned. Surface mount ferrite beads are employed in many electronic devices such as laptops, TV tuners, mobiles, and gaming systems for achieving signal integrity and decreasing interference. Ferrite beads (Fig. 1.28) have many advantages. They are very small, cost-effective, and light weight. They can be used effectively within the frequency range of 1 MHz to 1 GHz. The selection of approach for offsetting the noise is different depending on the source of the electromagnetic interference, frequency range, the required value of attenuation, and the environmental and electrical parameters such as temperature, maximum operating current, and the strength of field, DC voltage, and DC bias current. Note that ferrite bead is designed to work up to a certain DC current, above which they are abruptly affected by heat (Kabbur et al., 2021).

Arai and colleagues proposed that through deposition of a metallic coil on the epitaxial YIG, followed by providing a suitable pattern, the inductor in an integrated circuit can be developed. The second ferrite layer was deposited on the coil to increase the effective inductance of the device (Valenzuela, 2005).

1.8.6.4 Linear response application

In a separate application in modern technology, spinel ferrites have the capability to convert alternating current (AC) signals with small amplitude into large variations of magnetic flux. Since the variation in permeability with the magnetic field is linear in small fields, the action is called linear response application or small-signal application. In such practice, the permeability and working frequency of ferrite should be high, and the temperature dependence of permeability should be almost constant.

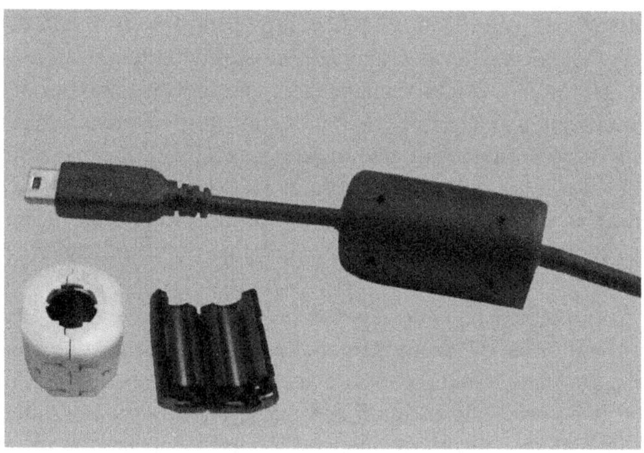

FIGURE 1.28

Ferrite beads used in electronic circuits.

From Ferrite bead. (n.d.). https://en.wikipedia.org/wiki/Ferrite_bead.

The Mn–Zn ferrite worked well to 1 MHz, and Ni–Zn ferrite is suitable for higher frequencies (Valenzuela, 2005).

1.8.6.5 Magnetoelectric sensors

The next generation of novel microelectronic devices, such as information components and sensors, could be fabricated by synthesizing the composite of ferrite (magnetic phase) and piezoelectric (electric phase) to provide magnetoelectronic components (Duong et al., 2006, 2010; Etier et al., 2012). The magnetoelectric sensors based on magnetic and piezoelectric thin films are illustrated in Fig. 1.29. Nanostructures of ferrite-piezoelectric in various configurations, including wire, pillar, and thin films, are considered to enhance the functionality of miniature devices. Among the whole series of ferrites, cobalt and nickel-zinc ferrites have the largest impact on increasing the magnetoelectric coefficient. In such composites, the electric field induces magnetization, and the magnetic field leads to electrical polarization. Converting magnetic field into electrical voltage is processed in two steps. First, the magnetic field creates a mechanical strain, and the mechanical strain causes voltage in the piezoelectric. Sensors for AC and DC magnetic fields and electric currents can be designed and fabricated by such nanocomposite, as discussed in more detail in Chapter 4.

1.8.7 Applications of ferrites in microwave devices

Microwave technology requires higher frequencies and bandwidths. The high-frequency application of ferrite in radar systems and telecommunications is rapidly increasing. Ferrites with nonconduction characteristics could be employed in several microwave devices, including miniature antennas, isolators, phase shifters, and circulators.

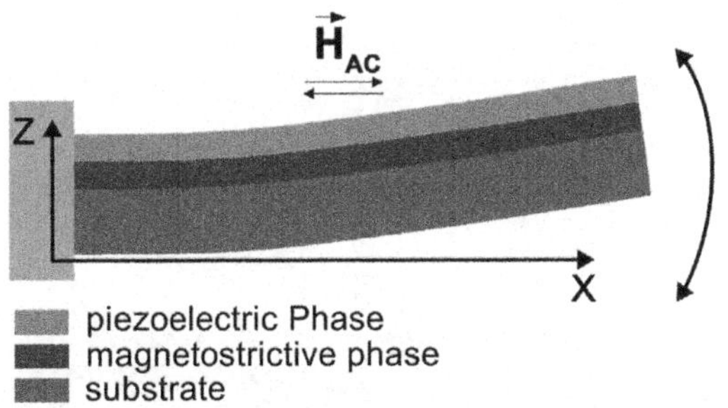

FIGURE 1.29

Schematics of magnetoelectric sensor contains magnetostrictive and piezoelectric layers.

From Gröttrup, J., Kaps, S., Carstensen, J., Smazna, D., Mishra, Y. K., Piorra, A., Kirchhof, C., Quandt, E., & Adelung, R. (2016). Piezotronic-based magnetoelectric sensor: fabrication and response. Physica Status Solidi (a), 213*(8), 2208–2215. https://doi.org/10.1002/pssa.201532924.*

1.8.7.1 Miniature antennas

The general function of the antenna is to convert the wave propagation; for example, the guided wave in a transmission line can be converted to a plane wave in free space, which is the transmission, and vice versa, which is reception. There are many types of antennas, including wire antennas, printed antennas such as printed dipoles, printed slots, and microstrip patches (for use within the frequency range of 400 MHz to 100 GHz), aperture antennas with open-ended waveguides, rectangular and circular horns, reflectors, and array antennas. The unique feature of the ferrite antenna was recognized as early as the late 1960s (Özgür et al., 2009b). Since ferrites contain nonreciprocal characteristics, by applying an external magnetic field, they can be electronically tuned, and thus the use of biased ferrite layers in the novel antenna is frequently investigated (Brown et al., 1999; Kramer et al., 2006; Maheri et al., 1988; Yang, 1996; Zaitsev et al., 1994). Ferrites with various configurations are used for fabricating antennas with tunable resonance frequencies. Array antenna with a phase array function and with variable ferrite phase shifters is employed to scan the main beam electrically without moving parts for agility and compactness (Özgür et al., 2009a). The typical array antenna is shown in Fig. 1.30. To prevent the eddy current losses, the resistivity of ferrite should be high. The bias magnetic field generated from a coil or an electromagnet can be applied in different directions, thereby making the antenna tunable.

1.8.7.2 Ferrite isolators

One of the important microwave devices is an isolator, which is a two-port device that transfers the forward power from an RF microwave generator to a load. The

FIGURE 1.30

Phase array antenna.

From (n.d.-a). https://en.wikipedia.org/wiki/Antenna_array#/media/File:55Zh6M_Nebo-M_mobile_multiband_radar_system_-02.jpg.

isolator has important applications in microwave engineering involving high-gain traveling wave-tube amplifiers. The typical ferrite isolator is shown in Fig. 1.31.

The advantage of the isolator is transferring the entire power from the generator, while the reflected electromagnetic wave from the mismatched load does not flow back into the RF generator. Without employing an isolator, the unwanted electromagnetic wave returning back to the generator can disturb the equipment by signal distortions. For obtaining double isolation in practice, using two isolators in the transmission line is preferred. The ferrite isolator involves a transmission line, a ferrite part, and a DC biasing magnetic field. Three types of isolators—resonance isolator, bias placement isolator, and Faraday isolator—are frequently used in microwave systems (Adam et al., 2002). The resonance isolator has a simple design compared with other isolators, since a small ferrite disc is used. A stub is connected to the main line, and a ferrite disc is placed in the center of the junction. A large external magnetic field should be applied to the ferrite, which is the main disadvantage of this kind of isolator.

The bias placement isolator consists of a wide strip conductor on a ferrite substrate. The operation of the isolator is based on the field displacement induced by the gyromagnetic feature of a normally magnetized ferrite substrate. To explain the Faraday isolator, it is essential to briefly define Faraday rotation. By passing an electromagnetic wave through ferrites, the plane of polarization continuously rotates to angle θ in one specific direction. This rotation can be done either clockwise or anticlockwise. The polarization plane changes in the same direction and could be directed in the axis of electromagnetic wave propagation. This is called Faraday

FIGURE 1.31

Ferrite isolator used in microwave systems.

From (n.d.-b). https://en.wikipedia.org/wiki/Isolator_(microwave)#/media/File:Isolator-(resonance-absorption-type)-in-WG16.jpg.

rotation. Faraday isolator with reciprocal function includes a rectangular waveguide and mechanical bend, while Faraday isolator with nonreciprocal function involves a circular waveguide containing a rod of ferrite.

A ferrite isolator with a helical transmission line is also used in microwave engineering. In this type of isolator, since the closed ring of soft ferrite with a square-loop magnetization curve is used, no external magnetic field is required. The main application of this isolator is usage in traveling wave tubes (Lax & Button, 1962; Özgür et al., 2009a).

1.8.7.3 Ferrite phase shifter

Phase shifters are employed for controlling the relative phase of each element in a phased array antenna. A large number of phase shifters employ ferrites to control the transmission of energy and perform a number of functions in microwave systems. The typical ferrite phase shifter is shown in Fig. 1.32. The high resistance, tailoring of the magnetic properties by adding various kinds of cations, ease of preparation, and performance considerations make ferrites the first-choice materials for these applications (Adkins et al., 1984; Fetisov & Srinivasan, 2005). Systems use one of three phase shifters: (1) The axial toroidal phase shifter is a nonreciprocal phase shifter consisting of a ferrite toroid centered axially along a rectangular waveguide. (2) The dual-mode phase shifter consists of two nonreciprocal channels composed of a ferrite rod and yoke. (3) The rotary field phase shifter has linearly polarized RF energy that is propagated in the input and output rectangular waveguides and converted to circular polarization in a ferrite half-wave plate (Lax & Button, 1962).

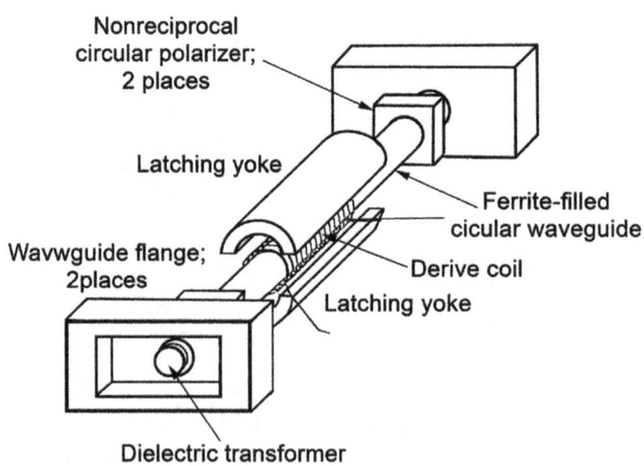

FIGURE 1.32

Typical ferrite phase shifter.

From Hord, W. E. (2001). Ferrite phase shifters. Wiley Encyclopedia of Electrical and Electronics Engineering. *https://doi.org/10.1002/047134608X.W4914.*

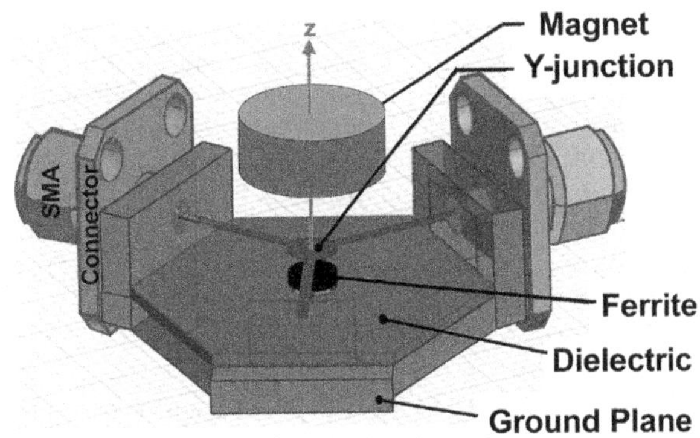

FIGURE 1.33

Schematics of the internal components of a circulator.

From Harris, V. G., & Sokolov, A. S. (2019). The self-biased circulator: ferrite materials design and process considerations. Journal of Superconductivity and Novel Magnetism, *32(1), 97–108. https://doi.org/10.1007/s10948-018-4928-9.*

1.8.7.4 Ferrite circulators

A circulator is a passive three- or four-port device in which a microwave or radiofrequency signal entering any port is transmitted to the next port in rotation. One of the important types of circulators is ferrite circulators (Lax & Button, 1962; Pozar, 1992; Soohoo, 1968).

Ferrite circulators are composed of magnetized ferrite materials. Fig. 1.33 displays the schematics of a ferrite circulator. They fall into two main classes: 4-port waveguide circulators based on Faraday rotation of waves propagating in a magnetized material and 3-port "Y-junction" circulators based on cancellation of waves propagating over two different paths.

References

Adam, J., Davis, L., Dionne, G., Schloemann, E., & Stitzer, S. N. (2002). Ferrite devices and materials. *IEEE Transactions on Microwave Theory and Techniques, 50*(3), 721–737. https://doi.org/10.1109/22.989957

Adelskold, V. (1938). X-ray studies on magneto-plumbite, $PbO_{.6}Fe_2O_3$, and other substances resembling "beta-alumina," $Na_2O.11Al_2O_3$. *Arkiv for Kemi, Mineralogioch Geologi, 29*, 1–9.

Adkins, L., Glass, H., Stearns, F., Carter, R., Chang, K. W., & Owens, J. (1984). Electronically variable time delays using cascaded magnetostatic delay lines. *Journal of Applied Physics, 55*(6), 2518–2520. https://doi.org/10.1063/1.333714

Braun, P., Hornstra, J., & Leenhouts, J. (1957). Philips Res. *Forestry Report, 12*, 491–548.

Brown, A. D., Volakis, J. L., Kempel, L. C., & Botros, Y. (1999). Patch antennas on ferromagnetic substrates. *IEEE Transactions on Antennas and Propagation, 47*(1), 26–32. https://doi.org/10.1109/8.752980

Chikazumi, S., & Graham, C. D. (2009). *Physics of ferromagnetism 2e (No. 94)*. Oxford University Press on Demand.

Chou, S. Y. (1997). Patterned magnetic nanostructures and quantized magnetic disks. *Proceedings of the IEEE, 85*(4), 652–671. https://doi.org/10.1109/5.573754

Coey, J. M. (2010). *Magnetism and magnetic materials*. Cambridge University Press.

Cullity, B. D., & Graham, C. D. (2011). *Introduction to magnetic materials*. John Wiley & Sons.

Cunha, I. T., Teixeira, I. F., Albuquerque, A. S., Ardisson, J. D., Macedo, W. A., Oliveira, H. S., ... Lago, R. M. (2016). Catalytic oxidation of aqueous sulfide in the presence of ferrites (MFe_2O_4, M = Fe, Cu, Co). *Catalysis Today, 259*, 222–227. https://doi.org/10.1016/j.cattod.2015.07.023

Dobin, A. Y., & Richter, H. (2006). Domain wall assisted magnetic recording. *Applied Physics Letters, 89*(6), 062512–062515. https://doi.org/10.1063/1.2335590

Duong, G. V., Groessinger, R., & Turtelli, R. S. (2006). Magnetoelectric properties of $CoFe_2O_4$-$BaTiO_3$ core-shell structure composite. *IEEE Transactions on Magnetics, 42*(10), 3611–3613. https://doi.org/10.1109/TMAG.2006.879748

Duong, G. V., Turtelli, R. S., & Groessinger, R. (2010). Magnetoelectric properties of $CoFe_2O_4$–$BaTiO_3$ core-shell structure composite studied by a magnetic pulse method. *Proceedings of the Joint European Magnetic Symposia, 322*(9), 1581–1584. https://doi.org/10.1016/j.jmmm.2009.09.022

Etier, M., Gao, Y., Shvartsman, V. V., Elsukova, A., Landers, J., Wende, H., & Lupascu, D. C. (2012). Cobalt ferrite/barium titanate core/shell nanoparticles. *Ferroelectrics, 438*(1), 115–122. https://doi.org/10.1080/00150193.2012.743773

Fetisov, Y. K., & Srinivasan, G. (2005). Electrically tunable ferrite-ferroelectric microwave delay lines. *Applied Physics Letters, 87*(10), 103502–103505. https://doi.org/10.1063/1.2037860

Goldman, A. (2006). *Modern ferrite technology*. Springer Science & Business Media.

Gomi, M., Satoh, K., & Abe, M. (1989). *New garnet films with giant Faraday rotation*. Oxford and IBH.

Hanke, I., & Neusser, P. (1984). Power ferrites for high frequencies. *IEEE Transactions on Magnetics, 20*(5), 1512–1514. https://doi.org/10.1109/TMAG.1984.1063243

Hoagland, A. S. (2003). History of magnetic disk storage based on perpendicular magnetic recording. *IEEE Transactions on Magnetics, 39*(4), 1871–1875. https://doi.org/10.1109/TMAG.2003.813786

Kabbur, S. M., Nadargi, D. Y., Kambale, R. C., Ghodake, U. R., & Suryavanshi, S. S. (2021). Microstructure and magnetic interactions of Co^{2+} substituted NiCuZn ferrites. *Journal of Magnetism and Magnetic Materials, 517*, 167376. https://doi.org/10.1016/j.jmmm.2020.167376

Khot, V., Salunkhe, A., Ruso, J., & Pawar, S. (2015). Improved magnetic induction heating of nanoferrites for hyperthermia applications: Correlation with colloidal stability and magneto-structural properties. *Journal of Magnetism and Magnetic Materials, 384*, 335–343. https://doi.org/10.1016/j.jmmm.2015.03.039

Knott, E. F., Schaeffer, J. F., & Tulley, M. T. (2004). *Radar cross section*. SciTech Publishing.

Kobayashi, K.-I., Morinaga, H., Araki, T., Naka, Y., & Oomura, T. (1992). Low-loss Ni-Zn-Cu ferrite for deflection yoke. *Journal of Magnetism and Magnetic Materials, 104*, 413–414. https://doi.org/10.1016/0304-8853(92)90857-K

Kobayashi, T. (2011). Cancer hyperthermia using magnetic nanoparticles. *Biotechnology Journal, 6*(11), 1342–1347. https://doi.org/10.1002/biot.201100045

Kohn, J., Eckart, D., & Cook, C. F. (1971). Crystallography of the hexagonal ferrites. *Science, 172*(3983), 519–525. https://doi.org/10.1126/science.172.3983.519

Kramer, B. A., Koulouridis, S., Chen, C.-C., & Volakis, J. L. (2006). A novel reflective surface for an UHF spiral antenna. *IEEE Antennas and Wireless Propagation Letters, 5*, 32–34. https://doi.org/10.1109/LAWP.2005.863613

Kugimiya, K. (1990). Ceramic materials for magnetic heads. *American Ceramic Society Bulletin, 69*(4), 696–702.

Kugimiya, K., Hirota, E., & Bando, Y. (1974). Magnetic heads made of a crystal-oriented spinel ferrite. *IEEE Transactions on Magnetics, 10*(3), 907–909. https://doi.org/10.1109/TMAG.1974.1058471

Lax, B., & Button, K. J. (1962). *Microwave ferrites and ferrimagnetics*. McGraw-Hill.

Liu, C., Zhang, Y., Zhang, Y., Fang, G., Zhao, X., Peng, K., & Zou, J. (2019). Multiple nature resonance behavior of $BaFe_xTiO_{19}$ controlled by Fe/Ba ratio and its regulation on microwave absorption properties. *Journal of Alloys and Compounds, 773*, 730–738. https://doi.org/10.1016/j.jallcom.2018.09.278

Maheri, H., Tsutsumi, M., & Kumagai, N. (1988). Experimental studies of magnetically scannable leaky-wave antennas having a corrugated ferrite slab/dielectric layer structure. *IEEE Transactions on Antennas and Propagation, 36*(7), 911–917. https://doi.org/10.1109/8.7195

Morrish, A. H. (2001). *The physical principles of magnetism*. IEEE Press.

O'Handley, R. C. (2000). *Modern magnetic materials: Principles and applications*. Wiley.

Özgür, Ü., Alivov, Y., & Morkoç, H. (2009a). Microwave ferrites, part 2: Passive components and electrical tuning. *Journal of Materials Science: Materials in Electronics, 20*(10), 911–952. https://doi.org/10.1007/s10854-009-9924-1

Özgür, Ü., Alivov, Y., & Morkoç, H. (2009b). Microwave ferrites, part 1: Fundamental properties. *Journal of Materials Science: Materials in Electronics, 20*(9), 789–834. https://doi.org/10.1007/s10854-009-9923-2

Pozar, D. M. (1992). Microstrip antennas. *Proceedings of the IEEE, 80*(1), 79–91. https://doi.org/10.1109/5.119568

Pullar, R. C. (2012). Hexagonal ferrites: A review of the synthesis, properties and applications of hexaferrite ceramics. *Progress in Materials Science, 57*(7), 1191–1334. https://doi.org/10.1016/j.pmatsci.2012.04.001

Salman, G., Bohan, A., & Jaed, G. (2017). Use of nano-magnetic material for removal of heavy metals from wastewater. *Engineering and Technology Journal, 35*(9), 903–908.

Smit, J. (1959). *Ferrites*. Philips Technical Laboratory.

Soohoo, R. (1968). Microwave ferrite materials and devices. *IEEE Transactions on Magnetics, 4*(2), 118–133. https://doi.org/10.1109/TMAG.1968.1066207

Spaldin, N. A. (2010). *Magnetic materials: Fundamentals and applications*. Cambridge University Press.

Valenzuela, R. (2005). *Magnetic ceramics*. Cambridge University Press.

Valenzuela, R. (2012). Novel applications of ferrites. *Physics Research International, 2012*, 1–9. Article ID 591839.

Victora, R. H., & Shen, X. (2005). Composite media for perpendicular magnetic recording. *IEEE Transactions on Magnetics, 41*(2), 537–542. https://doi.org/10.1109/TMAG.2004.838075

White, R. L., Newt, R., & Pease, R. F. W. (1997). Patterned media: A viable route to 50 Gbit/in/sup 2/and up for magnetic recording. *IEEE Transactions on Magnetics, 33*(1), 990–995. https://doi.org/10.1109/20.560144

Yang, H. Y. D. (1996). Characteristics of switchable ferrite microstrip antennas. *IEEE Transactions on Antennas and Propagation, 44*(8), 1127–1132. https://doi.org/10.1109/8.511821

Zaitsev, E. F., Yavon, Y. P., Komarov, Y. A., Guskov, A. B., & Kanivets, A. Y. (1994). MM-wave integrated phased arrays with ferrite control. *IEEE Transactions on Antennas and Propagation, 42*(3), 304–310. https://doi.org/10.1109/8.280716

CHAPTER

Ferrite characterization techniques 2

2.1 Introduction

As a fast-emerging and growing class of magnetic materials, ferrites have received tremendous attention and have numerous practical uses. So far, many characterization techniques have been employed to determine the phase, crystal structure, homogeneity, morphology, size distribution, band structure, and elemental composition of magnetic ferrites. In this regard, X-ray diffraction (XRD) technique and electron microscopes integrated with energy-dispersive spectroscopy can perform phase identification and elemental analysis, respectively. Physical features of ferrites, including thermal behavior, are also necessary to determine the temperature dependence of structural properties. The site occupancy of cations substituted in ferrites needs to be precisely determined to calculate the magnetic moment per formula unit and discuss the magnetic data obtained. The variation of magnetization and magnetic flux density of ferrites as a function of the magnetic field can also be measured using several reliable and reproducible techniques. For ferrite absorbers, the microwave attenuation constant and reflection loss value can be measured by different techniques, each with its strengths and limitations. In turn, this complicates the choice of the most appropriate technique. Therefore, combinatory characterization methods have been proposed to address the need for efficient and reliable analysis of ferrite properties. The main objective of this chapter is to introduce conventional experimental characterization techniques and summarize their merits and shortcomings. The operating principles of these techniques designated for characterizing ferrites in bulk, thin-film, powder, and composite forms are also discussed.

2.2 X-ray diffraction

XRD analysis was first used by W. H. Bragg and his son W. L. Bragg to study structural properties of crystalline materials. As a rapid and powerful nondestructive method, XRD can provide researchers with a wide range of information on phase composition and structural parameters such as crystallinity, lattice constant, preferred crystal orientation (texture), ordered-disordered transformation, average crystallite size, peak density, and residual stress (Lee, 2019; Zolotoyabko, 2014).

Magnetic Ferrites and Related Nanocomposites. https://doi.org/10.1016/B978-0-12-824014-4.00002-0

2.2.1 Fundamental principles of X-ray diffraction

The foundation of XRD is based on the constructive diffraction between monochromatic X-rays and a crystalline structure. The constructive diffraction occurs only when Bragg's law is satisfied:

$$n\lambda = 2d\sin\theta,\ n\text{: an integer} \tag{2.1}$$

Bragg's law indicates the general relationship between the incident X-ray wavelength (λ), the angle between the incident ray and crystalline planes (θ), and the distance between atomic planes in a crystal (d).

Since the X-ray wavelength is constant, and the angles at which the diffraction occurs are also known in the experiments, one can calculate the distance between atomic planes to determine phase composition or crystal structure (Lee, 2019; Sardela, 2014).

The relationship between the d-spacing and lattice parameter (a) of a cubic structure is expressed as follows:

$$d_{hkl} = \frac{a}{\sqrt{h^2 + k^2 + l^2}} \tag{2.2}$$

The Miller indices (hkl) indicate a series of parallel planes in a crystalline structure with d_{hkl} spacing. By combining Eqs. (2.1) and (2.2), the relationship between diffraction data and crystal parameters for a cubic crystal structure is obtained as follows:

$$\sin^2\theta = \frac{\lambda^2}{4a^2}\left(h^2 + k^2 + l^2\right) \tag{2.3}$$

For a given λ, the relationship between θ and the Miller indices of most crystalline materials has been determined and published by the International Center for Diffraction Data (Leng, 2013). If a unit cell has 1, 2, 3, …, N atoms with coordinates of $u_1v_1w_1$, $u_2v_2w_2$, …, $u_nv_nw_n$, the structure factor of the (hkl) plane (F_{hkl}) is given by

$$F_{hkl} = \sum_{n=1}^{N} f_n e^{2\pi i(hu_n + kv_n + lw_n)} \tag{2.4}$$

where F_{hkl} is a complex number representing the amplitude and phase of the resultant wave, and f is the atomic form factor. For example, consider a body-centered cubic (BCC) unit cell having two atoms located at (0,0,0) and (1/2,1/2,1/2) planes. Based on Eq. (2.4), F_{hkl} is expressed as

$$F_{hkl} = fe^{2\pi i(0)} + fe^{2\pi i\left(\frac{h}{2}+\frac{k}{2}+\frac{l}{2}\right)} = f\left[1 + e^{\pi i(h+k+l)}\right] \tag{2.5}$$

$$\text{If } (h+k+l) \text{is odd: } F_{hkl} = 0$$

$$\text{If } (h+k+l) \text{is even: } F_{hkl} = 2f$$

Therefore, the BCC lattice planes whose Miller indices sum to an even number are reflected. Conversely, reflections whose Miller indices sum to an odd number are forbidden.

A face-centered cubic (FCC) unit cell has the same four atoms located at (0,0,0), (1/2,1/2,0), (1/2,0,1/2), and (0,1/2,1/2) planes. In this case, F_{hkl} is as follows (Cullity & Stock, 2014; Leng, 2013):

$$F_{hkl} = fe^{2\pi i(0)} + fe^{2\pi i\left(\frac{h}{2}+\frac{k}{2}\right)} + fe^{2\pi i\left(\frac{k}{2}+\frac{l}{2}\right)} + fe^{2\pi i\left(\frac{h}{2}+\frac{l}{2}\right)} \tag{2.6}$$

$$F_{hkl} = f\left[1 + e^{\pi i(h+k)} + e^{\pi i(h+l)} + e^{\pi i(k+l)}\right] \tag{2.7}$$

If (h,k,l) indices are unmixed, i.e., all are even or all are odd, the three sets $(h + k)$, $(h + l)$, and $(k + l)$ will be even integers, and each exponential term in Eq. (2.7) will equal 1. For unmixed indices, F_{hkl} is obtained as follows:

$$F_{hkl} = 4f$$

If (h,k,l) indices are mixed, the sum of the three exponential terms in Eq. (2.7) equals -1. For mixed indices, F_{hkl} is zero. The allowed and forbidden diffraction conditions for different Bravais lattices are summarized in Table 2.1.

2.2.2 X-ray diffraction components and performance

An X-ray diffractometer basically detects diffracted rays from a sample and records the diffraction intensity for a range of the diffraction angle (2θ). A simplified schematic of an X-ray diffractometer system is presented in Fig. 2.1, comprising five main parts: the X-ray tube, rotation equipment, sample holder, X-ray detector, and primary and secondary optics. The instrument consists of a diffraction circle with the X-ray tube and detector on its perimeter and a sample at the center. Based on the sample rotation and the arms holding the tube and the detector, two geometries can be adopted for the XRD instrument: θ–2θ and θ–θ. In the θ–2θ geometry,

Table 2.1 The allowed and forbidden diffraction conditions for different Bravais lattices.

Bravais lattice	Allowed diffraction	Forbidden diffraction
Simple cubic	All	None
Base-centered cubic	h and k unmixed[a]	h and k mixed[a]
Body-centered cubic	$(h + k + l)$ even	$(h + k + l)$ odd
Face-centered cubic	(hkl) unmixed	(hkl) mixed

[a] *Applies to a cell with (001) face-centered.*

From Leng, Y. (2013). Materials characterization: introduction to microscopic and spectroscopic methods. *Wiley. https://books.google.com/books?id=URdwAAAAQBAJ.*

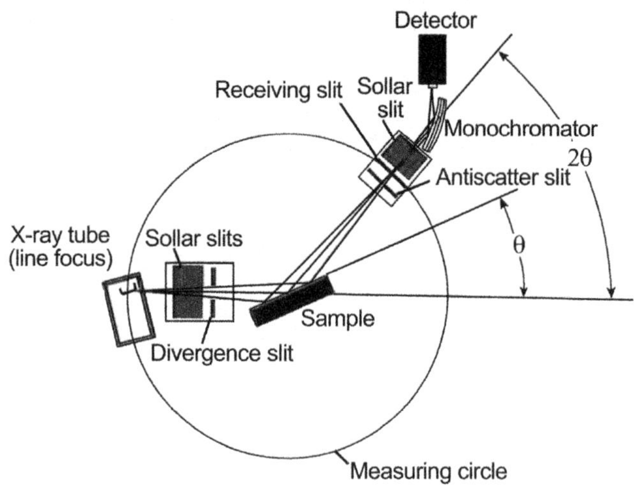

FIGURE 2.1

A simplified schematic of an X-ray diffractometer.

From Leng, Y. (2013). Materials characterization: introduction to microscopic and spectroscopic methods. *Wiley. https://books.google.com/books?id=URdwAAAAQBAJ.*

the sample rotates at an angle of θ, while the X-ray tube is fixed and the detector is located at an angle of 2θ. In the $\theta-\theta$ geometry, the sample is fixed, and both the tube and the detector rotate at an angle of θ.

X-ray rocking curve analysis is a common method for evaluating the texture of hard ferrite thin films used in high-density recording media. For this purpose, a $\omega-2\theta$ scan is carried out to identify the Bragg peak, followed by scanning the rocking curve (the ω scan). With the fixed detector at the $2\theta_B$ position (θ_B: Bragg angle), the ferrite thin film is tilted up to ω about an axis normal to the incident beam. In this way, the intensity of the diffracted beam versus plots would show typical Gaussian profiles. Moreover, the full width at the half-maximum value obtained from the rocking curve is employed to identify the preferred orientation of thin films.

X-rays are produced in an evacuated X-ray tube by bombarding the target surface with accelerated electrons. These are emitted from a heated filament (cathode) and accelerated by applying a high voltage between the filament and a target (anode). The produced X-ray spectrum contains continuous and characteristic radiation. Notably, Kα and Kβ are the most common and strongest characteristic X-rays, whose emissions are due to the electron transition from the L shell to the K shell, and the M shell to the K shell, respectively. Kα radiation consists of Kα_1 and Kα_2. The difference in wavelength between Kα_1 and Kα_2 is so small that they are not considered separate radiation. Nevertheless, in high-resolution XRD (HRXRD), Kα_2 radiation is separated from Kα_1, in turn increasing the spectrum resolution. The HRXRD technique is widely used to investigate the microstructure of epitaxially grown thin layers. In XRD, the monochromatic radiation must come from the characteristic X-rays. To this end, $K\beta$ radiation is removed from the spectrum by special filters. The resulting monochromatic X-rays are then collimated and concentrated by the primary optics and directed to the sample. The diffracted X-rays are passed through the secondary optics to converge and remove the background radiation in X-rays. Finally, the diffracted X-rays are detected by the detector, and their intensity versus 2θ is plotted by the software (Leng, 2013).

2.2.3 Applications

The most important applications of XRD are listed below:

- phase identification;
- determining structural parameters such as lattice constant and crystallite size;
- determining the degree of crystallinity;
- Texture analysis: Anisotropic distribution of grain orientations in a material is the preferred orientation or texture and greatly affects mechanical, optical, electric, and magnetic properties. By comparing the relative intensity (or the peak area) of various (*hkl*) orientations and the corresponding relative intensities from the nontextured sample of the same material, it is possible to achieve anisotropic distribution of crystalline planes;
- Residual stress: Due to its nondestructive nature, XRD is one of the most widely used techniques for studying residual stress. Determined by evaluating and analyzing XRD patterns, residual stress can influence the interplanar spacing, diffraction angle, and width of the peaks;
- Expansion tensor: Complementary equipment can be attached to the XRD instrument to determine the characteristics of the sample as a function of temperature or pressure (Leng, 2013).

2.2.4 Strengths and limitations

The strengths of XRD analysis include the following:

- a nondestructive, rapid, and powerful tool for identifying unknown materials;
- easy sample preparation;
- relatively simple data interpretation;
- ease of accessibility and availability of the XRD instrument.

The limitations of XRD analysis can be summarized as follows:

- the need to access a standard reference file for mineral compounds (*d*-spacing and *hkl*);
- low detection accuracy of $\sim 2\%$ for mixed materials;
- possibility of overlapping some peaks with others;
- the complexity of indexing the patterns of nonisometric crystal systems to determine a unit cell.

2.3 Scanning electron microscopy

Restrictions on the magnification and resolution of optical microscopes in the microstructural characterization of materials have led to the development of electron microscopes with high resolutions and magnifications of up to one million times. Scanning electron microscopy (SEM) is one of the most popular types of electron microscopic analyses and is widely used in various fields nowadays, allowing for the creation of high-magnification images in microscopic-scale characterizations of size, shape, composition, and crystallography, among other physical and chemical properties. The basic principles of SEM were first established by M. Knoll (1935) and M. Ardenne (1938). The first commercially available SEM instrument resulted from extensive development in 1965 by Professor C. Oatley and his colleagues at the University of Cambridge (Goldstein et al., 2017; Stokes, 2008).

The operating mechanism of SEM is based on sample bombardment with an electron beam (e-beam) and the detection of electrons and photons emitted from the sample and converted into a signal. The movement of the beam on the sample provides a set of signals, enabling the microscope to instantly display a reciprocal image of the sample surface on the screen. Accordingly, SEM performance is based on the following three principles: (a) interaction of the e-beam with the sample; (b) the possibility of producing and controlling the e-beam in electric and magnetic fields; and (c) the possibility of detecting the rays emitted from the sample following the interaction of the incoming e-beam with the sample.

2.3.1 The basic principles of scanning electron microscopy

A schematic representation of the processes occurring during the impingement of the e-beam on a sample is shown in Fig. 2.2. In the SEM instrument, the e-beam entering the sample usually contains electrons with energies of 1–50 keV. Generally, two main types of interactions between the e-beam and the sample can be considered: elastic scattering and inelastic scattering. The former causes the formation of backscattered electrons (BSEs), and the latter creates a variety of signals such as secondary electrons (SEs), auger electrons, photons (i.e., characteristic X-rays and continuum X-rays), and visible light (cathodoluminescence). These signals are obtained from the specific diffusion volume in the sample and can be used to study its various properties (e.g., surface topography, crystallography, composition, and spectrum). In fact, the variety of interactions between the e-beam and the sample may lead to different phenomena with potentially valuable applications (Hawkes & Spence, 2019; Stokes, 2008).

SEs and BSEs are commonly used for imaging samples. SEs originate from sample atoms by inelastic scattering of the beam electrons along their pathways in the sample surface regions (1–10 nm), providing a high-value signal for illustrating surface morphology and topography. The origin of BSEs is the multiple scattering of beam electrons within the sample. They come from deeper regions of the sample (typically ~2 μm), giving rise to information on the composition of multiphase

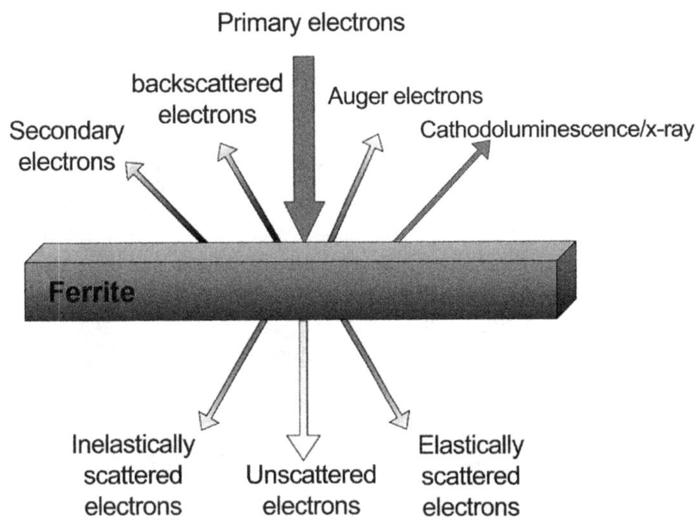

FIGURE 2.2

Schematic drawing of the processes when an electron beam impinges on a sample.

samples. For this reason, BSE images show high sensitivity to different atomic numbers of materials. In other words, a material with a higher atomic number has a brighter contrast in the corresponding BSE image. Characteristic X-rays are also used for elemental analysis (Egerton, 2005; Goldstein et al., 2017).

2.3.2 Scanning electron microscopy components and performance

In general, the SEM instrument comprises three major parts: the electron column, sample chamber, and computer/electronic control system. A simplified schematic of a conventional SEM instrument is presented in Fig. 2.3. The electron column has a long cylindrical body above the sample chamber, consisting of the electron gun, electromagnetic lenses (condenser and objective lens), scan coils, stigmator, and apertures. The main function of the electromagnetic lenses in the optical column of the microscope is to reduce the diameter of the e-beam and increase its density. After forming a parallel beam with an appropriate diameter, the scanner system uses two coils to move the e-beam on the sample surface. Therefore, the main performance of the electron column is to form a fine electron probe, scanning the desired area on the sample. To prevent the emitted electrons from scattering, the electron column is kept under a vacuum at all times (Ul-Hamid, 2018).

Electronic guns are divided into two categories based on their operating mechanisms: thermionic and field emission. In a thermionic system, the gun is heated at very high temperatures; a given percentage of electrons can reach a certain level of energy, thus leaving the surface of the gun. On the other hand, the tunneling

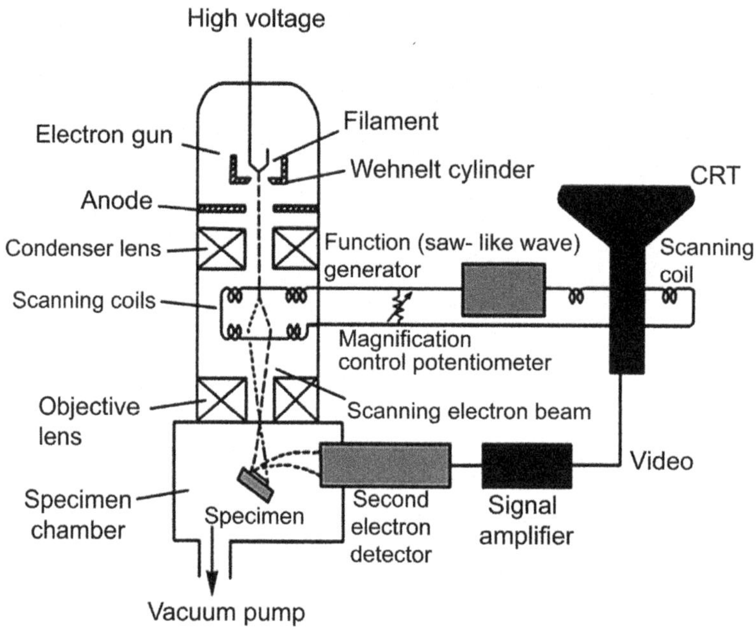

FIGURE 2.3

Schematic diagram of the basic components of a conventional scanning electron microscope.

From Kogure, T., Bergaya, F., & Lagaly, G. (2013). Chapter 2.9 – Electron microscopy. In Handbook of clay science *(Vol. 5, pp. 275–317). Elsevier. https://doi.org/10.1016/B978-0-08-098259-5.00011-1.*

phenomenon is used to produce electrons in a field emission gun. Currently, the commercial types of electron guns are tungsten filament, lanthanum hexaboride (LaB_6) emitter, Schottky field emission, and cold field emission. An ideal electron gun produces a stable e-beam with a fine source size, high brightness, and small energy dissipation. The production capacity of field emission filaments is much higher than that of thermionic ones; they are commonly used to achieve high magnification and resolution in modern SEMs (Dehm et al., 2012; Ul-Hamid, 2018).

The sample chamber is located at the bottom of the electron column and kept under a vacuum during SEM operation. It contains important components such as the sample stage, sample holder, CCD camera, and several detectors. After the interaction of the e-beam with the sample, a variety of functional signals such as SEs, BSEs, and characteristic X-rays are generated. The detectors specifically collect and convert the resultant signals emanating from the sample into an electrical pulse for processing; these are displayed as secondary and backscattered images or energy-dispersive X-ray spectra. The most commonly employed detector, the

Everhart–Thornley detector, can detect both SEs and BSEs. However, it is mostly used to generate SE images of the specimen surface. Other detectors such as solid-state, Robinson, and through-the-lens are used for detecting BSEs. In addition to detecting the number of BSEs, one can obtain their diffraction pattern using an electron backscattered diffraction (EBSD) detector. This is a powerful instrument for crystallographic analysis, determining grain orientations, and phase identification. The origin of EBSD patterns is similar to that of Kikuchi lines in transmission electron microscopy (TEM) (Hawkes & Spence, 2019; Ul-Hamid, 2018).

The computer/electronic control system consists of electric power supplies for the electron gun and the electromagnetic lenses, the scan generator, and electronic amplifiers to amplify various signals. In addition, the control system consists of image display/data output devices and special equipment to control the e-beam and selected signals. Other important requirements of the SEM include the vacuum system, water chiller, heater, and antivibration table. The SEM instrument also requires a high vacuum environment. Therefore, after placing the sample in the chamber, the microscope column and sample chamber are evacuated using combined prevacuum rotary and high-vacuum turbomolecular pumps. The main reason for a vacuum environment in electron microscopes is to prevent beam electrons, SEs, and BSEs from unwanted scattering following their collisions with atoms and molecules in the gas environment.

2.3.3 Energy- and wavelength-dispersive spectroscopy

By measuring the amount of X-ray energy released during the electron bombardment, it is possible to determine the type of atom in a sample, thereby forming the basis of the qualitative and quantitative analyses using energy-dispersive X-ray spectroscopy (EDS). The characteristic X-rays produced by beam–sample interactions also have a unique wavelength for the elements in the sample, forming the basis of wavelength-dispersive X-ray spectroscopy (WDS).

In SEM, the characteristic X-rays generated can be detected using either EDS or WDS, thus determining the type and number of elements in the sample. These spectroscopic techniques can also provide the element line scan and elemental distribution map. EDS and WDS analyses are powerful tools for quickly and nondestructively determining the microchemical components of a sample. In EDS and WDS, all elements with an atomic number greater than 5 can be detected. In other words, low atomic number elements (i.e., H, He, Li, and Be) cannot be analyzed using these techniques. A WDS system is beneficial for analyzing light elements (e.g., B, O, N, C, etc.). The WDS system has advantages over EDS such as better resolution at low concentrations, improved trace element detection limit, less clutter in the background, and better separation of overlapping peaks. However, the WDS system is more time-consuming and expensive than the EDS system (Ul-Hamid, 2018).

2.3.4 Applications

SEM is a very powerful tool for research investigations in a wide range of fields such as materials science, magnetic materials, nanodevices, semiconductor design, microelectronics, life sciences, geology, mineralogy, archeology, and forensic science. In materials science, SEM is used for a wide range of materials, including metals, ceramics, polymers, glasses, plastics, alloys, semiconductors, and composite powders, coatings, corrosion products, etc. The most common use of SEM is to observe and image surface details at micrometer and nanometer scales. The most important information obtained from SEM and related techniques are the following:

- morphology: size and shape of particles; size, shape, and distribution of precipitates, defects, voids, and cracks inside the sample; grain size and shape; microroughness; and fractured cross sections (using the SE mode)
- compositional contrast; grain boundaries; magnetic domains; and interfaces (using the BSE mode)
- elemental composition: elemental identification and quantification; line scans; and X-ray maps (using EDS or WDS)
- crystallographic analysis: grain orientations; characterization of grain boundaries; and phase identification (using EBSD)
- others: fabrication of nanodevices and printing of circuit boards (using SEM combined with electron and ion beams); and design validation of microelectronic and semiconductor devices (using SE or BSE modes).

2.3.5 Strengths and limitations

In the study of solids, no research devices can compete with SEM instruments for application scope and useful output information. In all research fields involving the characterization of solid materials, SEM is an important and critical tool. However, it has its strengths and limitations (Hawkes & Spence, 2019; Ul-Hamid, 2018).

SEM strengths are the following:

- ability to investigate a wide range of materials;
- ability to provide three-dimensional (3-D) detailed images (due to the large SEM depth of field);
- relatively straightforward interpretation of the images;
- ability to achieve high spatial resolution (1 nm) in modern models;
- capability of microchemical analysis (from the element beryllium to uranium);
- relatively easy and fast preparation of the sample (nonconductive samples such as ceramics should be coated, making them electrically conductive);
- nondestructive analysis;
- ability to perform several modes of imaging, spectroscopy, and diffraction analysis.

SEM limitations include the following:

- the need for a skilled and trained operator to work with the instrument;
- limited type (solid) and size (typically 1–150 mm) of the sample;
- the need for a vacuum environment when analyzing the sample;
- limited chemical analysis of elements with atomic numbers below 5 (H, He, Li, and Be);
- low sensitivity for elemental analysis with EDS compared with wet analytical methods (especially light elements);
- the need to use conductive coating for nonconductive samples;
- restrictions on imaging wet and oily samples (low vacuum and environmental SEMs can be used to overcome this limitation);
- relatively expensive to purchase and run (Dehm et al., 2012).

2.4 Transmission electron microscopy

TEM can be used to study the structural symmetry, orientation, and defects of crystals. The operating mechanism of TEM instruments involves an e-beam that can pass through the sample, which leads to a reduction in its energy. This is displayed on a phosphor plate or sent to a computer for processing. By examining and analyzing the images taken by TEM, desired information about the material structure can be obtained. The resolution of TEM ranges between 0.1 and 1 nm, and its magnification can be increased up to 1,500,000 times by increasing the electron accelerating voltage. Depending on the density of the material, electrons may pass through parts of the body and collide with the fluorescence plate, creating a blurred sample image (Reimer, 2013).

The main components of the TEM instrument are the electron gun (accelerating electrons in the range of 40–200 kV), lenses, sample chamber, illustration system, and vacuum pumps (see Fig. 2.4). Electromagnetic lenses generate a magnetic field parallel to the motion of electrons, thereby avoiding their collision with each other. As a result, parallel e-beams with a narrower and smaller spiral path enter the lens, converging at one point.

Compared with SEM, sample preparation of TEM is much more challenging. The sample must be thinned to a thickness of several tens of nanometers to observe its internal structure. Also, it must be dry and have low dimensions to be held by the sample holders. TEM sample preparation methods are divided into two general categories. The samples are thinned up to a certain thickness in the first category using mechanical or chemical processes. The samples are cut to a minimal thickness along crystalline planes in the second category.

Two types of images, bright-field and dark-field, can be formed in TEM. In bright-field imaging, some electrons that hit the sample can pass through it, whereas

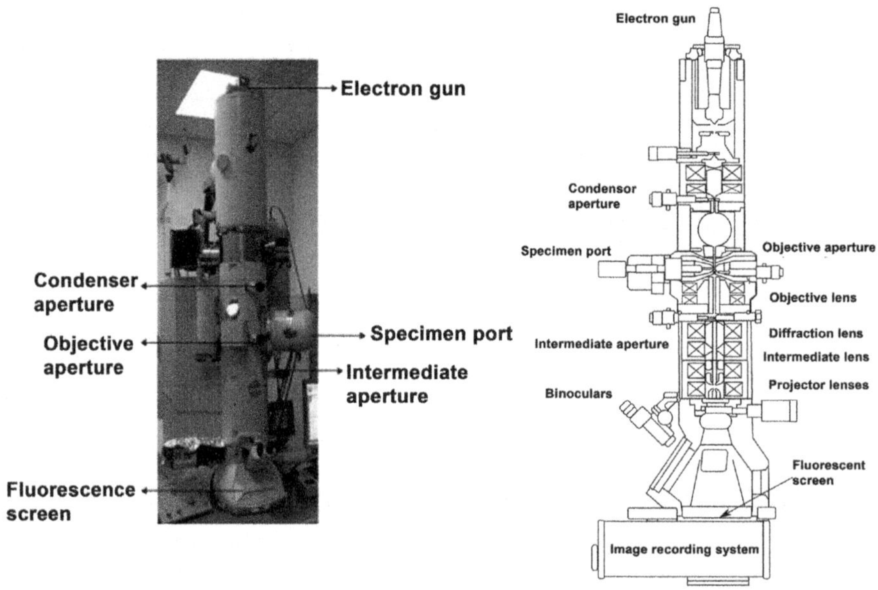

FIGURE 2.4

The key components of a transmission electron microscope.

From wikipedia.org. https://en.wikipedia.org/wiki/Transmission_electron_microscopy#/media/File:Scheme_TEM_en.svg.

others are reflected in elastic or nonelastic modes. Bright-field images are usually used to evaluate the crystal structure, grain morphology, and crystal defects. For example, Fig. 2.5 shows a TEM bright-field micrograph obtained from ferrite nanoparticles. The lattice fringe and particle size distribution can be seen in the micrograph.

By changing the size of the objective lens apertures, it is possible to transmit the diffracted rays and prevent the passage of the main rays, resulting in an image with a dark background. This low-resolution image reveals the areas of the sample through which the e-beams have passed, providing information about contrast distribution while identifying crystal defects (e.g., interface structure) (Zipare et al., 2018).

2.4.1 Selected area diffraction patterns

Since the wavelength of high-energy electrons is a few thousandths of a nanometer (thus being about 100 times larger than the distance between atoms in a solid), some electrons are diffracted at certain angles. Therefore, the crystal structure of the sample can be determined from diffracted high-energy electrons. In this case, the images obtained from the TEM instrument are seen as a set of points, each of which represents a suitable diffraction pattern of the sample crystal structure. The selected area

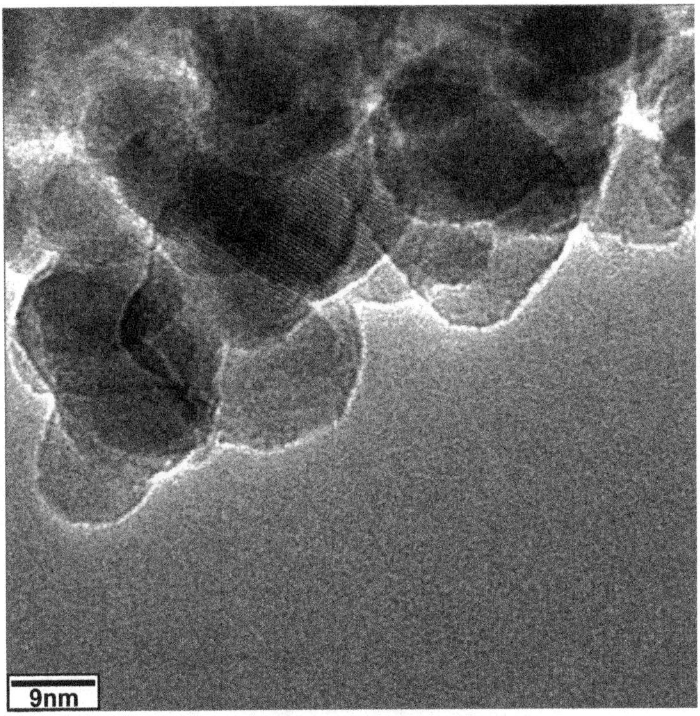

FIGURE 2.5

TEM bright-field micrograph of typical ferrite nanoparticles.

diffraction (SAD) pattern is thus a great way to identify crystal structure and defects; it is similar to the XRD. Electron diffraction patterns of gases, liquids, and amorphous solids appear as scattered halos. In the case of dense phases, there may be a slight atomic order in the form of crystalline arrays in areas with dimensions of 10–20 Å, so that part of the SAD pattern appears as a loop. Individual rings may be so wide that they overlap, making the appearance of the pattern nearly the same as the random order of the atoms. Circular diffraction patterns occur when many crystals are placed in different directions relative to the irradiated e-beam, coinciding with the diffracted electrons and giving rise to the crystallography of unknown materials.

The *d*-spacing of a specimen is calculated by the following expression (Biosciencenotes.com):

$$Rd = \lambda L \tag{2.8}$$

where R, d, L, and λ are the SAD ring radius, lattice distance, camera length, and electron wavelength, respectively.

For the diffraction of electrons, the atomic planes of a crystal must be placed almost parallel to the e-beam. Therefore, if several sets of crystal surfaces are oriented parallel to the e-beam, a diffraction pattern consisting of a regular arrangement of light points will form. The corresponding diffraction pattern would be complicated for ferrite nanoparticles containing several crystals with a different orientation. In this case, since the points of light are very close together, continuously connected rings are formed in the patterns, as shown in Fig. 2.6.

Over the past 40 years, various methods have been developed to overcome limitations of SAD analysis. Among them, the convergent beam electron diffraction (CBED) method has proved very simple and comprehensive. CBED patterns are formed by a small e-beam (below 100 nm in size) that must be created with a convergent half-angle of α. While a large α angle exhibits a blurred CBED spot, medium and small α angles have apparent diffraction spots.

High-resolution TEM (HRTEM) and CBED analyses of $NiFe_2O_4$ ferrite nanoparticles synthesized by a liquid–solid solution-assisted hydrothermal route are shown in Fig. 2.7. The SAD pattern obtained from the $NiFe_2O_4$ nanoparticles reveals an FCC structure. Furthermore, the XRD pattern of the $NiFe_2O_4$ nanoparticles

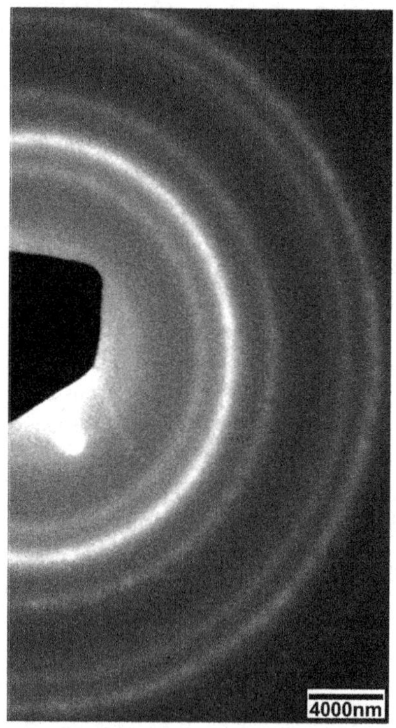

FIGURE 2.6

Selected area diffraction of typical ferrite nanoparticles.

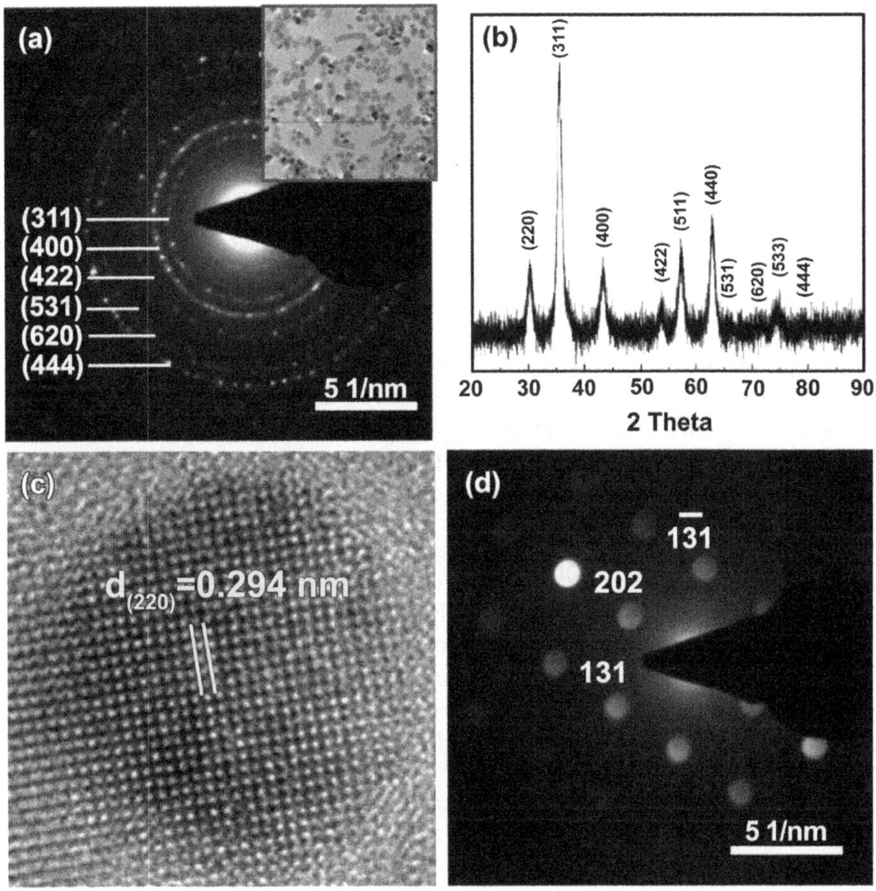

FIGURE 2.7

(A) Selected area diffraction (SAD) pattern obtained from $NiFe_2O_4$ nanoparticles. The inset shows the selected $NiFe_2O_4$ nanoparticles for the SAD analysis. (B) X-ray diffraction spectrum of $NiFe_2O_4$ nanoparticles. (C) High-resolution transmission electron microscopy image of a single $NiFe_2O_4$ particle. (D) Convergent beam electron diffraction pattern of a single $NiFe_2O_4$ nanoparticle.

From Li, X., Tan, G., Chen, W., Zhou, B., Xue, D., Peng, Y., Li, F., & Mellors, N. J. (2012). Nanostructural and magnetic studies of virtually monodispersed $NiFe_2O_4$ nanocrystals synthesized by a liquid–solid-solution assisted hydrothermal route. Journal of Nanoparticle Research, *14(3), 751. https://doi.org/10.1007/s11051-012-0751-6.*

confirms the crystal characterization obtained by the SAD pattern while also predicting a crystallite size of 9.5 nm and symmetry of Fd3m, according to JCPDS 10–0325. The bright-field HRTEM image of individual ferrite nanoparticles depicted in Fig. 2.7C indicates the formation of a single crystal. The CBED method with a spot size configuration of 0.5 nm was also used to investigate the crystal structure of a single $NiFe_2O_4$ nanoparticle. The converged beam was incident on the analyzed particle, reflecting a <101> oriented single crystal spinel ferrite structure indexed to (131), (202), and (131) planes.

An essential feature of TEM analysis is the ability to study grains and grain boundaries. For example, the microstructural development of cobalt ferrite has been investigated by Kim and colleagues using grain boundary observation with TEM (Kim et al., 2013), clarifying abnormal grain growth in $CoFe_2O_4$. Fig. 2.8 presents a TEM micrograph of a faceted grain boundary obtained from $CoFe_2O_4$. Marked by the arrow, one can see the presence of irregular curvature in the boundary. A faceted boundary is also observed in the inset of Fig. 2.8A; it is terminated with a {100}-plane in the upper grain. HRTEM images depicted in Fig. 2.8B show the occurrence of long-range faceted boundary in $CoFe_2O_4$. The left and right insets of Fig. 2.8B show typical faceted and singular {111}-terminated plane grain boundaries, respectively. Consequently, the grain boundary structure with abnormal grain growth is energetically anisotropic due to the faceted grain boundaries in $CoFe_2O_4$ systems.

2.4.2 Kikuchi diffraction lines

In crystalline samples, some nonelastically scattered electrons may be rediffracted elastically, causing Kikuchi lines to form. As the thickness of the crystal increases, the background of the scattering pattern tends to become nonelastic. The intensity of electrons scattered nonstatically depends on the scattering angle, and the maximum scattering angle is in the direct axis. The dominant role of Kikuchi lines and bands is mostly seen in the electronic distribution patterns of single crystals with relatively large thicknesses. While increasing sample thickness increases Kikuchi line intensity, it decreases the intensity of scattered light points. Kikuchi lines are placed symmetrically on either side of the scattered pages, moving in the scattering pattern width due to sample rotation. However, the location of the scattering points remains constant. Kikuchi lines are used to determine the grain direction on both sides of the subgrain boundary. Fig. 2.9 shows the Kikuchi patterns of a nickel–zinc ferrite [(Ni, Zn)Fe_2O_4] sample with a polycrystalline structure indexed with only four bands.

2.4.3 Electron energy-loss spectroscopy

Electron energy-loss spectroscopy (EELS) analysis is a new method for analyzing the microscopic properties of low-thickness specimens. It can be employed for high-precision elemental analysis, and imaging and mapping of surface elements. The resolution and detection power of EELS are about 10 times higher than those of the EDS analysis. Qualitative analysis of all elements with atomic numbers greater than 17 by the ionization edges of the inner layer of the elements is also the power of this spectroscopic technique. On the contrary, the EELS sample preparation is very time-consuming as the size of specimens must be very small. Moreover, spatial resolution is limited by the TEM instrument.

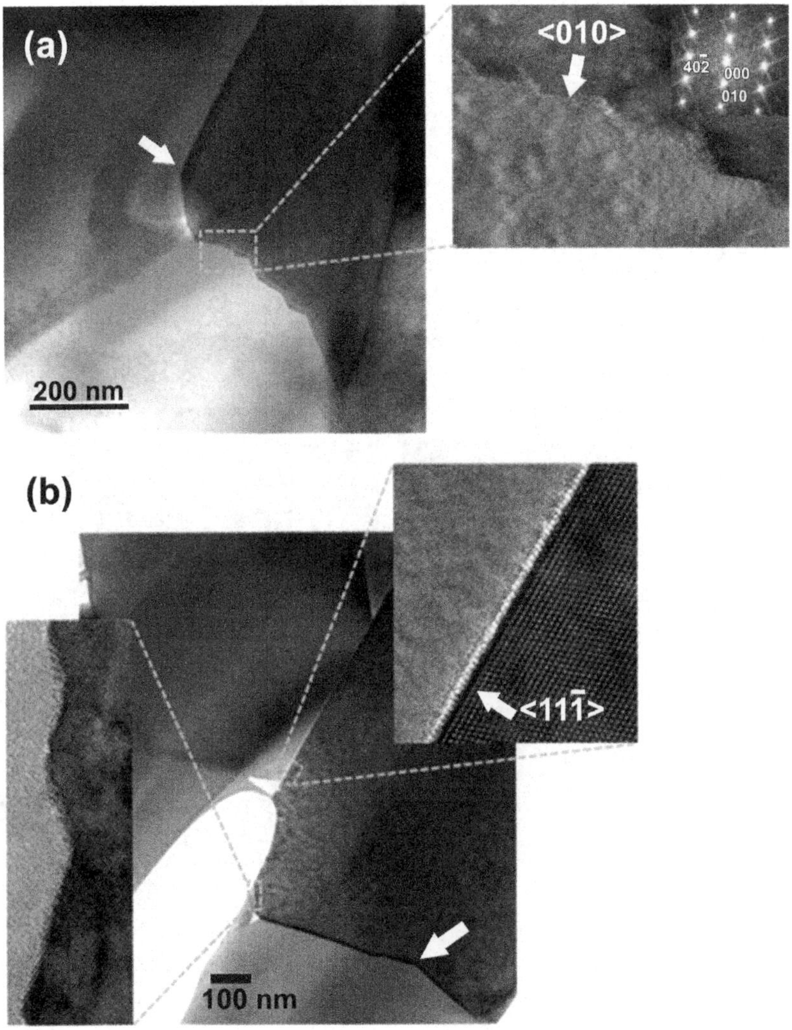

FIGURE 2.8

High-resolution transmission electron microscopy images of faceted grain boundaries in $CoFe_2O_4$: (A) projection of faceted grain boundaries marked by the *arrow* and shown in the magnified inset and (B) projection of singular and faceted grain boundaries.

From Kim, G.-Y., Jeon, J.-H., Kim, M.-H., Suvorov, D., & Choi, S.-Y. (2013). Microstructural development of cobalt ferrite ceramics and its influence on magnetic properties. Metals and Materials International, 19*(6), 1209–1213. https://doi.org/10.1007/s12540-013-6009-8.*

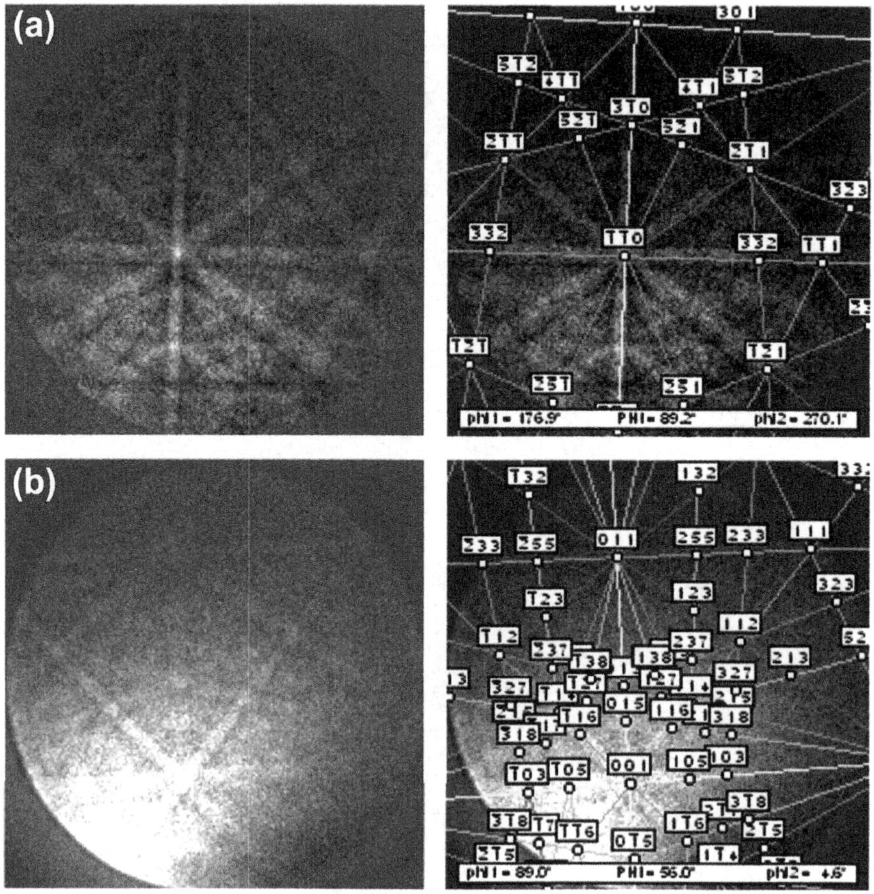

FIGURE 2.9

Kikuchi patterns and their indexation obtained from nickel-zinc ferrite $(Ni{-}Zn)Fe_2O_4$.

From Koblischka, M. R., & Koblischka-Veneva, A. (2013). Applications of the electron backscatter diffraction technique to ceramic materials. Phase Transitions, 86*(7), 651–660. https://doi.org/10.1080/01411594.2012.726729.*

The EELS analysis is based on nonelastic collisions occurring during the interaction between the electron and sample surface. Some electrons interacting nonelastically lose part of their energy without a significant change in their path.

The amount of energy loss of electrons during the nonelastic scattering is a function of the operation applied. These electrons are called loss electrons and can be used for imaging and chemical microanalysis. Fig. 2.10 shows oxygen K-edge (onset ~532 eV) and iron-L edge (onset ~708 eV) obtained from iron oxide nanoparticles. The iron L_3 to L_2 ratio of transitions in the core–shell structure is

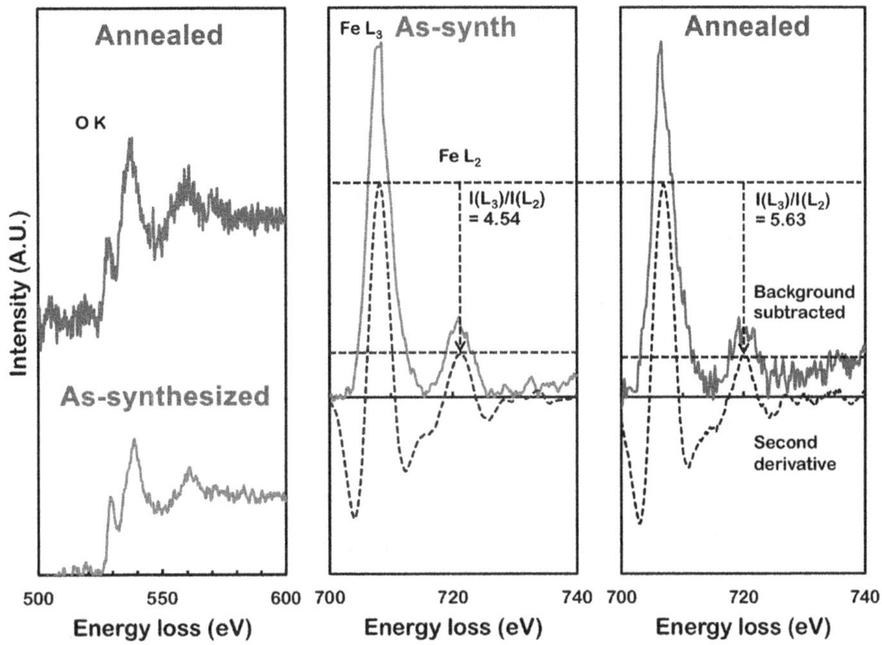

FIGURE 2.10

Electron energy-loss spectroscopy spectra of iron oxide nanoparticles synthesized from FeOOH before and after annealing at 100°C for 12 h.

From Hufschmid, R., Arami, H., Ferguson, R. M., Gonzales, M., Teeman, E., Brush, L. N., Browning, N. D., & Krishnan, K. M. (2015). Synthesis of phase-pure and monodisperse iron oxide nanoparticles by thermal decomposition. Nanoscale, *7(25), 11142–11154. https://doi.org/10.1039/C5NR01651G.*

proportional to the iron valence, leading to a phase transformation from magnetite to maghemite during the annealing process at 100°C for 12 h.

2.5 Atomic force microscopy

The atomic force microscopy (AFM) instrument is at the forefront of equipment for imaging the topography and surface roughness of nanomagnets, especially in thin-film configurations. It has the power to provide a true 3-D surface profile with atomic resolution. The AFM instrument comprises a cantilever with a sharp tip (made of silicon or silicon nitride) at its end that is employed to scan the surface of thin films. By scanning the surface with the tip at a constant height, forces applied between the tip and the sample lead to a deflection at the cantilever, which can be measured by a laser spot based on Hooke's law. The topographic map of thin films can be obtained by detecting the cantilever position in constant-height or constant-force modes. The specimen is mounted on a piezoelectric to move in the z direction to keep the force constant, whereas the scanning process is carried out along the x and y axes.

2.6 Magnetic force microscopy

Magnetic force microscopy (MFM) allows us to obtain useful information regarding the magnetization distribution while identifying magnetic domains in ferrite thin films. Furthermore, naturally occurring and artificially written domain configurations can be developed by MFM. Such a capability is provided by measuring the magnetic interaction between the sample and the MFM probe tip coated with a ferromagnetic thin film. The magnetic force between the sample and the tip is given by Jiles (2015), Bonnell (2000):

$$F = \mu_0 (m \cdot \nabla) H \tag{2.9}$$

where m is the magnetic moment of the tip, H is the stray magnetic field of the specimen, and μ_0 is the magnetic permeability of free space. The range of typically measured forces is in the order of 10^{-14} N.

It must be noted that the magnetization of the sample and tip can be changed by the magnetic field, thus imposing nonlinear interactions between them. In turn, this can complicate the explanation of magnetic domains. Thus, both the type of the tip and the ferromagnetic coating on it play important roles in the resultant magnetization. The sample does not need to be coated by a conducting layer To observe domain features, whereas it is essential for SEM sample preparation.

2.7 Fourier transform infrared spectroscopy

The Fourier transform infrared (FTIR) spectroscopic method involves passing an infrared (IR) beam through the sample so that part of it is absorbed by the sample and the rest passes through it.

2.7.1 Fundamental principles of Fourier transform infrared spectroscopy

FTIR spectra reveal the properties of molecules based on the absorption and transmission of IR. Similar to fingerprints, the molecular structure of the material is unique, thereby making FTIR spectroscopy a useful analytical tool for evaluating different types of materials (Hyperphysics, 2000).

To better understand the oscillations of molecular bonds, consider the bond as a spring with two masses attached to its two ends. In the IR spectral region, the absorption of light by a substance is due to the interaction between the vibrations of the sample's chemical bonds and radiation from the light source. More precisely, if the atoms on either side of a chemical bond (consider a simple two-atom bond) are different, they can create an electric dipole oscillating at a particular frequency. If such an asymmetric bond is irradiated by a single source of light with a frequency equal to the bipolar frequency of the bond, there will be an interaction between the bond and the radiation from the source. As a result of this interaction, the electric component of the radiation will be able to transfer its energy to the molecular bond, which in turn leads to a change in the vibration of the bond.

Polar molecules have positive and negative partial charges resulting in a bipolar torque. This is equal to the amount of the vector output obtained by multiplying the charge "q" by the distance of the bipolar torque "d" going from a negative to a positive charge (Nikolic, 2011). The condition for the energy absorption of IR waves by molecules is realized when their dipole moment changes during the vibration. In other words, molecules with a bipolar torque can absorb specific amounts of energy to change their bipolar torque vector, thus producing particular peaks in the corresponding IR spectrum. Based on these brief descriptions, two general conditions for absorbing IR radiation by a chemical bond can be considered as follows (Nikolic, 2011):

1. There is an electric dipole around the chemical band.
2. The frequencies of the electrical polarity of the bond and radiation should be exactly equal to each other.

Typically, molecules can have tensile and flexural vibrations. The former is realized in both symmetrical and asymmetrical modes. For an asymmetric vibration, the bipolar torque oscillates with the vibration frequency of the molecule, promoting it to the IR absorption band. In this case, the molecule is called "IR-active."

In the 8–25 μm spectral region, known as the "fingerprint" region, the spectrum of each substance is used to identify its functional groups and bonds formed. Therefore, when molecules are exposed to IR light, they absorb radiation at the frequencies of their vibrational modes. The IR radiation absorption of samples is measured as a function of frequency, producing a spectrum used for identifying functional groups and material structure (Mohamed et al., 2017).

In the FTIR instrument, the spectrum is emitted by an optical system consisting of several mirrors, known as a Michelson interferometer. As the mirror moves, each wavelength of light reflected in the mirror is intermittently interrupted by an interferometer, thereby producing wavelength interference. The output of this process results from two fixed and moving mirror beams at any given moment, indicating a specific wavelength at any given time. This continues until all wavelengths are generated, traversing the entire wavelength range. Finally, a computer system is needed to process the received information, thus converting the raw data (i.e., the absorption amount per position of the moving mirror) into useable information (i.e., the absorption per wavelength) (Bacsik et al., 2004). The most critical parts of the FTIR spectrometer are the laser light, signal source analyzer, sample holder, and opaque solids, as schematically shown in Fig. 2.11.

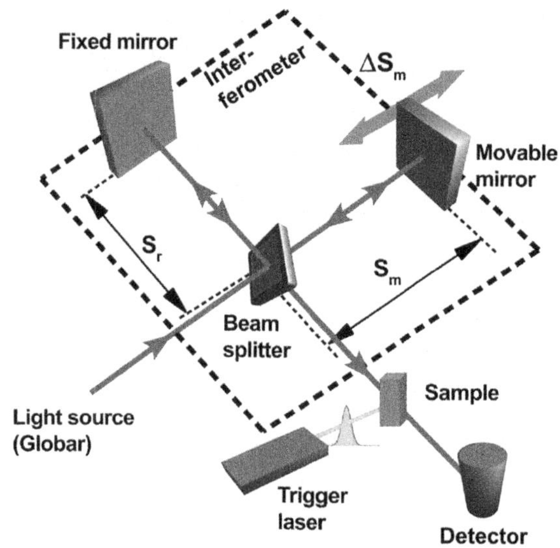

FIGURE 2.11

General schematic of the Fourier transform infrared spectrometer.

From Ritter, E., Puskar, L., Bartl, F. J., Aziz, E. F., Hegemann, P., & Schade, U. (2015). Time-resolved infrared spectroscopic techniques as applied to channel rhodopsin. Frontiers in Molecular Biosciences, 2, *38. https://doi.org/10.3389/fmolb.2015.00038.*

2.7.2 Typical applications of Fourier transform infrared analysis of ferrite nanoparticles

Various chemical methods have been used to synthesize ferrite nanoparticles in the solution phase, such as reverse micelle, hydrothermal, solvothermal, and sol-gel. Some of these methods use surfactants and capping or chelating agents (e.g., triethanolamine and citric acid) (Ghasemi et al., 2019).

FTIR analysis can detect the presence of functional groups in precursors (before the calcination process) and their removal (after the calcination) in the synthesis of ferrite nanoparticles. Another application of FTIR analysis involves the study of drug release by ferrite nanoparticles. For example, in research work, anticancer drugs such as doxorubicin have been loaded on cobalt ferrite nanoparticles. The existence of drugs on nanoparticles can be approved using the FTIR analysis (Dey et al., 2017; Wang et al., 2016). With the help of FTIR analysis, the location of cations in octahedral and tetrahedral positions of spinel ferrite nanoparticles can also be determined. The cations located in tetrahedral positions show a smaller wavenumber from the corresponding FTIR spectra than that in octahedral positions due to their shorter and stronger bonds. Notably, for cobalt ferrite nanoparticles, the Co−O bond at octahedral positions shows absorption at 400 cm^{-1}, whereas the Fe−O bond at the tetrahedral sites results in absorption at 600 cm^{-1}.

Lohar and colleagues have calculated the cation distribution in $Ni_{1-x}Cd_xFe_2O_4$ nanoparticles using FTIR analysis (Lohar et al., 2013). They reported that the incidence of ν_1 (600 cm^{-1}) and ν_2 (400 cm^{-1}) absorption bands is related to the intrinsic vibrations of tetrahedral and octahedral groups, respectively (see Fig. 2.12). These bands are mostly dependent on Fe—O distances and the nature of the cations occupied.

The force constant of the tetrahedral site (K_t) and octahedral site (K_O) can be considered based on employing the method suggested by Waldron (1955).

According to the Waldron theory, the force constants K_t and K_O for respective sites are calculated by

$$K_t = 7.62 \times M_1 \times \nu_1^2 \times 10^{-3}$$

$$K_O = 10.62 \times \frac{M_2}{2} \times \nu_2^2 \times 10^{-3} \quad (2.10)$$

where M_1 and M_2 are the respective molecular weight of cations on A and B sites measured using the cation distribution data. The bond lengths (R_A and R_B) have also been calculated using the formula given by Gorter (Gorter, 1954). The calculated values of R_A and R_B and the force constants K_t and K_O are listed in Table 2.2. With an increase in the Cd^{2+} substitution, force constants (K_t and K_O) and bond lengths (R_A and R_B) decrease and increase, respectively.

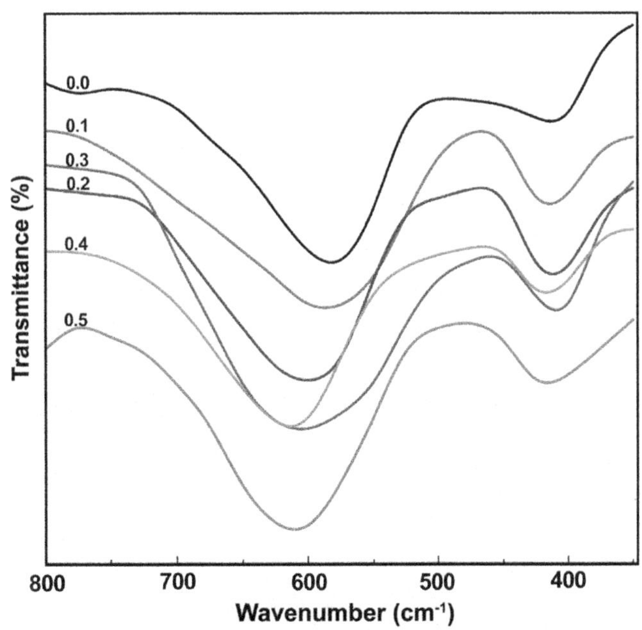

FIGURE 2.12

Fourier transform infrared spectroscopy of $Ni_{1-x}Cd_xFe_2O_4$ ($x = 0$–0.5) nanoparticles.

From Lohar, K., Patange, S., Mane, M., & Shirsath, S. E. (2013). Cation distribution investigation and characterizations of $Ni_{1-x}Cd_xFe_2O_4$ nanoparticles synthesized by citrate gel process. Journal of Molecular Structure, 1032, *105–110. https://doi.org/10.1016/j.molstruc.2012.07.055.*

Table 2.2 Absorption bands (ν_1 and ν_2), force constants (K_o and K_t), and bond lengths (R_A and R_B) of the $Ni_{1-x}Cd_xFe_2O_4$ system.

x	$\nu_1(cm^{-1})$	$\nu_2(cm^{-1})$	K_o(dyne/cm)	K_t(dyne/cm)	Bond length $R_A(\mathring{A})$	$R_B(\mathring{A})$
0.0	408.33	586.85	253511.16	113310.50	1.8936	1.8716
0.1	415.65	587.32	253510.16	112108.26	1.8946	1.8725
0.2	408.79	587.18	253499.72	110374.38	1.8989	1.8764
0.3	398.91	579.19	246502.55	106767.30	1.9107	1.8884
0.4	401.91	579.12	246452.25	108086.26	1.9112	1.8983
0.5	397.21	579.25	246295.33	108894.71	1.9296	1.8991

From Lohar, K., Patange, S., Mane, M., & Shirsath, S. E. (2013). Cation distribution investigation and characterizations of $Ni_{1-x}Cd_xFe_2O_4$ nanoparticles synthesized by citrate gel process. Journal of Molecular Structure, 1032, *105–10. https://doi.org/10.1016/j.molstruc.2012.07.055.*

Table 2.3 Frequencies ν_1 and ν_2 and force constants K_t and K_O for the *A* and *B* sites, respectively.

x	ν_1 (cm^{-1})	ν_2 (cm^{-1})	K_t ×10^5 (dyne/cm)	K_O ×10^5 (dyne/cm)
0.00	562.88	355.71	2.306	0.921
0.025	555.39	354.88	2.246	0.917
0.05	555.67	352.38	2.248	0.904
0.075	555.67	379.56	2.248	1.048
0.10	550.95	378.45	2.210	1.043

From Yadav, R. S., Havlica, J., Masilko, J., Kalina, L., Wasserbauer, J., Hajdúchová, M., Enev, V., Kuřitka, I., & Kozáková, Z. (2016). Impact of Nd^{3+} in $CoFe_2O_4$ spinel ferrite nanoparticles on cation distribution, structural and magnetic properties. Journal of Magnetism and Magnetic Materials, 399, *109–117. https://doi.org/10.1016/j.jmmm.2015.09.055.*

In a separate research study, an Nd cation has been added to a cobalt ferrite structure to synthesize $CoFe_{2-x}Nd_xO_4$ ($x = 0.0, 0.025, 0.05, 0.075$ and 0.1) composition (Yadav et al., 2016). They observed a shift in characteristic absorption bands due to the change in the lattice constant and cation redistribution. A higher force constant was also found for the band stretching of tetrahedral sites compared with the band stretching of the octahedral site. The force constants K_t and K_O calculated for the *A* and *B* sites, respectively, are listed in Table 2.3 using Eq. (2.11):

$$K = 4\pi^2 c^2 \nu^2 m \tag{2.11}$$

where K is the force constant, c is the light velocity ($c = 2.99 \times 10^{10}$ cm/s), ν denotes the vibration frequency of the *A* and *B* sites, and m is the reduced mass of Fe^{3+} and O^{2-} ions, respectively.

2.8 Thermal analysis methods

Thermal analysis methods refer to techniques for evaluating the temperature dependence of physical and mechanical properties of materials. In many cases, these methods can be used as a complementary analysis for XRD, SEM, and TEM investigations. Different methods of thermal analysis have been used for various purposes, including (Haines, 2012):

- thermal gravimetric analysis (TGA) for finding the thermal feature of materials subjected to heating under controlled heating conditions;
- simultaneous thermal analysis (STA) for investigating the role of temperature in the properties of materials;
- differential scanning calorimetry (DSC) for measuring heat flux of materials;

- differential thermal analysis (DTA) to find the reaction type in terms of thermodynamics;
- thermomechanical analysis for measuring dimensional changes of the sample;
- dynamic mechanical analysis to find viscoelastic features of polymers;
- dielectric thermal analysis for determining temperature dependence of dielectric properties;
- cooling curve thermal analysis to study the freezing process and its relationships with other factors such as structural changes, phase characteristics, and mechanical properties.

In the following, some of the abovementioned techniques are briefly discussed.

2.8.1 Thermal gravimetric analysis

TGA is based on measuring the weight changes of a sample with temperature under controlled atmospheres such as argon (Ar), preventing it from adverse reactions and oxidation. The sample is hung inside a furnace by a wire, and the end of it is connected to a magnetic core (Coats & Redfern, 1963). Depending on the motion of the core, an electric current proportional to the change in the sample weight is induced in the coil and sent to the recorder after amplification. The temperature of the sample is also measured by a thermocouple. The beginning (T_i) and end (T_f) temperatures of each heat event are measured at each stage of the weight change while also determining the amount of weight loss per event.

The most important factors affecting the T_i and T_f of the process are the heating rate, furnace environment, sample features, and sample holder type. The thermal equilibrium of the sample is mainly dependent on the exchange of heat with the atmosphere. Therefore, specimen shape and atmosphere can affect the onset and end temperatures of the reaction. As an effective practical factor, the sample holder is generally a flat tray or cylinder made of platinum (Pt) or alumina (Al_2O_3). It should be noted that the properties of an unknown sample (e.g., the shape, size, density, and sample amount) also play an essential role in the results of the TGA, influencing the thermal equilibrium of the material while also changing the onset and end temperatures of the reaction. The important components of the TGA instrument are the scale, heating furnace, and sample placement chamber, as shown in Fig. 2.13.

If the temperatures of two thermal events are close to each other on the TGA curves, it will be difficult to separate them in the corresponding weight change curves in terms of temperature. On the other hand, it is challenging to quickly determine the T_i and T_f of a thermal event in TGA curves. To overcome these difficulties, one can obtain the temperature curve derivative in terms of temperature by adding an electronic system to the TGA instrument, leading to differential TGA. In this case, the maximum point of the resultant curves indicates a transformation point in the thermal weight analysis curve.

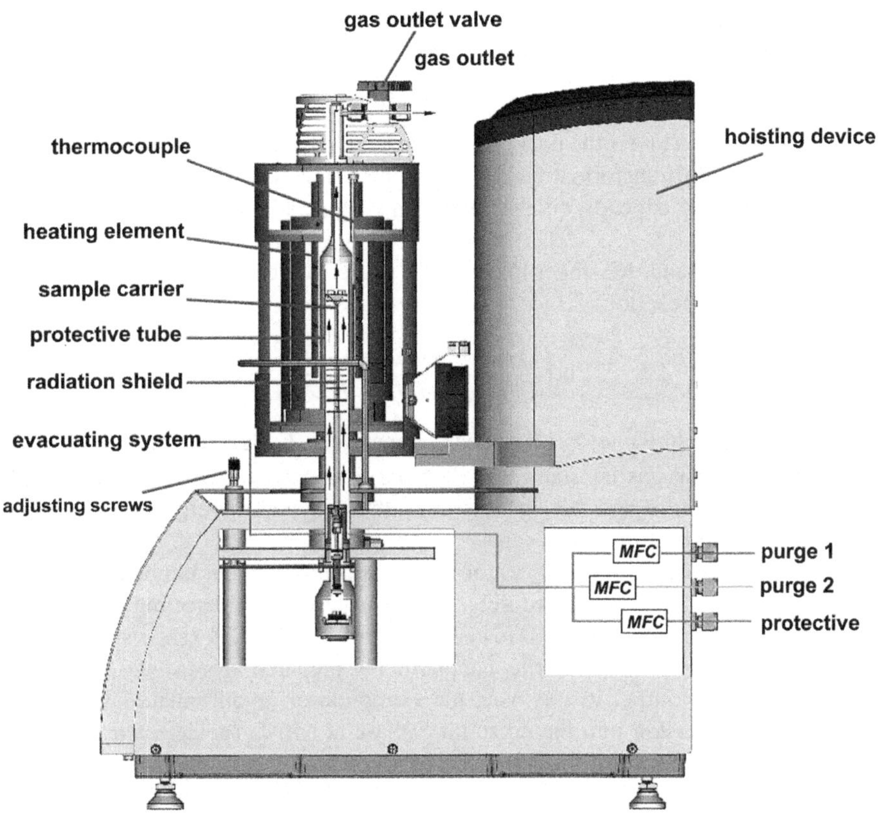

FIGURE 2.13

Thermal gravimetric analysis instrument components.

From Dudziak, T., Medvedovski, E., & Homa, M. (2018). Multilayered coatings for high-temperature steam oxidation: TGA studies up to 1000°C. Journal of Materials Engineering and Performance, *27(8), 4317–4335. https://doi.org/10.1007/s11665-018-3507-3.*

2.8.2 Differential thermal analysis

DTA is a thermoanalytical technique in which the material under consideration and a reference sample must be subjected to the same thermal cycles (having the same heating or cooling program), while also recording any temperature difference between them. This differential temperature is plotted relative to time or temperature, resulting in thermogram curves. Any release or absorption of heat from the sample is accompanied by a deviation from the reference material temperature, providing information about the transformation temperature.

The area under the peak in the DTA curves is proportional to the heat of the reaction, which can be assessed using the following equation (Charsley et al., 2019):

$$A = Km\Delta H \tag{2.12}$$

where A is the area under the peak, K is a constant, m is the sample mass, and ΔH is the enthalpy of the transformation. To determine ΔH, it is necessary to find K and A. The amount of K depends on the instrument type and the specifications of the sample.

On the other hand, Kissinger relationship is used to determine the activation energy of chemical reactions:

$$\ln\left(\frac{\beta}{T_p^2}\right) = \ln\left(\frac{AR}{E_a}\right) - \frac{E_a}{R}\cdot\frac{1}{T_p} \tag{2.13}$$

where β is the heating rate, T_p is the peak temperature, E_a denotes the activation energy, R is the total gas constant, and A is the frequency factor (Holba & Šesták, 2014). It is therefore possible to determine the activation energy by performing DTA at different heating rates.

One of the essential applications of DTA is to study the thermal behavior of ceramics, including ferrites. Accordingly, DTA is beneficial for detecting the transformation temperature of amorphous to crystalline ferrites. Fig. 2.14 shows an example of a DTA curve of nickel ferrite nanoparticles prepared from a gel containing triethanolamine-NiZnFe. In this way, the extraction of organic material at 523°C and the gel conversion into the crystalline phase at 656°C are detected. Note that even before thermal transformation, the baseline is not smooth, arising from the

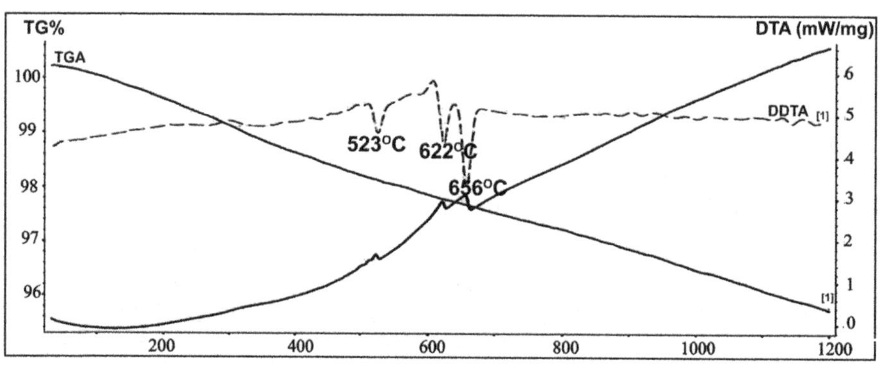

FIGURE 2.14

The differential thermal analysis curve of the gel containing triethanolamine-NiZnFe.

From Ghasemi, A., Loghman-Estarki, M. R., Torkian, S., & Tavoosi, M. (2019). The microstructure and magnetic behavior of spark plasma sintered iron/nickel zinc ferrite nanocomposite synthesized by the complex sol-gel method. Composites Part B: Engineering, 175, *107179. https://doi.org/10.1016/j.compositesb.2019.107179.*

difference between thermal capacity and thermal conductivity of the ferrite nanoparticles and the reference sample.

2.8.3 Simultaneous thermal analysis

STA is one of the most widely used methods to investigate the role of temperature in the properties of materials. This thermal analysis is "simultaneous" as it combines the results of DTA and TGA simultaneously. The conditions of STA are similar to those of TGA and DTA, measuring the effects of heat (temperature and enthalpy changes) and weight changes of the specimen simultaneously (Haines, 2012).

2.8.4 Differential scanning calorimetry

DSC involves the thermal decomposition of a substance whose heat capacity changes are continuously measured as a function of temperature. The specimen and reference sample are kept at the same temperature in the DSC method. The difference in energy required to keep the temperature constant is plotted as a function of temperature changes. Accordingly, the heat capacity changes in a heated (or cooled) specimen with a specific weight follow the changes in its heat flow. The melting point and melting behavior, heat transfer temperature, specific heat capacity, phase changes, reaction kinetics, metal oxidation and degradation, thermal decomposition and pyrolysis of ferrites, crystallization and amorphous behavior of ferrites, and Curie temperature could be determined by DSC.

The Curie temperature of magnetic materials can be obtained by placing a strong magnet above or below the specimen pan during analysis. As the Curie temperature is crossed, a transformation from ferrimagnetic or ferromagnetic to paramagnetic state occurs. This is signaled by a sharp change in specimen weight, as shown in Fig. 2.15.

2.9 Mössbauer spectroscopy of ferrites

In 1957, Rudolph Mössbauer observed the effect of recoilless gamma-ray emission and absorption, which is known as the Mössbauer effect. He received the Nobel Prize in Physics in 1961 for his achievement. As a versatile technique, Mössbauer spectroscopy provides researchers with very precise information regarding the site occupancies of cations in magnetic materials. A radioactive source is oscillated with a velocity of a few mm/s, recording the Mössbauer spectrum in discrete velocity steps with a resolution of 1 in 10^{12}. The information obtained from the Mössbauer spectroscopy is very useful for better understanding the nature of magnetic materials, especially ferrimagnetic ferrites. Thus, it is easily possible to calculate magnetic moments of ferrites by finding the site preference of atoms or cations, interpreting the magnetic data obtained from magnetometer devices such as the vibrating sample magnetometer (VSM), and superconducting quantum interference

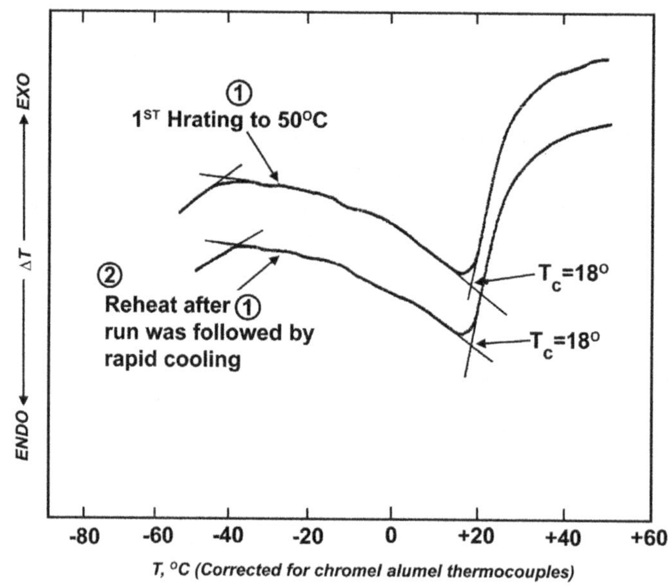

FIGURE 2.15

The differential scanning calorimetry curve of a ferromagnetic material used to determine Curie temperature.

From Williams, H. W., & Chamberland, B. L. (1969). Determination of Curie, Neel, or crystallographic transition temperatures via differential scanning calorimetry. Analytical Chemistry, 41*(14), 2084–2086. https://doi.org/10.1021/ac50159a056.*

device (SQUID). The magnetic sample must be sufficiently thin to allow the gamma rays to pass through it. The power of this technique can be realized by considering some theoretical aspects. It is well known that nuclei in atoms have different energy level transitions from the emission or absorption of gamma rays. The surrounding electronic and magnetic environments mainly affect changing or splitting of these energy levels. Variations in energy levels can be observed by employing the resonance-fluorescence method, providing information about an atom's surrounding environment within a system. It must be noted that the hyperfine interactions between the nucleus and its environment are very small, so the recoil of the nucleus prevents the occurrence of resonance as the gamma ray is emitted or absorbed (Cranshaw et al., 1985). If the emitting and absorbing nuclei exist in solid materials, the emitted and absorbed gamma ray will have the same energy, thereby characterizing the resonance condition. The energy levels in nuclei can be identified by isomer shift (IS), quadrupole splitting (QS), and magnetic splitting. Therefore, these represent the essential components in the identification of Mössbauer spectra.

The IS is required to find valence states and electron shielding in electronegative groups, originating as a consequence of the nonzero volume of the electron charge density as well as the nucleus due to s-electrons within it. Any difference in the s-electron environment between the source and absorber causes a shift in the resonance transition energy that is generally called an IS.

Consider Fe^{2+} and Fe^{3+} cations with respective electron configurations of $(3d)^6$ and $(3d)^5$. Because of its greater screening of the d-electrons, the Fe^{2+} cation has fewer s-electrons at the nucleus, leading to larger positive ISs than those in the Fe^{3+} cation. On the other hand, nuclei in states with an angular momentum quantum number of $I > 1/2$ have a nonspherical charge distribution, producing a nuclear quadrupole moment while also causing the splitting of the nuclear energy levels through the interaction of the nuclear quadrupole moment with the electric field gradient (Dyar et al., 2006). By applying an external magnetic field ($B_{applied}$), Zeeman splitting results from the nuclear spin moment dipolar interaction with the magnetic field, yielding a sextet pattern in the simplest case (see Fig. 2.16). As known, the effective magnetic field at the nucleus, B_{eff}, is expressed by

$$B_{eff} = \left(B_{contact} + B_{orbital} + B_{dipolar}\right) + B_{applied} \tag{2.14}$$

in which $B_{applied}$ denotes the applied field, and $B_{contact}$ is related to the spin on those electrons polarizing the spin density at the nucleus. $B_{orbital}$ and $B_{dipolar}$ are the orbital fields originating from the orbital moment, and the dipolar field due to the spin of the electrons, respectively (Gibb, 2013).

The intensity or area of the Mössbauer spectrum is related to the bonding strength of atoms based on the Debye-Waller factor (the f factor). It could be

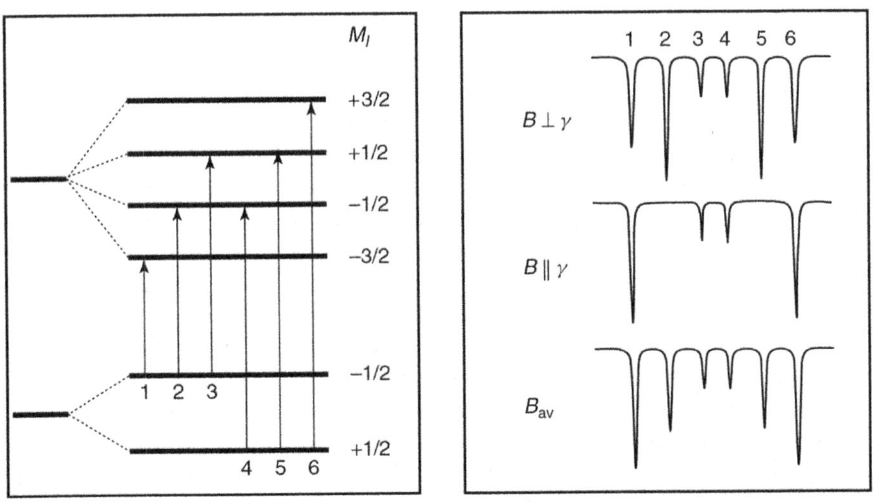

FIGURE 2.16

Sextet pattern in Mössbauer spectra.

From Schünemann, V., & Paulsen, H. (2011). Mössbauer spectroscopy. Encyclopedia of Inorganic and Bioinorganic Chemistry. *https://doi.org/10.1002/9781119951438.eibc0138.*

determined by the percentage of γ rays emitted or absorbed without energy loss (Long & Grandjean, 2013). Based on the Debye-Waller model, the f factor is expressed by

$$f = \exp\left\{ -\frac{3E_R}{2k_B\theta_D}\left[1 + 4\left(\frac{T}{\theta_D}\right)^2 \int_0^{\frac{\theta_D}{T}} \frac{x}{e^x - 1}dx\right]\right\} \tag{2.15}$$

where k_B is Boltzman constant, E_R denotes the recoil energy of the nuclide, and θ_D is the Debye temperature. By considering $T = 0$, the f factor still depends on recoil energy and Debye temperature by the following:

$$f_{T=0} = exp\left(\frac{-3E_R}{2k_B\theta_D}\right) \tag{2.16}$$

At room temperature and below, the local magnetic field at Mössbauer nuclei of ferrite materials is on the order of 50 T, originating from the nature of superexchange interaction between nuclear spins and transferring the spin density via the oxygen anions (Sawatzky et al., 1969). To realize the capability and power of the Mössbauer spectroscopic technique while also interpreting the practical results, an example of hexagonal ferrite is given in the following.

Strontium ferrite particles with micron-size configurations have been synthesized in my research group (Ghasemi & Šepelák, 2011). A small amount of hematite (α-Fe_2O_3) with the following characteristics: IS = 0.24 mm/s, QS = −0.1 mm/s, and B = 51.5 T, has also been formed along with the strontium ferrite. The ^{57}Fe Mössbauer spectroscopic measurements were carried out in transmission geometry at room temperature. The γ-ray source was isotope ^{57}Co in the Rh matrix, and the velocity scale in the experiment was calibrated relative to ^{57}Fe in Rh. Recoil spectral analysis software (Lagarec & Rancourt, 1998) was employed for the quantitative interpretation of the obtained results. A Lorentzian line width of Γ = 0.229(6) mm/s resulted from the fit of the spectrum. By assuming the same recoilless fractions of ferric cations defined by crystal sites of $12k$, $4f_1$, $4f_2$, $2a$, and $2b$ in the strontium ferrite particles, it was possible to determine the position of the Fe^{3+} cation from the Mössbauer subspectral intensities (Fig. 2.17).

By considering the IS values of the spectral components (Table 2.4), the absence of the ferrous cation in the samples is confirmed (Menil, 1985). Sextets with ISs of 0.24, 0.22, and 0.20 mm/s are assigned to octahedrally coordinated Fe^{3+} ion occupancies of $4f_2$, $12k$, and $2a$ sites, respectively, whereas IS = 0.14 mm/s is typical for Fe^{3+} cations in the tetrahedral ($4f_1$) position (Menil, 1985). Fe^{3+} ions in the trigonal bipyramidal ($2b$) site are represented by a sextet with IS = 0.17 mm/s and a large QS of 1.1 mm/s. A comparison of the relative intensities of the sextets indicates that the number of iron cations located in $4f_2$, $12k$, $2a$, $4f_1$, and $2b$ sites is 2, 6, 1, 2, and 1 [per f.u. of strontium ferrite], respectively, giving rise to the following crystal chemical formula: $Sr(Fe_2)_{4f_2}(Fe_6)_{12k}(Fe_1)_{2a}(Fe_2)_{4f_1}(Fe_1)_{2b}O_{19}$.

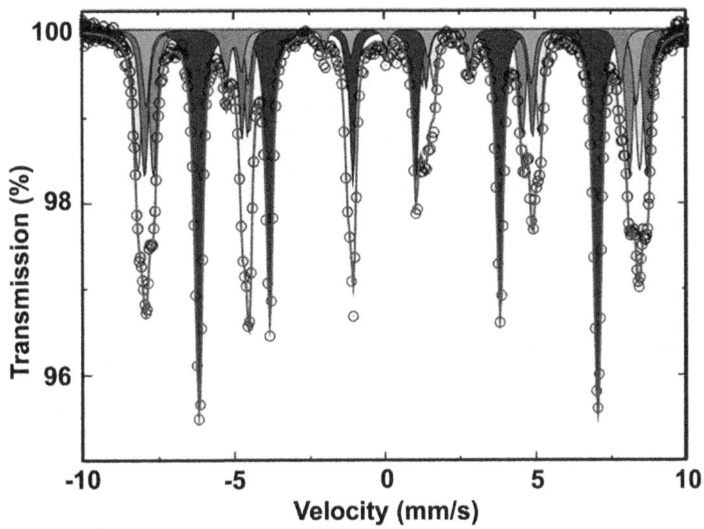

FIGURE 2.17

^{57}Fe Mössbauer spectrum of strontium ferrite particles.

From Ghasemi, A., & Šepelák, V. (2011). Correlation between site preference and magnetic properties of substituted strontium ferrite thin films. Journal of Magnetism and Magnetic Materials, 323*(12), 1727–33. https://doi.org/10.1016/j.jmmm.2011.02.010.*

Table 2.4 Hyperfine parameters determined by fitting Mössbauer spectrum of strontium ferrite particles.

Sublattice	Coordination of Fe^{3+} ions	Isomer shift (mm/s)	Quadrupole splitting (mm/s)	*B* (T)	Number of Fe^{3+} ions per f.u.
$4f_2$	Octahedral	0.235(5)	0.142(8)	51.62(5)	2
$12k$	Octahedral	0.222(2)	0.204(1)	40.96(8)	6
$2a$	Octahedral	0.203(1)	0.021(2)	50.31(2)	1
$4f_1$	Tetrahedral	0.140(2)	0.098(8)	48.89(7)	2
$2b$	Bipyramidal	0.170(3)	1.115(1)	40.59(1)	1

From Ghasemi, A., & Šepelák, V. (2011). Correlation between site preference and magnetic properties of substituted strontium ferrite thin films. Journal of Magnetism and Magnetic Materials, 323*(12), 1727–33. https://doi.org/10.1016/j.jmmm.2011.02.010.*

It is also possible to calculate the magnetic moment (m) per f.u. by finding the site occupancy of cations in the strontium ferrite crystal structure. It was reported that one Fe^{3+} cation has a magnetic moment of $m = 5\mu_B$. If spins of Fe^{3+} ions are assumed to be aligned collinearly in the five sublattices of strontium ferrite determined by the ^{57}Fe Mössbauer spectroscopy (Morel et al., 2002; Zi, Sun, Zhu, Yang, & Song, 2008), the resulting magnetic moment will be $m = 20\mu_B$/f.u., according to Table 2.5.

Table 2.5 Hyperfine parameters along with the spin orientation and calculation of magnetic moment per f.u. for strontium ferrite.

Sublattice	Coordination of Fe^{3+} ions	Number of Fe^{3+} ions per f.u.	Spin orientation	Magnetic moment Lper f.u.
$4f_2$	Octahedral	2	↓	$m = -2 \times 5\mu_B + 6 \times 5\mu_B + 1 \times 5\mu_B - 2 \times 5\mu_B + 1 \times 5\mu_B = 4 \times 5\mu_B = 20\mu_B$
$12k$	Octahedral	6	↑	
$2a$	Octahedral	1	↑	
$4f_1$	Tetrahedral	2	↓	
$2b$	Bipyramidal	1	↑	

From Ghasemi, A., & Šepelák, V. (2011). Correlation between site preference and magnetic properties of substituted strontium ferrite thin films. Journal of Magnetism and Magnetic Materials, 323*(12), 1727−1733. https://doi.org/10.1016/j.jmmm.2011.02.010.*

2.10 Magnetic anisotropy

Magnetocrystalline anisotropy, shape anisotropy, induced magnetic anisotropy, magnetostriction, and surface and interface anisotropies are the most important magnetic anisotropies investigated in materials. Depending on the nature of the specimen and its fabrication history, one or two of the aforementioned anisotropies play the dominant role in interpreting the magnetic features of ferrites. The direction of magnetization and the domain configuration are determined by competition between the magnetic anisotropies. The preferred magnetization directions providing the system with minimum energy are called the magnetic easy axes. Conversely, the directions inducing the highest level of energy are hard axes. To evaluate the magnetic properties of ferrites, understanding the role of various kinds of magnetic anisotropy is essential.

2.10.1 Magnetocrystalline anisotropy

The magnetocrystalline anisotropy originates from the electron spin-orbit coupling since the electron orbitals are linked to crystallographic orientation. This is the most important anisotropy in hexagonal ferrites with uniaxial growth, such as M-type ferrites. It must be noted that spinel ferrites have no strong preferred orientation of grains compared with hexagonal ferrites, leading to a negligible role for magnetocrystalline anisotropy in determining the direction of magnetization and magnetic properties. Generally, magnetization direction is defined by a direction cosine, as illustrated in Fig. 2.18.

The relationship between the direction cosine α_i, and θ and ϕ angles is as follows:

$$\alpha_1 = \sin\theta \cos\varphi \tag{2.17}$$

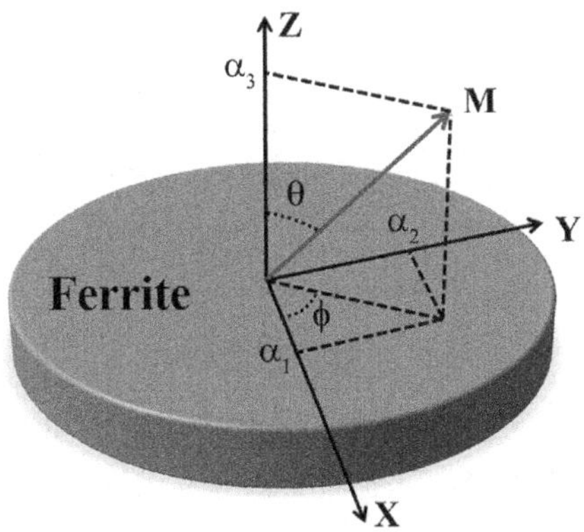

FIGURE 2.18

Direction of magnetization vector for Cartesian axes.

$$\alpha_2 = \sin\theta \sin\varphi \tag{2.18}$$

$$\alpha_3 = \cos\theta \tag{2.19}$$

The magnetocrystalline anisotropy energy per volume, E_{crys}, can be described by the following expression (Getzlaff, 2007):

$$E_{crys} = E_0 + \sum_{ij} b_{ij}\alpha_i\alpha_j + \sum_{ijkl} b_{ijkl}\alpha_i\alpha_j\alpha_k\alpha_l \tag{2.20}$$

where α and b are constants, and E_0 is the initial energy of the magnet. The general equations of E_{crys} for cubic, hexagonal, and tetragonal crystal structures can be rewritten as follows:

$$E_{cubic\ crys} = K_0 + K_1\left(\alpha_1^2\alpha_2^2 + \alpha_1^2\alpha_3^2 + \alpha_2^2\alpha_3^2\right) + K_2\alpha_1^2\alpha_2^2\alpha_3^2 + \cdots \tag{2.21}$$

$$E_{tetra\ crys} = K_0 + K_1\alpha_3^2 + K_2\alpha_3^4 + K_3\left(\alpha_1^4 + \alpha_2^4\right) + \cdots \tag{2.22}$$

$$E_{hex\ crys} = K_0 + K_1\left(\alpha_1^2 + \alpha_2^2\right) + K_2\left(\alpha_1^2 + \alpha_2^2\right)^2 + K_3\left(\alpha_1^2 + \alpha_2^2\right)^3 + K_4\left(\alpha_1^2 - \alpha_2^2\right) \times \left(\alpha_1^4 - 14\alpha_1^2\alpha_2^2 + \alpha_2^4\right) \tag{2.23}$$

The dependence of E_{crys} on the angle between the magnetization vector and the direction of the crystallographic axis θ for hexagonal structure can be given by

$$E_{hex\ crys} = K_0 + K_1\sin^2\theta + K_2\sin^4\theta + K_3\sin^6\theta + K_4\sin^6\theta\cos6\varphi \tag{2.24}$$

where K_0, K_1, K_2, … are anisotropy constants. In hexagonal ferrites, the K_1 is the most important anisotropy constant.

2.10.2 Shape anisotropy

If the ferrite morphology is not spherical, one or more specific directions will represent the easy magnetization axes. This phenomenon is known as shape anisotropy. To find the effect of morphology and obtain deeper insights into the shape anisotropy, the stray field energy density, E_{str}, of a sample must be defined as given below (Getzlaff, 2007):

$$E_{str} = -\frac{1}{2}\int \mu_0 M \cdot H_{demag} dV \tag{2.25}$$

where M and V are the magnetization and volume of the sample. H_{demag} is the demagnetization field given by

$$H_{demag} = -NM \tag{2.26}$$

where N is the demagnetization tensor. The magnetostatic energy is induced by a demagnetizing field correlated with the shape factor of magnets.

If the morphology of the sample is a sphere, the elements of the demagnetization tensor will be as follows:

$$N = \begin{bmatrix} \frac{1}{3} & 0 & 0 \\ 0 & \frac{1}{3} & 0 \\ 0 & 0 & \frac{1}{3} \end{bmatrix} \tag{2.27}$$

Moreover, E_{str} of the spherical morphology can be given by

$$E_{str} = \frac{1}{6}\mu_0 M^2 \tag{2.28}$$

From the above equation, all directions are energetically equivalent, giving rise to the isotropic behavior of the sample. This can be used for ferrite nanoparticles or micron-size particles with spherical morphology synthesized easily by a wet chemical method.

For specimens in the form of infinitely long cylinders with dimensions of $a = b$ and infinite c, the tensor amounts to the following:

$$N = \begin{bmatrix} \frac{1}{2} & 0 & 0 \\ 0 & \frac{1}{2} & 0 \\ 0 & 0 & 0 \end{bmatrix} \tag{2.29}$$

In this case, the corresponding E_{str} is obtained as follows:

$$E_{str} = \frac{1}{2}\mu_0 M^2 \cdot \frac{1}{2}\left(\sin^2\theta\cos^2\phi + \sin^2\theta\sin^2\phi\right) = \frac{1}{4}\mu_0 M^2 \sin^2\theta \tag{2.30}$$

Accordingly, based on the discussion above, one can interpret the magnetic properties of ferrite nanorods synthesized using a hydrothermal approach or grown in alumina templates by an electrodeposition technique. For thin films of ferrite prepared by a pulsed laser deposition or sputtering technique, the tensor is expressed by (Getzlaff, 2007)

$$N = \begin{bmatrix} 0 & 0 & 0 \\ 0 & 0 & 0 \\ 0 & 0 & 1 \end{bmatrix} \tag{2.31}$$

E_{str} of thin films is also as follows:

$$E_{str} = \frac{1}{2}\mu_0 M^2 \cos^2\theta \tag{2.32}$$

2.10.3 Induced magnetic anisotropy

After performing an annealing process under an external magnetic field on some metallic magnets such as FeCoCr and Alnico, spinodal decomposition with an ordered magnetic phase can be achieved, providing induced magnetic anisotropy. The annealing process should be carried out below the Curie temperature. The induced magnetic anisotropy energy, E_{ind}, is expressed by (Getzlaff, 2007)

$$E_{ind} \cdot \frac{1}{a} V = -F\left(\alpha_1^2\beta_1^2 + \alpha_2^2\beta_2^2 + \alpha_3^2\beta_3^2\right) - G(\alpha_1\alpha_2\beta_1\beta_2 + \alpha_2\alpha_3\beta_2\beta_3 + \alpha_1\alpha_3\beta_1\beta_3) \tag{2.33}$$

where α_i and β_i are direction cosines of magnetization measured during the annealing under an external magnetic field, and F and G are material constants. Since magnetically induced anisotropy is not often considered important in ferrites and related composites, further insights are not presented here.

2.10.4 Magnetostriction anisotropy

The elastic deformation in magnetic materials is caused by applying stress in the elastic region and the displacement of solid atoms. Alternatively, applying a magnetic field to soft magnets can lead to the same elastic deformation. This interplay between the deformation and magnetic field results in stress anisotropy or magnetostriction, playing the main role in soft magnets, including soft ferrites. Magnetostriction must be considered very carefully when dealing with soft ferrite thin films such as Mn–Zn ferrite thin films, whereas it can be neglected for hard ferrites whose magnetocrystalline anisotropy is much stronger than their magnetostriction anisotropy. Explanation of the deformation tensor is beyond the scope of this chapter, but it has been discussed in several works. By considering the elements of deformation tensor ε_{ij} and elastic constant c_{ij}, the elastic energy density for cubic $(E_{cubic\ el})$

and hexagonal ($E_{\text{hex el}}$) structures is given by Eqs. (2.34) and (2.35), respectively (Getzlaff, 2007):

$$E_{\text{cubic el}} = \frac{1}{2}c_{11}\left(\varepsilon_{11}^2 + \varepsilon_{22}^2 + \varepsilon_{33}^2\right) + c_{12}(\varepsilon_{11}\varepsilon_{22} + \varepsilon_{11}\varepsilon_{33} + \varepsilon_{22}\varepsilon_{33}) + \frac{1}{2}c_{44}\left(\varepsilon_{12}^2 + \varepsilon_{13}^2 + \varepsilon_{23}^2\right) \tag{2.34}$$

$$E_{\text{hex el}} = \frac{1}{2}c_{11}\left(\varepsilon_{11}^2 + \varepsilon_{22}^2\right) + \frac{1}{2}c_{33}\varepsilon_{33}^2 + c_{12}\varepsilon_{11}\varepsilon_{22} + c_{13}(\varepsilon_{11} + \varepsilon_{22})\varepsilon_{33} + \frac{1}{2}c_{44}\left(\varepsilon_{13}^2 + \varepsilon_{23}^2\right) + (c_{11} - c_{12})\varepsilon_{12}^2 \tag{2.35}$$

2.10.5 Magnetic surface and interface anisotropies

Due to the broken symmetry at the interface of low-dimensional ferrites, including nanoparticles, nanowires, nanodots, thin films, and multilayers with a high surface-to-volume ratio, the surface and interface anisotropies are very important parameters in understanding their magnetic properties. The effective anisotropy constant, K^{eff}, has two parts: one determining the volume and the other describing the surface contribution, according to the following equation:

$$K^{\text{eff}} = K^V + 2K^S/d \tag{2.36}$$

where d is the size of the magnet, K^V is the volume dependence of magnetocrystalline anisotropy, and K^S is the surface dependence of magnetocrystalline anisotropy. After finding the critical size d_c, one can determine perpendicular and in-plane magnetization orientations for $d < d_c$ and $d > d_c$, respectively.

2.11 Magnetic domains

Magnetic materials comprise domains magnetically saturated in different directions. Without applying a strong magnetic field, a random arrangement of domains is induced, leading to zero net magnetization of the material, as shown in Fig. 2.19.

Domain walls are distinct boundaries between adjacent domains, reducing the energy of the magnetic system. The domain wall width is determined by balancing the competing energy contributions. Magnetocrystalline anisotropy and exchange interaction strength play the main roles in determining the width of domain walls. Normally, a strong magnetocrystalline anisotropy will induce a narrow domain wall, whereas a strong exchange interaction will form a wider wall. The real image of magnetic domains, taken by optical microscopy, is shown in Fig. 2.20.

Consider a magnetized specimen with free poles at the interface that can generate a demagnetizing field and large magnetostatic energy, as schematically illustrated in Fig. 2.21A. In this case, the demagnetizing field opposes the specimen magnetization. By breaking down the single domain specimen into N domains, the total

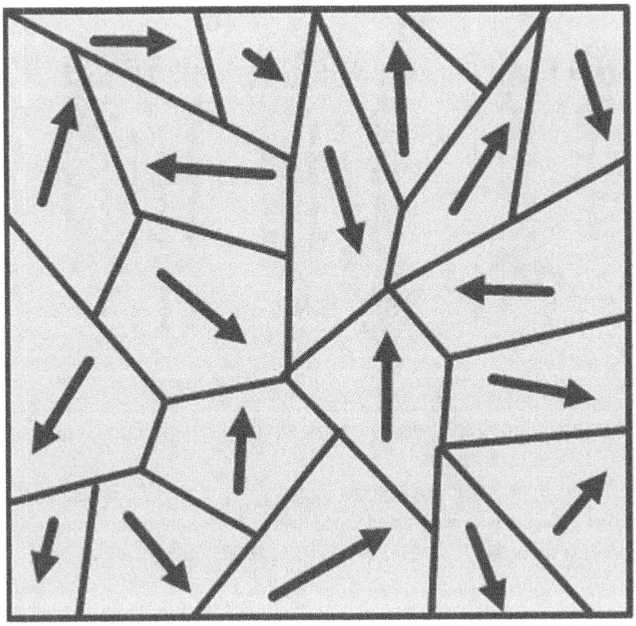

FIGURE 2.19

Random arrangement of magnetic domains.

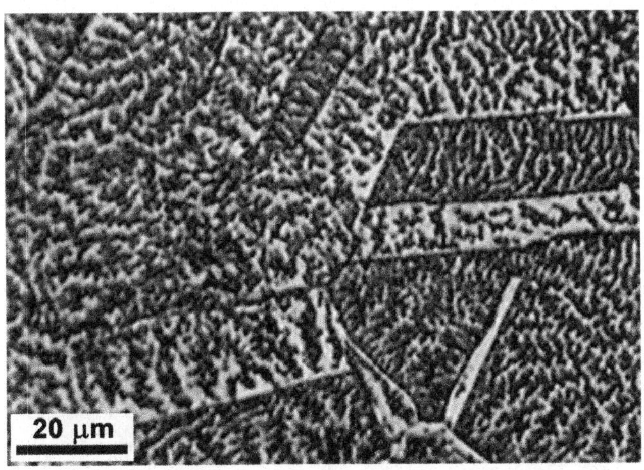

FIGURE 2.20

Magnetic domains of a bulk magnet obtained using optical microscopy.

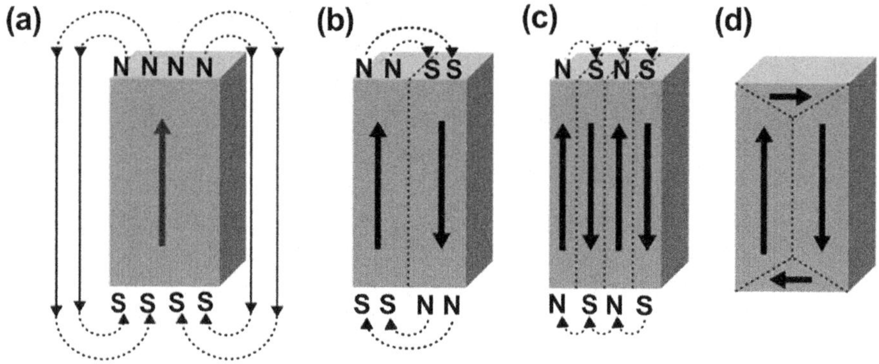

FIGURE 2.21

Schematic illustration of magnetic domains: (A) single domain, (B) two domains, (C) four domains, and (D) closure domains.

From Harris, I. R., Jewell, G. W., Kilner, J. A., Skinner, S. J., Irvine, S. J. C., & Edwards, P. P. (2012). Chapter 19 – Rare-earth magnets: properties, processing and applications. In Woodhead publishing series in energy *(pp. 600–639). Woodhead Publishing. https://doi.org/10.1533/9780857096371.4.600.*

magnetostatic energy is reduced by a factor $1/N$. Hence, the magnetic specimen with four domains depicted in Fig. 2.21C has a quarter of the magnetostatic energy shown in Fig. 2.21A. Dividing the specimen into several blocks increases the surface anisotropy energy, magnetostriction, and exchange energy. Finally, the magnetostatic energy becomes zero after forming closure domains, maintaining the system in the minimum of free energy.

The formation of closure domains is shown in Fig. 2.22. It is perceived that the closure domain is formed by minimizing the anisotropy energies. The orientation of the magnetization vector in the domain is shown as different colors.

FIGURE 2.22

Formation of the closure domain in a magnetic nanowire.

2.12 Vibrating sample magnetometer

Various machinery and techniques have been devised to measure the magnetization versus magnetic field for ferrites. Notably, the Faraday balance and the alternating gradient force magnetometer operate with force techniques, measuring the force exerted on a magnetized sample in a magnetic field gradient. On the other hand, the hysteresis meter and VSM are based on inductive techniques, measuring the voltage induced by a changing flux (Speliotis, 2005). The power of the VSM instrument is considered in more detail since it has numerous applications for measuring the magnetic characteristics of materials, especially ferrites. The VSM can measure the magnetic moment of a specimen vibrating in a uniform magnetic field. Many sources of error existing in other techniques can be eliminated using VSM. This measurement technique can also provide information about the variation of magnetization versus magnetic field under different temperatures and crystallographic orientations. Using VSM, it is possible to precisely detect magnetization variation as small as 10^{-5}–10^{-6} emu while having a stability of one part in 10^4 (Foner, 1959).

A typical schematic representation of the VSM instrument is given in Fig. 2.23. The sample is initially located in a uniform external magnetic field situated between a pair of pick-up coils. It is then vibrated by a vibrator in a sinusoidal fashion with a small amplitude ($\sim$1 mm) and a frequency of 50 Hz. The magnet's magnetization provides a magnetic stray field, changing the magnetic flux (Niazi et al., 2000). An AC voltage proportional to the magnetic moment is induced in the coils with the same frequency. A lock-in amplifier is used to control the frequency of vibrations and detect the signal. The output values are in millivolts (mV), which are normally calibrated to emu using a nickel standard sample (1 mV = 0.0847 emu). A Hall probe is employed to determine the applied magnetic field intensity. By utilizing a computer interface, the magnetic field and the signal proportional to the magnetic moment can be simultaneously recorded, resulting in a hysteresis loop.

2.12.1 Low-temperature vibrating sample magnetometer

A VSM instrument equipped with a cryogenic system for measuring magnetic properties at a low temperature is shown in Fig. 2.24. In this way, magnetization versus magnetic field can be traced in the range of 77–298 K. The hysteresis loops can be plotted very precisely in a temperature step of 1 K. The power of the low-temperature VSM is to determine the ferrimagnetic to superparamagnetic transition temperature in ferrite nanoparticles. The samples used can be powders, small bulks, and thin films. One of the main capabilities of low-temperature VSM is plotting zero-field cooling (ZFC) and field cooling (FC) curves. In ZFC, the ferrite must be cooled without applying a magnetic field to the desired temperature. During the heating of the specimen, the magnetic data should be collected under the application of a magnetic field. In FC, the ferrite must be cooled by applying a

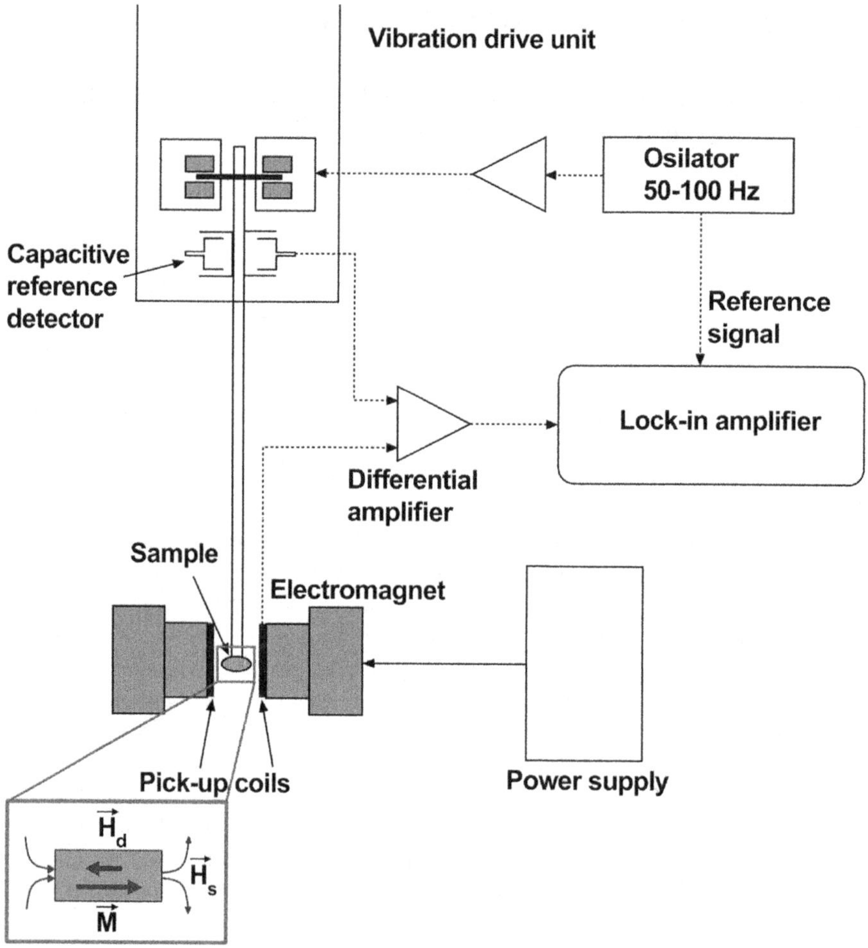

FIGURE 2.23

A schematic representation of the vibrating sample magnetometer instrument.

From Thomson, T. (2014). Magnetic properties of metallic thin films. In Metallic films for electronic, optical and magnetic applications *(pp. 454–546). Elsevier. https://doi.org/10.1533/9780857096296.2.454.*

magnetic field to the desired temperature, and the data can be collected in two ways, including field-cooled cooling or field-cooled warming. By intercepting ZFC and FC curves, one can estimate the blocking temperature.

2.12.2 High-temperature vibrating sample magnetometer

It is well known that the magnetization of magnetic materials is reduced when their temperature increases, arising from the effect of thermal energy. Understanding the

FIGURE 2.24

The low-temperature vibrating sample magnetometer (VSM) (left) and high-temperature VSM (right) for measuring hysteresis loops at elevated temperatures.

magnetic properties of ferrites at an elevated temperature is important, as it can be utilized in exchange spring magnets and high-frequency applications at high temperatures. By flowing hot inert gases in a ceramic tube and inserting ferrites into such an environment, one can measure hysteresis loops at high temperatures. The VSM instrument shown in Fig. 2.24 can plot magnetization versus magnetic field at different temperatures up to 1223K in a step of 278K, thereby measuring the Curie temperature. Temperature dependence of coercivity and magnetization of ferrites can also be determined by the high-temperature VSM.

2.13 *B–H* tracer

Many experimental arrangements can measure the *B* versus *H* curve of bulk samples. Consider a magnetic material in a toroid shape with a primary and secondary winding. If the current *I* flows in the primary winding, an average magnetic field of $H = 4\pi nI$ (in which n is the number of turns per unit length) will be generated. The direction of the field is parallel to the toroid axis. Increasing the toroid diameter and length also increases the uniformity of the magnetic field. When the current is changed, a voltage is induced on the secondary winding according to Faraday's law. By representing ε as the induced electromagnetic force, N as the number of turns of the secondary winding, and A as the effective cross section, one can obtain B as follows:

$$B = -\frac{1}{NA}\int \varepsilon dt \tag{2.37}$$

The variation in the magnetic flux density can be determined by the deflection of the ballistic galvanometer connected to the secondary winding when changing the

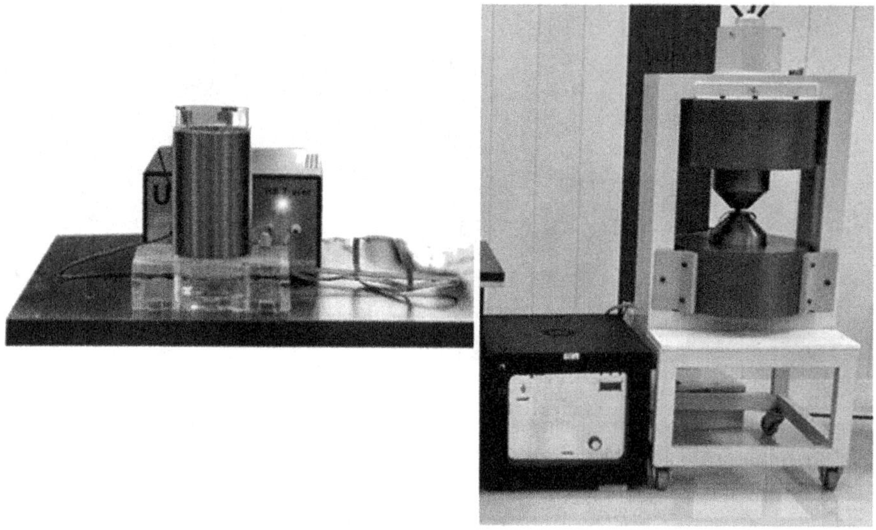

FIGURE 2.25

$B-H$ tracer used for measuring magnetic properties of soft (left) and hard (right) magnets.

current quickly. If an AC current is applied to the primary winding and the integrator is inserted in the secondary winding, the hysteresis loop can be displayed on the oscilloscope screen (Morrish, 2001). The secondary winding often consists of a small coil, called a pickup coil, that is inserted inside the solenoid (the primary winding). Several applicable parameters, including the maximum energy product, coercive field, and residual flux density B_r, can be extracted from $B-H$ curves. The coercivity of very soft magnets such as Mn–Zn ferrite, permalloy, and Mu-metals can be precisely determined using a solenoid with suitable accuracy. For hard ferrites, including strontium and barium ferrites, large magnets with an accuracy of 1 Oe are employed (Fig. 2.25).

2.14 Interpretation of the *M–H* loop

The magnetic properties of a sample can be evaluated by a plot of magnetization M as a function of an applied magnetic field H, providing the characteristic $M-H$ hysteresis loop (see Fig. 2.26). During one cycle of a hysteresis loop, the potential energy returns to its original value, consuming the hysteresis loss as heat. The hysteresis loss is measured based on the area surrounded by the hysteresis loop. The hysteresis loss, W, is expressed by

$$W = \oint H \cdot dM \tag{2.38}$$

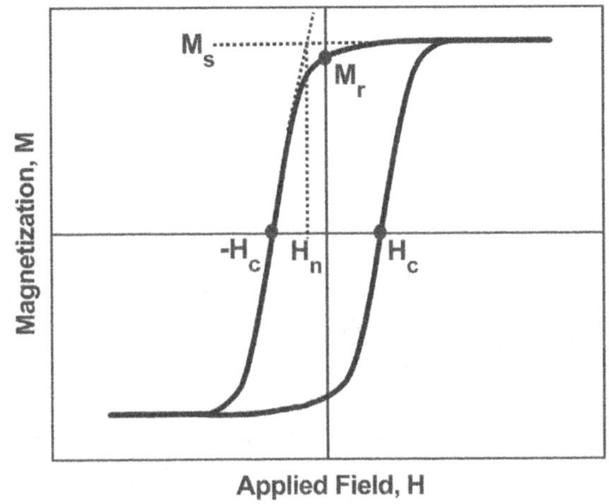

FIGURE 2.26

The *M–H* hysteresis loop and its corresponding magnetic parameters.

In the demagnetizing state (i.e., the origin of the hysteresis loop), magnetization vectors are isotropically distributed, reflecting no magnetic features of the sample. When the external field is applied in the plus (positive) direction, the magnetization vectors with the minus direction are reversed by domain wall displacement. Further increasing the magnetic field causes all magnetization vectors to be aligned parallel to the applied field, giving rise to a saturation state. Decreasing the intensity of the magnetic field from the saturation state causes the vectors to turn back to the nearest easy axes, thereby reaching the remanence magnetization. Further decreasing the applied field in the minus direction reverses the magnetization vectors from the plus direction, reaching the coercive field and forming a different polar distribution in the coercivity point (Chikazumi, 1984).

The magnetic parameters which can be extracted, either directly or indirectly, from the hysteresis loop are explained below:

1. Saturation magnetization, M_s, is the maximum magnetization of magnetic material when fully saturated. The strength of the dipole moments and the packing density of atoms in unit volume greatly influence this parameter. The atomic dipole moment is affected by the nature of the atom and the electronic structure of the specimen. The packing density is determined by the crystal structure and the presence of any nonmagnetic secondary phase, inclusions, and other defects in the specimen. In other words, M_s depends on the magnitude of alignment between magnetic moments. Thermal effects cause the moments to be misaligned, reducing magnetization saturation. It is well known that soft ferrites (e.g., Mn–Zn and Ni–Zn ferrites) have high saturation magnetization values.
2. H_c is the magnetic field at zero magnetization, indicating the level of difficulty to demagnetize the material. It should be noted that although the magnetization

value at both the demagnetizing state and coercivity is zero, the corresponding magnetization of domains is completely different. Hard ferrites, such as barium and strontium ferrites, have a high coercivity and moderate saturation magnetization, whereas soft spinel ferrites show low coercivity. So far, several coercivity mechanisms have been proposed, as tabulated in Table 2.6. Notably, Brown's theorem, suggested for an ideal magnet, indicates that the magnetocrystalline anisotropy field governs the magnetization reversal. Coherent rotation of magnetization that is nearly independent of the size of particles is a well-known type of coercivity mechanism. Alternatively, the exchange energy plays an important role in the nonuniform curling reversal mode. The coercivity formula of an infinite cylinder and a sphere is also expressed. It has been found that the domain wall pinning is a very important mechanism affecting coercivity. In an extreme case with no defects in the crystals, perfectly flat walls can be formed, minimizing both the wall energy and magnetostatic energy. The coercivity depends explicitly on L_xL_y (Bertotti, 1998).

Basically, the domain wall acts as a one-dimensional (1-D) object, tailoring the 1-D defect density of $\rho LxLy$. The impact of pinning increases with an increase in the degree of bowing (1-D and 2-D). This is because the coercivity is related to $(\rho E_p^2)^\alpha$ through increasing exponent 1/2, 2/3, and 1. The possible pinning source in polycrystalline magnets is the presence of defects, including point defects (e.g., vacancies and interstitial and substantial atoms), line defects (e.g., edge, screw, and mixed dislocations), and surface and volume defects (e.g., twin boundaries, stacking faults, voids, porosities, and unwanted secondary phases). Specifically, the dislocations induce stronger pinning effects through magnetoelastic coupling. The presence of a nonmagnetic phase in the form of a matrix, secondary phases grown in a magnetic matrix, and residual stress imposed in the fabrication method may also act as pinning sites, playing important roles in determining the coercivity. As well, grain boundaries, phase boundaries, and dislocation tangles influence coercivity. Intrinsic fluctuations in the local exchange and anisotropy constants, formation of short-range chemical order, and time-dependent structural rearrangement in amorphous magnets can also lead to domain wall pinning.

The role of particle size in coercivity is complicated. The given general formula of coercivity implies that it has an opposite trend as a function of size for large diameter particles. Since the density of grain boundaries increases with decreasing the size of particles and grains, the pinning site density increases, thus enhancing the coercivity. It is worth mentioning that the limit of superparamagnetism must always be considered. Brown's equation is not satisfied in real materials since it does not consider several important factors, such as possible inhomogeneity in magnetic properties, particle interaction, shape factors, thermal activation, and vestigial domains acting as domain nucleation sites under a reverse field. Consequently, Brown's paradox is a correction to include such effects in real materials. In ferrites, the aforementioned

Table 2.6 Coercivity mechanisms and related expressions in magnetic materials.

Coercivity mechanisms	Corresponding formula	Definitions and supplementary explanations
Brown's equation	$H_C = \frac{2K_1}{\mu_0 M_s} - N_{\|} M_S \equiv H_a - N_{\|} M_S$	H_a is magnetocrystalline anisotropy field. K_1 is magnetocrystalline anisotropy constant. $N\|$ is parallel demagnetization factor (equations are not met in real materials)
Coherent rotation of magnetization	$H_C = \frac{2K_1}{\mu_0 M_s} + \left(N_{\perp} - N_{\|}\right) M_S$	$N\perp$ is perpendicular demagnetization factor independent of particle size
Curling mode of magnetization	$H_C = \frac{2K_1}{\mu_0 M_s} - N_{\|} M_S + \frac{L_{Ex}^2}{R^2} q_C^2 M_S$	$L_{Ex} = \left(\frac{2A}{\mu_0 M_s^2}\right)^{\frac{1}{2}}$: exchange length, A: exchange stiffness constant, $q_c =$ 1.8412 for infinite cylinder and $q_c = 2.0816$ for sphere
Rigid wall	$H_C = \frac{C}{\mu_0 M_s}\left(\frac{\rho}{L_x L_z}\right)^{\frac{1}{2}} \frac{E_P}{\delta_P^{\frac{1}{2}}}$	E_p is pinning energy. $\rho L_x L_y$ is one-dimensional defect density, C is constant and $\delta_P = \sqrt{\frac{A}{K_1}}$
Wall bowing: one dimensional	$H_C = \frac{C}{\mu_0 M_s}\left(\frac{\rho}{L_x}\right)^{\frac{2}{3}} \frac{E_P^{\alpha 3}}{\gamma_w^{\gamma 3} \delta_p}$	$\gamma_W = 4(AK_1)^{\frac{1}{2}}$
Wall bowing: two dimensional	$H_C = \frac{C}{\mu_0 M_s} \frac{\rho E_P^2}{\gamma_w \delta_P^2}$	$\delta_P = \sqrt{\frac{A}{K_1}}$ and ρ is average number of defects
Dislocations	$H_C = C \frac{G \lambda_s b}{\mu_0 M_s}\left(\frac{\rho_d}{L_x L_z}\right)^{\frac{1}{2}} (l\delta_P)^{\frac{1}{2}}$	G is shear modulus, b amplitudes of Burgers vector, λ_s is saturation magnetostriction, l is typical length of dislocation, and ρ_d is dislocation density
Residual stress	$H_C \cong \frac{3}{5\pi} \frac{\nu}{\mu_0 M_s} \frac{\lambda_s^2 \sigma^2}{K_1}\left(1.386 - \frac{1}{2}\ln k\right)$	ν is volume fraction of materials, σ is residual stress, k is hardness parameter $\lambda_S \sigma \ll K_1, k = \frac{2K_1}{\mu_0 M_s} = \frac{H}{M_s}, H = \frac{2K_1}{\mu_0 M_s}$
Nonmagnetic inclusions	$H_C \cong \frac{2}{\pi} \frac{\nu}{\mu_0 M_s} K_1 \left(0.386 - \frac{1}{2}\ln k\right)$	$k = \frac{2K_1}{\mu_0 M_s} = \frac{H_a}{M_s}$
Particle size	$H_C = \frac{2K}{\mu_0 M_s}\left(1 - \sqrt{\frac{25k_B T}{\alpha K D^3}}\right) = \frac{2K}{\mu_0 M_s}\left[1 - \left(\frac{D_B}{D}\right)^{\frac{3}{2}}\right]$	$\alpha K D_B^3 = 25k_B T$, α is dependent on the particle geometry, and D_B is critical diameter
Brown's paradox	$H_C = \alpha \frac{2K_1}{\mu_0 M_s} - N_{eff} M_S$	α and N_{eff} (effective demagnetization factor) are phenomenological parameters measuring deviation from the ideal condition

mechanisms are very important, and normally, several of them are involved simultaneously.

3. Remanent magnetization, M_r, is magnetization after the applied field has been removed. Basically, high remanent magnetization is desired in ferrite-based high-density recording media. Several formulas of M_r could be found in the literature. For example, the remanence of material with an isotropic distribution of easy axes and uniaxial magnetic anisotropy, such as cobalt, is given by

$$M_r = \int_0^{\frac{\pi}{2}} M_s \cos\theta \sin\theta d\theta = \frac{M_s}{2} \tag{2.39}$$

The remanent magnetization in most highly stressed magnets is about one-half of their saturation magnetization (Chikazumi, 1984).

4. Remanent squareness ratio, S, is defined as the ratio of the remanent magnetization to the saturation magnetization (M_r/M_s), ranging from 0 to 1. It must be noted that a high S value (nearly equal to 1) along with a high nucleation field (discussed below) is effective in reducing the effect of disturbing external and internal magnetic fields and spin fluctuations imposed by thermal relaxation.

5. Coercivity squareness, S^*, and the hysteresis loop slope parameter, α, are defined by Eqs. (2.40) and (2.41), respectively. Both equations represent the slope of the hysteresis loop at H_c, indicative of the degree of intergranular exchange coupling in magnets:

$$\left.\frac{dM}{dH}\right|_{H_c} = \frac{M_r}{H_C(1 - S^*)} \tag{2.40}$$

$$\alpha = 4\pi \left(\frac{dM}{dH}\right)_{(H=H_c)} \tag{2.41}$$

The degree of intergranular exchange coupling significantly affects the features of hysteresis loops. The demagnetizing field reduces the slope of the hysteresis loop, whereas the intergranular exchange coupling has an opposite effect on it (Campbell, 1996). Assuming only the demagnetizing effect with completely exchange decoupled grains exists, α becomes unity. Nevertheless, the intergranular exchange coupling is present in practical cases, thereby reducing the demagnetization effects, which in turn results in $\alpha > 1$.

6. The anisotropy field, H_a, is an internal field and equivalent to the torque applied to the spins to make them aligned with the easy axis direction, thereby minimizing the energy (Spaldin & Mathur, 2003).

7. The nucleation field, H_n, is the magnetic field obtained via the tangent to the hysteresis loop at H_c. Moreover, H_n represents the field at which some grains start to reverse their magnetization, indicating the stability of oriented spins against thermal relaxation and external and internal magnetic fields (O'Hendley, 1999).

2.15 Henkel plot

Magnetic interactions in ferrites can be determined by the Henkel plot method (or δM measurement). For this purpose, two different remanent magnetizations should be provided. The first one is $M_r(H)$, which can be reached after applying and removal of the external field subsequently. The second one is demagnetization remanence $M_d(H)$, which can be obtained by saturating the ferrite in one direction, followed by applying and removing a direct external field in the reverse direction. Note that $M_r(H)$ and $M_d(H)$ are normalized by $M_r(\infty)$: $M_r = M_r(H)/M_r(\infty)$ and $M_d = M_d(H)/M_d(\infty)$. The variation of $\delta M = (M_d(H)\text{-}[1-2M_r(H)])$ with the applied field is an alternative to the exploration of the interaction types (Wohlfarth, 1958).

2.16 Alternating current magnetic susceptibility

AC magnetic susceptibility measurement is a standard technique to study the dynamical magnetic properties of bulk and nanostructured materials (Rostamnejadi et al., 2009). Depending on the material and magnetic field, it presents a complicated response function, being used to explore static and dynamic magnetic properties. According to the first law of thermodynamics, the free energy, F, of a magnetic system in the presence of a magnetic field, H, is given by the following equation:

$$dF = -SdT - PdV - \mu_0 MdH \tag{2.42}$$

where T is the temperature, V is the volume, and M is the magnetization. In turn, F is related to the partition function, Z, as follows:

$$F = -k_B T \ln Z \tag{2.43}$$

where k_B is the Boltzmann constant. The partition function of a canonical ensemble is obtained from the system Hamiltonian H by

$$Z = tr\left(e^{-\beta \hat{H}}\right) \tag{2.44}$$

where $tr()$ is the trace of a matrix and β is the thermodynamic beta defined as $1/k_B T$. The magnetization is obtained from F using Eq. (2.45):

$$M = -\frac{1}{\mu_0}\frac{\partial F}{\partial H} \tag{2.45}$$

Consequently, the magnetic susceptibility, χ, can be obtained from magnetization through the following relation:

$$\chi = \frac{\partial M}{\partial H} = -\frac{1}{\mu_0}\frac{\partial^2 F}{\partial H^2} \tag{2.46}$$

It is assumed that an AC magnetic field is applied to the magnetic system as follows (Boca, 1999):

$$H_{AC} = H_0 \cos(\omega t) \tag{2.47}$$

Usually, the magnetization cannot follow the applied field, giving rise to a time-delayed response as

$$M_{AC} = M_0 \cos(\omega t - \theta) = M_0 \cos(\theta)\cos(\omega t) + M_0 \sin(\theta)\sin(\omega t) \tag{2.48}$$

$$= \chi' H_0 \cos(\omega t) + \chi'' H_0 \sin(\omega t) \tag{2.49}$$

in which $\chi' = M_0/H_0 \cos(\theta)$ and $\chi'' = M_0/H_0 \sin(\theta)$. Therefore, in the presence of the AC magnetic field, χ is a complex quantity as given by

$$\chi = \chi' + i\chi'' \tag{2.50}$$

In general, the real part of the complex susceptibility, χ', represents the storage of energy. The imaginary part, χ'', is indicative of the energy dissipation due to the dynamic magnetic processes in a magnetic material. In fact, the dynamic magnetic response of a nonlinear magnetic material has a complicated function, as given by higher-order harmonic terms according to the following equation:

$$M_{AC} = H_0 \sum_n \left[\chi'_n \cos(n\omega t) + \chi''_n \sin(n\omega t)\right] \tag{2.51}$$

in which the nth-order AC magnetic susceptibility, χ_n, is defined as

$$\chi_n = \left[\chi'_n \cos(n\omega t) + \chi''_n \sin(n\omega t)\right];\ n = 1, 2, 3, \ldots \tag{2.52}$$

The real part of χ_n is given by

$$\chi'_n = \frac{1}{\pi H_0} \int_0^{\pi} M(t)\cos(n\omega t)d(\omega t) \tag{2.53}$$

Alternatively, the imaginary part of χ_n is expressed by

$$\chi''_n = \frac{1}{\pi H_0} \int_0^{\pi} M(t)\sin(n\omega t)d(\omega t) \tag{2.54}$$

The magnetic field acting on a magnetic material consists of an external magnetic field H_{ext} and an internal demagnetization field $H_d = -NM$ as follows:

$$H = H_d + H_{ext} = H_{ext} - NM \tag{2.55}$$

where N $(0 \le N \le 1)$ is the demagnetization factor. The external, internal, and total magnetic susceptibilities are defined as $\chi_{ext} = dM/(dH_{ext})$, $\chi_{int} = dM/(dH_d)$, and $\chi = dM/dH$, respectively. The relation between internal and external magnetic susceptibilities is given by

$$\chi_{\text{int}} = \frac{\chi_{\text{ext}}}{1 - N\chi_{\text{ext}}} \tag{2.56}$$

The real and imaginary parts of the internal AC susceptibility in Eq. (2.56) are obtained based on external susceptibility as follows:

$$\chi'_{int} = \frac{\left[\chi'_{ext} - N\left(\chi'^{2}_{ext} + \chi''^{2}_{ext}\right)\right]}{\left[N^2\left(\chi'^{2}_{ext} + \chi''^{2}_{ext}\right) - 2N\chi'_{ext} + 1\right]} \tag{2.57}$$

$$\chi''_{int} = \frac{\chi''_{ext}}{\left[N^2\left(\chi'^{2}_{ext} + \chi''^{2}_{ext}\right) - 2N\chi'_{ext} + 1\right]} \tag{2.58}$$

Therefore, in the case of magnetic nanostructures, it may be necessary to correct the demagnetization field contribution to the magnetic susceptibility.

The AC magnetic susceptibility measurement could distinguish between the superparamagnetism and spin glass and superspin glass behaviors in ferrite nanoparticles. In this regard, Fig. 2.27 shows the variation of real and imaginary parts of the effective magnetic susceptibility versus temperature at different frequencies for typical ferrite nanoparticles. The temperature at which susceptibility reaches a maximum value is the blocking temperature (Rostamnejadi et al., 2009).

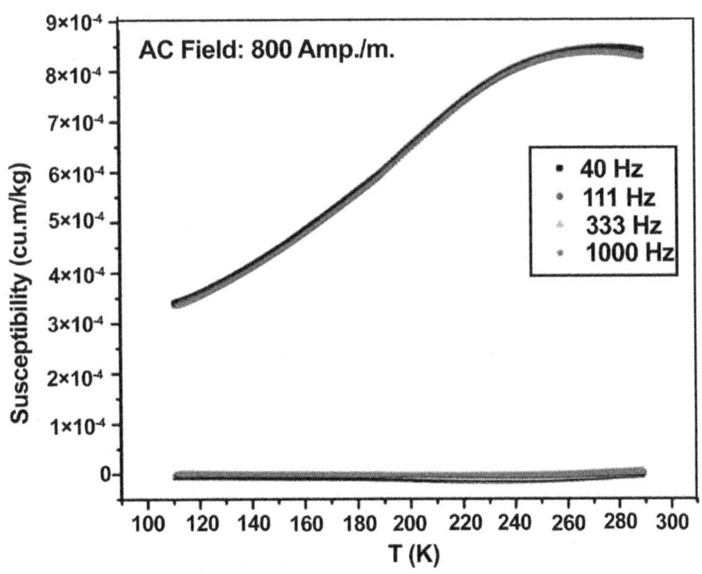

FIGURE 2.27

Variation of magnetic susceptibility versus temperature at different frequencies for typical ferrite nanoparticles.

Table 2.7 provides useful information regarding the phenomenological models and related formulas for interacting and noninteracting nanoparticles. Generally, the frequency dependence of the magnetization relaxation for noninteracting ferrite nanoparticles can be explained by the Néel-Brown model. The blocking temperature, T_B, is defined as a temperature at which the potential barrier (effective magnetic anisotropy energy) is equal to the thermal energy. When $T > T_B$, the magnetization vector is aligned in the direction of an external field if nanoparticles are in the superparamagnetic state. When $T < T_B$, the thermal energy becomes less than the potential barrier energy. In the case of interacting magnetic nanoparticles, the Vogel-Fulcher law describes the relaxation of magnetization in superparamagnets (Dormann et al., 1997, 1999).

The frequency dependence of T_B obtained by plotting the AC magnetic susceptibility provides two empirical parameters c_1 and c_2. These are employed to distinguish between the superparamagnetic behavior and the spin freezing in spin glass systems (Dormann et al., 1997, 1999), according to the following equations:

$$c_1 = \frac{\Delta T_f}{T_f \Delta(\log_{10} f)} \tag{2.59}$$

$$c_2 = \frac{T_f - T_0}{T_f} \tag{2.60}$$

where T_f reflects the mean value of T_B, and ΔT_f is the difference between blocking temperatures measured in the frequency interval of $\Delta(\log_{10} f)$. Based on c_1 and c_2 values presented in Table 2.8, it is possible to determine the types of magnetic interactions in magnetic nanoparticles (Dormann et al., 1997, 1999).

2.17 First-order reversal curve measurement

It is worth noting that first-order reversal curve (FORC) analysis was initially employed to investigate the magnetic properties of mineral species. Conventionally, hysteresis curves provide researchers with overall information about the average behavior of magnetic systems, such as coercive field value and remanence ratio. Remarkably, FORC measurements reveal the interplay between average and individual processes occurring in arrays of magnetic nanostructures, outperforming other characterization techniques, including δH plots, isothermal remanent magnetization, and direct current demagnetization curves (Pike & Fernandez, 1999).

2.17.1 First-order reversal curve analysis background

As an advanced and powerful method for analyzing detailed magnetic properties such as coercive field distribution (CFD), interaction field distribution (IFD), magnetic states (single-domain, multidomain, pseudo-single-domain, and superparamagnetic), magnetization reversal processes (vortex domain wall, transverse

Table 2.7 Phenomenological models and related formulas for interpretation of magnetic interactions in nanoparticles.

Phenomenological model	Formula	Description
Néel-Brown (noninteracting superparamagnetic behavior)	$\tau = \tau_0 \exp\left(\frac{E_a}{k_B T}\right)$	E_a: the effective magnetic anisotropy energy, $E_a = k_{eff} V sin^2\theta$ V: particle volume θ: angle between the magnetic moment and the easy axis τ: characteristic relaxation time$\tau = 1/f$ f: measurement frequency τ_0: relaxation time for individual magnetic moments, ranging between 10^{-9}–10^{-13}s
Vogel-Fulcher (interacting magnetic nanoparticles)	$\tau = \tau_0 \exp\left(\frac{E_a}{k_B (T - T_0)}\right)$	T: characteristic temperature, showing the onset of the blocking process T_0: effective temperature, indicating the strength of the interparticle interaction
Critical slowing down (superspin glass behavior)	$\tau = \tau_0 \left(T/T_g - 1\right)^{-z\nu}$	T_g: spin glass transition temperature ν: correlation length critical exponent ξ : correlation length : $\xi \approx \left(T/T_g - 1\right)^{-\nu}$ $z\nu$ is a critical exponent in the range of 7–8 for spin glass nanoparticles

Table 2.8 The values of c_1 and c_2 for different types of interaction.

Type of interaction	c_1	c_2
Non-interacting magnetic particles	0.1–0.13	1
Weak interaction (inhomogeneous freezing)	0.03–0.06	0.3–0.6
Medium to strong interaction (superspin glass state)	0.005–0.02	0.07–0.3

From Dormann, J., Fiorani, D., & Tronc, E. (1999). On the models for interparticle interactions in nanoparticle assemblies: comparison with experimental results. Journal of Magnetism and Magnetic Materials, *202(1), 251–267. https://doi.org/10.1016/S0304-8853(98)00627-1.*

domain wall, and single vortex), FORC analysis has shown promising results for the past 2 decades. In this regard, many-body interacting systems such as nanowire, nanopillar, nanotube, nanodot, and multilayer film arrays have been investigated by the FORC analysis to reach further insights into their magnetic properties (Béron et al., 2008; Proenca et al., 2013). This approach is quite helpful for deeply analyzing the magnetic characteristics of soft ferrites. However, very few studies thus far have used FORC analysis for this purpose. Therefore, other soft magnetic systems, including metallic nanowires, are herein explained in detail.

2.17.2 First-order reversal curve analysis measurements

The FORC method is calculated through a mathematical model called the classical Preisach model (Preisach, 1935). A set of elementary processes known as hysterons represent hysteresis mathematically, characterizing the parameters coercivity (H_c) and interaction or bias field (H_u), according to Fig. 2.28.

Experimentally speaking, FORC analysis is obtained as follows. Initially, the magnetic sample (consisting of bulk or nanostructured materials) is saturated with a maximum magnetic field labeled as $H_{\max}$. This strong magnetic field is then reduced to a reversal field labeled as H_r, in such a way that $H_r < H_{\max}$. Afterward, the magnetic field is reversed and swept back to the saturation state of the sample. In turn, this results in a series of remanence field steps (Fig. 2.29). The aforementioned procedure is repeated for numerous levels of H_r while also measuring the magnetization M (H_r, H) at each step (Béron et al., 2008; Proenca et al., 2013).

The FORC distribution or density is then defined as follows (Fanny Béron et al., 2010):

$$\rho_{\text{FORC}}(H, H_r) = -\frac{1}{2}\frac{\partial^2 M(H, H_r)}{\partial H \partial H_r}(H > H_r) \tag{2.61}$$

A FORC diagram can be plotted in a 45-degree rotated $\{H_c, H_u\}$ coordinate system, in which $H_c = (H - H_r)/2$ and $H_u = (H + H_r)/2$. For instance, a FORC diagram obtained from a nanostructured system is shown in Fig. 2.29.

Mathematically speaking, the polynomial surface $a_1 + a_2H + a_3H^2 + a_4H_r + a_5H_r^2 + a_6HH_r$ is fitted over a local grid centered on a point to calculate

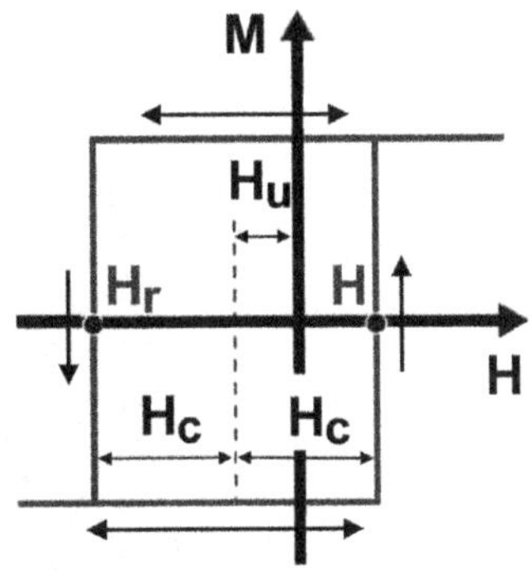

FIGURE 2.28

The classical Preisach model of hysteron, representing it mathematically. Here, the magnetization abruptly switches down at $H_r = -(H_c + H_u)$, while switching back up at a certain value of $H = (H_c - H_u)$. Note: H_c and H_u parameters are the coercivity and bias fields relating to the hysteron mathematically.

From Béron, F., Clime, L., Ciureanu, M., Ménard, D., Cochrane, R. W., & Yelon, A. (2007). Reversible and quasireversible information in first-order reversal curve diagrams. Journal of Applied Physics, 101*(9), 09J107.*

$r(H_r, H)$ at that point (Muxworthy & Roberts, 2007). It should be noted that the polynomial surface mixed second derivative is a_6 and is scaled by a factor of -0.5 based on Eq. (2.61). Thus, the $-0.5a_6$ value represents $r(H_r, H)$, which would be the desired point on the grid. The $r(H_r, H)$ is assessed at all points (located on the grid) within the FORC diagram boundaries. After contouring the corresponding data, they are plotted through a varying color map ranging from red (maximum ρ_{FORC}) to blue (minimum ρ_{FORC}), thereby representing the FORC distribution (see Fig. 2.29B). It is worth mentioning that the term $(2SF + 1)^2$ is known as a fitted data point number, where SF is considered a smoothing factor. Increasing SF decreases noise contribution and increases FORC diagram smoothness (Muxworthy & Roberts, 2007). For example, Fig. 2.30A–C illustrates FORC diagrams of FeCo nanowire arrays plotted using various SF values (ranging between 2 and 6). In the case of well-behaved magnetic samples, SF is selected to be 2. Alternatively, the SF of magnetic samples with a small signal-to-noise ratio is selected to be 9.

Measurement instruments for obtaining hysteresis data with high sensitivity include the VSM, SQUID magnetometer, and alternating gradient magnetometer (AGM). While the sensitivity of the VSM is smaller than that of AGM, it benefits from the advantage of holding larger magnetic samples (up to several grams). Therefore, the VSM is preferred over AGM and is commonly used for FORC measurements (Roberts et al., 2014).

Several effective parameters (arising from the magnetic properties of the sample) must be selected to measure FORCs. These parameters include the saturating field

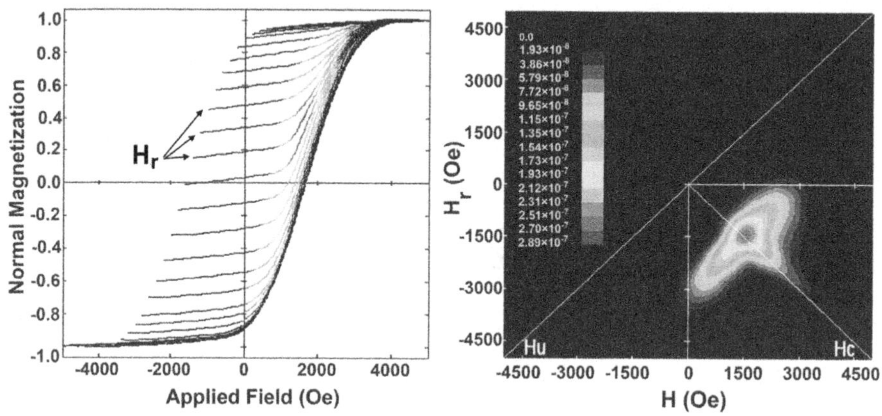

FIGURE 2.29

(A) Set of experimental first-order reversal curves and (B) First-order reversal curve diagram of Fe nanowire arrays.

(H_{sat} or H_{max}), ΔH_r range, the averaging time (t_{avg}), the field increment (ΔH), and the number of FORCs to be measured (N). To optimize the experimental FORC, ΔH and ΔH_r ranges can be varied based on the coercivity and H_{sat} of the sample. However, it is required to keep the $\Delta H_r/\Delta H$ ratio between 1 and 2.5 (high and low hysteresis susceptibility, respectively). While decreasing ΔH allows for better resolution of FORC diagram features, it considerably increases both FORC measurement time and FORC noise (Roberts et al., 2014).

On the other hand, the measurement time of each data point (t_{avg}) can vary between 0.1 and 1 s. By increasing t_{avg}, the overall time needed to measure FORCs will increase while also improving the signal-to-noise ratio. It should be noted that an increase in t_{avg} would not remove the noise induced by the electromagnet, thereby increasing undesired effects involved with the drift of the instrument (Roberts et al., 2014). Consequently, depending on the sample, t_{avg} should be set at its optimum amount, thus improving the signal-to-noise ratio. Another important parameter is H_{sat}, meaning that the sample should be magnetically saturated, and applying magnetic fields larger than H_{sat} will increase the measurement time without providing any extra information. The parameter N is also important, as it specifies the time required for completing a measurement sequence. This is dependent on the sample. Nevertheless, based on the study by Pike and Roberts (Alikhani et al., 2016; Pike & Fernandez, 1999; Roberts et al., 2014), 99 FORCs are adequate for many magnetic samples. Other parameters involved in the FORC measurements are the pause time (for being settled at H_{sat}) and the pause at H_r (Roberts et al., 2014).

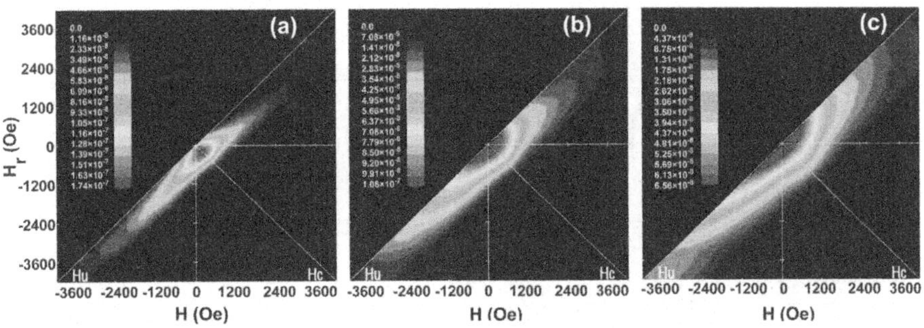

FIGURE 2.30

Illustration of first-order reversal curve diagrams of FeCo nanowire arrays plotted using SF values of (A) 2, (B) 4, and (C) 6.

2.17.3 First-order reversal curve analysis applications

Beyond conventional hysteresis curve measurements, a FORC diagram provides important characteristics such as CFD and IFD. The FORC behavior strongly depends on the type of hysterons, so an antiparallel IF (being opposite to the magnetization) elongates the distribution along the H_u axis. Moreover, a parallel IF causes the FORC distribution to be elongated along the H_c axis.

Alternatively, depending on the magnetization reversal processes occurring in the sample, FORC diagrams can manifest two discernible parts, according to Fig. 2.31. In the case of reversible processes (having $H_c = 0$), the FORC distribution elongates along the interaction axis (H_u). In the case of irreversible processes (i.e., $H_c \neq 0$), the FORC distribution would appear elsewhere in the diagrams (Béron et al., 2008).

Reversible processes are characterized by calculating the reversible indicator (η). In other words, the slope (χ_{FORC}) of minor curves at each H_r is calculated to obtain η. In turn, this reflects the reversibility of the magnetization process, being normalized to susceptibility (χ_{Hyst}) of the upper branch of the major hysteresis curve at the same H_r (Fig. 2.31B). Thus, one can conclude the following relation for obtaining η (Béron et al., 2010):

$$\eta(H = H_r) = \frac{\chi_{\mathrm{FORC}}(H = H_r)}{\chi_{\mathrm{Hyst}}(H = H_r)} \tag{2.62}$$

It is worth noting that single-domain states contribute to irreversible switching of magnetization, whereas multidomain and vortex states result from reversible components (Alikhani et al., 2016). This capability (i.e., distinguishing between reversible and irreversible components as provided by the FORC analysis) is crucial for interpreting magnetization reversal mechanisms in magnetic nanostructured materials.

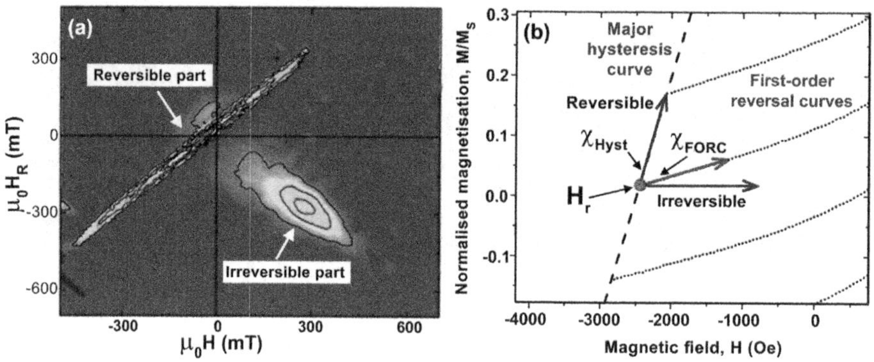

FIGURE 2.31

(A) A typical first-order reversal curve diagram comprising two discernible parts. (B) calculation of the reversible indicator, η, at a reversal field of H_r.

From Gilbert, D. A., Murray, P. D., De Rojas, J., Dumas, R. K., Davies, J. E., & Liu, K. (2021). Reconstructing phase-resolved hysteresis loops from first-order reversal curves. Scientific Reports, 11*(1), 4018. https://doi.org/10.1038/s41598-021-83349-z.*

Mathematically, an irreversible change, $dM^{\text{irreversible}}$, is the difference between the major hysteresis curve and the FORC magnetization (Winklhofer et al., 2008), as given by

$$dM^{\text{irreversible}} = \lim_{H \to H_r}[M(H) - M(H, H_r)] \tag{2.63}$$

The reversible change, $dM^{\text{reversible}}$ is expressed by

$$dM^{\text{reversible}} = \lim_{H \to H_r}[M(H, H_r) - M(H_r)] \tag{2.64}$$

Notably, reversible and irreversible components of a nanowire array system extracted using the FORC diagram (Fig. 2.32A) are depicted in Fig. 2.32B and Fig. 2.32C, respectively.

By employing the cross-sectional view of CFD (Fig. 2.32D), one can obtain the H_c of the irreversible peak at $H_u = 0$, thereby estimating the individual coercivity (H_c^{FORC}) of the nanowires. Defined as the half-width of IFD (ΔH_u^{FORC}) (Béron et al., 2010), one can also quantitatively investigate magnetostatic interactions of the system (Fig. 2.32E).

From another perspective, FORC analysis can identify the domain structure of magnetic materials (Egli et al., 2010; Leonhardt et al., 2004; Roberts et al., 2014). Generally, single-domain states are known with closed contours broadening along the H_c axis, whereas multidomain systems show contours broadened along the $H_u = 0$ axis. Analytically, FORC functions of single-domain nanoparticles have been modeled, manifesting symmetric CFDs around the H_c axis with teardrop contours. Multidomain nanoparticles have shown oval contours around the H_u axis of FORC diagrams. Pseudo-single-domain nanoparticles result in oval-like contours

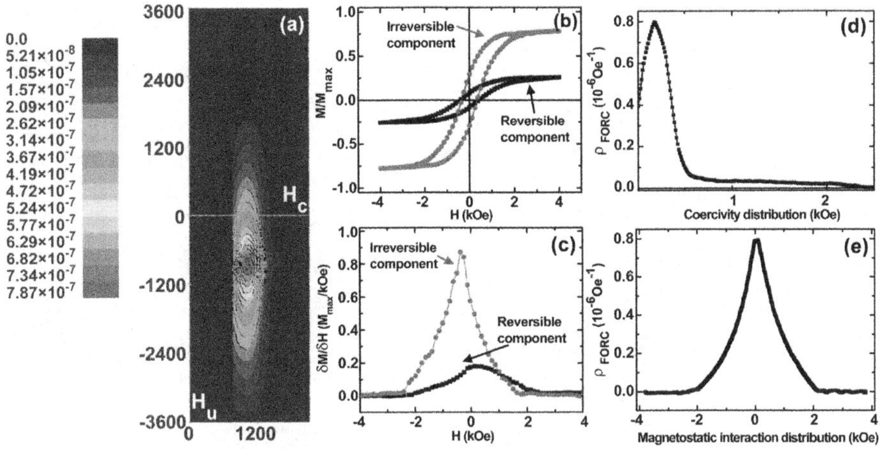

FIGURE 2.32

(A) First-order reversal curve diagram, (B) reversible and irreversible hysteresis curves, (C) reversible and irreversible magnetization components, (D) cross-section of coercive field distribution, and (E) cross-section of interaction field distribution of the CoFeNi nanowire array system.

with a FORC peak moving toward the H_u axis. Note that the transitional state between single-domain and multidomain states is a pseudo-single-domain state comprising magnetic nanoparticle mixtures with single domain-like and multidomain-like properties (Leonhardt et al., 2004; Muxworthy & Dunlop, 2002). Similar results have been obtained in the case of experimental FORC diagrams of magnetic nanowires. In this regard, Fig. 2.33A–C shows single-domain, pseudo-single-domain, and multidomain nanowires, respectively.

Importantly, FORC analysis has been recently associated with popular characterization techniques such as XRD, SEM, and TEM. In this respect, a progressive shift from horizontal to vertical distributions parallel to the H_u axis has been observed when increasing the grain size of single-domain particle systems (Roberts, Pike, & Verosub, 2000). Moreover, comparing XRD peaks and FORC diagrams of Co nanowire arrays interestingly reveals a direct correlation between Co XRD peak intensities and FORC features (Ramazani et al., 2014). Uniform and nonuniform length distributions characterized by SEM analysis have also been involved with CFD and IFD of magnetic nanowires (Dobrotă & Stancu, 2013; Samanifar et al., 2015). Superparamagnetic fractions of magnetic nanoparticles characterized by TEM and FORC analyses have found similar changing trends with each other (Almasi-Kashi et al., 2016; Kumari et al., 2014). Fig. 2.34A shows a FORC diagram obtained from hollow Co_2FeAl nanoparticles, indicating the presence of a mixture of single-domain and superparamagnetic grains.

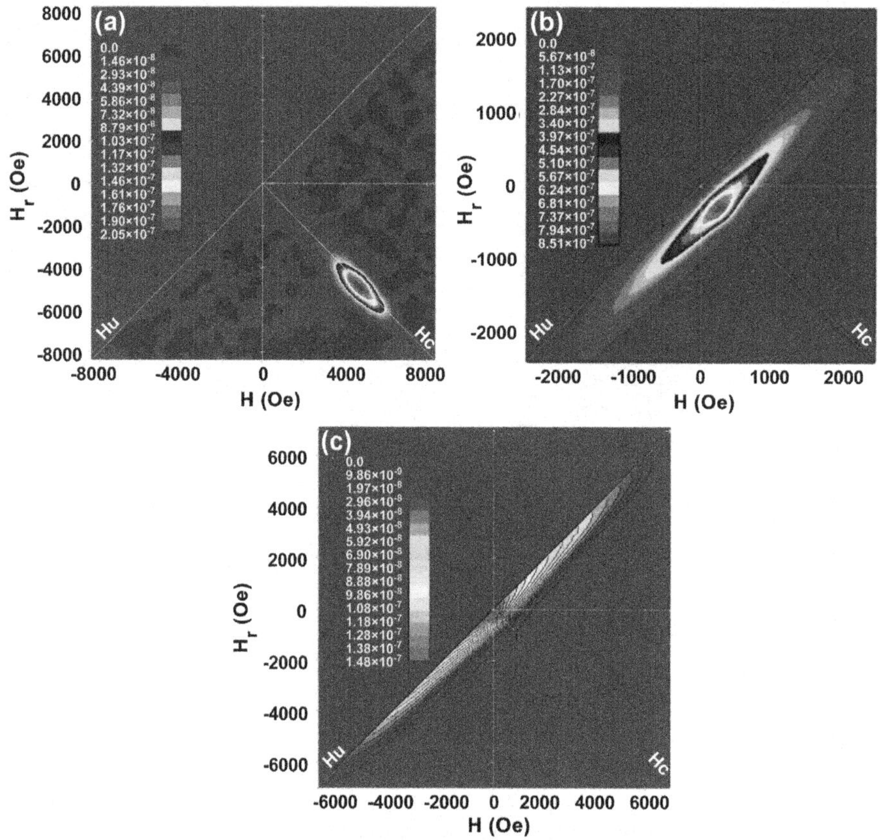

FIGURE 2.33

First-order reversal curve diagrams of (A) single-domain, (B) pseudo-single-domain, and (C) multidomain magnetic nanowires.

From Ramazani, A., Almasi Kashi, M., & Montazer, A. H. (2014). Fabrication of single crystalline, uniaxial single domain co nanowire arrays with high coercivity. Journal of Applied Physics, *115(11), 113902. https://doi.org/10.1063/1.4868582.*

FORC analysis can also be used to test novel magnetic phenomena and possibilities. Notably, micromagnetic simulations have indicated that the nonuniform magnetization of fine particles may fan into vortex or flower states (Roberts et al., 2014). In this way, researchers have considered the vortex or flower states to provide new explanations, outperforming many traditional ones for magnetic behavior observed in previous studies. Fig. 2.34B demonstrates, for the first time, the capturing of the occurrence of a single vortex state in magnetic (FeNi) nanowires using the FORC analysis experimentally.

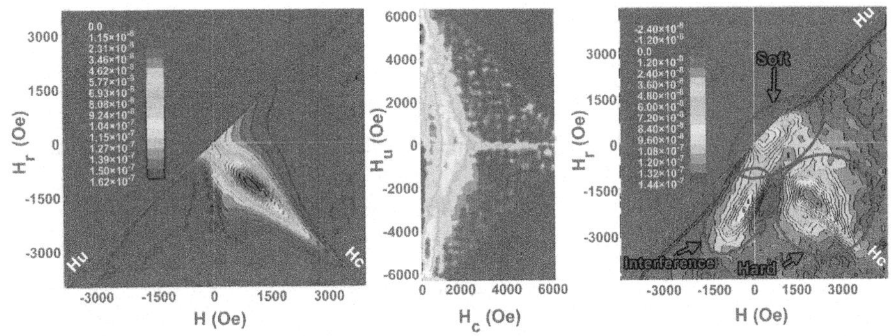

FIGURE 2.34

First-order reversal curve diagrams of (A) a mixture of single-domain and superparamagnetic grains in hollow Co_2FeAl nanoparticles, (B) FeNi nanowires with a single vortex state, and (C) CoFe/Cu multilayer nanowires with soft/hard magnetic phases.

From Almasi-Kashi, M., Ramazani, A., Alikhanzadeh-Arani, S., Pezeshki-Nejad, Z., & Hassan Montazer, A. (2016). Synthesis, characterization and magnetic properties of hollow Co_2FeAl nanoparticles: the effects of heating rate. New Journal of Chemistry, 40*(6), 5061–70. https://doi.org/10.1039/C6NJ00646A.*

For magnetic phases, FORC analysis has provided detailed information about soft, hard, and interference regions taking place in a wide variety of nanostructured and bulk materials. Fig. 2.34C shows a FORC diagram of novel CoFe/Cu multilayer nanowires (Jafari-Khamse et al., 2016). Interestingly, besides confirming the presence of soft/hard magnetic phases, an interference region (depicted as a ridge along the H_u axis) is revealed by the FORC analysis, indicating magnetic interactions between the two phases.

2.18 Measurement of permeability and Curie temperature

Permeability involves the interaction of a material with an applied magnetic field and is defined as the ratio of the induction B to the field H. The initial permeability, μ_i, is the relative permeability at low field strength, constituting the most important parameter for determining the characteristics of soft magnetic materials.

The correlation between permeability, flux density, and magnetizing force of a wound core is determined by Ampere's and Faraday's laws, as given below (Van Valkenburg, 2001):

$$H = \frac{0.4\mu NI}{\ell} \text{(Ampere's law)} \tag{2.65}$$

$$B_{\max} = \frac{V_{\text{rms}} \times 10^8}{4.44 \times f \times A \times N} \text{(Faraday's law)} \tag{2.66}$$

$$\mu = \frac{B}{H} \tag{2.67}$$

where H is the external field in Oe, N denotes the number of turns, μ is the magnetic permeability, I is the peak magnetizing current in amperes, A is the effective cross-sectional area in cm^2, B_{max} is the maximum flux density in Gauss, V_{rms} is the voltage (V) across the coil, and f is the frequency in Hz. Moreover, ℓ is the effective magnetic path length (in cm), which can be calculated by employing Ampere's law and averaging the magnetizing force according to Eq. (2.68):

$$\ell = \frac{\pi(\mathrm{OD} - \mathrm{ID})}{\ln\left(\frac{\mathrm{OD}}{\mathrm{ID}}\right)} \tag{2.68}$$

where OD and ID denote the respective outer and inner diameters of the core depicted in Fig. 2.35.

The correlation between effective permeability and inductance measurement is given by

$$\mu_e' = \frac{\ell(L_{eff} - L_w)}{\mu_0 N^2 A} \tag{2.69}$$

$$\mu_e'' = \frac{\ell(L_{eff} - L_w)}{\mu_0 N^2 \omega A} \tag{2.70}$$

where L_{eff} is the inductance of a toroidal coil, L_w is the inductance of an air-core coil, and ω is the angular frequency.

On the other hand, the Curie temperature can be measured by plotting the inductance of a magnet versus temperature. Typical inductance-temperature curves of $Li_{0.3}Zn_{0.4}Mo_xFe_{2.3-x}O_4$ ($x = 0$–0.03) ferrite are shown in Fig. 2.36. The slight increase observed in the inductance confirms that the temperature coefficient of the initial permeability ($\mu_i/\Delta T$) is positive, followed by an abrupt reduction at the Curie temperature.

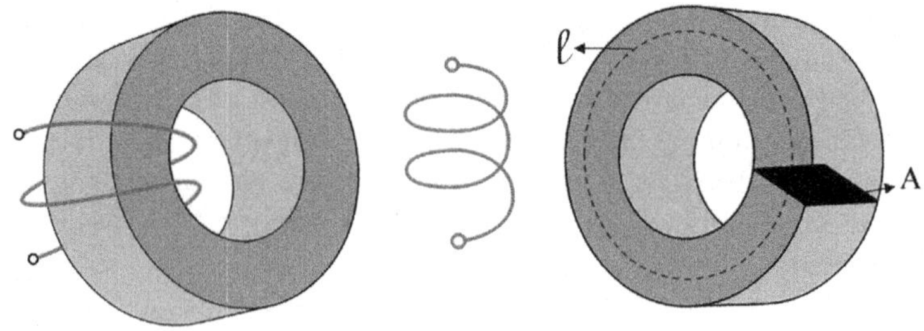

FIGURE 2.35

Illustration of the effective cross-sectional area (A) and the effective magnetic path length (ℓ).

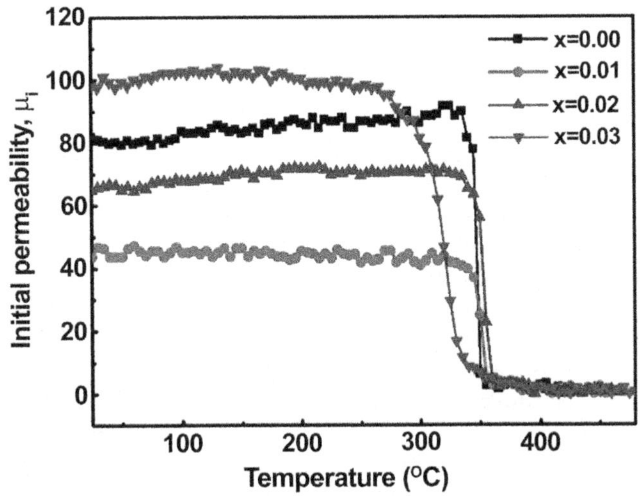

FIGURE 2.36

Variation of inductance versus temperature for a typical $Li_{0.3}Zn_{0.4}Fe_{2.3-x}Mo_xO_4$ ferrites, indicating Curie temperatures in the range of 330–360°C.

From Gao, Y., & Wang, Z. (2019). Effect of Mo substitution on the structural and soft magnetic properties of Li–Zn ferrites. Journal of Sol-Gel Science and Technology, 91*(1), 111–6. https://doi.org/10.1007/s10971-019-05008-0.*

2.19 Measurement of permittivity

Permittivity is indicative of the electronic properties of a material exposed to an electric field. According to Eq. (2.71), the dielectric constant (κ) is defined as the ratio of complex permittivity to the permittivity of free space ($\varepsilon_0 = 8.85 \times 10^{-12}$ F/m) (Agilent Basics of Measuring the Dielectric Properties of Materials).

$$\kappa^* = \varepsilon_r^* = \frac{\varepsilon^*}{\varepsilon_0} = \varepsilon_r' - j\varepsilon_r'' = \left(\frac{\varepsilon'}{\varepsilon_0}\right) - j\left(\frac{\varepsilon''}{\varepsilon_0}\right) \tag{2.71}$$

Using a contacting electrode method (see Fig. 2.37), one can determine permittivity, involving the capacitance of the electrodes, as follows:

$$\varepsilon_r = \frac{t_m \times C_p}{A \times \varepsilon_0} = \frac{t_m \times C_p}{\pi\left(\frac{d}{2}\right)^2 \times \varepsilon_0} \tag{2.72}$$

where C_p is the equivalent parallel capacitance of the sample in Farad, t_m is the average thickness of the sample, A denotes the electrode surface area, d is the guarded electrode diameter, and ε_0 is the permittivity of free space.

The main problem of this method is the formation of an air gap between the sample and electrodes. The measurement errors involve the thickness of the sample and

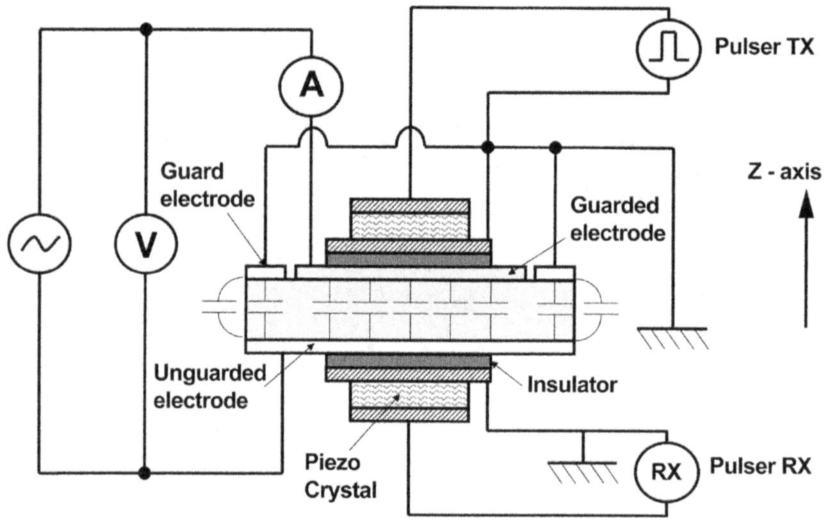

FIGURE 2.37

The measuring method of permittivity by using contacting electrodes.

From Josh, M. (2017). A combination dielectric and acoustic laboratory instrument for petrophysics. Measurement Science and Technology, 28*(12), 125904.*

the thickness of the air gap (t_a). The effect of the air gap on the measurement can be eliminated by coating the surface of the dielectric material with thin films while using a noncontact electrode method. In this case, two capacitances are measured with and without the presence of the dielectric material. Accordingly, the air gap ($t_a = t_g - t_m$) should be extremely small compared with t_m. The permittivity formula is given by

$$\varepsilon_r' = \frac{1}{1 - \left(1 - \frac{C_{s1}}{C_{s2}}\right) \times \frac{t_g}{t_m}} \tag{2.73}$$

where C_{s1} is the capacitance without dielectric insertion, C_{s2} is the capacitance with dielectric insertion, t_g is the gap between the guarded and unguarded electrodes, and t_m is the average thickness of the dielectric material (Literature, 2014).

2.20 Definition of intrinsic impedance

The complex permittivity and permeability of a ferrite are expressed by the following equations:

$$\varepsilon_r = \varepsilon_r' + i\varepsilon_r'' \tag{2.74}$$

$$\mu = \mu_r' + i\mu_r'' \tag{2.75}$$

The real and imaginary parts indicate the storage and loss of electromagnetic energy, respectively. Based on Eq. (2.76), the electrical conductivity, σ, has an impact on the dielectric loss:

$$\varepsilon_r'' = \frac{\sigma}{\omega\varepsilon_0} \tag{2.76}$$

where ω is the angular frequency, and ε_0 is the permittivity of free space. The complex permittivity and permeability in a polar system can be rewritten as follows:

$$\varepsilon_r = |\varepsilon_r|e^{i\delta} \tag{2.77}$$

$$\mu_r = |\mu_r|e^{i\delta_m} \tag{2.78}$$

where δ and δ_m represent the electrical and magnetic losses, respectively, as given below:

$$\tan\delta = \frac{\varepsilon_r''}{\varepsilon_r'} \tag{2.79}$$

$$\tan\delta_m = \frac{\mu_r''}{\mu_r'} \tag{2.80}$$

The refractive index, n, is expressed by

$$n = \frac{k}{k_0} = \sqrt{\mu_r\varepsilon_r} \tag{2.81}$$

where k and k_0 are the wavenumbers of the magnet and free space, respectively. The latter is given by

$$k_0 = \omega\sqrt{\mu_0\varepsilon_0} \tag{2.82}$$

The intrinsic impedance of a ferrite absorber, Z, is related to the permeability and permittivity, according to the following expression:

$$Z = Z_0\sqrt{\frac{\mu_r}{\varepsilon_r}} \tag{2.83}$$

where Z_0 is the intrinsic impedance of free space ($Z_0 = 377\ \Omega$) (Collin, 2007).

2.21 Microwave reflection loss (*R*) measurements

The transmission line theory could be employed to calculate normalized input impedance Z_{in} of a ferrite absorber with a thickness of d, as follows (Knott et al., 2004):

$$Z_{in} = \sqrt{\frac{\mu_r}{\varepsilon_r}}\tan h\left(-ik_0 d\sqrt{\mu_r\varepsilon_r}\right) \tag{2.84}$$

The input impedance is used for measuring reflection loss, R (in dB), of a microwave absorber as given below:

$$R = \frac{Z_{in} - 1}{Z_{in} + 1} \tag{2.85}$$

The practical values of the reflection loss are summarized in the following:

$R = 0$ dB: electromagnetic wave energy will be completely reflected.

$R = -3$ dB: half of the electromagnetic wave energy will be reflected.

$R = -10$ dB: 1/10th of the electromagnetic wave energy will be reflected.

$R = -20$ dB: 1/100th of the electromagnetic wave energy will be reflected.

$R \leq -30$ dB: 1/1000th or less of the electromagnetic wave energy will be reflected.

Amplitudes and phase angles of waves reflected or scattered from a junction relative to the incident waves are directly measurable quantities (Collin, 2007). In other words, the scattered wave amplitudes are related to the incident wave amplitudes linearly. The scattering matrix (also called the Collin matrix) can describe this linear relationship. Consider the two-port junction of Fig. 2.38. The reflected waves of b_1 and b_2 are related to the incident waves of a_1 and a_2 by the following scattering matrix:

$$\begin{bmatrix} b_1 \\ b_2 \end{bmatrix} = [s]\begin{bmatrix} a_1 \\ a_2 \end{bmatrix} = \begin{bmatrix} s_{11} & s_{12} \\ s_{21} & s_{22} \end{bmatrix}\begin{bmatrix} a_1 \\ a_2 \end{bmatrix} \tag{2.86}$$

Vector network analysis (VNA) is an essential microwave measurement approach to characterizing the scattering matrix of an absorber, thereby allowing for the calculation of the real and imaginary parts of permeability and permittivity (Baker-Jarvis et al., 1990; Chen et al., 2004; Marks & Williams, 1991; Vittoria, 1998; Wu et al., 2007).

Fig. 2.39 shows the VNA instrument that can be connected to the sample holder by a coaxial line or a waveguide.

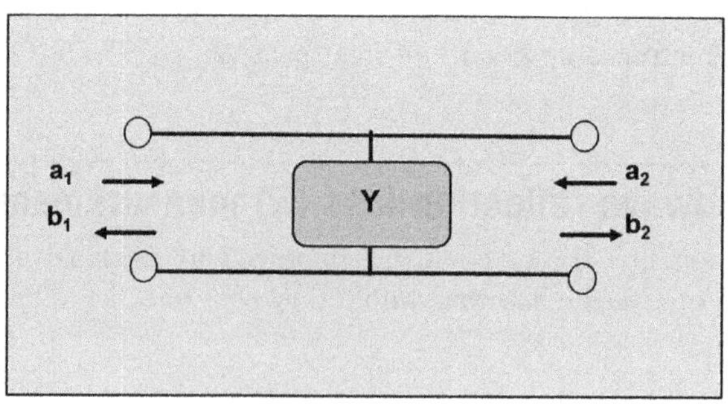

FIGURE 2.38

A two-port junction for measuring scattering matrix elements.

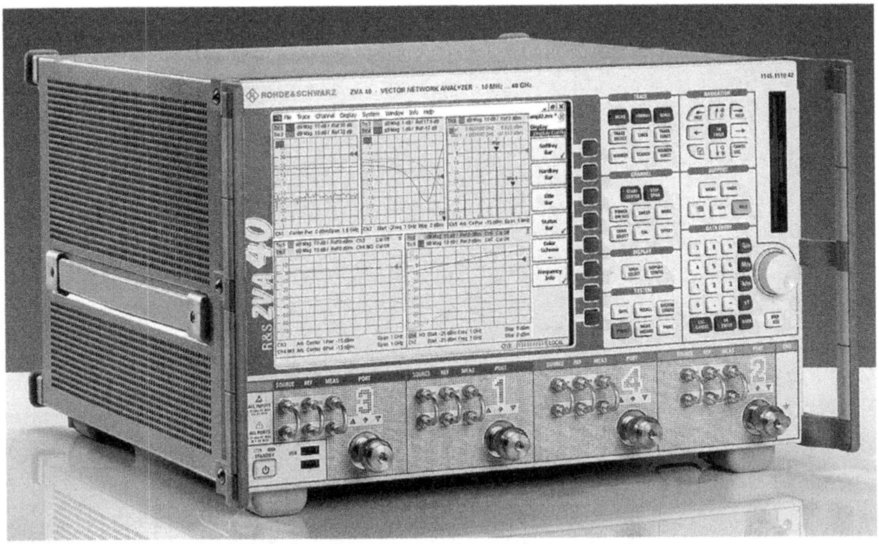

FIGURE 2.39

Schematic of a vector network analysis instrument for measuring the reflection loss.

From wikipedia.org. https://en.wikipedia.org/wiki/Network_analyzer_(electrical)#/media/File:Netzwerkanalysator_ZVA40_RSD.jpg.

A ring-shaped ferrite absorber is used for measuring scattering parameters in a coaxial cable. However, the absorber should be shaped in the form of a slab in waveguides (Fig. 2.40).

Consider a ring-shaped ferrite with a thickness of t connected by a coaxial line to the ports (Fig. 2.41).

The parameters Z_2 and k_2 could be related to permeability and permittivity by the following expressions:

$$Z_2 = Z_0\sqrt{\frac{\mu_r}{\varepsilon_r}} \tag{2.87}$$

$$k_2 = k_0\sqrt{\mu_r\varepsilon_r} \tag{2.88}$$

The normalized input impedance could be calculated by measuring the reflection coefficient in short circuit (SC) and open circuit (OC) conditions:

$$Z_{SC} = Z_r\frac{(1-w^2)}{(1+w^2)} \tag{2.89}$$

$$Z_{OC} = Z_r\frac{(1+w^2)}{(1-w^2)} \tag{2.90}$$

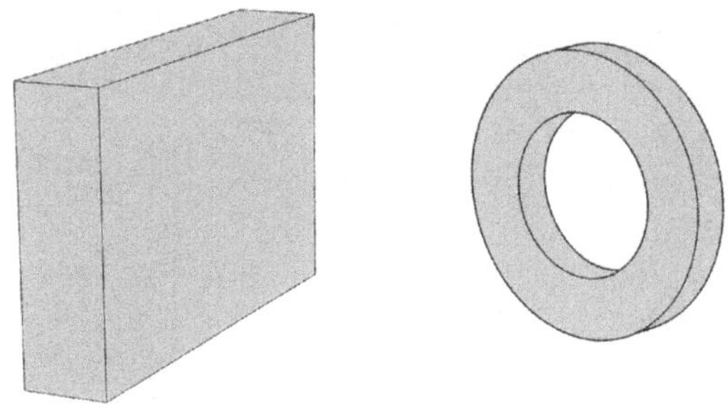

FIGURE 2.40

Shapes of a ferrite absorber for measuring scattering parameters in a coaxial cable and waveguides.

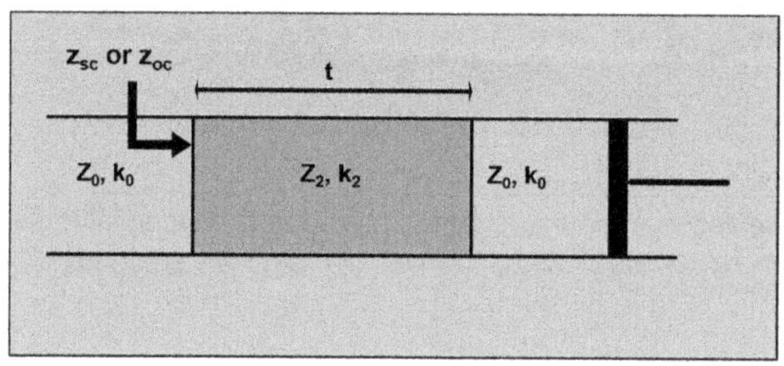

FIGURE 2.41

A ring-shaped ferrite with a thickness of t connected by a coaxial line for measuring complex permeability and permittivity.

in which w and Z_r are as follows:

$$w = \exp(ik_2 t) \tag{2.91}$$

$$Z_r = \frac{Z_2}{Z_0} \tag{2.92}$$

By substituting, one can easily obtain the following expression:

$$\sqrt{\frac{\mu_r}{\varepsilon_r}} = \sqrt{Z_{SC} Z_{OC}} \tag{2.93}$$

$$\sqrt{\mu_r \varepsilon_r} = \frac{1}{i2k_0 t} \ln \frac{\sqrt{Z_{OC}} - \sqrt{Z_{SC}}}{\sqrt{Z_{OC}} + \sqrt{Z_{SC}}} \tag{2.94}$$

After solving the above equations, permeability and permittivity can be calculated as follows (Knott et al., 2004):

$$\mu_r = \frac{\sqrt{Z_{OC} Z_{SC}}}{i2k_0 t} \ln \frac{\sqrt{Z_{OC}} - \sqrt{Z_{SC}}}{\sqrt{Z_{OC}} + \sqrt{Z_{SC}}} \tag{2.95}$$

$$\varepsilon_r = \frac{1}{i2k_0 t\sqrt{Z_{OC}Z_{SC}}} \ln \frac{\sqrt{Z_{OC}} - \sqrt{Z_{SC}}}{\sqrt{Z_{OC}} + \sqrt{Z_{SC}}} \tag{2.96}$$

As another example, consider a ferrite absorber with a specific thickness of $t < \frac{\lambda}{2}$, where λ is the wavelength of the electromagnetic wave. The parameters Z_0 and k_0 are the characteristic impedance and propagation constants of a coaxial line without ferrite, respectively. The directions of microwave propagation are identified by A_1, A_2, and A_3, according to Fig. 2.42. Moreover, the reflection directions are identified by B_1, B_2, and B_3. Taking into account that $B_3 = 0$ (i.e., a perfect absorber), and B_3/A_3 and A_1/A_3 represent the scattering matrix parameters with and without the presence of the absorber, respectively.

In this way, the scattering parameters are as follows:

$$S_{11} = \frac{u(1 - w^2)}{u^2 - w^2} \tag{2.97}$$

$$S_{21} = \frac{w(u^2 - 1)}{u^2 - w^2} \tag{2.98}$$

where u and w are given by

$$u = \frac{1 + y_r}{1 - y_r} \tag{2.99}$$

$$y_r = \sqrt{\frac{\varepsilon_r}{\mu_r}} \tag{2.100}$$

$$w^2 = \frac{u(1 - uS_{11})}{u - S_{11}} \tag{2.101}$$

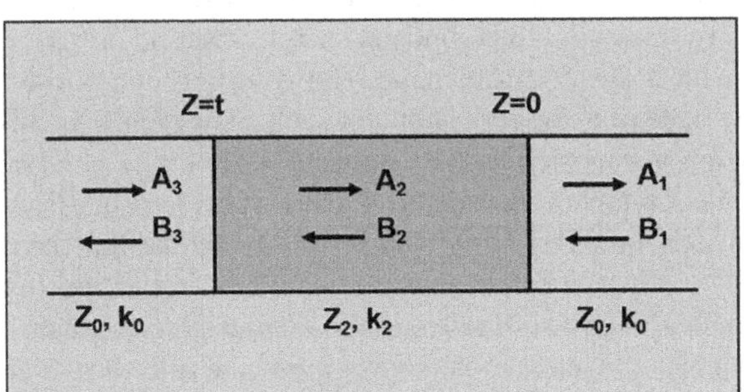

FIGURE 2.42

The representative circuit for measuring complex permeability and permittivity.

After measuring S_{11} and S_{21} by the VNA instrument, it is possible to calculate u and w. By using some simple mathematical substitutions, one has the following equation:

$$u^2 + 2uQ + 1 = 0 \tag{2.102}$$

where Q is a complex parameter given by

$$Q = \left[(S_{21})^2 - (S_{11})^2 - 1\right]/2S_{11} \tag{2.103}$$

Thus, u is determined as follows:

$$u = -Q \pm (Q^2 - 1)^{\frac{1}{2}} \tag{2.104}$$

Now, the parameter U can be introduced by the following expression:

$$U = \frac{u - 1}{u + 1} \tag{2.105}$$

Thus, $U = y_r$. As a result, one obtains

$$\sqrt{\mu_r \varepsilon_r} = \frac{\ln w}{i2\pi k_0 t} \tag{2.106}$$

Finally, permeability and permittivity can be calculated by measuring the parameters in the SC and OC conditions, as follows:

$$\varepsilon_r = \frac{U \ln w}{i2k_0 t} \tag{2.107}$$

$$\mu_r = \frac{\ln w}{i2k_0 t} \tag{2.108}$$

In a nonmagnetic sample (e.g., pure dielectric or conductive specimens), the only requirement is that electrical permittivity is measured without finding S_{11} and S_{21} simultaneously (Balanis, 1966). From a practical standpoint, reflection loss measurements of composite specimens fabricated by mixing ferrite and polymer could be carried out. To this end, the composite sample must be pressed in a ring or slab configuration with a specific thickness. The resulting composite should then be coated on the surface of a perfect conductor such as oxygen-free copper. The reflection loss variation versus frequency within the GHz range can be measured by a VNA instrument. A typical reflection loss (the vertical axis in dB) versus frequency (the horizontal axis in GHz) curve for a substituted barium ferrite composite is depicted in Fig. 2.43. It is found that the maximum reflection loss is about −27 dB, and the bandwidth covering reflection loss values greater than −10 dB is about 2 GHz. The nanocomposite sample shows moderate microwave absorption in the X-band frequency (8–12 GHz).

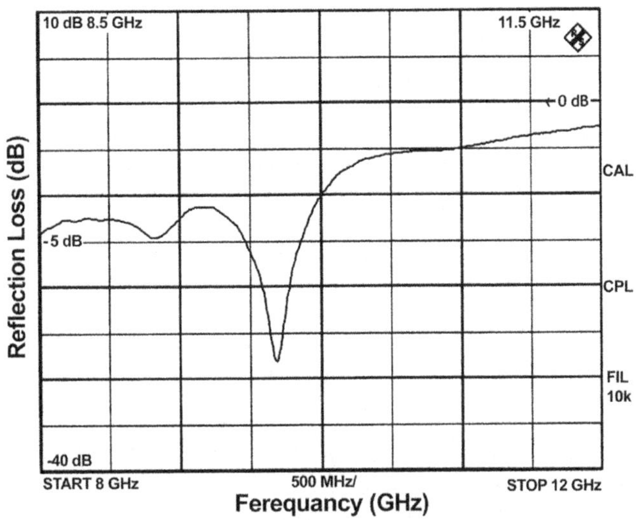

FIGURE 2.43

Reflection loss versus frequency in the microwave X band (8–12 GHz) for a typical barium ferrite nanocomposite.

References

http://academy.cba.mit.edu/classes/input_devices/meas.pdf, 2006.

Alikhani, M., Ramazani, A., Almasi Kashi, M., Samanifar, S., & Montazer, A. H. (2016). Irreversible evolution of angular-dependent coercivity in $Fe_{80}Ni_{20}$ nanowire arrays: Detection of a single vortex state. *Journal of Magnetism and Magnetic Materials, 414*, 158–167. https://doi.org/10.1016/j.jmmm.2016.04.035

Almasi-Kashi, M., Ramazani, A., Alikhanzadeh-Arani, S., Pezeshki-Nejad, Z., & Hassan Montazer, A. (2016). Synthesis, characterization and magnetic properties of hollow Co_2FeAl nanoparticles: The effects of heating rate. *New Journal of Chemistry, 40*(6), 5061–5070. https://doi.org/10.1039/C6NJ00646A

Bacsik, Z., Mink, J., & Keresztury, G. (2004). FTIR spectroscopy of the atmosphere. I. Principles and methods. *Applied Spectroscopy Reviews, 39*(3), 295–363. https://doi.org/10.1081/ASR-200030192

Baker-Jarvis, J., Vanzura, E. J., & Kissick, W. A. (1990). Improved technique for determining complex permittivity with the transmission/reflection method. *IEEE Transactions on Microwave Theory and Techniques, 38*(8), 1096–1103. https://doi.org/10.1109/22.57336

Balanis, C. A. (1966). *Investigation of a proposed technique for measurements of dielectric constants and losses at V-band using the Fabry-Perot principle.*

Béron, F., Clime, L., Ciureanu, M., Ménard, D., Cochrane, R., & Yelon, A. (2008). Magnetostatic interactions and coercivities of ferromagnetic soft nanowires in uniform length arrays. *Journal of Nanoscience and Nanotechnology, 8*(6), 2944–2954. https://doi.org/10.1166/jnn.2008.159

Béron, F., Carignan, L.-P., Ménard, D., & Yelon, A. (2010). Extracting individual properties from global behaviour: First-order reversal curve method applied to magnetic nanowire arrays. *Electrodeposited Nanowires and Their Applications*, 167–188.

Bertotti, G. (1998). *Hysteresis in magnetism: For physicists, materials scientists, and engineers*. Academic Press.

http://www.biosciencenotes.com/transmission-electron-microscope-tem/, 2018.

Boca, R. (1999). *Theoretical foundations of molecular magnetism*. Elsevier.

Bonnell, D. (2000). *Scanning probe microscopy and spectroscopy: Theory, techniques, and applications*. John Wiley & Sons.

Campbell, P. (1996). *Permanent magnet materials and their application*. Cambridge University Press.

Charsley, E., Price, D., Hunter, N., Gabbott, P., Kett, V., Gaisford, S., Priestley, I., Duncan, J., Royall, P., & Scowen, I. (2019). *Principles of thermal analysis and calorimetry*. Royal Society of Chemistry.

Chen, L.-F., Ong, C., Neo, C., Varadan, V., & Varadan, V. K. (2004). *Microwave electronics: Measurement and materials characterization*. John Wiley & Sons.

Chikazumi, S. (1984). *Physics of magnetism*. Kyoujiseitai No Butsuri.

Coats, A., & Redfern, J. (1963). Thermogravimetric analysis. A review. *Analyst, 88*(1053), 906–924. https://doi.org/10.1039/AN9638800906

Collin, R. E. (2007). *Foundations for microwave engineering*. John Wiley & Sons.

Cranshaw, T. E., Dale, B. W., Longworth, G. O., & Johnson, C. E. (1985). *Mössbauer spectroscopy and its applications*. Cambridge University Press.

Cullity, B. D., & Stock, S. R. (2014). Elements of X-ray diffraction. In *Prentice Hall* (3rd ed.). Pearson.

Dehm, G., Howe, J. M., & Zweck, J. (2012). *In-situ electron microscopy: Applications in physics, chemistry and materials science*. John Wiley & Sons.

Dey, C., Baishya, K., Ghosh, A., Goswami, M. M., Ghosh, A., & Mandal, K. (2017). Improvement of drug delivery by hyperthermia treatment using magnetic cubic cobalt ferrite nanoparticles. *Journal of Magnetism and Magnetic Materials, 427*, 168–174. https://doi.org/10.1016/j.jmmm.2016.11.024

Dobrotă, C.-I., & Stancu, A. (2013). What does a first-order reversal curve diagram really mean? A study case: Array of ferromagnetic nanowires. *Journal of Applied Physics, 113*(4), 043928–043939. https://doi.org/10.1063/1.4789613

Dormann, J., Fiorani, D., & Tronc, E. (1999). On the models for interparticle interactions in nanoparticle assemblies: Comparison with experimental results. *Journal of Magnetism and Magnetic Materials, 202*(1), 251–267. https://doi.org/10.1016/S0304-8853(98)00627-1

Dormann, J. L., Fiorani, D., & Tronc, E. (1997). *Magnetic relaxation in fine-particle systems*. John Wiley & Sons.

Dyar, M. D., Agresti, D. G., Schaefer, M. W., Grant, C. A., & Sklute, E. C. (2006). Mössbauer spectroscopy of earth and planetary materials. *Annual Review of Earth and Planetary Sciences, 34*, 83–125. https://doi.org/10.1146/annurev.earth.34.031405.125049

Egerton, R. F. (2005). *Physical principles of electron microscopy, 56*. Springer.

Egli, R., Chen, A. P., Winklhofer, M., Kodama, K. P., & Horng, C.-S. (2010). Detection of noninteracting single domain particles using first-order reversal curve diagrams. *Geochemistry, Geophysics, Geosystems, 11*(1), 1–22. https://doi.org/10.1029/2009GC002916

Foner, S. (1959). Versatile and sensitive vibrating-sample magnetometer. *Review of Scientific Instruments, 30*(7), 548–557. https://doi.org/10.1063/1.1716679.

Getzlaff, M. (2007). *Fundamentals of magnetism*. Springer Science & Business Media.

Ghasemi, A., Loghman-Estarki, M. R., Torkian, S., & Tavoosi, M. (2019). The microstructure and magnetic behavior of spark plasma sintered iron/nickel zinc ferrite nanocomposite synthesized by the complex sol-gel method. *Composites Part B: Engineering, 175*, 107179–107194. https://doi.org/10.1016/j.compositesb.2019.107179

Ghasemi, A., & Šepelák, V. (2011). Correlation between site preference and magnetic properties of substituted strontium ferrite thin films. *Journal of Magnetism and Magnetic Materials, 323*(12), 1727–1733. https://doi.org/10.1016/j.jmmm.2011.02.010

Gibb, T. C. (2013). *Principles of Mössbauer spectroscopy*. Springer.

Goldstein, J. I., Newbury, D. E., Michael, J. R., Ritchie, N. W., Scott, J. H. J., & Joy, D. C. (2017). *Scanning electron microscopy and X-ray microanalysis*. Springer.

Gorter, E. W. (1954). *Saturation magnetization and crystal chemistry of ferrimagnetic oxides. Philips Research Report*.

Haines, P. J. (2012). *Thermal methods of analysis: Principles, applications and problems*. Springer Science & Business Media.

Hawkes, P. W., & Spence, J. C. (2019). *Springer handbook of microscopy*. Springer Nature.

Holba, P., & Šesták, J. (2014). Imperfections of Kissinger evaluation method and crystallization kinetics. *Glass Physics and Chemistry, 40*(5), 486–495. https://doi.org/10.1134/S1087659614050058

Hyperphysics. (2020). http://hyperphysics.phy-astr.gsu.edu/hbase/mod3.html.

Jafari-Khamse, E., Almasi Kashi, M., & Ramazani, A. (2016). First-order-reversal-curve (FORC) diagrams of alternative chain of soft/hard magnetic CoFe/Cu multilayer nanowires. *Current Applied Physics, 16*(4), 486–496. https://doi.org/10.1016/j.cap.2015.12.001

Jiles, D. (2015). *Introduction to magnetism and magnetic materials*. CRC Press.

Kim, G.-Y., Jeon, J.-H., Kim, M.-H., Suvorov, D., & Choi, S.-Y. (2013). Microstructural development of cobalt ferrite ceramics and its influence on magnetic properties. *Metals and Materials International, 19*(6), 1209–1213. https://doi.org/10.1007/s12540-013-6009-8

Knott, E. F., Schaeffer, J. F., & Tulley, M. T. (2004). *Radar cross section*. Institution of Engineering and Technology.

Kumari, M., Widdrat, M., Tompa, É., Uebe, R., Schüler, D., Pósfai, M., Faivre, D., & Hirt, A. M. (2014). Distinguishing magnetic particle size of iron oxide nanoparticles with first-order reversal curves. *Journal of Applied Physics, 116*(12), 124304. https://doi.org/10.1063/1.4896481

Lagarec, K., & Rancourt, D. G. (1998). *RECOIL, Mössbauer spectral analysis software for windows (version 1.0)*. Canada: Department of Physics, University of Ottawa.

Lee, M. (2019). *X-ray diffraction for materials research*. Apple Academic Press.

Leng, Y. (2013). *Materials characterization: Introduction to microscopic and spectroscopic methods*. Wiley.

Leonhardt, R., Krása, D., & Coe, R. S. (2004). Multidomain behavior during Thellier paleointensity experiments: A phenomenological model. *Paleo and Archeointensity: Methods, Techniques and New Results, 147*(2), 127–140. https://doi.org/10.1016/j.pepi.2004.01.009

Literature. (2014). http://literature.cdn.keysight.com/litweb/pdf/16451-90020.pdf.

Lohar, K., Patange, S., Mane, M., & Shirsath, S. E. (2013). Cation distribution investigation and characterizations of $Ni_{1-x}Cd_xFe_2O_4$ nanoparticles synthesized by citrate gel process. *Journal of Molecular Structure, 1032*, 105–110. https://doi.org/10.1016/j.molstruc.2012.07.055

Long, G. J., & Grandjean, F. (2013). *Mössbauer spectroscopy applied to magnetism and materials science, 1*. Springer Science & Business Media.

Marks, R. B., & Williams, D. F. (1991). Characteristic impedance determination using propagation constant measurement. *IEEE Microwave and Guided Wave Letters, 1*(6), 141–143. https://doi.org/10.1109/75.91092

Menil, F. (1985). Systematic trends of the ^{57}Fe Mössbauer isomer shifts in (FeO_n) and (FeF_n) polyhedra. Evidence of a new correlation between the isomer shift and the inductive effect of the competing bond TX (→ Fe)(where X is O or F and T any element with a formal positive charge). *Journal of Physics and Chemistry of Solids, 46*(7), 763–789. https://doi.org/10.1016/0022-3697(85)90001-0

Mohamed, M. A., Jaafar, J., Ismail, A., Othman, M., & Rahman, M. (2017). Fourier transform infrared (FTIR) spectroscopy. In *Membrane characterization* (pp. 3–29). Elsevier.

Morel, A., Le Breton, J., Kreisel, J., Wiesinger, G., Kools, F., & Tenaud, P. (2002). Sublattice occupation in $Sr_{1-x}LaxFe_{12-x}Co_xO_{19}$ hexagonal ferrite analyzed by Mössbauer spectrometry and Raman spectroscopy. *Journal of Magnetism and Magnetic Materials, 242*, 1405–1407. https://doi.org/10.1016/S0304-8853(01)00962-3

Morrish, A. H. (2001). *The physical principles of magnetism*. IEEE Magnetics Society.

Muxworthy, A. R., & Dunlop, D. J. (2002). First-order reversal curve (FORC) diagrams for pseudo-single-domain magnetites at high temperature. *Earth and Planetary Science Letters, 203*(1), 369–382. https://doi.org/10.1016/S0012-821X(02)00880-4

Muxworthy, A. R., & Roberts, P. (2007). *First-order reversal curve (FORC) diagrams, 1* pp. 266–272). Springer. https://doi.org/10.1007/978-1-4020-4423-6_99

Niazi, A., Poddar, P., & Rastogi, A. (2000). A precision, low-cost vibrating sample magnetometer. *Current Science, 79*(1), 99–109.

Nikolic, G. (2011). *Fourier transforms: New analytical approaches and FTIR strategies*. BoD–Books on Demand.

O'Hendley, R. (1999). *Modern magnetic materials. Principles and applications*. Johns Wiley & Sons.

Pike, C., & Fernandez, A. (1999). An investigation of magnetic reversal in submicron-scale Co dots using first order reversal curve diagrams. *Journal of Applied Physics, 85*(9), 6668–6676. https://doi.org/10.1063/1.370177

Preisach, F. (1935). Über die magnetische Nachwirkung. *Zeitschrift Für Physik, 94*(5), 277–302. https://doi.org/10.1007/BF01349418

Proenca, M. P., Sousa, C. T., Escrig, J., Ventura, J., Vazquez, M., & Araujo, J. P. (2013). Magnetic interactions and reversal mechanisms in Co nanowire and nanotube arrays. *Journal of Applied Physics, 113*(9), 093907. https://doi.org/10.1063/1.4794335

Ramazani, A., Almasi Kashi, M., & Montazer, A. H. (2014). Fabrication of single crystalline, uniaxial single domain Co nanowire arrays with high coercivity. *Journal of Applied Physics, 115*(11), 113902. https://doi.org/10.1063/1.4868582

Reimer, L. (2013). *Transmission electron microscopy: Physics of image formation and microanalysis, 36*. Springer.

Roberts, A. P., Heslop, D., Zhao, X., & Pike, C. R. (2014). Understanding fine magnetic particle systems through use of first-order reversal curve diagrams. *Reviews of Geophysics, 52*(4), 557–602. https://doi.org/10.1002/2014RG000462

Roberts, A. P., Pike, C. R., & Verosub, K. L. (2000). First-order reversal curve diagrams: A new tool for characterizing the magnetic properties of natural samples. *Journal of Geophysical Research: Solid Earth, 105*(12), 28461–28475. https://doi.org/10.1029/2000JB900326

Rostamnejadi, A., Salamati, H., Kameli, P., & Ahmadvand, H. (2009). Superparamagnetic behavior of $La_{0.67}Sr_{0.33}MnO_3$ nanoparticles prepared via sol–gel method. *Journal of Magnetism and Magnetic Materials, 321*(19), 3126–3131. https://doi.org/10.1016/j.jmmm.2009.05.035

Samanifar, S., Almasi Kashi, M., Ramazani, A., & Alikhani, M. (2015). Reversal modes in FeCoNi nanowire arrays: Correlation between magnetostatic interactions and nanowires length. *Journal of Magnetism and Magnetic Materials, 378*, 73–83. https://doi.org/10.1016/j.jmmm.2014.10.155

Sardela, M. (2014). *Practical materials characterization*. Springer.

Sawatzky, G., Van Der Woude, F., & Morrish, A. H. (1969). Mössbauer study of several ferrimagnetic spinels. *Physical Review, 187*(2), 747–757. https://doi.org/10.1103/PhysRev.187.747

Spaldin, N. A., & Mathur, N. D. (2003). Magnetic materials: Fundamentals and device applications. *Physics Today, 56*(12), 62–63.

Speliotis, D. (2005). *Getting the most from your vibrating sample magnetometer*. MA, USA: ADE Technologies Inc.

Stokes, D. (2008). *Principles and practice of variable pressure/environmental scanning electron microscopy (VP-ESEM)*. John Wiley & Sons.

Ul-Hamid, A. (2018). *A beginners' guide to scanning electron microscopy*. Springer.

Van Valkenburg, M. E. (2001). *Reference data for engineers: Radio, electronics, computers and communications*. Elsevier.

Vittoria, C. (1998). *Elements of microwave networks: Basics of microwave engineering*. World Scientific.

Waldron, R. (1955). Infrared spectra of ferrites. *Physical Review, 99*(6), 1727–1735. https://doi.org/10.1103/PhysRev.99.1727

Wang, G., Ma, Y., Wei, Z., & Qi, M. (2016). Development of multifunctional cobalt ferrite/graphene oxide nanocomposites for magnetic resonance imaging and controlled drug delivery. *Chemical Engineering Journal, 289*, 150–160. https://doi.org/10.1016/j.cej.2015.12.072

Winklhofer, M., Dumas, R. K., & Liu, K. (2008). Identifying reversible and irreversible magnetization changes in prototype patterned media using first- and second-order reversal curves. *Journal of Applied Physics, 103*(7), 07C518–07C521. https://doi.org/10.1063/1.2837888

Wohlfarth, E. P. (1958). Relations between different modes of acquisition of the remanent magnetization of ferromagnetic particles. *Journal of Applied Physics, 29*(3), 595–596. https://doi.org/10.1063/1.1723232

Wu, Y., Tang, Z.-X., Zhang, B., & Xu, Y. (2007). Permeability measurement of ferromagnetic materials in microwave frequency range using support vector machine regression. *Progress in Electromagnetics Research, 70*, 247–256. https://doi.org/10.2528/PIER07012801

Yadav, R. S., Havlica, J., Masilko, J., Kalina, L., Wasserbauer, J., Hajdúchová, M., … Kožáková, Z. (2016). Impact of Nd^{3+} in $CoFe_2O_4$ spinel ferrite nanoparticles on cation distribution, structural and magnetic properties. *Journal of Magnetism and Magnetic Materials, 399*, 109–117. https://doi.org/10.1016/j.jmmm.2015.09.055

Zi, Z., Sun, Y., Zhu, X., Yang, Z., & Song, W. (2008). Structural and magnetic properties of $SrFe_{12}O_{19}$ hexaferrite synthesized by a modified chemical co-precipitation method. *Journal of Magnetism and Magnetic Materials, 320*(21), 2746–2751. https://doi.org/10.1016/j.jmmm.2008.06.009

Zipare, K., Bandgar, S., & Shahane, G. (2018). Effect of Dy-substitution on structural and magnetic properties of MnZn ferrite nanoparticles. *Journal of Rare Earths, 36*(1), 86–94. https://doi.org/10.1016/j.jre.2017.06.011

Zolotoyabko, E. (2014). *Basic concepts of X-ray diffraction.* John Wiley & Sons.

CHAPTER

Magnetic ferrites

3

3.1 Introduction

Hard magnetic ferrites are frequently used in permanent magnets, magnetic recording media, and microwave-absorbing materials. Moreover, multilayer chip inductors and electromagnetic suppressors can be made from soft magnetic ferrites. Garnet thin films and bulk materials have found important uses in devices operating in the gigahertz frequency range. The structural and magnetic properties of hexagonal ferrites are attributed mostly to their substituted cations. Substituting hexaferrites with various cations introduces strains to the crystal lattice, thus changing the magnetic ordering. The spinel ferrites, such as Mn–Zn, Ni–Zn, and Ni–Cu–Zn, can be used in many devices at low and intermediate frequencies. The ferrimagnetic and superparamagnetic features of soft ferrite can be determined by varying magnetic susceptibility versus temperature at different frequencies. The biomedical application of iron oxide nanoparticles is also studied. Garnets, which can be used in tunable microwave devices, circulators, isolators, phase shifters, magneto-optical devices, and tunable filters, are also considered.

3.2 Important definitions

The current chapter includes an investigation of the magnetic features of ultrafine particles. Since the magnetic state of nanoparticles with a single-domain configuration differs completely from those of their bulk counterparts, some elementary explanations of single-domain and superparamagnetic features are necessary. Then, Snoek's law for finding the frequency-dependent permeability value of magnetic materials is explained.

3.2.1 Particles with single-domain structure

Experimental investigations on the magnetic properties of ultrafine particle systems have revealed the variation of coercivity as a function of particle size, as schematically depicted in Fig. 3.1. By considering the minimization of energy, domain walls (DWs) are favorably formed in large particles while involving the reversal of

Magnetic Ferrites and Related Nanocomposites. https://doi.org/10.1016/B978-0-12-824014-4.00003-2

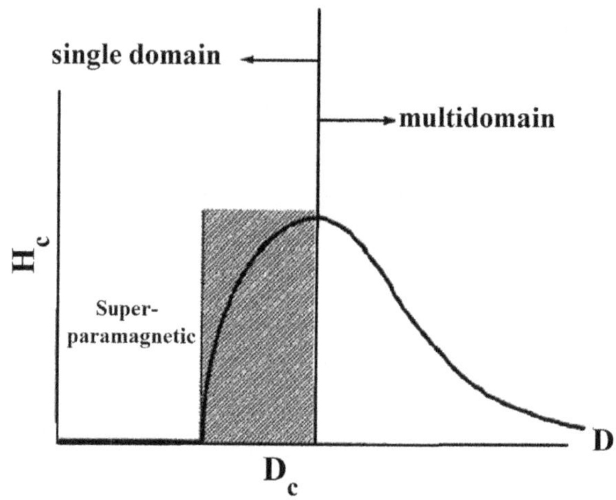

FIGURE 3.1

Schematic representation of the variation of coercivity (H_c) as a function of diameter for ultrafine particle systems.

From Philippova, O., Barabanova, A., Molchanov, V., & Khokhlov, A. (c. 2011). Magnetic polymer beads: Recent trends and developments in synthetic design and applications. European Polymer Journal, *47(4), 542–559. https://doi.org/10.1016/j.eurpolymj.2010.11.006.*

magnetization by DW nucleation and motion processes. Energy requirements prevent DW formation when the size of particles is reduced to a critical diameter (D_c), leading to a single-domain structure. DW nucleation and motion are then not responsible for magnetization reversal. Therefore, spins of the single-domain structure are simultaneously rotated by a coherent mode, giving rise to higher coercive field (H_c) values. For particles with diameters smaller than the single-domain size, thermal fluctuations can significantly affect the magnetic system, leading to a superparamagnetic state.

As noted, magnetic materials in particles are expected to have a single-domain structure when their size is reduced to below a critical value, as first predicted by Frenkel and Doefman (1930). Kittel (1946) was the first worker to estimate critical particle sizes. In the case of typical spherical ferromagnetic particles, the critical radii have been estimated to be approximately 10 to 1000 nm.

Proportionality exists between magnetic moment magnitude and particle volume. The total magnetic moment of each unit could be as high as thousands of μ_B. Normally, the particles are assumed to be ellipsoidal. In this way, the magnetic moments tend to be aligned along the longest axis of the ellipsoid, describing the direction for the lowest energy of the shape anisotropy (Stoner & Wohlfarth, 1991).

Eq. (3.1) gives the critical radius (r_c) of a particle, below which it behaves as a single-domain particle (O'Handley, 2000):

$$r_c \approx 9\frac{(AK_u)^{\frac{1}{2}}}{\mu_0 M_s^2} \tag{3.1}$$

where A and K_u are the exchange and uniaxial anisotropy constants, respectively. Moreover, μ_0 and M_s are the vacuum permeability and saturation magnetization, respectively. For Fe, Co, and γ-Fe_2O_3 particles, the corresponding r_c values are 15, 35, and 30 nm, whereas the r_c of $SmCo_5$ particles has been reported as high as 750 nm (Givord et al., 1991). Particles with a single-domain structure can have magnetic moments ranging from 10^3 to 10^5 μ_B, depending on their size and composition (Blundell, 2003).

3.2.2 Time variation of magnetization

Understanding the time variation of magnetization in a magnetic system is important for tuning practical applications related to the magnetization reversal (Néel, 1949, 1951) while investigating related mechanisms from a fundamental point of view.

Regardless of the magnetic system type, the time variation of magnetization can be expressed by the following equation:

$$\frac{dM(t)}{dt} = -\frac{M(t) - M(t=\infty)}{\tau} \tag{3.2}$$

where $M\ (t=\infty)$ and τ are the equilibrium magnetization and characteristic relaxation time, respectively. τ is related to the relaxation with an energy barrier, as given below:

$$\tau^{-1} = f_0 e^{-\Delta E/k_B T} \tag{3.3}$$

where ΔE is the energy barrier, k_B is the Boltzmann constant, and f_0 is a constant ($f_0 = 10^9\ s^{-1}$) (Brown Jr, 1959; Kneller, 1964). In the case of uniaxial anisotropy, ΔE is obtained from the product of the anisotropy constant and the magnetic system volume. The particle volume and anisotropy constant magnitude can influence f_0 (Aharoni, 1969; Brown Jr, 1963). In this respect, more appropriate f_0 values in the range of 10^{12} to $10^{13} s^{-1}$ have been suggested by some experimental investigations (Dickson et al., 1993; Xiao et al., 1986). Nevertheless, since the variation of τ is governed by the exponential argument, it is not necessary to set the f_0 value exactly. ΔE with heights from 25 to 32 $k_B T$ may play a role in the measurement times of magnetization in a range of 1 to 1000 s. It should be emphasized that the particle size (D) can critically affect τ [$\tau \approx \exp(D^3)$]. Notably, with $f_0 = 10^9\ s^{-1}$, $K = 10^6$ erg/cm^3, $T = 300$ K, and $D = 11.4$ nm, τ equals 0.1 s.

3.2.3 Superparamagnetic state

Based on the theoretical calculations by Néel (1949), thermal fluctuations can lead to an H_c of near zero for ultrafine particles, preventing stable magnetization of the magnetic system. By considering Eq. (3.3), the critical ΔE (ΔE_{crit}) for an arbitrary measurement time ($\tau = 100$ s) with $f_0 = 10^9 s^{-1}$ can be obtained as follows:

$$\Delta E_{crit} = \ln(\tau f_0) k_B T = 25\ k_B T \tag{3.4}$$

When the magnetic particle has a uniaxial anisotropy with an anisotropy constant of K, ΔE will equal KV (at $H = 0$), thus realizing the superparamagnetic state:

$$KV = 25\ k_B T \tag{3.5}$$

Again, as can be inferred, since τ depends strongly on the exponential argument, defining the exact value of f_0 is of little importance. In the case of $f_0 = 10^{12}\ s^{-1}$, the relation 3.5 becomes $KV = 32\ k_B T$. For particles with $\Delta E > \Delta E_{crit}$ at a measurement time scale of 100 s, relaxation does not occur, thus blocking them during the measurement period. Accordingly, one can define a blocking temperature (T_B) as follows:

$$T_B = \Delta E / 25\ k_B \tag{3.6}$$

Increasing the measurement field decreases the T_B of a superparamagnetic system. In this case, T_B is proportional to $H^{2/3}$ and H^2 at high and low magnetic fields, respectively (Chantrell et al., 1991; Wohlfarth, 1980). At $T = T_B$, the variation of magnetization as a function of temperature, $M(T)$, evidences the presence of a cusp in zero-field-cooled (ZFC) susceptibility. Alternatively, the particles can be freely aligned with the magnetic field during the measurement time when $T > T_B$, giving rise to a superparamagnetic state. In this way, the particles behave similarly to spins in a paramagnetic state.

It is possible to model magnetic nanoparticles as two-phase systems. One phase is composed of a highly crystalline core, and the other is a disordered surface layer surrounding the core. The atomic disordering of the surface layer is caused by vacancies, lattice strain, and broken bonds. The spin arrangement of the surface layer is also destabilized due to spin frustration, leading to the formation of canted spin structures (see Fig. 3.2). Theoretical and experimental investigations have evidenced the existence of such complex intrinsic spin structures in fine particle systems, resulting in a reduction in M_s. These spin structures are not completely saturated even under high magnetic fields (Ahmed et al., 2002; Sohn et al., 1998; Zhang et al., 1997, 2001). Moreover, in the case of antiferromagnetic nanoparticles, shifts may be observed in the hysteresis loops, owing to exchange bias between the crystalline core and surface layer magnetization (Park et al., 2007).

3.2.4 Snoek's law

Highly permeable magnetic materials have important potential applications in various radio frequency and microwave-engineered devices. One can evaluate the permeability of a material at high frequency using the ferromagnetic resonance (FMR) frequency, f_{res}, and static permeability, μ_s. Reaching a high dynamic permeability value requires high parameters for both μ_s and f_{res}. The product of μ_s and f_{res} parameters is limited for the majority of magnetic materials, based on the Snoek's law given below (Snoek, 1948):

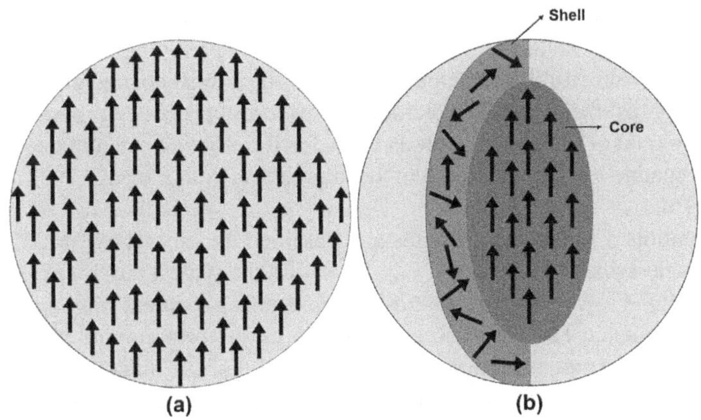

FIGURE 3.2

(A) A schematic illustrating the uniform magnetization in a spherical particle, resulting in the collinear spins. (B) A schematic illustrating a two-phase magnetic particle system, leading to different magnetization for the core and surface structure. The spins existing on the surface are canted with respect to the magnetization of the particle core, thus being subjected to spin excitations with low energy.

$$(\mu_s - 1)f_{\text{res}} = \frac{2}{3}\gamma 4\pi M_0 \tag{3.7}$$

where $4\pi M_0$ is the saturation magnetization, and γ represents the gyromagnetic ratio.

Thereby, the achievement of high-frequency permeability values is limited according to Eq. (3.7). One can also rewrite Eq. (3.7) as follows:

$$(\mu_s - 1)\cdot f_{\text{res}}^2 \leq (\gamma 4\pi M_0) \tag{3.8}$$

As can be inferred, the limitation on magnetic properties at high frequencies corresponds to the equality in Eq. (3.8), as predicted for magnetic materials with planar anisotropy magnetized uniformly. In this respect, ferromagnetic thin films (Perrin et al., 1996), flake-shaped composites (Walser et al., 1998), typical ferrites (Adenot et al., 2002; Rozanov et al., 2005), and amorphous microwires (Torrejon et al., 2009) can be considered the most important examples of such magnetic systems. For other materials, the left side of Eq. (3.8) is less than its right side. Thus, the resultant magnetic properties at a high frequency will be well below the limiting value.

3.3 Hexagonal ferrites

Ferrite materials with hexagonal structures, usually known as hexaferrites, show beneficial magnetic properties such as a high coercivity, large magnetocrystalline anisotropy constant, moderate saturation magnetization and Curie temperature, and high resonance frequency. It is possible to synthesize hexaferrite materials using various techniques, including solid-state reactions, coprecipitation, sol–gel, hydrothermal, and water-in-oil microemulsions. It has been well documented that various factors such as the synthesis technique, effective processing parameters, and type and amount of substituted cations in the ferrite structure can influence the structural and magnetic characteristics of hexaferrites (Drmota et al., 2012; Ebrahimi et al., 2016).

As mentioned earlier, hexaferrites are classified into six types—M, Y, W, Z, X, and U—as described in Chapter 1. Fig. 3.3 shows the typical ternary phase diagram of BaO-MeO–Fe_2O_3 systems, where Me represents a divalent ion among the first transition elements. The chemical composition in the formation of M-, Y-, W-, Z-, X- and U-type ferrites can be determined using the phase diagram. In this regard, one will find that only two oxides (BaO and Fe_2O_3) are required to prepare M-type barium ferrites, whereas stoichiometric amounts of MeO should be considered when synthesizing other types of hexaferrites.

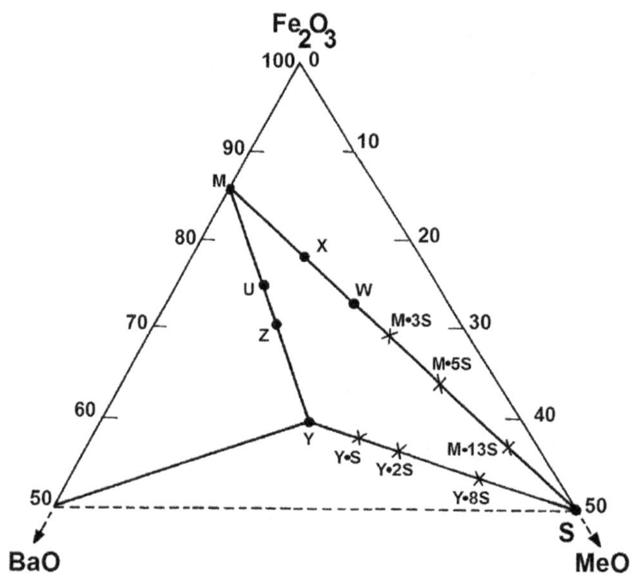

M : $BaFe^{III}_{12}O_{19}$
X : $Ba_2Me^{II}_2Fe^{III}_{28}O_{46}$
Z : $Ba_3Me^{II}_2Fe_{24}O_{41}$
S : $MeFe_2O_4$
W : $BaMe^{II}_2Fe^{III}_{16}O_{27}$
Y : $Ba_2Me^{II}_2Fe^{III}_{12}O_{22}$
U : $Ba_4Me^{II}_2Fe^{III}_{36}O_{60}$

FIGURE 3.3

The ternary phase diagram of BaO–MeO–Fe_2O_3 systems.

From Sudakar, C., Subbanna, G., & Kutty, T. (c. 2004). Nanoparticle composites having structural intergrowths of hexaferrite and spinel ferrites prepared by gel-to-crystallite conversion and their magnetic properties. Journal of Magnetism and Magnetic Materials, 268*(1-2), 75–88. https://doi.org/10.1016/S0304-8853(03)00476-1 76.*

In the following, different kinds of hexagonal ferrites are discussed, and the role of substituted cations in their magnetic characteristics is explained.

3.3.1 M-type hexagonal ferrites

The $BaFe_{12}O_{19}$ and $SrFe_{12}O_{19}$ ferrites have been highlighted as the most important groups of M-type ferrites. In this respect, Table 3.1 presents the magnetic properties of $BaFe_{12}O_{19}$ and $SrFe_{12}O_{19}$ hexaferrites (Luo et al., 2012).

Many researchers acknowledge that the orientation of the magnetization vector along the *c*-axis can provide the ferrites with large magnetocrystalline anisotropy, resulting in a high coercivity and resonance frequency. The substituted cations may be distributed in various crystallographic positions, allowing for the tuning of the magnetic anisotropy. Five crystallographic occupancy sites, including $4f_1$(tetrahedral) and $4f_2$(octahedral) with a spin-down orientation, and 12*k* (octahedral), 2*a* (octahedral), and 2*b* (bipyramidal) with a spin-up orientation, could exist in M-type ferrites.

By incorporating various substitutions in the aforementioned sites, the magnetization can be reoriented from the *c*-axis (the perpendicular direction) to the *c*-plane (the in-plane or basal plane), thereby changing the magnetocrystalline anisotropy, resonance frequency, and coercivity (Ghasemi, 2015; Ghasemi et al., 2007, 2010b). The role of substituted cations on the magnetic feature is strongly dependent on the site preferences of cations in the ferrite crystal lattice. Thus, a powerful instrument such as Mössbauer spectroscopy is required for precise information about cation site occupancy.

3.3.1.1 Typical Mössbauer spectroscopic analysis

Over the last decade, several M-type hexagonal ferrites have been synthesized by my research group. The site preference of substituted cations in the crystal structure of ferrites was also evaluated and interpreted in great detail. The Mössbauer spectroscopy results discussed in the following were obtained from divalent–four valence cations of Zn^{2+}–Sn^{4+} in $SrFe_{12-x}(Sn_{0.5}Zn_{0.5})_xO_{19}$, and trivalent–trivalent Cr^{3+}–Al^{3+} in $SrFe_{12-x}(Cr_{0.5}Al_{0.5})_xO_{19}$ compounds are explained. The magnetic moment per formula unit (f.u.) of Zn–Sn-substituted ferrites has been calculated based on the data obtained from the Mössbauer spectroscopic measurements, thereby predicting the variation behavior of saturation magnetization.

Table 3.1 The magnetic properties of $BaFe_{12}O_{19}$ and Sr $Fe_{12}O_{19}$ hexaferrites.

Hexaferrites	M_s (A m^2/kg)	H_c (kA/m)	M_r/M_s isotropic	H_a (kA/m)	K_1 ($\times 10^5$ J/m^3)	T_c (°C)
$BaFe_{12}O_{19}$	72	477.4	0.5	1353	3	450
$SrFe_{12}O_{19}$	74–92	533.16	0.5	1592	3.5	460

3.3.1.1.1 Mössbauer spectroscopic measurements of $SrFe_{12-x}(Sn_{0.5}Zn_{0.5})_xO_{19}$

In a research study, $SrFe_{12-x}(Sn_{0.5}Zn_{0.5})_xO_{19}$ ($x = 0–5$) powder' comprising randomly oriented particles have been synthesized (Ghasemi & Šepelák, 2011). The site occupancy of substituted cations was precisely determined using ^{57}Fe Mössbauer spectroscopic measurements for $SrFe_{12-x}(Sn_{0.5}Zn_{0.5})_xO_{19}$ ($x = 0–5$) in a transmission geometry at 293 K. The γ-ray sources were provided by isotopes ^{57}Co in *Rh*. The velocity scale was then calibrated relative to ^{57}Fe in *Rh*. Recoil spectral analysis software was used to evaluate the resultant Mössbauer spectra quantitatively (Lagarec & Rancourt, 1998). The distribution of ferric cations was investigated over nonequivalent cation sublattices of the hexagonal ferrite structure to determine Mössbauer subspectral intensities. In this case, recoilless fractions of Fe^{3+} cations were assumed to be the same for all available crystal sites in the structure.

Fig. 3.4 compares ^{57}Fe Mössbauer spectra of the $SrFe_{12-x}(Sn_{0.5}Zn_{0.5})_xO_{19}$ structure with various x in the range of 0–5. Increasing the iron substitution amount (x) results in the formation of five sextets (a typical ferrimagnetic hexagonal structure) with higher asymmetry inside each line, followed by their slow collapse and lower intensity. Finally, a doublet structure located in the center of the spectra replaces the sextets, leading to the emergence of a paramagnetic state. According to the fitted Mössbauer data, the maximum hyperfine magnetic field, B_{max} (acting on the iron nuclei), is reduced from 51.6 to 48 T with an increasing x from 0 to 5 within the $SrFe_{12-x}(Sn_{0.5}Zn_{0.5})_xO_{19}$ structure.

The typical ^{57}Fe Mössbauer spectrum of the ferrite structure with $x = 2$ is shown in Fig. 3.5; it involves some constraints in its fitting procedure. For each sextet in the spectrum of the $SrFe_{10}(Sn_{0.5}Zn_{0.5})_2O_{19}$ structure, the six absorption lines are assumed to have intensity ratios of 3:2:1:1:2:3. It should be noted that the isomer shift (IS) and quadrupole splitting (QS) values of $SrFe_{10}(Sn_{0.5}Zn_{0.5})_2O_{19}$ were fixed during the fitting procedure. In addition, since the Fe^{3+}(12*k*) cations were sensitive to their nearest magnetic neighbors, hyperfine fields with a broad distribution were considered for the corresponding subspectrum.

Only four subspectra were needed to obtain a good fit for the $SrFe_{10}(Sn_{0.5}Zn_{0.5})_2O_{19}$ structure, with respective relative intensity ratios of 1:6:1:2 in $4f_2$, 12*k*, 2*a*, and $4f_1$ sites. As the $4f_2$ site can be occupied by only one Fe^{3+} cation, an interesting observation is that the relative intensity of the $Fe^{3+}(4f_2)$ cations in the $SrFe_{10}(Sn_{0.5}Zn_{0.5})_2O_{19}$ structure subspectrum is half the size than that in the $SrFe_{12}O_{19}$ one.

Table 3.2 presents the Mössbauer hyperfine parameters of $SrFe_{10}(Sn_{0.5}Zn_{0.5})_2O_{19}$, indicating a decrease in the average hyperfine magnetic fields after the substitution of Fe^{3+} cations by Sn and Zn ions in the ferrite material. A comparison of Table 3.2 and Table 2.4 (Chapter 2) shows a significant decrease in the local fields for the Fe^{3+} cations located in the 12*k* and 2*a* sites of $SrFe_{10}(Sn_{0.5}Zn_{0.5})_2O_{19}$ relative to $SrFe_{12}O_{19}$.

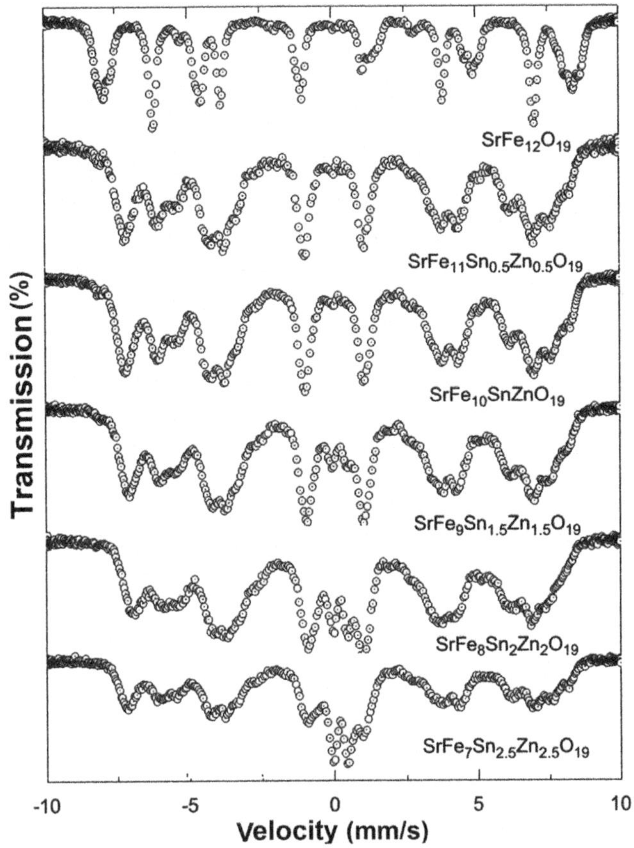

FIGURE 3.4

^{57}Fe Mössbauer spectra obtained from $SrFe_{12-x}(Sn_{0.5}Zn_{0.5})_xO_{19}$ structure with different x (ranging from 0 to 5).

From Ghasemi, A., & Šepelák, V. (c. 2011). Correlation between site preference and magnetic properties of substituted strontium ferrite thin films. Journal of Magnetism and Magnetic Materials, 323*(12), 1727–1733. https://doi.org/10.1016/j.jmmm.2011.02.010. 1729.*

Thus, it can be concluded that the $4f_2$ and $2b$ sites are preferentially occupied by Sn and Zn ions in $SrFe_{12-x}(Sn_{0.5}Zn_{0.5})_xO_{19}$ structure, being in agreement with previously published data on Mössbauer hyperfine parameters of $BaFe_{12-x}(Zr_{0.5}Zn_{0.5})_xO_{19}$ (Z. Li et al., 2000).

Obradors et al. (1985) have also proposed the possibility of the occupation of the $2b$ site by Zn^{2+} ions, indicating a pseudotetrahedral character.

According to the ^{57}Fe Mössbauer analysis results, Sn and Zn ions occupy the $4f_2$ and 2b sites of $SrFe_{12-x}(Sn_{0.5}Zn_{0.5})_xO_{19}$ preferentially. The uncompensated ferric ion magnetic moments give rise to the total magnetic moment of $SrFe_{12-x}(Sn_{0.5}Zn_{0.5})_xO_{19}$ as the Zn^{2+} and Sn^{4+} ions have no magnetic moment.

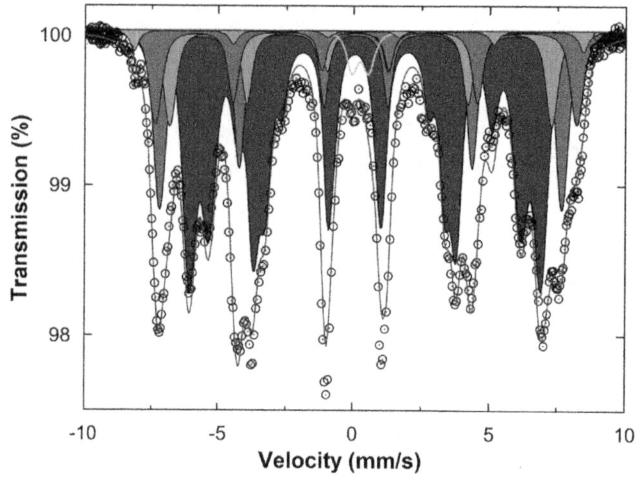

FIGURE 3.5

^{57}Fe Mössbauer spectrum obtained from $SrFe_{10}(Sn_{0.5}Zn_{0.5})_2O_{19}$ structure.

From Ghasemi, A., & Šepelák, V. (c. 2011). Correlation between site preference and magnetic properties of substituted strontium ferrite thin films. Journal of Magnetism and Magnetic Materials, 323*(12), 1727–1733. https://doi.org/10.1016/j.jmmm.2011.02.010.*

Table 3.2 Mössbauer hyperfine parameters extracted from the corresponding fitted spectrum of $SrFe_{10}(Sn_{0.5}Zn_{0.5})_2O_{19}$ structure at room temperature.

Sublattice	Coordination of Fe^{3+} ions	IS[a] (mm/s)	QS[a] (mm/s)	*B* (T)	Number of Fe^{3+} ions per f.u.
$4f_2$	Octahedral	0.24	0.14	48.32(7)	1
$12k$	Octahedral	0.22	0.2	36.93(5)	6
$2a$	Octahedral	0.2	0.02	43.79(1)	1
$4f_1$	Tetrahedral	0.14	0.1	46.09(9)	2

IS, isomer shift; QS, quadrupole splitting.
[a] *Fixed parameter.* $\Gamma = 0.386(4)$ *mm/s.*

From Table 3.3, accounting for collinear spin alignment, it is evident the magnetic moment is enhanced (25 μ_B/f.u.) when 1 Fe^{3+} cations are substituted by Zn^{2+} and Sn^{4+} ones in the $4f_2$ sites. Likewise, the resulting magnetic moment is further enhanced to 30 μ_B when 2 Fe^{3+} cations are substituted by diamagnetic Zn^{2+} and Sn^{4+} ones in the $4f_2$ site, as presented in Table 3.4.

When the nonmagnetic cations substitute Fe^{3+} ones in $SrFe_{12-x}(Sn_{0.5}Zn_{0.5})_xO_{19}$, the canted magnetic structures formed as the Fe^{3+} cation spins with a relatively high substitution degree (x) are no longer collinear. In this respect, both the occupation

Table 3.3 Calculation of magnetic moment per formula unit for $SrFe_{12-x}(Sn_{0.5}Zn_{0.5})_xO_{19}$ ($x = 1$) according to the Mössbauer spectroscopy results.

Sublattice	Coordination of Fe^{3+} ions	Number of Fe^{3+} ions per f.u.	Spin orientation	Magnetic moment per formula unit
$4f_2$	Octahedral	1	↓	$m = -1 \times 5\ \mu_B + 6 \times 5\ \mu_B + 1 \times 5\ \mu_B - 2 \times 5\ \mu_B + 1 \times 5\ \mu_B = 5 \times 5\ \mu_B = \mathbf{25}\ \mu_B$
$12k$	Octahedral	6	↑	
$2a$	Octahedral	1	↑	
$4f_1$	Tetrahedral	2	↓	
$2b$	Bipyramidal	1	↑	

Table 3.4 Calculation of magnetic moment per formula unit for $SrFe_{12-x}(Sn_{0.5}Zn_{0.5})_xO_{19}$ ($x = 2$) according to the Mössbauer spectroscopy results.

Sublattice	Coordination of Fe^{3+} ions	Number of Fe^{3+} ions per f.u.	Spin orientation	Magnetic moment per formula unit
$4f_2$	Octahedral	0	↓	$m = 0 \times 5\ \mu_B + 6 \times 5\ \mu_B + 1 \times 5\ \mu_B - 2 \times 5\ \mu_B + 1 \times 5\ \mu_B = 6 \times 5\ \mu_B = \mathbf{30}\ \mu_B$
$12k$	Octahedral	6	↑	
$2a$	Octahedral	1	↑	
$4f_1$	Tetrahedral	2	↓	
$2b$	Bipyramidal	1	↑	

factors of the Fe^{3+} cation sublattice and the average spin canting angles [Ψ_i ($i = 4f_2$, $12k$, $2a$, $4f_1$, $2b$)] can influence the resulting magnetic moment of each sublattice.

Accordingly, the noncollinear magnetic structure can have a magnetic moment as given below:

$$m = -m_{4f_2}\cos\Psi_{4f_2} + m_{12k}\cos\Psi_{12k} + m_{2a}\cos\Psi_{2a} - m_{4f_1}\cos\Psi_{4f_1} + m_{2b}\cos\Psi_{2b} \quad (3.9)$$

It is inferred that an enhancement in the $2b$ site substitution of Fe^{3+} cations by the diamagnetic ones reduces the magnetic moment under collinear spin alignment. However, one can expect the magnetic moment to increase for different spin canting angles Ψ_i.

3.3.1.1.2 ^{57}Fe Mössbauer analysis of $SrFe_{12-x}(Cr_{0.5}Al_{0.5})_xO_{19}$

In a separate research study, ^{57}Fe Mössbauer analysis of $SrFe_{12-x}(Cr_{0.5}Al_{0.5})_xO_{19}$ ($x = 0$–2.5) has been carried out (Ghasemi et al., 2010a). The ^{57}Fe Mössbauer spectra of $SrFe_{12-x}(Cr_{0.5}Al_{0.5})_xO_{19}$ ferrite samples with typical x ranged between 0 and 2.5, as shown in Fig. 3.6. Increasing x leads to the broadening of the five sextets when comparing them with the spectral Gaussian width.

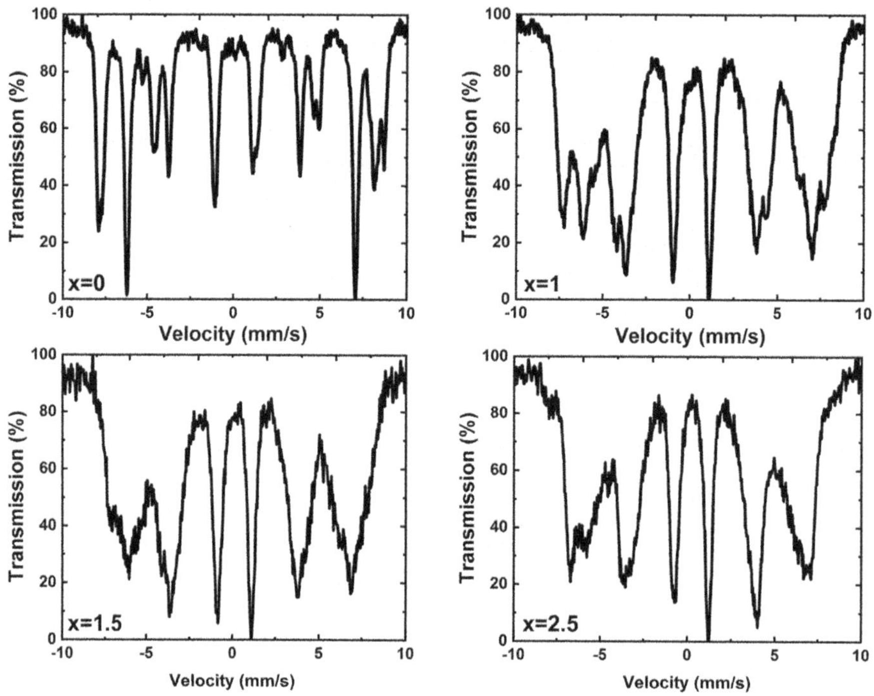

FIGURE 3.6

^{57}Fe Mössbauer spectra obtained from $SrFe_{12-x}(Cr_{0.5}Al_{0.5})_xO_{19}$ structure.

The results indicate that a broad distribution of the magnetic hyperfine field exists in the substituted samples. Thus, the average magnetic hyperfine fields, B, are reduced with increasing the Fe substitution amount (x) in the five $SrFe_{12-x}(Cr_{0.5}Al_{0.5})_xO_{19}$ sublattices. In particular, this stands true for Fe ions located in the $4f_2$ and $4f_1$ sites.

Notably, the occurrence of a reduction in the super-transferred field may be generally responsible for the decrease observed in the local magnetic hyperfine fields. In fact, the magnetic ions surrounding the Fe^{3+} ion in a lattice site can lead to the generation of the super-transferred field. The spin density transfer in complex oxides may take place from one lattice site (e.g., tetrahedral) to another site (e.g., octahedral), being considerably more effective than the reverse transfer of the sites (Van der Woude & Sawatzky, 1971). Note that the substitution of Fe^{3+} ions can modify the spin density transfer. Consequently, the detailed magnetic ion distribution in the structure may not affect the local magnetic fields of Fe^{3+} ion in some lattice sites (e.g., Fe^{3+} ion in the $2a$ site). From Fig. 3.6, the Mössbauer spectra indicate a reduction in the $12k$ subspectrum relative intensity with increasing x.

It is possible to detect a superparamagnetic state in the Mössbauer spectrum of a material whose particles have a shorter relaxation time of magnetization reversal than the Larmor precession time of nuclear magnetic moment (which is sometimes also called the "time window" of Mössbauer spectroscopy, ranging from 10^{-8} to 10^{-9} s). Since $SrFe_{12-x}(Cr_{0.5}Al_{0.5})_xO_{19}$ particles have sizes ranging from 1 to 5 μm, the Al and Cr cation site preference evaluated is not expected to show superparamagnetic behavior.

3.3.1.2 Magnetic properties of substituted strontium ferrite

Using a sol–gel technique, strontium ($SrFe_{12}O_{19}$) and substituted strontium [$SrFe_9(Mn_{0.5}Co_{0.5}Zr)_{3/2}O_{19}$] ferrites have been prepared (Ghasemi & Morisako, 2008). Figs. 3.7A and 3.7B show respective hysteresis loops ($J = 4\pi M$ versus H, and B versus H) of strontium and substituted strontium ferrite (SrM) at room

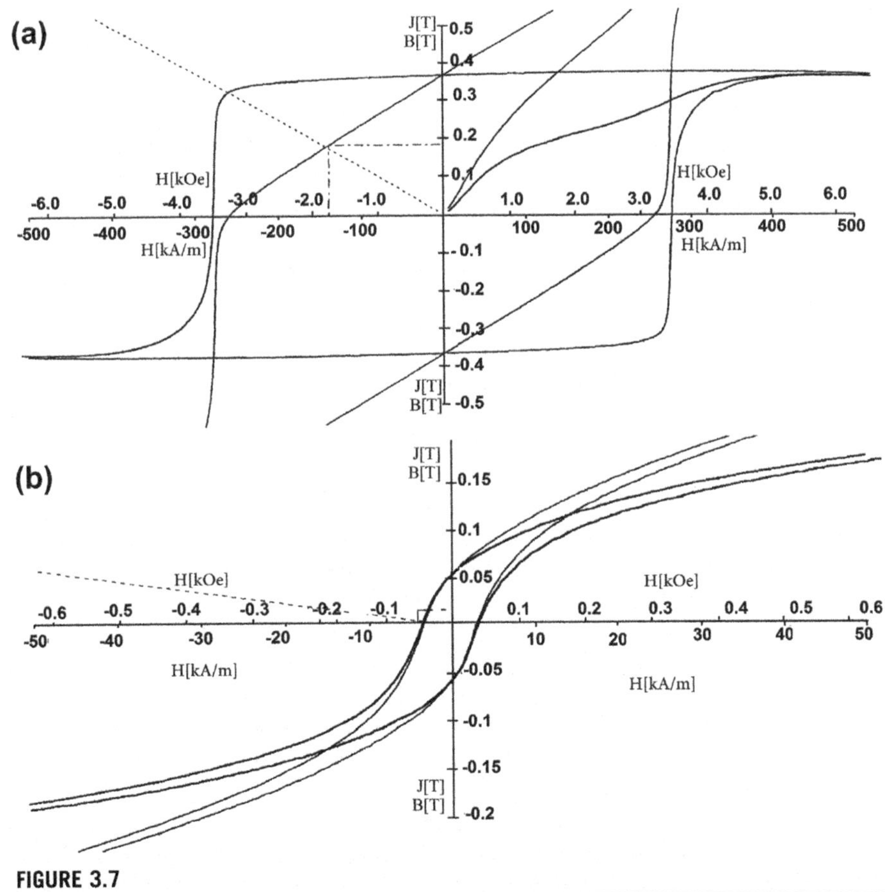

FIGURE 3.7

Hysteresis loops obtained from: (A) strontium ferrite, and (B) substituted strontium ferrite.

From Ghasemi, A., & Morisako, A. (c. 2008). Structural and electromagnetic characteristics of substituted strontium hexaferrite nanoparticles. Journal of Magnetism and Magnetic Materials, 320*(6), 1167–1172. https://doi.org/10.1016/j.jmmm.2007.11.004 1169.*

temperature. As can be seen, a large coercivity and hysteresis loop area are obtained for the $SrFe_{12}O_{19}$ ferrite. As well, the remanent magnetization of $SrFe_{12}O_{19}$ is higher than that of the substituted ferrite. It is worth noting that the large hysteresis loop area, along with the higher remanent magnetization, may lead to a larger hysteresis loss for the $SrFe_{12}O_{19}$.

The strong uniaxial anisotropy of the SrM leads to a large H_c of about 258.7 kA/m. However, the H_c is sharply reduced from 258.7 to 3.4 kA/m after the substitution of SrM with Mn, Co, and Zr, arising from a decrease in the magnetocrystalline anisotropy.

To investigate the role of substitution in the anisotropy constant variation, it is necessary to briefly understand the law of approach to saturation (LAS), as expressed below:

$$M(H) = M_s\left(1 - \frac{A}{H} - \frac{B}{H^2} - \frac{C}{H^3}\right) + \chi_d H \tag{3.10}$$

where M_s is the saturation magnetization, and χ_d is the high field differential susceptibility. Moreover, the presence of crystal inhomogeneities can be indicated by the term A/H. Theoretically speaking, the magnetic energy required to saturate the material would be infinite if the term A/H does not vanish at sufficiently high magnetic fields. The magnetic anisotropy is related to the terms B/H^2 and C/H^3. In the case of a uniaxial hexagonal compound, the terms B/H^2 and C/H^3 can be written as follows:

$$\frac{B}{H^2} + \frac{C}{H^3} = \left(\frac{1}{15}\right)\left(\frac{H_a}{H}\right)^2 + \left(\frac{2}{105}\right)\left(\frac{H_a}{H}\right)^3 \tag{3.11}$$

$$H_a = \frac{2K_1}{M_s} \tag{3.12}$$

where H_a and K_1 are the anisotropy field and the first anisotropy constant, respectively. The magnetic moment of Fe, Mn, and Zr can be assumed as $\mu(\text{Fe}) = 5\ \mu_B$, $\mu(\text{Mn}) = 3\ \mu_B$, and $\mu(\text{Zr}) = 0\ \mu_B$. For Co^{2+} ions located in tetrahedral and octahedral sites, the corresponding magnetic moments are 3 and 3.7 μ_B, respectively. The partial quenching of Co^{2+} ion angular magnetic moment in the octahedral sites may be responsible for the slightly higher magnetic moment of the latter case. The first anisotropy constant (K_1) is reduced from 5.36×10^{-2} to 1.09×10^{-2} kJ/kg after the substitution of $SrFe_{12}O_{19}$ with Mn, Co, and Zr, confirming the rotation of the easy axis from the perpendicular to the in-plane direction. As a result, nearly planar magnetic anisotropy can be obtained.

Elsewhere, using a coprecipitation technique, Mg–Co–Ti-substituted $SrFe_{12}O_{19}$ hexaferrite nanoparticles with the chemical formula of $SrFe_{12-2x}(MgCo)_{x/2}Ti_xO_{19}$ ($x = 0$–2.5) have been synthesized at 900°C (Gordani et al., 2014). The Fourier transform infrared spectra of the resulting ferrites with different x are shown in Fig. 3.8. The infrared spectroscopy absorption behavior of the ferrite samples remains almost the same by changing the substitution amount. The absorption peaks

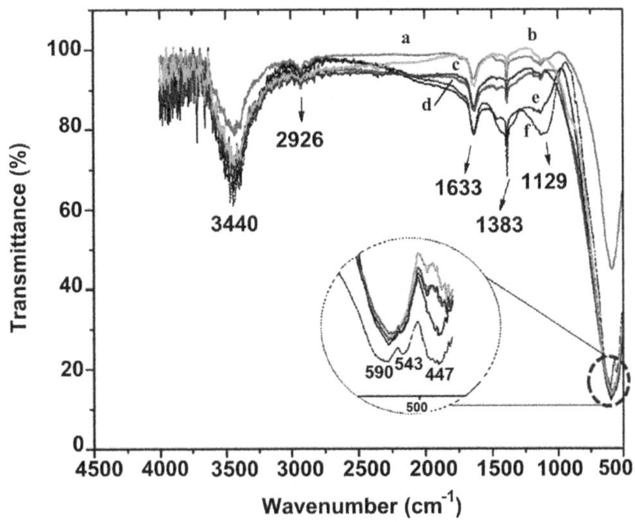

FIGURE 3.8

Fourier transform infrared spectra obtained from $SrFe_{12-2x}(MgCo)_{x/2}Ti_xO_{19}$ hexaferrite nanoparticles with: (A) $x=0$, (B) $x=0.5$, (C) $x=1$, (D) $x=1.5$, (E) $x=2$ and (F) $x=2.5$.

From Gordani, G. R., Ghasemi, A., & Saidi, A. (c. 2014). Enhanced magnetic properties of substituted Sr-hexaferrite nanoparticles synthesized by coprecipitation method. Ceramics International, 40*(3), 4945–4952. https://doi.org/10.1016/j.ceramint.2013.10.096 4949.*

that appeared at 436, 465, 543, and 590 cm^{-1} are assigned to the asymmetric stretching and out-of-plane bending vibrations taking place in the octahedral and tetrahedral sites of the Mg-, Co-, and Ti-substituted Sr-hexaferrites. The asymmetric stretching band of $-CH_2-$ is responsible for the emergence of the peak at 2926 cm^{-1}, indicating the presence of an sp^3 hybridized C—H bond. The stretching vibration and hydroxyl group (—OH) deformation vibration lead to the relatively broad peaks at approximately 1630 and 3440 cm^{-1}, arising from the wet atmosphere. Peaks ranging from 1100 to 1500 cm^{-1} refer to metal—oxygen—metal bands, including Co—O—Co or Fe—O—Fe.

Alternatively, Fig. 3.9 shows hysteresis loops of the hexaferrites for various compositions. Increasing x decreases the saturation magnetization (M_s) from 39.55 to 19.27 emu/g, except for $x = 0.5$ (40 emu/g), coinciding with the reduction of the remanence magnetization (M_r) in all hexaferrite samples. An enhancement in the nonmagnetic phases (e.g., α-Fe_2O_3) of the substituted Sr-hexaferrites along with the occupation of the substituted ions could be responsible for the decreasing trends of M_s and M_r. The increase in M_s from $x = 0$ to 0.5 could occur when the nonmagnetic (Mg^{2+} and Ti^{4+}) and magnetic (Co^{2+}) ions replace the spin-down sites ($4f_1$ and $4f_2$). The substituted cations with $x > 0.5$ can also replace the spin-up sites, especially at the $12k$ octahedral site.

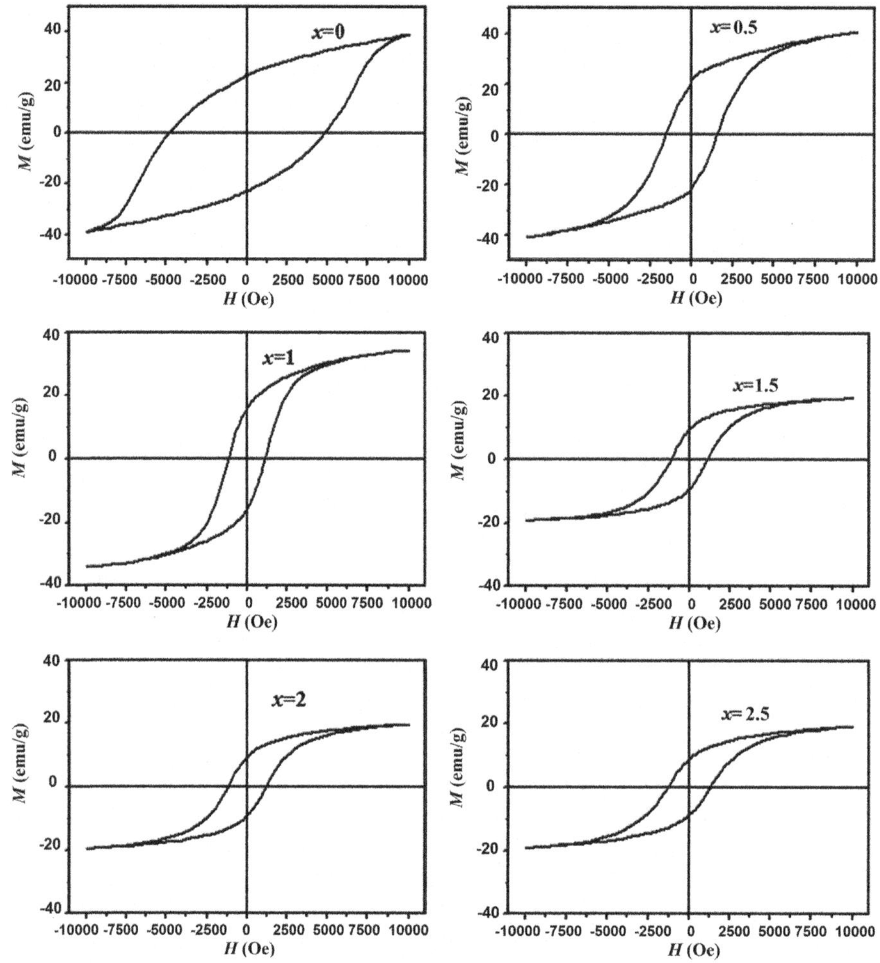

FIGURE 3.9

Vibrating sample magnetometry graphs of substituted Sr-hexaferrite nanoparticles synthesized by coprecipitation method.

From Gordani, G. R., Ghasemi, A., & Saidi, A. (c. 2014). Enhanced magnetic properties of substituted Sr-hexaferrite nanoparticles synthesized by coprecipitation method. Ceramics International, 40*(3), 4945–4952. https://doi.org/10.1016/j.ceramint.2013.10.096 4949.*

3.3.1.3 Design criteria for perpendicular magnetic recording media

Perpendicularly oriented M-type ferrites can be used in high-density recording media. To this end, hexagonal ferrites with high purity and uniaxial anisotropy have been epitaxially grown on appropriate substrates. In order to realize high-density magnetic recording media, the following criteria need to be fulfilled (Bertram,

1994; Fullerton et al., 2000; Moser & Weller, 1999; Plumer et al., 2001; Weller & Moser, 1999):

Squareness ratio of 1
High coercivity (H_c)
Negative nucleation field (H_n)
Moderate saturation magnetization (M_s)
Nanosized grains
Magnetically isolated grains with reduced exchange interaction
High thermal stability
High corrosion and wear resistance
Good mechanical properties
Minimal production cost

Notably, strontium and barium ferrites have the following advantages, making them promising candidates for magnetic recording media applications (Kojima, 1982, Chapter 1; Morisako et al., 1999; Speliotis, 1987):

Narrow switching field distribution
Negligible exchange interaction in dot array configuration
Perpendicular magnetic anisotropy
High anisotropy field
Excellent chemical stability and mechanical strength
Low thermal instability
High corrosion immunity
Enhanced tribological interface, avoiding the need of having overcoats

In this direction, taking into consideration the aforementioned requirements, high-density recording media have been prepared based on SrM thin films using a DC magnetron sputtering system (Kaewrawang et al., 2009). The SrM thin films were deposited on silicon wafer substrate oxidized under thermal treatment (SiO_2/Si) and sapphire substrate with a single-crystal structure oriented along the (00*l*) direction (Al_2O_3 (00*l*)) using a Pt underlayer at various temperatures. The typical X-ray diffraction (XRD) patterns of the resulting SrM/Pt films deposited on the SiO_2/Si and Al_2O_3 substrates at room temperature (T_u = r.t.) and T_u= 400°C are shown in Fig. 3.10. As observed, all the thin films show fcc (111) preferential orientation for the Pt underlayer and hexagonal c-plane for the SrM layer. Moreover, increasing T_u from r.t. to 400°C increases the Pt(111) diffraction line intensity. Thereby, the SrM phase emerges with a stronger (00*l*) diffraction line deposited on the Pt underlayer. At the same T_u, a diffraction line with stronger intensity is observed for the SrM/Pt thin films deposited on the SiO_2/Si substrate compared with the diffraction line intensity of the SrM/Pt thin films deposited on the Al_2O_3 substrate. In addition, the SrM films are promoted along the c-axis, arising from the (111) oriented Pt underlayer. On the other hand, it is possible to investigate the dispersion angle ($\Delta\theta_{50}$) by the full-width at half-maximum of the rocking curve of Pt(111) and SrM(008). In this case, increasing T_u decreases the corresponding

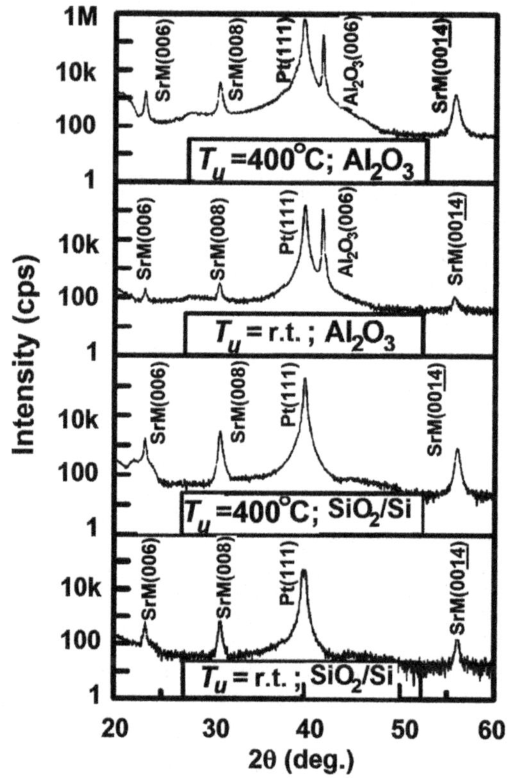

FIGURE 3.10

X-ray diffraction patterns of strontium ferrite (SrM)/Pt thin films deposited on SiO_2/Si and Al_2O_3 substrates at room temperature (T_u = r.t.) and T_u = 400°C.

From Kaewrawang, A., Ghasemi, A., Liu, X., & Morisako, A. (c. 2009). Crystallographic and magnetic properties of SrM thin films on Pt underlayer prepared at various substrate temperatures. Journal of Magnetism and Magnetic Materials, 321*(13), 1939–1942. https://doi.org/10.1016/j.jmmm.2008.12.023 1940.*

$\Delta\theta_{50}$. The relation between $\Delta\theta_{50}$ values of SrM(008) and Pt(111) deposited on the SiO_2/Si and Al_2O_3 substrates is shown in Fig. 3.11. The $\Delta\theta_{50}$ of SrM(008) is observed to depend on the $\Delta\theta_{50}$ of Pt(111). In other words, decreasing the $\Delta\theta_{50}$ of the Pt(111) underlayer reduces the $\Delta\theta_{50}$ value of SrM(008). It is worth noting that the $\Delta\theta_{50}$ value of Pt(111) deposited on the Al_2O_3 substrate with a single-crystal structure is approximately 0.771 smaller than that deposited on the SiO_2/Si substrate, indicating the possibility of improving the SrM crystallographic structure.

Perpendicular and in-plane M–H loops measured at T_u = r.t., 400, and 600°C are depicted in Fig. 3.12, indicating perpendicular magnetic anisotropy for all thin films. The occurrence of the magnetic anisotropy in the perpendicular direction agrees with the results obtained from the XRD analysis. The anisotropy field of SrM/Pt thin films is obtained to be about 18.5 kOe, being almost the same as that of its

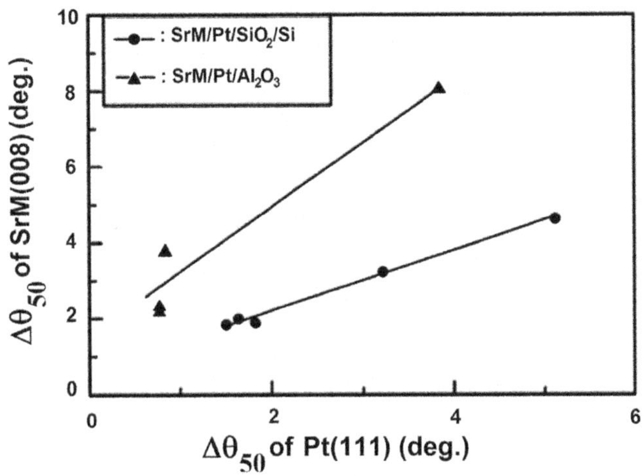

FIGURE 3.11

The relation between $\Delta\theta_{50}$ of SrM(008) and Pt(111) deposited on SiO_2/Si and Al_2O_3 substrates.

From Kaewrawang, A., Ghasemi, A., Liu, X., & Morisako, A. (c. 2009). Crystallographic and magnetic properties of SrM thin films on Pt underlayer prepared at various substrate temperatures. Journal of Magnetism and Magnetic Materials, 321*(13), 1939–1942. https://doi.org/10.1016/j.jmmm.2008.12.023 1940.*

bulk value. Furthermore, M_s is higher for the SrM/Pt deposited on the SiO_2/Si substrate than that on the Al_2O_3 substrate. Increasing T_u gradually decreases M_s of SrM/Pt/SiO_2/Si from 260 to 225 emu/cm^3. Coincidentally, it increases M_s of SrM/Pt/Al_2O_3 from 160 to 215 emu/cm^3. Further increasing T_u reduces M_s of SrM/Pt/Al_2O_3 from 215 to 190 emu/cm^3.

The perpendicular coercivity, $H_{c_\perp}$, is higher than the in-plane coercivity, $H_{c||}$, indicating the formation of the SrM thin film with perpendicular anisotropy. In the case of SrM films deposited on the Al_2O_3 substrate, H_c increases gradually with increasing T_u. Further increasing T_u maintains the H_c constant to 4.2 kOe, higher than the corresponding H_c of SrM films deposited on the SiO_2/Si substrate. For all T_u values, the SrM layers deposited on Pt/SiO_2/Si have a larger grain size than those deposited on Pt/Al_2O_3, according to scanning electron microscopy (SEM) investigations (not shown here).

Alternatively, $H_{c||}$ is seen to decrease from 0.8 to 0.5 kOe, and 0.4 to 0.1 kOe for SrM/Pt/Al_2O_3 and SrM/Pt/SiO_2/Si when increasing T_u from r.t. to 400°C, respectively. Further incasing T_u increases $H_{c||}$ of SrM/Pt/SiO_2/Si from 0.1 to 0.9 kOe. Increasing T_u decreases squareness (S) of SrM/Pt/SiO_2/Si and SrM/Pt/Al_2O_3 from 0.82 to 0.77, and 0.75 to 0.68, respectively. A decreasing trend is also observed for $S_{||}$, being the same as that for $H_{c||}$. Notably, $S_{||}$ is minimized at $T_u = 400$°C, reaching 0.05 and 0.08 for SrM/SiO_2/Si and SrM/Al_2O_3 thin films, respectively.

In a separate research study, SrM thin films were deposited on Au nano-dots using a DC magnetron sputtering system (Kaewrawang et al., 2010). Improved Au

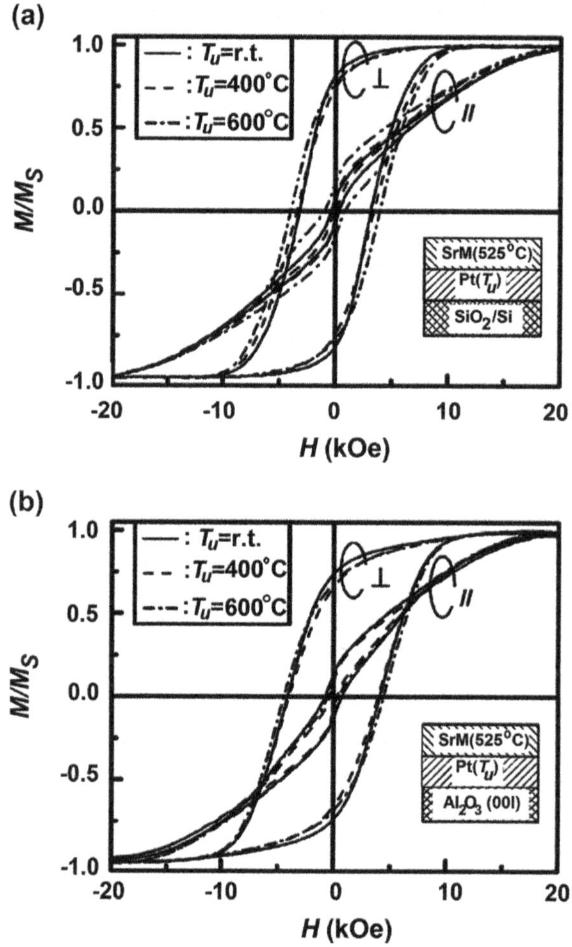

FIGURE 3.12

M—H loops measured at perpendicular and in-plane directions for strontium ferrite (SrM)/ Pt thin films deposited on: (A) SiO_2/Si, and (B) Al_2O_3 substrates using T_u = r.t., 400 and 600°C.

From Kaewrawang, A., Ghasemi, A., Liu, X., & Morisako, A. (c. 2009). Crystallographic and magnetic properties of SrM thin films on Pt underlayer prepared at various substrate temperatures. Journal of Magnetism and Magnetic Materials, 321*(13), 1939–1942. https://doi.org/10.1016/j.jmmm.2008.12.023 1940.*

nano-dot crystalline orientation was attempted by employing single-crystal sapphire [Al_2O_3 (00*l*)], single-crystal MgO (111), and Corning glass substrates. In fact, the orientation of the c-axis of the hexagonal ferrite layer may be promoted by the resulting Au film underlayer.

The fcc Au (111) plane and c-plane of hcp SrM have a misfit ratio of 2.1%. It is possible to grow Au island structures on SiO_2/Si (Kaewrawang et al., 2008) and glass substrates using $T_u > 100°C$ (A Kaewrawang et al., 2009), whereas continuous films have been reported to form at T_u = r.t. Furthermore, Au particles can absorb the infrared (Suzuki et al., 2004), contributing to the crystallization of SrM. In this direction, using the Au underlayer, isolated SrM thin films have been prepared. Fig. 3.13 shows a schematic cross-sectional representation of the SrM/Au structure.

Transmission electron microscopy (TEM) investigations were carried out in order to detail the morphology of the Au island structures, and the results are shown in Fig. 3.14. As can be seen, while the Au structure is in the form of a strip at T_u = r.t. (Fig. 3.14A), it has an island structure at $T_u = 400°C$, according to Fig. 3.14B. The size of the Au dot ranges from 10 to 20 nm, with an average of approximately 15 nm and a spacing of 10 nm.

Perpendicular and in-plane hysteresis loops of the SrM/Au thin films deposited on the Al_2O_3, MgO, and Corning glass substrates are shown in Fig. 3.15. Regardless

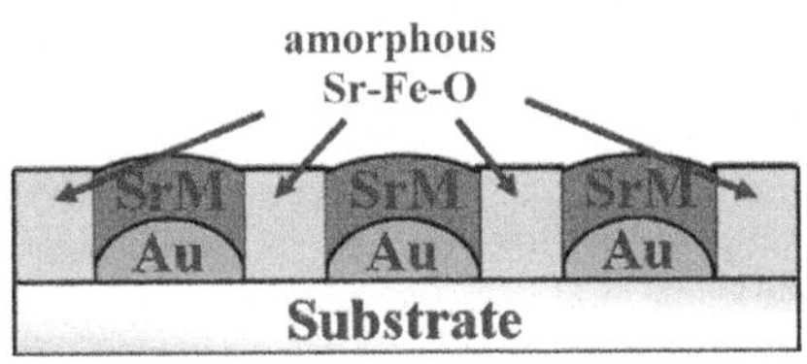

FIGURE 3.13

A schematic cross-sectional representation of strontium ferrite (SrM)/Au structure.

From Kaewrawang, A., Ghasemi, A., Liu, X., & Morisako, A. (c. 2010). Fabrication, crystallographic and magnetic properties of SrM perpendicular films on Au nano-dot arrays. Journal of Alloys and Compounds, *492(1–2), 44–47. https://doi.org/10.1016/j.jallcom.2009.11.174 47.*

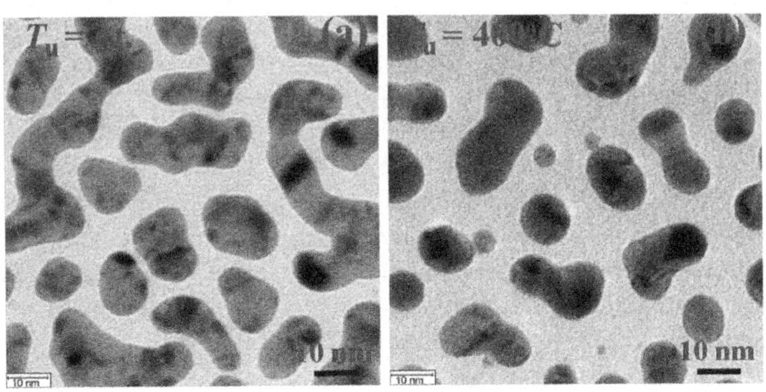

FIGURE 3.14

Transmission electron miscroscopy images obtained from Au island structures prepared at (A) T_U= r.t., and (B) T_U= 400°C.

From Kaewrawang, A., Ghasemi, A., Liu, X., & Morisako, A. (c. 2010). Fabrication, crystallographic and magnetic properties of SrM perpendicular films on Au nano-dot arrays. Journal of Alloys and Compounds, *492(1–2), 44–47. https://doi.org/10.1016/j.jallcom.2009.11.174 47.*

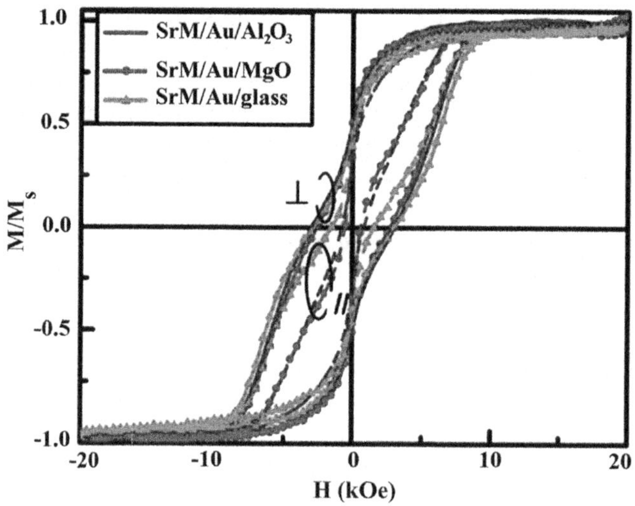

FIGURE 3.15

Hysteresis loops of strontium ferrite (SrM)/Au/Al_2O_3, SrM/Au/MgO and SrM/Au/glass thin films measured in perpendicular and in-plane directions.

From Kaewrawang, A., Ghasemi, A., Liu, X., & Morisako, A. (c. 2010). Fabrication, crystallographic and magnetic properties of SrM perpendicular films on Au nano-dot arrays. Journal of Alloys and Compounds, *492(1–2), 44–47. https://doi.org/10.1016/j.jallcom.2009.11.174 47.*

of the substrate type, all the SrM/Au thin films show magnetic anisotropy in the perpendicular direction. Moreover, the presence of in-plane orientation is evidenced by the hysteresis loop measured in the in-plane magnetization, which can be justified based on the study carried out by Sui et al. (1993). They proposed that platelet and acicular grains have a c-axis orientation perpendicular to and in the film plane, respectively. Hexagonal platelets are known to grow slowly and rapidly along the c-axis and in the ab plane, respectively (Smit & Wign, 1959, p. 143). The growth of grains is not restricted in the lateral direction when the initial nucleus has a c-axis perpendicular to the plane of the film. As a result, a considerably higher film volume is expected to be occupied by the platelet grains than with acicular ones, being in agreement with the hysteresis loop measurements depicted in Fig. 3.15. Since no preferential orientation is observed in the basal planes of the grains with the *c*-axis in the film plane, platelet grains (having a *c*-axis perpendicular to the plane) mainly give rise to the signature detection by the XRD. M_s, $H_{c_\perp}$, and $S_\perp$ of the SrM/Au thin film deposited on the single-crystal Al_2O_3 (00*l*) substrate are higher than those deposited on the single-crystal MgO and glass substrates. The maximum M_s value of SrM/Au/Al_2O_3 is approximately 125 emu/cm^3. It should be noted that the theoretical M_s value of standard SrM film (360 emu/cm^3) is higher than that of all the prepared SrM/Au thin films, arising from the presence of nonmagnetic phases between the magnetic particle spaces in the SrM thin films, as well as the substitution of Fe atoms by Au ones in the SrM lattice. The SrM film magnetic moment is reduced when Au atoms occupy unidirectional Fe sites with the spin-up orientation.

Fig. 3.16 compares δ*m(H)* curves of the SrM/Au (deposited on the different substrates) and SrM/Pt thin films. As inferred, magnetostatic interactions are present in all the thin films. However, SrM/Au thin films have weaker interactions than the continuous SrM/Pt film due to their smaller δ*m* curve peak (Kaewrawang et al., 2010). In fact, the nonmagnetic phases existing between the magnetic particle spaces in the SrM thin films may reduce the intergranular interactions.

3.3.1.4 Magnetic properties of barium ferrites

As mentioned, barium ferrite is an important type of hexagonal ferrite whose magnetic feature can be tuned by incorporating substituted cations. In research work, $BaFe_{10}CrAlO_{19}$/Au thin films were deposited on a Corning glass substrate using a sputtering technique and then annealed via a rapid thermal process. By keeping the ferrite thickness and the heating time constant to 50 nm and 15 min, respectively, the structural and magnetic properties of the resulting thin films with annealing temperatures ranging between 500 and 600°C (25°C in each step) have been investigated (Ghasemi, 2016a).

To study the mechanism of magnetization reversal involved in the thin films, the angular dependence of coercivity was obtained and compared with theoretical magnetization reversal models, including coherent magnetization rotation and DW motion based on the Stoner–Wohlfarth (SW) and Kondorsky theories,

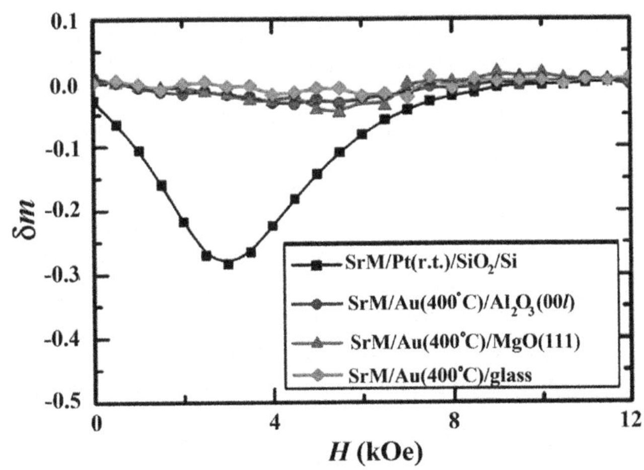

FIGURE 3.16

Comparison of δ*m*(*H*) curves of strontium ferrite (SrM)/Au thin films (deposited on the different substrates) and SrM/Pt thin film.

From Kaewrawang, A., Ghasemi, A., Liu, X., & Morisako, A. (c. 2010). Fabrication, crystallographic and magnetic properties of SrM perpendicular films on Au nano-dot arrays. Journal of Alloys and Compounds, *492(1–2), 44–47. https://doi.org/10.1016/j.jallcom.2009.11.174 47.*

respectively. By applying a magnetic field to multidomain materials, magnetic domains with lower energy can be expanded while constricting domains with higher energy. In turn, this causes the occurrence of DW movement. On the contrary, the magnetization reversal can occur without moving the DWs when the magnetization rotates in the direction of the magnetic field applied to single-domain materials.

The Kondorsky law defines the relationship between the coercive field and magnetic field angle, H_c (θ_H), when $0 < \theta_H < \pi/2$, as given below (Kondorski, 1937):

$$H_c(\theta_H) = \frac{H_c(0)}{\cos\theta_H} \tag{3.13}$$

where θ_H is the angle between the magnetic field and the easy axis. From Eq. (3.13), one can illustrate the pinning mechanism involved in the magnetization reversal.

The SW model is expected to be realized in single-domain magnets, reversing the magnetization via coherent rotation without moving the DWs. In this case, H_c (θ_H) can be expressed as follows (Stoner & Wohlfarth, 1948):

$$H_c(\theta_H) = \begin{cases} \dfrac{H_c(0)}{\left[(\cos\theta_H)^{2/3} + (\sin\theta_H)^{2/3}\right]^{3/2}}; & 0 \le \theta \le \dfrac{\pi}{4} \\ \dfrac{H_c(0)\sin(2\theta_H)}{2}; & \dfrac{\pi}{4} \le \theta \le \dfrac{\pi}{2} \end{cases} \tag{3.14}$$

Importantly, the SW model should dominate the magnetization process in ideally designed recording media. This is because magnetic moments store the information in each bit magnetically. If DW pinning is the dominant mechanism involved in the magnetization reversal, the magnetic domains may be expanded and shrunk, thus varying the magnetization vector orientation.

Angular dependence of coercivity [$H_c(\theta°)/H_c(0°)$] measured for $BaFe_{10}CrAlO_{19}$/Au/Corning glass thin films with different annealing temperatures is depicted in Fig. 3.17. For comparison, $H_c(\theta°)/H_c(0°)$ calculated for SW and DW motion modes are included as well. While they differ in the magnitude of their angular dependence of coercivity, all the thin film samples show similar variation behavior of $H_c(\theta°)/H_c(0°)$, being in agreement with the SW model. Note that although $H_c(\theta°)/H_c(0°)$ behavior does not completely follow the coherent rotation, its deviation from the DW motion is quite large. In this way, the $BaFe_{10}CrAlO_{19}$/Au/Corning glass thin films annealed at different temperatures may be suitable for magnetic recording media applications.

Elsewhere, using a coprecipitation technique, substituted barium hexaferrite [$BaFe_{12-x}(MnMgTiZr)_{x/4}O_{19}$ ($x = 0-2.5$, in a step of $x = 0.5$)] nanoparticles have been synthesized (Amirabadizade et al., 2016). By applying a maximum magnetic field of 20 kOe using vibrating sample magnetometry (VSM) at room temperature, the resulting nanoparticles were investigated in terms of magnetic properties. Fig. 3.18 shows the corresponding hysteresis loops of the barium hexaferrite samples with different compositions. Increasing x from 0 to 2 decreases the coercivity from 4.8 to 0.8 kOe. Based on previous reports, magnetocrystalline anisotropy and particle size can influence the coercivity of ferrite nanoparticles. In this regard,

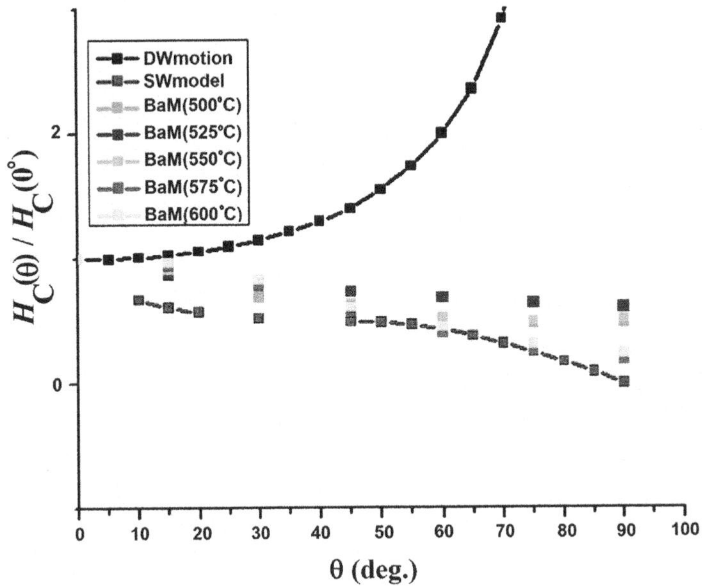

FIGURE 3.17

Angular dependence of coercivity [$H_c(\theta°)/H_c(0°)$] measured for $BaFe_{10}CrAlO_{19}$ ferrite thin films annealed at different temperatures, along with $H_c(\theta°)/H_c(0°)$ calculated for Stoner–Wohlfarth (SW) and domain wall (DW) motion modes.

From Ghasemi, A. (c. 2016). Coherent rotation of magnetization or Kondorsky model of exchange-decoupled $BaFe_{10}CrAlO_{19}$ dot array. Journal of Alloys and Compounds, 684, *245–253. https://doi.org/10.1016/j.jallcom. 2016.05.190 251.*

the reduction in the coercivity of the substituted barium hexaferrites can be attributed to an increase in particle size and a decrease in magnetocrystalline anisotropy.

On the other hand, increasing x from 0 to 1 increases the saturation magnetization from 32.3 to 61 emu/g. Further increasing x to 2.5 decreases the saturation magnetization to 38.9 emu/g as the Mn^{2+}, Mg^{2+}, Ti^{4+}, and Zr^{4+} cations are preferentially substituted in the lattice sites. Basically, the Mn^{2+} ion has a magnetic moment of 3 μ_B which is smaller than the magnetic moment of Fe^{3+} ions (5 μ_B). Moreover, Mg^{2+}, Ti^{4+}, and Zr^{4+} are considered nonmagnetic ions. Thus, the replacement of spin-down ($4f_1$ and $4f_2$ sites) Fe^{3+} ions by the nonmagnetic and the magnetic ions might be responsible for the enhancement observed in the saturation magnetization of the hexaferrite nanoparticles. At $x = 2.5$, the saturation magnetization reduction can be explained by the replacement of spin-up ($2a$, $2b$, and $12k$ sites) Fe^{3+} ions by the Mg^{2+}, Ti^{4+}, Zr^{4+} and Mn^{2+} ones.

3.3.1.5 Microwave absorption characteristics of hexagonal ferrites

It is well documented that unsubstituted ferrites show an unsatisfactorily low reflection loss value and bandwidth, demonstrating ineffective absorbing characteristics by their high FMR frequency ($\sim$ 48 GHz). As well, it has been evidenced that composites consisting of substituted M-type ferrites can provide considerably higher electromagnetic absorption effects.

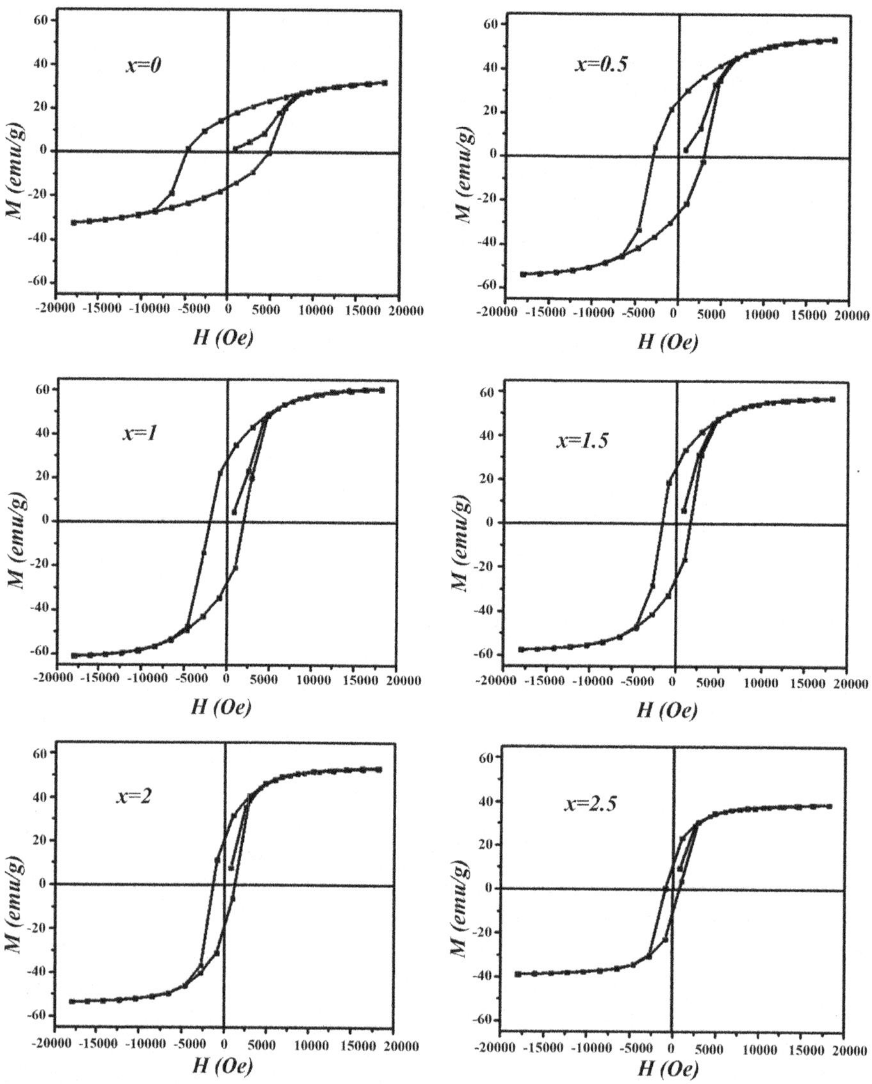

FIGURE 3.18

Room-temperature hysteresis loops of $BaFe_{12-x}(MnMgTiZr)_{x/4}O_{19}$ nanoparticles with different compositions ($x = 0$–2.5).

From Amirabadizade, A., Shiri, N., & Ghasemi, A. (c. 2016). The Role of Mn–Mg–Ti–Zr Substitution on Structural and Magnetic Features of $BaFe_{12-x}(MnMgTiZr)_{x/4}O_{19}$ Nanoparticles. Journal of Superconductivity and Novel Magnetism, *29(2), 515–520. https://doi.org/10.1007/s10948-015-3332-y 518.*

The variation of reflection loss as a function of frequency for typical substituted hexagonal ferrites is shown in Figs. 3.19–3.21. As inferred, the resonance frequency of substituted ferrites with a higher substituted cation content shifts to lower frequencies while increasing the resulting absorption bandwidth. Moreover, the presence of two resonance frequencies is evidenced for all thicknesses. In fact, the reflection dip results in the DW motion and the spin resonance at the lower and higher frequencies, respectively. When the absorber thickness was increased, the resonance frequency shifted lower.

3.3.2 W-type hexagonal ferrites

The W-type hexaferrite has the chemical composition of $SrMe_2Fe_{16}O_{27}$ (Me = any divalent element). The ions in the W-type hexaferrite structure can be distributed at seven different lattice sites (i.e., $4f_{vi}$, $2d$, $12k$, $6g$, $4f$, $4f_{iv}$, and $4e$) with different spin coordination and orientation (Pullar, 2012).

In a research study, magnetic properties of $SrZn_xFe_{2-x}Fe_{16}O_{27}$ ($x = 0.0–2.0$), SrW hexaferrites synthesized at different temperatures with a reduced oxygen atmosphere of 10^{-3} atm have been investigated (You & Yoo, 2018). In this regard, the

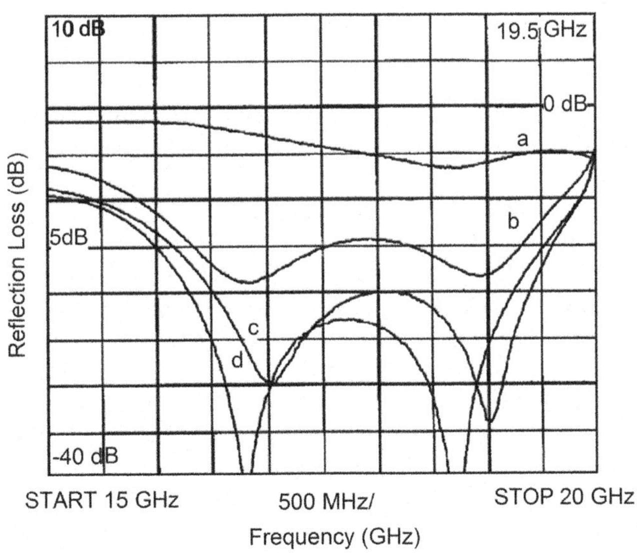

FIGURE 3.19

Absorption characteristics of the composite consisting of: (A) barium ferrite, (B) $BaFe_{11}(Mn_{0.5}Co_{0.5}Zr)_{1/2}O_{19}$, (C) $BaFe_{10}(Mn_{0.5}Co_{0.5}Zr)_{2/2}O_{19}$, and (d) $BaFe_9(Mn_{0.5}Co_{0.5}Zr)_{3/2}O_{19}$.

From Ghasemi, A., Liu, X., & Morisako, A. (c. 2007). Magnetic and microwave absorption properties of $BaFe_{12-x}(Mn_{0.5}Cu_{0.5}Zr)_{x/2}O_{19}$ synthesized by sol–gel processing. Journal of Magnetism and Magnetic Materials, 316*(2), e105–e108. https://doi.org/10.1016/j.jmmm.2007.02.043 108.*

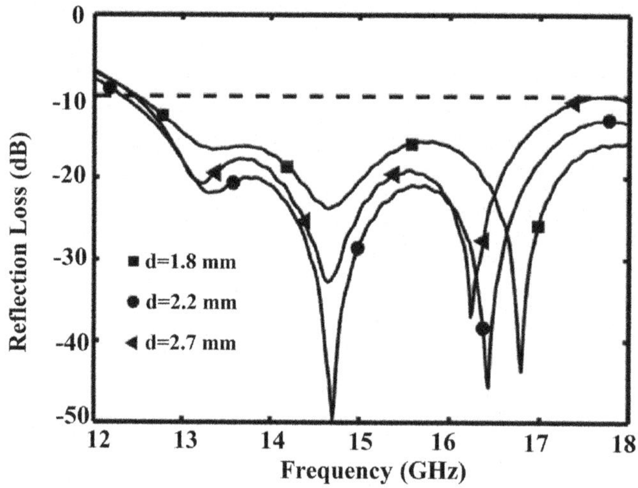

FIGURE 3.20

Absorption characteristics of rubber composite consisting of $BaFe_9Mg_{1.5}Ti_{1.5}O_{19}$ with thicknesses of 1.8, 2.2 and 2.7 mm.

From Shams, M. H., Salehi, S. M. A., & Ghasemi, A. (c. 2008). Electromagnetic wave absorption characteristics of Mg–Ti substituted Ba-hexaferrite. Materials Letters, *62(10-11), 1731–1733. https://doi.org/10.1016/j.matlet.2007.09.073 1733.*

influence of the substitution of Zn^{2+} for Fe^{2+} on the saturation magnetization (M_s) was emphasized. Fig. 3.22A shows hysteresis loops of pure and Zn-substituted ferrite samples.

The coercivity of the ferrite samples ranges from 90 to 120 Oe. The variations of M_s and H_a as a function of x are shown in Fig. 3.22B. The M_s reaches its maximum value at $x = 1$. As can be seen, increasing x from 0.0 to 2.0 continuously decreases H_a from 19468 to 14640 Oe, which can be assigned to the weakening of the super-exchange interactions among magnetic ions as well as the spin–orbit coupling interaction after introducing the nonmagnetic Zn^{2+} ions into the ferrite structure.

Co_2-W is considered an interesting material among different W-type Sr ferrites due to its high permeability and magnetic loss properties (Stergiou & Litsardakis, 2011). To efficiently tune the magnetic properties (including saturation magnetization, coercive field, and FMR frequency) of ferrites, one can change the ferrite composition, preparation method, and calcination temperature while substituting magnetic and nonmagnetic ions (Khan et al., 2011). According to the literature (Ahmad, Grössinger et al., 2013; Khan et al., 2011; Pullar, 2012; Stergiou & Litsardakis, 2011a,b), magnetic properties of the $SrCo_2$ W-type hexagonal ferrite are improved after substituting it with different ions. It is worth noting that without changing the iron content, the W-type hexaferrites can be widely tuned in terms of magnetic properties by choosing appropriate substitution divalent cations (Özgür et al., 2009).

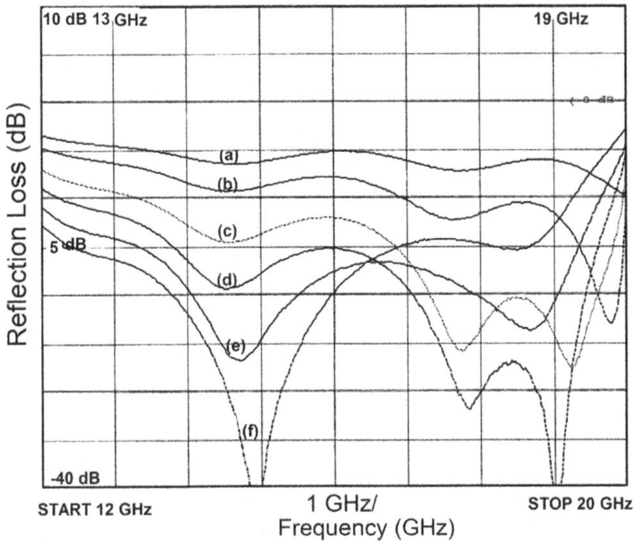

FIGURE 3.21

Absorption characteristics of the $BaCo_{0.5}Mn_{0.5}Ti_{1.0}Fe_{10}O_{19}$ ferrite composite with different thicknesses: (A) $t = 1.35$ mm, (B) $t = 1.70$ mm, (C) $t = 1.80$ mm, (D) $t = 2.00$ mm, (E) $t = 2.10$ mm, and (F) $t = 2.20$ mm.

From Choopani, S., Keyhan, N., Ghasemi, A., Sharbathi, A., Maghsoudi, I., & Eghbali, M. (c. 2009). Static and dynamic magnetic characteristics of $BaCo_{0.5}Mn_{0.5}Ti_{1.0}Fe_{10}O_{19}$. Journal of Magnetism and Magnetic Materials, *321(13), 1996–2000. https://doi.org/10.1016/j.jmmm.2008.12.030 1999.*

In an investigation, using a sol–gel autocombustion method, $SrCo_2$ W-type hexaferrites with the composition of $SrCo_{2-x}Zn_xCe_yFe_{16-y}O_{27}$ ($x = 0.0–1.0$; and $y = 0.0–0.1$) have been synthesized by sintering them at a temperature of 1050°C (Akhtar & Khan, 2018b). Fig. 3.23 shows XRD patterns of $SrCo_2$ W-type hexaferrites with different Zn and Ce contents. The (hkl) index has been clarified for each sample. Evidently, the formation of single-phase W-type ferrite is detected for the whole series of samples without the presence of secondary phases.

On the other hand, hysteresis loops of the $SrCo_2$ W-type hexaferrites measured at room temperature are shown in Fig. 3.24, indicating the single-phase magnetic behavior of the Ce-Zn-substituted $SrCo_2$ hexagonal ferrites. This also agrees with the structural results obtained for them. Details of the hysteresis loops are depicted in the inset of Fig. 3.24. The saturation magnetization and remanence of the resulting hexaferrites decrease with increased Ce-Zn concentration, whereas the corresponding coercivity increases.

Elsewhere, using a sol–gel autoignition method, Mn- and Ti-substituted W-type hexagonal ferrites with the composition of $Ba_{0.5}Sr_{0.5}Co_2Mn_xTi_xFe_{16-2x}O_{27}$ ($x = 0.00–2.50$) have been synthesized (Akhtar, Javed, et al., 2020). Hysteresis loops of the hexaferrites were measured at room temperature, and the results are presented

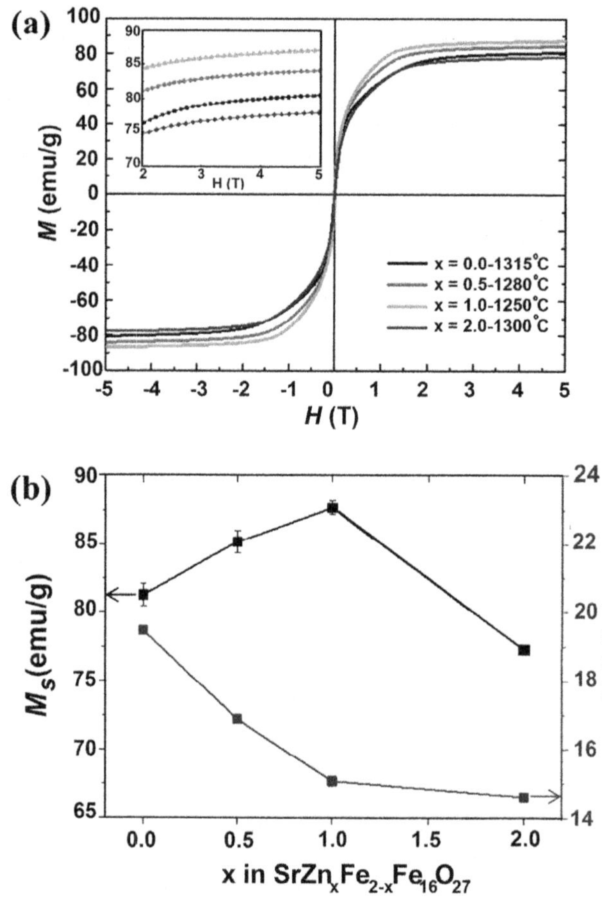

FIGURE 3.22

(A) Hysteresis loops obtained from $SrZn_xFe_{2-x}Fe_{16}O_{27}$ ($x = 0.0–2.0$) SrW hexaferrites. The inset demonstrates the loops in a lower magnetic field range (2–5 T). (B) The variations of M_S and H_a as a function of x.

From You, J., & Yoo, S. (c. 2018). Improved magnetic properties of Zn-substituted strontium W-type hexaferrites. Journal of Alloys and Compounds, 763, *459–465. https://doi.org/10.1016/j.jallcom.2018.05.296 17.*

in Fig. 3.25. As observed, the formation of S-shaped hysteresis loops indicates the single-phase magnetic behavior. The XRD analysis also confirmed the magnetic single-phase behavior of the W-type ferrites.

The variations of M_r, M_s, squareness ratio (M_r/M_s), and H_c as a function of Mn-Ti concentration are shown in Fig. 3.26. As can be seen, increasing the Mn-Ti substitution content decreases M_r, M_s, and H_c, whereas it increases the squareness ratio in the range of 0.67–0.75. Moreover, the variations of Bohr magneton, magnetocrystalline anisotropy constant, initial permeability, and magnetic anisotropy field as a

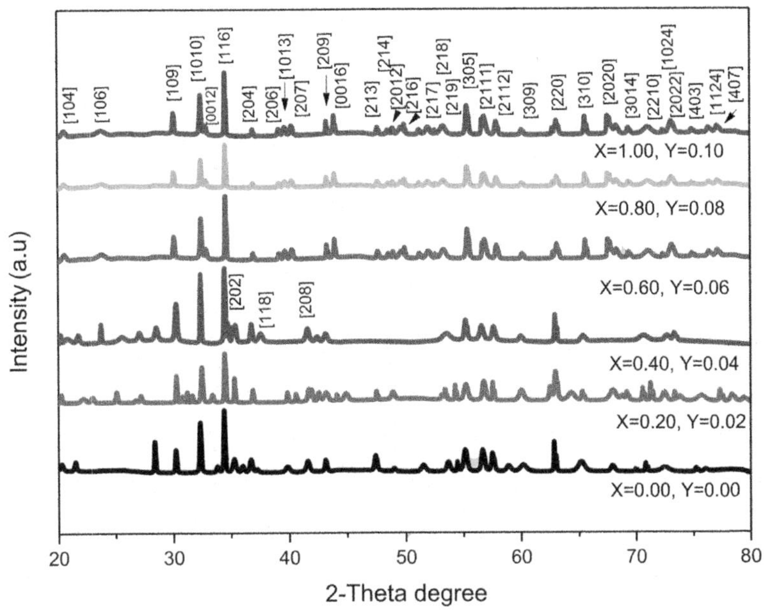

FIGURE 3.23

X-ray diffraction patterns obtained from $SrCo_{2-x}Zn_xCe_yFe_{16-y}O_{27}$ ($x = 0.0–1.0$; and $y = 0.0–0.1$) W-type hexaferrites.

From Akhtar, M. N., & Khan, M. A. (c. 2018). Structural, physical and magnetic evaluations of Ce-Zn substituted $SrCo_2$ W-type hexaferrites prepared via sol gel auto combustion route. Ceramics International, *44(11), 12921–12928. https://doi.org/10.1016/j.ceramint.2018.04.104.*

function of Mn-Ti concentration are depicted in Fig. 3.27. The mentioned parameters decreased with an increase in substitutions.

To calculate μ_B, initial permeability (μ_i), H_a, and K, the following relations can be used (Akhtar & Khan, 2018a; Akhtar et al., 2020; Hossain & Rahman, 2011; Ikram et al., 2018; Kabbur et al., 2018; Yousaf et al., 2020):

$$\text{Bohr magneton}(\mu_B) = \frac{M \times M_s}{5585 \times \rho_{\text{Xray}}} \tag{3.15}$$

$$\text{Initial permeability}(\mu_i) = \frac{M_s^2 \times D}{\text{K}} \tag{3.16}$$

$$\text{Anisotropy field}(H_a) = \frac{2\text{K}}{\mu_0 M_s} \tag{3.17}$$

$$\text{Anisotropy constant}(\text{K}) = \frac{H_c \times M_s}{0.96} \tag{3.18}$$

where M is the molecular weight of the substituted hexaferrite, ρ_{Xray} is the density calculated from the XRD pattern, and D is the grain size.

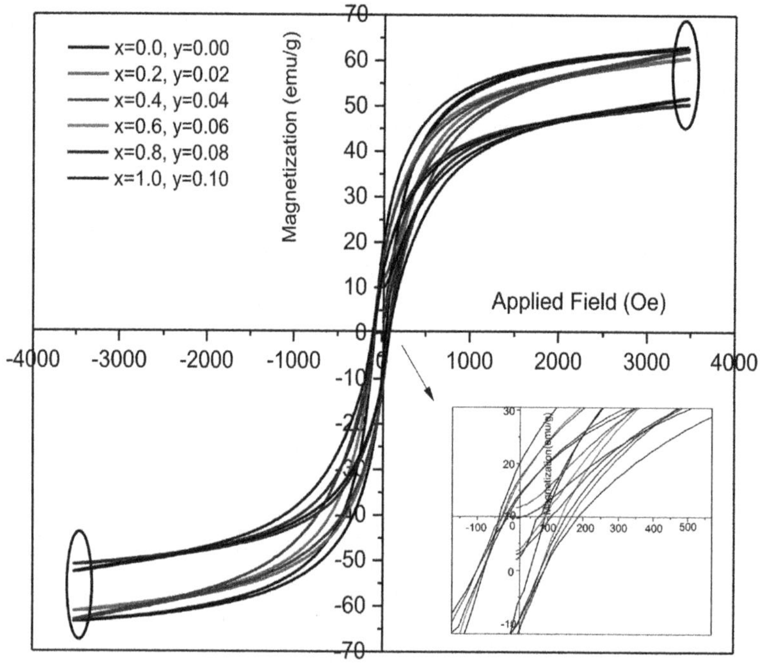

FIGURE 3.24

Hysteresis loops of $SrCo_{2-x}Zn_xCe_yFe_{16-y}O_{27}$ ($x = 0.0–1.0$; and $y = 0.0–0.1$) hexaferrites measured at room temperature. The inset shows details of the hysteresis loops.

From Akhtar, M. N., & Khan, M. A. (c. 2018). Structural, physical and magnetic evaluations of Ce-Zn substituted $SrCo_2$ W-type hexaferrites prepared via sol gel auto combustion route. Ceramics International, *44(11), 12921–12928. https://doi.org/10.1016/j.ceramint.2018.04.104.*

The Heisenberg superexchange interactions and the spin canting mechanism can play important roles in adjusting magnetic characteristics. As well, the octahedral and tetrahedral sites are preferably occupied by the cations, leading to the reduction in the μ_B, μ_i, H_a, and K of the hexaferrites.

In another research study, using a high-energy ball milling process and subsequent sintering at 1300°C, $BaCo_2Fe_{16-x}Ga_xO_{27}$ ($x = 0.0$, 0.2, 0.4, 0.6, and 0.8) powders have been prepared, and their magnetic properties have been investigated (Mahmood et al., 2018). Fig. 3.28 shows hysteresis loops of the resulting W-type hexaferrites measured at room temperature. The S-shaped loops indicate that all the ferrite samples are magnetically soft in nature. Furthermore, the samples are not completely saturated at the applied magnetic field of 10 kOe.

The variations of magnetization and temperature derivative of magnetization as a function of temperature for samples measured at an applied magnetic field of 100 Oe are shown in Fig. 3.29. Increasing the temperature initially increases the magnetization to a peak value for all the ferrite samples. Further increasing the temperature progressively decreases the magnetization to a plateau. For $T > 450°C$, the

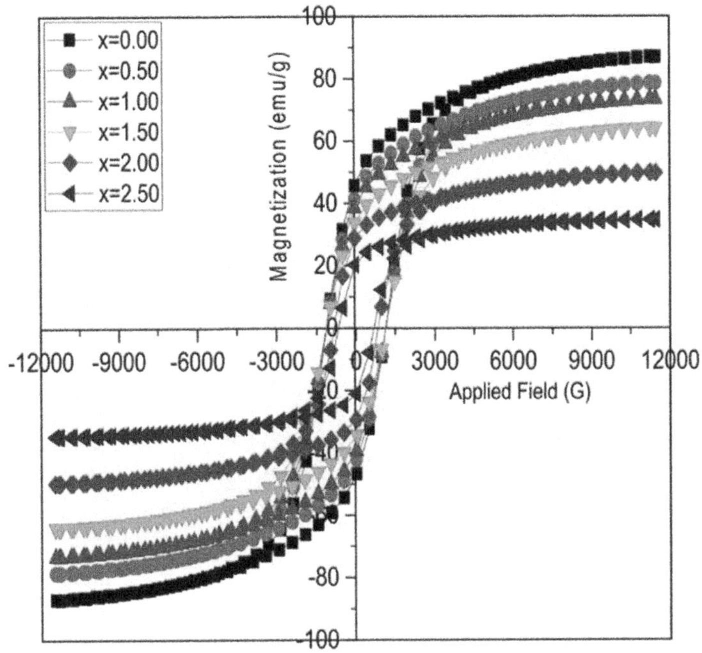

FIGURE 3.25

Hysteresis loops of $Ba_{0.5}Sr_{0.5}Co_2Mn_xTi_x Fe_{16-2x}O_{27}$ ($x = 0.00$–2.50) W-type hexaferrites measured at room temperature.

Akhtar, M. N., Javed, S., Ahmad, M., Sulong, A., & Khan, M. A. (c. 2020). Sol gel derived MnTi doped Co$_2$ W-type hexagonal ferrites: Structural, physical, spectral and magnetic evaluations. Ceramics International, *46(6), 7842–7849. https://doi.org/10.1016/j.ceramint.2019.12.003.*

magnetization is sharply reduced, reflecting a double-step pattern. In turn, this leads to the formation of two peaks in the temperature derivative of the magnetization curve, corresponding to two magnetic phases. The sharper negative peak formed in the temperature range of $451°C < T < 485°C$ can be assigned to the Curie temperature (T_c) of the W-type hexagonal ferrite phase. Accordingly, the T_c of the hexaferrite phase for each sample can be extracted, as presented in Table 3.5. Note that the second phase might arise from the presence of a minority magnetic phase.

Elsewhere, using a coprecipitation method and subsequent calcination at 1100°C for 3 h, W-type hexagonal ferrites with the composition of $SrCo_{2-x}(MnZnCa)_{x/3}$ $Fe_{16}O_{27}$ ($0.0 \leq x \leq 0.5$) have been prepared (Ghasemi, 2016b). XRD patterns indicated that single-phase W-type hexaferrites were formed without the involvement of any impurity phases. Moreover, the crystallite size increased from 50 to 85 nm with an increase in x from 0.0 to 0.5.

To obtain the elemental composition of the resulting hexagonal ferrites, energy-dispersive X-ray spectroscopy (EDS) analysis was carried out, and the results are tabulated in Table 3.6, and the corresponding spectra are shown in Fig. 3.30.

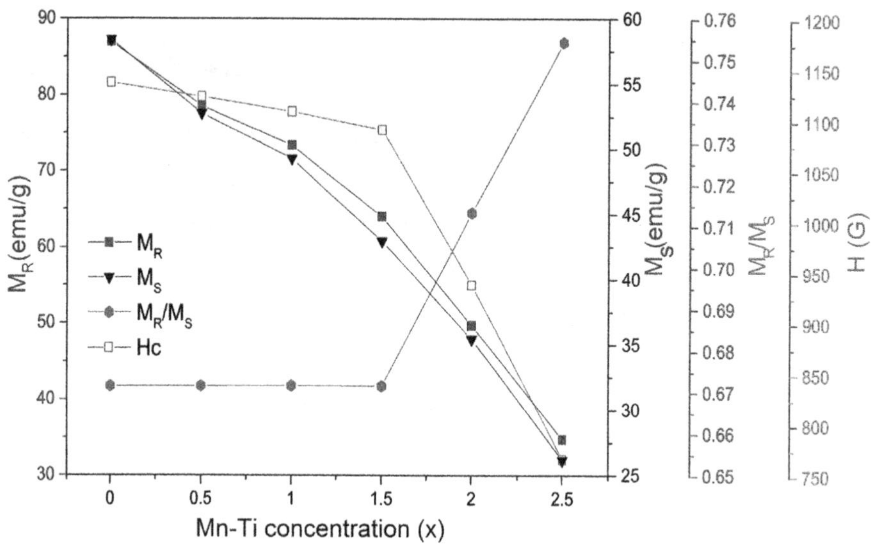

FIGURE 3.26

The variations of remanence (M_r), saturation magnetization (M_s), squareness ratio (M_r/M_s) and coercivity (H_c) as a function of Mn-Ti concentration (x) for $Ba_{0.5}Sr_{0.5}Co_2Mn_xTi_xFe_{16-2x}O_{27}$ W-type hexaferrites.

From Akhtar, M. N., Javed, S., Ahmad, M., Sulong, A., & Khan, M. A. (c. 2020). Sol gel derived MnTi doped Co_2 W-type hexagonal ferrites: Structural, physical, spectral and magnetic evaluations. Ceramics International, *46(6), 7842–7849. https://doi.org/10.1016/j.ceramint.2019.12.003.*

From the EDS spectra, one can reach an agreement between the Sr, Co, Mn, Zn, Ca, Fe, and O contents and the compositions of the dissolved reactants. In turn, this confirms the replacement of cobalt ions by the cations in the hexagonal lattice structure.

VSM hysteresis loops of $SrCo_{2-x}(MnZnCa)_{x/3}Fe_{16}O_{27}$ nanoparticles are shown in Fig. 3.31. Evidently, the saturation magnetization decreases from 78.5 to 72.3 emu/g with an increase in the substitution amount from $x = 0.0$ to 0.5. The lowest saturation magnetization is obtained for the hexaferrite nanoparticles with the substitution amount of $x = 0.4$. In fact, the octahedral positions are occupied by the magnetic ions in the magnetoplumbite structure, whereas the tetrahedral positions are filled by the nonmagnetic ions. This reduces the superexchange interaction between the tetrahedral and octahedral positions, thus decreasing the saturation magnetization (Iqbal, Khan, Mizukami, & Miyazaki, 2011; Ri et al., 2012). Alternatively, the hexaferrite nanoparticles with the substitution amount of $x = 0.3$ result in the highest coercivity (1650 Oe).

Using a chemical coprecipitation method, $SrCo_{2-x}Mn_xFe_{16}O_{27}$ ($0.0 \leq x \leq 0.5$) W-type hexaferrite nanoparticles have also been prepared, followed by pressing their dried powder via a uniaxial hot-press system (Ghasemi, 2016e). The formation of

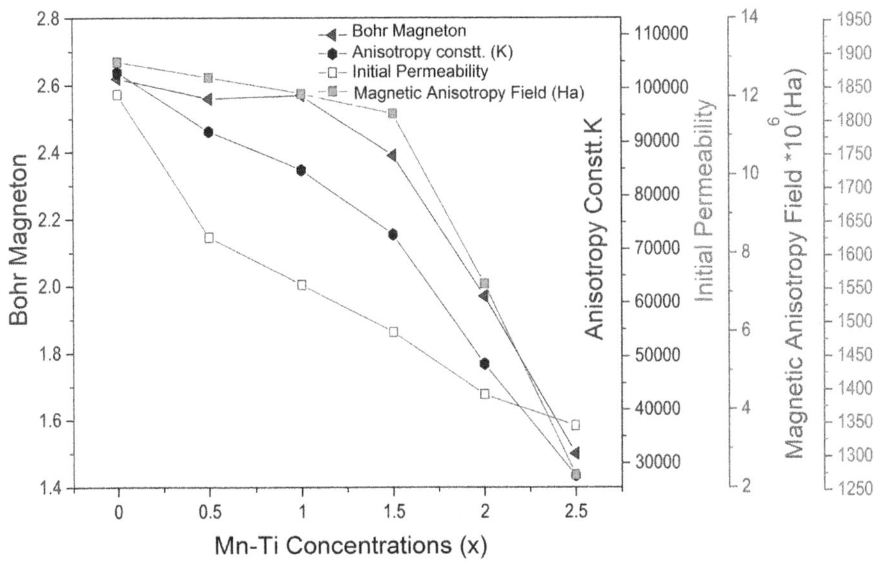

FIGURE 3.27

The variations of Bohr magneton, magnetocrystalline anisotropy constant (K), initial permeability, and magnetic anisotropy field (H_a) as a function of Mn-Ti concentration (x) for $Ba_{0.5}Sr_{0.5}Co_2Mn_xTi_x Fe_{16-2x}O_{27}$ W-type hexaferrites.

From Akhtar, M. N., Javed, S., Ahmad, M., Sulong, A., & Khan, M. A. (c. 2020). Sol gel derived MnTi doped Co_2 W-type hexagonal ferrites: Structural, physical, spectral and magnetic evaluations. Ceramics International, *46(6), 7842–7849. https://doi.org/10.1016/j.ceramint.2019.12.003.*

the single-phase hexaferrite was indicated by XRD analysis while evidencing the absence of any secondary phases. Fig. 3.32 shows perpendicular and in-plane VSM hysteresis loops of $SrCo_{2-x}Mn_xFe_{16}O_{27}$ bulks. For $x \leq 0.3$, the samples show perpendicular anisotropy since the coercivity of the in-plane direction is smaller than that of the perpendicular one. The coercive field values measured in the in-plane and perpendicular directions are almost equal for $x = 0.4$ and 0.5. Moreover, the saturation magnetization values measured in the in-plane and perpendicular directions are the same for all the hexaferrite samples.

Fig. 3.33 shows angular-dependent coercivity and initial magnetization curves measured for the W-type hexaferrite bulks with $x = 0.1$, 0.3, 0.4, and 0.5. As can be seen, the magnetization reversal mechanism of samples with $x \leq 0.3$ roughly obeys the SW mechanism. For $x = 0.4$ and 0.5, the DW motion model is more responsible for the reversal process of magnetization. In fact, the DW motion may be restricted due to the presence of defects, impurity, grain boundary, and other defects in the ferrites, thus reducing the wall energy. On the other hand, based on the corresponding initial magnetization curves, the pinning model is expected to occur for $x = 0.4$ and 0.5.

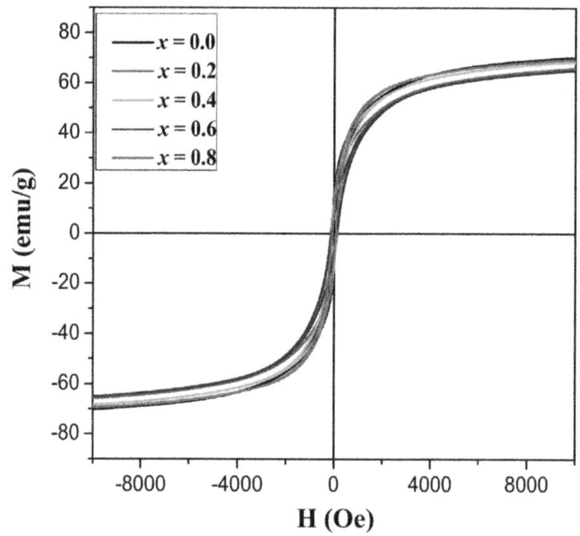

FIGURE 3.28

Hysteresis loops of $BaCo_2Fe_{16-x}Ga_xO_{27}$ ($0.0 \leq x \leq 0.8$) W-type hexagonal ferrites measured at room temperature.

From Mahmood, S. H., Al Sheyab, Q., Bsoul, I., Mohsen, O., & Awadallah, A. (c. 2018). Structural and magnetic properties of Ga-substituted Co_2–W hexaferrites. Current Applied Physics, *18(5), 590–598. https://doi.org/10.1016/j.cap.2018.02.013.*

In a separate research study, W-type hexaferrites consisting of $SrCo_{2-x}(NiZnMg)_{x/3}Fe_{16}O_{27}$ ($x = 0$–0.5) micron particles have been synthesized using a coprecipitation method (Ghasemi, 2016c). Fig. 3.34 depicts SEM images of the resulting W-type hexaferrite particles with different compositions. In terms of morphology, it is seen that the particles have mainly the hexagonal configuration. It has been known that the pH of precursors used in the coprecipitation methods can influence the morphology of the resultant particles (Goldman, 2006). In this direction, increasing the substitution content varies the pH of the solvent, requiring a high NaOH amount in order to obtain the appropriate pH value for the precipitation. As a result, the different morphologies of the particles arise from their varied nucleation rates.

SEM images with larger magnifications were taken from $SrCo_{2-x}(NiZnMg)_{x/3}Fe_{16}O_{27}$ ($x = 0.2$) particles are shown in Fig. 3.35, indicating the occurrence of heterogeneous nucleation on the surface of large particles. In other words, the small-size particles are formed preferentially on the outer surface of large particles, giving rise to the coexistence of large and small particles. From Fig. 3.35, the hexagonal platelet morphology of the particles is also magnified. This indicates that the appropriate use of the effective processing parameter in the synthesis method can form uniform hexagonal particles.

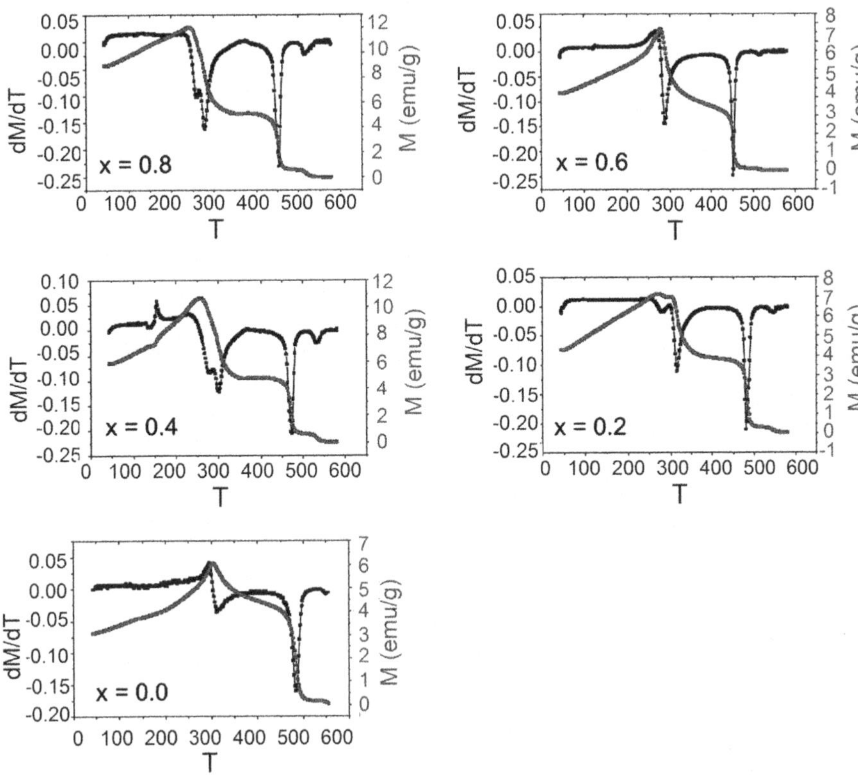

FIGURE 3.29

The variations of magnetization and temperature derivative of magnetization as a function of temperature for $BaCo_2Fe_{16-x}Ga_xO_{27}$ ($0.0 \leq x \leq 0.8$) W-type hexagonal ferrites measured at an applied magnetic field of 100 Oe.

From Mahmood, S. H., Al Sheyab, Q., Bsoul, I., Mohsen, O., & Awadallah, A. (c. 2018). Structural and magnetic properties of Ga-substituted Co_2–W hexaferrites. Current Applied Physics, 18*(5), 590–598. https://doi.org/10.1016/j.cap.2018.02.013.*

Fig. 3.36 shows hysteresis loops of the $SrCo_{2-x}(NiZnMg)_{x/3}Fe_{16}O_{27}$ ($x = 0$–0.5) particles, along with the corresponding M_s, H_c and M_r values. The formation of a large hysteresis loop area indicates that all the hexaferrite particle samples possess hard magnetic features.

From Fig. 3.36, increasing the substitution content reduces the saturation magnetization of all the ferrite samples, reaching a minimum M_s of 11.6 emu/g at $x = 0.5$. The magnetic moment of Co^{2+} cation is higher than that of Mg^{2+}, $Ni^{2+}$and Zn^{2+} cations (Chauhan et al., 2013; Goldman, 2006; Stergiou & Litsardakis, 2011). As well, the Co^{2+} cation has a tendency to occupy $4f_{VI}$ and $6g$ lattice sites, as discussed in the literature (Iqbal et al., 2012; Iqbal, Khan, Mizukami, & Miyazaki, 2011). The former has a down spin, and the latter has an up spin in the crystallographic planes

Table 3.5 Curie temperature (T_c) of $BaCo_2Fe_{16-x}Ga_xO_{27}$ W-type hexagonal ferrites with different Ga contents (x).

x	0	0.2	0.4	0.6	0.8
T_c (°C)	485	480	473	451	452

Table 3.6 Elemental composition of $SrCo_{2-x}(MnZnCa)_{x/3}Fe_{16}O_{27}$ ($0.0 \leq x \leq 0.5$) W-type hexagonal ferrites extracted from EDS spectra.

Sample	Chemical formula	Mn (wt.%)	Zn (wt.%)	Ca (wt.%)	Co (wt.%)
$x = 0.0$	$SrCo_2Fe_{16}O_{27}$	0	0	0.5	8.7
$x = 0.1$	$SrCo_{1.9}Mn_{0.033}Zn_{0.033}Ca_{0.033}Fe_{16}O_{27}$	0.2	0.4	0.4	9.4
$x = 0.2$	$SrCo_{1.8}Mn_{0.066}Zn_{0.066}Ca_{0.066}Fe_{16}O_{27}$	0.5	0.4	0.3	8.8
$x = 0.3$	$SrCo_{1.7}Mn_{0.1}Zn_{0.1}Ca_{0.1}Fe_{16}O_{27}$	0.5	0.6	0.8	8.3
$x = 0.4$	$SrCo_{1.6}Mn_{0.133}Zn_{0.133}Ca_{0.133}Fe_{16}O_{27}$	0.5	0.8	0.6	7.9
$x = 0.5$	$SrCo_{1.5}Mn_{0.166}Zn_{0.166}Ca_{0.166}Fe_{16}O_{27}$	0.8	1.2	0.7	8.7

(Herzer, 1990). The tetrahedral sites of spinel structures are preferably occupied by nonmagnetic metal cations (Iqbal, Khan, Takeda, et al., 2011; Ri et al., 2012; Sadiq et al., 2012). Thus, the net values of magnetic moments and saturation magnetization are reduced since the cobalt and iron cations are replaced by ones with lower magnetic moments.

A direct relationship also exists between ferrite composition and coercivity. For $x = 0, 0.1, 0.2, 0.3, 0.4$ and 0.5, the respective H_c values are 2555, 3145, 3968, 4663, 4923 and 5107 Oe. An enhancement in the anisotropy constant of the hexaferrite samples could be responsible for the increasing trend of H_c.

3.3.3 Y-type hexagonal ferrites

While most hexagonal ferrites are magnetically hard in nature, Y-type hexaferrite is a soft magnetic material with easy magnetization planes being perpendicular to the c-axis (Murtaza et al., 2014). Electrical and magnetic properties of Y-type hexaferrites substituted with divalent and trivalent cations such as Al-Ga-In (Lim et al., 2012), Zn (El Ata & Attia, 2003), Co–Cu (Bai et al., 2006), Mg (Elahi et al., 2013), Mn-Tb (Ali et al., 2013), Pb (Júnior et al., 2010), Sr (Ahmad, Ali, et al., 2013), and Cr (Iqbal & Liaqat, 2010), have been featured in various studies.

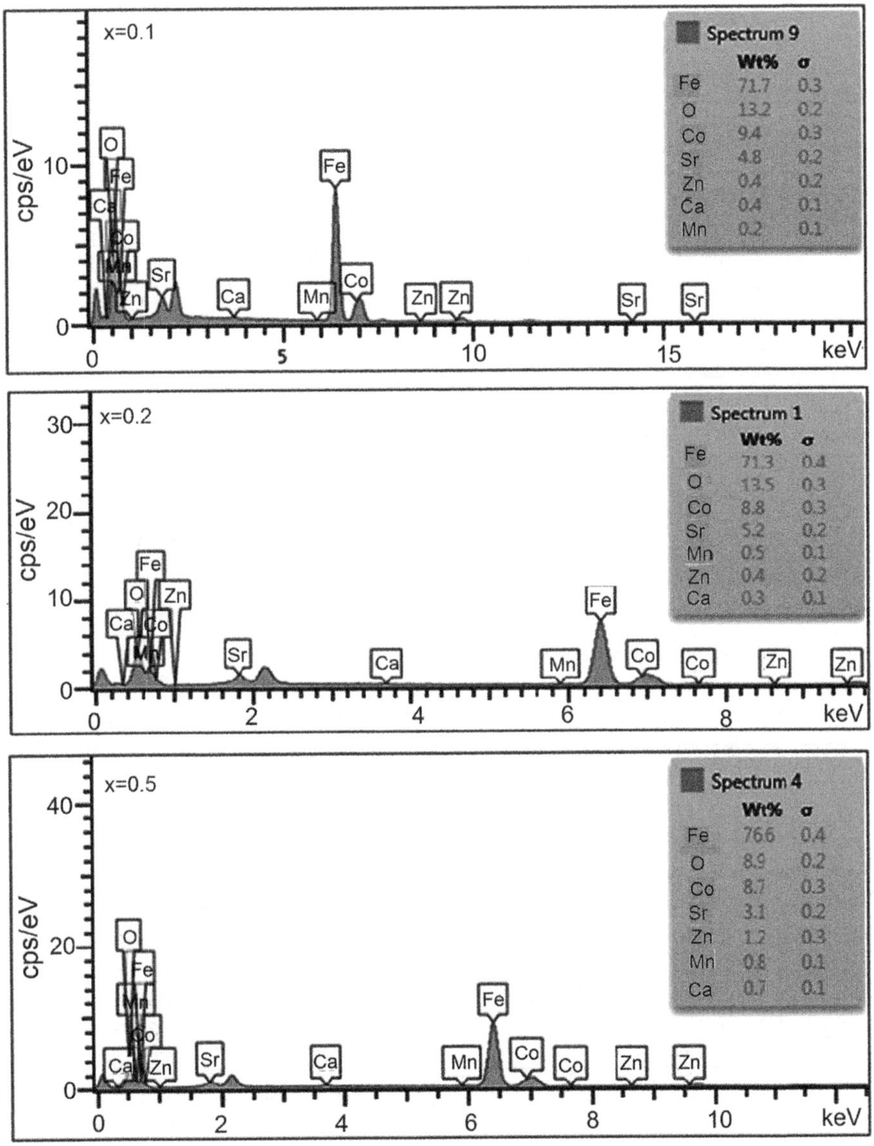

FIGURE 3.30

EDS spectra of typical MnZnCa-substituted W-type hexagonal ferrites.

From Ghasemi, A. (c. 2016). Effects of divalent ion substitution on the microstructure and magnetic properties of $SrCo_{2-x}(MnZnCa)_{x/3}Fe_{16}O_{27}$ Nanoparticles. Journal of Superconductivity and Novel Magnetism, *29(7), 1943–1952. https://doi.org/10.1007/s10948-016-3503-5 1949.*

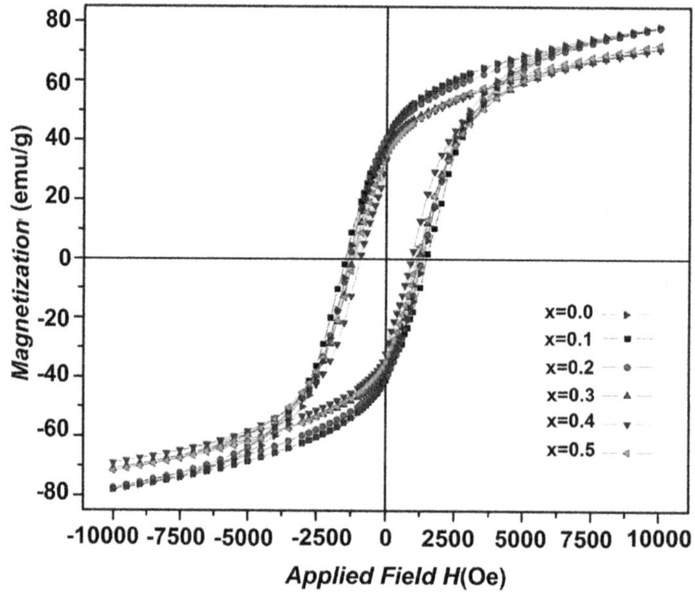

FIGURE 3.31

Vibrating sample magnetometry hysteresis loops of $SrCo_{2-x}(MnZnCa)_{x/3}Fe_{16}O_{27}$ W-type hexaferrite nanoparticles.

From Ghasemi, A., Gordani, G. R., & Ghasemi, E. (c. 2019). Co_2W hexaferrite nanoparticles-carbon nanotube microwave absorbing nanocomposite. Journal of Magnetism and Magnetic Materials, 469, *391–397. https://doi.org/10.1016/j.jmmm.2018.09.010 395.*

In a study, $Ba_2Fe_2^{2+}Fe_{12}^{3+}O_{22}(Fe_2Y)$ Y-type hexaferrite polycrystalline ceramics were synthesized using a solid-state reaction method, followed by calcination and sintering in a nitrogen atmosphere (Zhang & Zhang, 2020). Hysteresis loops of the resulting Fe_2Y hexaferrites measured at temperatures of 5, 150, and 300 K are shown in Fig. 3.37A. As observed, the hexaferrites have planar anisotropy, indicating that they are magnetically soft. By applying a magnetic field of 40 kOe, the Fe_2Y hexaferrites become saturated, resulting in M_s values of 39.9, 42.3, and 38.5 emu/g at temperatures of 5, 150, and 300 K, respectively. In other words, the M_s of hexaferrites changes with a variation in temperature, which is abnormal behavior for the temperature dependence of magnetization (M–T) of ferrite materials. The spin magnetic structure of the Y-type hexaferrite polycrystalline ceramics might cause the occurrence of abnormal M–T behavior.

M-T curves of the Fe_2Y hexaferrite were measured under different applied fields, and the results are shown in Fig. 3.37B and C. In this regard, the hexaferrite sample was initially cooled to 5 K under a zero magnetic field in order to obtain the ZFC measurement. In the following process, the M–T measurement of the sample was carried out under a specific magnetic field when heating it up to 300 K. On the contrary, the field-cooled (FC) measurement of the sample was performed by cooling it

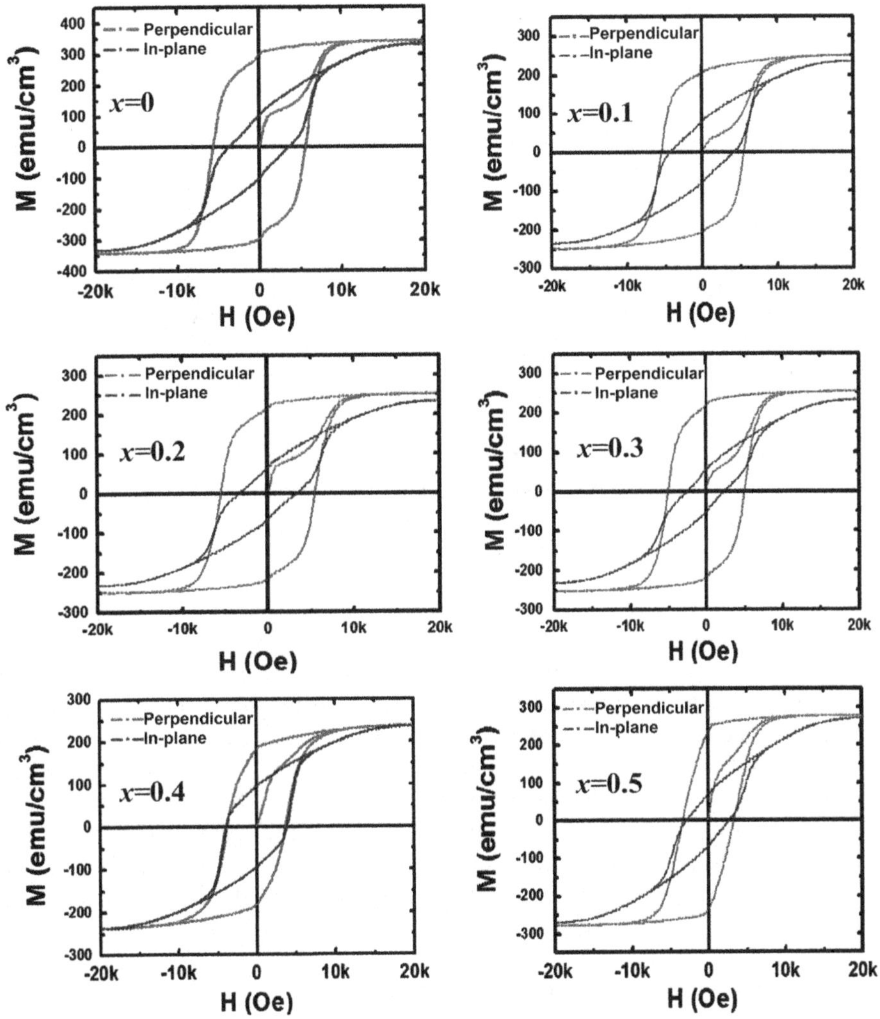

FIGURE 3.32

Perpendicular and in-plane vibrating sample magnetometry hysteresis loops of $SrCo_{2-x}Mn_xFe_{16}O_{27}$ ($0.0 \le x \le 0.5$) W-type hexaferrites.

From Ghasemi, A. (c. 2016). Stoner–Wohlfarth rotation or domain wall motion mechanism in W-type magnetic hexaferrite nanoparticles. Ceramics International, *42(3), 4143–4149. https://doi.org/10.1016/j.ceramint.2015. 11.087 6.*

under a nonzero magnetic field. All the M—T curves observed in Fig. 3.37 indicate the irreversibility of the ZFC/FC measurements. For temperatures above room temperature, the M-T curve depicted in Fig. 3.37D shows that Fe_2Y hexaferrite magnetization is sharply reduced at 634 K, being indicative of its Curie temperature (T_c).

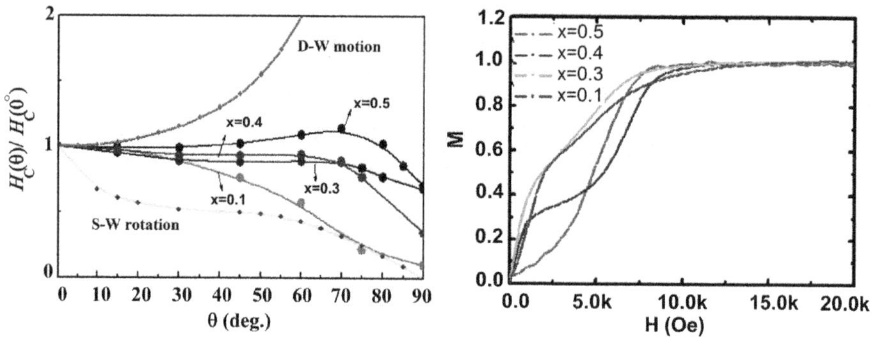

FIGURE 3.33

Angular-dependent coercivity and initial magnetization curves of $SrCo_{2-x}Mn_xFe_{16}O_{27}$ W-type hexaferrites.

From Ghasemi, A. (c. 2016). Stoner–Wohlfarth rotation or domain wall motion mechanism in W-type magnetic hexaferrite nanoparticles. Ceramics International, *42(3), 4143–4149. https://doi.org/10.1016/j.ceramint.2015. 11.087 6.*

To shed light on its complex magnetic behavior, the variation of ac magnetic susceptibility as a function of frequency for the Fe_2Y hexaferrite is shown in Fig. 3.38A and B. In this respect, two prominent rises are found in the in-phase component (χ') when increasing the temperature to ~50 and 130 K, being in perfect agreement with the dc M–T curves. The aforementioned changes are reflected in the out-of-phase part (χ'') in the form of two peaks. The peak that appeared at 50 K highly depends on the frequency value, whereas the peak at ~130 K is independent of frequency. Thus, increasing the frequency shifts the temperature of the peak at 50 K, reflecting typical spin glass behavior. Moreover, using the relation $p = \delta T_f / (T_f \, \delta \log_{10} f)$ (in which f and T_f are the frequency and the temperature of the peak, respectively), one can explore the frequency dependence of the shift of the peak. The parameter p is ~ 0.12 for the Fe_2Y hexaferrite, being typical with regard to canonical spin glass systems with p in the range of 0.0045–0.28.

It is also possible to characterize the magnetic state with the glass-like behavior based on the temperature dependence and time dependence of thermoremanent magnetization (TRM), as shown in Fig. 3.38C and D. To this end, under magnetic fields of 100 Oe, 1 kOe and 10 kOe, the ferrite sample was initially cooled from room temperature to 5 K, followed by cutting off the field and collecting data in the temperature range from 5 to 300 K. Increasing the temperature decreases the TRM value while leading to the appearance of two kinks at 50 and 130 K. This variation behavior is more prominent for larger magnetic fields, further confirming the two magnetic transitions. Fig. 3.38D indicates that TRM experiences a slow relaxation after performing the field-cooling process from room temperature to 50 K. Accordingly, the relaxation property of the Fe_2Y ceramic is evidenced, having logarithmic time dependence as a spin glass behavior.

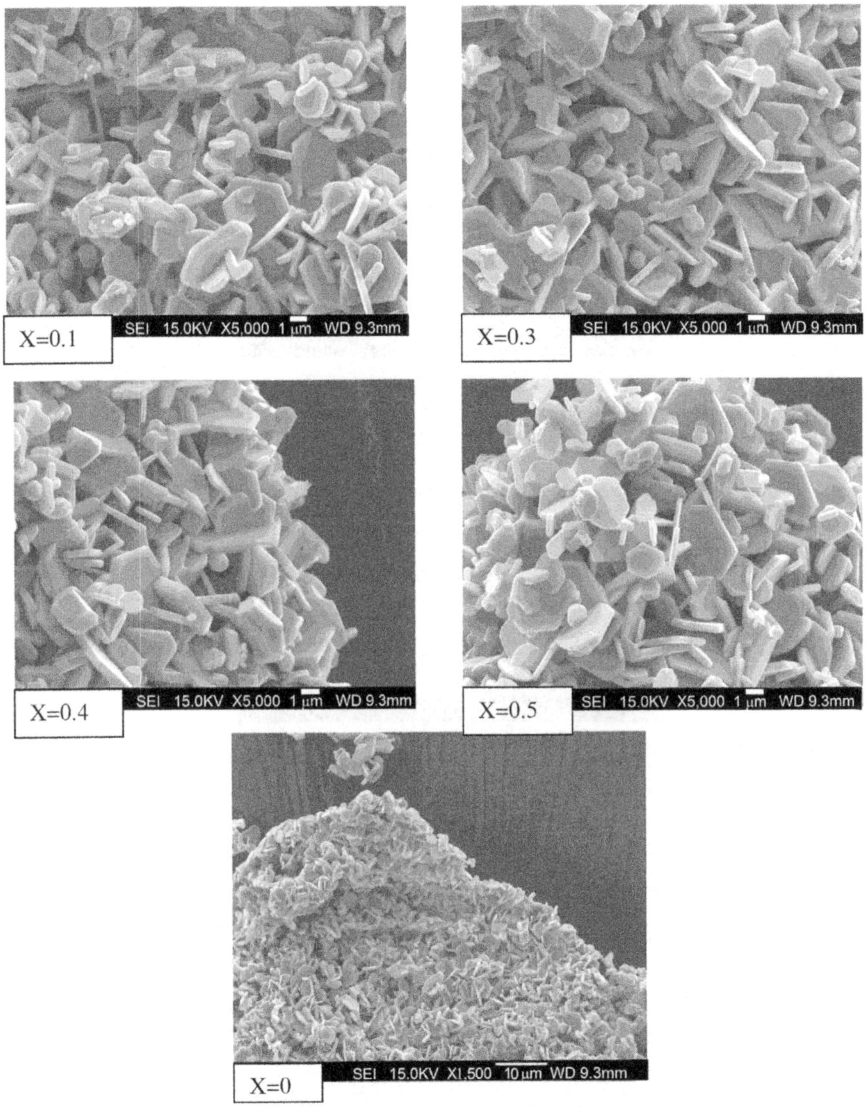

FIGURE 3.34

Scanning electron microscopy images of $SrCo_{2-x}(NiZnMg)_{x/3}Fe_{16}O_{27}$ ($x = 0$–0.5) W-type hexaferrite particles.

From Ghasemi, A. (c. 2016). On the magnetic analysis of external magnetic field annealing of Co$_2$-W hexagonal shape particles. Journal of Superconductivity and Novel Magnetism, 29*(6), 1601–1610. https://doi.org/10.1007/s10948-016-3435-0 6.*

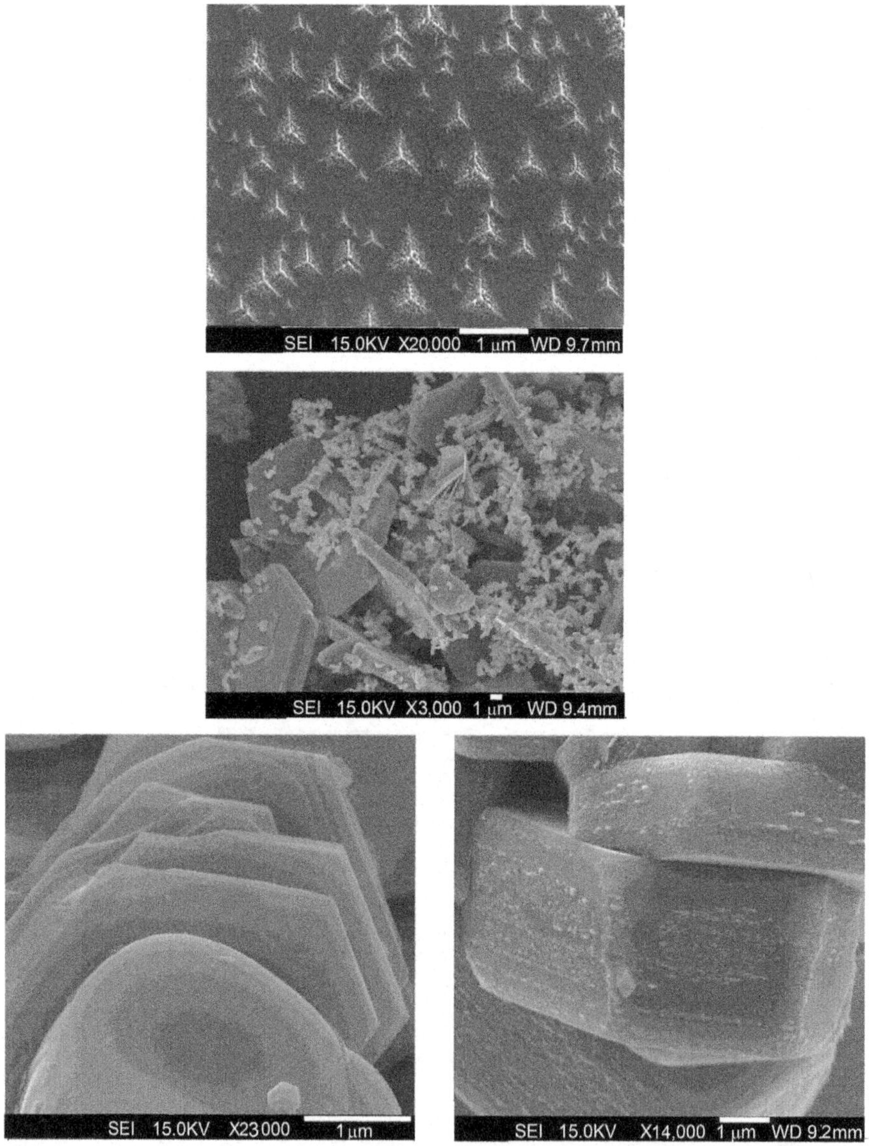

FIGURE 3.35

Scanning electron microscopy images of sample $x = 0.2$ illustrating the formation of hexagonal crystals, along with the nucleation of small particles on the surface of large particles.

From Ghasemi, A. (c. 2016). On the magnetic analysis of external magnetic field annealing of co$_2$-W hexagonal shape particles. Journal of Superconductivity and Novel Magnetism, *29(6), 1601–1610. https://doi.org/10.1007/s10948-016-3435-0 6.*

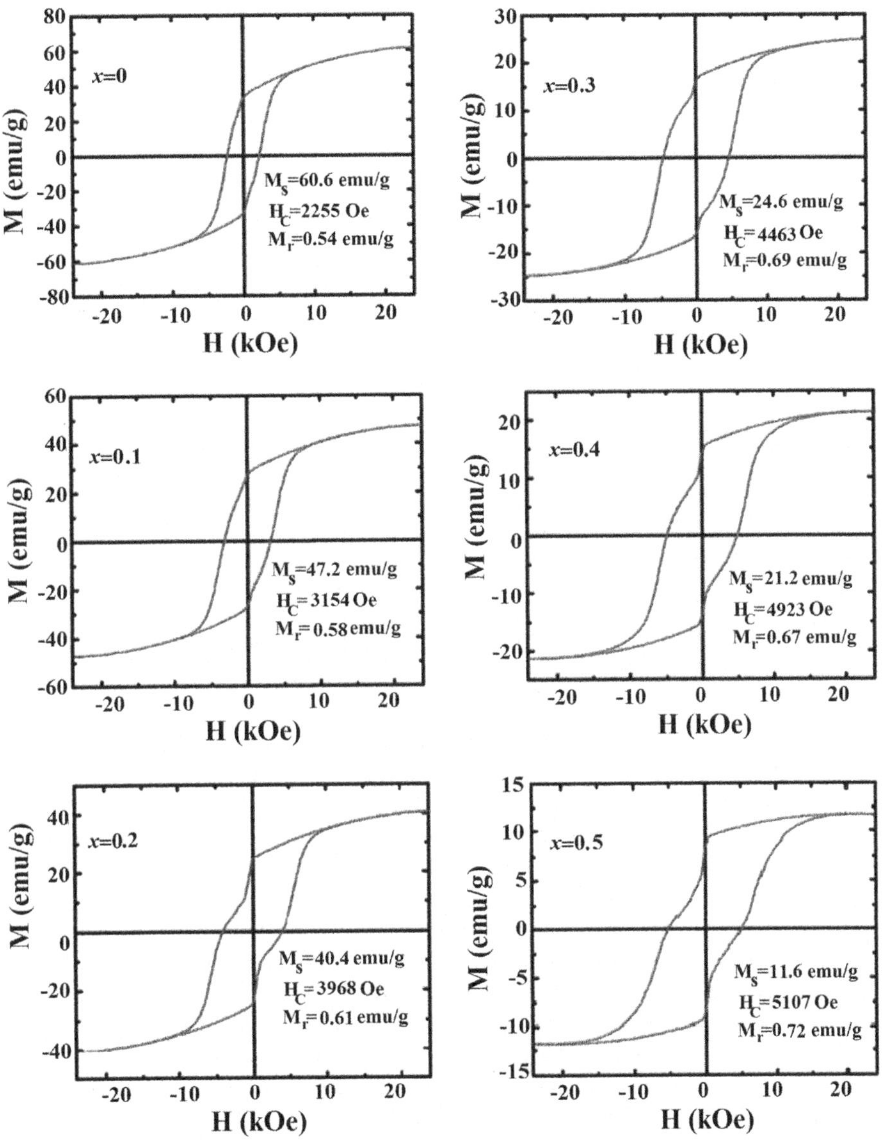

FIGURE 3.36

Hysteresis loops of $SrCo_{2-x}(NiZnMg)_{x/3}Fe_{16}O_{27}$ particles with $x = 0$, 0.1, 0.2, 0.3, 0.4 and 0.5, along with the corresponding M_s, H_c and M_r values.

From Ghasemi, A. (c. 2016). On the magnetic analysis of external magnetic field annealing of Co_2-W hexagonal shape particles. Journal of Superconductivity and Novel Magnetism, *29(6), 1601–1610. https://doi.org/10.1007/s10948-016-3435-0 6.*

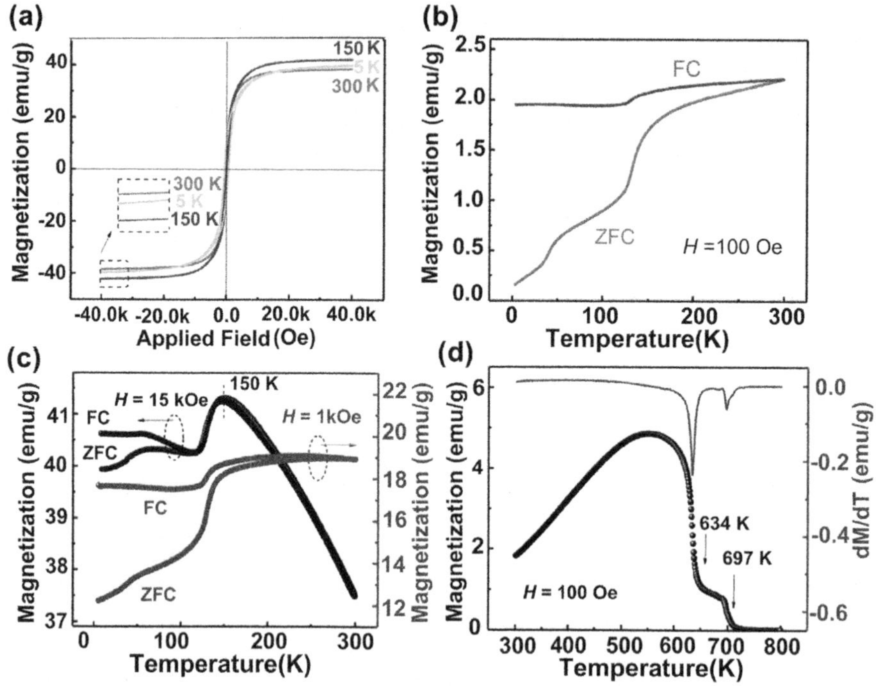

FIGURE 3.37

(A) Hysteresis loops of Fe_2Y hexaferrite ceramic measured at different temperatures (5, 150 and 300 K), and (B–D) the variation of magnetization as a function of temperature for the Fe_2Y hexaferrite ceramic under different magnetic fields.

From Zhang, X., & Zhang, J. (c. 2020). Synthesis and magnetic properties of $Ba_2Fe_{14}O_{22}$ Y-type hexaferrite. Materials Letters, 269, *127642. https://doi.org/10.1016/j.matlet.2020.127642 3.*

Elsewhere, using citric acid sol–gel auto-combustion and sonochemistry methods, $Ba_{0.5}Sr_{1.5}Zn_2Fe_{12}O_{22}$ powders have been synthesized, and the effect of the synthesis method on structural and magnetic properties have been investigated (Georgieva et al., 2019). Fig. 3.39A and B show hysteresis loops measured at 4.2 K and room temperature (300 K) for the powder synthesized by the sol–gel auto-combustion method, respectively. Moreover, hysteresis loops measured at 4.2 K and room temperature (300 K) for the powder synthesized by the sonochemistry method are shown in Fig. 3.40A and B, respectively. As observed, the hysteresis loops of the $Ba_{0.5}Sr_{1.5}Zn_2Fe_{12}O_{22}$ powders are very narrow, regardless of the synthesis method.

The initial magnetization curves measured for powders synthesized by the auto-combustion and sonochemistry methods are shown in Figs. 3.39C and 3.40C, respectively. As inferred, by applying magnetic fields higher than 30 kOe, the magnetization curves of all the powders become saturated. In the magnetic field range of 2–6 kOe, the ferrimagnetic behavior of the powder synthesized by the

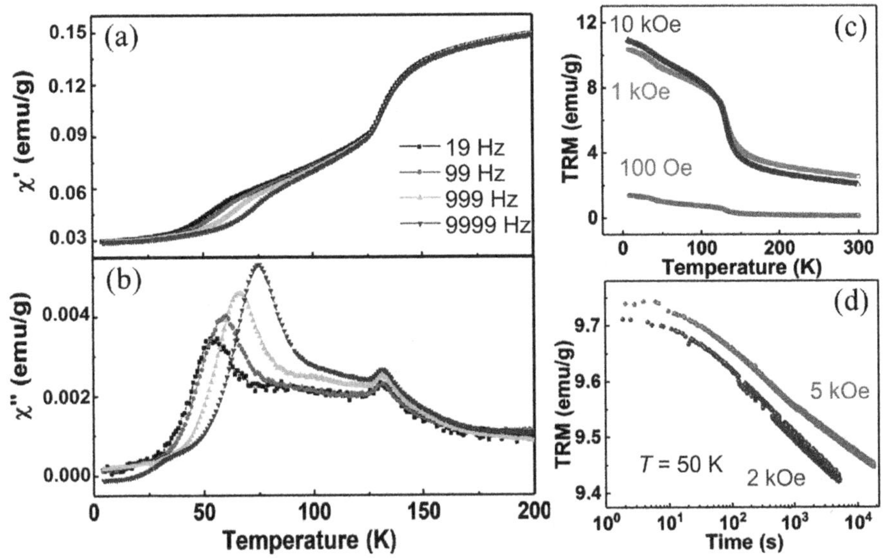

FIGURE 3.38

The variations of: (A) in-phase (χ'), and (B) out-of-phase (χ'') components as a function of temperature for Fe_2Y hexaferrite ceramic under different frequencies. (C) TRM curves for magnetic fields of 100 Oe, 1 kOe, and 10 kOe. (D) The variation of TRM as a function of time for magnetic fields of 2 and 5 kOe at 50 K.

From Zhang, X., & Zhang, J. (c. 2020). Synthesis and magnetic properties of $Ba_2Fe_{14}O_{22}$ Y-type hexaferrite. Materials Letters, 269, *127642. https://doi.org/10.1016/j.matlet.2020.127642 3.*

sonochemistry method changes, arising from a variation in the $M(H)$ curve slope at 300 K. The presence of intermediate phases, as well as the occurrence of metamagnetic transitions could be responsible for the change observed in the ferrimagnetic behavior at 300 K.

ZFC-FC magnetization curves measured under an applied magnetic field of 100 Oe for the powders synthesized by the sol–gel auto-combustion and sonochemistry methods are shown in Fig. 3.41A and B, respectively. As can be seen, increasing the temperature monotonically decreases all the FC curves, whereas it increases the ZFC curves together with some changes in the magnetization. In order to determine the temperature of these magnetization changes, the first derivatives of the ZFC curves were also obtained. Two temperature maxima are observed at 74 and 232 K in the ZFC curve of the ferrite sample synthesized by the auto-combustion method. In the case of the sample synthesized by the sonochemistry method, two temperature maxima appear at 71 and 232 K.

A maximum temperature of 139 K is also observed in the first derivatives of the ZFC curves for both samples, leading to an inflection point. In turn, this indicates the beginning of a magnetic phase transition, which ends at the temperature of 232 K. The magnetization behavior of the ZFC-FC curves was also studied for the sample

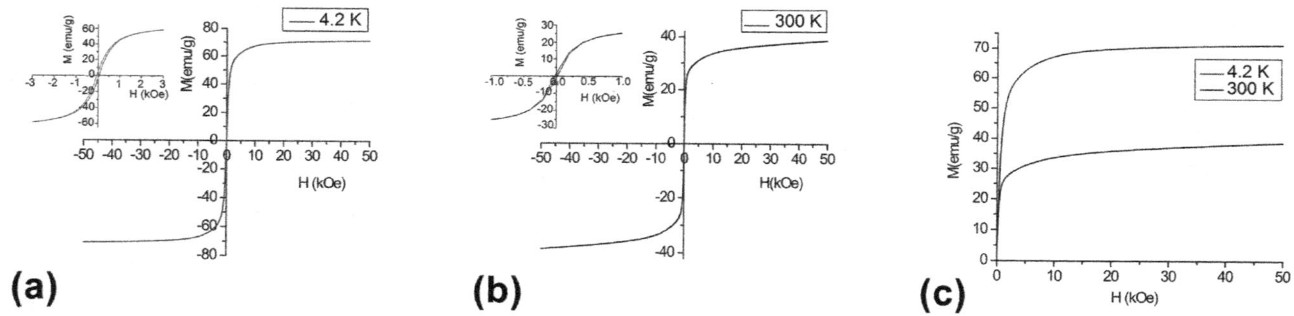

FIGURE 3.39

Hysteresis loops measured at (A) 4.2 K and (B) 300 K for $Ba_{0.5}Sr_{1.5}Zn_2Fe_{12}O_{22}$ powder synthesized by the auto-combustion method. (C) The corresponding initial magnetization curves.

From Georgieva, B., Kolev, S., Krezhov, K., Ghelev, C., Kovacheva, D., Vertruyen, B., Closset, R., Tran, L. M., Babij, M., & Zaleski, A. J. (c. 2019). Structural and magnetic characterization of Y-type hexaferrite powders prepared by sol-gel auto-combustion and sonochemistry. Journal of Magnetism and Magnetic Materials, 477, *131–135. https://doi.org/10.1016/j.jmmm.2019.01.033 133.*

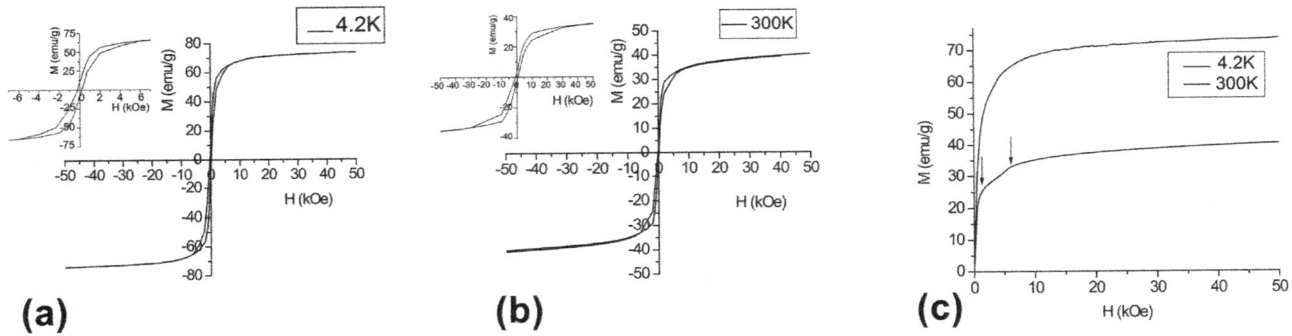

(a) (b) (c)

FIGURE 3.40

Hysteresis loops measured at (A) 4.2 K and (B) 300 K for $Ba_{0.5}Sr_{1.5}Zn_2Fe_{12}O_{22}$ powder synthesized by the sonochemistry method. (C) The corresponding initial magnetization curves.

From Georgieva, B., Kolev, S., Krezhov, K., Ghelev, C., Kovacheva, D., Vertruyen, B., Closset, R., Tran, L. M., Babij, M., & Zaleski, A. J. (c. 2019). Structural and magnetic characterization of Y-type hexaferrite powders prepared by sol-gel auto-combustion and sonochemistry. Journal of Magnetism and Magnetic Materials, 477, *131–135. https://doi.org/10.1016/j.jmmm.2019.01.033 133.*

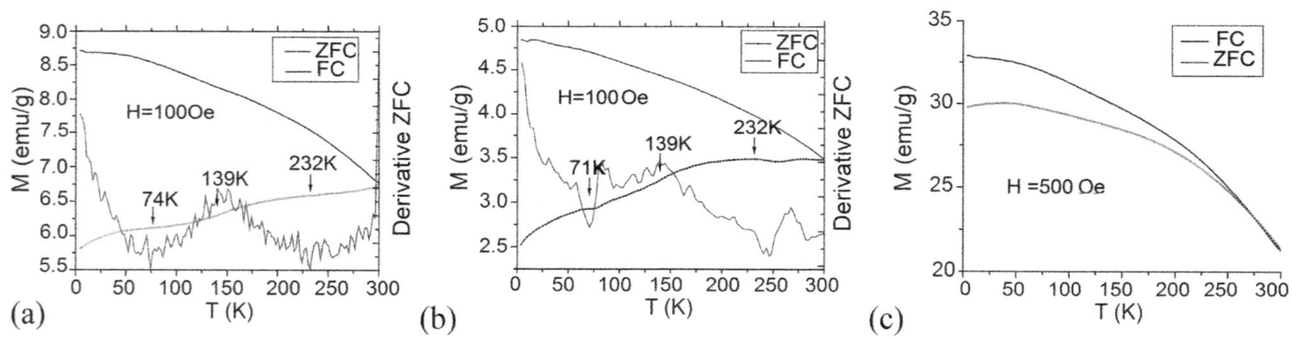

FIGURE 3.41

Zero-field-cooled–field-cooled curves measured for $Ba_{0.5}Sr_{1.5}Zn_2Fe_{12}O_{22}$ powders synthesized using: (A and C) sol–gel auto-combustion, and (B) sonochemistry methods. The applied magnetic fields in parts (A and B) and (C) are 100 and 500 Oe, respectively.

From Georgieva, B., Kolev, S., Krezhov, K., Ghelev, C., Kovacheva, D., Vertruyen, B., Closset, R., Tran, L. M., Babij, M., & Zaleski, A. J. (c. 2019). Structural and magnetic characterization of Y-type hexaferrite powders prepared by sol-gel auto-combustion and sonochemistry. Journal of Magnetism and Magnetic Materials, 477, *131–135. https://doi.org/10.1016/j.jmmm.2019.01.033 133.*

synthesized by the sol–gel auto-combustion method at a higher magnetic field (500 Oe), and the results are shown in Fig. 3.41C. In this case, increasing the temperature monotonically reduces both the ZFC and the FC curves, with separation below 275 K.

In another investigation, Y-type hexaferrites ($Ba_{0.5}Sr_{1.5}Co_2Fe_{12}O_{22}$) with polycrystalline structure have been synthesized at different temperatures in the range of 900–1250°C (Hernández-Gómez et al., 2019). Fig. 3.42 shows hysteresis loops of calcined (900°C) and sintered (1050, 1150, and 1250°C) Y-type hexaferrite samples. As can be seen, compared with the sintered samples, the calcination of the Y-type hexaferrite sample leads to a relatively wide hysteresis loop. Increasing the sintering temperature induces the formation of a mixed phase consisting of M- and Y-type hexaferrites and $SrFe_2O_4$ and $CoFe_2O_4$. By performing the sintering process at 1050°C, a pure Y-type hexaferrite is obtained whose hysteresis loop is narrow, which is in accordance with ferroxplana compounds with soft magnetic characteristics. At the sintering temperatures of 1050, 1150, and 1250°C, the coercive field values are 157.5, 102, and 59 Oe, respectively. Therefore, increasing the sintering temperature causes the coercive field to decrease due to higher densification. The LAS was used to investigate the value of saturation magnetization in the sintered samples, resulting in M_s of 20.4, 21.6, and 33.2 emu/g at the sintering temperatures of 1050, 1150, and 1250°C, respectively.

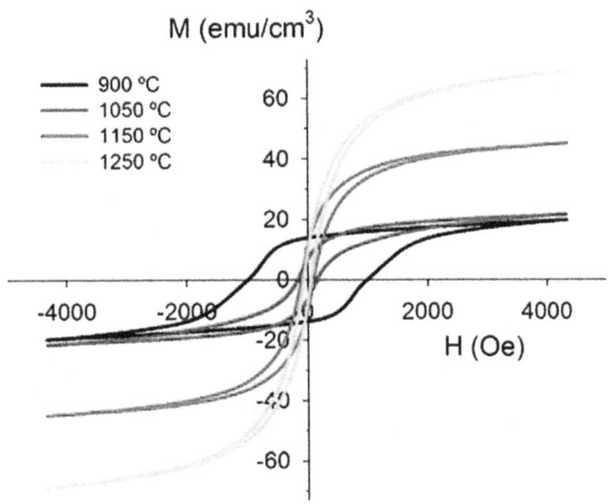

FIGURE 3.42

Hysteresis loops of calcined (900°C) and sintered (1050, 1150, and 1250°C) $Ba_{0.5}Sr_{1.5}Co_2Fe_{12}O_{22}$ Y-type hexaferrites.

From Hernández-Gómez, P., Martín-González, D., Torres, C., & Muñoz, J. (c. 2019). Broadband transverse susceptibility in multiferroic Y-type hexaferrite $Ba_{0.5}Sr_{1.5}Co_2Fe_2O_{22}$. Journal of Magnetism and Magnetic Materials, 476, *478–482. https://doi.org/10.1016/j.jmmm.2019.01.035 479.*

Using a solid-state reaction method, $Ba_{0.5}Sr_{1.5}Zn_{2-x}Mg_xFe_{11}AlO_{22}$ ($x = 0$, 0.4, 0.8, 1.2, 1.6, and 2.0) Y-type hexaferrites have also been synthesized, and the Mg substitution effect on structural, magnetic, and dielectric properties has been studied (Zhang et al., 2017). ZFC curves of the hexaferrite samples measured under $H = 0.01$ T are depicted in Fig. 3.43A–F. As observed, the occurrence of two prominent phase transitions is evidenced at the temperatures of T_1 and T_2.

Hysteresis loops of the hexaferrites measured at room temperature under a magnetic saturation field of 4.5 T are shown in Fig. 3.44. Depending on the magnetic field and the Mg substitution content, all samples behave differently in terms of magnetization measured at room temperature. In the case of $x \leq 0.4$, the corresponding hysteresis loops indicate the existence of a typical ferrimagnetic state arising from the transition temperature of $T_2 < 300$ K. For $x \geq 0.8$, simple ferrimagnetic characteristics are not reflected in the hysteresis loops. At low magnetic fields, the magnetization sharply increases with increasing the field. Further increasing the field up to the saturation state leads to a slow increase in the magnetization, together with the emergence of several kinks. Accordingly, it could be stated that by applying the magnetic field, the magnetic phase of the hexaferrite samples may experience transitions into a ferrimagnetic state successively. In this case, some intermediate magnetic phases between conical and proper-screw spin states are involved in the transitions.

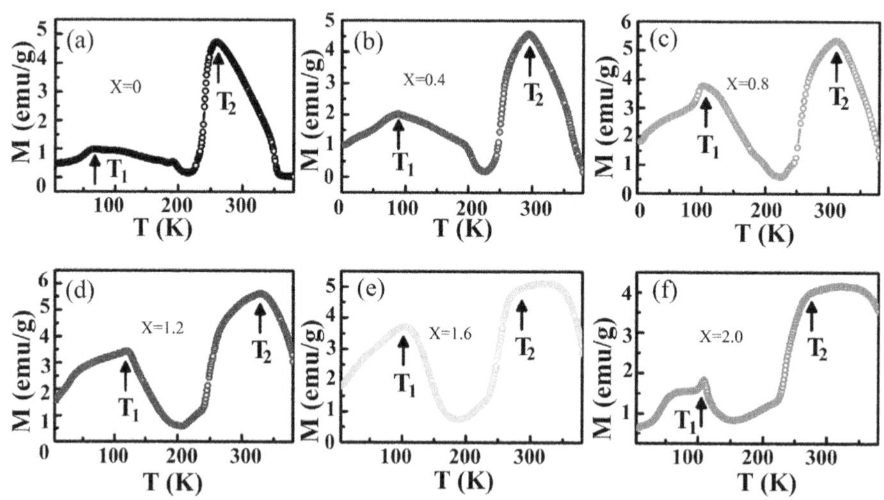

FIGURE 3.43

Zero-field-cooled curves of $Ba_{0.5}Sr_{1.5}Zn_{2-x}Mg_xFe_{11}AlO_{22}$ Y-type hexaferrites measured under $H = 0.01$ T for: (A) $x = 0$, (B) $x = 0.4$, (C) $x = 0.8$, (D) $x = 1.2$, (E) $x = 1.6$, and (F) $x = 2.0$.

From Zhang, M., Yin, L., Liu, Q., Kong, X., Zi, Z., Dai, J., & Sun, Y. (c. 2017). Magnetic properties and magnetodielectric effect in Y-type hexaferrite $Ba_{0.5}Sr_{1.5}Zn_{2-x}Mg_xFe_{11}AlO_{22}$. Journal of Alloys and Compounds, *725, 1252–1258. https://doi.org/10.1016/j.jallcom.2017.07.286 1257.*

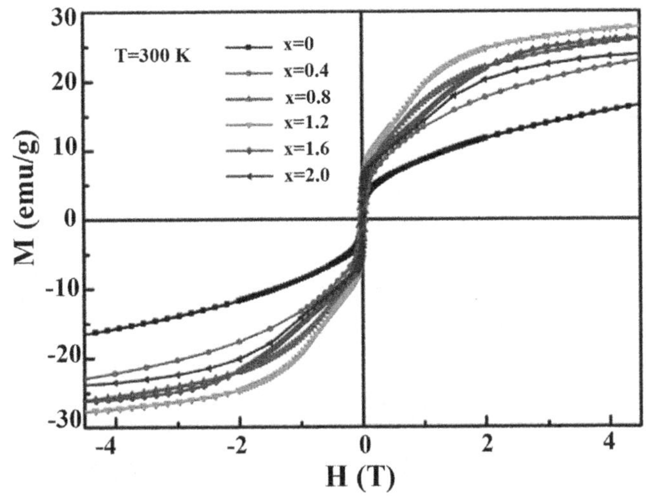

FIGURE 3.44

Hysteresis loops of $Ba_{0.5}Sr_{1.5}Zn_{2-x}Mg_xFe_{11}AlO_{22}$ hexaferrites measured at T = 300 K. *From Zhang, M., Yin, L., Liu, Q., Kong, X., Zi, Z., Dai, J., & Sun, Y. (c. 2017). Magnetic properties and magnetodielectric effect in Y-type hexaferrite $Ba_{0.5}Sr_{1.5}Zn_{2-x}Mg_xFe_{11}AlO_{22}$.* Journal of Alloys and Compounds, 725, *1252–1258. https://doi.org/10.1016/j.jallcom.2017.07.286 1257.*

The variations of dielectric constant as a function of magnetic field (i.e., the magnetodielectric effect) for $Ba_{0.5}Sr_{1.5}Zn_{2-x}Mg_xFe_{11}AlO_{22}$ ($x = 0$, 0.4 and 0.8) hexaferrites measured at different temperatures (ranged between 30 and 250 K) are shown in Fig. 3.45. The magnetodielectric effect taking place in the hexaferrites is indicated by the presence of two distinct peaks below a temperature of 250 K. One of the peaks is sharp at the nearly zero magnetic field, and the other is broad at high magnetic fields. The magnetodielectric coefficient is maximized at the temperature of 200 K, giving rise to values approximating 8%, 6%, and 4% for $x = 0$, 0.4, and 0.8, respectively. Since the octahedral lattice sites can be occupied by Mg^{2+} ions preferentially (thus leading to the absence of ferrous and ferric ions), the substituted hexaferrite samples show lower dielectric constants compared with the pure sample. For $T > 200$ K, a gradual decrease is observed in the intensity of the sharp peak around the zero magnetic field, according to Fig. 3.45B. Concurrently, the broad peak intensity increases at the high magnetic fields, as shown in Fig. 3.45C. Notably, the single peak at the nearly zero magnetic field disappears completely, indicating the gradual vanishing of the transverse conical spin state. Additionally, the double peaks emerge around the magnetic field of ~ 1.5 T. The variation behavior of the dielectric constant is evidently different for pure and Mg-substituted hexaferrite samples at 250 K. In this way, the peak fields are observed to be continuously evolved from the low to high temperatures, giving rise to ferroelectric transitions up to 250 K induced by the magnetic field for the Mg-substituted hexaferrites. Since the measurement temperature (250 K) and transition temperature (259 K) are similar, the ferroelectric phase

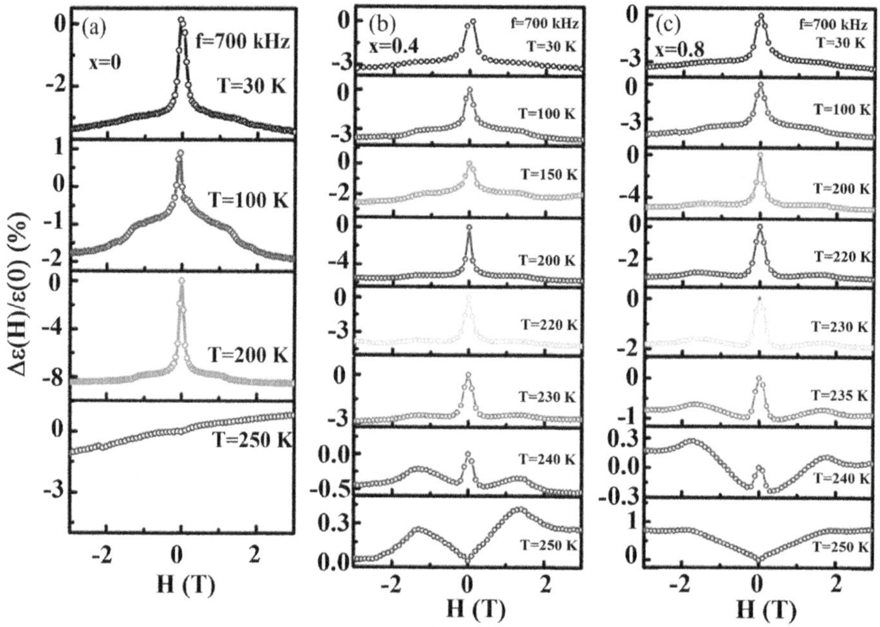

FIGURE 3.45

The variation of dielectric constant as a function of magnetic field for $Ba_{0.5}Sr_{1.5}Zn_{2-x}Mg_xFe_{11}AlO_{22}$ hexaferrites with: (A) $x = 0$, (B) $x = 0.4$, and (C) $x = 0.8$ measured at a temperature range of 30–250 K.

From Zhang, M., Yin, L., Liu, Q., Kong, X., Zi, Z., Dai, J., & Sun, Y. (c. 2017). Magnetic properties and magnetodielectric effect in Y-type hexaferrite $Ba_{0.5}Sr_{1.5}Zn_{2-x}Mg_xFe_{11}AlO_{22}$. Journal of Alloys and Compounds, *725, 1252–1258. https://doi.org/10.1016/j.jallcom.2017.07.286 1257.*

has disappeared for the pure hexaferrite sample. Thereby, measurement temperatures below the spin ordering temperature are only expected to reflect the ferroelectric phase induced by the magnetic field. Furthermore, the magnetodielectric effect is opposite for the measurement temperatures below and above 240 K, likely owing to the variation in the magnetic structures. It is worth noting that increasing the Mg content increases the magnetodielectric coefficient for measurement temperatures less than 200 K, arising from an increase in the transition temperature. In turn, this causes the longitudinal conical spin state to appear. The negative magnetodielectric effect may exist in the $Ba_{0.5}Sr_{1.5}Zn_{2-x}Mg_xFe_{11}AlO_{22}$ system located in the ferroelectric transition region whose dielectric constant is reduced as a function of magnetic field.

In a separate study, $Ba_{0.5}Sr_{1.5}Zn_2(Fe_{1-x}In_x)_{12}O_{22}$ ($0 \leq x \leq 0.1$) Y-type hexaferrites have been synthesized using a solid-state reaction method. The effect of In substitution on structural and magnetic properties of the resulting hexaferrites has also been studied (Zhang et al., 2018). Fig. 3.46 shows ZFC curves of the Y-type hexaferrites with $x = 0$, 0.02, 0.04, 0.06, 0.08, and 0.1 measured under $H = 100$ Oe.

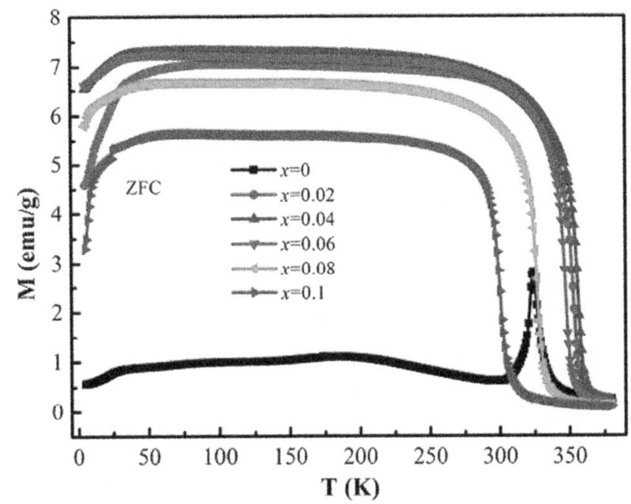

FIGURE 3.46

Zero-field-cooled curves of $Ba_{0.5}Sr_{1.5}Zn_2(Fe_{1-x}In_x)_{12}O_{22}$ ($0 \leq x \leq 0.1$) Y-type hexaferrites measured under $H = 100$ Oe.

From Zhang, M., Yin, L., Liu, Q., Zi, Z., Dai, J., & Sun, Y. (c. 2018). Indium doping effect on the magnetic properties of Y-type hexaferrite $Ba_{0.5}Sr_{1.5}Zn_2(Fe_{1-x}In_x)_{12}O_{22}$. Current Applied Physics, 18*(9), 1001–1005. https://doi.org/10.1016/j.cap.2018.05.017 13.*

Increasing the temperature leads to a perceptible phase transition from proper-screw spin to collinear ferrimagnetic spin at a transition temperature of T_S. A sharp peak is observed for the pure sample ($x = 0$) at a temperature of ~ 324 K. As well, T_S increases to a maximum value when the In content increases to $x = 0.04$. Increasing the In content beyond $x = 0.04$ decreases the corresponding T_S. In this regard, increasing x from 0 to 0.04 increases T_S from 324 to 356 K, which can be assigned to the occurrence of strong superexchange interaction.

Hysteresis loops of samples measured at room temperature under a saturation magnetic field of 4.5 T are shown in Fig. 3.47. As can be seen, the magnetic field and the In content influence the magnetization behavior of all samples. For pure ($x = 0$) and substituted ($x \leq 0.08$) hexaferrite samples, simple ferrimagnetic behavior is not observed in the hysteresis loops. At low magnetic fields, increasing the field sharply increases the magnetization. Further increasing the field up to the saturation state gradually increases the magnetization. Nevertheless, a typical ferrimagnetic behavior without the occurrence of the gradual magnetization increase is seen in the hysteresis loop of $x = 0.1$, accompanied by T_S of ~ 300 K. Alternatively, increasing the In content decreases M_s, which can be attributed to the metal ion distribution in different lattice sites. As a result, $18h_{VI}$, $3b_{VI}$, and $3a_{VI}$ lattice sites are expected to be occupied by the In ions.

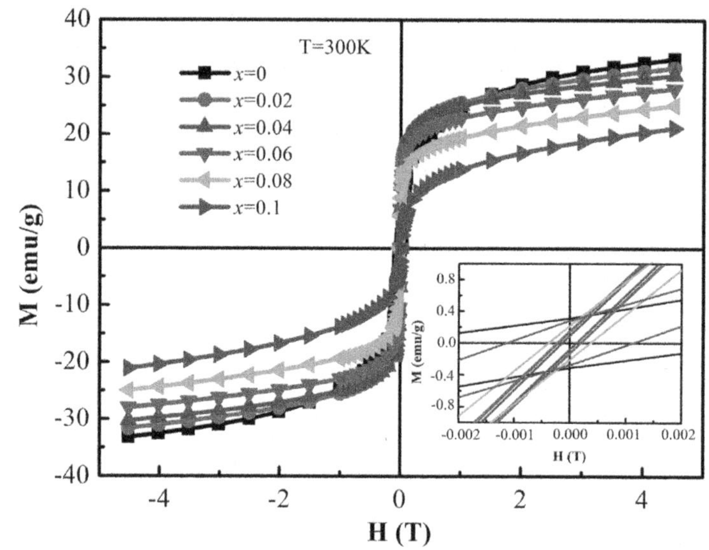

FIGURE 3.47

Hysteresis loops of $Ba_{0.5}Sr_{1.5}Zn_2(Fe_{1-x}In_x)_{12}O_{22}$ ($0 \leq x \leq 0.1$) Y-type hexaferrites measured at $T = 300$ K.

From Zhang, M., Yin, L., Liu, Q., Zi, Z., Dai, J., & Sun, Y. (c. 2018). Indium doping effect on the magnetic properties of Y-type hexaferrite $Ba_{0.5}Sr_{1.5}Zn_2(Fe_{1-x}In_x)_{12}O_{22}$. Current Applied Physics, *18(9), 1001–1005. https://doi.org/10.1016/j.cap.2018.05.017 13.*

3.3.4 Z-type hexagonal ferrites

The Z-type hexaferrite is a complex compound with a hexagonal structure, comprising the simple M and Y hexaferrites. Structural and magnetic properties of Z-type hexaferrites have been found to change with the substitution of different magnetic and nonmagnetic ions such as Ni^{2+}, Zn^{2+}, Cu^{2+}, La^{3+}, Dy^{3+}, and Al^{3+} (Kamishima et al., 2007; Mu et al., 2011; Wang et al., 2004, 2003; Wu et al., 2013; Xu et al., 2009; Zhang et al., 2002).

The Z-type hexaferrite consisting of $Ba_3Co_{1.8}Mn_{0.066}Ni_{0.066}Cu_{0.066}Fe_{24-2y}Al_yCr_yO_{41}$ ($0.0 \leq y \leq 0.8$) nanoparticles has been synthesized using a coprecipitation-assisted solid-state method, followed by calcination at 1000°C for 3 h (Ghasemi & Gordani, 2016). The XRD pattern exhibited the formation of the Z-type hexaferrite, being accompanied by a small amount of M- and Y-type hexaferrites. Fig. 3.48 displays VSM hysteresis loops of the resulting Z-type hexaferrite nanoparticles. It is clearly observed that the ferrite sample with the composition of $y = 0.4$ has the highest coercivity and magnetization. The coercivity and magnetization of the samples with $y = 0$ and 0.4 are obtained to be 1306 and 1850 Oe, and 60.73 and 75 emu/g, respectively. For $y > 0.4$, the coercivity of the corresponding ferrite samples decreases because of the grain growth. Moreover, while the preferred distribution of Al^{3+} and Cr^{3+} ions (replacing Fe^{3+} ions) at spin-down positions may

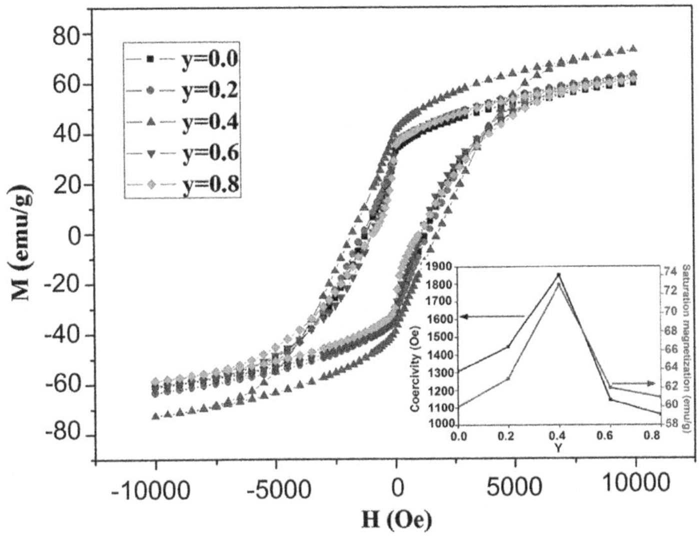

FIGURE 3.48

Vibrating sample magnetometry hysteresis loops obtained from Z-type hexaferrite consisting of $Ba_3Co_{1.8}Mn_{0.066}Ni_{0.066}Cu_{0.066}Fe_{24-2y}Al_yCr_yO_{41}$ ($0.0 \leq y \leq 0.8$) nanoparticles. The inset shows the variations of coercivity and saturation magnetization as a function of *y*.

From Ghasemi, A., & Gordani, G. R. (c. 2016). Characterization and investigation of magnetic and microwave properties of Al–Cr-substituted Z-type barium hexaferrite nanoparticles. Journal of Superconductivity and Novel Magnetism, 29*(3), 795–801. https://doi.org/10.1007/s10948-015-3337-6 5.*

enhance the saturation magnetization for $y \leq 0.4$, the ion occupation changes to the $12k_{VI}$ and $2a_{VI}$ octahedral positions when $y > 0.4$.

Elsewhere, barium Z-type hexaferrites ($Ba_3Co_2Cr_{2x}Fe_{24-2x}O_{41}$; $x = 0$, 0.3, 0.6 and 0.9) have been prepared by a solid-state reaction process (Magham et al., 2017). Fig. 3.49 shows XRD patterns of ferrites sintered at 1300°C, indicating the formation of pure single-phase Z-type hexaferrite structure without the presence of impurities or secondary phases. It can be inferred that since the Z-type hexaferrite is a mixture of M- and Y-type structures, a high temperature is required to form it.

VSM hysteresis loops of ferrites are shown in Fig. 3.50. Evidently, the saturation magnetization decreases with increasing the Cr^{3+} ion concentration. Notably, the saturation magnetization is reduced from 54.62 emu/g ($x = 0$) to 38.99 emu/g ($x = 0.9$). As the Fe^{3+} ions are substituted by Cr^{3+} ions at octahedral lattice sites, the saturation magnetization decreases with increasing the Cr^{3+} ion concentration. Moreover, the coercivity increases with the substitution of Cr^{3+} ions. For example, the coercivity increases from 28.21 Oe ($x = 0$) to 245.01 Oe ($x = 0.9$). As mentioned previously, the coercivity of hexaferrites is directly dependent on the anisotropy field. Accordingly, the anisotropy field shows an increase from 11.25 to 14.7 kOe when x is increased from 0 to 0.9. The inset reflects the variation of remanent

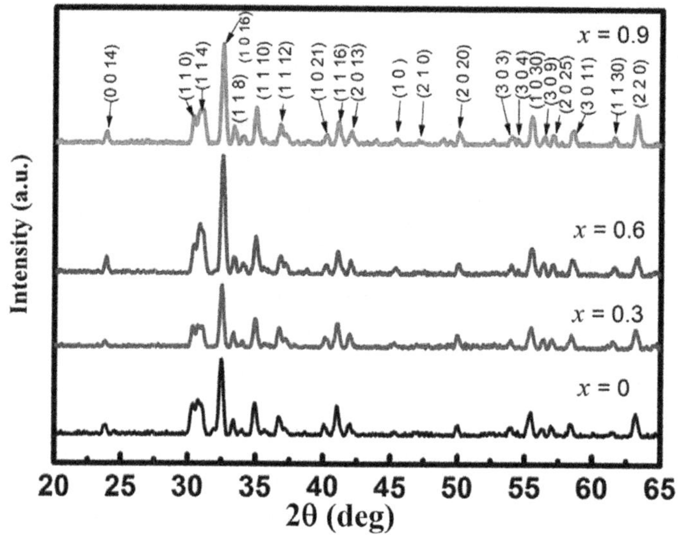

FIGURE 3.49

X-ray diffraction patterns of $Ba_3Co_2Cr_{2x}Fe_{24-2x}O_{41}$ ($x = 0$, 0.3, 0.6 and 0.9) Z-type hexagonal ferrites.

From Magham, S., Sharma, M., Shannigrahi, S., Tan, H. R., Sharma, V., Meng, Y. S., Idapalpati, S., Ramanujan, R., & Repaka, D. (c. 2017). Development of Z-type hexaferrites for high frequency EMI shielding applications. Journal of Magnetism and Magnetic Materials, 441, *303–309. https://doi.org/10.1016/j.jmmm.2017.05.066 308.*

magnetization with mole percentage of chromium content. It is observed that at a chromium content of 0.3, the maximum value of M_r is obtained.

Fig. 3.51 shows the variation of reflection loss as a function of frequency for the ferrites. A reflection loss of −28.42 dB is obtained for the $Ba_3Co_2Fe_{24}O_{41}$ sample at the frequency of 5.2 GHz. The $Ba_3Co_2Cr_{2x}Fe_{24-2x}O_{41}$ hexaferrite with the composition of $x = 0.6$ results in the maximum reflection loss (−35.20 dB) at 5.34 GHz. It is seen that the microwave absorption of the hexaferrite samples is enhanced from 95% for $x = 0$ to 99.8% for $x = 0.6$. Furthermore, no considerable variations were observed in the resonance frequencies of the samples when increasing the Cr^{3+} concentration, although the microwave absorption properties of the samples improved.

In another study, magnetic properties of $Sr_{3-x}Ba_xCo_2Fe_{24}O_{41}$ ($0.0 \leq x \leq 3.0$) Z-type hexaferrites prepared with a conventional solid-state reaction method have been investigated (Tang et al., 2016). XRD patterns confirmed the formation of a pure single phase of hexaferrite. In addition, all the peaks were slightly left-shifted with an increasing Ba^{2+}/Sr^{2+} ratio owing to the larger radius of Ba^{2+} over that of Sr^{2+}. In addition, both lattice parameters (a and c) increased with an increase in the Ba^{2+} concentration to $x = 2$. In other words, the cell volume of the Z-type hexaferrite structure increased with an increase in the Ba^{2+} concentration.

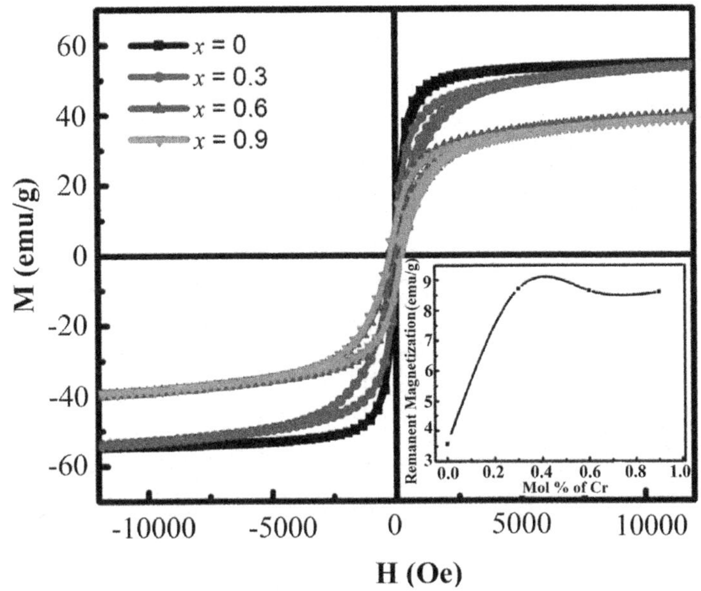

FIGURE 3.50

Vibrating sample magnetometry hysteresis loops of $Ba_3Co_2Cr_{2x}Fe_{24-2x}O_{41}$ ($x = 0$, 0.3, 0.6 and 0.9) Z-type hexagonal ferrites.

From Magham, S., Sharma, M., Shannigrahi, S., Tan, H. R., Sharma, V., Meng, Y. S., Idapalpati, S., Ramanujan, R., & Repaka, D. (c. 2017). Development of Z-type hexaferrites for high frequency EMI shielding applications. Journal of Magnetism and Magnetic Materials, 441, *303–309. https://doi.org/10.1016/j.jmmm.2017.05.066 308.*

Fig. 3.52 shows the variations of saturation magnetization and coercivity as a function of Ba^{2+}content for ferrites at different temperatures (50–350 K). As can be seen, M_s of the Z-type hexaferrites is rapidly reduced with an increase in the temperature. In general, M_s has a decreasing trend when the Ba^{2+} concentration increases. In fact, the substitution of Ba^{2+} with Sr^{2+} leads to an alteration in the site preference of cobalt ions in the $Sr_{3-x}Ba_xCo_2Fe_{24}O_{41}$ Z-type hexagonal ferrite crystal. Moreover, the superexchange interaction between the magnetic ions may be reduced with the substitution of Ba^{2+} ions, having a larger ionic radius than that of Sr^{2+} ones. Therefore, the weakened superexchange interactions can cause a reduction in the M_s. Similarly, H_c decreases rapidly with an increase in the temperature. Additionally, while H_c initially increases with an increase in the Ba^{2+} content, it decreases with further enhancing the substitution content. The Ba^{2+}/Sr^{2+} ratio can affect the bond angles between Fe^{3+} ions due to their different radii while changing the magnetocrystalline anisotropy energy.

In a separate study, the effects of Bi/Co co-substitution on the microstructure, magnetic and dielectric properties of $Ba_{3-x}Bi_xCo_{2+x}Fe_{24-x}O_{41}$ (x = 0.05, 0.1, 0.15, 0.2, and 0.4) Z-type hexagonal ferrites synthesized by a conventional ceramic method have been evaluated (Li et al., 2019). The XRD patterns illustrated that the

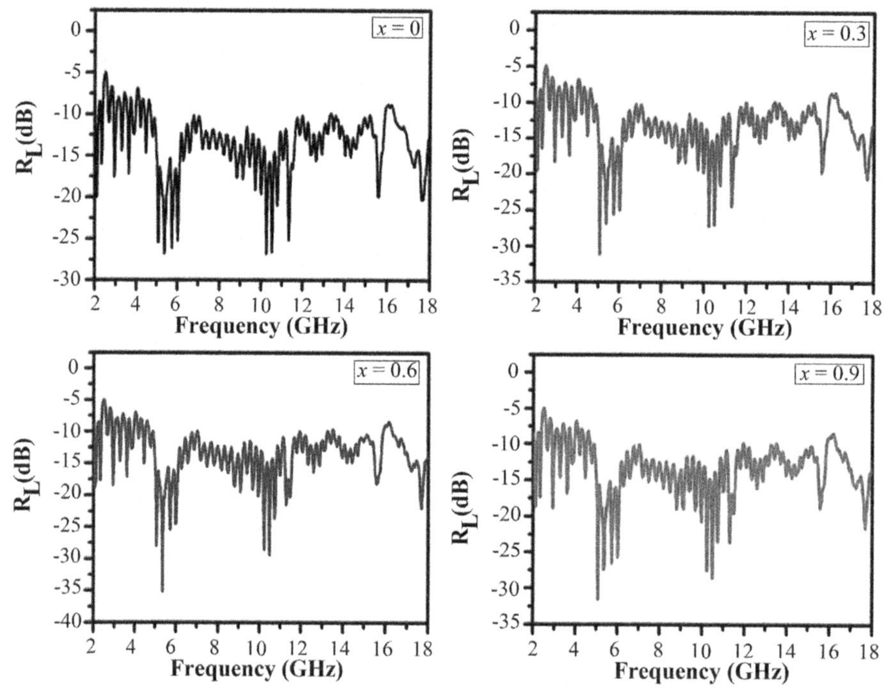

FIGURE 3.51

The variation of reflection loss as a function of frequency for $Ba_3Co_2Cr_{2x}Fe_{24-2x}O_{41}$ ($x =$ 0, 0.3, 0.6, and 0.9) Z-type hexagonal ferrites.

From Magham, S., Sharma, M., Shannigrahi, S., Tan, H. R., Sharma, V., Meng, Y. S., Idapalpati, S., Ramanujan, R., & Repaka, D. (c. 2017). Development of Z-type hexaferrites for high frequency EMI shielding applications. Journal of Magnetism and Magnetic Materials, 441, *303–309. https://doi.org/10.1016/j.jmmm.2017.05.066 308.*

ferrite sample with $x = 0.05$ resulted in a Z-type hexaferrite structure accompanied by small M- and Y-type structural content as the secondary phases. The samples with high Bi/Co codoping concentrations ($x \geq 0.15$) contained a Z-type hexaferrite phase with small contents of W- and Y-type hexaferrites as the secondary phases.

The variations of complex permeability and permittivity as a function of frequency are shown in Fig. 3.53. The permeability of the ferrite samples with $x =$ 0.1–0.4 is high at a frequency greater than 1.83 GHz. The sample with $x = 0.05$ has the maximum permeability (18.3 ± 0.43). Increasing the co-substitution content initially reduces the value of the real part of permeability. However, at higher frequencies, it increases the permeability real part of the ferrite samples. Since sample $x = 0.05$, contains large intragranular pore, the significant dielectric dispersion was observed in Fig. 3.53C and D, while for $x > 0.05$, no dispersion phenomenon is seen.

Fig. 3.54 exhibits VSM hysteresis loops of the Z-type hexagonal ferrites. All of the samples show soft magnetic properties, possessing low coercivity and high saturation magnetization. The maximum saturation magnetization (55.77 emu/g) and

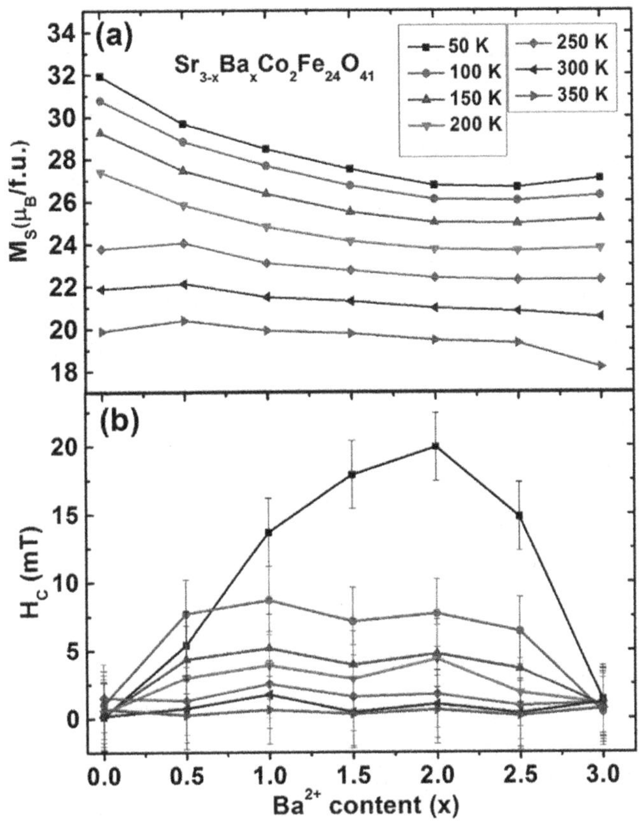

FIGURE 3.52

The variations of: (A) saturation magnetization, and (B) coercivity as a function of Ba^{2+} content for $Sr_{3-x}Ba_xCo_2Fe_{24}O_{41}$ Z-type hexagonal ferrites at different temperatures (50–350 K).

From Tang, R., Jiang, C., Zhou, H., & Yang, H. (c. 2016). Effects of composition and temperature on the magnetic properties of (Ba, Sr) $3Co_2Fe_{24}O_{41}$ Z type hexaferrites. Journal of Alloys and Compounds, 658, *132–138. https://doi.org/10.1016/j.jallcom.2015.10.207 137.*

minimum coercivity (22.1 Oe) values are obtained for the sample with $x = 0.2$. Moreover, the magnetic moment and the saturation magnetization increase with an increase in the co-substitution concentration. The coercivity and remanent magnetization of the sample with $x = 0.05$ are obtained to be 62.3 Oe and 4.9 emu/g, respectively. Increasing the co-substitution concentration up to $x = 0.15$ reduces both coercivity and remanent magnetization. In fact, the presence of the hard magnetic phase (the M-type hexaferrite) in the structure is responsible for the high coercivity of the samples with low co-substitution levels.

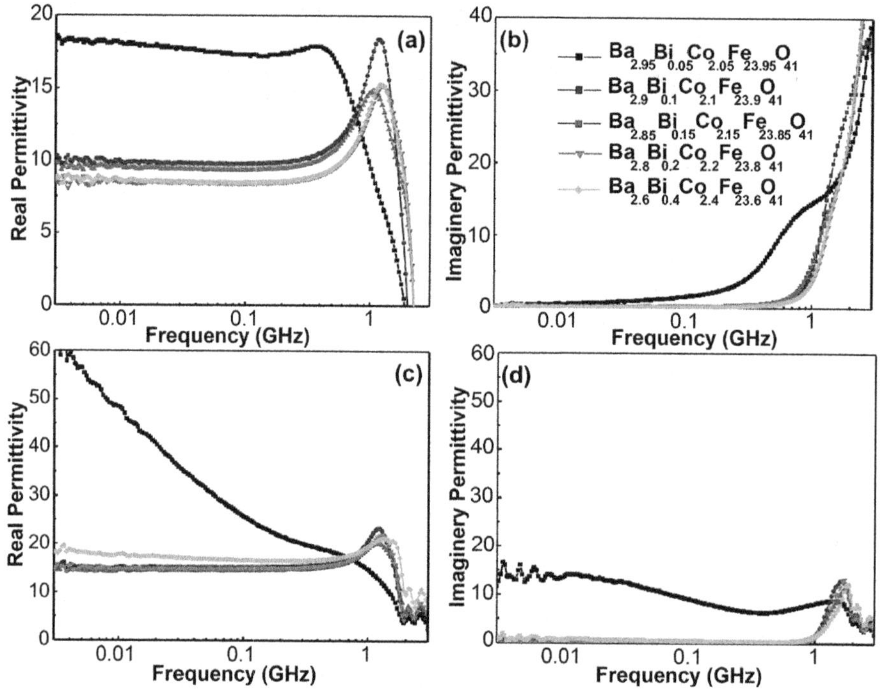

FIGURE 3.53

Variation of (A–B) complex permeability and (C–D) complex permittivity of $Ba_{3-x}Bi_xCo_{2+x}Fe_{24-x}O_{41}$ Z-type hexagonal ferrites.

From Li, J., Li, S., Fu, T., Shao, N., & Che, S. (c. 2019). Preparation and properties of $Ba_{3-x}Bi_xCo_{2+x}Fe_{24-x}O_{41}$ Z-type hexaferrites. Journal of Magnetism and Magnetic Materials, 488, *165366. https://doi.org/10.1016/j.jmmm.2019.165366 4.*

3.3.5 X-type hexagonal ferrites

The X-type hexagonal ferrites are an important family of magnetic materials, having high saturation magnetization, low coercivity, high electrical resistivity, high Curie temperature, and low eddy current loss. As permanent magnet materials, X-type hexagonal ferrites can play a remarkable role in many technological and industrial applications such as refrigerators, magnetic recording media, and microwave absorbers.

In a research study, $Ba_{2-x}Dy_xCu_2Fe_{28-y}Co_yO_{46}$ ($x = 0.0, 0.02, 0.06, 0.1, y = 0.0, 0.1, 0.3, 0.5$) X-type hexagonal ferrites have been synthesized using a sol–gel method, followed by a sintering process at 1300°C (Asif et al., 2019). Fig. 3.55 shows XRD patterns of the resulting ferrites. All the peaks in the patterns indicate the formation of a pure X-type hexagonal ferrite structure with no secondary phases. The values of lattice parameters and c/a ratio vary with the substitution amount of x and y in the composition, arising from the difference between the ionic radii of Dy^{3+}

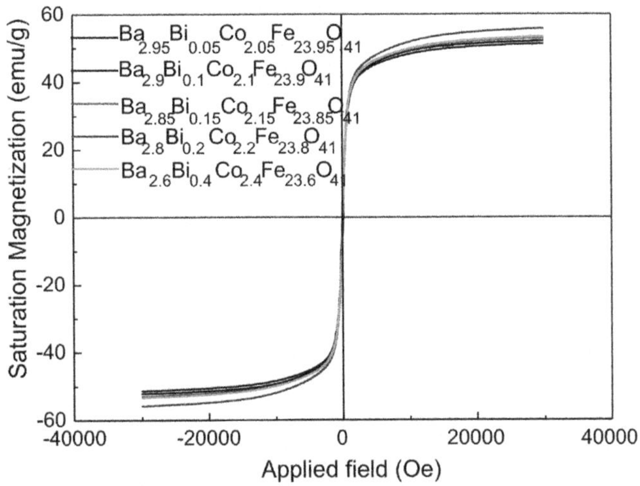

FIGURE 3.54

Vibrating sample magnetometry hysteresis loops of $Ba_{3-x}Bi_xCo_{2+x}Fe_{24-x}O_{41}$ ($x = 0.05$, 0.1, 0.15, 0.2 and 0.4) Z-type hexagonal ferrites.

From Li, J., Li, S., Fu, T., Shao, N., & Che, S. (c. 2019). Preparation and properties of $Ba_{3-x}Bi_xCo_{2+x}Fe_{24-x}O_{41}$ Z-type hexaferrites. Journal of Magnetism and Magnetic Materials, 488, *165366. https://doi.org/10.1016/j.jmmm.2019.165366 4.*

and Co^{2+} with those of Ba^{2+} and Fe^{3+}. Notably, the ionic radius of Ba^{2+} and Dy^{3+} is 1.49 and 0.912 Å, respectively. Moreover, some alterations occur in the lattice parameters due to the substitution of Co^{2+} with Fe^{3+}, having an ionic radius of 0.745 and 0.645 Å, respectively.

VSM hysteresis loops of ferrites are displayed in Fig. 3.56. The saturation magnetization is seen to decrease from 22.43 to 14.53 emu/g with increasing x, $y = 0$, 0 to x, $y = 0.02$, 0.1. Further increasing the substitution amount to x, $y = 0.1$, 0.5 enhances the saturation magnetization to 48.33 emu/g. The variations observed in the saturation magnetization can be attributed to the distribution of cations in different sites of the unit cell. The Co^{2+} occupies octahedral sites in the R block, which leads to an enhancement in the saturation magnetization. On the other hand, as the Co^{2+} ions prefer to be distributed in the octahedral sites, Fe^{2+} ions occupying the octahedral sites in the S block may be transferred to tetrahedral sites with a down spin in the S block, thereby decreasing the magnetization. It is worth noting that all ferrite samples have squareness ratios smaller than 0.5, indicating the presence of multimagnetic domains. In addition, coercivity decreases from 317 G (x, $y = 0$, 0) to 158 G (x, $y = 0.1$, 0.5). The grain size of ferrites increases with the substitution of Dy-Co ions, leading to a decrease in the resultant coercivity.

Elsewhere, dielectric and magnetic properties of $Ba_{2-x}Mg_xCo_{2-y}Sr_yFe_{28}O_{46}$ ($x = y = 0$–0.5) X-type hexagonal ferrites synthesized by a sol–gel autocombustion route have been investigated (Ejaz et al., 2018). The variations of real and imaginary

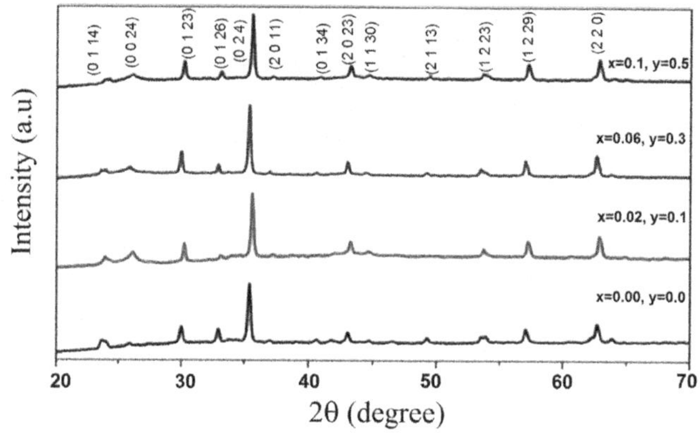

FIGURE 3.55

X-ray diffraction patterns of $Ba_{2-x}Dy_xCu_2Fe_{28-y}Co_yO_{46}$ X-type hexagonal ferrites.

From Asif, S. U., Rizwan, S., Awan, M. Q., Khan, M. W., Sadiq, I., Mazhar, M. E., Ahmad, A., Hussain, S. S., Khan, E. U., & Hu, W. (c. 2019). Effect of Dy-Co on physical and magnetic properties of X-type hexaferrites ($Ba_{2-x}Dy_xCu_2Fe_{28-y}Co_yO_{46}$). Chinese Journal of Physics, 61, *47–54. https://doi.org/10.1016/j.cjph.2019. 07.007 52.*

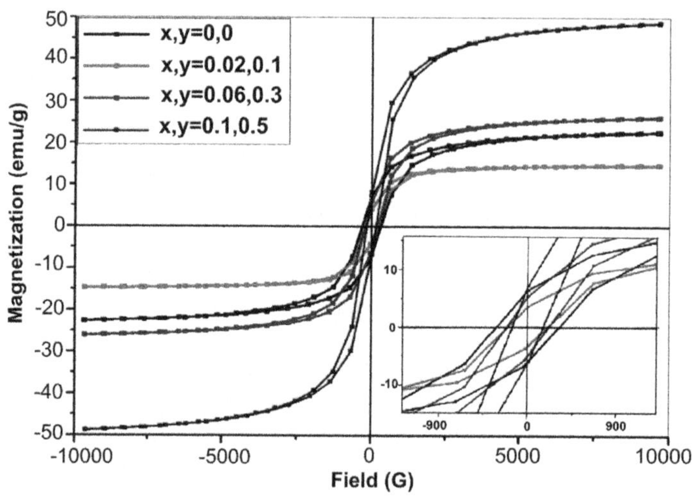

FIGURE 3.56

Vibrating sample magnetometry hysteresis loops of $Ba_{2-x}Dy_xCu_2Fe_{28-y}Co_yO_{46}$ X-type hexagonal ferrites. The inset shows the hysteresis loops in a smaller magnetic field range with higher resolution.

From Asif, S. U., Rizwan, S., Awan, M. Q., Khan, M. W., Sadiq, I., Mazhar, M. E., Ahmad, A., Hussain, S. S., Khan, E. U., & Hu, W. (c. 2019). Effect of Dy-Co on physical and magnetic properties of X-type hexaferrites ($Ba_{2-x}Dy_xCu_2Fe_{28-y}Co_yO_{46}$). Chinese Journal of Physics, 61, *47–54. https://doi.org/10.1016/j.cjph.2019. 07.007 52.*

parts of permittivity as a function of frequency for the resulting ferrites are shown in Figs. 3.57 and 3.58, respectively. As can be seen, the real and imaginary parts of permittivity decrease with increasing frequency.

The resonance peaks are observed to occur above a frequency of 2 GHz, which can be caused by the interfacial polarization based on the Maxwell-Wagner theory. As well, the substitution of Mg-Sr ions ($x = y = 0.1$) in the hexagonal ferrites can decrease the real part of permittivity. Further increasing the substitution amount increases the corresponding dielectric constant, reaching a maximum value at $x = y = 0.4$. However, the dielectric constant is sharply reduced at $x = y = 0.5$. The high value of the permittivity real part at the low frequency is attributed to dislocations and other defects. Alternatively, the permittivity imaginary part increases with an increase in the substitution amount to $x = y = 0.4$.

Fig. 3.59 shows VSM hysteresis loops of ferrites. It is seen that the saturation magnetization decreases from 69.569 to 65.785 emu/g with increasing the Mg-Sr substituting ions from $x = y = 0$ to $x = y = 0.5$. Meanwhile, the remanent magnetization decreases from 35.53 emu/g ($x = y = 0$) to 33.099 emu/g ($x = y = 0.5$). Basically, the saturation magnetization of the Mg-Sr-substituted hexagonal ferrites depends on the superexchange interaction among octahedral $3a_{v1}$ and tetrahedral

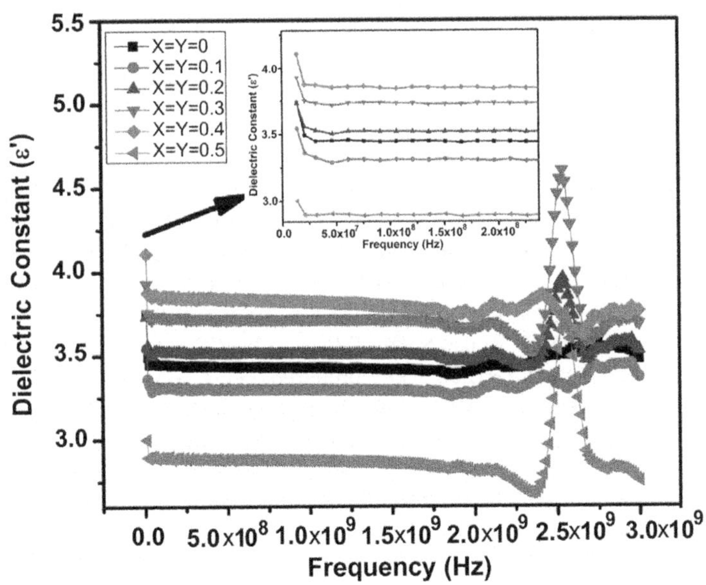

FIGURE 3.57

The variation of real part of permittivity as a function of frequency for $Ba_{2-x}Mg_xCo_{2-y}Sr_yFe_{28}O_{46}$ ($x = y = 0$–0.5) X-type hexagonal ferrites.

From Ejaz, S. R., Khan, M. A., Warsi, M. F., Akhtar, M. N., & Hussain, A. (c. 2018). Study of structural transformation and hysteresis behavior of Mg-Sr substituted X-type hexaferrites. Ceramics International, *44(15), 18903–18912. https://doi.org/10.1016/j.ceramint.2018.07.126.*

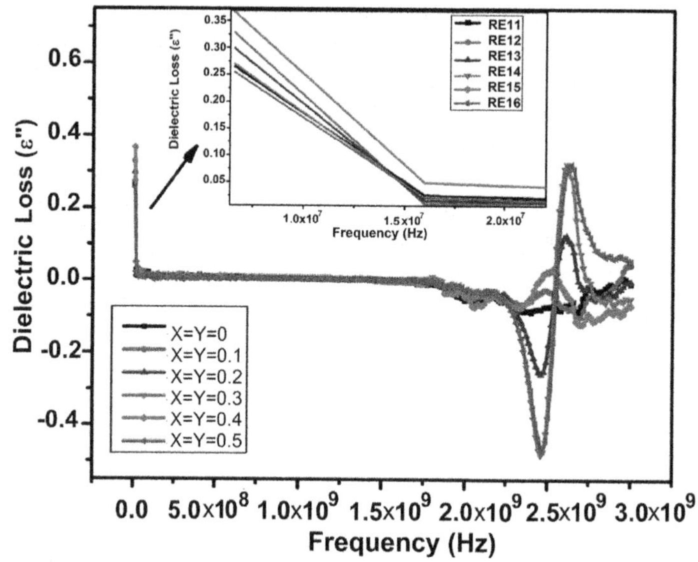

FIGURE 3.58

The variation of imaginary part of permittivity as a function of frequency for $Ba_{2-x}Mg_xCo_{2-y}Sr_yFe_{28}O_{46}$ ($x = y = 0$–0.5) X-type hexagonal ferrites.

From Ejaz, S. R., Khan, M. A., Warsi, M. F., Akhtar, M. N., & Hussain, A. (c. 2018). Study of structural transformation and hysteresis behavior of Mg-Sr substituted X-type hexaferrites. Ceramics International, *44(15), 18903–18912. https://doi.org/10.1016/j.ceramint.2018.07.126.*

$6c_{IV}$ sites of the S blocks. Since the magnetic moments of Mg (0.85 μ_B) and Sr (2.7 μ_B) are lower than those of Ba (2.3 μ_B) and Co (3.7 μ_B), the substitution of Mg–Sr ions in the $Ba_{2-x}Mg_xCo_{2-y}Sr_yFe_{28}O_{46}$ hexaferrites reduces A–B superexchange magnetic interactions. It should be noted that the value of coercivity increases from 2143.82 for $x = y = 0$ to 2889.47 Oe for $x = y = 0.1$.

In a separate research study, $Sr_2Ni_2T_xFe_{28-x}O_{46}$ (T = Cr, Bi, Al, and In; x = 0.25) X-type hexagonal ferrites were prepared using a sol–gel route (Majeed et al., 2020). Figs. 3.60 and 3.61 show the variations of real and imaginary parts of permittivity as a function of frequency for the resulting hexagonal ferrites. The permittivity real part of all the ferrite samples increases with increasing the frequency in the range of 0.5–1.5 GHz, arising from the polarization at the grain boundaries. The resonance peaks are also observed in the frequency range of 1–2 GHz.

In fact, the occurrence of resonance is related to the matching of the hopping frequency of electrons (between Fe^{2+} and Fe^{3+} ions) with the applied field frequency. The substitution of Fe^{3+} ions by Cr^{3+}, Bi^{3+}, Al^{3+}, and In^{3+} ones increases the corresponding dielectric constant because of the enhancement in polarization, as well as the large resistance of grain boundary at low frequencies. For the samples substituted with Cr^{3+} and Bi^{3+} cations, the dielectric constant is seen to decrease at frequencies

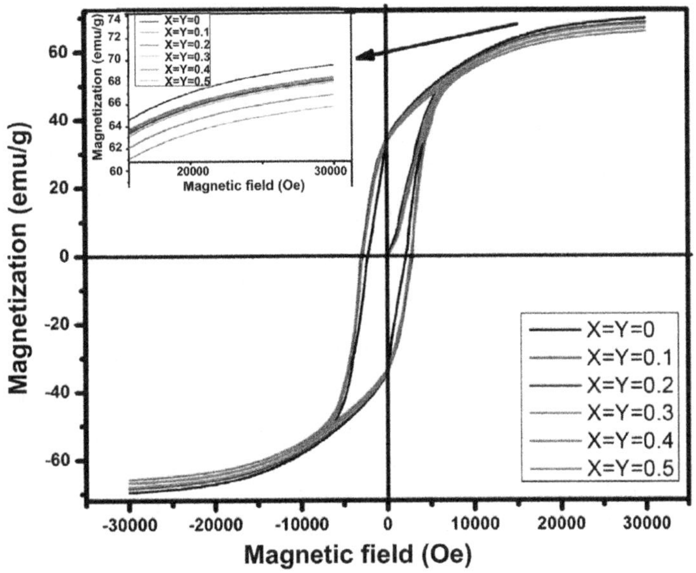

FIGURE 3.59

Vibrating sample magnetometry hysteresis loops of $Ba_{2-x}Mg_xCo_{2-y}Sr_yFe_{28}O_{46}$ X-type hexagonal ferrites.

From Ejaz, S. R., Khan, M. A., Warsi, M. F., Akhtar, M. N., & Hussain, A. (c. 2018). Study of structural transformation and hysteresis behavior of Mg-Sr substituted X-type hexaferrites. Ceramics International, *44(15), 18903–18912. https://doi.org/10.1016/j.ceramint.2018.07.126 8.*

of 1.21 and 1.23 GHz. This can be attributed to the reduction in the grain boundary resistance at such frequencies.

The values of permittivity imaginary part for all of the samples are extremely small in the frequency ranges of 0.001–1 GHz and 2–3 GHz. Notably, the resonance peaks are observed in the frequency range of 1–2 GHz, being shifted to a lower frequency range following the substitution of ions due to the reduction in the hopping or jumping probability. Essentially, a decrease in the Fe^{3+} cations at octahedral sites may be responsible for the reduction in the hopping probability. The values of tangent loss are obtained to be higher than 1.0 for the ferrite samples substituted with Cr^{3+} and Bi^{3+} cations. In the case of the samples substituted with Al^{3+} and In^{3+} cations, the corresponding tangent loss values are less than 1.0.

3.3.6 U-type hexagonal ferrites

The high intrinsic magnetocrystalline anisotropy of the U-type hexaferrite has made it a suitable candidate for microwave absorbers working in the GHz frequency range (Lisjak et al., 2004; Lisjak & Drofenik, 2003, 2004; Pullar & Bhattacharya, 2001). Some research has focused on the magnetic and microwave absorption properties of the U-type hexaferrites (Dimri et al., 2006; Kamishima et al., 2015; Lisjak et al.,

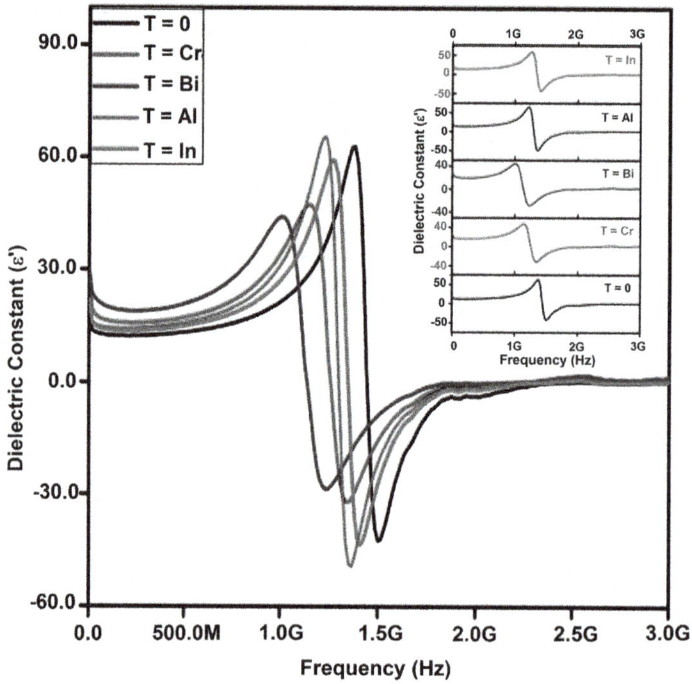

FIGURE 3.60

The variation of real part of permittivity as a function of frequency for $Sr_2Ni_2T_xFe_{28-x}O_{46}$ ($T=$ Cr, Bi, Al, and In; $x=0.25$) X-type hexagonal ferrites.

From Majeed, A., Khan, M. A., Lodhi, M. Y., Ahmad, R., & Ahmad, I. (c. 2020). Structural, microwave permittivity, and complex impedance studies of cation (Cr, Bi, Al, In) substituted SrNi-X hexagonal nano-sized ferrites. Ceramics International, *46(2), 1907–1915. https://doi.org/10.1016/j.ceramint.2019.09.168.*

2006; Shannigrahi et al., 2013). Notably, $Ba_4Co_{2-3x}Cr_{2x}Fe_{36}O_{60}$ ($0.00 \leq x \leq 0.60$) U-type hexagonal ferrites have been prepared by a solid-state reaction technique, followed by investigating the magnetic and microwave absorption properties (Kumar et al., 2016). Fig. 3.62 shows VSM hysteresis loops of the $Ba_4Co_{2-x}Cr_{2x}Fe_{36}O_{60}$ U-type hexagonal ferrites with different compositions. As can be seen, increasing the Cr^{3+} ion concentration from $x=0.0$ to 0.60 decreases the saturation magnetization from 59.1 to 50.5 emu/g. Since the Cr^{3+} cations fill the octahedral sites of hexaferrites, a reduction occurs in the number of Fe^{3+} ions. In other words, the saturation magnetization decreased with increasing the Cr^{3+} content because of the partial conversion of Fe^{3+} ($5\ \mu_B$) ions into Fe^{2+} ($4\ \mu_B$) ones at the octahedral sites. On the other hand, the coercivity increases from 48.5 to 419.9 Oe with increasing the Cr^{3+} concentration from $x=0.00$ to 0.60 due to the alteration of the magnetocrystalline anisotropy direction from the c-plane to the c-axis.

Fig. 3.63 shows the variations of complex permittivity and permeability as a function of frequency for hexagonal ferrites. As mentioned earlier, the number of Fe^{3+} ions

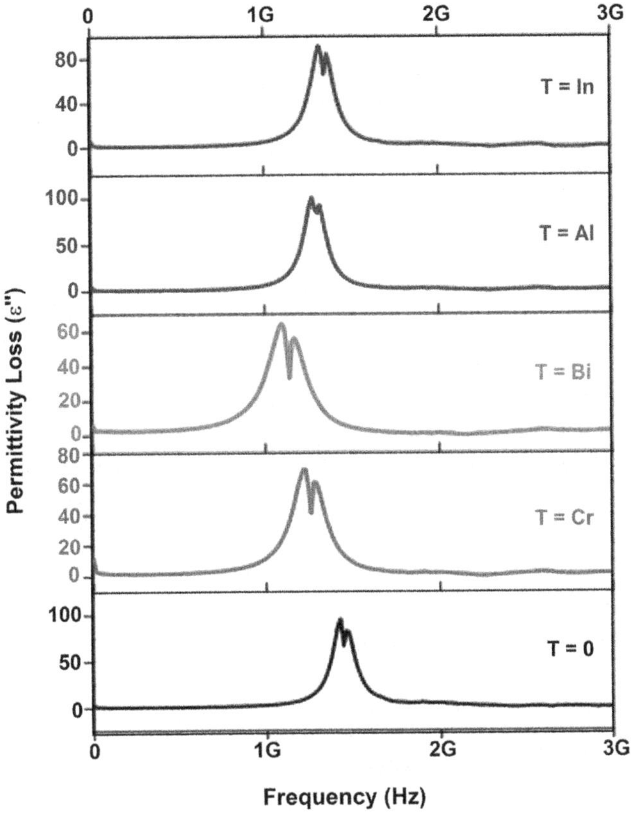

FIGURE 3.61

The variation of imaginary part of permittivity as a function of frequency for $Sr_2Ni_2T_xFe_{28-x}O_{46}$ (T = Cr, Bi, Al, and In; $x = 0.25$) X-type hexagonal ferrites.

From Majeed, A., Khan, M. A., Lodhi, M. Y., Ahmad, R., & Ahmad, I. (c. 2020). Structural, microwave permittivity, and complex impedance studies of cation (Cr, Bi, Al, In) substituted SrNi-X hexagonal nano-sized ferrites. Ceramics International, *46(2), 1907–1915. https://doi.org/10.1016/j.ceramint.2019.09.168.*

decreases with an increase in the Cr^{3+} content, followed by the Fe^{3+} ion conversion into Fe^{2+} ions as the Cr^{3+} cations occupy the octahedral sites in hexaferrites. In this way, the motion of charge carriers between the Fe^{3+} and Fe^{2+} ions decreases, leading to an increase in the electrical resistivity. As a result, both the real (ε') and the imaginary (ε'') parts of permittivity are reduced when increasing the Cr^{3+} content. Likewise, the magnetization at octahedral sites decreases with an increase in the Cr^{3+} content, giving rise to a reduction in the real part of permeability (μ'). The presence of peaks in the imaginary part of permeability (μ'') curves confirms the maximum loss of magnetic energy. It is interesting to find that the peak in the μ'' curve shifts from 7.54 to 10.75 GHz with increasing the Cr^{3+} concentration from $x = 0.0$ to 0.60, arising from the enhancement in the magnetocrystalline anisotropy.

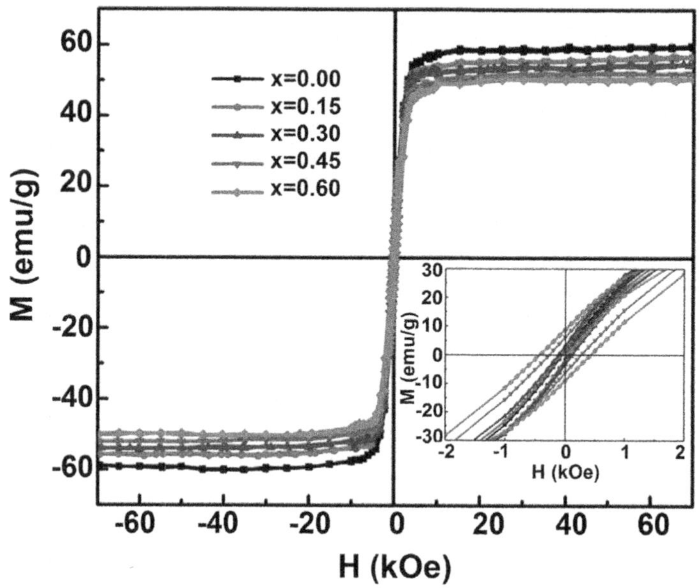

FIGURE 3.62

Vibrating sample magnetometry hysteresis loops of $Ba_4Co_{2-3x}Cr_{2x}Fe_{36}O_{60}$ ($0.00 \leq x \leq 0.60$) U-type hexagonal ferrites.

From Kumar, S., Meena, R. S., & Chatterjee, R. (c. 2016). Microwave absorption studies of Cr-doped Co–U type hexaferrites over 2–18 GHz frequency range. Journal of Magnetism and Magnetic Materials, 418, *194–199. https://doi.org/10.1016/j.jmmm.2016.02.084.*

Fig. 3.64 shows the variation of reflection loss versus frequency for the $Ba_4Co_{2-3x}Cr_{2x}Fe_{36}O_{60}$ ($0.00 \leq x \leq 0.60$) U-type hexagonal ferrites with different thicknesses (1.7–1.9 mm). The maximum reflection loss value is obtained to be −34.90 dB at a frequency of 8.2 GHz for ferrite with the composition of $Ba_4Co_{1.1}Cr_{0.6}Fe_{36}O_{60}$, having a thickness of 1.7 mm. While increasing the Cr^{3+} content shifts the peak of the maximum reflection loss to a higher frequency (from the C-band to X-band), increasing the sample thickness shifts it to a lower frequency. The $Ba_4Co_{2-3x}Cr_{2x}Fe_{36}O_{60}$ U-type hexaferrites with the compositions of $x =$ 0.30, 0.45 and 0.60 exhibit also considerable absorption in the Ku-band.

In a separate study, $Ba_4Zn_{2-x}Co_xFe_{36}O_{60}$ ($0.4 \leq x \leq 1.2$) U-type hexagonal ferrites have been evaluated in terms of magnetic characteristics (Chen et al., 2020). Fig. 3.65 shows VSM hysteresis loops of the ferrite samples. It is seen that the coercivity decreases from 195 to 110 Oe with increasing the Co^{2+} content from $x = 0.4$ to 1.2. Meanwhile, the anisotropy field is reduced from 7528 Oe ($x = 0.4$) to 4833 Oe ($x = 1.2$). The magnetic parameters obtained for ferrites are listed in Table 3.7.

Fig. 3.66 shows the variations of anisotropy field and total resonance linewidth (ΔH) as a function of Co^{2+} content for the U-type hexagonal ferrites.

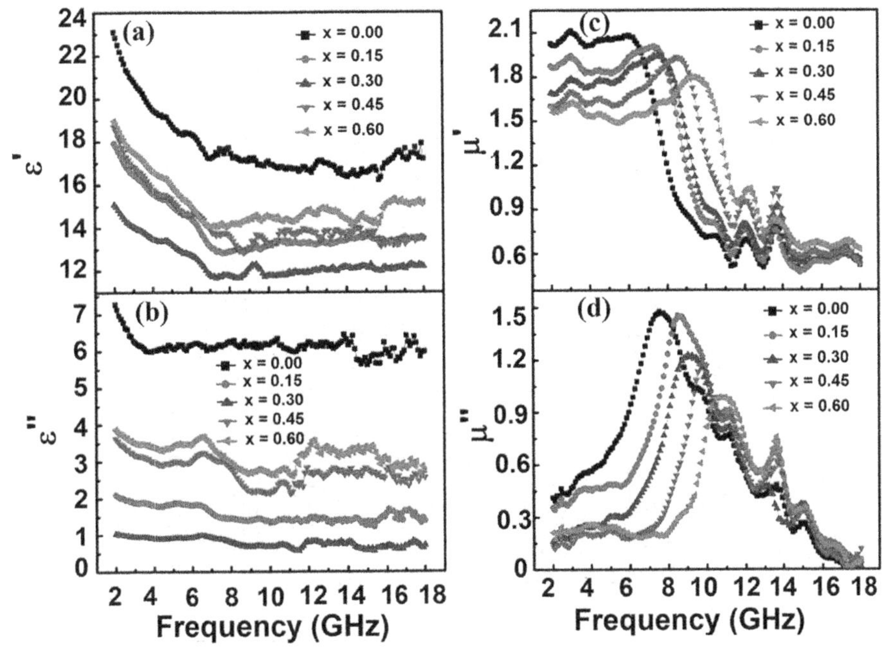

FIGURE 3.63

The variations of: (A and B) complex permittivity (ε' and ε''), and (C and D) complex permeability (μ' and μ'') as a function of frequency for $Ba_4Co_{2-3x}Cr_{2x}Fe_{36}O_{60}$ ($0.00 \leq x \leq 0.60$) U-type hexagonal ferrites.

From Kumar, S., Meena, R. S., & Chatterjee, R. (c. 2016). Microwave absorption studies of Cr-doped Co–U type hexaferrites over 2–18 GHz frequency range. Journal of Magnetism and Magnetic Materials, 418, *194–199. https://doi.org/10.1016/ j.jmmm.2016.02.084.*

In fact, ΔH depends on four significant contributions, according to the following relation:

$$\Delta H = \Delta H_i + \Delta H_a + \Delta H_p + \Delta H_s \tag{3.19}$$

where ΔH_i is the intrinsic anisotropy, ΔH_a is the random anisotropy, ΔH_p is the porosity, and ΔH_s is the surface flatness part. Thus, it can be inferred that ΔH is mostly influenced by the anisotropy field as the chemical composition, porosity, and surface condition of the four different samples remain approximately the same, leading to the similar levels of ΔH_i, ΔH_p, and ΔH_s. In this direction, it was seen that the anisotropy field decreased with an increase in the Co^{2+} content, thereby reducing the ΔH.

The synthesis of $Ba_4Me_2Fe_{36}O_{60}$ (Me = Cu, Co, Fe, Mg, and Mn) U-type hexagonal ferrites has also been carried out by a chemical citrate solution method (Dimri et al., 2011). VSM hysteresis loops of the resulting U-type hexagonal ferrites are shown in Fig. 3.67. The $Ba_4Fe_2Fe_{36}O_{60}$ hexaferrite is seen to possess the highest

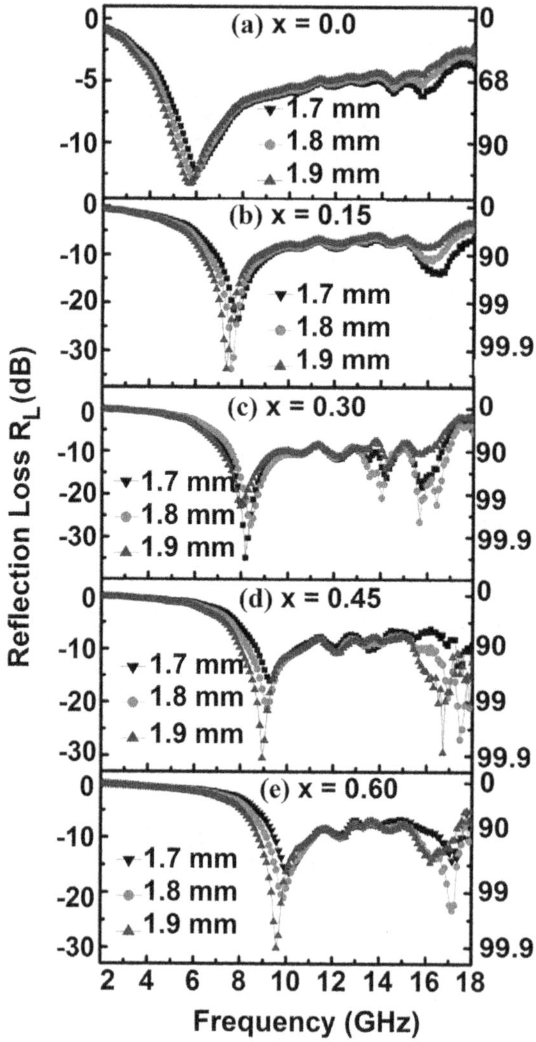

FIGURE 3.64

The variation of reflection loss versus frequency for $Ba_4Co_{2-3x}Cr_{2x}Fe_{36}O_{60}$ ($0.00 \leq x \leq 0.60$) U-type hexagonal ferrites with different thicknesses.

From Kumar, S., Meena, R. S., & Chatterjee, R. (c. 2016). Microwave absorption studies of Cr-doped Co–U type hexaferrites over 2–18 GHz frequency range. Journal of Magnetism and Magnetic Materials, 418, *194–199. https://doi.org/10.1016/j.jmmm.2016.02.084.*

coercivity (470 Oe) among the other U-type hexaferrites. The lowest coercivity (50 Oe) is obtained for the $Ba_4Mg_2Fe_{36}O_{60}$ hexaferrite. Alternatively, the saturation magnetization is maximized (70 emu/g) and minimized (53 emu/g) for the $Ba_4Cu_2Fe_{36}O_{60}$ and $Ba_4Mg_2Fe_{36}O_{60}$ hexaferrites, respectively. The reduction in

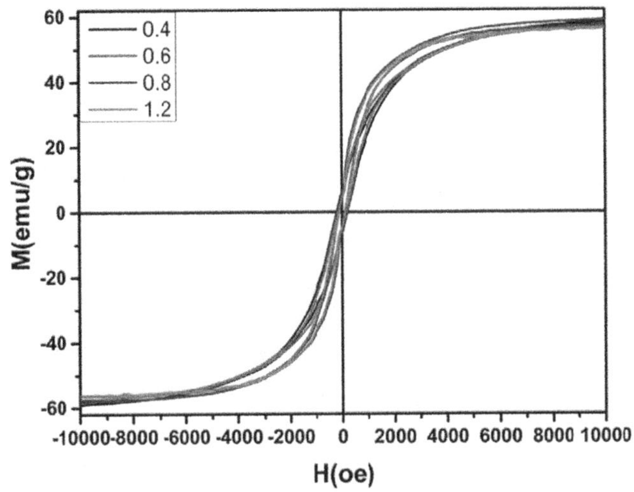

FIGURE 3.65

Vibrating sample magnetometry hysteresis loops of $Ba_4Zn_{2-x}Co_xFe_{36}O_{60}$ ($0.4 \le x \le 1.2$) U-type hexagonal ferrites.

From Chen, J., Liu, Y., Wang, Y., Yin, Q., Liu, Q., Wu, C., & Zhang, H. (c. 2020). Magnetic and microwave properties of polycrystalline U-type hexaferrite Ba4Zn2– xCoxFe36060. Journal of Magnetism and Magnetic Materials, 496, *165948. https://doi.org/10.1016/j.jmmm.2019.165948.*

Table 3.7 The magnetic parameters of $Ba_4Zn_{2-x}Co_xFe_{36}O_{60}$ ($0.4 \le x \le 1.2$) U-type hexagonal ferrites.

x	$4\pi M_s$ (G)	H_c (Oe)	M_r/M_s (%)	H_a (Oe)	K_1 (J/m^3)
0.4	3405	195	9.80	7528	103.563
0.6	3327	163	9	6917	92.256
0.8	3427	126	10	6044	82.437
1.2	3239	110	8	4833	62.299

the saturation magnetization of the U-type hexaferrites can be attributed to the formation of intermediate phases such as Z-type hexaferrite and $BaFe_2O_4$ with lower magnetization. It is worth noting that the maximum squareness ratio (M_r/M_s) is 0.2, belonging to the $Ba_4Fe_2Fe_{36}O_{60}$ hexaferrite.

Elsewhere, $(Ba_{1-3x}Bi_{2x})_4Co_2Fe_{36}O_{60}$ ($x = 0.05$, 0.10 and 0.15) U-type hexagonal ferrites have been prepared by a solid-state reaction route (Kumar & Chatterjee, 2018). VSM hysteresis loops of the resulting ferrites are shown in Fig. 3.68. The substitution of Bi^{3+} ions increases the saturation magnetization from 59.1 emu/g ($x =$ 0.0) to 63.5 emu/g ($x = 0.10$). The unit cell of U-type hexagonal ferrites has a

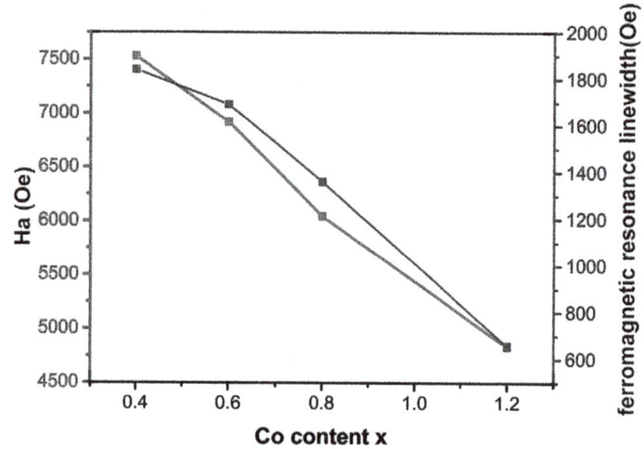

FIGURE 3.66

The variations of anisotropy field and resonance linewidth (ΔH) as a function of Co^{2+} content for $Ba_4Zn_{2-x}Co_xFe_{36}O_{60}$ U-type hexagonal ferrites.

From Chen, J., Liu, Y., Wang, Y., Yin, Q., Liu, Q., Wu, C., & Zhang, H. (c. 2020). Magnetic and microwave properties of polycrystalline U-type hexaferrite $Ba_4Zn_{2-x}Co_xFe_{36}O_{60}$. Journal of Magnetism and Magnetic Materials, 496, *165948. https://doi.org/10.1016/j.jmmm.2019.165948.*

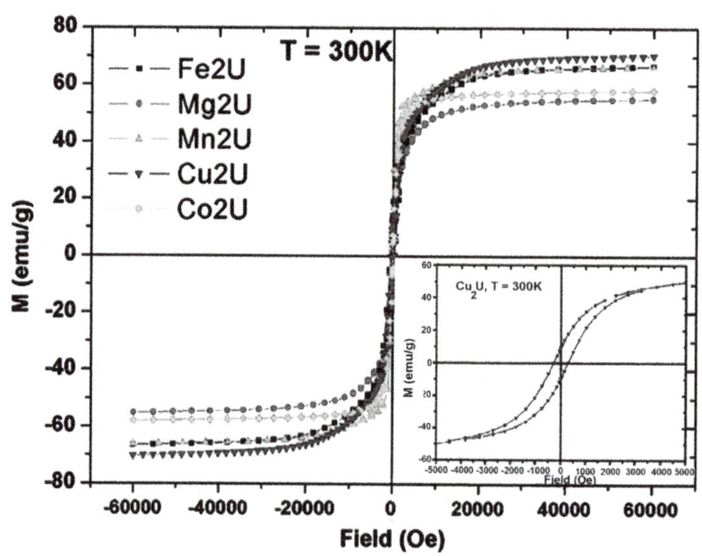

FIGURE 3.67

Vibrating sample magnetometry hysteresis loops of $Ba_4Me_2Fe_{36}O_{60}$ (Me = Cu, Co, Fe, Mg, and Mn) U-type hexagonal ferrites. The inset shows large magnification hysteresis loop of the Cu-substituted U-type ferrite.

From Dimri, M. C., Khanduri, H., Kooskora, H., Heinmaa, I., Joon, E., & Stern, R. (c. 2011). Magnetic properties and 57Fe NMR studies of U-type hexaferrites. Journal of Magnetism and Magnetic Materials, 323*(16), 2210–2213. https://doi.org/10.1016/j.jmmm.2011.03.033 2212.*

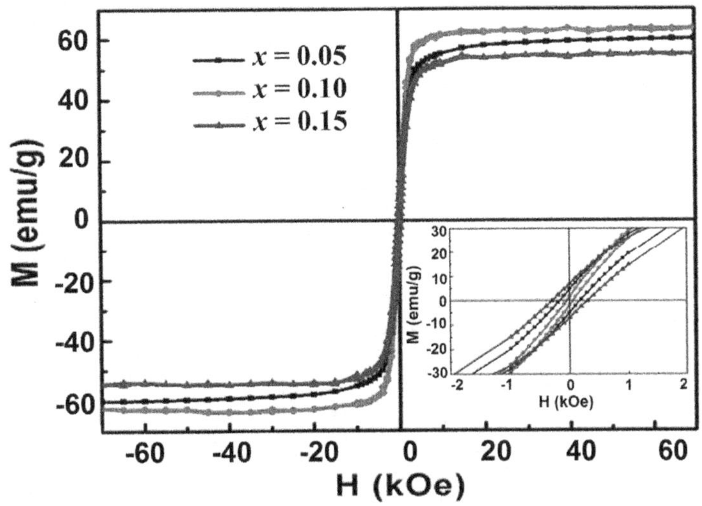

FIGURE 3.68

Vibrating sample magnetometry hysteresis loops of $(Ba_{1-3x}Bi_{2x})_4Co_2Fe_{36}O_{60}$ ($x = 0.05$, 0.10 and 0.15) U-type hexagonal ferrites.

From Kumar, S., & Chatterjee, R. (c. 2018). Complex permittivity, permeability, magnetic and microwave absorbing properties of Bi^{3+} substituted U-type hexaferrite. Journal of Magnetism and Magnetic Materials, 448, *88–93. https://doi.org/10.1016/j.jmmm.2017.06.123 16.*

stacking sequence of RSR*S*T*S* blocks, in which large cations of $4Ba^{2+}$ are distributed on the oxygen lattice, whereas $2Co^{2+}$ and $36Fe^{3+}$ cations occupy five different crystallographic sites, i.e., three octahedral sites ($12k$, $4a$, and $4f_{VI}$), one tetrahedral site ($8f_{IV}$), and one trigonal bipyramidal site ($2b$). It has been demonstrated that the superexchange interaction can couple three parallel spin ($12k$, $4a$, and $4f_{VI}$) and two anti-parallel spin ($4f_{VI}$ and $8f_{IV}$) sites through the O^{2-} ions. Since the substitution of Bi^{3+} with Ba^{2+} can transform the Fe^{3+} ions into Fe^{2+} ones at the octahedral sites, the magnetocrystalline anisotropy field increases with an increase in the Bi^{3+} ions, thereby enhancing the coercivity. In fact, the increase in the magnetocrystalline anisotropy field may be attributed to the existence of anisotropic Fe^{2+} ions at the octahedral lattice sites.

Fig. 3.69 shows the variations of complex permittivity and complex permeability as a function of frequency for ferrites. Evidently, Fig. 3.69A and B indicate that ε' and ε'' decrease with increasing the Bi^{3+} ions. In fact, the mechanism $Ba^{2+} + Fe^{3+} = Bi^{3+} + Fe^{2+}$ can prevent the movement of charge carriers between Fe^{3+} and Fe^{2+} ions, leading to an enhancement in the electrical resistivity. Thus, the real and imaginary parts of permittivity are reduced.

In addition, the permittivity value decreases as the Ba^{2+}cation with higher polarizability is replaced by the Bi^{3+} cation with lower polarizability. Likewise, the presence of pores in the microstructure can block the paths of electrical conduction, thereby reducing the permittivity. From Fig. 3.69C, the maximum value of μ' is

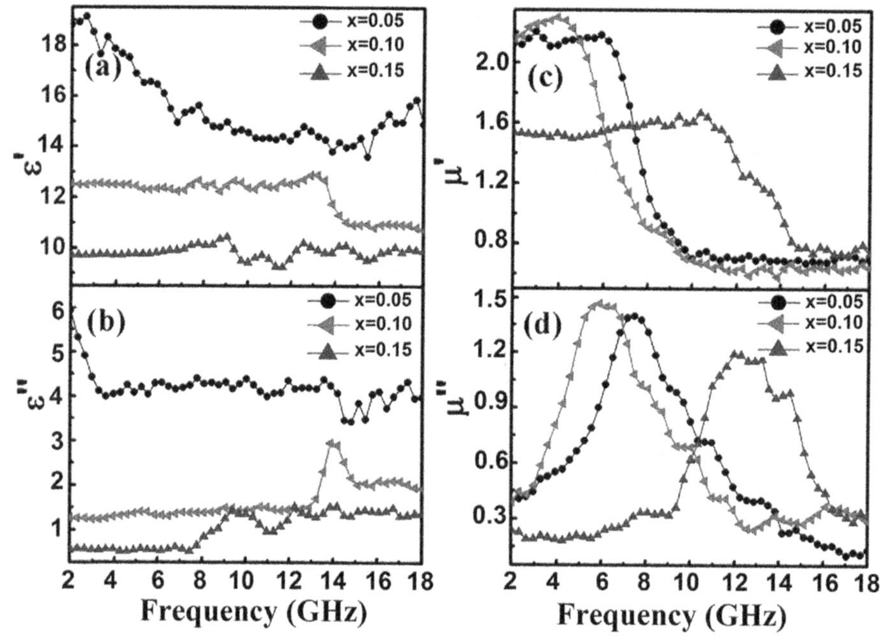

FIGURE 3.69

The variations of: (A and B) complex permittivity, and (C and D) complex permeability as a function of frequency for $(Ba_{1-3x}Bi_{2x})_4Co_2Fe_{36}O_{60}$.

From Kumar, S., & Chatterjee, R. (c. 2018). Complex permittivity, permeability, magnetic and microwave absorbing properties of Bi^{3+} substituted U-type hexaferrite. Journal of Magnetism and Magnetic Materials, 448, *88–93. https://doi.org/10.1016/j.jmmm.2017.06.123 16.*

seen to increase with an increase in the Bi^{3+} ions from $x = 0.05$ to $x = 0.10$. In fact, the stability of the Fe^{3+}-O-Fe^{3+} superexchange interaction can increase the hyperfine fields at $12k$ and $2b$ sites, arising from the higher permeability real part. Moreover, the peak value of μ'' increases with an increase in the Bi^{3+} ions to $x = 0.1$. Note that the reduction in the maximum value of permeability for the higher Bi^{3+}content ($x = 0.15$) can be attributed to the magnetic dilution.

Fig. 3.70 shows the variation of reflection loss versus frequency for ferrites. Interestingly, the substitution of Bi^{3+} ions increases the relative permeability to permittivity ratio, being excellent for enhanced impedance matching at the absorber interface. A maximum reflection loss of −35.5 dB is achieved for the $(Ba_{0.55}Bi_{0.30})_4Co_2Fe_{36}O_{60}$ hexaferrite at 10.6 GHz, having an absorber thickness of 1.7 mm. This hexaferrite also exhibits a broad bandwidth of 8.4 GHz (over X- and Ku- bands) for a reflection loss of higher than −10 dB. In addition, the matching frequency increases from 5.2 to 10.6 GHz with increasing the Bi^{3+} ions.

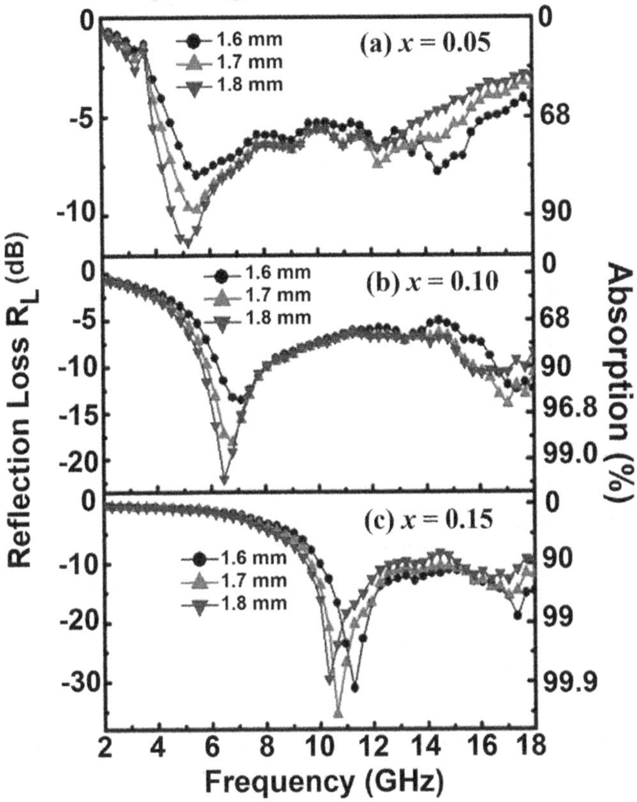

FIGURE 3.70

The variation of reflection loss versus frequency for $(Ba_{1-3x}Bi_{2x})_4Co_2Fe_{36}O_{60}$ hexagonal ferrites.

From Kumar, S., & Chatterjee, R. (c. 2018). Complex permittivity, permeability, magnetic and microwave absorbing properties of Bi^{3+} substituted U-type hexaferrite. Journal of Magnetism and Magnetic Materials, 448, 88–93. *https://doi.org/10.1016/j.jmmm.2017.06.123 16.*

3.4 Spinel ferrites

It is normally feasible to recognize differences between soft and hard magnetic ferrites. Soft ferrite has a small coercivity and is saturated by applying a relatively low magnetic field. Among many factors determining magnetic characteristics, relatively large permeability is extremely important for understanding the soft ferrite nature.

3.4.1 Simple spinel ferrites

Spinel ferrites contain two main groups, including simple and mixed ferrites. In simple ferrites, only one divalent cation, such as Ni^{2+}, Mg^{2+}, Co^{2+}, Mn^{2+}, Cu^{2+}, Cd^{2+}, and Zn^{2+}, is incorporated into the crystal structure of ferrites. Table 3.8 presents the site preference of ions and corresponding magnetic moment of spinel ferrites (Kotnala & Shah, 2015).

In the following, some typical simple spinel ferrite is studied. Lithium ferrite is considered a very important magnetic material because of its high Curie temperature, mechanical hardness, chemical stability, and high resistivity. Thus far, the lithium ferrite has been used in several electronic devices operating in both low- and high-frequency systems, including power transformers, rod antennas, microwave absorbers, and rechargeable lithium batteries.

In a research study, the role of annealing temperature and particle size on the magnetic properties of $Li_{0.5}Fe_{2.5}O_4$ nanoferrite was evaluated (Shirsath et al., 2011). Fig. 3.71 shows the variations of coercivity and saturation magnetization as a function of annealing temperature for $Li_{0.5}Fe_{2.5}O_4$ nanoparticles. In this respect, a maximum coercivity of 151 Oe is obtained at an annealing temperature of 1075 K. Further increasing the temperature reduces the coercivity of the nanoparticles while increasing their size. Thus, the magnetic structure of the $Li_{0.5}Fe_{2.5}O_4$ nanoparticles changes from a single domain to a multidomain when increasing the annealing temperature. Alternatively, the saturation magnetization of the nanoparticles ranges from 34 to 52 emu/g. In other words, it is seen that the saturation magnetization increases with increasing the annealing temperature.

Fig. 3.72 shows the variation of blocking temperature (T_B) as a function of particle size for the $Li_{0.5}Fe_{2.5}O_4$ nanoparticles. As observed, increasing the particle size increases T_B. In fact, nanoparticles with larger sizes show enhanced anisotropy energy, thus reducing the possibility of the occurrence of a leap across the magnetic anisotropy barrier. For this reason, the blocking temperature shifts to a higher temperature.

Copper ferrite is another type of spinel ferrite that has several potential technological applications. In a research study, copper ferrites consisting of nanoparticles were synthesized and then annealed at different temperatures (Naseri et al., 2013). VSM hysteresis loops of $CuFe_2O_4$ nanoparticles annealed at temperatures in the range of 673−823 K are shown in Fig. 3.73. The coercivity and remanence ratio (M_r/M_s) increase with increases in the annealing temperature. In this case, the coercivity increases from 391 to 1010 Oe with increasing the annealing temperature from 673 to 823 K. Meanwhile, it is observed that the saturation magnetization and the remanent magnetization increase with increasing the annealing temperature. The maximum saturation magnetization of the $CuFe_2O_4$ nanoparticles is obtained to be 21.63 emu/g. It is worth noting that the saturation magnetization of bulk $CuFe_2O_4$ is 33.4 emu/g. The difference between saturation magnetization values of the nanoscale and bulk $CuFe_2O_4$ may be attributed to the surface effects of nanoparticles. In fact, the spins existing on the surface of the nanoparticles are disordered. Hence, the saturation magnetization and remanent magnetization of larger particles are higher than those of smaller ones. In addition, the existence of impurity and secondary phases such as α-Fe_2O_3 and CuO can lead to a reduction in the magnetization.

Table 3.8 Ionic distribution along with the magnetic moment of some ferrites with spinel structure.

Ferrite	Ionic distribution		Magnetic moment of tetrahedral ions (μ_B)	Magnetic moment of octahedral ions (μ_B)	Magnetic moment per molecule (μ_B)	
	Tetrahedral site	Octahedral site			Theory	Experiment
Fe_3O_4	Fe^{3+}	$Fe^{2+} + Fe^{3+}$	5	4+5	4	4.1
$NiFe_2O_4$	Fe^{3+}	$Ni^{2+} + Fe^{3+}$	5	5+2	2	2.3
$Li_{0.5}Fe_{2.5}O_4$	Fe^{3+}	$Fe^{3+}_{1.5} + Li^{+}_{0.5}$	5	7.5+0	2.5	2.6
$MgFe_2O_4$	Fe^{3+}	$Mg^{2+} + Fe^{3+}$	5	0+5	0	1.1
$CoFe_2O_4$	Fe^{3+}	$Co^{2+} + Fe^{3+}$	5	3+5	3	3.7
$MnFe_2O_4$	Fe^{3+}	$Mn^{2+} + Fe^{3+}$	5	5+5	5	4.6
$CuFe_2O_4$	Fe^{3+}	$Cu^{2+} + Fe^{3+}$	5	1+5	1	1.3
$CdFe_2O_4$	Cd^{2+}	$Fe^{3+} + Fe^{3+}$	0	5−5	0	1
$ZnFe_2O_4$	Zn^{2+}	$Fe^{3+} + Fe^{3+}$	0	5−5	0	1

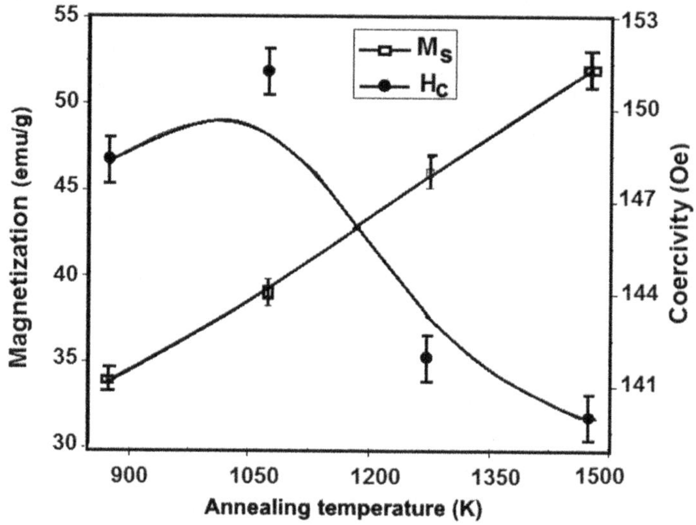

FIGURE 3.71

The variations of coercivity and saturation magnetization as a function of annealing temperature for $Li_{0.5}Fe_{2.5}O_4$ nanoparticles.

From Shirsath, S. E., Kadam, R., Gaikwad, A. S., Ghasemi, A., & Morisako, A. (c. 2011). Effect of sintering temperature and the particle size on the structural and magnetic properties of nanocrystalline $Li_{0.5}Fe_{2.5}O_4$. Journal of Magnetism and Magnetic Materials, *323(23), 3104–3108. https://doi.org/10.1016/j.jmmm.2011.06.065 3108.*

Likewise, increasing the particle size of the $CuFe_2O_4$ nanoparticles is expected to increase their M_s and M_r.

Magnesium ferrite ($MgFe_2O_4$) has a wide range of technological applications, such as heterogeneous catalysts, hyperthermia agents, humidity, and oxygen sensors (Maehara et al., 2002; Oliver et al., 1995; Shimizu et al., 1985; Verma et al., 2004). In research, low-temperature magnetic properties and Mössbauer spectroscopy analysis of $MgFe_2O_4$ were investigated (Liu et al., 2000). ZFC curves of $MgFe_2O_4$ nanoparticles with different particle sizes are shown in Fig. 3.74. At specific temperatures, the magnetization of $MgFe_2O_4$ ferrites with particle sizes in the range of 4–45 nm exhibits a maximum value. The magnetization of ferrite nanoparticles is then reduced, reflecting a paramagnetic behavior. From the inset of Fig. 3.74, the blocking temperature increases with increasing the particle size.

The temperature dependence of Mössbauer spectra of $MgFe_2O_4$ nanoparticles with a particle size of 20 nm is shown in Fig. 3.75. As can be seen, a notable doublet absorption component is formed at 410 K, indicating the superparamagnetic nature of the $MgFe_2O_4$ nanoparticles. Moreover, decreasing the measurement temperature to 300 K leads to the formation of the sextet pattern along with the doublet signature.

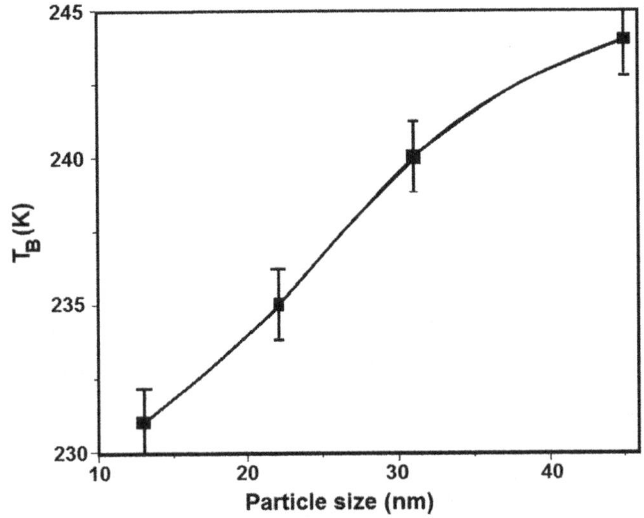

FIGURE 3.72

The variation of blocking temperature (T_B) as a function of particle size for $Li_{0.5}Fe_{2.5}O_4$ nanoparticles.

From Shirsath, S. E., Kadam, R., Gaikwad, A. S., Ghasemi, A., & Morisako, A. (c. 2011). Effect of sintering temperature and the particle size on the structural and magnetic properties of nanocrystalline $Li_{0.5}Fe_{2.5}O_4$. Journal of Magnetism and Magnetic Materials, 323*(23), 3104–3108. https://doi.org/10.1016/j.jmmm.2011.06.065 3108.*

3.4.2 Mixed spinel ferrites

The spinel ferrite could be synthesized with more than one divalent cation (e.g., Mn^{2+}, Zn^{2+}, Mg^{2+}, and Cu^{2+}). Accordingly, a group of ferrites such as Mn–Zn, Ni–Zn, Mn–Mg–Zn, Mg–Cu–Zn, and Ni–Cu–Zn with spinel structures can be prepared and used for the low and intermediate frequency range applications. The applications of several mixed soft ferrite materials, including the Mn–Zn and Ni–Zn ferrites, along with the operating frequency range, are listed in Table 3.9.

It has been found that the initial permeability of ferrite materials changes with frequency. In this regard, Fig. 3.76 shows the correlation between the operating frequency and permeability of several soft magnetic ferrites. Interestingly, the permeability of Mn–Zn ferrite is the highest compared with that of other ferrite materials. However, increasing the frequency considerably increases enhancement of the loss parameter. This is because the resistivity of Mn–Zn ferrite is the lowest among the ferrite group with the spinel structure, making it less appropriate for applications at higher frequencies. Therefore, the Mn–Zn ferrite materials with high permeability are mostly employed as filters and signal transformers (Müller et al., 2018).

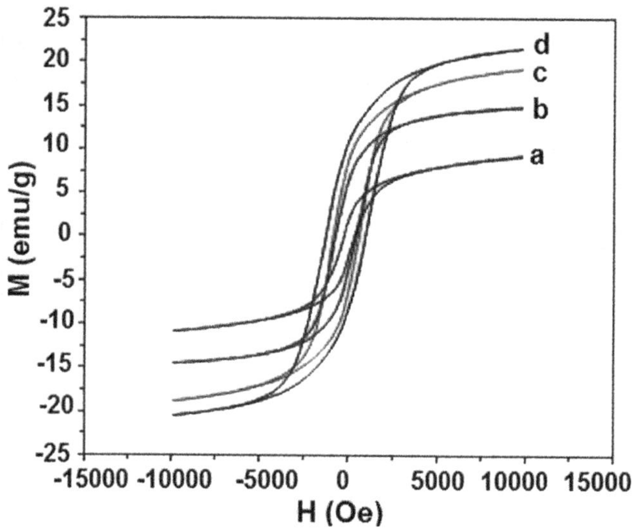

FIGURE 3.73

Vibrating sample magnetometry hysteresis loops of $CuFe_2O_4$ nanoparticles annealed at different temperatures: (A) 673 K, (B) 723 K, (C) 773 K, and (D) 823 K.

From Naseri, M. G., Saion, E. B., Ahangar, H. A., & Shaari, A. H. (c. 2013). Fabrication, characterization, and magnetic properties of copper ferrite nanoparticles prepared by a simple, thermal-treatment method. Materials Research Bulletin, 48*(4), 1439–1446. https://doi.org/10.1016/j.materresbull.2012.12.039 1444.*

3.4.2.1 Mn–Zn ferrite

The field strength, as well as the temperature, can influence the magnetic permeability of Mn–Zn ferrite. In a research study, the magnetic permeability of Mn–Zn ferrite was investigated (Kachniarz & Salach, 2016). Fig. 3.77 shows the variation of magnetic permeability as a function of the magnetic field measured at different temperatures (−20, 20, and 60°C). It can be observed that the permeability initially increases with increasing the field, reaching a maximum value. The permeability value then begins to decrease with further increasing the magnetic field. On the other hand, the maximum permeability is enhanced by increasing the temperature. Notably, the Mn–Zn ferrite has a maximum permeability value within the range of 4000–5000 at 60°C.

The sintering temperature has an important role in controlling the size of nanoparticles. In a study, the role of particle size on the magnetic properties of Mn–Zn ferrite is studied (Mathur et al., 2008). Fig. 3.78 shows the variation of particle size (D) versus sintering temperature for $Mn_{0.4}Zn_{0.6}Fe_2O_4$ nanoparticles. This result displays that increasing the sintering temperature continuously increases the size of nanoparticles.

Figs. 3.79 and 3.80 show the variation of saturation magnetization and coercivity as a function of particle size (*D*) for $Mn_{0.4}Zn_{0.6}Fe_2O_4$ nanoparticles, respectively. As observed, the saturation magnetization is enhanced with an increase in the particle

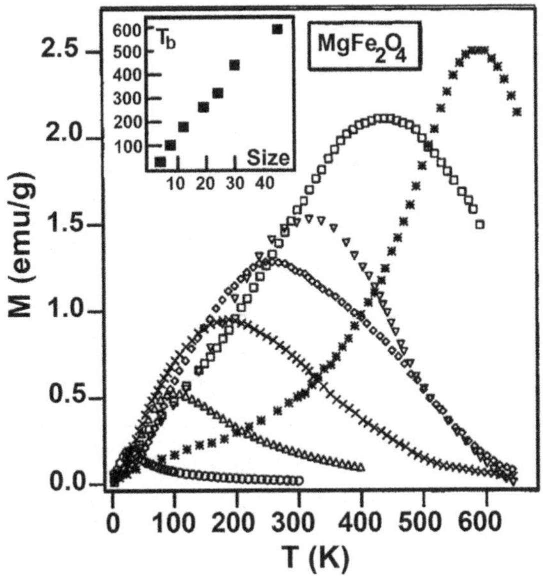

FIGURE 3.74

Zero-field-cooled curves of $MgFe_2O_4$ nanoparticles with different particle sizes (○ = 4 nm, Δ = 7.5 nm, × = 12 nm, ◇ = 19 nm, □ = 24 nm, ∇ = 30 nm, and * = 45 nm). The inset shows the variation of blocking temperature as a function of particle size.

From Liu, C., Zou, B., Rondinone, A. J., & Zhang, Z. J. (c. 2000). Chemical control of superparamagnetic properties of magnesium and cobalt spinel ferrite nanoparticles through atomic level magnetic couplings. Journal of the American Chemical Society, 122*(26), 6263–6267. https://doi.org/10.1021/ja000784g 6266.*

size. This variation behavior of saturation magnetization could be ascribed to the surface effects. In other words, increasing the size of the nanoparticles reduces the surface/volume ratio, which causes the saturation magnetization to increase. From Fig. 3.80, the coercivity has a maximum value at $D = 25.8$ nm. In fact, $Mn_{0.4}Zn_{0.6}Fe_2O_4$ nanoparticles with $D < 25.8$ nm show a single-domain behavior, whereas a multidomain configuration occurs for the nanoparticles having $D >$ 25.8 nm.

3.4.2.2 Ni–Zn ferrite

Ni–Zn ferrite is also a suitable candidate for practical application in the midfrequency range. The variations of μ' and μ'' as a function of frequency for Ni–Zn ferrite composites and sintered ferrite are shown in Fig. 3.81A and B, respectively (Nakamura et al., 1994). At the low-frequency region, the μ' of the sintered ferrite is approximately 1400. Increasing the frequency from 1 to 100 MHz decreases μ' to 10. Alternatively, a maximum of about 800 is obtained for the μ'' of the sintered ferrite at a frequency of around 1 MHz.

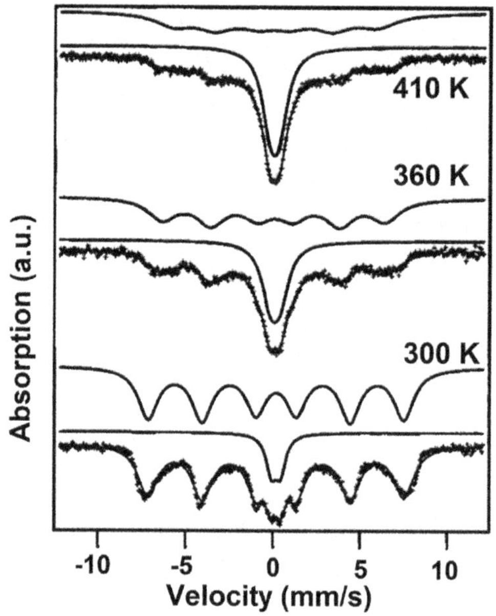

FIGURE 3.75

The temperature dependence of Mössbauer spectra of $MgFe_2O_4$ nanoparticles with a particle size of 20 nm.

From Liu, C., Zou, B., Rondinone, A. J., & Zhang, Z. J. (c. 2000). Chemical control of superparamagnetic properties of magnesium and cobalt spinel ferrite nanoparticles through atomic level magnetic couplings. Journal of the American Chemical Society, 122*(26), 6263–6267. https://doi.org/10.1021/ja000784g 6266.*

Fig. 3.81 shows dispersion curves of complex permeability (μ^*) for composite samples A, B, C, D, E, and F with ferrite contents of 50.0, 68.0, 72.0, 95.0, 96.5, and 98.0 vol%, respectively. To prepare these samples, the sintered cores were cracked and bonded together. Fig. 3.81 shows that the real and imaginary parts of μ^* are reduced at the low-frequency region when ferrite content decreases. Furthermore, μ' of the ferrite composite is higher than that of the sintered ferrite at the high-frequency region (e.g., ~100 MHz). Decreasing the ferrite content shifts the μ'' shoulder and μ'' peak frequencies to higher values. Similarly, the permeability dispersion of samples D, E, and F shows significant reductions in μ' and μ'' in the low-frequency region while shifting the μ'' peak frequency toward a higher frequency.

By superpositioning the two components involved in the magnetizing mechanisms through the spin rotation and DW movement, one can describe the permeability spectrum of polycrystalline ferrites as follows (Rado, 1953):

$$\mu(\omega) = 1 + \chi_{spin}(\omega) + \chi_{dw}(\omega) \tag{3.20}$$

Table 3.9 The applications of soft ferrite materials along with the operating frequency range.

Application	Frequency	Ferrite material
Communication coils	1 kHz–1 MHz	Mn–Zn
	0.5–80 MHz	Ni–Zn
Pulse transformers	–	Mn–Zn
		Ni–Zn
Transformers	~300 kHz	Mn–Zn
Flyback transformers	15.75 kHz	Mn–Zn
Deflection yoke cores	15.75 kHz	Mn–Zn
		Mn–Mg–Zn
		Ni–Zn
Antenna	0.4–50 MHz	Ni–Zn
Intermediate frequency transformers	0.4–200 MHz	Ni–Zn
Magnetic heads	1 kHz–10 MHz	Mn–Zn
Isolators	30 MHz–30 GHz	MnMgAl
Circulators		YIG
Splitters		YIG
Temperature responsive switches	–	MnCuZn

where χ_{spin} and χ_{dw} represent the spin rotation and DW components, respectively. When χ_{spin} and χ_{dw} are of relaxation and resonance types, respectively, one can have the following equations:

$$\chi_{\text{spin}}(\omega) = \frac{K_s}{1 + i(\omega/\omega^{\text{res}})} \tag{3.21}$$

$$\chi_{\text{dw}}(\omega) = \frac{K_{\text{dw}}\omega_{\text{dw}}^2}{\omega_{\text{dw}}^2 - \omega^2 + i\beta\omega} \tag{3.22}$$

where K_s is the static spin susceptibility, ω^{res} is the spin resonance frequency, K_{dw} is the static susceptibility of the DW movement, ω_{dw} is the DW resonance frequency, and β is the DW damping factor. By numerically fitting the permeability curve of the sintered ferrite (Fig. 3.81) to the above-mentioned equations, one can obtain the parameters $K_s = 839$ and $\omega^{\text{res}} = 9.17$ MHz for the spin rotation and $K_{\text{dw}} = 603$, $\omega_{\text{dw}} =$ 3.62 MHz, and $\beta = 5.34$ MHz for the DW movement components. By inserting these parameters into Eqs. (3.20)–(3.22), it is possible to calculate the permeability spectrum. In this regard, Fig. 3.82 shows the permeability spectrum calculated for the sintered Ni-Zn ferrite with the spin rotation and DW movement components. As can be inferred, the experimental dispersion and the calculated curve agree well. Moreover, at the high-frequency region ($>$ 10 MHz), the spin rotation component determines the complex permeability.

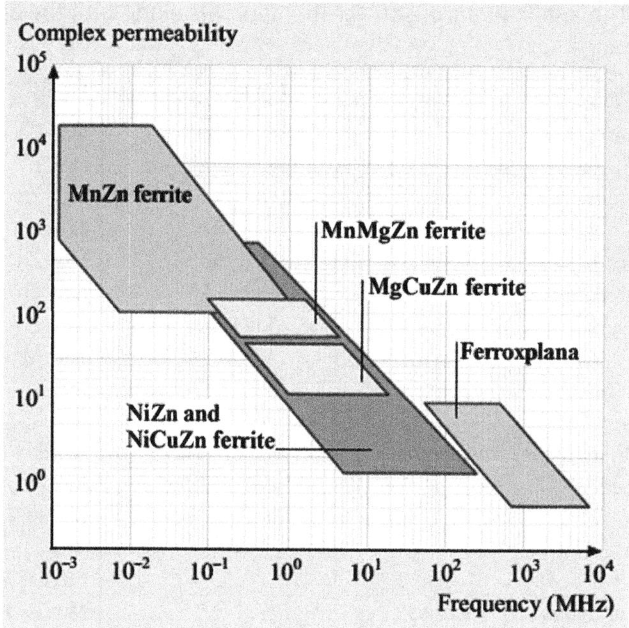

FIGURE 3.76

The correlation between the operating frequency and permeability of several soft magnetic ferrites.

From Müller, M., Harada, H., & Warlimont, H. (c. 2018). Magnetic materials. In Springer Handbook of Materials Data *(pp. 757–811). Springer. https://doi.org/10.1007/978-3-319-69743-7_22.*

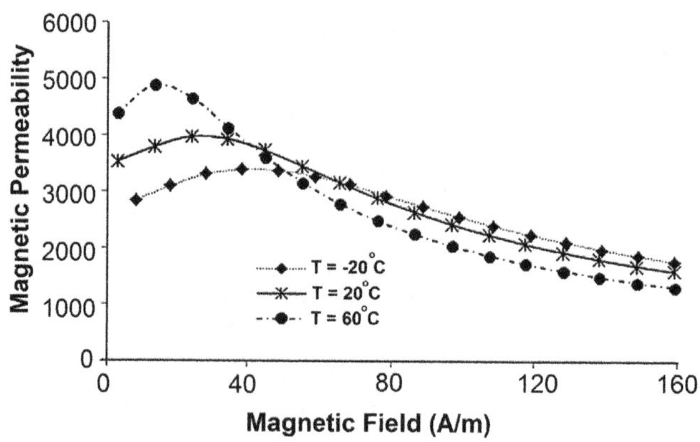

FIGURE 3.77

The variation of magnetic permeability as a function of magnetic field measured at different temperatures for Mn-Zn ferrite.

From Kachniarz, M., & Salach, J. (c. 2016). Influence of temperature and magnetizing field on the magnetic permeability of soft ferrite materials. In International Conference on Systems, Control and Information Technologies *2016 (pp. 698–704). Springer.*

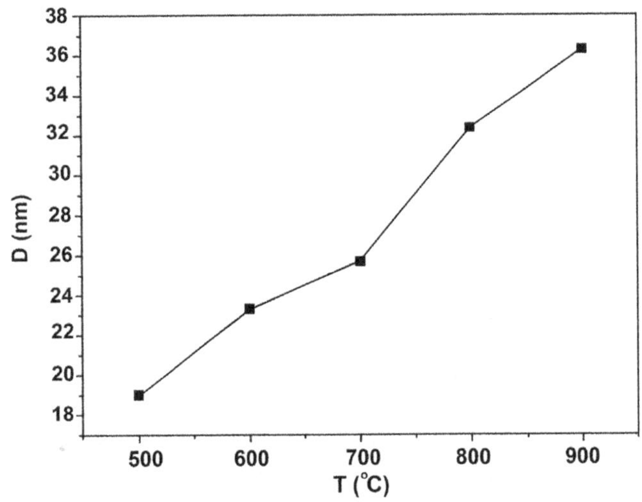

FIGURE 3.78

The variation of particle size (D) versus sintering temperatures for $Mn_{0.4}Zn_{0.6}Fe_2O_4$ nanoparticles.

From Mathur, P., Thakur, A., & Singh, M. (c. 2008). Effect of nanoparticles on the magnetic properties of Mn–Zn soft ferrite. Journal of Magnetism and Magnetic Materials, 320*(7), 1364–1369. https://doi.org/10.1016/j.jmmm.2007.11.008.*

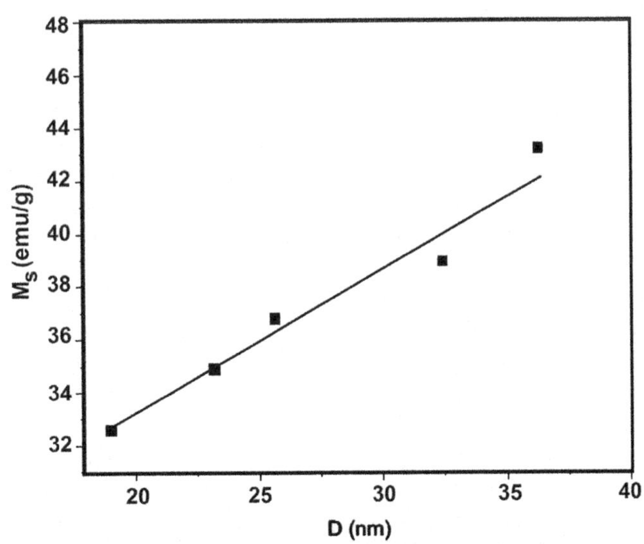

FIGURE 3.79

The variation of saturation magnetization (σ_S) as a function of particle size (D) for $Mn_{0.4}Zn_{0.6}Fe_2O_4$ nanoparticles.

From Mathur, P., Thakur, A., & Singh, M. (c. 2008). Effect of nanoparticles on the magnetic properties of Mn–Zn soft ferrite. Journal of Magnetism and Magnetic Materials, 320*(7), 1364–1369. https://doi.org/10.1016/j.jmmm.2007.11.008.*

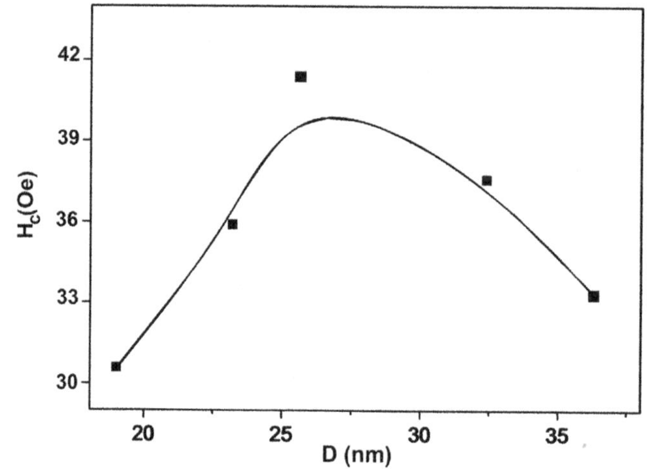

FIGURE 3.80

The variation of coercivity (H_c) as a function of particle size (D) for $Mn_{0.4}Zn_{0.6}Fe_2O_4$ nanoparticles.

From Mathur, P., Thakur, A., & Singh, M. (c. 2008). Effect of nanoparticles on the magnetic properties of Mn–Zn soft ferrite. Journal of Magnetism and Magnetic Materials, 320*(7), 1364–1369. https://doi.org/10.1016/j.jmmm.2007.11.008.*

The magnetic properties of Ni-Zn ferrites with a composition of $Ni_{0.5-x}Mn_xZn_{0.5}Fe_2O_4$ ($0.0 \leq x \leq 0.5$) have also been evaluated (Shirsath et al., 2010). Fig. 3.83 shows variations in saturation magnetization and coercivity as a function of Mn content (x). The variation in saturation magnetization can be dependent on the distribution of cations and the exchange interaction between ions. By substituting manganese ions with nickel ones, some of the iron ions migrate from tetrahedral to octahedral lattices sites, thus increasing the concentration of iron ions at the octahedral sites. In fact, the magnetic moment of the B sublattice is enhanced for $x \leq 0.3$. Moreover, the A–B interaction is reduced with an increase in the manganese concentration ($x > 0.3$). The B–B interaction also increases with an increase in the number of iron ions at the octahedral site, giving rise to a spin canting state. Accordingly, the magnetization of the Ni-Zn ferrites decreases when $x > 0.3$.

On the other hand, the coercivity decreases with an increase in the manganese concentration ($x < 0.3$), according to Fig. 3.83. In this case, the lowest value of coercivity occurs at $x = 0.3$. Further increasing the manganese concentration ($x > 0.3$) increases the resultant coercivity. In fact, the presence of porosity in the ferrites can act as a demagnetizing field. By increasing the manganese concentration, the porosity was observed to decrease, thereby reducing the coercivity. In addition, the coercivity is directly correlated with the magnetocrystalline anisotropy constant. It was illustrated that the number of Ni^{2+} ions is reduced with an increase in the

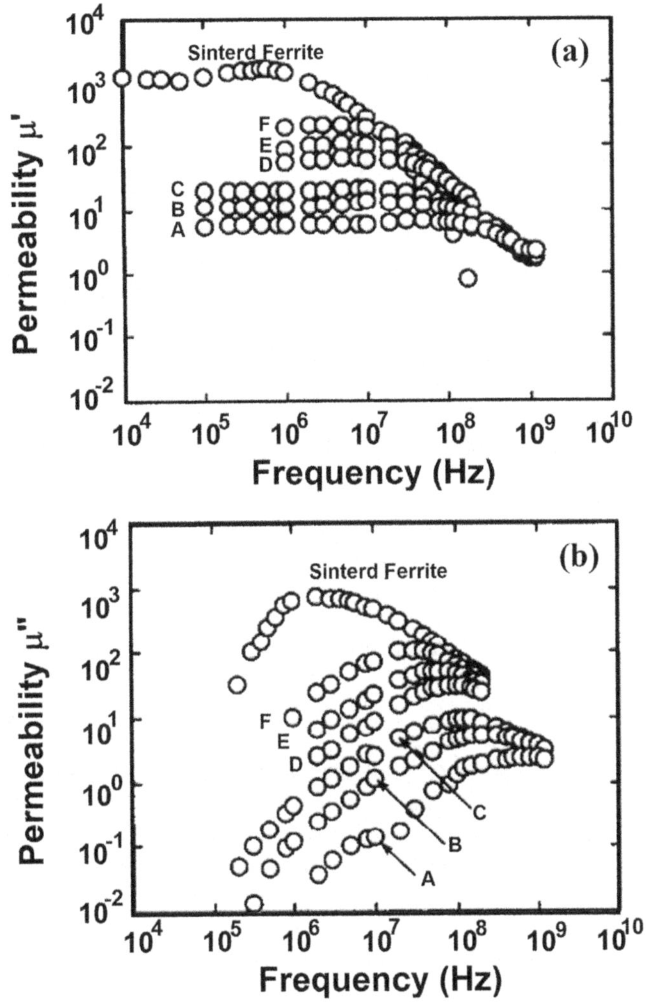

FIGURE 3.81

The variations of: (A) permeability real part (μ'), and (B) permeability imaginary part (μ'') as a function of frequency for Ni-Zn ferrite composites and sintered ferrite. (A, B, C, D, E, and F curves refer to composites with ferrite contents of 50.0, 68.0, 72.0, 95.0, 96.5, and 98.0 vol%, respectively).

From Nakamura, T., Tsutaoka, T., & Hatakeyama, K. (c. 1994). Frequency dispersion of permeability in ferrite composite materials. Journal of Magnetism and Magnetic Materials, 138*(3), 319–328. https://doi.org/10.1016/0304-8853(94)90054-X 322.*

Mn^{2+} content, leading to a decrease in the magnetocrystalline anisotropy constant. Thus, the coercivity of the Ni–Zn ferrites decreases when $x \leq 0.3$.

Fig. 3.84 shows the variation of initial permeability (μ_i) as a function of Mn content (x) in the $Ni_{0.5-x}Mn_xZn_{0.5}Fe_2O_4$ composition. Several parameters such as the

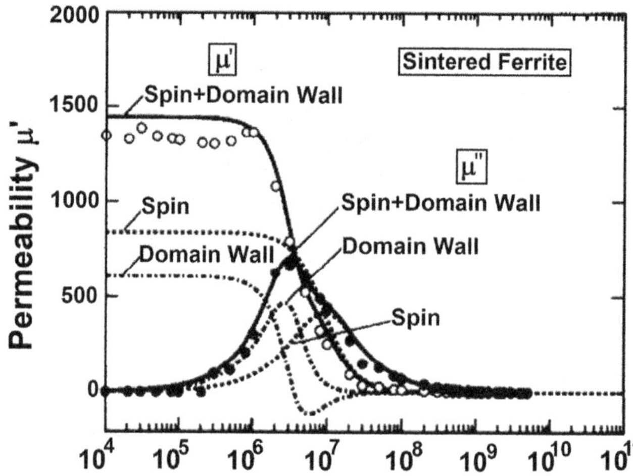

FIGURE 3.82

Comparison of experimental and calculated permeability spectra of a sintered ferrite. The open (μ') and solid (μ'') circles represent the experimental values. The total permeability comprising the spin rotation (*broken lines*) and domain wall movement (*dash-dotted lines*) components is shown by the solid lines.

From Nakamura, T., Tsutaoka, T., & Hatakeyama, K. (c. 1994). Frequency dispersion of permeability in ferrite composite materials. Journal of Magnetism and Magnetic Materials, 138*(3), 319–328. https://doi.org/10.1016/0304-8853(94)90054-X 322.*

composition, impurity, porosity, grain size, and magnetic anisotropy can affect the initial permeability of ferrite materials. Permeability has been shown to increase with decreases in magnetocrystalline anisotropy. Moreover, it increases with an increase in grain size. The permeability arising from the DW movement is expressed as follows:

$$\mu_i - 1 = \frac{3\pi M_s^2 D}{4\gamma} \tag{3.23}$$

where M_s is the saturation magnetization, D is the average particle size, and γ is the magnetic DW energy. It has been evidenced that the magnetic DW energy has a direct relation with the first magnetocrystalline anisotropy constant (Globus & Guyot, 1972). It is interesting to find that the permeability increases with an increase in the manganese concentration ($x < 0.3$). In fact, the magnetocrystalline anisotropy constant is reduced with an increase in the Mn^{2+} content, thus enhancing the initial permeability. The permeability of the $Ni_{0.5-x}Mn_xZn_{0.5}Fe_2O_4$ ferrite has a maximum value at $x = 0.3$. Further increasing the manganese concentration ($x > 0.3$) decreases the resultant permeability, arising from the role of particle size and porosity.

Fig. 3.85 shows the variation of Curie temperature (T_C) as a function of Mn content in the $Ni_{0.5-x}Mn_xZn_{0.5}Fe_2O_4$ composition for experimental and calculated data. As can be seen, the experimental and calculated Curie temperatures are reduced with

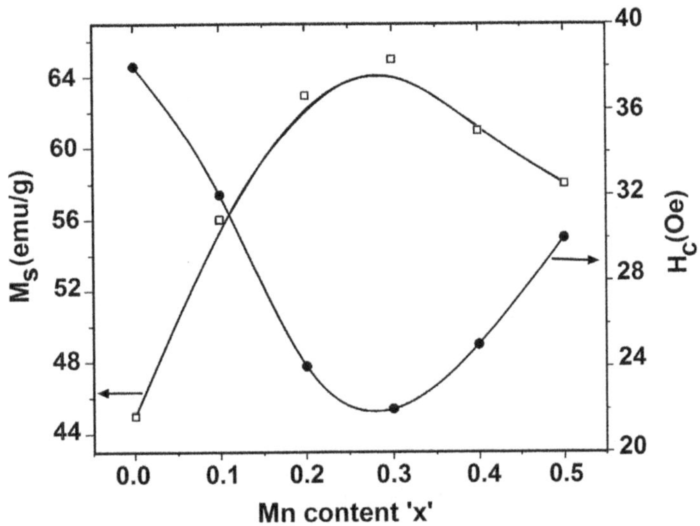

FIGURE 3.83

The variations of saturation magnetization (M_S) and coercivity (H_c) as a function of Mn content (x) in $Ni_{0.5-x}Mn_xZn_{0.5}Fe_2O_4$ composition.

From Shirsath, S. E., Toksha, B., Kadam, R., Patange, S., Mane, D., Jangam, G. S., & Ghasemi, A. (c. 2010). Doping effect of Mn^{2+} on the magnetic behavior in Ni–Zn ferrite nanoparticles prepared by sol–gel auto-combustion. Journal of Physics and Chemistry of Solids, *71(12), 1669–1675. https://doi.org/10.1016/j.jpcs.2010.08.016 1675.*

an increase in the manganese concentration. The Curie temperature of the ferrites can be influenced by the strength of the exchange interactions. The lattice parameter and the distances between the ions increase when increasing the Mn concentration, causing the A–B superexchange interactions to decrease. Consequently, the Curie temperature of the $Ni_{0.5-x}Mn_xZn_{0.5}Fe_2O_4$ composition is reduced with an increase in the manganese concentration.

In a separate research study, using a reverse micelle process, $Ni_{0.6}Zn_{0.4}Fe_{2-x}Cr_xO_4$ ($x = 0–0.5$) ferrite nanoparticles were synthesized (Ghasemi, 2016d). By measuring the AC magnetic susceptibility of typical ferrites ($x = 0$) as a function of temperature at different frequencies, one can study the magnetic dynamic behavior taking place in the nanoparticles. In this direction, Fig. 3.86 shows the variations of real and imaginary parts of AC susceptibility as a function of temperature for $Ni_{0.6}Zn_{0.4}Fe_2O_4$ nanoparticles measured at different frequencies (ranging between 40 and 1000 Hz) under an AC magnetic field ($H = 115$ Oe).

In this regard, a frequency-dependent peak is observed near the temperature of 105 K, being indicative of the blocking/freezing process occurring in superparamagnetic/spin glass magnetic systems. As well, increasing the frequency shifts this peak to a higher temperature (Pachpinde et al., 2014). In fact, one can define a threshold point for thermal activation based on the blocking temperature. Above the blocking

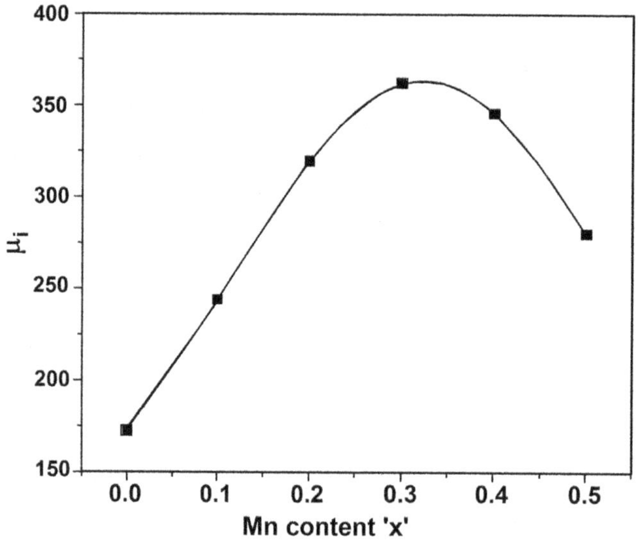

FIGURE 3.84

The variation of initial permeability (μ_i) as a function of Mn content (x) in the $Ni_{0.5-x}Mn_xZn_{0.5}Fe_2O_4$ composition.

From Shirsath, S. E., Toksha, B., Kadam, R., Patange, S., Mane, D., Jangam, G. S., & Ghasemi, A. (c. 2010). Doping effect of Mn^{2+} on the magnetic behavior in Ni–Zn ferrite nanoparticles prepared by sol–gel auto-combustion. Journal of Physics and Chemistry of Solids, *71(12), 1669–1675. https://doi.org/10.1016/j.jpcs.2010.08.016 1675.*

temperature, the thermal activation dominates the magnetocrystalline anisotropy, so the direction of the nanoparticle magnetization follows that of the magnetic field applied. In this way, the nanoparticles show superparamagnetic properties above the blocking temperature. Due to the random orientation of nanoparticles, decreasing the temperature reduces the overall susceptibility in the $Ni_{0.6}Zn_{0.4}Fe_2O_4$ composition, according to Fig. 3.86.

To evaluate the noninteracting and interacting behaviors of fine powders, Neél-Brown and Vogel-Fulcher models can be employed, respectively. Essentially, the blocking temperature (T_B) is related to the working frequency based on the Neél-Brown model as described in Chapter 2.

By matching the relaxation time with the working frequency at the maximum value of the out-of-phase susceptibility component, it is then possible to obtain T_B (Dormann et al., 1999). The attempt time ranges between $\tau_0 = 10^{-23}$ and 10^{-21} s. Nevertheless, the τ_0 values from the calculations are considered unphysical, as they are several orders of magnitude smaller than the values typically obtained for a noninteracting superparamagnetic nanoparticle assembly ($\tau_0 = 10^{-8}$–10^{-11} s). In other words, the Neél-Brown model is not capable of describing the dynamic behavior occurring in the $Ni_{0.6}Zn_{0.4}Fe_2O_4$ nanoparticle system. The existence of

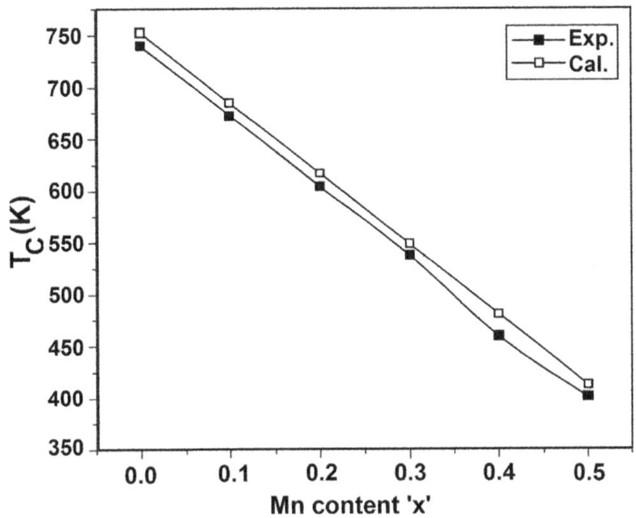

FIGURE 3.85

The variation of Curie temperature (T_C) as a function of Mn content (x) in the $Ni_{0.5-x}Mn_xZn_{0.5}Fe_2O_4$ composition for experimental and calculated data.

From Shirsath, S. E., Toksha, B., Kadam, R., Patange, S., Mane, D., Jangam, G. S., & Ghasemi, A. (c. 2010). Doping effect of Mn^{2+} on the magnetic behavior in Ni–Zn ferrite nanoparticles prepared by sol–gel auto-combustion. Journal of Physics and Chemistry of Solids, *71(12), 1669–1675. https://doi.org/10.1016/j.jpcs.2010.08.016 1675.*

strong interactions between the nanoparticles may be responsible for the experimental deviations from the Neél-Brown model, influencing the blocking temperatures by changing the potential barrier.

In the case of interacting nanoparticle systems, the Vogel-Fulcher model sheds light on the frequency dependence of T_B as expressed in Chapter 2. The experimental susceptibility data was fitted for the ferrite sample, leading to attempt time values ranging between 5.8×10^{-9} and 3.3×10^{-10}s. As a result, the presence of strong interactions between nanoparticles was confirmed, as the experimental data agreed well with the Vogel-Fulcher model. Moreover, for the ferrite samples synthesized by the reverse micelle method, T_B increases nearly linearly with increasing the chromium concentration, arising from the nanoparticle size effect. In fact, T_B is well known to be linearly proportional to the particle volume and the anisotropy constant. Since the particle size increases with increasing the Cr content, the aforementioned variation behavior of T_B versus chromium concentration can be justified accordingly.

In the last years, Ni–Cu–Zn ferrites have been employed in several electronic devices, such as chip inductors, owing to their high permeability in the RF region, high electrical resistivity, hard mechanical characteristics, and high Curie temperature.

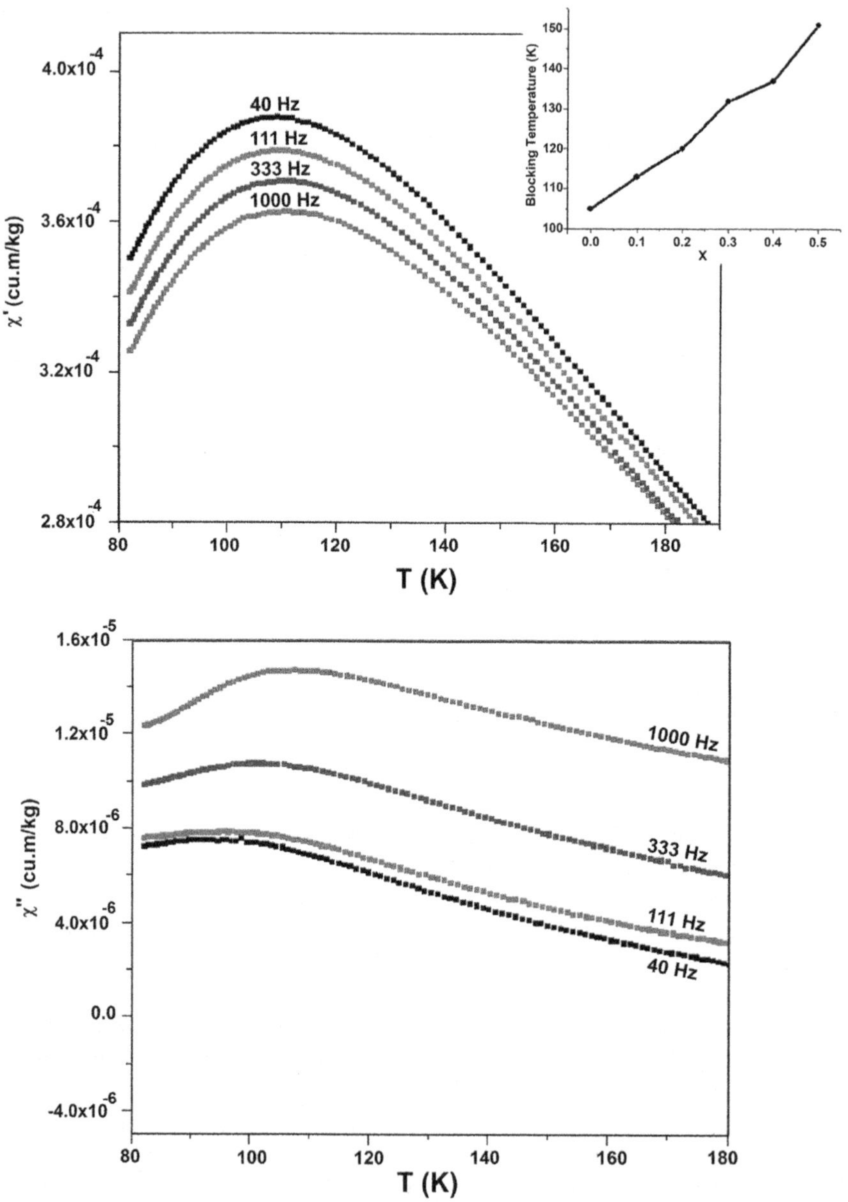

FIGURE 3.86

The variations of real and imaginary parts of AC susceptibility as a function of temperature for $Ni_{0.6}Zn_{0.4}Fe_2O_4$ nanoparticles measured at different frequencies. The variation of blocking temperature as a function of Cr content (x) is demonstrated in the inset.

From Ghasemi, A. (c. 2016). Real and imaginary parts of magnetic susceptibility of fine dispersed nanoparticles synthesized by reverse micelle: from superparamagnetic trend to ferrimagnetic state. Journal of Cluster Science, *27(3), 979–992. https://doi.org/10.1007/s10876-016-0978-y 12.*

In a research study, $Ni_{0.6-x}Cu_xZn_{0.4}Fe_2O_4$ ($x = 0$–0.5) nanoparticles were synthesized using a reverse micelle technique (Ghasemi, Ghasemi, & Paimozd, 2011). VSM hysteresis loops of the resulting nanoparticles measured at room temperature are shown in Fig. 3.87. Basically, the ferrite nanoparticle magnetization initially increases by increasing the magnetic field up to a specific value, thereby aligning the magnetic moments of the core with the field applied. Further increasing the magnetic field would influence the particle's shell, thus slowing down the slope in the increasing trend of magnetization. All the nanoparticles are superparamagnetic, since their coercivity and remanent magnetization are approximately zero. Moreover, they are not magnetically saturated after applying a field of 10 kOe. The saturation magnetization (M_s) value can be expressed by the following equation:

$$M = M_s\left(\frac{1-\alpha}{H}\right) \tag{3.24}$$

where M is the magnetization at the high magnetic field (H), and α is the fitting parameter.

From Fig. 3.87, the maximum magnetization decreases with increasing the copper content. While zinc and nickel cations occupy tetrahedral and octahedral sites, respectively, iron and copper ones fill both sites. Nevertheless, the iron and copper cations prefer to occupy the octahedral sites. For this reason, the B sublattice has a higher magnetization than the A sublattice. Since nickel has a higher magnetic

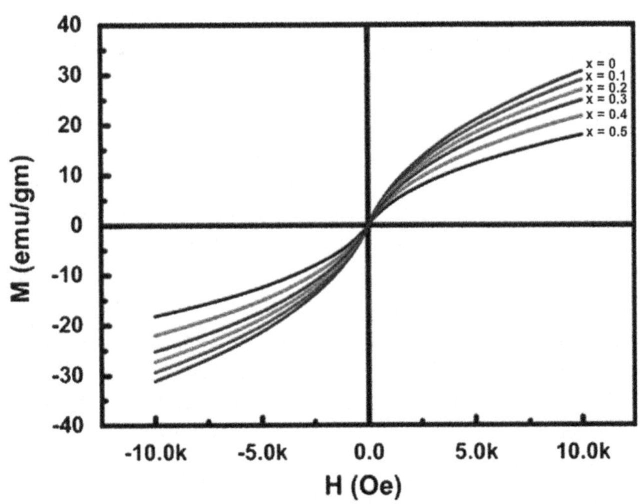

FIGURE 3.87

Vibrating sample magnetometry hysteresis loops of $Ni_{0.6-x}Cu_xZn_{0.4}Fe_2O_4$ ($x = 0$–0.5) nanoparticles measured at room temperature.

From Ghasemi, A., Ghasemi, E., & Paimozd, E. (c. 2011). Influence of copper cations on the magnetic properties of NiCuZn ferrite nanoparticles. Journal of Magnetism and Magnetic Materials, 323*(11), 1541–1545. https://doi.org/10.1016/j.jmmm.2011.01.014 1543.*

moment than copper, the saturation magnetization is reduced when nickel is substituted by copper. Owing to surface effects, nanocrystalline ferrites have lower saturation magnetization than their bulk counterparts. The core structure of nanoparticles comprises ferrimagnetically aligned spins, whereas the magnetic moments existing on the shell structure are disordered according to the magnetic core–shell model of nanoparticles. In fact, the occurrence of disordering in the shell structure is caused by the local chemical disorder and broken exchange bonds (Puli et al., 2015).

It should be noted that in the case of superparamagnetic particles, it is possible to estimate an upper bound for the average size (d_{max}) by taking into account the major contribution near the zero-field caused by the size distribution of the largest particles. This is represented by the following equation:

$$d_{max} = \left[\frac{18 k_B T}{\pi} \frac{(dM/dH)_{H=0}}{\rho M_s^2} \right]^{1/3} \tag{3.25}$$

where k_B, T, ρ, dM/dH, and M_s are the Boltzmann constant, temperature, the sample density, the slope of magnetization near the zero-field, and saturation magnetization of nanoparticles, respectively.

3.4.2.3 Zn–Co ferrite

In a research study using a sol–gel process, $Zn_{1-x}Co_xFe_2O_4$ ($0 \leq x \leq 1$) ferrite nanoparticles were synthesized, and their ^{57}Fe Mössbauer spectra were measured in the transmission mode at room temperature (Ghasemi et al., 2011b). Based on the relation $I_{(A)}/I_{[B]} = (f_{(A)}/f_{[B]})(\lambda/(2-\lambda))$, in which $I_{(A)}/I_{[B]}$ is the ratio of the intensities of subspectra for Fe^{3+} ions coordinated in tetrahedral (A) and octahedral [B] sites, $f_{[B]}/f_{(A)}$ is the ratio of the recoilless fractions, the degree of inversion (i.e., λ) has been calculated for the $Zn_{1-x}Co_xFe_2O_4$ ($x = 1$) ferrite nanoparticles by assuming that $f_{[B]}/f_{(A)} = 1$.

Fig. 3.88 shows a Mössbauer spectrum obtained from the $Zn_{1-x}Co_xFe_2O_4$ ($x = 0$) nanoparticles at room temperature. In this case, a paramagnetic doublet with an isomer shift (IS) of 0.19(6) mm/s along with a quadrupole splitting (QS) of 34(8) mm/s is observed in the Mössbauer spectrum, corresponding to Fe^{3+} ions located in the octahedral coordination of oxygen ions (Menil, 1985). The presence of the doublet component characterizes the preparation of the $ZnFe_2O_4$ compound (Šepelák et al., 1998). It should be noted that the ferrite sample contains a small amount (~5 wt.%) of the α-Fe_2O_3 (hematite) phase, giving rise to the sextet with the following hyperfine parameters: $IS = 0.23(5)$ mm/s, $QS = -0.08(7)$ mm/s, and $H = 51.28(5)$ T. For better clarity, Fig. 3.89 shows the central part of the Mössbauer spectrum.

By superpositioning of two magnetically split sextets with the hyperfine parameters $IS(A) = 0.13(3)$ mm/s, $H(A) = 48.46(9)$ T, $IS[B] = 0.22(6)$ mm/s, and $H[B] = 49.93(6)$ T, one can fit the second end-member spectrum of $Zn_{1-x}Co_xFe_2O_4$ ($0 \leq x \leq 1$) series, i.e., $CoFe_2O_4$. Accordingly, the local magnetic fields (H) at the Mössbauer nucleus positions can be detected via the splitting of the characteristic sextet spectra. The local magnetic fields of magnetic oxides are on the order of 50 T at room

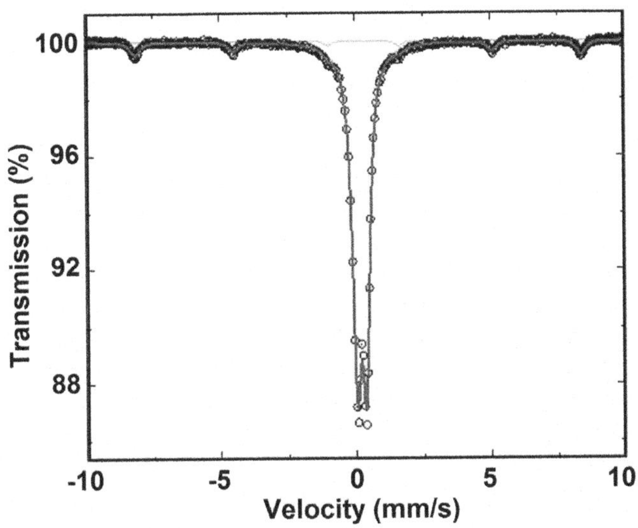

FIGURE 3.88

The Mössbauer spectrum obtained from $Zn_{1-x}Co_xFe_2O_4$ ($x = 0$) ferrite nanoparticles at room temperature.

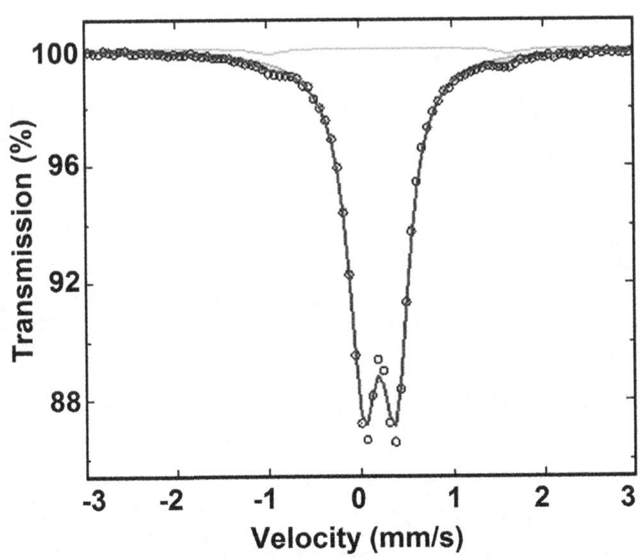

FIGURE 3.89

The Mössbauer spectrum central part of the $Zn_{1-x}Co_xFe_2O_4$ ($x = 0$) ferrite nanoparticles.

temperature and lower temperatures. In the case of spinel ferrites, the emergence of high local fields can be attributed to indirect-exchange (superexchange) interactions taking place between nuclear spins, transferring the spin density through oxygen anions.

The hyperfine parameters for the $CoFe_2O_4$ structure characterize ferric ions located in the (A) and [B] coordination of oxygen ions (Sawatzky et al., 1969). The fraction of Fe^{3+} cations located in (A) $CoFe_2O_4$ sites is 0.86(2) for $I_{(A)}/I_{[B]} = 43.13/56.87$. In other words, it is possible to have $(Co_{0.14}Fe_{0.86})[Co_{0.86}Fe_{1.14}]O_4$ as the structural chemical formula for $CoFe_2O_4$, featuring the cation site occupancy of the compound. In this way, the synthesized $CoFe_2O_4$ nanoparticles have a partial inverse spinel structure based on the obtained results of Fig. 3.90. In turn, this accords with a previous investigation into the $CoFe_2O_4$ material, indicating its cation distribution dependency on the thermal process history (Sawatzky et al., 1969).

In summary, a transition occurs from paramagnetic (represented by the doublet component) to ferrimagnetic behavior when the Co content increases. This behavior can be justified, as diamagnetic Zn^{2+} ions are substituted by magnetic Co^{2+} cations.

3.4.3 Cobalt ferrites

Cobalt ferrite is an important hard magnetic material, having a cubic spinel structure. It has attracted extreme attention due to its outstanding magnetic characteristics, including relatively high coercivity, large magnetocrystalline anisotropy, and

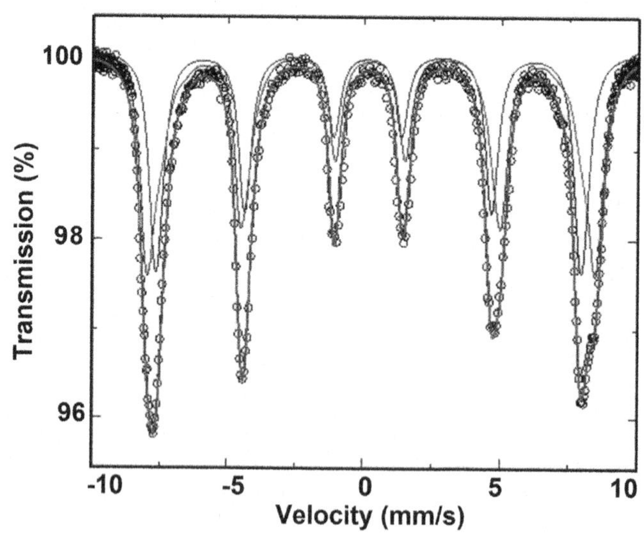

FIGURE 3.90

^{57}Fe Mössbauer spectrum obtained from $Zn_{1-x}Co_xFe_2O_4$ ($x = 1$) ferrite nanoparticles at room temperature. The subspectra of Fe^{3+} cations located in (A) and (B) $CoFe_2O_4$ sublattices are represented by *blue* and *red* colors, respectively.

intermediate saturation magnetization (Maaz et al., 2009). Moreover, cobalt ferrite is considered one of the most useful materials for various applications such as high-density magnetic recording media, microwave devices, high-sensitivity sensors, and biomedical industries (Caltun et al., 2008; Mata-Zamora et al., 2007; Shu & Qiao, 2009; Tomitaka et al., 2010).

3.4.3.1 Rare earth-substituted cobalt ferrites

In the last years, many studies have focused on the synthesis and magnetic properties of rare-earth cation-substituted cobalt ferrite nanoparticles. Although rare-earth elements are expensive, they can have a magnetic moment per atom exceeding that of iron, making them useful materials for technological applications.

In a research study carried out by my group, terbium-substituted cobalt ferrite nanoparticles with the composition of $CoFe_{2-x}Tb_xO_4$ ($x = 0-0.5$) were prepared by reverse micelle at room temperature (Sodaee et al., 2013). Fig. 3.91 shows typical Mössbauer spectra obtained from the resulting nanoparticles. As well, Table 3.10 presents Mössbauer parameters extracted from the spectra of the $CoFe_{2-x}Tb_xO_4$ ferrite nanoparticles. A six-line pattern indicating the ferrimagnetic behavior of the ferrite nanoparticles is observed in the Mössbauer spectra. The appearance of the two sextets in the spectra is attributed to iron distribution in the tetrahedral (1) and octahedral (2) sites of the cobalt ferrites.

Fig. 3.92 shows room-temperature VSM hysteresis loops of $CoFe_{2-x}Tb_xO_4$ ($x = 0-0.5$) ferrite nanoparticles. The coercive field and saturation magnetization values were extracted from the hysteresis loops, and the results are summarized in Table 3.11. As can be seen, increasing the terbium substitution content from $x = 0.1$ to 0.5 decreases the coercive field from 41 to 19 Oe.

The existence of the large magnetocrystalline anisotropy in the cobalt ferrite nanoparticles can be ascribed to the Co^{2+} ions located in the octahedral sites of the cubic spinel lattice structure. Due to the substitution of terbium cations, Co^{2+} ions migrate from octahedral to tetrahedral sites, leading to a reduction in the magnetocrystalline anisotropy constant as well as in the coercive field. In addition, increasing the terbium content increases the size of particles. When the particle size increases, DWs become less pinned due to a decrease in the volume fraction of grain boundaries. Alternatively, increasing the terbium content enhances the saturation magnetization of the ferrite nanoparticles. These results show that although terbium cations are diffused in both octahedral and tetrahedral sites, they prefer the latter. Thus, the substitution of iron cations with the terbium ones causes the saturation magnetization to increase from 75 emu/g ($x = 0.1$) to 90 emu/g ($x = 0.5$).

Recently, magnetic materials with one-dimensional structures have become important for performing research on advanced nanomagnets. This is because of the possibility of tuning the magnetic easy axis of a particle through its crystal structure.

Using a hydrothermal process without employing any template or surfactant, $CoFe_{2-x}Gd_xO_4$ ($x = 0-0.05$) nanorods have been prepared, and their morphological and magnetic properties have been investigated (Sodaee et al., 2016, 2017). Fig. 3.93

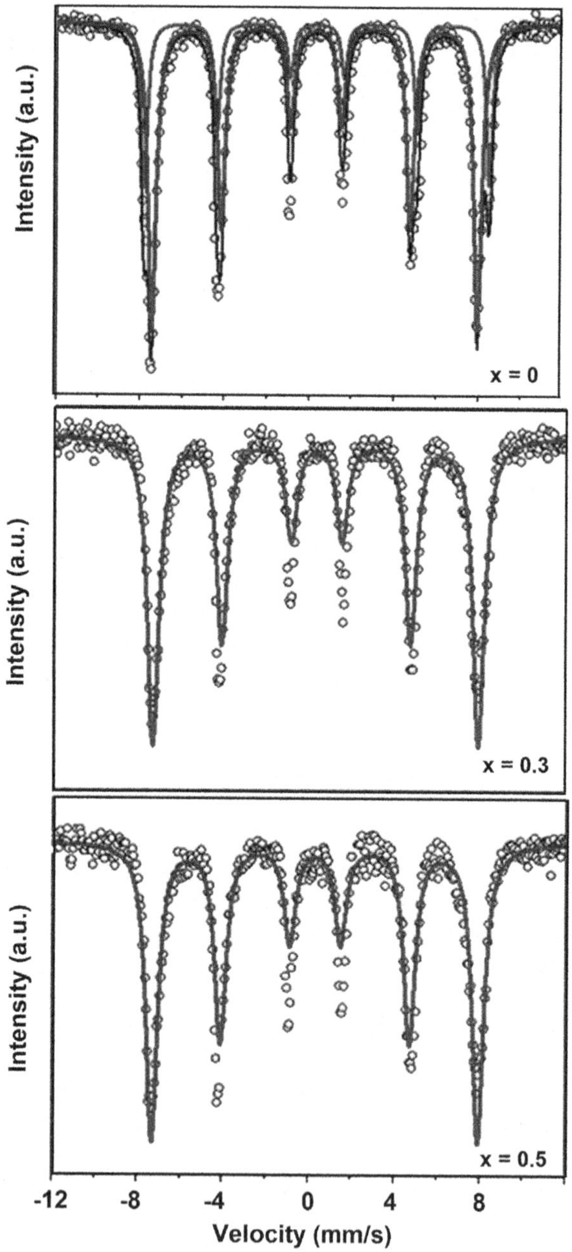

FIGURE 3.91

Mössbauer spectra obtained from $CoFe_{2-x}Tb_xO_4$ ferrite nanoparticles with $x = 0$, 0.3 and 0.5.

From Sodaee, T., Ghasemi, A., Paimozd, E., Paesano Jr, A., & Morisako, A. (c. 2013). The role of terbium cation substitution on the magnetic properties of cobalt ferrite nanoparticles. Journal of Magnetism and Magnetic Materials, 330, *169–173. https://doi.org/10.1016/j.jmmm.2012.10.050 4.*

Table 3.10 Mössbauer parameters extracted from the spectra of $CoFe_{2-x}Tb_xO_4$ ferrite nanoparticles with $x = 0$, 0.3 and 0.5.

Ferrite	Site	δ (mm/s)	QS (mm/s)	Γ (mm/s)	B_{hf} (T)	A (%)
$CoFe_2O_4$	Sextet 1	0.28	−0.02	0.44	47.4	70.3
	Sextet 2	0.34	0.01	0.35	50.6	28.7
$CoFe_{1.7}Tb_{0.3}O_4$	Sextet 1	0.3	0	0.64	47.4	100
$CoFe_{1.5}Tb_{0.5}O_4$	Sextet 1	0.31	0.01	0.65	47.2	100

shows field-emission scanning electron microscopy (FE-SEM) images of the resulting nanorods. As can be seen, arrays of ferrite nanorods have been synthesized over a large area by the hydrothermal process. During the formation of the nanorods, the morphology can be affected by the growth rate in different directions. It has been found that the average lengths of nanorods are reduced with increasing the gadolinium content.

Normally, crystallographic facets with lower energy surround a particle with a single-crystal structure, providing it with a particular shape. Upon the formation of a nucleus, it starts to grow in different directions. The growth of the nucleus is

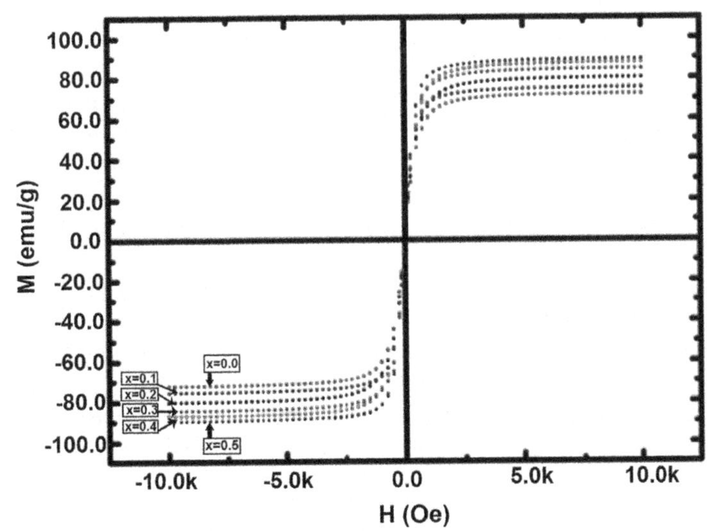

FIGURE 3.92

Room-temperature vibrating sample magnetometry hysteresis loops of $CoFe_{2-x}Tb_xO_4$ ($x = 0$–0.5) ferrite nanoparticles.

From Sodaee, T., Ghasemi, A., Paimozd, E., Paesano Jr, A., & Morisako, A. (c. 2013). The role of terbium cation substitution on the magnetic properties of cobalt ferrite nanoparticles. Journal of Magnetism and Magnetic Materials, 330, *169–173. https://doi.org/10.1016/j.jmmm.2012.10.050.*

Table 3.11 The coercive field and saturation magnetization of $CoFe_{2-x}Tb_xO_4$ ($x = 0.1–0.5$).

	x = 0.1	x = 0.2	x = 0.3	x = 0.4	x = 0.5
H_c (Oe)	41	36	29	24	19
M_s (emu/g)	75	81	84	88	90

known to be limited by the following factors: the atom addition kinetics in the interface, the capillarity, and the heat and mass diffusion.

For some materials, atomic attachment kinetics can play an important role in the growth process. When the atoms are transited from the liquid to the crystal, the transition kinetics is ignored due to its rapid occurrence. Accordingly, nonfaceted growth morphology is obtained, being typical for metal compounds. In the case of nonmetallic intermetallic or compounds such as ferrites, the faceted growth mode may take place, involving a large kinetic term. Nevertheless, this term may not necessarily dominate the growth process.

A nonfaceted substance is solidified when its atoms easily attach to any point on the surface. Under this condition, although the interplay between the capillarity factors and heat/solute diffusion processes mostly determines the crystal shape, the small anisotropy remaining in the physical properties (including the interface energy) leads to the formation of dendrite arms directed toward specific crystallographic directions. In the case of faceted materials, the added atoms grow rapidly, comprising high-index crystallographic planes. Consequently, the higher index planes disappear because of their quicker growth, giving rise to the formation of a characteristic crystal bounded by faces with the lowest growth rates. Frequently, impurities are capable of changing the growth process of specific planes, so that different growth forms can be induced for the same crystal structure.

Fig. 3.94 shows VSM hysteresis loops of $CoFe_{2-x}Gd_xO_4$ ($x = 0–0.05$) nanorods. As can be seen, the coercivity is enhanced from 590.35 to 826.10 Oe with increasing the gadolinium content from $x = 0$ to 0.05. Meanwhile, the saturation magnetization decreases from 72.36 to 53.61 emu/g with the substitution of gadolinium cations. Since the morphology of samples is in rod shapes, finding the importance of shape anisotropy on the magnetic analysis of substituted ferrites is required. Eq. (3.26) defines the shape anisotropy constant (K_s) as follows:

$$K_s = \frac{1}{2}\mu_0(N_a - N_c)M^2 \tag{3.26}$$

where N_c and N_a are the respective demagnetizing factors in c and a hexagonal directions. For a rod, N_c, N_b, and N_a demagnetizing factors can be defined as follows:

$$N_c = \frac{C_3}{m^2}(\ln(2m) - 1) \tag{3.27}$$

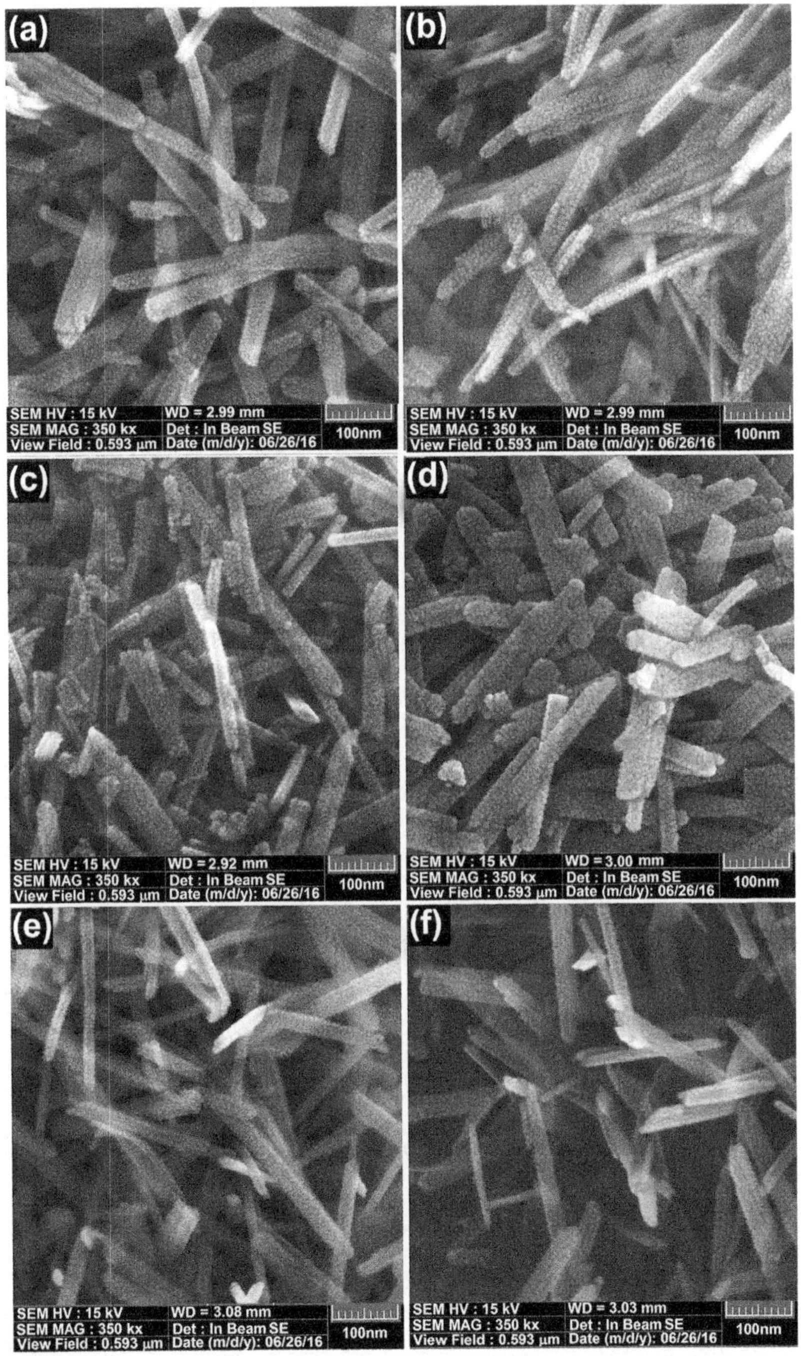

FIGURE 3.93

Field-emission scanning electron microscopy images obtained from $CoFe_{2-x}Gd_xO_4$ nanorods with: (A) $x = 0$, (B) $x = 0.01$, (C) $x = 0.02$, (D) $x = 0.03$, (E) $x = 0.04$, and (F) $x = 0.05$.

From Sodaee, T., Ghasemi, A., & Razavi, R. S. (c. 2016). Controlled growth of large-area arrays of gadolinium-substituted cobalt ferrite nanorods by hydrothermal processing without use of any template. Ceramics International, *42(15), 17420–17428. https://doi.org/10.1016/j.ceramint.2016.08.042 7.*

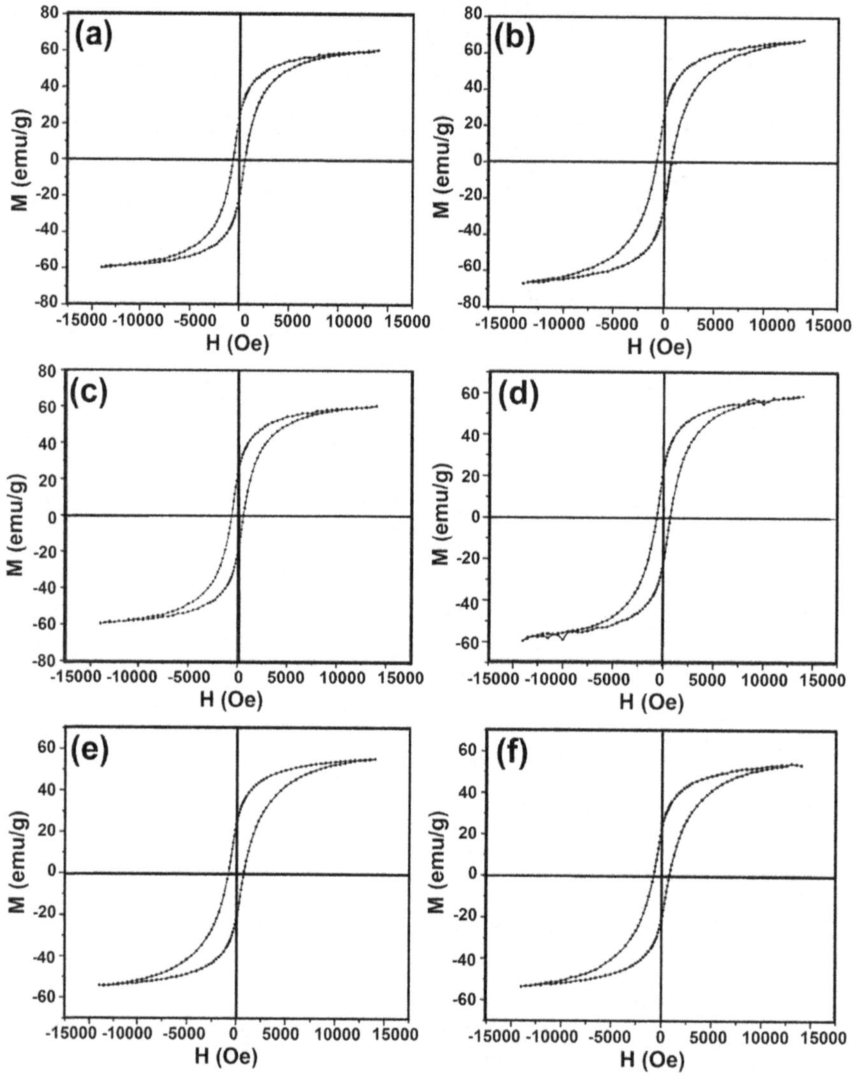

FIGURE 3.94

Vibrating sample magnetometry hysteresis loops obtained from $CoFe_{2-x}Gd_xO_4$ nanorods with: (A) $x = 0$, (B) $x = 0.01$, (C) $x = 0.02$, (D) $x = 0.03$, (E) $x = 0.04$, and (F) $x = 0.05$.
From Sodaee, T., Ghasemi, A., & Razavi, R. S. (c. 2016). Controlled growth of large-area arrays of gadolinium-substituted cobalt ferrite nanorods by hydrothermal processing without use of any template. Ceramics International, *42(15), 17420–17428. https://doi.org/10.1016/j.ceramint.2016.08.042 7.*

$$N_b = N_a \approx \frac{C_3}{2} \tag{3.28}$$

where $C_3 = 1$, and $m = c/a$. The easy direction of magnetization is along the c-axis. For any axis normal to c, the direction of magnetization is equally hard.

The shape anisotropy strength is determined by the axial ratio (c/a) of the object and the magnetization value (M). It has been evidenced that the shape anisotropy is less effective than the intrinsic magnetocrystalline anisotropy at a maximum c/a of $\sim$20. Thereby, the shape anisotropy constant (K_s) is smaller than the first magnetocrystalline anisotropy constant (K_1) when $c/a \sim 20$, and it can be stated that the main effect of gadolinium substitution on the magnetic features of cobalt ferrite nanorods is tuning magnetocrystalline anisotropy constant.

Using a hydrothermal process, $CoFe_{1.98}Gd_{0.02}O_4$ ferrite nanocrystals have also been prepared at different reaction temperatures ranging between 100 and 200°C (Sodaee et al., 2017). FE-SEM images of the resulting ferrite nanocrystals with the different reaction temperatures are shown in Fig. 3.95. As can be seen, the reaction temperature influences the morphology of the ferrite grains. Notably, while the grains are mostly in the form of rods at 100°C, two distinct types of grains with different sizes and shapes are formed at 120°C.

The rod morphology comprises elongated grains, whereas spherical shapes are mostly observed for the smallest grains. At the reaction temperature of 160°C, the $CoFe_{1.98}Gd_{0.02}O_4$ ferrite grains mostly have spherical morphology. While the ferrite grains are in the form of cubes at 200°C, a mixture of nanospheres and nanocubes are observed at 180°C. From Fig. 3.95, it is inferred that the morphological characteristics of the nanoparticles synthesized by the hydrothermal process are considerably influenced by the reaction temperature.

Fig. 3.96 shows VSM hysteresis loops of the $CoFe_{1.98}Gd_{0.02}O_4$ ferrite nanocrystals with different reaction temperatures. Moreover, Table 3.12 presents magnetic parameters (H_c and M_s) of the ferrite nanocrystals extracted from the hysteresis loops. As observed, the coercivity increases from 656.25 to 965.48 Oe with increasing the reaction temperature from 100 to 160°C. At $T = 200$°C, the nanocrystals have a coercive field of 793.26 Oe. Furthermore, M_s increases from 60.76 to 66.42 emu/g with increasing the reaction temperature from $T = 100$ to 160°C.

3.4.3.2 Transition cation-substituted cobalt ferrites

Transition cation-substituted cobalt ferrite nanoparticles have also been prepared, and their magnetic properties have been evaluated. For example, using a sol–gel autocombustion method, $Co_xNi_{1-x}Fe_2O_4$ ($x = 0-1$) ferrite nanoparticles have been synthesized (Torkian et al., 2017). XRD patterns of the resulting ferrite nanoparticles calcined at 1200°C for 2 h are shown in Fig. 3.97A. All the nanoparticle samples are seen to have pure cubic spinel structure without any secondary phases, reflecting (220), (311), (222), (400), (422), (511), and (440) planes. Among the peaks observed in the XRD patterns, the (311) plane has the maximum intensity, which is magnified in Fig. 3.97B. The substitution of cobalt cation increases the

FIGURE 3.95

Field-emission scanning electron microscopy images obtained from $CoFe_{1.98}Gd_{0.02}O_4$ nanocrystals with different reaction temperatures: (A) $T = 100°C$, (B) $T = 120°C$, (C) $T = 140°C$, (D) $T = 160°C$, (E) $T = 180°C$, and (F) $T = 200°C$.

From Sodaee, T., Ghasemi, A., & Razavi, R. S. (c. 2017). Shape factors dependence of magnetic features of $CoFe_{2-x}Gd_xO_4$ nanocrystals. Journal of Alloys and Compounds, 693, *1231–1242. https://doi.org/10.1016/j.jallcom.2016.10.102 23.*

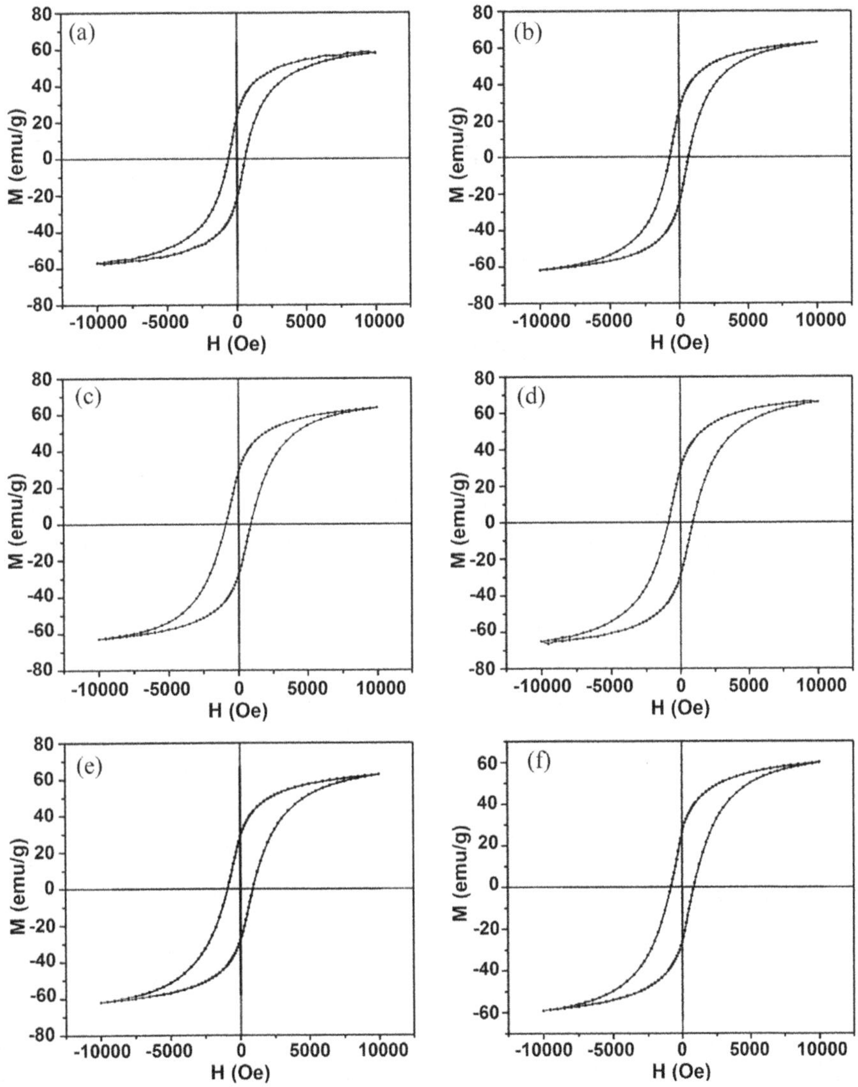

FIGURE 3.96

Vibrating sample magnetometry hysteresis loops obtained from $CoFe_{1.98}Gd_{0.02}O_4$ nanocrystals with different reaction temperatures: (A) $T = 100°C$, (B) $T = 120°C$, (C) $T = 140°C$, (D) $T = 160°C$, (E) $T = 180°C$, and (F) $T = 200°C$.

From Sodaee, T., Ghasemi, A., & Razavi, R. S. (c. 2017). Shape factors dependence of magnetic features of $CoFe_{2-x}Gd_xO_4$ nanocrystals. Journal of Alloys and Compounds, 693, *1231–1242. https://doi.org/10.1016/j.jallcom.2016.10.102 23.*

Table 3.12 Magnetic parameters of $CoFe_{1.98}Gd_{0.02}O_4$ nanocrystals with different reaction temperatures extracted from the hysteresis loop measurements.

	T = 100°C	*T* = 120°C	*T* = 140°C	*T* = 160°C	*T* = 180°C	*T* = 200°C
H_c (Oe)	656.25	691.1	859.37	965.48	907.45	793.26
M_s (emu/g)	60.76	62.21	62.29	66.42	61.51	59.03

lattice parameter from 8.34 to 8.38 Å, arising from the higher ionic radius of Co^{2+} (0.74 Å) than that of Ni^{+2} (0.69 Å).

After substituting, ferrite structure and magnetic properties, including saturation magnetization and coercivity, were enhanced. All the nanoparticle samples showed ferrimagnetic behavior at room temperature, being transformed from soft magnetic (for pure $NiFe_2O_4$) to relatively hard magnetic (for pure $CoFe_2O_4$) ferrites. The variations of saturation magnetization and coercivity as a function of cobalt content (x) are shown in Fig. 3.98. In this respect, substituting Ni^{2+} cations with Co^{2+} ones enhances saturation magnetization and coercivity. In other words, $NiFe_2O_4$ and

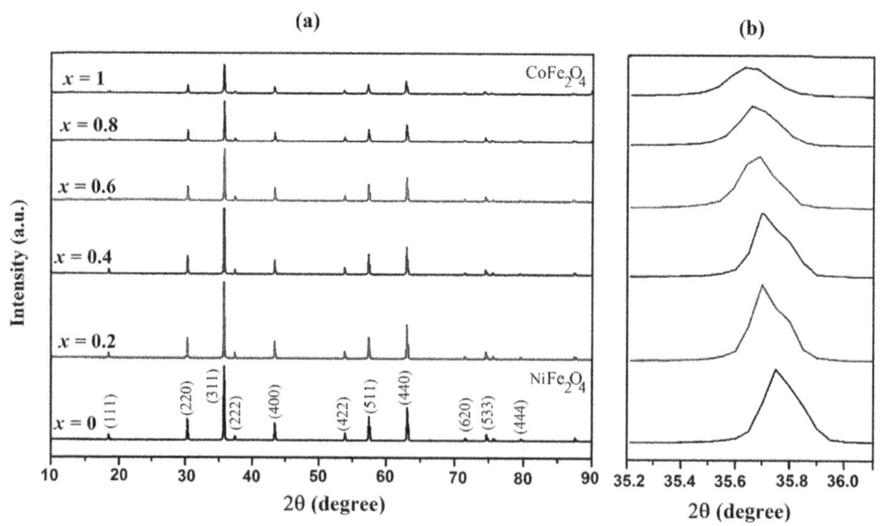

FIGURE 3.97

X-ray diffraction patterns obtained from $Co_xNi_{1-x}Fe_2O_4$ ($x = 0–1$) ferrite nanoparticles calcined at 1200°C for 2 h, and (B) the corresponding magnified view of the (311) peak.

From Torkian, S., Ghasemi, A., & Razavi, R. S. (c. 2017). Cation distribution and magnetic analysis of wideband microwave absorptive CoxNi1−xFe2O4 ferrites. Ceramics International, *43(9), 6987–6995. https://doi.org/10.1016/j.ceramint.2017.02.124 6994.*

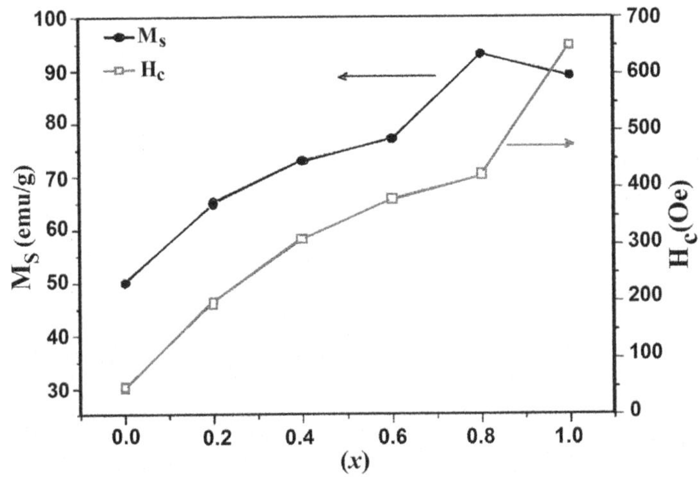

FIGURE 3.98

The variations of saturation magnetization and coercivity as a function of cobalt content (x) for $Co_xNi_{1-x}Fe_2O_4$ ferrite nanoparticles calcined at 1200°C for 2 h.

From Torkian, S., Ghasemi, A., & Razavi, R. S. (c. 2017). Cation distribution and magnetic analysis of wideband microwave absorptive $CoxNi_{1-x}Fe_2O_4$ *ferrites.* Ceramics International, *43(9), 6987–6995. https://doi.org/10.1016/j.ceramint.2017.02.124 6994.*

$CoFe_2O_4$ nanoparticles have saturation magnetization values of 33 and 89 emu/g, respectively. Moreover, the $Co_{0.8}Ni_{0.2}Fe_2O_4$ ferrite nanoparticles have respective saturation magnetization and coercivity of 93 emu/g and 420 Oe. Since Co^{2+} cations have a higher magnetic moment than Ni^{2+} ones, their substitution can increase the net magnetic moment. The enhancement in the coercivity of the substituted ferrites is attributed to the high Co^{2+} magnetic anisotropy, thus inducing a higher magnetocrystalline anisotropy constant. In addition, the particle size is reduced by incorporating Co^{2+}cations, which enhances coercivity.

Elsewhere, using a coprecipitation method, $Co_{1-x}Ni_xFe_2O_4$ ($0.0 \leq x \leq 1.0$ with a step of 0.2) nanoparticles with different sizes have been prepared (Mesbahinia et al., 2019). By employing a vibrating sample magnetometer instrument equipped with the FORC software, magnetic properties were measured at room temperature after applying a maximum field of $\pm$15 kOe. For better clarity, the step size was set to be the same for all the nanoparticle samples. Moreover, the smoothing factor was 2. Fig. 3.99 shows FORC diagrams of nanoparticles. The coercive field and interaction field distributions extracted from the diagrams are also depicted in Fig. 3.100.

Decreasing the Ni content widens the distribution of the coercive field significantly. Meanwhile, the distribution maximum point shifts to higher coercive fields, thus increasing the coercivity of the samples. Notably, in the case of the highest Ni content, two soft and hard magnetic segments are observed. The emergence of the soft phase can be attributed to an increase in the contribution of the

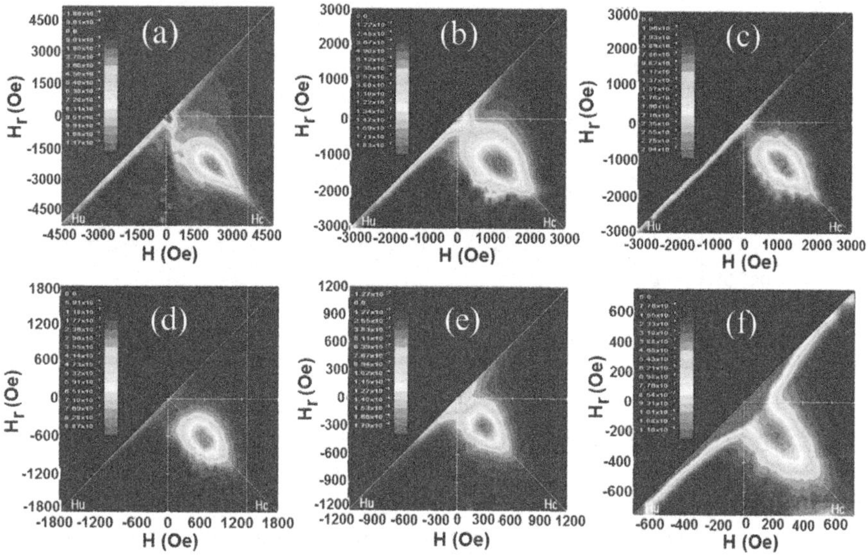

FIGURE 3.99

FORC diagrams obtained from $Co_{1-x}Ni_xFe_2O_4$ ferrite nanoparticles with (A) $x = 0$, (B) $x = 0.2$, (C) $x = 0.4$, (D) $x = 0.6$, (E) $x = 0.8$, and (F) $x = 1.0$.

From Mesbahinia, A., Almasi-Kashi, M., Ghasemi, A., & Ramezani, A. (c. 2019). First order reversal curve analysis of cobalt-nickel ferrite. Journal of Magnetism and Magnetic Materials, 473, *161–168. https://doi.org/10.1016/j.jmmm.2018.10.057 16.*

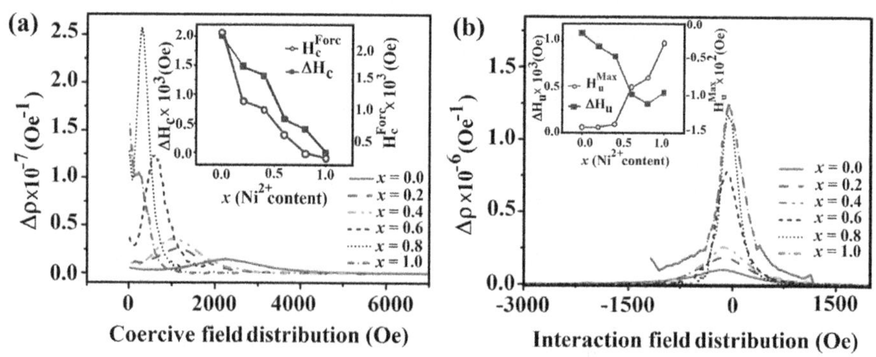

FIGURE 3.100

(A) Coercive field distribution, and (B) interaction field distribution of $Co_{1-x}Ni_xFe_2O_4$ ($0.0 \leq x \leq 1.0$) ferrite nanoparticles extracted from the FORC diagrams. The insets show FWHM values of coercive field (ΔH_c) and interaction field (ΔH_u) distributions, along with H_c^{FORC} and H_u^{max}.

From Mesbahinia, A., Almasi-Kashi, M., Ghasemi, A., & Ramezani, A. (c. 2019). First order reversal curve analysis of cobalt-nickel ferrite. Journal of Magnetism and Magnetic Materials, 473, *161–168. https://doi.org/10.1016/j.jmmm.2018.10.057 16.*

superparamagnetic phase of the sample, causing the coercivity to be reduced by almost 50%. Table 3.13 presents the magnetic parameters obtained from the FORC diagrams of ferrite powders.

According to Fig. 3.100B, the interaction field distribution is expanded by decreasing the Ni content while increasing the saturation magnetization. In this way, the correlation between the saturation magnetization and the interaction field distribution is evidently indicated, revealing the presence of magnetostatic-type interactions taking place between the nanoparticles.

It is inferred that the hard and soft phases are combined, so the magnetic behavior starts to be governed by the soft content when enhancing the Ni content of the samples. In turn, this leads to the formation of the superparamagnetic state. Increasing the Ni content decreases the magnetostatic interaction distribution width, which is compatible with the reduction observed in the magnetization of the nanoparticle samples. To investigate the magnetic characteristics of the nanoparticles in detail, it is possible to use the FORC analysis in order to estimate the superparamagnetic fraction by taking into consideration the reversible to irreversible components of magnetization. In other words, due to the increase in the Ni content, one can study the magnetic behavior of the nanoparticles when reducing their coercive field distribution to zero. In this regard, Fig. 3.101 shows the reversible and irreversible components extracted from the FORC analysis. The irreversible component is larger than the reversible one for $Co_{1-x}Ni_xFe_2O_4$ nanoparticles with $x = 0$, 0.2, 0.4, and 0.6. This can be attributed to the presence of large coercive fields along with the multiaxial crystalline anisotropy of ferrites with cubic structure. It should be noted that due to the nearly symmetric shape of the nanoparticles, the shape anisotropy contribution is negligible.

Due to the coercive field reduction, the contribution of superparamagnetic nanoparticles is enhanced. In turn, the reversible component percentage increases with increasing the Ni content. Further increasing the Ni content ($x = 0.8$ and 1.0) causes the reversible component to overcome the irreversible one, thus further reducing the coercive field of particles. Accordingly, the $Co_{1-x}Ni_xFe_2O_4$ nanoparticles with $x = 1.0$ appear to have more pronounced superparamagnetic behavior, being manifested in the FORC diagram as well as in the corresponding reversible component.

Using a spark plasma sintering (SPS) process, $Co_{1-x}Ni_xFe_2O_4$ ($0.0 \leq x \leq 1.0$) bulk samples have also been prepared (Mesbahinia et al., 2020). In this case, the powders were placed in a graphite die, followed by applying pressure of 50 MPa with a rate of 4.3 MPa/min. The samples were then heated from room temperature to 900°C having a heating rate of 50°C/min. The FORC analysis was carried out in order to investigate the irreversible magnetic characteristics of the samples (see Fig. 3.102).

To obtain high-resolution FORC diagrams, 60 reversal curves were at least plotted by choosing appropriate ΔH_r while applying a smoothing factor of 2. The coercive field and interaction field distributions extracted from the diagrams of $Co_{1-x}Ni_xFe_2O_4$ ($0.0 \leq x \leq 1.0$) bulk samples are shown in Fig. 3.103A and B, respectively.

Table 3.13 Magnetic parameters extracted from FORC diagrams of $Co_{1-x}Ni_xFe_2O_4$ ($0.0 \leq x \leq 1.0$) ferrite nanoparticles.

x	Nanoparticle sample	ΔH_c (Oe)	ΔH_u (Oe)	H_c^{FORC} (Oe)	H_u^{max} (Oe)	Irreversibility	Reversibility	Irreversibility/ reversibility
0	$CoFe_2O_4$	2000.47	1092.87	2268	−147	71	29	2.45
0.2	$Co_{0.8}Ni_{0.2}Fe_2O_4$	1485.29	939.43	1150	−147	70	30	2.33
0.4	$Co_{0.6}Ni_{0.4}Fe_2O_4$	1313.38	832.03	1001	−142	65	35	1.86
0.6	$Co_{0.4}Ni_{0.6}Fe_2O_4$	576.29	424.3	594	−90	64	36	1.78
0.8	$Co_{0.2}Ni_{0.8}Fe_2O_4$	398.55	321.46	282	−77	49	51	0.96
1	$NiFe_2O_4$	134	446.14	221	−28	32	68	0.47

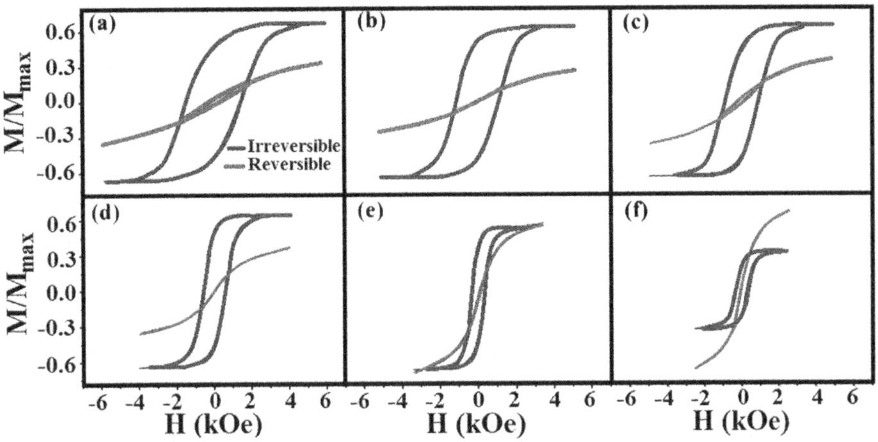

FIGURE 3.101

Comparison of irreversible and reversible hysteresis loops obtained from $Co_{1-x}Ni_xFe_2O_4$ ferrite nanoparticles with x: (A) $x = 0$, (B) $x = 0.2$, (C) $x = 0.4$, (D) $x = 0.6$, (E) $x = 0.8$, and (F) $x = 1$.

From Mesbahinia, A., Almasi-Kashi, M., Ghasemi, A., & Ramezani, A. (c. 2019). First order reversal curve analysis of cobalt-nickel ferrite. Journal of Magnetism and Magnetic Materials, 473, *161–168. https://doi.org/10.1016/j.jmmm.2018.10.057 16.*

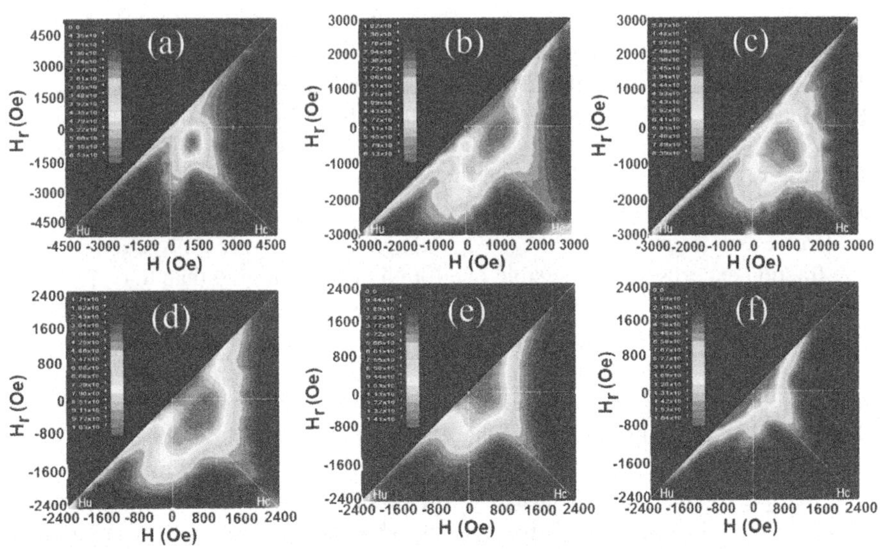

FIGURE 3.102

FORC diagrams obtained from $Co_{1-x}Ni_xFe_2O_4$ bulk samples with (A) $x = 0$, (B) $x = 0.2$, (C) $x = 0.4$, (D) $x = 0.6$, (E) $x = 0.8$, and (F) $x = 1$.

From Mesbahinia, A., Almasi-Kashi, M., Ghasemi, A., & Ramazani, A. (c. 2020). FORC investigation of Co-Ni bulk ferrite consolidated by spark plasma sintering technique. Journal of Magnetism and Magnetic Materials, 497*, 165976. https://doi.org/10.1016/j.jmmm.2019.165976 17.*

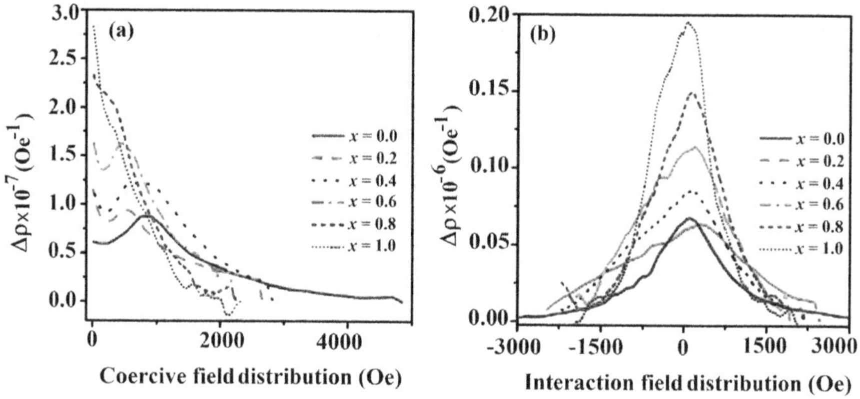

FIGURE 3.103

(A) Coercive field distribution, and (B) interaction field distribution of $Co_{1-x}Ni_xFe_2O_4$ ($0.0 \leq x \leq 1.0$) bulk ferrites prepared by the spark plasma sintering process.

From Mesbahinia, A., Almasi-Kashi, M., Ghasemi, A., & Ramazani, A. (c. 2020). FORC investigation of Co-Ni bulk ferrite consolidated by spark plasma sintering technique. Journal of Magnetism and Magnetic Materials, *497, 165976. https://doi.org/10.1016/j.jmmm.2019.165976 17.*

Moreover, Table 3.14 presents the magnetic parameters of the samples obtained using the FORC measurements. As well, reversible, irreversible, and total hysteresis loops of the samples are depicted in Fig. 3.104, allowing for the calculation of reversibility and irreversibility, according to Table 3.14. In this regard, the reversibility increases from 52% to 81% with increasing the Ni content from $x = 0.0$ to 1.0. Meanwhile, H_c^{FORC} is reduced from 839 to about 0 Oe, leading to an enhancement in the soft content. By comparing the FORC results obtained from the powder and bulk cobalt-nickel ferrite samples, one can justify their different magnetic behavior. In fact, increasing the substitution of Co with Ni decreases the coercive field values while enhancing the corresponding reversibility. In this case, the reversible fraction exceeds the irreversible one for Ni contents higher than 0.4, giving rise to the emergence of a superparamagnetic phase in the resulting Ni-rich ferrites. Therefore, the increase in the Ni content together with the decrease in the magnetocrystalline anisotropy leads to an increase in the superparamagnetic phase, thus reducing the corresponding coercivity from 1673 Oe (for $CoFe_2O_4$) to 134 Oe (for $NiFe_2O_4$). It has been reported that the crystalline anisotropy of the Co ferrite is almost two times larger than that of the Ni ferrite (Yasemian et al., 2019).

From Fig. 3.103A, FORC analysis of the bulk ferrites with the Co-rich content shows decreased coercive field distributions, whereas Ni-rich ferrites have relatively constant coercive field distributions on average. Alternatively, Fig. 3.103B reveals that the interaction field distribution of all the ferrites considerably increases compared with the powder state. Thus, the SPS process of the Co-rich ferrite powders (without having the superparamagnetic phase) sharply increases the interactions, thus leading to an increase in the ΔH_u and reversible fraction. In turn, a

Table 3.14 Magnetic parameters extracted from FORC diagrams of $Co_{1-x}Ni_xFe_2O_4$ ($0.0 \leq x \leq 1.0$) bulk ferrites.

x	Sample	ΔH_u (Oe)	H_c^{FORC} (Oe)	$H_c^{Irreversible}$ (Oe)	Irreversibility	Reversibility	Irreversibility/reversibility
0	$CoFe_2O_4$	1611.31	839	1083.98	48	52	0.92
0.2	$Co_{0.8}Ni_{0.2}Fe_2O_4$	2657.14	504	909.62	46	54	0.85
0.4	$Co_{0.6}Ni_{0.4}Fe_2O_4$	2106.09	672	899.37	53	47	1.12
0.6	$Co_{0.4}Ni_{0.6}Fe_2O_4$	1983.48	450	653.63	40	60	0.66
0.8	$Co_{0.2}Ni_{0.8}Fe_2O_4$	1440.24	0	498.02	31	69	0.45
1	$NiFe_2O_4$	1156.15	0	433.2	19	81	0.23

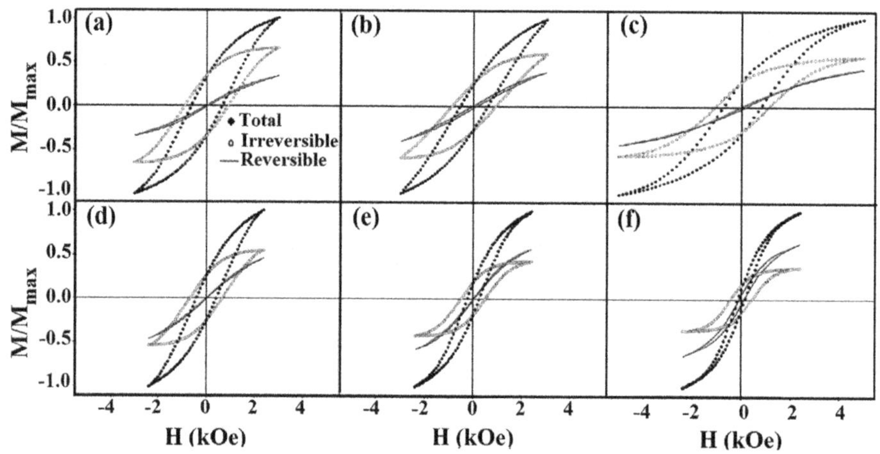

FIGURE 3.104

Reversible, irreversible, and total hysteresis loops of $Co_{1-x}Ni_xFe_2O_4$ bulk ferrites with: (A) $x = 0$, (B) $x = 0.2$, (C) $x = 0.4$, (D) $x = 0.6$, (E) $x = 0.8$, and (F) $x = 1$.

From Mesbahinia, A., Almasi-Kashi, M., Ghasemi, A., & Ramazani, A. (c. 2020). FORC investigation of Co-Ni bulk ferrite consolidated by spark plasma sintering technique. Journal of Magnetism and Magnetic Materials, *497, 165976. https://doi.org/10.1016/j.jmmm.2019.165976 17.*

multidomain state is formed compared with the single-domain state of the powders, reducing the corresponding coercivity by 50%. In the case of the Ni-rich ferrites, the SPS process reduces the superparamagnetic phase (thus increasing the coercivity) while forming the multidomain state with decreased coercivity. Therefore, the competition occurring between the increase and decrease in the coercivity leads to the constant coercive field distribution of the Ni ferrite. This is further confirmed by the variation behavior of the coercivity in the powder state compared with the bulk one when increasing the Ni content.

Elsewhere, a sol–gel process has been used to synthesize $Co_{1-x}Ni_{x/2}Sr_{x/2}Fe_2O_4$ ($x = 0, 0.4, 1$) nanoparticles (Ghasemi et al., 2012). Typical Mössbauer spectra of the resulting nanoparticles measured at room temperature are depicted in Fig. 3.105. The Mössbauer hyperfine parameters extracted from the spectra are presented in Table 3.15. When $x \leq 0.20$, one can fit the Mössbauer spectra with two distinct sextets. On the other hand, the fit procedure comprises a magnetic hyperfine field distribution and a doublet pattern when $x > 0.20$.

Two magnetic components are observed in the Mössbauer spectrum of the stoichiometric cobalt ferrite ($x = 0$). The hyperfine parameters of the cobalt ferrite nanoparticles agree well with those reported in previous investigations into the spinel cobalt ferrite compound. The relative areas in the spectrum show site occupancy expected for 1 (Site A): 2 (Site B). By introducing 20% of Ni-Sr into the structure, the corresponding Mössbauer pattern does not considerably change. However, a minimum variation is observed in the resultant hyperfine parameters as well as in the

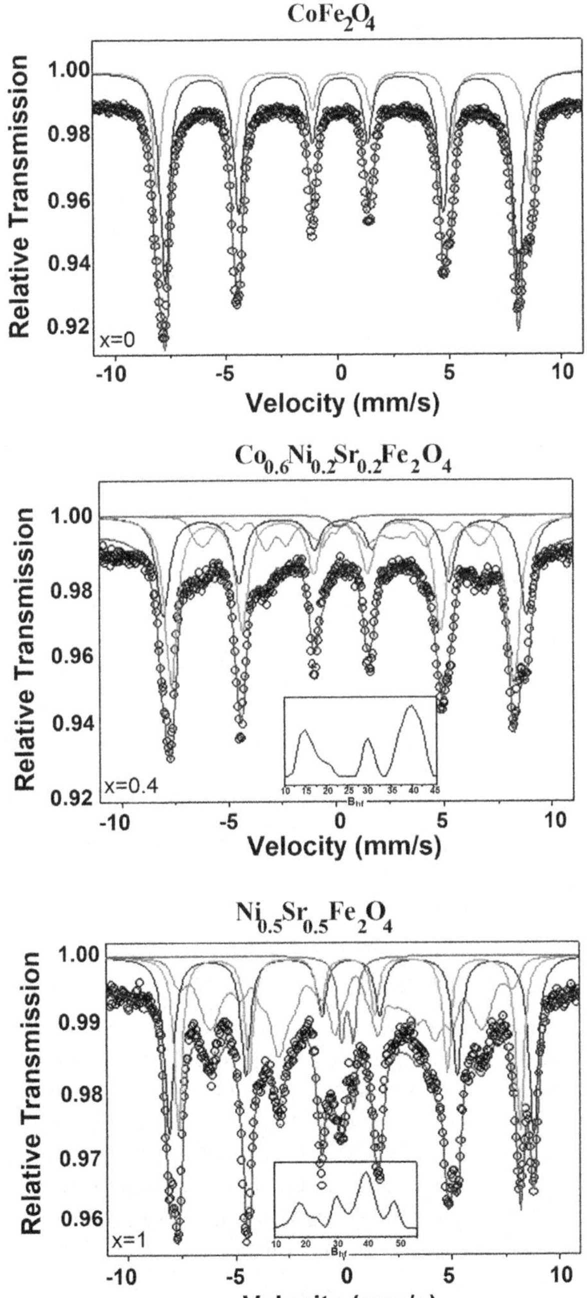

FIGURE 3.105

Typical Mössbauer spectra of $Co_{1-x}Ni_{x/2}Sr_{x/2}Fe_2O_4$ ($x = 0, 0.4, 1$) nanoparticles.

From Ghasemi, A., Paesano Jr, A., & Machado, C. F. C. (c. 2012). Structural and magnetic characteristics of $Co_{1-x}Ni_{x/2}Sr_{x/2}Fe_2O_4$ nanoparticles. Journal of Magnetism and Magnetic Materials, 324*(14), 2193–2198. https://doi.org/10.1016/j.jmmm.2012.02.019 2194.*

Table 3.15 Mössbauer hyperfine parameters of $Co_{1-x}(NiSr)_xFe_2O_4$ ($x = 0-0.5$) nanoparticles.

Ferrite	Site/ component	IS (mm/s)	QS (mm/s)	B_{hf} (T)	Γ (mm/s)	Area (%)
$Ni_{0.5}Sr_{0.5}Fe_2O_4$	Dist. B_{hf}	0.38	−0.5	–	0.41	43.5
	Sextet 1	0.37	−0.02	52.4	0.51	28.1
	Sextet 2	0.25	0.02	49	0.43	23.6
	Doublet	0.23	0.53	–	0.4	4.8
$Co_{0.6}(Ni_{0.2}Sr_{0.2})Fe_2O_4$	Dist. B_{hf}	0.35	−0.22	–	0.36	20.5
	Site A/Sextet 1	0.39	0	52.2	0.7	29.5
	Site B/Sextet 2	0.27	0.01	49.3	0.68	49
	Doublet 1	0.18	0.4	–	0.57	1
$CoFe_2O_4$	Site A/Sextet 1	0.37	0	51.9	0.45	32.5
	Site B/Sextet 2	0.28	−0.01	49.1	0.5	67.5

subspectral areas. Further increasing the Ni-Sr content (x) leads to the conversion of a part of the discrete components into a magnetic hyperfine field distribution, along with the appearance of minor nonmagnetic contribution in the form of a doublet pattern. In the case of $x = 1$, the summed area for both sites is ~50%, indicating the weakening of magnetic superexchange interactions due to the presence of Ni−Sr in the spinel ferrite structure. In fact, due to the slight variation in the Mössbauer fraction in site A, the evolution of the subspectral areas indicates that the relaxation of iron is mostly related to site B. It is worth noting that reducing the cobalt content does not considerably change the hyperfine magnetic fields with respect to the distinct sextet patterns.

3.5 Biomedical aspects of magnetite

Magnetite (Fe_3O_4) nanoparticles with a multidomain magnetic structure have been used for different applications because of their high Curie temperature ($T_C^{bulk} = 840$ K) and saturation magnetization ($M_s^{bulk} = 98$ emu/g). The multidomain magnetite with a high saturation magnetization value can be considered a promising candidate for fabricating exchange spring magnets, which will be discussed in Chapter 5. The magnetic properties of Fe_3O_4 nanoparticles are predominantly determined based on the size of the particles. Since large nanoparticles have relatively high remanence and coercive fields, they could be aggregated under the magnetic field. Thus, small-size magnetite nanoparticles with superparamagnetic behavior are suitable for the production of colloidally stable dispersions, providing high magnetization, good chemical stability, biocompatibility, and low toxicity. In this respect, the colloidally stable superparamagnetic Fe_3O_4 nanoparticles can be used for several applications, including magnetic resonance imaging, hyperthermia, catalysis, and drug delivery.

In a research study, hyperthermia properties of cobalt-doped magnetite ($Co_xFe_{3-x}O_4$; $0.0 \leq x \leq 1.0$) nanoparticles have been investigated (Fantechi et al., 2015). In this regard, hyperthermia performance has been evaluated by the specific absorption rate (SAR) parameter. SAR values can be calculated from the initial slope of the kinetic curves, $dT/dt(0)$, as follows:

$$\mathrm{SAR} = \frac{1}{m_{\mathrm{metal}}}\left(\sum_i c_i m_i\right)\frac{dT}{dt(0)} \tag{3.29}$$

where i represents each species (solvents, surfactants, and nanoparticles), c_i is the specific heat, m_i is the mass, and m_{metal} is the total mass of metal. The hyperthermia properties of nanoparticles depend on several parameters, such as the externally applied field (H_0), frequency (f), the saturation magnetization of the material (M_s), and the time of the magnetization reversal (τ), according to the following relations:

$$\mathrm{SAR} \propto f H_0 M_s^2 V\left[\frac{2\pi f\tau}{1+(2\pi f\tau)^2}\right] \tag{3.30}$$

and

$$\tau = \tau_0 \exp\frac{KV}{k_B T} \tag{3.31}$$

where V, K, and τ_0 are the average volume, magnetic anisotropy constant of nanoparticles, and length of time, respectively. The field amplitude and frequency were set to $H_0 = 12$ kA/m and $f = 183$ kHz, respectively, thereby avoiding any undesirable side effects on human health. Fig. 3.106 shows the variations in

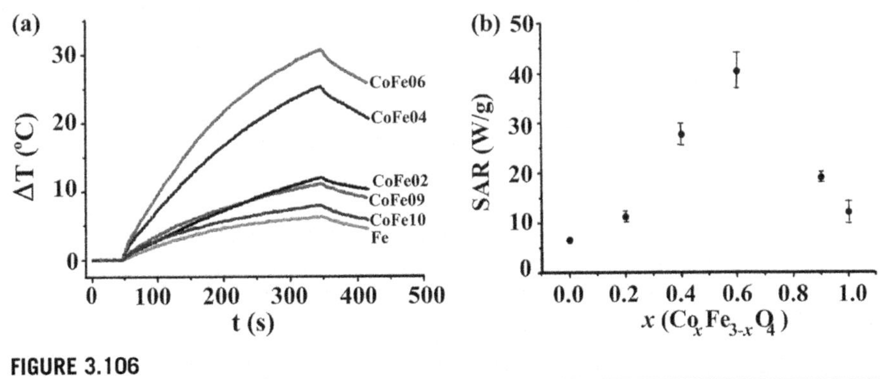

FIGURE 3.106

The variations of: (A) temperature rise (ΔT) versus time, and (B) Specific absorption rate (SAR) versus Co content (x) for $Co_xFe_{3-x}O_4$ nanoparticles (x *was* calculated from the Co:Fe ratio obtained from elemental analysis (ICP-AES)).

From Fantechi, E., Innocenti, C., Albino, M., Lottini, E., & Sangregorio, C. (c. 2015). Influence of cobalt doping on the hyperthermic efficiency of magnetite nanoparticles. Journal of Magnetism and Magnetic Materials, 380, *365–371. https://doi.org/10.1016/j.jmmm.2014.10.082 369.*

temperature rise (ΔT) versus time and SAR versus Co content (x) for nanoparticles. As can be seen, while pure magnetite nanoparticles with a particle size of 8 nm show poor heating efficiency, increasing the Co content increases SAR, reaching a maximum value for an intermediate Co content ($x = 0.6$). Therefore, the doping of the magnetite with Co may be a suitable strategy for improving the hyperthermia properties of ultrafine magnetite nanoparticles (<10 nm).

Fig. 3.107 shows the temperature dependence of ZFC/FC magnetization measured for $Co_xFe_{3-x}O_4$ ($0.0 \leq x \leq 1.0$) nanoparticles. The blocking temperature of the nanoparticles is maximum when $x = 0.6$.

VSM hysteresis loops and the variation in anisotropy constant (K) versus Co content (x) for nanoparticles are also shown in Fig. 3.108A and B, respectively. The saturation magnetization decreases with a decrease in the particle size, arising from the formation of a spin-disordered layer at the nanoparticle surface. Moreover, the coercivity and anisotropy constant increase with increasing the cobalt content, reaching maximum values for $x = 0.6$. Further increasing the cobalt content leads to a reduction in the coercivity and anisotropy constant of the nanoparticles. The evaluation of the magnetic properties clearly indicates that both magnetic moment and anisotropy can influence the SAR values, demonstrating the same variation behavior when increasing the Co content.

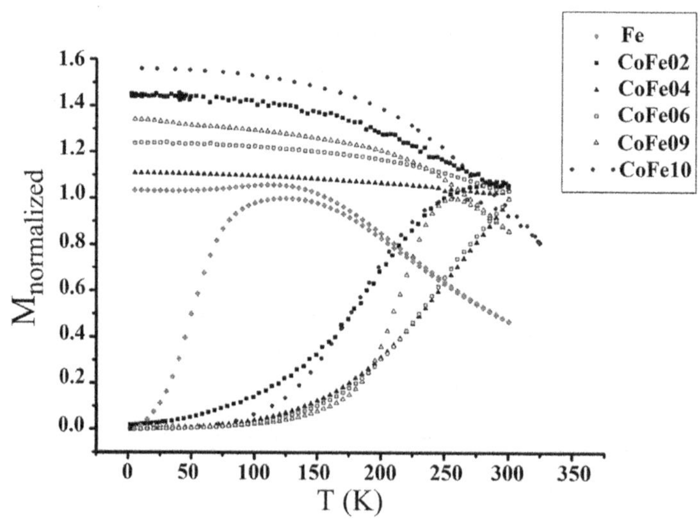

FIGURE 3.107

The temperature dependence of Zero-field-cooled/field-cooled magnetization measured for $Co_xFe_{3-x}O_4$ ($0.0 \leq x \leq 1.0$) nanoparticles.

From Fantechi, E., Innocenti, C., Albino, M., Lottini, E., & Sangregorio, C. (c. 2015). Influence of cobalt doping on the hyperthermic efficiency of magnetite nanoparticles. Journal of Magnetism and Magnetic Materials, 380, *365–371. https://doi.org/10.1016/j.jmmm.2014.10.082 369.*

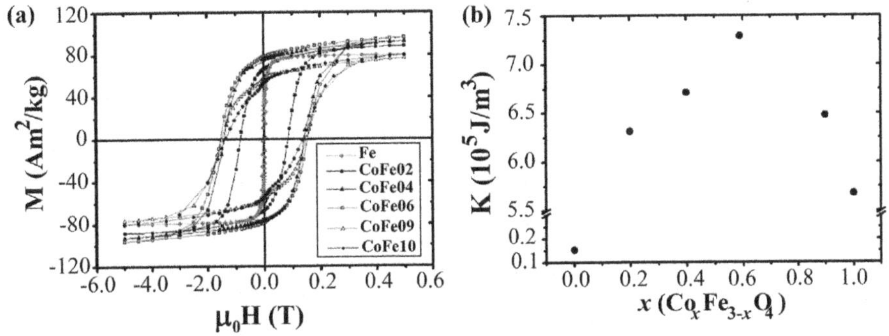

FIGURE 3.108

(A) Vibrating sample magnetometry hysteresis loops, and (B) variation of anisotropy constant (K) versus Co content (x) for $Co_xFe_{3-x}O_4$ ($0.0 \leq x \leq 1.0$) nanoparticles.

From Fantechi, E., Innocenti, C., Albino, M., Lottini, E., & Sangregorio, C. (c. 2015). Influence of cobalt doping on the hyperthermic efficiency of magnetite nanoparticles. Journal of Magnetism and Magnetic Materials, 380, *365–371. https://doi.org/10.1016/j.jmmm.2014.10.082 369.*

In another research study, Fe_3O_4/polyethylene glycol (PEG) nanocomposites and Fe_3O_4 nanoparticles were prepared using a coprecipitation method (Zhao et al., 2010). The saturation magnetization and coercivity were 67.06 emu/g and 7 Oe, and 64.11 emu/g and 14 Oe, for Fe_3O_4 nanoparticles and Fe_3O_4/PEG nanocomposites, respectively. Fig. 3.109 shows the time dependence of temperature for the resulting nanocomposite and nanoparticle samples. As can be seen, the temperatures of physiological suspensions comprising Fe_3O_4 nanoparticles and Fe_3O_4/PEG nanocomposites are 89.2 and 72.2°C after 100 s of AC magnetic field exposure, respectively. In fact, it is possible to kill tumor cells by heating the cancer area (including the implanted thermoseeds) up to elevated temperatures of 41–46°C under an external AC magnetic field. Consequently, the Fe_3O_4/PEG nanocomposites with a core–shell configuration are suitable candidates as thermoseeds for hyperthermia.

Elsewhere, using a sonochemical method, hematite/magnetite ferrofluid nanocomposites have been synthesized for hyperthermia purposes (Zayed et al., 2016). To prepare the nanocomposites, variable weight percentages of $(1-x)Fe_3O_4$ and $(x)Fe_2O_3$ ($0.0 \leq x \leq 1.0$) were physically mixed. High-resolution transmission electron microscopy (HRTEM) and FE-SEM micrographs exhibited the spherical and rod-like shape of the magnetite and hematite, respectively. Moreover, the average particle sizes were 12 and 11 nm for Fe_3O_4 and Fe_2O_3 nanoparticles, respectively. The HRTEM micrograph and selected area diffraction pattern obtained from $(1-x)Fe_3O_4/(x)Fe_2O_3$ nanocomposites with $x = 0.3$ are shown in Fig. 3.110. As observed, the morphology of the nanocomposites is mostly spherical due to the high magnetite content. The nanocomposites with $x = 0.7$ showed rod morphology because of the high hematite concentration.

Fig. 3.111 shows VSM hysteresis loops of ferrofluid nanocomposites, indicating the absence of remanence magnetization and coercive field at room temperature.

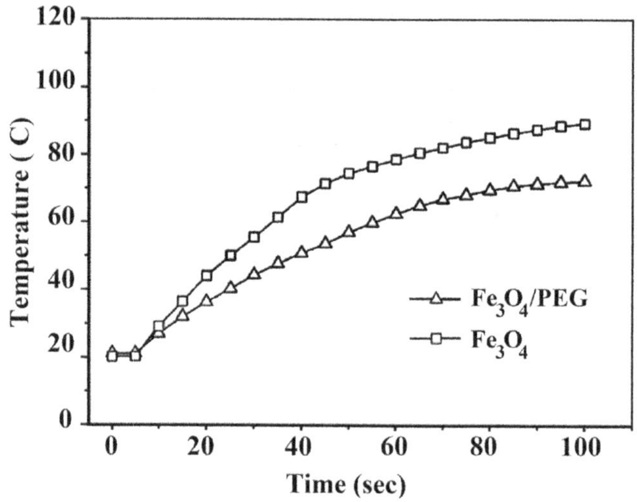

FIGURE 3.109

The time dependence of temperature for Fe_3O_4/PEG nanocomposites and Fe_3O_4 nanoparticles measured under a frequency of 80 kHz and an AC magnetic field of 30 kA/m.

From Zhao, D., Teng, P., Xu, Y., Xia, Q., & Tang, J. (c. 2010). Magnetic and inductive heating properties of Fe_3O_4/polyethylene glycol composite nanoparticles with core–shell structure. Journal of Alloys and Compounds, *502(2), 392–395. https://doi.org/10.1016/j.jallcom.2010.04.177 394.*

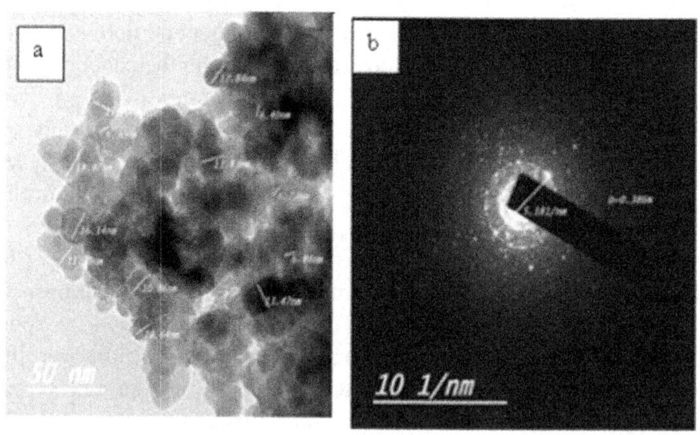

FIGURE 3.110

(A) High-resolution transmission electron microscopy micrograph and (B) Selected area diffraction pattern of $(1-x)Fe_3O_4/(x)Fe_2O_3$ nanocomposites with $x = 0.3$.

From Zayed, M., Ahmed, M., Imam, N., & El Sherbiny, D. H. (c. 2016). Preparation and structure characterization of hematite/magnetite ferro-fluid nanocomposites for hyperthermia purposes. Journal of Molecular Liquids, 222, *895–905. https://doi.org/10.1016/j.molliq.2016.07.082 904.*

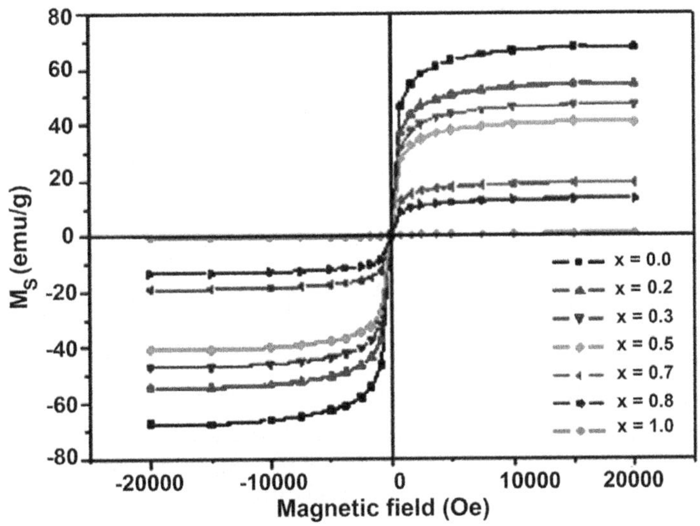

FIGURE 3.111

Vibrating sample magnetometry hysteresis loops of $(1-x)Fe_3O_4/(x)Fe_2O_3$ $(0.0 \leq x \leq 1.0)$ nanocomposites measured at room temperature.

From Zayed, M., Ahmed, M., Imam, N., & El Sherbiny, D. H. (c. 2016). Preparation and structure characterization of hematite/magnetite ferro-fluid nanocomposites for hyperthermia purposes. Journal of Molecular Liquids, *222, 895–905. https://doi.org/10.1016/j.molliq.2016.07.082 904.*

In addition, increasing the hematite concentration increases the saturation magnetization of the nanocomposites. The nanocomposites with $x = 0.2$ show magnetic properties suitable for hyperthermia applications.

The time dependence of temperature measured for the ferrofluid nanocomposites with $x = 0.2$ is shown in Fig. 3.112. As can be seen, the highest temperature rise is carefully measured to be $\Delta T = 43°C$ after 20 min of the magnetic field exposure.

In another study, Fe_3O_4–chitosan nanoparticles with a core–shell structure and an average particle size of 10.5 nm were synthesized, and their magnetic and hyperthermia features were evaluated (Qu et al., 2010). As a partially acetylated glucosamine biopolymer, chitosan has several beneficial properties such as hydrophilicity, biocompatibility, and biodegradability. Fig. 3.113 compares VSM hysteresis loops of Fe_3O_4, oleic acid-modified Fe_3O_4, and Fe_3O_4–chitosan nanoparticles. The authors claim that all the nanoparticle samples are superparamagnetic at room temperature due to their small size being less than the superparamagnetic critical size (25 nm). The saturation magnetization of the Fe_3O_4–chitosan nanoparticles (30.7 emu/g) is smaller than that of the Fe_3O_4 nanoparticles, which can be attributed to the presence of a large amount of diamagnetic chitosan.

The schematic illustration of the experimental setup used for the calorimetric measurements is shown in Fig. 3.114A. The frequency of the AC magnetic field was kept constant at 80 kHz. The AC magnetic field strength was controlled by

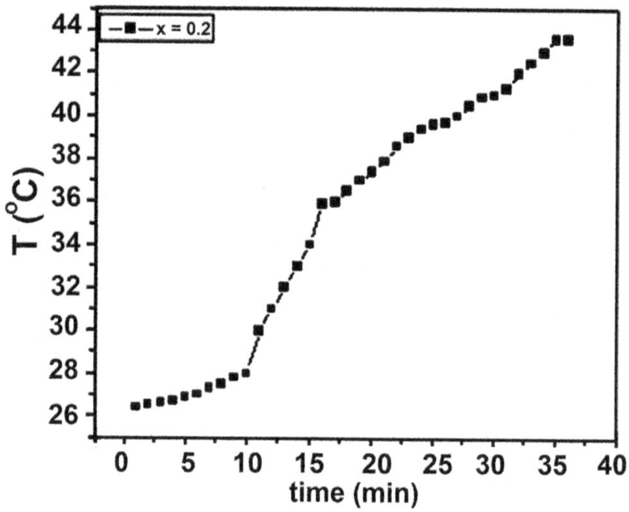

FIGURE 3.112

The time dependence of temperature measured for $(1-x)Fe_3O_4/(x)Fe_2O_3$ ferrofluid nanocomposites with $x = 0.2$.

From Zayed, M., Ahmed, M., Imam, N., & El Sherbiny, D. H. (c. 2016). Preparation and structure characterization of hematite/magnetite ferro-fluid nanocomposites for hyperthermia purposes. Journal of Molecular Liquids, 222, *895–905. https://doi.org/10.1016/j.molliq.2016.07.082 904.*

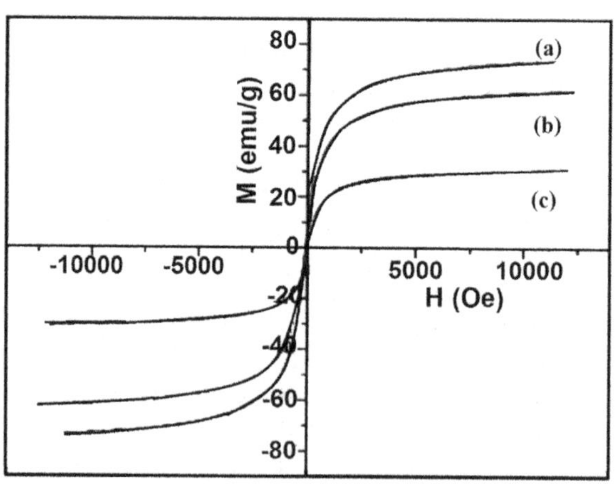

FIGURE 3.113

Vibrating sample magnetometry hysteresis loops of: (A) Fe_3O_4, (B) oleic acid-modified Fe_3O_4, and (C) Fe_3O_4–chitosan nanoparticles.

Qu, J., Liu, G., Wang, Y., & Hong, R. (2010). Advanced Powder Technology, 21*(4), 461–467. https://doi.org/10.1016/j.apt.2010.01.008.*

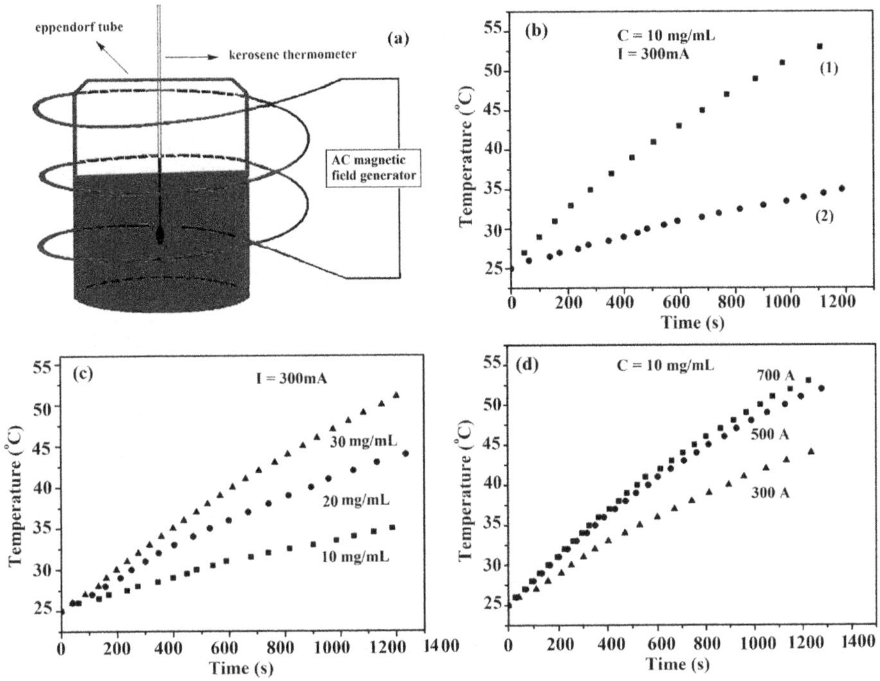

FIGURE 3.114

(A) Schematic illustration of the experimental setup for calorimetric measurement. (B) The time dependence of temperature for: (1) Fe_3O_4, and (2) Fe_3O_4–chitosan nanoparticles. (C) The time dependence of temperature measured for Fe_3O_4–chitosan nanoparticles at a fixed current of $I = 300$ A and different concentrations. (D) The time dependence of temperature measured for Fe_3O_4–chitosan nanoparticles at a fixed concentration of $C = 20$ mg/mL and different currents.

From Qu, J., Liu, G., Wang, Y., & Hong, R. (c. 2010). Preparation of Fe_3O_4–chitosan nanoparticles used for hyperthermia. Advanced Powder Technology, 21*(4), 461–467. https://doi.org/10.1016/j.apt.2010.01.008 465.*

applying apparent current in the range of 100–900 A. The time dependence of temperature measured at a fixed current of $I = 300$ A and concentration of 10 (mg/mL) for the Fe_3O_4 and Fe_3O_4–chitosan nanoparticles is shown in Fig. 3.114B. It is observed that the Fe_3O_4 nanoparticles outperform the Fe_3O_4–chitosan ones in terms of heating efficiency. In other words, the temperature of the Fe_3O_4 nanoparticles increases from 25°C to 53°C in 1200 s, whereas that of the Fe_3O_4–chitosan nanoparticles reaches 35°C at the same time. The reduction observed in the hyperthermia properties of Fe_3O_4–chitosan nanoparticles is due to the decrease in the magnetization of the Fe_3O_4 nanoparticles as a result of the chitosan coating. In addition, increasing the concentration from 10 to 30 mg/mL increases the temperature from 35 to 50°C in 1200 s (see Fig. 3.114C. Likewise, the temperature increases rapidly with increasing the apparent current. Notably, the temperature of the Fe_3O_4–

chitosan sample reaches 43.9°C under the apparent current of 300 A in 1200 s. Moreover, by applying the apparent currents of 500 and 700 A, the temperature of the sample reaches 51.4 and 53.1°C, respectively (Fig. 3.114D). Therefore, the apparent current has a great effect on the heating efficiency, and consequently, hyperthermia features of the nanoparticles. In this direction, the Fe_3O_4–chitosan nanoparticles have great potential in hyperthermia under the large applied apparent current.

$CuFe_2O_4$ and Fe_3O_4 nanoparticles have also been compared in terms of hyperthermia measurements (Fotukian et al., 2020). Fig. 3.115A and B show the time dependence of temperature measured for $CuFe_2O_4$ and Fe_3O_4 nanoparticles at different field amplitudes, respectively. As observed, increasing the field amplitude increases the temperature for both nanoparticle samples. Basically, the heat production induced by magnetic nanoparticles under a high-frequency magnetic field is defined by the three following mechanisms: (1) hysteresis loss, (2) Brownian relaxation, and (3) Néel relaxation. The hysteresis loss can be measured by integrating the hysteresis loop area. The relaxation of the magnetic moment takes place in a single-domain nanoparticle by Néel or Brownian relaxation mode. In the case of superparamagnetic nanoparticles, Néel and Brownian relaxation mechanisms are involved. It is interesting to state that the combination of the aforementioned three mechanisms is more desirable for heat production.

The SAR values obtained from the nanoparticle samples at different amplitudes are shown in Fig. 3.115C. The SAR values of $CuFe_2O_4$ nanoparticles are seen to increase from 24.8 to 44.9 W/g with increasing the magnetic field from 13 to 19 kA/m. Moreover, the SAR values of Fe_3O_4 nanoparticles are smaller than those of the $CuFe_2O_4$ ones under all the applied magnetic fields. This can be attributed to the lower magnetization of the Fe_3O_4 nanoparticles under the low applied magnetic fields compared with the $CuFe_2O_4$ nanoparticles. In addition, the anisotropy energy of Fe_3O_4 nanoparticles is higher than that of the $CuFe_2O_4$ nanoparticles. It has been evidenced that particles with lower magnetic anisotropy energy could lead to higher heating efficiency under lower applied magnetic fields during the hyperthermia experiments. In other words, the magnetic field applied to the Fe_3O_4 nanoparticles is not high enough to overcome their magnetic anisotropy energy barrier during the hyperthermia measurements. Consequently, the higher hyperthermia performance of the $CuFe_2O_4$ nanoparticles can be related to their higher magnetization and lower magnetic anisotropy energy compared with the Fe_3O_4 nanoparticles.

In another study, hyperthermia measurements of polyvinyl alcohol (PVA)-Fe_3O_4 nanocomposites were evaluated (Lao & Ramanujan, 2004). Fig. 3.116 shows the time dependence of temperature measured for the PVA-Fe_3O_4 nanocomposites at different contents of iron oxide under a constant magnetic field of $H = 2.5$ kA/m. As observed, increasing the iron oxide content increases the temperature.

The time dependence of temperature measured for PVA-Fe_3O_4 nanocomposites with 2 wt% iron oxide content under various magnetic field amplitudes is shown in Fig. 3.117. Evidently, increasing the magnetic field amplitude from 1.69 to 2.53 kA/m increases the temperature from 40 to 47°C. It is worth mentioning that healthy

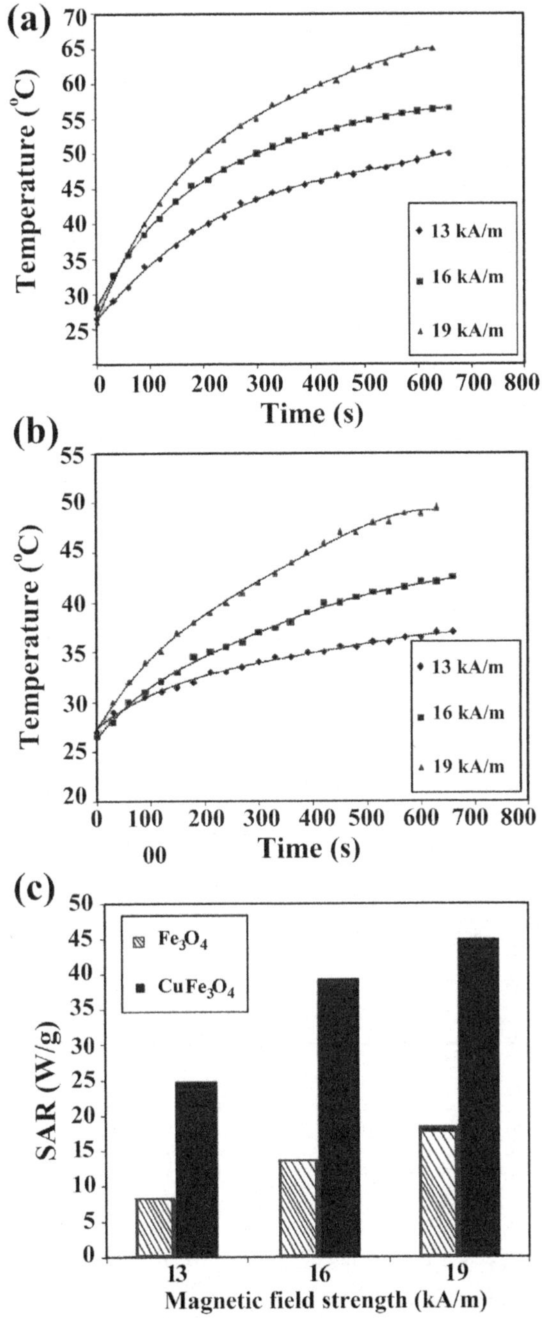

FIGURE 3.115

The time dependence of temperature measured for: (A) $CuFe_2O_4$, and (B) Fe_3O_4 nanoparticles. (C) Specific absorpton rate (SAR) values of the $CuFe_2O_4$ and Fe_3O_4 nanoparticle samples at different field amplitudes.

From Fotukian, S. M., Barati, A., Soleymani, M., & Alizadeh, A. M. (c. 2020). Solvothermal synthesis of $CuFe_2O_4$ and Fe_3O_4 nanoparticles with high heating efficiency for magnetic hyperthermia application. Journal of Alloys and Compounds, 816, *152548. https://doi.org/10.1016/j.jallcom.2019.152548 16.*

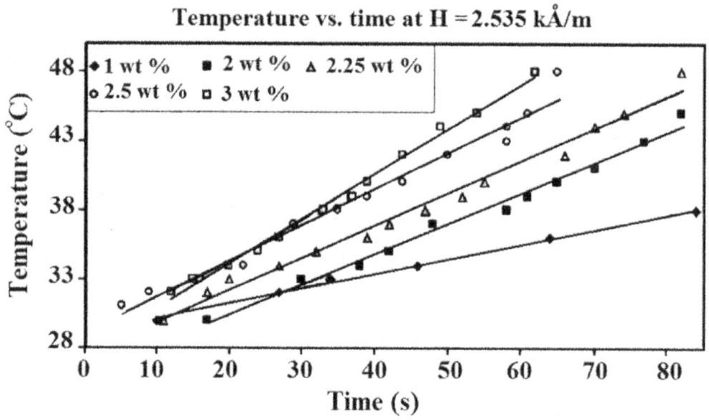

FIGURE 3.116

The time dependence of temperature measured for PVA-Fe_3O_4 nanocomposites with different iron oxide contents.

From Lao, L., & Ramanujan, R. (c. 2004). Magnetic and hydrogel composite materials for hyperthermia applications. Journal of Materials Science: Materials in Medicine, 15*(10), 1061–1064. https://doi.org/10.1023/B:JMSM.0000046386.78633.e5 1063.*

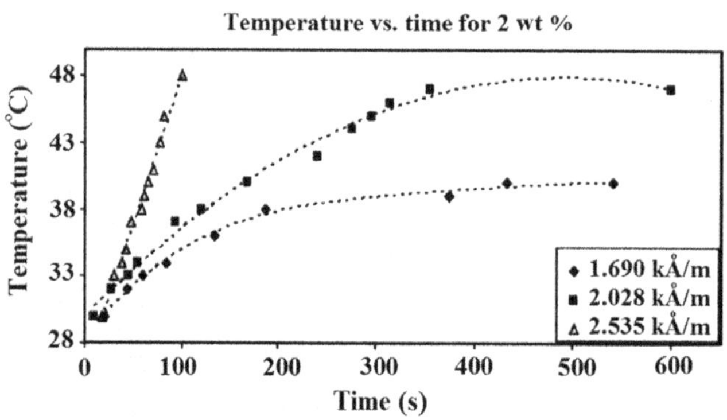

FIGURE 3.117

The time dependence of temperature measured for PVA-Fe_3O_4 nanocomposites with 2 wt % iron oxide content at various magnetic field amplitudes.

From Lao, L., & Ramanujan, R. (c. 2004). Magnetic and hydrogel composite materials for hyperthermia applications. Journal of Materials Science: Materials in Medicine, 15*(10), 1061–1064. https://doi.org/10.1023/B:JMSM.0000046386.78633.e5 1063.*

tissues surrounding the tumor can be burnt and damaged when exposed to higher temperatures ($T \geq 48°C$).

Fig. 3.118 shows the variation of SAR versus square magnetic field strength for PVA-Fe_3O_4 nanocomposites with 2 wt%. iron oxide. It is seen that the SAR values increase linearly with the square of the magnetic field.

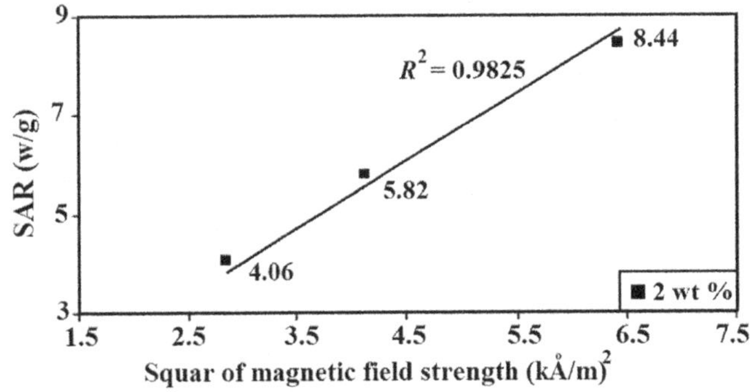

FIGURE 3.118

The variation in specific absorpton rate (SAR) versus square of magnetic field for PVA-Fe_3O_4 nanocomposites with 2 wt% iron oxide content.

From Lao, L., & Ramanujan, R. (c. 2004). Magnetic and hydrogel composite materials for hyperthermia applications. Journal of Materials Science: Materials in Medicine, 15*(10), 1061–1064. https://doi.org/10.1023/B:JMSM.0000046386.78633.e5 1063.*

Biosilicate materials are used for tissue engineering applications, improving cell adhesion, proliferation, spreading, and differentiation. Diopside ($CaMgSi_2O_6$) bioceramics with a slow degradation rate provide comparative advantages in bone tissue regeneration. In a research study, the hyperthermia performance of diopside-magnetite nanocomposites has been investigated (Abdellahi et al., 2018). Fig. 3.119 shows XRD patterns, VSM hysteresis loops, and hyperthermia measurements of the resulting nanocomposites consisting of 10–30% magnetite. The major peaks corresponding to Fe_3O_4, together with some peaks of the diopside structure, can be observed in the XRD patterns. The VSM hysteresis loops indicate that the magnetization increases with an increase in the magnetite concentration. It is inferred from the hyperthermia measurements that the temperature increases with increasing the magnetite concentration. The higher values of SAR at a high magnetite content enhance the safety factor in hyperthermia applications.

Elsewhere, $(Zn_xMn_{1-x}Fe_{0.6})Fe_2O_4$ nanoparticles with an average particle size of 13 nm have been prepared using a coprecipitation process, followed by the investigation of magnetic and hyperthermia properties (de Mello et al., 2019). Fig. 3.120 shows VSM hysteresis loops of nanoparticles. It is observed that all the nanoparticle samples show superparamagnetic behavior with remanence magnetization and coercivity values close to zero at room temperature. Moreover, the pure magnetite nanoparticles (Fe_3O_4) have the lowest saturation magnetization. The substitution of manganese and zinc cations in the magnetite composition increases the saturation magnetization, arising from the higher magnetic moment of Mn^{2+} ([Ar] $4s^2$ $3d^5$; 5 μ_B) compared with that of Fe^{2+}. However, the substitution of Zn^{2+} ([Ar] $4s^2$ $3d^{10}$, 0 μ_B) changes the site occupancy of cations in magnetite. This is because zinc cations preferably occupy tetrahedral sites, leading to the migration of some Fe^{3+} ions from tetrahedral to octahedral sites.

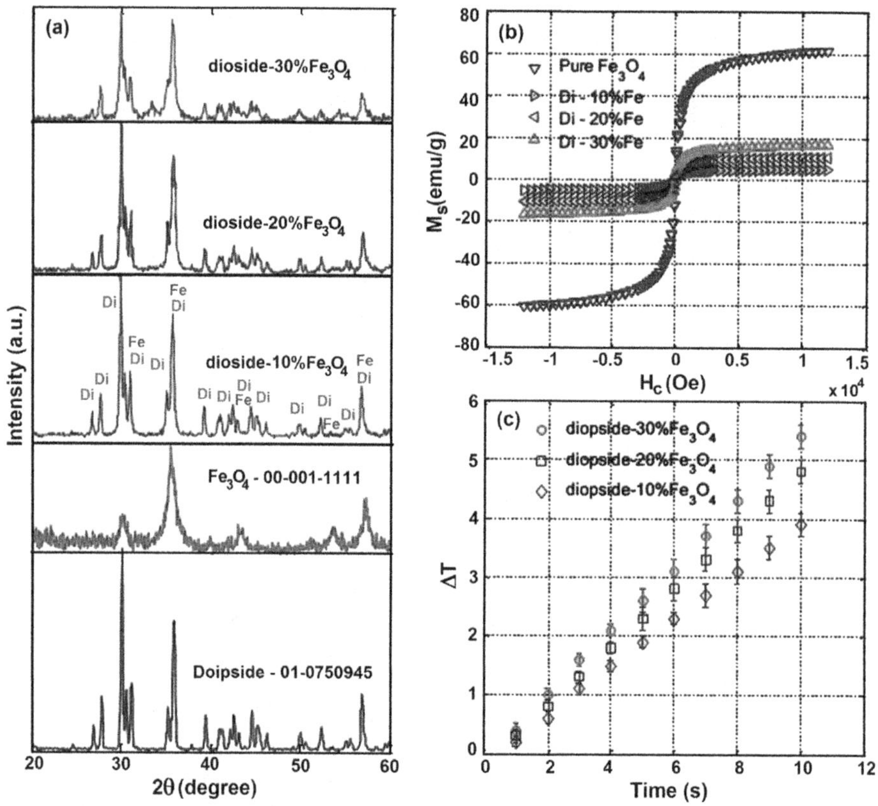

FIGURE 3.119

(A) X-ray diffraction patterns, (B) Vibrating sample magnetometry hysteresis loops, and (C) hyperthermia measurements of diopside-magnetite nanocomposites with different magnetite contents.

From Abdellahi, M., Karamian, E., Najafinezhad, A., Ranjabar, F., Chami, A., & Khandan, A. (c. 2018). Diopside-magnetite; A novel nanocomposite for hyperthermia applications. Journal of the Mechanical Behavior of Biomedical Materials, *77, 534–538. https://doi.org/10.1016/j.jmbbm.2017.10.015.*

The hyperthermia measurements of Fe_3O_4, $(Mn_{0.4}Fe_{0.6})Fe_2O_4$, $(Zn_{0.1}Mn_{0.3}Fe_{0.6})Fe_2O_4$, $(Zn_{0.2}Mn_{0.2}Fe_{0.6})Fe_2O_4$, $(Zn_{0.3}Mn_{0.1}Fe_{0.6})Fe_2O_4$ and $(Zn_{0.4}Fe_{0.6})Fe_2O_4$ nanoparticles are exhibited in Fig. 3.121. It is clear that the $(Mn_{0.4}Fe_{0.6})Fe_2O_4$ nanoparticles have the highest temperature rise over 600 s and at the first 30 s. The lowest temperature rise is obtained for $(Zn_{0.3}Mn_{0.1}Fe_{0.6})Fe_2O_4$ and $(Zn_{0.4}Fe_{0.6})Fe_2O_4$ nanoparticles over 600 s and at the first 30 s, respectively. The SAR values of the $(Mn_{0.4}Fe_{0.6})Fe_2O_4$ and $(Zn_{0.4}Fe_{0.6})Fe_2O_4$ nanoparticles are 46.0 and 4.2 W/g, respectively. Furthermore, the magnetic hyperthermia effect is enhanced with an increase in the Mn content in the magnetite composition. It is worth noting that the

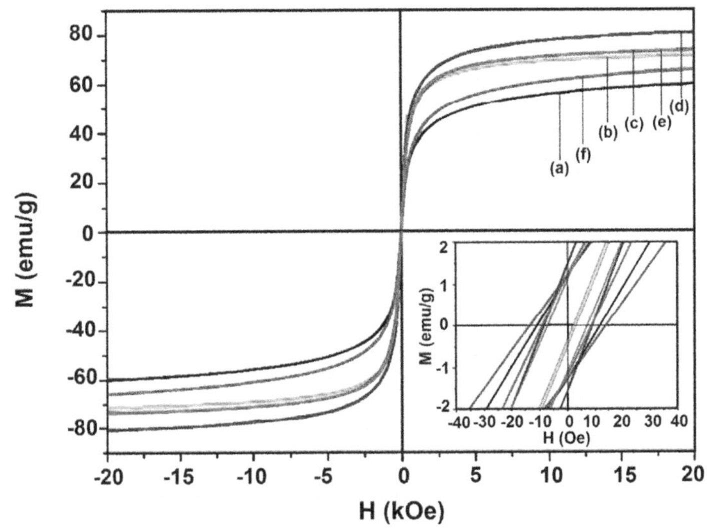

FIGURE 3.120

Vibrating sample magnetometry hysteresis loops of: (A) Fe_3O_4, (B) $(Mn_{0.4}Fe_{0.6})Fe_2O_4$, (C) $(Zn_{0.1}Mn_{0.3}Fe_{0.6})Fe_2O_4$, (D) $(Zn_{0.2}Mn_{0.2}Fe_{0.6})Fe_2O_4$, (E) $(Zn_{0.3}Mn_{0.1}Fe_{0.6})Fe_2O_4$, and (F) $(Zn_{0.4}Fe_{0.6})Fe_2O_4$ nanoparticles.

From de Mello, L. B., Varanda, L. C., Sigoli, F. A., & Mazali, I. O. (c. 2019). Coprecipitation synthesis of (Zn-Mn)-codoped magnetite nanoparticles and their application in magnetic hyperthermia. Journal of Alloys and Compounds, 779, *698–705. https://doi.org/10.1016/j.jallcom.2018.11.280.*

Néel-Brownian relaxation process is mainly responsible for the hyperthermia performance of the samples.

In a separate research study, the hyperthermia performance of novel bredigite-magnetite nanocomposites ($Ca_7MgSi_4O_{16}$-Fe_3O_4) has been evaluated for bone tissue engineering purposes (Khandan & Ozada, 2017). Fig. 3.122 shows the time dependence of temperature for the bredigite-magnetite nanocomposites with different magnetite concentrations (0–30 wt%). As can be seen, increasing the magnetite concentration leads to an increase in the temperature. The nanocomposites consisting of 30 wt% magnetite particles result in the highest temperature change (27°C) during 50 s. The hyperthermia treatment significantly increases the tumor tissue perfusion while enhancing the local oxygen content. This, in turn, leads to a suitable condition for γ-radiation, thereby killing the tumor cells.

Fig. 3.123 shows the heat absorption of the bredigite-magnetite nanocomposites with different magnetite concentrations. As inferred, the heat absorption of the nanocomposites continuously increases with increasing the Fe_3O_4 content, confirming that the bredigite-magnetite nanocomposites can be a good candidate for hyperthermia applications.

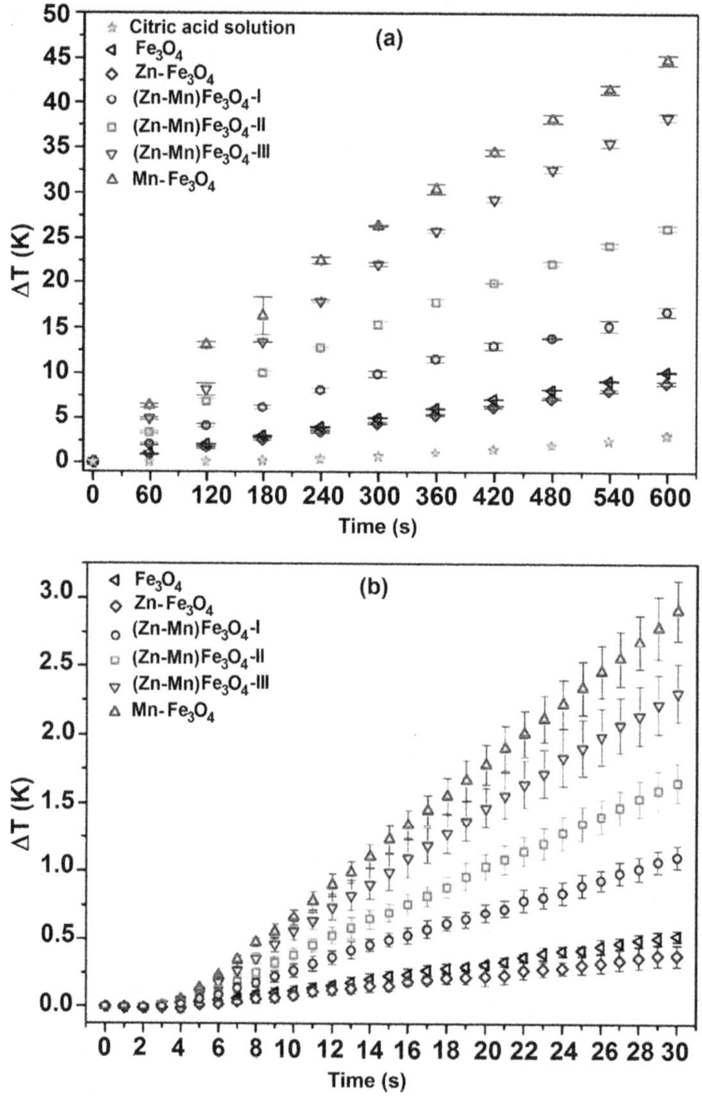

FIGURE 3.121

The time dependence of temperature measured for Fe_3O_4, $(Mn_{0.4}Fe_{0.6})Fe_2O_4$, $(Zn_{0.1}Mn_{0.3}Fe_{0.6})Fe_2O_4$, $(Zn_{0.2}Mn_{0.2}Fe_{0.6})Fe_2O_4$, $(Zn_{0.3}Mn_{0.1}Fe_{0.6})Fe_2O_4$, $(Zn_{0.4}Fe_{0.6})Fe_2O_4$ nanoparticles, and citric acid solution: (A) over 600 s, and (B) at the first 30 s.

From de Mello, L. B., Varanda, L. C., Sigoli, F. A., & Mazali, I. O. (c. 2019). Coprecipitation synthesis of (Zn-Mn)-codoped magnetite nanoparticles and their application in magnetic hyperthermia. Journal of Alloys and Compounds, 779, *698–705. https://doi.org/10.1016/j.jallcom.2018.11.280 14.*

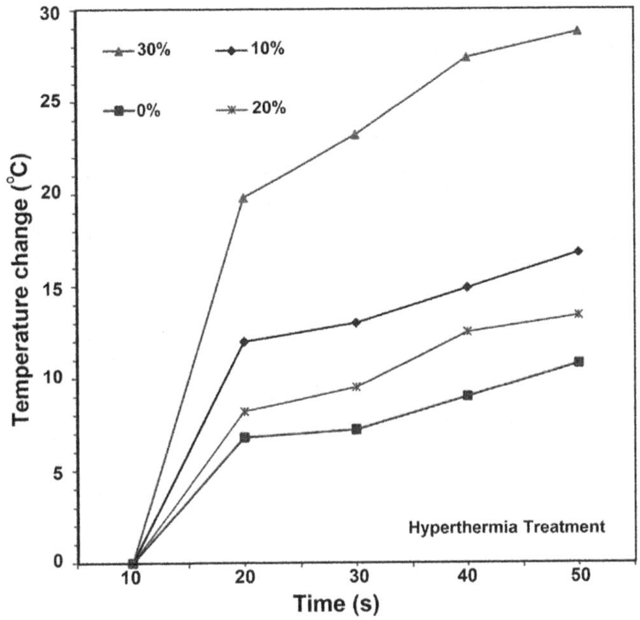

FIGURE 3.122

The time dependence of temperature measured for bredigite-magnetite nanocomposites with different magnetite concentrations.

From Khandan, A., & Ozada, N. (c. 2017). Bredigite-Magnetite ($Ca_7MgSi_4O_{16}$-Fe_3O_4) nanoparticles: A study on their magnetic properties. Journal of Alloys and Compounds, 726, *729–736. https://doi.org/10.1016/j.jallcom.2017.07.288 31.*

3.6 Garnets

Garnets are classified as a group of magnetic oxides, having a cubic structure with the chemical formula of $M_3Fe_5O_{12}$, in which M is usually yttrium (Y) or one of the rare-earth elements such as Ce, Gd, and Nd. From an application standpoint, the garnets can be used in many electronic devices, including tunable microwave devices (e.g., radar, telecommunication, and RF measurement systems) and filters, circulators, isolators, phase shifters, and various magneto-optical devices (Goldman, 2006).

3.6.1 Yttrium iron garnet

The most important magnetic garnet is the yttrium iron garnet (YIG), having many specific features, including high melting point, large resistivity, low dielectric loss, high thermal stability, low thermal expansion, high chemical stability, and high saturation magnetization. It is possible to tune the electrical and magnetic properties of YIG through the control of effective processing parameters and substitution content (Bahadur, 1992; Ganne et al., 2007).

It has been well documented that the magnetic properties of YIG nanoparticles depend on the particle size and annealing temperature. For example, YIG nanoparticles have been synthesized with different particle sizes of 9, 14, 25, and 60 nm (Rajendran et al., 2006). The blocking temperature and saturation magnetization decreased with decreasing the particle size due to the large surface-to-volume ratios of the nanoparticles. Fig. 3.124 shows hysteresis loops of the YIG nanoparticles with different particle sizes. The saturation magnetization and coercivity of the nanoparticle with an average particle size of 25 nm were obtained as 20.6 emu/g and 51 Oe, respectively, indicating the soft magnetic nature. However, the nanoparticle samples with the average particle sizes of 9 and 14 nm are not saturated at the moderate magnetic field strength, reflecting their almost superparamagnetic behavior.

Fig. 3.125 shows the variation of AC susceptibility as a function of temperature for the YIG nanoparticles with particle sizes of 9 and 14 nm. As observed, the blocking temperature increases with an increase in the particle size, being the threshold point of thermal activation. As noted in Chapter 2, the magnetization direction of each nanoparticle is aligned with the applied field direction for temperatures above the blocking temperature, exhibiting paramagnetic features. Alternatively, without

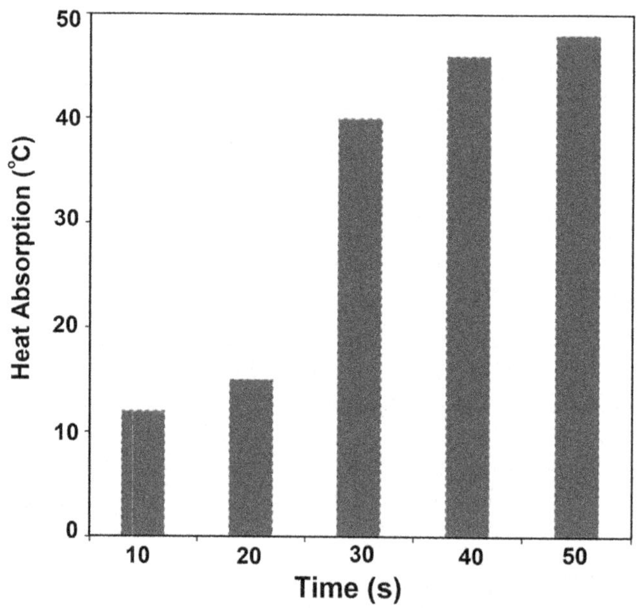

FIGURE 3.123

The heat absorption of bredigite-magnetite nanocomposites with different magnetite concentrations.

From Khandan, A., & Ozada, N. (c. 2017). Bredigite-Magnetite ($Ca_7MgSi_4O_{16}$-Fe_3O_4) nanoparticles: A study on their magnetic properties. Journal of Alloys and Compounds, 726, *729–736. https://doi.org/10.1016/j.jallcom.2017.07.288 31.*

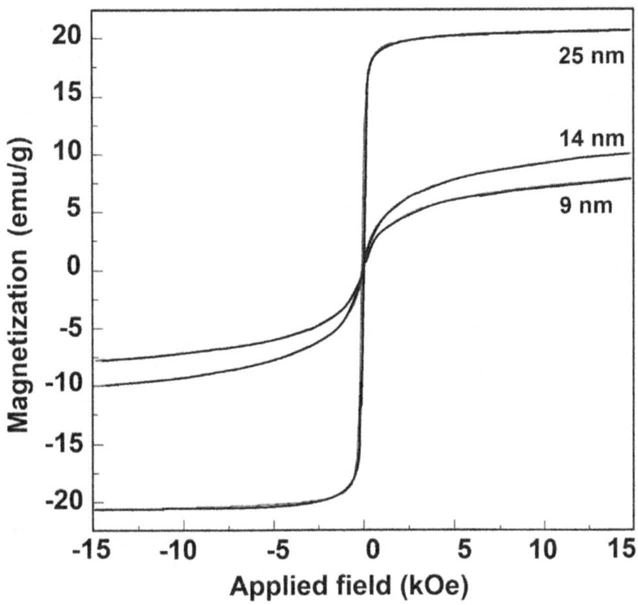

FIGURE 3.124

Hysteresis loops obtained from YIG nanoparticles with different particle sizes.

From Rajendran, M., Deka, S., Joy, P., & Bhattacharya, A. (c. 2006). Size-dependent magnetic properties of nanocrystalline yttrium iron garnet powders. Journal of Magnetism and Magnetic Materials, 301*(1), 212–219. https://doi.org/10.1016/j.jmmm.2005.06.027 216.*

any movement of the nanoparticles, the magnetization direction can rotate from the applied field direction back to its own easy axis for temperatures below the blocking temperature.

In a separate research study, YIG nanoparticles were synthesized using a citrate precursor gel (Nguyet et al., 2012). Hysteresis loops of the YIG nanoparticles were measured at different temperatures ranging between 5 and 560 K, and the results are shown in Figs. 3.126 and 3.127. The inset of Fig. 3.126 shows the hysteresis loops in a lower magnetic field range, providing more information regarding the variation of coercivity and saturation magnetization with temperature. As inferred, the coercivity decreases with increasing the temperature, which can be explained by the competition between the thermal and anisotropy energies. Evidently, the saturation magnetization also decreases with increasing the temperature, which can be attributed to nonmagnetic impurity and surface dead layers in the YIG nanoparticles. The inset of Fig. 3.127 reflects the thermomagnetic curve (M-T) measured in the applied field of 100 Oe. The curve displays that the Curie temperature is about 550 K. The larger difference in the obtained Curie temperature results between the hysteresis loops and the M–T curve relates to the strength of the applied field.

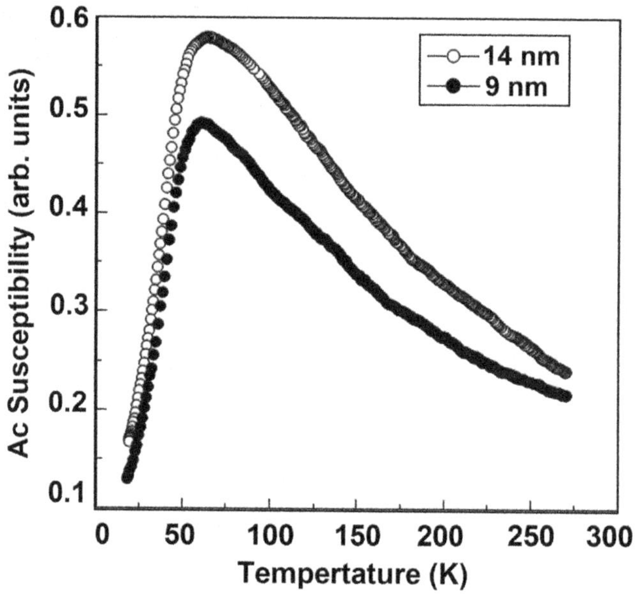

FIGURE 3.125

The variation of AC susceptibility as a function of temperature for YIG nanoparticles with particle sizes of 9 and 14 nm.

From Rajendran, M., Deka, S., Joy, P., & Bhattacharya, A. (c. 2006). Size-dependent magnetic properties of nanocrystalline yttrium iron garnet powders. Journal of Magnetism and Magnetic Materials, 301*(1), 212–219. https://doi.org/10.1016/j.jmmm.2005.06.027 216.*

3.6.2 Substituted yttrium iron garnet nanoparticles

In the following, the role of different cation substitutions on the structural and magnetic features of YIG nanoparticles is discussed.

Aluminum-substituted YIG nanoparticles with the composition of $Y_3Al_xFe_{5-x}O_{12}$ ($0.0 \leq x \leq 2.2$) have been synthesized via a sol–gel autocombustion method, and their structural and magnetic properties have been investigated (Musa et al., 2017). Fig. 3.128 shows the variations of saturation magnetization and coercivity as a function of Al concentration for the resulting nanoparticles. The saturation magnetization decreases from 20.721 to 3.7586 emu/g with increasing the Al concentration from 0.0 to 2.2. In addition, the substitution of paramagnetic Al^{3+} ions with ferromagnetic Fe^{3+} ones can reduce the net magnetic moment, thus weakening the superexchange interaction between a and d sites in the crystal structure. On the other hand, the coercivity of YIG nanoparticles decreases with an increase in the Al concentration to $x = 0.2$. Further increasing the Al concentration enhances the coercivity, reaching a maximum value at $x = 1.0$. The coercive field decreases for Al concentrations higher than $x = 1.0$. Consequently, the high Al content in the YIG nanoparticles can weaken the magnetic properties of the garnet.

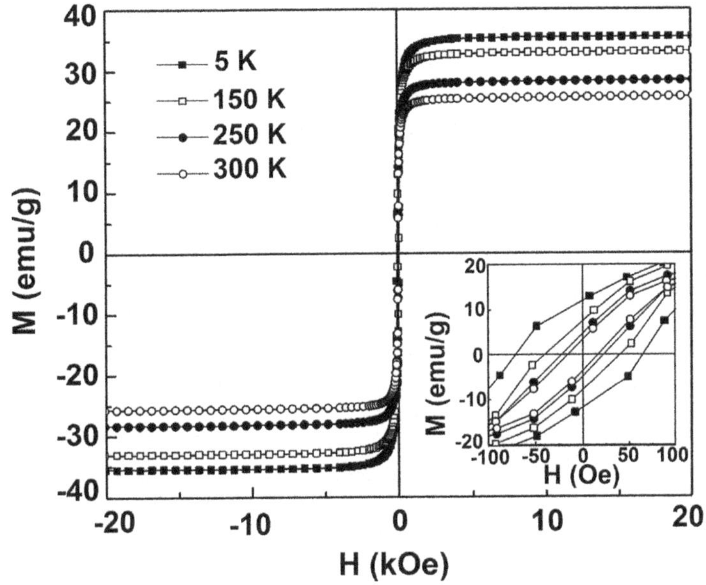

FIGURE 3.126

Hysteresis loops of YIG nanoparticles measured at low temperatures.

From Nguyet, D. T. T., Duong, N. P., Satoh, T., Anh, L. N., & Hien, T. D. (c. 2012). Temperature-dependent magnetic properties of yttrium iron garnet nanoparticles prepared by citrate sol–gel. Journal of Alloys and Compounds, 541, *18–22. https://doi.org/10.1016/j.jallcom.2012.06.122.*

In another research study, structural and magnetic properties of Co-substituted YIG nanoparticles with the composition of $Y_3(Co_yFe_{1-y})_5O_{12}$ have been investigated (Peña-Garcia, Guerra, Buitrago, et al., 2018). Rietveld refinement profiles of XRD patterns for the prepared samples with $y = 0.000$, 0.015, and 0.030 are shown in Fig. 3.129. The experimental data is shown with the black symbols (the light diagonal cross), whereas the calculated profile is represented by the solid red line. As observed, a single-phase garnet is obtained by the sol–gel method at an annealing temperature of 900°C. The Rietveld refinement reveals that the Co ions are located in the tetrahedral and octahedral sites of the YIG. It is expected that the lattice constant decreases with the substitution of Co ions in the YIG nanoparticle samples because of the smaller ionic radius of cobalt ions compared with that of Fe^{3+} ones. With the distribution of Co ions in the *a* and *d* sites of the YIG sample, the angles between $Co^{3+}(a)$-O^{2-}-$Y^{3+}(c)$ and $Co^{3+}(d)$-O^{2-}-$Y^{3+}(c)$ are reduced from 102.6 to 100.6 degree, and 124.3 to 123.6 degree, respectively.

Fig. 3.130 shows VSM hysteresis loops of the $Y_3(Co_yFe_{1-y})_5O_{12}$ nanoparticles. The formation of narrow loops indicates the soft magnetic nature of the Co-substituted YIG nanoparticles. The investigation of magnetic properties demonstrates that the saturation magnetization decreases from 26.63 to 24.92 emu/g with increasing the Co content from $y = 0.000$ to 0.030. This is attributed to the

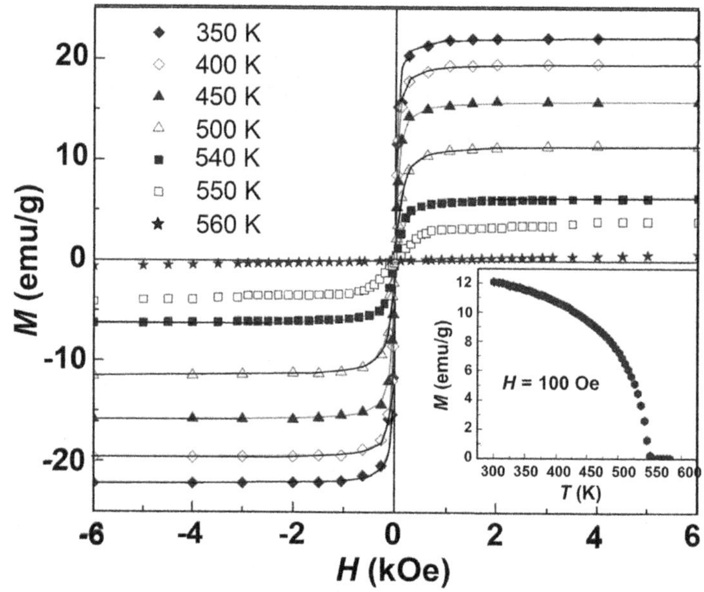

FIGURE 3.127

Hysteresis loops of YIG nanoparticles measured at relatively high temperatures.

From Nguyet, D. T. T., Duong, N. P., Satoh, T., Anh, L. N., & Hien, T. D. (c. 2012). Temperature-dependent magnetic properties of yttrium iron garnet nanoparticles prepared by citrate sol–gel. Journal of Alloys and Compounds, 541, *18–22. https://doi.org/10.1016/j.jallcom.2012.06.122.*

substitution of Fe^{3+} by Co^{3+} in the tetrahedral and octahedral sites of YIG. In fact, this substitution leads to a reduction in the superexchange interactions taking place in the crystalline lattice while reducing the total magnetic moment. Alternatively, the coercivity decreases from 82.21 to 75.68 Oe with increasing the Co content from 0.000 to 0.015. Further increasing the Co content from 0.025 to 0.030 increases the coercivity from 87.32 to 94.57 Oe. At the low substitution content, the average particle sizes are smaller than the critical diameter of the single-domain configuration, resulting in decreased coercivity, while the pinning positions of the magnetic moments increase at the higher concentration of Co ions. The enhancement in coercivity for $y > 0.015$ is assigned to the increased magnetocrystalline anisotropy.

In another study, the structural and magnetic properties of Cu-substituted YIG nanoparticles with the composition of $Y_3(Cu_yFe_{1-y})_5O_{12}$ ($0.00 \leq y \leq 0.05$) were investigated (Leal et al., 2019). XRD patterns (not shown here) exhibited the presence of Fe_2O_3 in the composition as a secondary phase for high copper content. Fig. 3.131 shows SEM micrographs of the nanoparticles with $y = 0.00$ and 0.05. The formation of rounded irregular shapes and elongated particles is evidenced. Accordingly, the substitution of Cu in the YIG nanoparticles increases the corresponding particle size.

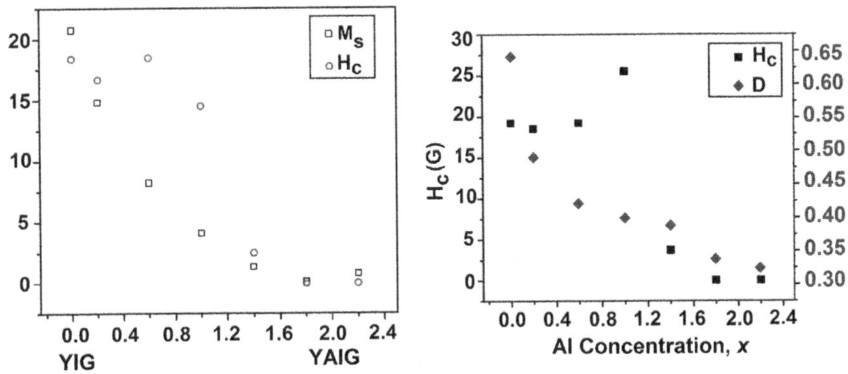

FIGURE 3.128

The variations of saturation magnetization and coercivity as a function of Al concentration for $Y_3Al_xFe_{5-x}O_{12}$ nanoparticles.

From Musa, M. A., Osman, N. H., Hassan, J., & Zangina, T. (c. 2017). Structural and magnetic properties of yttrium iron garnet (YIG) and yttrium aluminum iron garnet (YAIG) nanoferrite via sol-gel synthesis. Results in Physics, 7, *1135–1142. https://doi.org/10.1016/j.rinp.2017.02.038*

Fig. 3.132 displays VSM hysteresis loops of the $Y_3(Cu_yFe_{1-y})_5O_{12}$ nanoparticles. Increasing the Cu content from 0.00 to 0.01 increases the saturation magnetization from 24.04 to 29.49 emu/g. Further increasing the Cu content leads to a reduction in the saturation magnetization. For $y = 0.03$ and 0.05, the saturation magnetization is obtained to be 26.05 and 26.35 emu/g, respectively. The Cu ions prefer to occupy a-sites of the YIG nanoparticle sample, reducing the superexchange interaction due to the structural distortions.

The variation of saturation magnetization is indicative of the substitution of Fe/Cu cations in the octahedral sites of YIG, which in turn causes the saturation magnetization of YIG nanoparticles to be reduced. The coercive field values of the $Y_3(Cu_yFe_{1-y})_5O_{12}$ nanoparticles are 27.18, 19.09, 24.71, and 19.9 Oe for $y =$ 0.00, 0.01, 0.03 and 0.05, respectively.

The structural and magnetic properties of Ni-substituted YIG nanoparticles with the composition of $Y_3(Ni_xFe_{1-x})_5O_{12}$ ($0.00 \leq x \leq 0.05$) have also been investigated (Peña-Garcia et al., 2019). The Rietveld refinement performed on XRD patterns of the nanoparticles confirmed the substitution of Ni in tetrahedral (d) and octahedral (a) sites of the YIG structure. VSM hysteresis loops of the nanoparticles are shown in Fig. 3.133. Saturation magnetization increases with increased Ni content to $x =$ 0.01, resulting in a maximum value of 26.31 emu/g. The saturation magnetization decreases for higher contents of Ni ($x > 0.01$). The replacement of Fe ions by Ni ions in the octahedral and tetrahedral sites reduces the total magnetic moment. Meanwhile, the coercivity increases with an increase in the Ni content to $x =$ 0.01, giving rise to a maximum value of 66.45 Oe. Further increasing the Ni content ($x > 0.01$) decreases the coercivity.

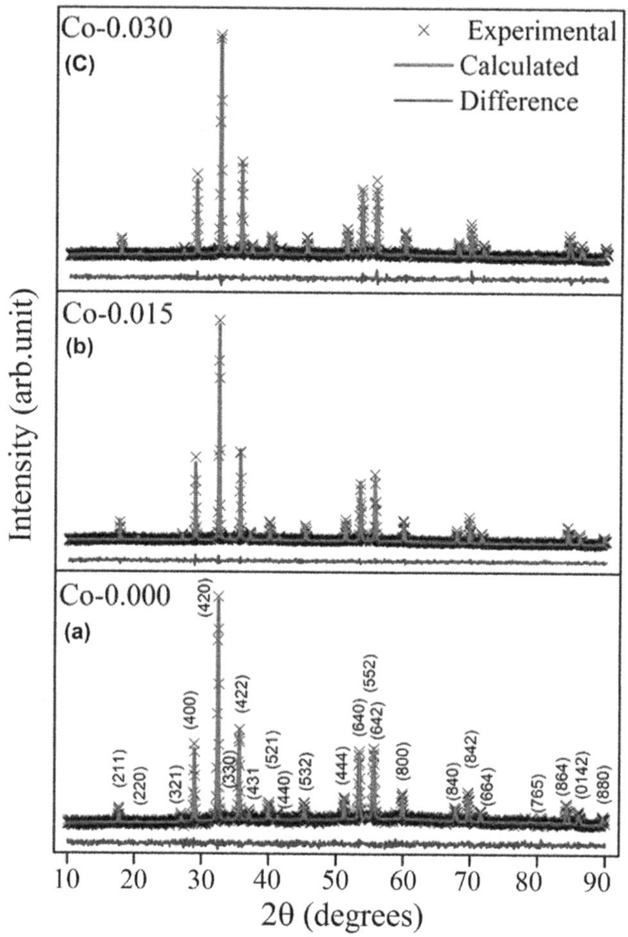

FIGURE 3.129

Rietveld refinement profiles of X-ray diffraction patterns for $Y_3(Co_yFe_{1-y})_5O_{12}$ nanoparticles with: (A) $y = 0.000$, (B) $y = 0.015$, and (C) $y = 0.030$.

From Peña-Garcia, R., Guerra, Y., Buitrago, D. M., Leal, L., Santos, F., & Padrón-Hernández, E. (c. 2018). Synthesis and characterization of yttrium iron garnet nanoparticles doped with cobalt. Ceramics International, *44(10), 11314–11319. https://doi.org/10.1016/j.ceramint.2018.03.179*

Elsewhere, the temperature dependence of saturation magnetization has been evaluated for Zn-substituted YIG nanoparticles (Peña-Garcia, Guerra, de Souza et al., 2018). The variation of saturation magnetization as a function of temperature $M(T)$ for nanoparticles was fitted by the extension of the Bloch's law, as suggested by Kobler and colleagues (Köbler & Hoser, 2014; Köbler et al., 2005) as follows:

$$M(T) = M_s(1 - \alpha T^{\varepsilon}) \tag{3.32}$$

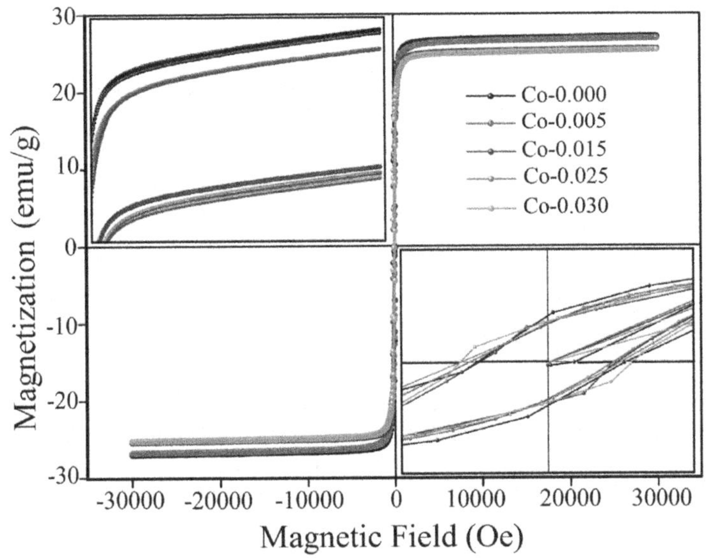

FIGURE 3.130

Vibrating sample magnetometry hysteresis loops obtained from $Y_3(Co_yFe_{1-y})_5O_{12}$ ($0.000 \leq y \leq 0.030$) nanoparticles.

From Peña-Garcia, R., Guerra, Y., Buitrago, D. M., Leal, L., Santos, F., & Padrón-Hernández, E. (c. 2018). Synthesis and characterization of yttrium iron garnet nanoparticles doped with cobalt. Ceramics International, *44(10), 11314–1319. https://doi.org/10.1016/j.ceramint.2018.03.179.*

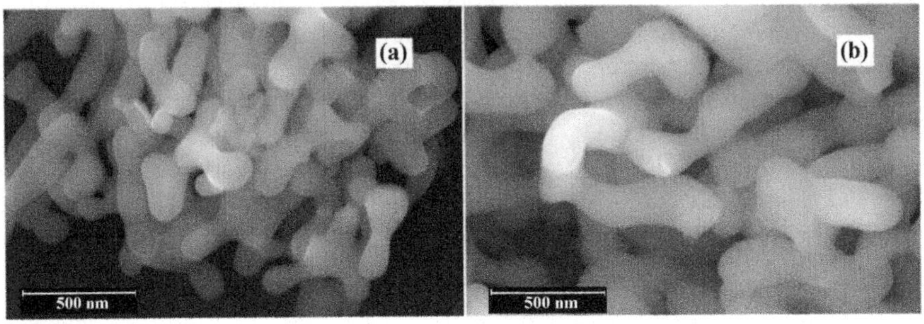

FIGURE 3.131

Scanning electron microscopy micrographs of $Y_3(Cu_yFe_{1-y})_5O_{12}$ nanoparticles with: (A) $y = 0$, and (B) $y = 0.05$.

From Leal, L., Guerra, Y., Padrón-Hernández, E., Rodrigues, A., Santos, F., & Peña-Garcia, R. (c. 2019). Structural and magnetic properties of yttrium iron garnet nanoparticles doped with copper obtained by sol gel method. Materials Letters, 236, *547–549. https://doi.org/10.1016/j.matlet.2018.11.004 7.*

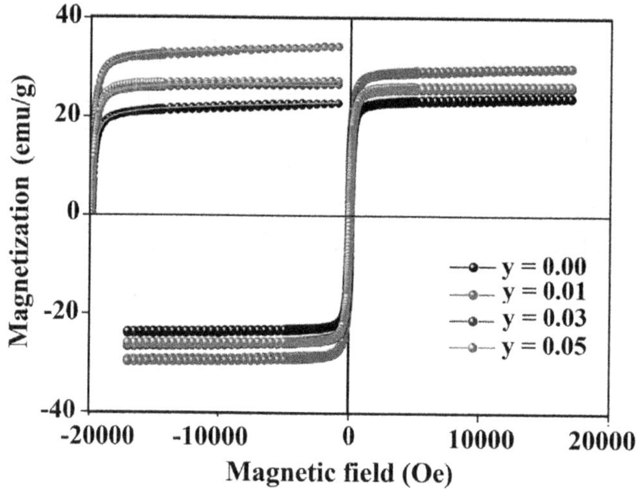

FIGURE 3.132

Vibrating sample magnetometry hysteresis loops of $Y_3(Cu_yFe_{1-y})_5O_{12}$ ($0.00 \le y \le 0.05$) nanoparticles.

From Leal, L., Guerra, Y., Padrón-Hernández, E., Rodrigues, A., Santos, F., & Peña-Garcia, R. (c. 2019). Structural and magnetic properties of yttrium iron garnet nanoparticles doped with copper obtained by sol gel method. Materials Letters, 236, *547–549. https://doi.org/10.1016/j.matlet.2018.11.004 7.*

where M_s is the saturation magnetization, and the coefficient α is related to the magnetocrystalline anisotropy and microstructural features. The exponent ε is determined by the dimensionality, predominant spin, and crystal structure. The ε values are 9/2, 2, 3/2, 3, and 5/2 for homogeneous magnetic materials. The fitted data is shown in Fig. 3.134. The pure YIG nanoparticles exhibit ferrimagnetic characteristics. In the temperature range of 50–300 K (ΔT_0), M_s varies from 37 to 25 emu/g. In the temperature range of 50 to 110 K (ΔT_1), the exponent values obtained from experimental data fitted by the extension of the Bloch law are $\varepsilon_1 = 5.47$, 5.35, 8.70, and 9.50, being different from theoretical values due to structural variation in the YIG nanoparticles at low temperatures. At higher temperatures (> 130 K), the variation in M_s accords with Kobler's equation. In the ΔT_0 range, no clear relation is found between the experimental findings and Kobler's equation. In the ΔT_1 range, the exponent ε_1 increases with increasing the Zn concentration. In this case, the exponent values of the $Y_3(Zn_xFe_{1-x})_5O_{12}$ nanoparticles are obtained to be $\varepsilon_1 =$ 8.70 and 9.51 for $x = 0.03$ and 0.05, respectively. Therefore, the fitted values are different from the expected ones ($\varepsilon = 9/2$, 2, 3/2, 3, and 5/2). It is assumed that the atomic disorder may gradually increase with increasing the dopant concentration, leading to such differences in ε.

The growth behavior and magnetic properties of Bi-substituted YIG nanoparticles have been investigated in another study (Kim et al., 2007). While the Bi-substituted YIG nanoparticles were amorphous at low annealing temperatures

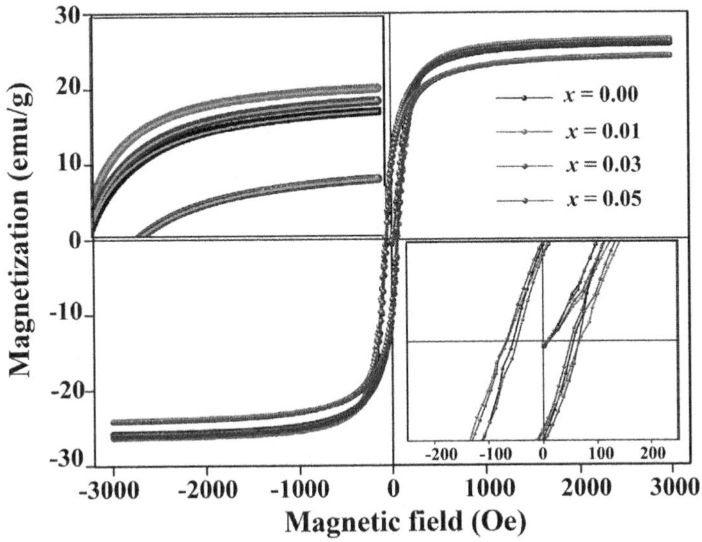

FIGURE 3.133

Vibrating sample magnetometry hysteresis loops of $Y_3(Ni_xFe_{1-x})_5O_{12}$ (0.00 ≤ x ≤ 0.05) nanoparticles.

From Peña-Garcia, R., Guerra, Y., Santos, F., Almeida, L., & Padrón-Hernández, E. (c. 2019). Structural and magnetic properties of Ni-doped yttrium iron garnet nanopowders. Journal of Magnetism and Magnetic Materials, *492, 165650. https://doi.org/10.1016/j.jmmm.2019.165650 9.*

(< 650°C), they were transformed to the garnet phase when increasing the annealing temperature above 700°C. Fig. 3.135A shows a TEM micrograph of the garnet nanoparticles annealed at 650°C. The average particle size is 50 nm. The lattice constant obtained from the electron diffraction pattern depicted in Fig. 3.135B is 1.260 nm. The diffraction pattern of the garnet nanoparticles also exhibits the garnet phase directed along the [012] axis, being indicative of single-phase nanoparticles.

The saturation magnetization and coercivity of $Y_3Fe_5O_{12}$ nanoparticles increased from 1.91 emu/g and 2 Oe to 16.1 emu/g and 45 Oe, respectively, after introducing the Bi content. It was interesting to realize that the substitution of Bi can enhance the superexchange interaction between the ferric ions at the octahedral and tetrahedral sites while increasing the Curie temperature and magnetic moments. This is because of the occurrence of lattice distortion in the crystal structure, leading to the enhancement in superexchange coupling. Fig. 3.136 shows the variation of saturation magnetization as a function of annealing temperature and time for $Y_3Fe_5O_{12}$ nanoparticles. Increasing the annealing temperature from 600 to 750°C increases the saturation magnetization due to the garnet phase transformation from amorphous to crystal. The saturation magnetization of the YIG nanoparticles is 23.9 emu/g, being smaller than that of bulk YIG (26.8 emu/g). This may arise from the spin disorder at the surface of nanoparticles. In addition, increasing the

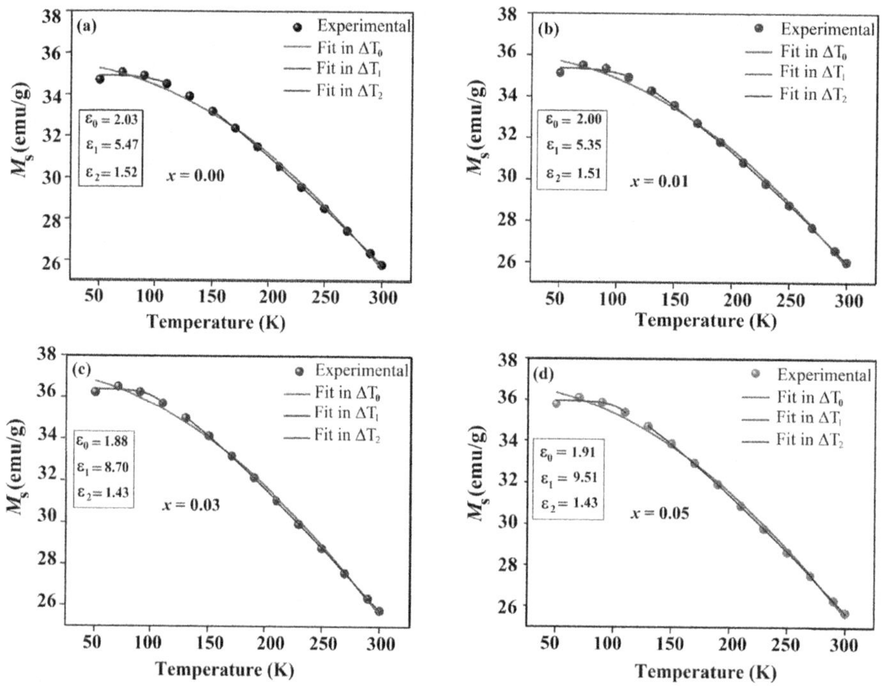

FIGURE 3.134

The temperature dependence of the saturation magnetization (M_S) for $Y_3(Zn_xFe_{1-x})_5O_{12}$ nanoparticles with the composition of: (A) $x = 0.0$, (*B*) $x = 0.01$, (C) $x = 0.03$, and (D) $x = 0.05$. Note: ΔT_0, ΔT_1, and ΔT_2 correspond to temperature ranges of 50–300 K, 50–110 K, and 110–300 K, respectively.

From Peña-Garcia, R., Guerra, Y., de Souza, F., Gonçalves, L., & Padrón-Hernández, E. (c. 2018). The extended Bloch's law in yttrium iron garnet doped with Zn, Ni and Co. Physica E: Low-dimensional Systems and Nanostructures, 103, *354–360. https://doi.org/10.1016/j.physe.2018.06.027 355.*

annealing time to 1 h increases the saturation magnetization. Further increasing the annealing time does not have an effect on the saturation magnetization Fig. 3.136B. The variation of magnetization as a function of annealing time can also be defined by the Avrami's theory for the phase transformation as follows:

$$\frac{M}{M_s} = 1 - \exp(-kt^n) \tag{3.33}$$

where n and k are constant parameters, and t is the annealing time. The calculated parameters of k and n are shown in Fig. 3.136C. It should be noted that parameter k reflects the rate of nucleation and growth, and parameter n represents the mode of crystal growth. At a temperature of 700°C, the parameters are $n = 2.5$ and $k = 0.63$ (Fig. 3.136C), while at 700°C, $n = 2.5$ and $k = 14.56$ (Fig. 3.136D). The value of $n = 2.5$ for both temperatures confirms that the nucleation of garnet takes place

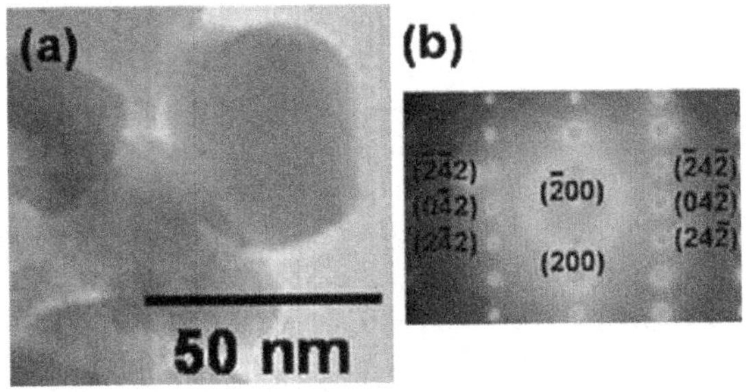

FIGURE 3.135

(A) Transmission electron microscopy micrograph, and (B) electron diffraction pattern of $Y_{1.2}Bi_{1.8}Fe_5O_{12}$ nanoparticles annealed at 650°C.

From Kim, T., Nasu, S., & Shima, M. (c. 2007). Growth and magnetic behavior of bismuth substituted yttrium iron garnet nanoparticles. Journal of Nanoparticle Research, *9(5), 737–743. https://doi.org/10.1007/s11051-6-9082-9 741.*

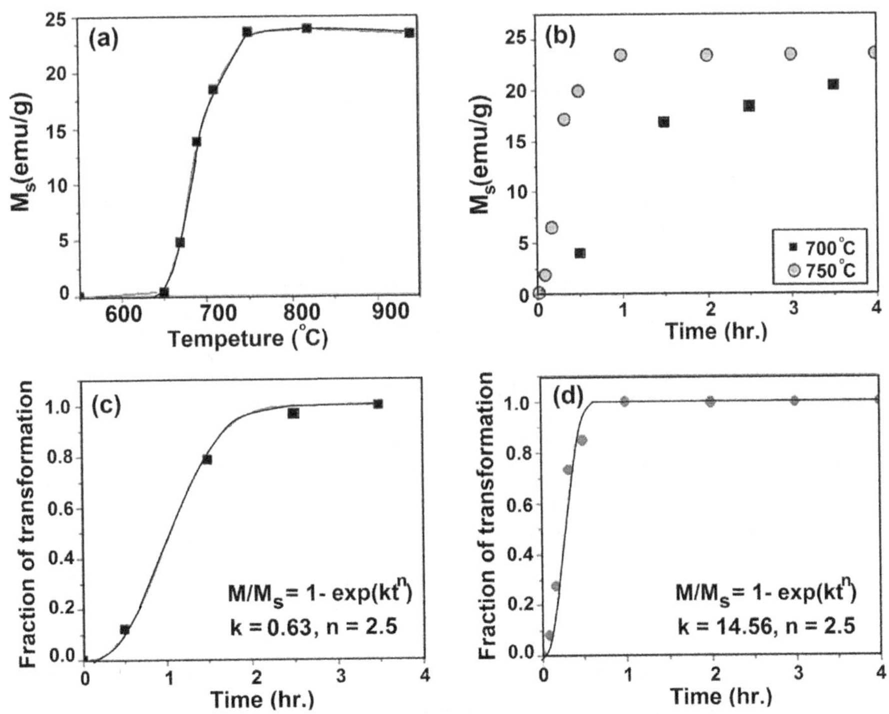

FIGURE 3.136

The variation of saturation magnetization as a function of annealing temperature and time for $Y_3Fe_5O_{12}$ nanoparticles.

From Kim, T., Nasu, S., & Shima, M. (c. 2007). Growth and magnetic behavior of bismuth substituted yttrium iron garnet nanoparticles. Journal of Nanoparticle Research, *9(5), 737–743. https://doi.org/10.1007/s11051-6-9082-9 741.*

homogeneously in particle configurations and could grow in three-dimensional mode.

Mössbauer spectra of $Y_3Fe_5O_{12}$ nanoparticles are shown in Fig. 3.137. As observed, the YIG nanoparticles annealed at 820°C have two sets of sextets related to the high-spin Fe^{3+} ions located on the tetrahedral and octahedral sites in the garnet. The respective hyperfine fields of ^{57}Fe at the tetrahedral and octahedral sites in the garnet are 39 and 48 T. The areal ratio of the absorption peaks is 1.5, which is attributed to the stoichiometric cation occupancy ratio of 24:16 in the tetrahedral and octahedral sites.

Elsewhere, $Y_{1.2}Bi_{1.8}Fe_5O_{12}$ nanoparticles have been prepared by microwave-assisted and conventional coprecipitation techniques (Hong et al., 2009). Fig. 3.138 shows VSM hysteresis loops of the nanoparticle samples prepared by the microwave-assisted (M1) and conventional coprecipitation (C2) techniques. The saturation magnetization values of M1 and C2 samples are 11.26 and 12.81 emu/g, respectively. The lower saturation magnetization of the microwave-assisted coprecipitated nanoparticles can be related to the higher surface-to-volume ratio of the particles.

3.6.3 Rare-earth-substituted yttrium iron garnet nanoparticles

The influence of Dy^{3+} substitution on magnetic, electrical, and dielectric properties of YIG nanoparticles has been investigated (Bhosale et al., 2020). The XRD patterns confirmed that the single phase of the nanoparticles was formed without any

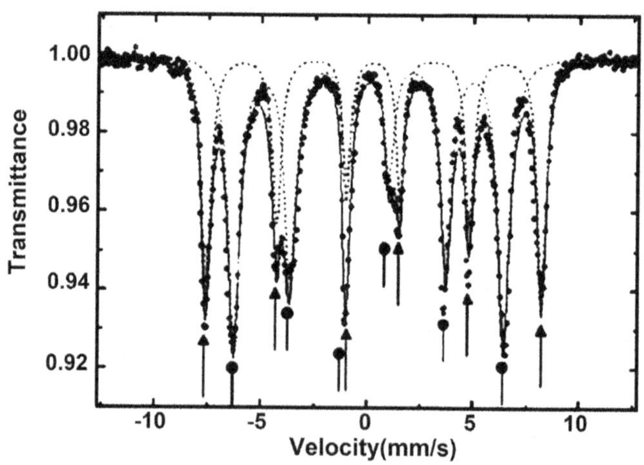

FIGURE 3.137

Mössbauer spectra obtained from $Y_3Fe_5O_{12}$ nanoparticles.

From Kim, T., Nasu, S., & Shima, M. (c. 2007). Growth and magnetic behavior of bismuth substituted yttrium iron garnet nanoparticles. Journal of Nanoparticle Research, *9(5), 737–743. https://doi.org/10.1007/s11051-6-9082-9.*

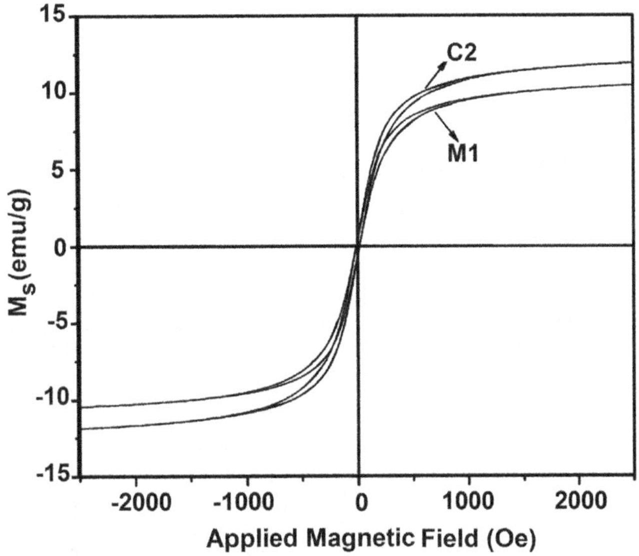

FIGURE 3.138

Vibrating sample magnetometry hysteresis loops of $Y_{1.2}Bi_{1.8}Fe_5O_{12}$ nanoparticles prepared by microwave-assisted (M1) and conventional coprecipitation (C2) techniques.
From Hong, R., Wu, Y., Feng, B., Di, G., Li, H., Xu, B., Zheng, Y., & Wei, D. (c. 2009). Microwave-assisted synthesis and characterization of Bi-substituted yttrium garnet nanoparticles. Journal of Magnetism and Magnetic Materials, 321*(8), 1106–1110. https://doi.org/10.1016/j.jmmm.2008.11.009.*

secondary phase and impurity. Fig. 3.139 shows VSM hysteresis loops of $Y_{3-x}Dy_xFe_5O_{12}$ ($0.0 \le x \le 1.0$) nanoparticles. The magnetic results illustrate that the saturation magnetization of the nanoparticles decreases from 24.33 to 19.09 emu/g with increasing the Dy concentration from $x = 0.0$ to 1.0. The reduction in the saturation magnetization is attributed to the alignment of the magnetic moments of Dy^{3+} ions (10 μ_B) opposite to the effective magnetic moments formed by Fe^{3+} ions.

The variations of dielectric constant and dielectric loss versus frequency for nanoparticles are shown in Fig. 3.140. The dielectric constant and dielectric loss decrease with increasing the Dy concentration. In addition, the dielectric constant and dielectric loss have been confirmed to initially decrease with an increase in frequency. At higher frequencies, the dielectric properties reach a constant value. The dielectric constant reduction at lower frequencies can be related to the relaxation of electric dipoles.

Elsewhere, magnetic properties of $Y_{2.9-y}Ce_{0.1}Gd_yFe_5O_{12}$ ($0.0 \le y \le 1.0$) nanoparticles have been evaluated (Xu et al., 2008). Fig. 3.141 shows the variation of M_s versus particle size for different Gd contents. In this case, while M_s decreases with an increase in the Gd concentration, it increases with increasing the particle size. This

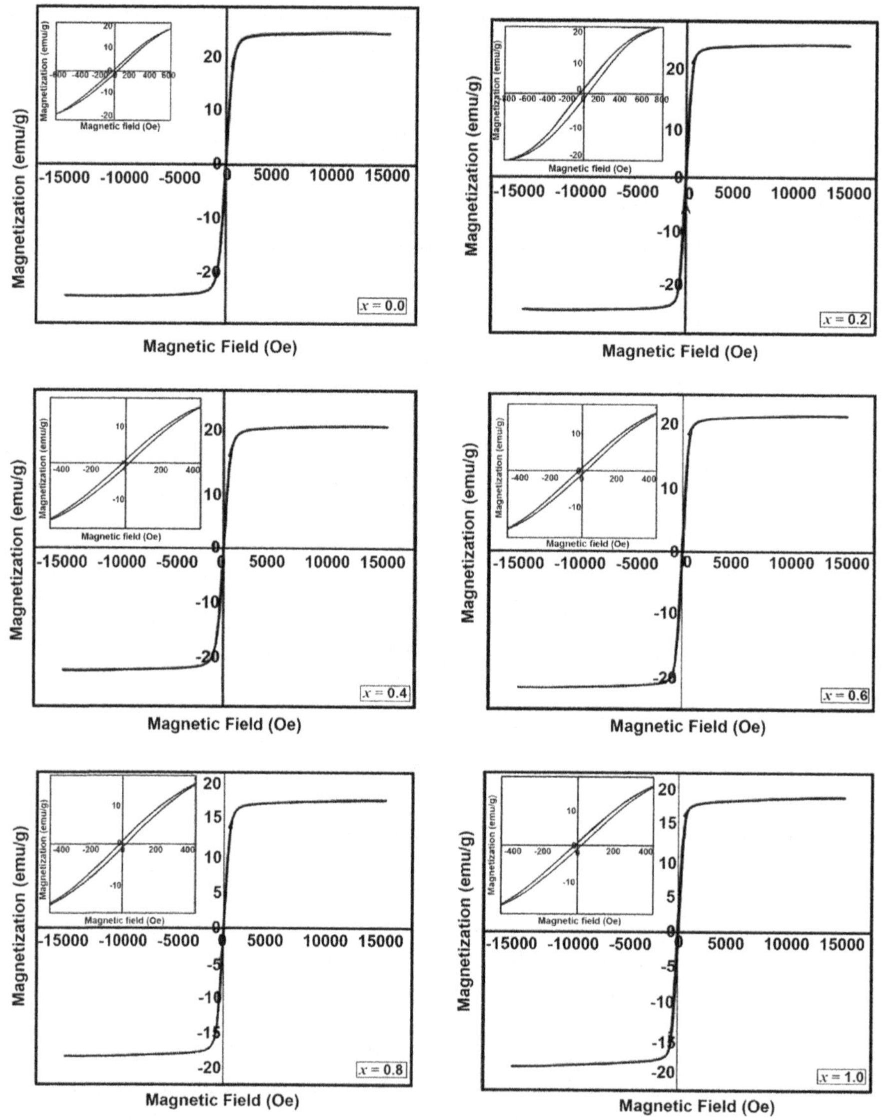

FIGURE 3.139

Vibrating sample magnetometry hysteresis loops of $Y_{3-x}Dy_xFe_5O_{12}$ ($0.0 \leq x \leq 1.0$) nanoparticles.

From Bhosale, A. B., Somvanshi, S. B., Murumkar, V., & Jadhav, K. (c. 2020). Influential incorporation of RE metal ion (Dy^{3+}) in yttrium iron garnet (YIG) nanoparticles: Magnetic, electrical and dielectric behaviour. Ceramics International, 46*(10), 15372–15378. https://doi.org/10.1016/j.ceramint.2020.03.081.*

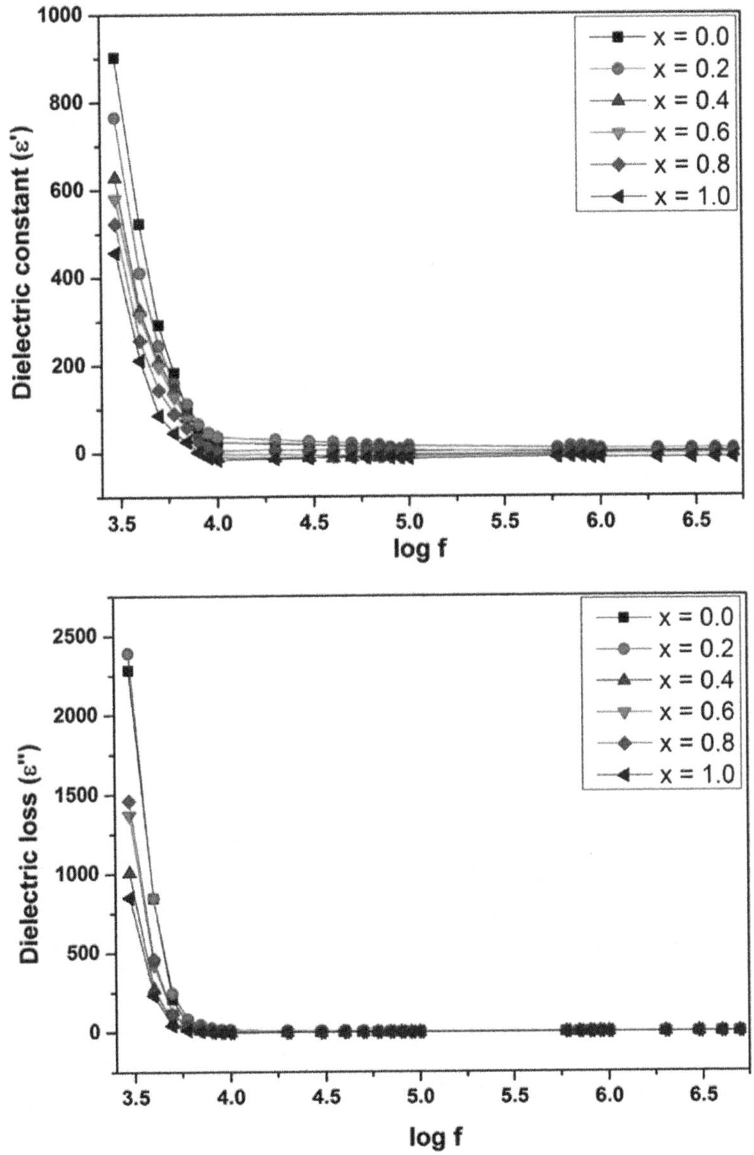

FIGURE 3.140

The variations of dielectric constant and dielectric loss versus frequency for $Y_{3-x}Dy_xFe_5O_{12}$ $(0.0 \leq x \leq 1.0)$ nanoparticles.

From Bhosale, A. B., Somvanshi, S. B., Murumkar, V., & Jadhav, K. (c. 2020). Influential incorporation of RE metal ion (Dy^{3+}) in yttrium iron garnet (YIG) nanoparticles: Magnetic, electrical and dielectric behaviour. Ceramics International, 46*(10), 15372–15378. https://doi.org/10.1016/j.ceramint.2020.03.081.*

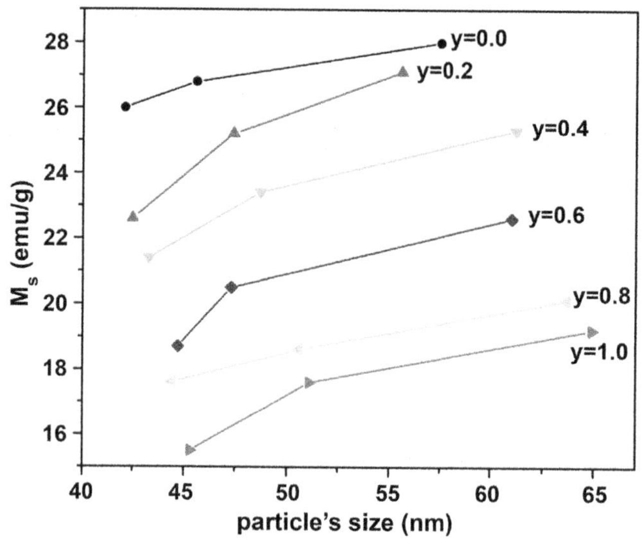

FIGURE 3.141

The variation of M_s versus particle size for $Y_{2.9-y}Ce_{0.1}Gd_yFe_5O_{12}$ ($0.0 \leq y \leq 1.0$) nanoparticles.

From Xu, H., Yang, H., Xu, W., & Feng, S. (c. 2008). Magnetic properties of Ce, Gd-substituted yttrium iron garnet ferrite powders fabricated using a sol–gel method. Journal of Materials Processing Technology, *197(1–3), 296–300. https://doi.org/10.1016/j.jmatprotec.2007.06.061.*

may be caused by the higher surface-to-volume ratio of the smaller particles, resulting in the emergence of a nonmagnetic surface dead-layer.

For microwave device applications, magnetic and crystallographic properties of other rare-earth (R = Ce, La, and Nd)-substituted YIG nanoparticles have also been evaluated (Sharma & Kuanr, 2018). In this respect, hysteresis loops, the plot of magnetization M versus $1/H^2$ at high magnetic fields, saturation magnetization, and magnetocrystalline anisotropy constant (K_1) of YIG and $Y_{2.85}R_{0.15}Fe_5O_{12}$ nanoparticles are shown in Fig. 3.142. The hysteresis loops show that the saturation magnetization increases after the substitution of the rare-earth cations. This arises from the structural distortion, influencing the a-d superexchange interaction. In fact, the distribution of lanthanide ions at the dodecahedral sites can enhance the superexchange interaction.

Based on the VSM data, K_1 can be calculated as follows (Chikazumi & Graham, 2009):

$$M = M_s\left[1 - \frac{8K_1^2}{105\mu_0^2M_s^2H^2}\right] + \chi H \qquad (3.34)$$

where M is the magnetization, H is the applied magnetic field, μ_0 is the free space permeability, and χH is the paramagnetic part. Based on Eq. (3.34), the first

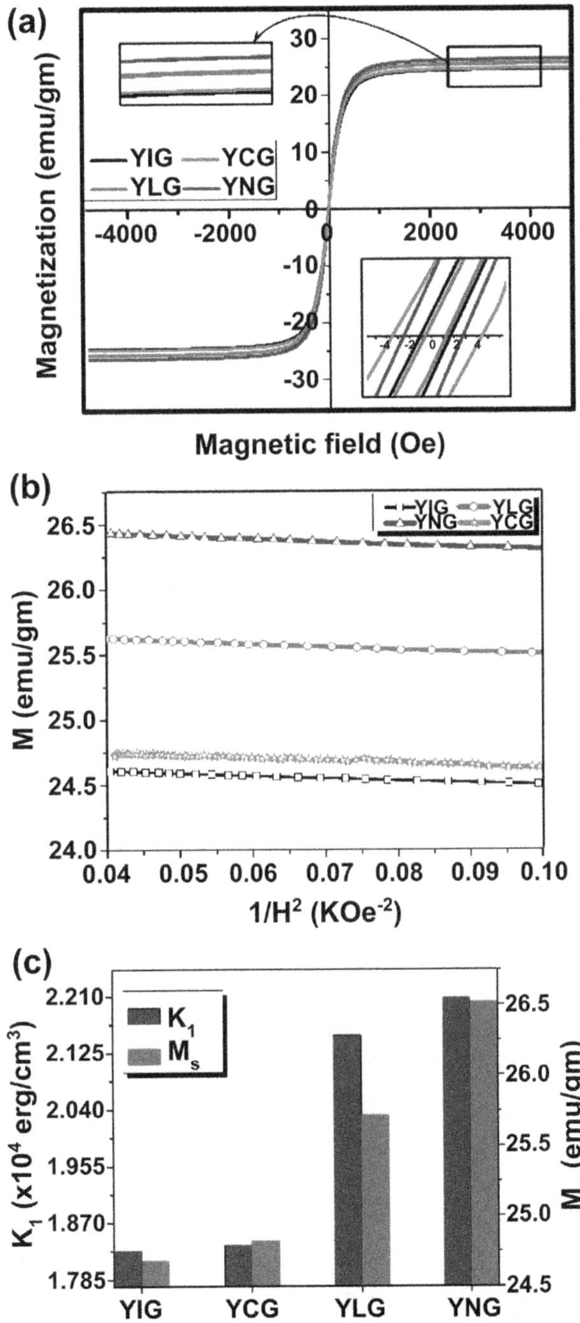

FIGURE 3.142

(A) Vibrating sample magnetometry hysteresis loops, (B) the plot of magnetization M versus $1/H^2$ at high magnetic fields, and (C) saturation magnetization (M_s) and magnetocrystalline anisotropy constant (K_1) of YIG and $Y_{2.85}R_{0.15}Fe_5O_{12}$ (R = Ce (YCG), La (YLG), and Nd (YNG)) nanoparticles.

From Sharma, V., & Kuanr, B. K. (c. 2018). Magnetic and crystallographic properties of rare-earth substituted yttrium-iron garnet. Journal of Alloys and Compounds, 748, *591–600. https://doi.org/10.1016/j.jallcom.2018.03.086 19.*

anisotropy constant can be determined from the slope of the straight line in Fig. 3.142B. The calculated values of K_1 and measured M_s are shown in Fig. 3.142C for the whole series of garnets. The anisotropy constant is enhanced from $K_1 = 1.830\times10^4$ to 2.207×10^4 erg/cm^3 after substituting Nd in the YIG nanoparticles. Based on the magnetic measurements, it is confirmed that the $Y_{2.85}Ce_{0.15}Fe_5O_{12}$ nanoparticles have the lowest value of saturation magnetization and highest value of coercivity among all the lanthanide-substituted YIG nanoparticles.

Fig. 3.143 displays the resonance field (H_r) and FMR linewidth (ΔH) of the $Y_{2.85}R_{0.15}Fe_5O_{12}$ nanoparticles at frequencies of 6 and 20 GHz. Evidently, the FMR absorption increases with the substitution of the rare-earth cations in the YIG nanoparticles, arising from the enhancement in the saturation magnetization and effective susceptibility of the doped garnets (Bowler, 2006; Suzuki et al., 2006). It is also observed that H_r increases with increasing the operating frequency (f_r).

The relation between f_r and H_r can be given by (Sharma et al., 2017):

$$f_r = \frac{\gamma}{2\pi}\left(H_r + H_{\text{eff}}\right) \tag{3.35}$$

$$H_{\text{eff}} = H_D + H_a + H_{\text{int}} \tag{3.36}$$

where γ is the gyromagnetic ratio, H_D is the demagnetizing field, H_a is the anisotropy field, and H_{int} is the internal field. It was reported that the gyromagnetic ratio

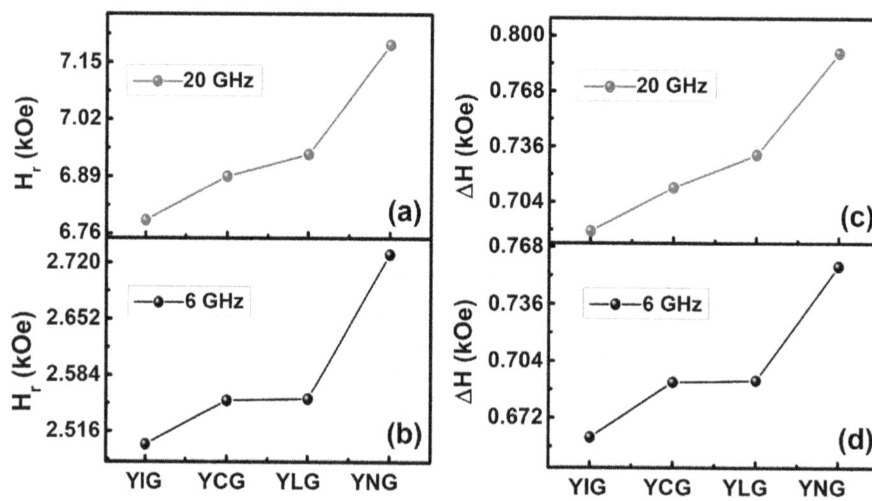

FIGURE 3.143

(A and B) The resonance field (H_r), and (C and D) FMR linewidth (ΔH) of YIG and $Y_{2.85}R_{0.15}Fe_5O_{12}$ (R = Ce, La, and Nd) nanoparticles at frequencies of 6 and 20 GHz.

From Sharma, V., & Kuanr, B. K. (c. 2018). Magnetic and crystallographic properties of rare-earth substituted yttrium-iron garnet. Journal of Alloys and Compounds, 748, *591–600. https://doi.org/10.1016/j.jallcom.2018.03.086 19.*

slightly increased with the substitution of rare-earth cations in the YIG nanoparticles. The reasoning behind the increase in the gyromagnetic ratio could be related to the enhancement in the superexchange interaction between Fe^{3+}– Fe^{2+} ions and the spin–orbit coupling after the substitution of the rare-earth cations (Fu et al., 2009; Youssef & Brosseau, 2006).

The values of the Gilbert damping parameter were also calculated by fitting ΔH versus frequency based on the following phenomenological equation:

$$\Delta H = \Delta H_0 + \frac{4\pi f \alpha}{\gamma} \tag{3.37}$$

where ΔH_0 is the extrinsic portion of the linewidth, and α is the intrinsic Gilbert damping parameter.

The Gilbert damping parameter and relaxation time of YIG and $Y_{2.85}R_{0.15}Fe_5O_{12}$ (R = Ce, La, and Nd) nanoparticles are shown in Fig. 3.144. As inferred, H_r and ΔH values increase with the substitution of the rare-earth cations in the YIG nanoparticles. Indeed, the coupling between spin phonons and ferric lattice is enhanced with the substitution of the rare-earth ions, resulting from their magnetic moments induced by the 4f electrons (Sekijima et al., 1999). Meanwhile, the Gilbert damping parameter increases from $\alpha = 2.45 \times 10^{-3}$ to 3.8×10^{-3} after substituting Nd in the YIG nanoparticles. In this case, the spin–orbit coupling is enhanced in the rare-earth-substituted nanoparticle samples. In addition, coupling between Fe^{3+}– Fe^{2+} magnetic ions increases with the substitution of the rare-earth cations, arising from the presence of defects in the YIG crystal structure. Consequently, the

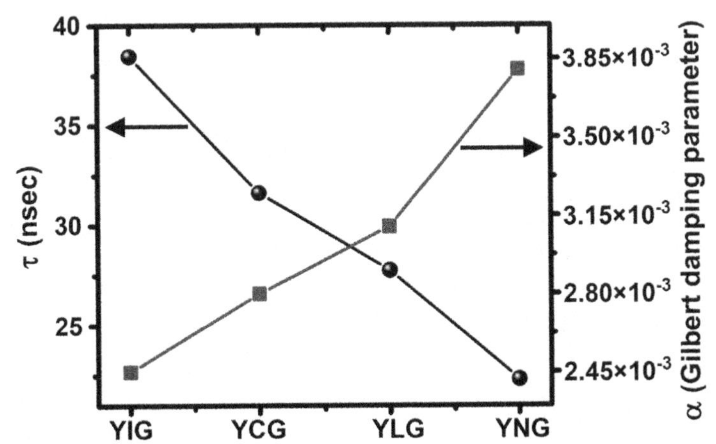

FIGURE 3.144

Gilbert damping parameter (α) and relaxation time (τ) for YIG and $Y_{2.85}R_{0.15}Fe_5O_{12}$ (R = Ce, La, and Nd) nanoparticles at a frequency of 10 GHz.

From Sharma, V., & Kuanr, B. K. (c. 2018). Magnetic and crystallographic properties of rare-earth substituted yttrium-iron garnet. Journal of Alloys and Compounds, 748, *591–600. https://doi.org/10.1016/j.jallcom.2018.03.086 19.*

Nd-substituted YIG nanoparticles can act as a suitable candidate for broadband-width microwave filter applications.

Alternatively, the relaxation time (τ) was calculated from Eq. (3.38) (Fähnle & Illg, 2011):

$$\tau = \frac{1}{\gamma \alpha H_r} \tag{3.38}$$

It is interesting to find that the substitution of the rare-earth cations in the YIG crystal structure decreases τ. Notably, τ is reduced from 38.45 to 22.3 ns after the substitution of Nd in the YIG nanoparticles. The relaxation time reduction may provide memory devices with low switching time and fast speed. Therefore, the Nd-substituted YIG nanoparticles are useful for designing high-speed memory devices (Sharma & Kuanr, 2018).

In another research study, microstructural and magnetic properties of holmium (Ho)-substituted YIG nanoparticles with the composition of $Y_{1.5}Ho_{1.5}Fe_5O_{12}$ have been investigated (Hapishah et al., 2019). $B-H$ hysteresis loops of nanoparticles sintered at different temperatures (1000–400°C) are shown in Fig. 3.145. The saturation induction decreases after the substitution of Ho, which can be related to the reduction in the superexchange interaction of metal cations at the tetrahedral,

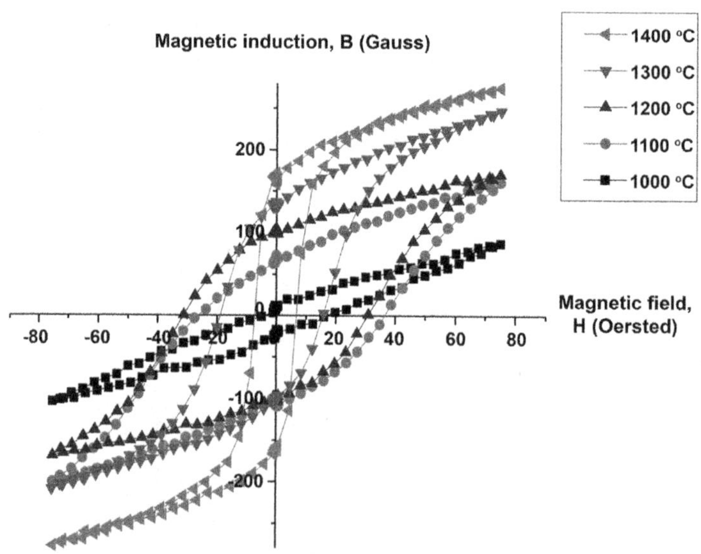

FIGURE 3.145

$B-H$ hysteresis loops obtained from $Y_{1.5}Ho_{1.5}Fe_5O_{12}$ nanoparticles sintered at different temperatures.

From Hapishah, A., Hamidon, M., Syazwan, M., & Shafiee, F. (c. 2019). Effect of grain size on microstructural and magnetic properties of holmium substituted yttrium iron garnets ($Y_{1.5}Ho_{1.5}Fe_5O_{12}$). Results in Physics, 14, *102391. https://doi.org/10.1016/j.rinp.2019.102391 7.*

octahedral, and dodecahedral sites. In fact, the substitution of yttrium ions with the rare-earth holmium ones can change the strength of magnetic interactions. In addition, the saturation induction increases with increasing the sintering temperature. On the other hand, the coercivity increases for all the Ho-YIG nanoparticle samples compared with the pure YIG. This may be attributed to the presence of segregation of rare-earth metal cations at the grain boundaries, as well as the involvement of other factors such as the secondary phase due to the crystal imperfection (Jahanbin et al., 2010; Tebble & Craik, 1969).

The variation of permeability as a function of frequency for the $Y_{1.5}Ho_{1.5}Fe_5O_{12}$ nanoparticles sintered at the different temperatures is shown in Fig. 3.146. The value of permeability is seen to increase with increasing the sintering temperature because of the enhancement in the density and grain size of the Ho-substituted YIG nanoparticles. The internal stress is also reduced by increasing the sintering temperature, giving rise to a decrease in magnetic anisotropy. Therefore, the resonance frequency can decrease with an increase in permeability, according to Snoek's law. In addition, the values of μ' and μ'' of Ho-YIG nanoparticle samples are smaller than those of pure YIG. The substitution of holmium ions in the garnet lattice may result in lattice distortion or crystalline field, thus generating internal stress. In turn, this can hinder the DW displacement, leading to the reduction in the real and imaginary parts of permeability (Ishino & Narumiya, 1987). Therefore, due to their enhanced soft magnetic properties, the holmium-substituted YIG nanoparticles are suitable for being employed in the microwave region of electronic devices.

3.6.4 Other types of garnet nanoparticles

Other garnet types also contain new developments and have many technological applications. Gadolinium-substituted iron garnet (GdIG) nanoparticles have been prepared using a high-energy ball milling method and evaluated for structural and magnetic properties (Joseyphus et al., 2006). In fact, by applying high-energy ball milling, the gadolinium iron garnet can be decomposed into $GdFeO_3$, Fe_2O_3, and Gd_2O_3. The first quadrant hysteresis loops at 300 K show that with an increase in milling time up to 25 h, the saturation magnetization increases, but this is reduced at a higher milling time (32 h). Fig. 3.147 shows the first quadrant hysteresis loops of the GdIG nanoparticles at 77 and 300 K for various milling times. The magnetization measured at 77 K is observed to decrease with increasing the milling time, owing to the enhancement in the volume fraction of the antiferromagnetic and nonmagnetic phases that appeared during the milling process. Moreover, the formation of the antiferromagnetic oxides and surface effects with a large volume fraction is responsible for reducing magnetization at 300 K for 32 h milling.

Fig. 3.148 shows Mössbauer spectra of the GdIG nanoparticles for various milling times at 16 K. The loss of oxygen is revealed by the IS values of the nanoparticles milled for longer times. The absence of the superparamagnetic doublet confirms that the blocking temperature of the nanoparticles is higher than 16 K. In the case of as-prepared GdIG nanoparticles, the hyperfine magnetic field is

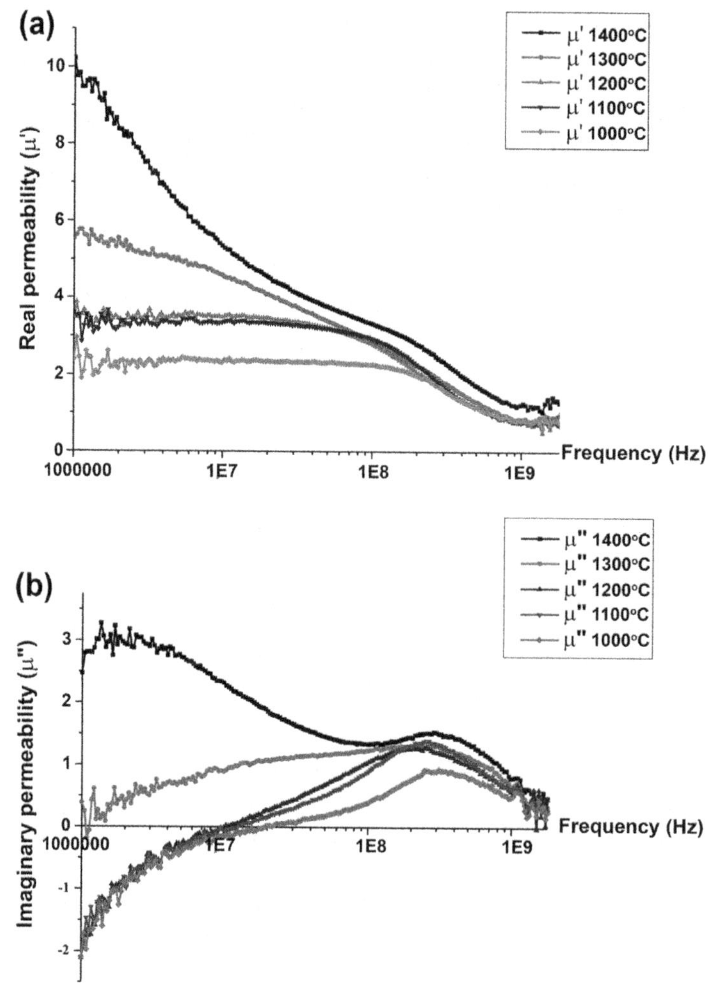

FIGURE 3.146

The variations of real and imaginary parts of permeability as a function frequency for $Y_{1.5}Ho_{1.5}Fe_5O_{12}$ nanoparticles sintered at different temperatures.

From Hapishah, A., Hamidon, M., Syazwan, M., & Shafiee, F. (c. 2019). Effect of grain size on microstructural and magnetic properties of holmium substituted yttrium iron garnets ($Y_{1.5}Ho_{1.5}Fe_5O_{12}$). Results in Physics, 14, 102391. https://doi.org/10.1016/j.rinp.2019.102391 7.

enhanced to 54.7 and 48.2 T for the octahedral and tetrahedral sites, respectively. Moreover, for the sample ball milled for 10 h, the two sextets of the octahedral and tetrahedral sites with the intensity ratio of 2:3 exhibit the respective hyperfine fields of 54.7 and 48.0 T. The formation of $GdFeO_3$ with cubic structure and α-Fe_2O_3 with hexagonal structure is also confirmed from the Mössbauer

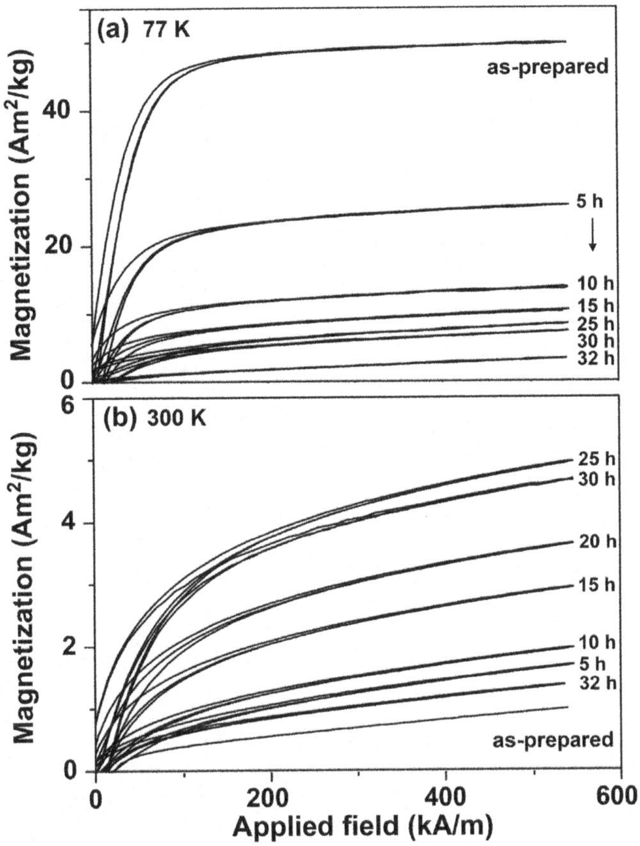

FIGURE 3.147

The first quadrant hysteresis loops of gadolinium-substituted iron garnet nanoparticles at (A) 77 K, and (B) 300 K for various milling times.

From Joseyphus, R., Narayanasamy, A., Nigam, A., & Krishnan, R. (c. 2006). Effect of mechanical milling on the magnetic properties of garnets. Journal of Magnetism and Magnetic Materials, 296*(1), 57–64. https://doi.org/10.1016/j.jmmm.2005.04.018 64.*

spectroscopy of sample milled for 10 h. The unidentified sextet with small relative intensity is also observed. The Mössbauer spectra for garnets fabricated at milling times of 25 and 30 h provide features similar to those of the 10 h milled garnet. It must be noted that with an increase in milling time, the intensity of orthoferrite and α-Fe_2O_3 phases increases. The hyperfine field values for garnet milled for 30 h are slightly lower than those for samples milled for 10 and 20 h. The ISs are also slightly more positive, which could be related to the loss of oxygen during milling, which caused the reduction of Fe^{3+} to Fe^{2+}.

In another research study, the influence of Gd-Ce substitution on structural and magnetic characteristics of garnet $Gd_3Ce_{3-x}Fe_5O_{12}$ ($0.0 \leq x \leq 3.0$) has been

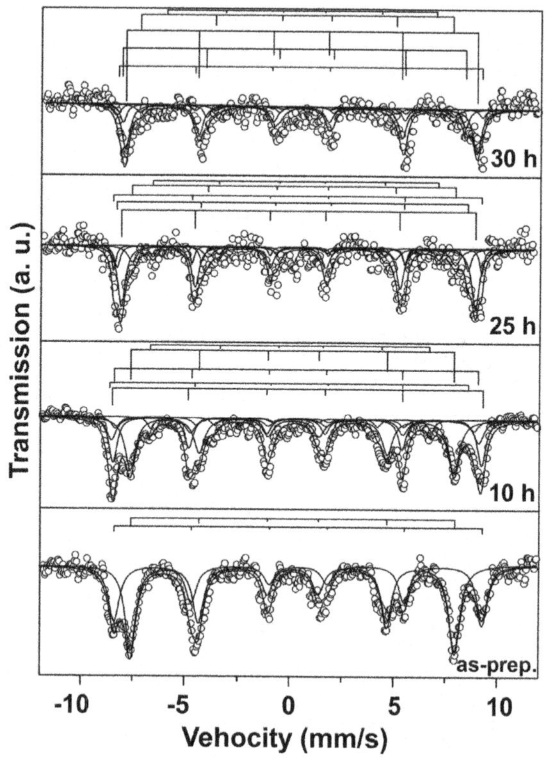

FIGURE 3.148

Mössbauer spectra of gadolinium-substituted iron garnet nanoparticles for various milling times at 16 K.

From Joseyphus, R., Narayanasamy, A., Nigam, A., & Krishnan, R. (c. 2006). Effect of mechanical milling on the magnetic properties of garnets. Journal of Magnetism and Magnetic Materials, 296*(1), 57–64. https://doi.org/10.1016/j.jmmm.2005.04.018 64.*

evaluated (Akhtar et al., 2016). The XRD patterns confirmed the presence of single-phase Gd-Ce garnet nanoparticles. The variations of M_r, M_s, M_r/M_s ratio, and H_c as a function of Ce concentration (x) for nanoparticles are shown in Fig. 3.149. Evidently, M_s and M_r decrease with increasing Ce concentration, which may be related to nanoparticle size and morphology. However, coercivity is enhanced by increasing the Ce^{3+} concentration from $x = 0$ to 2.5. A further increase in the Ce^{3+} concentration to $x = 3.0$ leads to reduced coercivity.

Variations in the magnetic moment and magnetocrystalline anisotropy constant as a function of Gd–Ce concentration are shown in Fig. 3.150. As can be seen, the magnetocrystalline anisotropy constant increases with an increase in concentration to $x = 2.5$, followed by its reduction when x is increased to 3.0. The varied behavior of the magnetic moments of the nanoparticles is related to their

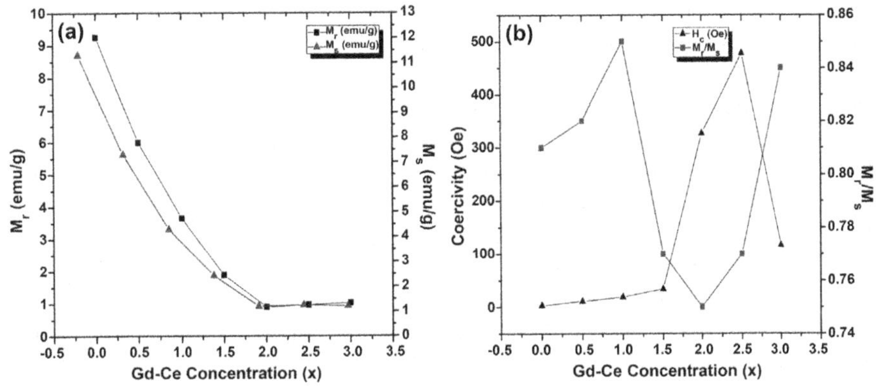

FIGURE 3.149

The variations of remanence magnetization (M_r), M_S, M_r/M_S ratio, and coercivity as a function of Gd–Ce concentration (x) for $Gd_3Ce_{3-x}Fe_5O_{12}$ nanoparticles.

From Akhtar, M. N., Sulong, A., Ahmad, M., Khan, M. A., Ali, A., & Islam, M. (c. 2016). Impacts of Gd–Ce on the structural, morphological and magnetic properties of garnet nanocrystalline ferrites synthesized via sol–gel route. Journal of Alloys and Compounds, 660, *486–495. https://doi.org/10.1016/j.jallcom.2015.11.146 24.*

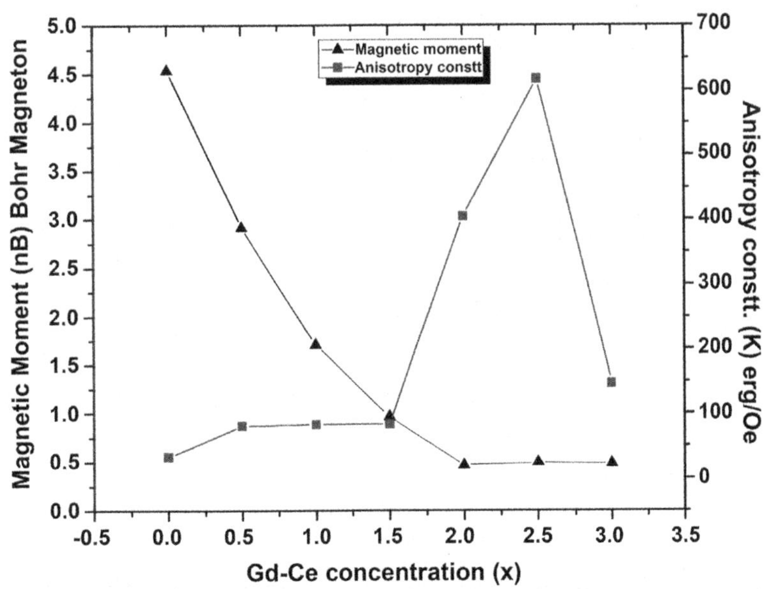

FIGURE 3.150

The variations of magnetic moment and magnetocrystalline anisotropy constant as a function of Gd-Ce concentration (x) for $Gd_3Ce_{3-x}Fe_5O_{12}$ nanoparticles.

From Akhtar, M. N., Sulong, A., Ahmad, M., Khan, M. A., Ali, A., & Islam, M. (c. 2016). Impacts of Gd–Ce on the structural, morphological and magnetic properties of garnet nanocrystalline ferrites synthesized via sol–gel route. Journal of Alloys and Compounds, 660, *486–495. https://doi.org/10.1016/j.jallcom.2015.11.146 24.*

superexchange interactions, which involve the net magnetic moments of Gd^{3+} and Ce^{3+} in tetrahedral, octahedral, and dodecahedral sites.

Elsewhere, structural and magnetic properties of several rare-earth iron garnet ($R_3Fe_5O_{12}$, where R=Sm, Eu, Gd, Tb, Dy, Ho, Er, Tm, Yb, and Lu) nanoparticles prepared by an aqueous sol–gel method have been studied (Opuchovic et al., 2017). The phase purity of the different lanthanide iron garnets depended on the final annealing temperature. Mössbauer spectra of $R_3Fe_5O_{12}$ nanoparticles are shown in Fig. 3.151, consisting of two magnetic subspectra (sextets). These correspond to the octahedral and tetrahedral sublattices. For "*a*" and "*d*" sites, the magnetic hyperfine fields are 49 and 39 T, respectively. Interestingly, the area ratio corresponding to 24*d* tetrahedral and 16*a* octahedral sites is 3:2 for the following single-phase garnet nanoparticles: $Gd_3Fe_5O_{12}$ annealed at 1000°C and $Lu_3Fe_5O_{12}$ annealed at 800°C.

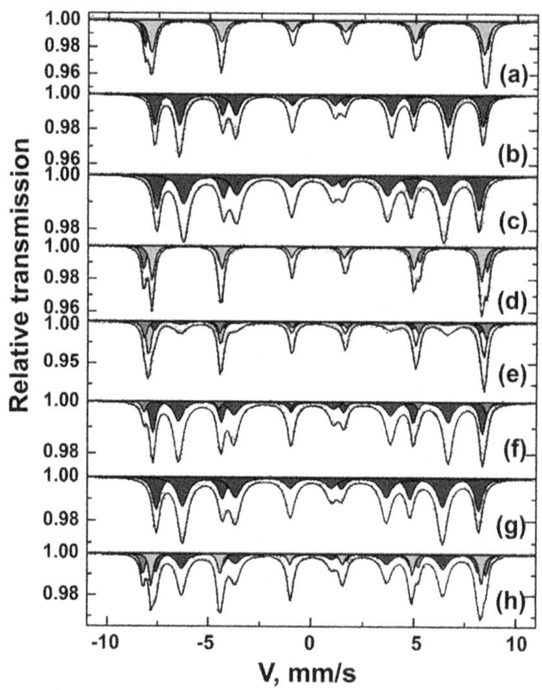

FIGURE 3.151

Mössbauer spectra of rare-earth iron garnet nanoparticles annealed at different temperatures: (A) $Gd_3Fe_5O_{12}$, 800°C, (B) $Gd_3Fe_5O_{12}$, 1000°C, (C) $Lu_3Fe_5O_{12}$, 800°C, (D) $Lu_3Fe_5O_{12}$, 1000°C, (E) $Sm_3Fe_5O_{12}$, 800°C, (F) $Sm_3Fe_5O_{12}$, 1000°C, (G) $Yb_3Fe_5O_{12}$, 800°C, and (H) $Yb_3Fe_5O_{12}$, 1000°C.

From Opuchovic, O., Kareiva, A., Mazeika, K., & Baltrunas, D. (c. 2017). Magnetic nanosized rare earth iron garnets R3Fe5O12: sol–gel fabrication, characterization and reinspection. Journal of Magnetism and Magnetic Materials, 422, *425–433. https://doi.org/10.1016/j.jmmm.2016.09.041 431.*

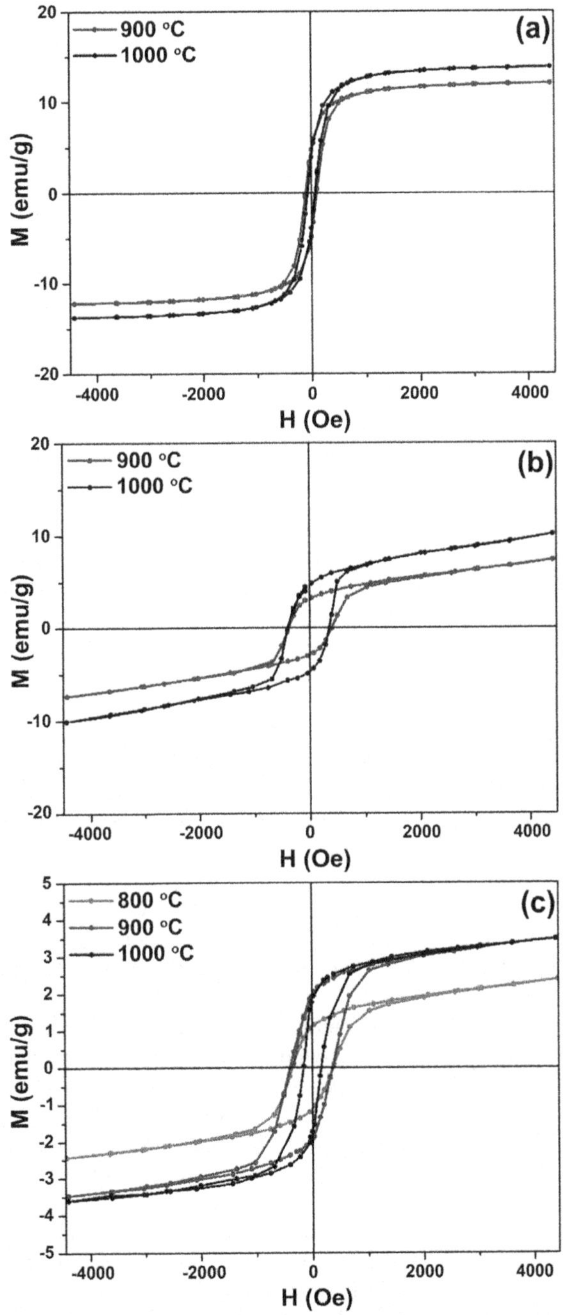

FIGURE 3.152

Vibrating sample magnetometry hysteresis loops of: (A) $Eu_3Fe_5O_{12}$, (B) $Gd_3Fe_5O_{12}$, and (C) $Tb_3Fe_5O_{12}$ nanoparticles annealed at different temperatures.

From Opuchovic, O., Kareiva, A., Mazeika, K., & Baltrunas, D. (c. 2017). Magnetic nanosized rare earth iron garnets R3Fe5012: sol–gel fabrication, characterization and reinspection. Journal of Magnetism and Magnetic Materials, 422, *425–433. https://doi.org/10.1016/j.jmmm.2016.09.041 431.*

For $Sm_3Fe_5O_{12}$ nanoparticles annealed at 800°C and $Yb_3Fe_5O_{12}$ nanoparticles annealed at 1000°C, the garnet subspectra area ratio is 1:7.

VSM hysteresis loops of $R_3Fe_5O_{12}$ (R = Eu, Gd, and Tb) nanoparticles annealed at different temperatures are exhibited in Fig. 3.152. Increasing the particle size and annealing temperature increases the saturation magnetization. The $Eu_3Fe_5O_{12}$ nanoparticles show lower coercivity and higher saturation magnetization than the $Gd_3Fe_5O_{12}$ and $Tb_3Fe_5O_{12}$ nanoparticles. As well, the lowest saturation magnetization and highest coercivity are obtained for the $Gd_3Fe_5O_{12}$ nanoparticles. Clearly, the annealing temperature imposes a significant variation in the magnetic characteristics of the synthesized garnets.

References

Abdellahi, M., Karamian, E., Najafinezhad, A., Ranjabar, F., Chami, A., & Khandan, A. (2018). Diopside-magnetite; A novel nanocomposite for hyperthermia applications. *Journal of the Mechanical Behavior of Biomedical Materials, 77*, 534–538. https://doi.org/10.1016/j.jmbbm.2017.10.015

Adenot, A., Acher, O., Taffary, T., & Longuet, L. (2002). Sum rules on the dynamic permeability of hexagonal ferrites. *Journal of Applied Physics, 91*(10), 7601–7603. https://doi.org/10.1063/1.1447505

Aharoni, A. (1969). Effect of a magnetic field on the superparamagnetic relaxation time. *Physical Review, 177*(2), 793–796. https://doi.org/10.1103/PhysRev.177.793

Ahmad, M., Ali, Q., Ali, I., Ahmad, I., Khan, M. A., Akhtar, M. N., ... Rana, M. (2013). Effects of Sr-substitution on the structural and magnetic behavior of Ba-based Y-type hexagonal ferrites. *Journal of Alloys and Compounds, 580*, 23–28. https://doi.org/10.1016/j.jallcom.2013.05.076

Ahmad, M., Grössinger, R., Kriegisch, M., Kubel, F., & Rana, M. (2013). Characterization of Sr-substituted W-type hexagonal ferrites synthesized by sol–gel autocombustion method. *Journal of Magnetism and Magnetic Materials, 332*, 137–145. https://doi.org/10.1016/j.jmmm.2012.12.039

Ahmed, S. R., Ogale, S., Papaefthymiou, G. C., Ramesh, R., & Kofinas, P. (2002). Magnetic properties of $CoFe_2O_4$ nanoparticles synthesized through a block copolymer nanoreactor route. *Applied Physics Letters, 80*(9), 1616–1618. https://doi.org/10.1063/1.1456258

Akhtar, M. N., Javed, S., Ahmad, M., Sulong, A., & Khan, M. A. (2020). Sol gel derived MnTi doped Co2 W-type hexagonal ferrites: Structural, physical, spectral and magnetic evaluations. *Ceramics International, 46*(6), 7842–7849. https://doi.org/10.1016/j.ceramint.2019.12.003

Akhtar, M. N., & Khan, M. A. (2018). Effect of rare earth doping on the structural and magnetic features of nanocrystalline spinel ferrites prepared via sol gel route. *Journal of Magnetism and Magnetic Materials, 460*, 268–277.

Akhtar, M. N., & Khan, M. A. (2018). Structural, physical and magnetic evaluations of Ce-Zn substituted SrCo2 W-type hexaferrites prepared via sol gel auto combustion route. *Ceramics International, 44*(11), 12921–12928. https://doi.org/10.1016/j.ceramint.2018.04.104

Akhtar, M. N., Nazir, M. S., Tahir, Z., Qamar, S., & Khan, M. A. (2020). Impact of Co doping on physical, structural, microstructural and magnetic features of MgZn nanoferrites for high frequency applications. *Ceramics International, 46*(2), 1750–1759. https://doi.org/10.1016/j.ceramint.2019.09.149

Akhtar, M. N., Sulong, A., Ahmad, M., Khan, M. A., Ali, A., & Islam, M. (2016). Impacts of Gd–Ce on the structural, morphological and magnetic properties of garnet nanocrystalline ferrites synthesized via sol–gel route. *Journal of Alloys and Compounds, 660*, 486–495. https://doi.org/10.1016/j.jallcom.2015.11.146

Ali, I., Islam, M., Ashiq, M. N., Iqbal, M. A., Khan, H. M., & Karamat, N. (2013). Effect of Tb–Mn substitution on DC and AC conductivity of Y-type hexagonal ferrite. *Journal of Alloys and Compounds, 579*, 576–582. https://doi.org/10.1016/j.jallcom.2013.06.182

Amirabadizade, A., Shiri, N., & Ghasemi, A. (2016). The role of Mn–Mg–Ti–Zr substitution on structural and magnetic features of $BaFe_{12-x}$ $(MnMgTiZr)_{x/4}O_{19}$ nanoparticles. *Journal of Superconductivity and Novel Magnetism, 29*(2), 515–520. https://doi.org/10.1007/s10948-015-3332-y

Asif, S. U., Rizwan, S., Awan, M. Q., Khan, M. W., Sadiq, I., Mazhar, M. E., … Hu, W. (2019). Effect of Dy-Co on physical and magnetic properties of X-type hexaferrites $(Ba_{2-x}Dy_xCu_2Fe_{28-y}Co_yO_{46})$. *Chinese Journal of Physics, 61*, 47–54. https://doi.org/10.1016/j.cjph.2019.07.007

Bahadur, D. (1992). Current trends in applications of magnetic ceramic materials. *Bulletin of Materials Science, 15*(5), 431–439. https://doi.org/10.1007/BF02745292

Bai, Y., Zhou, J., Gui, Z., & Li, L. (2006). Phase formation process, microstructure and magnetic properties of Y-type hexagonal ferrite prepared by citrate sol–gel auto-combustion method. *Materials Chemistry and Physics, 98*(1), 66–70. https://doi.org/10.1016/j.matchemphys.2005.08.067

Bertram, H. N. (1994). *Theory of magnetic recording*. Cambridge University Press.

Bhosale, A. B., Somvanshi, S. B., Murumkar, V., & Jadhav, K. (2020). Influential incorporation of RE metal ion (Dy^{3+}) in yttrium iron garnet (YIG) nanoparticles: Magnetic, electrical and dielectric behaviour. *Ceramics International, 46*(10), 15372–15378. https://doi.org/10.1016/j.ceramint.2020.03.081

Blundell, S. (2003). *Magnetism in condensed matter*. American Association of Physics Teachers.

Bowler, N. (2006). Designing dielectric loss at microwave frequencies using multi-layered filler particles in a composite. *IEEE Transactions on Dielectrics and Electrical Insulation, 13*(4), 703–711. https://doi.org/10.1109/TDEI.2006.1667727

Brown, W. F., Jr. (1959). Relaxational behavior of fine magnetic particles. *Journal of Applied Physics, 30*(4), 130–132. https://doi.org/10.1063/1.2185851

Brown, W. F., Jr. (1963). Thermal fluctuations of a single-domain particle. *Physical Review, 130*(5), 1677–1686. https://doi.org/10.1103/PhysRev.130.1677

Caltun, O., Dumitru, I., Feder, M., Lupu, N., & Chiriac, H. (2008). Substituted cobalt ferrites for sensors applications. *Journal of Magnetism and Magnetic Materials, 320*(20), 869–873. https://doi.org/10.1016/j.jmmm.2008.04.067

Chantrell, R. W., El-Hilo, M., & O'Grady, K. (1991). Spin-glass behavior in a fine particle system. *IEEE Transactions on Magnetics, 27*(4), 3570–3578. https://doi.org/10.1109/20.102929

Chauhan, S., Arora, M., Sati, P., Chhoker, S., Katyal, S., & Kumar, M. (2013). Structural, vibrational, optical, magnetic and dielectric properties of $Bi_{1-x}Ba_xFeO_3$ nanoparticles. *Ceramics International, 39*(6), 6399–6405. https://doi.org/10.1016/j.ceramint.2013.01.066

Chen, J., Liu, Y., Wang, Y., Yin, Q., Liu, Q., Wu, C., & Zhang, H. (2020). Magnetic and microwave properties of polycrystalline U-type hexaferrite $Ba_4Zn_{2-x}CoxFe_{36}O_{60}$. *Journal of*

Magnetism and Magnetic Materials, 496, 165948. https://doi.org/10.1016/j.jmmm.2019.165948

Chikazumi, S., & Graham, C. D. (2009). *Physics of Ferromagnetism 2e, 94*. Oxford University Press on Demand.

Dickson, D., Reid, N., Hunt, C., Williams, H., El-Hilo, M., & O'Grady, K. (1993). Determination of f0 for fine magnetic particles. *Journal of Magnetism and Magnetic Materials, 125*(3), 345–350. https://doi.org/10.1016/0304-8853(93)90109-F

Dimri, M. C., Kashyap, S. C., & Dube, D. C. (2006). Complex permittivity and permeability of $Co_2U(Ba_4Co_2Fe_{36}O_{60})$ hexaferrite bulk and composite thick films at radio and microwave frequencies. *IEEE Transactions on Magnetics, 42*(11), 3635–3640. https://doi.org/10.1109/TMAG.2006.882378

Dimri, M. C., Khanduri, H., Kooskora, H., Heinmaa, I., Joon, E., & Stern, R. (2011). Magnetic properties and ^{57}Fe NMR studies of U-type hexaferrites. *Journal of Magnetism and Magnetic Materials, 323*(16), 2210–2213. https://doi.org/10.1016/j.jmmm.2011.03.033

Dormann, J. L., Fiorani, D., & Tronc, E. (1999). On the models for interparticle interactions in nanoparticle assemblies: comparison with experimental results. *Journal of Magnetism and Magnetic Materials, 202*(1), 251–267. https://doi.org/10.1016/S0304-8853(98)00627-1

Drmota, A., Drofenik, M., & Žnidaršič, A. (2012). Synthesis and characterization of nanocrystalline strontium hexaferrite using the co-precipitation and microemulsion methods with nitrate precursors. *Ceramics International, 38*(2), 973–979. https://doi.org/10.1016/j.ceramint.2011.08.018

Ebrahimi, F., Bakhshi, S. R., Ashrafizadeh, F., & Ghasemi, A. (2016). Synthesis and optimization of the magnetic properties of aligned strontium ferrite nanowires. *Materials Research Bulletin, 76*, 240–246. https://doi.org/10.1016/j.materresbull.2015.11.061

Ejaz, S. R., Khan, M. A., Warsi, M. F., Akhtar, M. N., & Hussain, A. (2018). Study of structural transformation and hysteresis behavior of Mg-Sr substituted X-type hexaferrites. *Ceramics International, 44*(15), 18903–18912. https://doi.org/10.1016/j.ceramint.2018.07.126

El Ata, A. A., & Attia, S. (2003). Dielectric dispersion of Y-type hexaferrites at low frequencies. *Journal of Magnetism and Magnetic Materials, 257*(2–3), 165–174. https://doi.org/10.1016/S0304-8853(02)00446-8

Elahi, A., Ahmad, M., Ali, I., & Rana, M. (2013). Preparation and properties of sol–gel synthesized Mg-substituted Ni_2Y hexagonal ferrites. *Ceramics International, 39*(2), 983–990. https://doi.org/10.1016/j.ceramint.2012.07.016

Fähnle, M., & Illg, C. (2011). Electron theory of fast and ultrafast dissipative magnetization dynamics. *Journal of Physics: Condensed Matter, 23*(49), 493201–493218. https://doi.org/10.1088/0953-8984/23/49/493201

Fantechi, E., Innocenti, C., Albino, M., Lottini, E., & Sangregorio, C. (2015). Influence of cobalt doping on the hyperthermic efficiency of magnetite nanoparticles. *Journal of Magnetism and Magnetic Materials, 380*, 365–371. https://doi.org/10.1016/j.jmmm.2014.10.082

Fotukian, S. M., Barati, A., Soleymani, M., & Alizadeh, A. M. (2020). Solvothermal synthesis of $CuFe_2O_4$ and Fe_3O_4 nanoparticles with high heating efficiency for magnetic hyperthermia application. *Journal of Alloys and Compounds, 816*, 152548–152560. https://doi.org/10.1016/j.jallcom.2019.152548

Fullerton, E. E., Margulies, D., Schabes, M. E., Carey, M., Gurney, B., Moser, A., … Rosen, H. (2000). Antiferromagnetically coupled magnetic media layers for thermally stable high-density recording. *Applied Physics Letters, 77*(23), 3806–3808. https://doi.org/10.1063/1.1329868

Frenkel, J., & Doefman, J. (1930). Spontaneous and induced magnetisation in ferromagnetic bodies. *Nature, 126*(3173), 274–275. https://doi.org/10.1038/126274a0

Fu, Y., Sun, L., Wang, J., Bai, X., Kou, Z., Zhai, Y., … Lu, H. (2009). Magnetic properties of $(Ni_{83}Fe_{17})_{1-x}$-Gd thin films with diluted Gd doping. *IEEE Transactions on Magnetics, 45*(10), 4004–4007. https://doi.org/10.1109/TMAG.2009.2024164

Ganne, J.-P., Lebourgeois, R., Paté, M., Dubreuil, D., Pinier, L., & Pascard, H. (2007). The electromagnetic properties of Cu-substituted garnets with low sintering temperature. *Journal of the European Ceramic Society, 27*(8–9), 2771–2777. https://doi.org/10.1016/j.jeurceramsoc.2006.11.054

Georgieva, B., Kolev, S., Krezhov, K., Ghelev, C., Kovacheva, D., Vertruyen, B., Closset, R., Tran, L. M., Babij, M., & Zaleski, A. J. (2019). Structural and magnetic characterization of Y-type hexaferrite powders prepared by sol-gel auto-combustion and sonochemistry. *Journal of Magnetism and Magnetic Materials, 477*, 131–135. https://doi.org/10.1016/j.jmmm.2019.01.033

Ghasemi, A. (2015). Tuning static and high frequency magnetic properties of $SrFe_{12-x}(Mn_{0.5}Co_{0.5}Ti)_{x/2}O_{19}$ nanoparticles and thin films via chemical control. *Journal of Magnetism and Magnetic Materials, 378*, 340–344. https://doi.org/10.1016/j.jmmm.2014.10.162

Ghasemi, A. (2016a). Coherent rotation of magnetization or Kondorsky model of exchange-decoupled $BaFe_{10}CrAlO_{19}$ dot array. *Journal of Alloys and Compounds, 684*, 245–253. https://doi.org/10.1016/j.jallcom.2016.05.190

Ghasemi, A. (2016b). Effects of divalent ion substitution on the microstructure and magnetic properties of $SrCo_{2-x}$ (MnZnCa) x/3$Fe_{16}O_{27}$ nanoparticles. *Journal of Superconductivity and Novel Magnetism, 29*(7), 1943–1952. https://doi.org/10.1007/s10948-016-3503-5

Ghasemi, A. (2016c). On the magnetic analysis of external magnetic field annealing of Co 2-W hexagonal shape particles. *Journal of Superconductivity and Novel Magnetism, 29*(6), 1601–1610. https://doi.org/10.1007/s10948-016-3435-0

Ghasemi, A. (2016d). Real and imaginary parts of magnetic susceptibility of fine dispersed nanoparticles synthesized by reverse micelle: From superparamagnetic trend to ferrimagnetic state. *Journal of Cluster Science, 27*(3), 979–992. https://doi.org/10.1007/s10876-016-0978-y

Ghasemi, A. (2016e). Stoner–Wohlfarth rotation or domain wall motion mechanism in W-type magnetic hexaferrite nanoparticles. *Ceramics International, 42*(3), 4143–4149. https://doi.org/10.1016/j.ceramint.2015.11.087

Ghasemi, A., Ghasemi, E., & Paimozd, E. (2011). Influence of copper cations on the magnetic properties of NiCuZn ferrite nanoparticles. *Journal of Magnetism and Magnetic Materials, 323*(11), 1541–1545. https://doi.org/10.1016/j.jmmm.2011.01.014

Ghasemi, A., & Gordani, G. R. (2016). Characterization and investigation of magnetic and microwave properties of Al–Cr-substituted Z-type barium hexaferrite nanoparticles. *Journal of Superconductivity and Novel Magnetism, 29*(3), 795–801. https://doi.org/10.1007/s10948-015-3337-6

Ghasemi, A., Liu, X., & Morisako, A. (2007). Magnetic and microwave absorption properties of $BaFe_{12-x}(Mn_{0.5}Cu_{0.5}Zr)_{x/2}O_{19}$ synthesized by sol–gel processing. *Journal of Magnetism and Magnetic Materials, 316*(2), 105–108. https://doi.org/10.1016/j.jmmm.2007.02.043

Ghasemi, A., & Morisako, A. (2008). Structural and electromagnetic characteristics of substituted strontium hexaferrite nanoparticles. *Journal of Magnetism and Magnetic Materials, 320*(6), 1167–1172. https://doi.org/10.1016/j.jmmm.2007.11.004

Ghasemi, A., Paesano, A., Jr., & Machado, C. F. C. (2012). Structural and magnetic characteristics of $Co_{1-x}Ni_{x/2}Sr_{x/2}Fe_2O_4$ nanoparticles. *Journal of Magnetism and Magnetic Materials, 324*(14), 2193–2198. https://doi.org/10.1016/j.jmmm.2012.02.019

Ghasemi, A., & Šepelák, V. (2011). Correlation between site preference and magnetic properties of substituted strontium ferrite thin films. *Journal of Magnetism and Magnetic Materials, 323*(12), 1727–1733. https://doi.org/10.1016/j.jmmm.2011.02.010

Ghasemi, A., Šepelák, V., Liu, X., & Morisako, A. (2010a). Mössbauer spectroscopy and magnetic susceptibility studies of Cr–Al substituted strontium ferrite particles. *Journal of Applied Physics, 107*(9), 09A743.

Ghasemi, A., Šepelák, V., Liu, X., & Morisako, A. (2010b). The role of cations distribution on magnetic and reflection loss properties of ferrimagnetic $SrFe_{12-x}(Sn_{0.5}Zn_{0.5})_xO_{19}$. *Journal of Applied Physics, 107*(9), 09A734.

Ghasemi, A., Šepelák, V., Shirsath, S. E., Liu, X., & Morisako, A. (2011). Mössbauer spectroscopy and magnetic characteristics of Zn_{1-x} Cox Fe_2O_4 ($x = 0-1$) nanoparticles. *Journal of Applied Physics, 109*(7), 07A512.

Givord, D., Lu, Q., & Rossignol, M. (1991). In G. C. Hadjipanayis, & G. A. Prinz (Eds.), *Science and Technology of nanostructured magnetic materials*. Plenum.

Globus, A., & Guyot, M. (1972). Wall displacement and bulging in magnetization mechanisms of the hysteresis loop. *Physica Status Solidi, 52*(2), 427–431. https://doi.org/10.1002/pssb.2220520211

Goldman, A. (2006). *Modern ferrite technology*. Springer Science & Business Media.

Gordani, G. R., Ghasemi, A., & Saidi, A. (2014). Enhanced magnetic properties of substituted Sr-hexaferrite nanoparticles synthesized by co-precipitation method. *Ceramics International, 40*(3), 4945–4952. https://doi.org/10.1016/j.ceramint.2013.10.096

Hapishah, A., Hamidon, M., Syazwan, M., & Shafiee, F. (2019). Effect of grain size on microstructural and magnetic properties of holmium substituted yttrium iron garnets ($Y_{1.5}Ho_{1.5}Fe_5O_{12}$). *Results in Physics, 14*, 102391–102399. https://doi.org/10.1016/j.rinp.2019.102391

Hernández-Gómez, P., Martín-González, D., Torres, C., & Muñoz, J. (2019). Broadband transverse susceptibility in multiferroic Y-type hexaferrite $Ba_{0.5}Sr_{1.5}Co_2Fe_2O_{22}$. *Journal of Magnetism and Magnetic Materials, 476*, 478–482. https://doi.org/10.1016/j.jmmm.2019.01.035

Herzer, G. (1990). Grain size dependence of coercivity and permeability in nanocrystalline ferromagnets. *IEEE Transactions on Magnetics, 26*(5), 1397–1402. https://doi.org/10.1109/20.104389

Hong, R., Wu, Y., Feng, B., Di, G., Li, H., Xu, B., Zheng, Y., & Wei, D. (2009). Microwave-assisted synthesis and characterization of Bi-substituted yttrium garnet nanoparticles. *Journal of Magnetism and Magnetic Materials, 321*(8), 1106–1110. https://doi.org/10.1016/j.jmmm.2008.11.009

Hossain, A. A., & Rahman, M. (2011). Enhancement of microstructure and initial permeability due to Cu substitution in $Ni_{0.50-x}Cu_xZn_{0.50}Fe_2O_4$ ferrites. *Journal of Magnetism and Magnetic Materials, 323*(15), 1954–1962. https://doi.org/10.1016/j.jmmm.2011.02.031

Ikram, S., Arshad, M. I., Mahmood, K., Ali, A., Amin, N., & Ali, N. (2018). Structural, magnetic and dielectric study of La^{3+} substituted $Cu_{0.8}Cd_{0.2}Fe_2O_4$ ferrite nanoparticles synthesized by the co-precipitation method. *Journal of Alloys and Compounds, 769*, 1019–1025. https://doi.org/10.1016/j.jallcom.2018.08.065

Iqbal, M. J., Khan, R. A., Mizukami, S., & Miyazaki, T. (2011). Mössbauer and magnetic study of Mn, Zr and Cd substituted W-type hexaferrites prepared by co-precipitation. *Materials Research Bulletin, 46*(11), 1980–1986. https://doi.org/10.1016/j.materresbull.2011.07.014

Iqbal, M. J., Khan, R. A., Mizukami, S., & Miyazaki, T. (2012). Mössbauer, magnetic and microwave absorption characteristics of substituted W-type hexaferrites nanoparticles. *Ceramics International, 38*(5), 4097–4103. https://doi.org/10.1016/j.ceramint.2012.01.067

Iqbal, M. J., Khan, R. A., Takeda, S., Mizukami, S., & Miyazaki, T. (2011). W-type hexaferrite nanoparticles: A consideration for microwave attenuation at wide frequency band of 0.5–10 GHz. *Journal of Alloys and Compounds, 509*(28), 7618–7624. https://doi.org/10.1016/j.jallcom.2011.04.103

Iqbal, M. J., & Liaqat, F. (2010). Physical and electrical properties of nanosized Mn-and Cr-doped strontium Y-type hexagonal ferrites. *Journal of the American Ceramic Society, 93*(2), 474–480. https://doi.org/10.1111/j.1551-2916.2009.03385.x

Ishino, K., & Narumiya, Y. (1987). Development of magnetic ferrites: Control and application of losses. *American Ceramic Society Bulletin, 66*(10), 1469–1474.

Jahanbin, T., Hashim, M., & Mantori, K. A. (2010). Comparative studies on the structure and electromagnetic properties of Ni–Zn ferrites prepared via co-precipitation and conventional ceramic processing routes. *Journal of Magnetism and Magnetic Materials, 322*(18), 2684–2689. https://doi.org/10.1016/j.jmmm.2010.04.008

Joseyphus, R., Narayanasamy, A., Nigam, A., & Krishnan, R. (2006). Effect of mechanical milling on the magnetic properties of garnets. *Journal of Magnetism and Magnetic Materials, 296*(1), 57–64. https://doi.org/10.1016/j.jmmm.2005.04.018

Júnior, G. P., Rodrigues, H., Almeida, J., Sancho, E., Góes, J., Costa, M., ... Sombra, A. (2010). Study of the dielectric and magnetic properties of Co_2Y, Y-type hexaferrite ($Ba_2Co_2Fe_{12}O_{22}$) added with PbO and Bi_2O_3 in the RF frequency range. *Journal of Alloys and Compounds, 493*(1–2), 326–334. https://doi.org/10.1016/j.jallcom.2009.12.094

Kabbur, S., Ghodake, U., Nadargi, D., Kambale, R. C., & Suryavanshi, S. (2018). Effect of Dy^{3+} substitution on structural and magnetic properties of nanocrystalline Ni-Cu-Zn ferrites. *Journal of Magnetism and Magnetic Materials, 451*, 665–675. https://doi.org/10.1016/j.jmmm.2017.12.006

Kachniarz, M., & Salach, J. (2016). Influence of temperature and magnetizing field on the magnetic permeability of soft ferrite materials. In *International conference on systems, control and information Technologies 2016*. Springer.

Kaewrawang, A., Ghasemi, A., Liu, X., & Morisako, A. (2009). Structural and magnetic characteristics of SrM thin films on Au dot arrays. *Journal of physics: Conference series, 191*(1), 012026. IOP Publishing.

Kaewrawang, A., Ghasemi, A., Liu, X., & Morisako, A. (2009). Crystallographic and magnetic properties of SrM thin films on Pt underlayer prepared at various substrate temperatures. *Journal of Magnetism and Magnetic Materials, 321*(13), 1939–1942. https://doi.org/10.1016/j.jmmm.2008.12.023

Kaewrawang, A., Ghasemi, A., Liu, X., & Morisako, A. (2010). Fabrication, crystallographic and magnetic properties of SrM perpendicular films on Au nano-dot arrays. *Journal of Alloys and Compounds, 492*(1–2), 44–47. https://doi.org/10.1016/j.jallcom.2009.11.174

Kaewrawang, A., Ishida, G., Liu, X., & Morisako, A. (2008). Epitaxial growth of SrM (001) film on Au (111). *IEEE Transactions on Magnetics, 44*(11), 2899–2902. https://doi.org/10.1109/TMAG.2008.2002584

Kamishima, K., Ito, C., Kakizaki, K., Hiratsuka, N., Shirahata, T., & Imakubo, T. (2007). Improvement of initial permeability for Z-type ferrite by Ti and Zn substitution. *Journal of Magnetism and Magnetic Materials, 312*(1), 228–233. https://doi.org/10.1016/j.jmmm.2006.10.004

Kamishima, K., Tajima, R., Watanabe, K., Kakizaki, K., Fujimori, A., Sakai, M., & Abe, H. (2015). Crystallographic and magnetic properties of Cu_2U-type hexaferrite. *Journal of Magnetism and Magnetic Materials, 375*, 54–60. https://doi.org/10.1016/j.jmmm.2014.09.055

Khan, I., Sadiq, I., & Ashiq, M. N. (2011). Role of Ce–Mn substitution on structural, electrical and magnetic properties of W-type strontium hexaferrites. *Journal of Alloys and Compounds, 509*(31), 8042–8046. https://doi.org/10.1016/j.jallcom.2011.05.013

Khandan, A., & Ozada, N. (2017). Bredigite-magnetite ($Ca_7MgSi_4O_{16}$-Fe_3O_4) nanoparticles: A study on their magnetic properties. *Journal of Alloys and Compounds, 726*, 729–736. https://doi.org/10.1016/j.jallcom.2017.07.288

Kim, T., Nasu, S., & Shima, M. (2007). Growth and magnetic behavior of bismuth substituted yttrium iron garnet nanoparticles. *Journal of Nanoparticle Research, 9*(5), 737–743. https://doi.org/10.1007/s11051-006-9082-9

Kittel, C. (1946). Theory of the structure of ferromagnetic domains in films and small particles. *Physical Review, 70*(11–12), 965–971. https://doi.org/10.1103/PhysRev.70.965

Kneller, E. (1964). Magnetic interactions in super paramagnetic particle. In *Proceedings of the international conference on magnetism, Nottingham, September 1964*. Proceedings of the Physical Society.

Köbler, U., & Hoser, A. (2014). Magnetism of powder samples and of single crystals. *Journal of Magnetism and Magnetic Materials, 349*, 88–94. https://doi.org/10.1016/j.jmmm.2013.07.059

Köbler, U., Hoser, A., & Schäfer, W. (2005). On the temperature dependence of the magnetic excitations. *Physica B: Condensed Matter, 364*(1–4), 55–70. https://doi.org/10.1016/j.physb.2005.03.035

Kojima, H. (1982). *Ferromagnetic materials (3)*. North-Holland Publishing Company (Chapter 1).

Kondorski, E. (1937). *Physikalische Zeitschrift der Sowjetunion, 11*, 597.

Kotnala, R., & Shah, J. (2015). Ferrite materials: Nano to spintronics regime. In *Handbook of magnetic materials, 23*Elsevier.

Kumar, S., & Chatterjee, R. (2018). Complex permittivity, permeability, magnetic and microwave absorbing properties of Bi^{3+} substituted U-type hexaferrite. *Journal of Magnetism and Magnetic Materials, 448*, 88–93. https://doi.org/10.1016/j.jmmm.2017.06.123

Kumar, S., Meena, R. S., & Chatterjee, R. (2016). Microwave absorption studies of Cr-doped Co–U type hexaferrites over 2–18 GHz frequency range. *Journal of Magnetism and Magnetic Materials, 418*, 194–199. https://doi.org/10.1016/j.jmmm.2016.02.084

Lagarec, K., & Rancourt, D. G. (1998). *Recoil-Mössbauer spectral analysis software for Windows*. University of Ottawa.

Lao, L., & Ramanujan, R. (2004). Magnetic and hydrogel composite materials for hyperthermia applications. *Journal of Materials Science: Materials in Medicine, 15*(10), 1061–1064. https://doi.org/10.1023/B:JMSM.0000046386.78633.e5

Leal, L., Guerra, Y., Padrón-Hernández, E., Rodrigues, A., Santos, F., & Peña-Garcia, R. (2019). Structural and magnetic properties of yttrium iron garnet nanoparticles doped with copper obtained by sol gel method. *Materials Letters, 236*, 547–549. https://doi.org/10.1016/j.matlet.2018.11.004

Li, J., Li, S., Fu, T., Shao, N., & Che, S. (2019). Preparation and properties of $Ba_{3-x}Bi_xCo_{2+x}Fe_{24-x}O_{41}$ Z-type hexaferrites. *Journal of Magnetism and Magnetic Materials, 488*, 165366–165370. https://doi.org/10.1016/j.jmmm.2019.165366

Li, Z., Ong, C., Yang, Z., Wei, F., Zhou, X., Zhao, J., & Morrish, A. (2000). Site preference and magnetic properties for a perpendicular recording material: $BaFe\ 12_{-x}Zn\ _{x/2}Zr_{x/2}O_{19}$ nanoparticles. *Physical Review B, 62*(10), 6530–6537. https://doi.org/10.1103/PhysRevB.62.6530

Lim, J. T., Kim, C. M., Lee, B. W., & Kim, C. S. (2012). Investigation of magnetic properties of non-magnetic ion (Al, Ga, In) doped $Ba_2Mg_{0.5}Co_{1.5}Fe_{12}O_{22}$. *Journal of Applied Physics, 111*(7), 7A518–7A520. https://doi.org/10.1063/1.3679023

Lisjak, D., & Drofenik, M. (2003). Synthesis and characterization of Zn_2U ($Ba_4Zn_2Fe_{36}O_{60}$) hexaferrite powder. *Journal of Applied Physics, 93*(10), 8011–8013. https://doi.org/10.1063/1.1540159

Lisjak, D., & Drofenik, M. (2004). The thermal stability range and magnetic properties of U-type hexaferrites. *Journal of Magnetism and Magnetic Materials, 272*, 1817–1819. https://doi.org/10.1016/j.jmmm.2003.12.879

Lisjak, D., Makovec, D., & Drofenik, M. (2004). Formation of U-type hexaferrites. *Journal of Materials Research, 19*(8), 2462–2470. https://doi.org/10.1557/JMR.2004.0317

Lisjak, D., McGuiness, P., & Drofenik, M. (2006). Thermal instability of Co-substituted barium hexaferrites with U-type structure. *Journal of Materials Research, 21*(2), 420–427. https://doi.org/10.1557/jmr.2006.0048

Liu, C., Zou, B., Rondinone, A. J., & Zhang, Z. J. (2000). Chemical control of superparamagnetic properties of magnesium and cobalt spinel ferrite nanoparticles through atomic level magnetic couplings. *Journal of the American Chemical Society, 122*(26), 6263–6267. https://doi.org/10.1021/ja000784g

Luo, H., Rai, B., Mishra, S., Nguyen, V., & Liu, J. (2012). Physical and magnetic properties of highly aluminum doped strontium ferrite nanoparticles prepared by auto-combustion route. *Journal of Magnetism and Magnetic Materials, 324*(17), 2602–2608. https://doi.org/10.1016/j.jmmm.2012.02.106

Maaz, K., Usman, M., Karim, S., Mumtaz, A., Hasanain, S., & Bertino, M. (2009). Magnetic response of core-shell cobalt ferrite nanoparticles at low temperature. *Journal of Applied Physics, 105*(11), 113917–113921. https://doi.org/10.1063/1.3139293

Maehara, T., Konishi, K., Kamimori, T., Aono, H., Naohara, T., Kikkawa, H., … Kawachi, K. (2002). Heating of ferrite powder by an AC magnetic field for local hyperthermia. *Japanese Journal of Applied Physics, 41*(3), 1620. https://doi.org/10.1143/jjap.41.1620

Magham, S., Sharma, M., Shannigrahi, S., Tan, H. R., Sharma, V., Meng, Y. S., Idapalpati, S., Ramanujan, R., & Repaka, D. (2017). Development of Z-type hexaferrites for high frequency EMI shielding applications. *Journal of Magnetism and Magnetic Materials, 441*, 303–309. https://doi.org/10.1016/j.jmmm.2017.05.066

Mahmood, S. H., Al Sheyab, Q., Bsoul, I., Mohsen, O., & Awadallah, A. (2018). Structural and magnetic properties of Ga-substituted Co_2-W hexaferrites. *Current Applied Physics, 18*(5), 590–598. https://doi.org/10.1016/j.cap.2018.02.013

Majeed, A., Khan, M. A., Lodhi, M. Y., Ahmad, R., & Ahmad, I. (2020). Structural, microwave permittivity, and complex impedance studies of cation (Cr, Bi, Al, In) substituted SrNi-X hexagonal nano-sized ferrites. *Ceramics International, 46*(2), 1907–1915. https://doi.org/10.1016/j.ceramint.2019.09.168

Mata-Zamora, M., Montiel, H., Alvarez, G., Saniger, J., Zamorano, R., & Valenzuela, R. (2007). Microwave non-resonant absorption in fine cobalt ferrite particles. *Journal of Magnetism and Magnetic Materials, 316*(2), 532–534. https://doi.org/10.1016/j.jmmm.2007.03.011

Mathur, P., Thakur, A., & Singh, M. (2008). Effect of nanoparticles on the magnetic properties of Mn–Zn soft ferrite. *Journal of Magnetism and Magnetic Materials, 320*(7), 1364–1369. https://doi.org/10.1016/j.jmmm.2007.11.008

de Mello, L. B., Varanda, L. C., Sigoli, F. A., & Mazali, I. O. (2019). Co-precipitation synthesis of (Zn-Mn)-co-doped magnetite nanoparticles and their application in magnetic hyperthermia. *Journal of Alloys and Compounds, 779*, 698–705. https://doi.org/10.1016/j.jallcom.2018.11.280

Menil, F. (1985). Systematic trends of the ^{57}Fe Mössbauer isomer shifts in (FeO_n) and (FeF_n) polyhedra. Evidence of a new correlation between the isomer shift and the inductive effect of the competing bond TX (→ Fe)(where X is O or F and T any element with a formal positive charge). *Journal of Physics and Chemistry of Solids, 46*(7), 763–789. https://doi.org/10.1016/0022-3697(85)90001-0

Mesbahinia, A., Almasi-Kashi, M., Ghasemi, A., & Ramazani, A. (2020). FORC investigation of Co-Ni bulk ferrite consolidated by spark plasma sintering technique. *Journal of Magnetism and Magnetic Materials, 497*, 165976–165995. https://doi.org/10.1016/j.jmmm.2019.165976

Mesbahinia, A., Almasi-Kashi, M., Ghasemi, A., & Ramezani, A. (2019). First order reversal curve analysis of cobalt-nickel ferrite. *Journal of Magnetism and Magnetic Materials, 473*, 161–168. https://doi.org/10.1016/j.jmmm.2018.10.057

Morisako, A., Matsumoto, M., & Naoe, M. (1999). Properties of c-axis oriented Ba-ferrite sputtered films. *Journal of Magnetism and Magnetic Materials, 193*(1–3), 110–113. https://doi.org/10.1016/S0304-8853(98)00498-3

Moser, A., & Weller, D. (1999). Thermal effect limits in ultrahigh-density magnetic recording. *IEEE Transactions on Magnetics, 35*(6), 2808–2813. https://doi.org/10.1109/20.809134

Mu, C., Liu, Y., Song, Y., Wang, L., & Zhang, H. (2011). Improvement of high-frequency characteristics of Z-type hexaferrite by dysprosium doping. *Journal of Applied Physics, 109*(12), 123925–123929. https://doi.org/10.1063/1.3600068

Müller, M., Harada, H., & Warlimont, H. (2018). Magnetic materials. In *Springer Handbook of materials data*. Springer.

Murtaza, G., Ahmad, R., Hussain, T., Ayub, R., Ali, I., Khan, M. A., & Akhtar, M. N. (2014). Structural and magnetic properties of Nd–Mn substituted Y-type hexaferrites synthesized by microemulsion method. *Journal of Alloys and Compounds, 602*, 122–129. https://doi.org/10.1016/j.jallcom.2014.02.156

Musa, M. A., Osman, N. H., Hassan, J., & Zangina, T. (2017). Structural and magnetic properties of yttrium iron garnet (YIG) and yttrium aluminum iron garnet (YAlG) nanoferrite via sol-gel synthesis. *Results in Physics, 7*, 1135–1142. https://doi.org/10.1016/j.rinp.2017.02.038

Nakamura, T., Tsutaoka, T., & Hatakeyama, K. (1994). Frequency dispersion of permeability in ferrite composite materials. *Journal of Magnetism and Magnetic Materials, 138*(3), 319–328. https://doi.org/10.1016/0304-8853(94)90054-X

Naseri, M. G., Saion, E. B., Ahangar, H. A., & Shaari, A. H. (2013). Fabrication, characterization, and magnetic properties of copper ferrite nanoparticles prepared by a simple, thermal-treatment method. *Materials Research Bulletin, 48*(4), 1439–1446. https://doi.org/10.1016/j.materresbull.2012.12.039

Néel, L. (1949). Théorie du traînage magnétique des ferromagnétiques en grains fins avec applications aux terres cuites. *Annales Geophysicae, 5*, 99–136.

Néel, L. (1951). Le traînage magnétique. *Journal de Physique et Le Radium, 12*(3), 339–351. https://doi.org/10.1051/jphysrad:01951001203033900

Nguyet, D. T. T., Duong, N. P., Satoh, T., Anh, L. N., & Hien, T. D. (2012). Temperature-dependent magnetic properties of yttrium iron garnet nanoparticles prepared by citrate sol–gel. *Journal of Alloys and Compounds, 541*, 18–22. https://doi.org/10.1016/j.jallcom.2012.06.122

Obradors, X., Collomb, A., Pernet, M., Samaras, D., & Joubert, J. (1985). X-ray analysis of the structural and dynamic properties of $BaFe_{12}O_{19}$ hexagonal ferrite at room temperature. *Journal of Solid State Chemistry, 56*(2), 171–181. https://doi.org/10.1016/0022-4596(85)90054-4

O'handley, R. C. (2000). *Modern magnetic materials: Principles and applications*. Wiley.

Oliver, S., Willey, R., Hamdeh, H., Oliveri, G., & Busca, G. (1995). Structure and magnetic properties of magnesium ferrite fine powders. *Scripta Metallurgica et Materialia, 33*(10–11), 1695–1701. https://doi.org/10.1016/0956-716X(95)00412-O

Opuchovic, O., Kareiva, A., Mazeika, K., & Baltrunas, D. (2017). Magnetic nanosized rare earth iron garnets $R_3Fe_5O_{12}$: Sol–gel fabrication, characterization and reinspection. *Journal of Magnetism and Magnetic Materials, 422*, 425–433. https://doi.org/10.1016/j.jmmm.2016.09.041

Özgür, Ü., Alivov, Y., & Morkoç, H. (2009). Microwave ferrites, part 1: Fundamental properties. *Journal of Materials Science: Materials in Electronics, 20*(9), 789–834. https://doi.org/10.1007/s10854-009-9923-2

Pachpinde, A., Langade, M., Lohar, K., Patange, S., & Shirsath, S. E. (2014). Impact of larger rare earth Pr_{3+} ions on the physical properties of chemically derived $Pr_xCoFe_{2-x}O_4$ nanoparticles. *Chemical Physics, 429*, 20–26. https://doi.org/10.1016/j.chemphys.2013.11.018

Park, T.-J., Papaefthymiou, G. C., Viescas, A. J., Moodenbaugh, A. R., & Wong, S. S. (2007). Size-dependent magnetic properties of single-crystalline multiferroic $BiFeO_3$ nanoparticles. *Nano Letters, 7*(3), 766–772. https://doi.org/10.1021/nl063039w

Peña-Garcia, R., Guerra, Y., Buitrago, D. M., Leal, L., Santos, F., & Padrón-Hernández, E. (2018). Synthesis and characterization of yttrium iron garnet nanoparticles doped with cobalt. *Ceramics International, 44*(10), 11314–11319. https://doi.org/10.1016/j.ceramint.2018.03.179

Peña-Garcia, R., Guerra, Y., de Souza, F., Gonçalves, L., & Padrón-Hernández, E. (2018). The extended Bloch's law in yttrium iron garnet doped with Zn, Ni and Co. *Physica E: Low-Dimensional Systems and Nanostructures, 103*, 354–360. https://doi.org/10.1016/j.physe.2018.06.027

Peña-Garcia, R., Guerra, Y., Santos, F., Almeida, L., & Padrón-Hernández, E. (2019). Structural and magnetic properties of Ni-doped yttrium iron garnet nanopowders. *Journal of Magnetism and Magnetic Materials, 492*, 165650. https://doi.org/10.1016/j.jmmm.2019.165650

Perrin, G., Acher, O., Peuzin, J., & Vukadinovic, N. (1996). Sum rules for gyromagnetic permeability of ferromagnetic thin films: Theoretical and experimental results. *Journal of Magnetism and Magnetic Materials*, 157–158, 289–290. https://doi.org/10.1016/0304-8853(96)80019-9

Plumer, M. L., Van Ek, J., & Weller, D. (2001). *The physics of ultra-high-density magnetic recording, 41*. Springer Science & Business Media.

Pullar, R., & Bhattacharya, A. (2001). The synthesis and characterisation of Co_2X ($Ba_2Co_2Fe_{28}O_{46}$) and Co_2U ($Ba_4Co_2Fe_{36}O_{60}$) ferrite fibres, manufactured from a sol-gel process. *Journal of Materials Science, 36*(19), 4805–4812. https://doi.org/10.1023/A:1017947625940

Pullar, R. C. (2012). Hexagonal ferrites: A review of the synthesis, properties and applications of hexaferrite ceramics. *Progress in Materials Science, 57*(7), 1191–1334. https://doi.org/10.1016/j.pmatsci.2012.04.001

Puli, V. S., Adireddy, S., & Ramana, C. (2015). Chemical bonding and magnetic properties of gadolinium (Gd) substituted cobalt ferrite. *Journal of Alloys and Compounds, 644*, 470–475. https://doi.org/10.1016/j.jallcom.2015.05.031

Qu, J., Liu, G., Wang, Y., & Hong, R. (2010). Preparation of Fe_3O_4–chitosan nanoparticles used for hyperthermia. *Advanced Powder Technology, 21*(4), 461–467. https://doi.org/10.1016/j.apt.2010.01.008

Rado, G. T. (1953). Magnetic spectra of ferrites. *Reviews of Modern Physics, 25*(1), 81. https://doi.org/10.1103/RevModPhys.25.81

Rajendran, M., Deka, S., Joy, P., & Bhattacharya, A. (2006). Size-dependent magnetic properties of nanocrystalline yttrium iron garnet powders. *Journal of Magnetism and Magnetic Materials, 301*(1), 212–219. https://doi.org/10.1016/j.jmmm.2005.06.027

Ri, C., Li, L., & Qi, Y. (2012). Anisotropy of the electrical conductivity in W-type hexagonal ferrites $BaFe_{18}O_{27}$ and $BaCo_2Fe_{16}O_{27}$ from first principles. *Journal of Magnetism and Magnetic Materials, 324*(8), 1498–1502. https://doi.org/10.1016/j.jmmm.2011.11.041

Rozanov, K., Li, Z., Chen, L., & Koledintseva, M. (2005). Microwave permeability of Co_2Z composites. *Journal of Applied Physics, 97*(1), 13905–13911. https://doi.org/10.1063/1.1827911

Sadiq, I., Khan, I., Aen, F., Islam, M., & Rana, M. (2012). Influence of rare earth Ce^{3+} on structural, electrical and magnetic properties of Sr^{2+} based W-type hexagonal ferrites. *Physica B: Condensed Matter, 407*(8), 1256–1261. https://doi.org/10.1016/j.physb.2012.01.115

Sawatzky, G., Van der woude, F., & Morrish, A. H. (1969). *Physical Review, 187*(2), 747–757. https://doi.org/10.1103/PhysRev.187.747

Sekijima, T., Kishimoto, H., Fujii, T., Wakino, K., & Okada, M. (1999). Magnetic, optical and microwave properties of rare-earth-substituted fibrous yttrium iron garnet single crystals grown by floating zone method. *Japanese Journal of Applied Physics, 38*(10), 5874–5878. https://doi.org/10.1143/jjap.38.5874

Šepelák, V., Wißmann, S., & Becker, K. (1998). A temperature-dependent Mössbauer study of mechanically activated and non-activated zinc ferrite. *Journal of Materials Science, 33*(11), 2845–2850. https://doi.org/10.1023/A:1017589920725

Shannigrahi, S., Au, W., Kumar, V. S., Liu, L., Yang, Z., Cheng, C., … Ramanujan, R. (2013). Synthesis and electromagnetic properties of U-type hexaferrites $Ba_4B_2Fe_{36}O_{60}$ (B: Co, Ni, Cu). *Journal of Magnetism and Magnetic Materials, 325*, 63–68. https://doi.org/10.1016/j.jmmm.2012.08.019

Sharma, V., & Kuanr, B. K. (2018). Magnetic and crystallographic properties of rare-earth substituted yttrium-iron garnet. *Journal of Alloys and Compounds, 748*, 591–600. https://doi.org/10.1016/j.jallcom.2018.03.086

Sharma, V., Saha, J., Patnaik, S., & Kuanr, B. K. (2017). YIG based broad band microwave absorber: A perspective on synthesis methods. *Journal of Magnetism and Magnetic Materials, 439*, 277–286. https://doi.org/10.1016/j.jmmm.2017.04.098

Shimizu, Y., Arai, H., & Seiyama, T. (1985). Theoretical studies on the impedance-humidity characteristics of ceramic humidity sensors. *Sensors and Actuators, 7*(1), 11–22. https://doi.org/10.1016/0250-6874(85)87002-5

Shirsath, S. E., Kadam, R., Gaikwad, A. S., Ghasemi, A., & Morisako, A. (2011). Effect of sintering temperature and the particle size on the structural and magnetic properties of nanocrystalline $Li_{0.5}Fe_{2.5}O_4$. *Journal of Magnetism and Magnetic Materials, 323*(23), 3104–3108. https://doi.org/10.1016/j.jmmm.2011.06.065

Shirsath, S. E., Toksha, B., Kadam, R., Patange, S., Mane, D., Jangam, G. S., & Ghasemi, A. (2010). Doping effect of Mn^{2+} on the magnetic behavior in Ni–Zn ferrite nanoparticles prepared by sol–gel auto-combustion. *Journal of Physics and Chemistry of Solids, 71*(12), 1669–1675. https://doi.org/10.1016/j.jpcs.2010.08.016

Shu, C., & Qiao, H. (2009). Tuning magnetic properties of magnetic recording media cobalt ferrite nano-particles by co-precipitation method. In *2009 Symposium on photonics and optoelectronics*. IEEE.

Smit, J., & Wign, H. (1959). *Ferrites*. John Willy & Sons.

Snoek, J. (1948). Dispersion and absorption in magnetic ferrites at frequencies above one Mc/s. *Physica, 14*(4), 207–217. https://doi.org/10.1016/0031-8914(48)90038-X

Sodaee, T., Ghasemi, A., Paimozd, E., Paesano, A., Jr., & Morisako, A. (2013). The role of terbium cation substitution on the magnetic properties of cobalt ferrite nanoparticles. *Journal of Magnetism and Magnetic Materials, 330*, 169–173. https://doi.org/10.1016/j.jmmm.2012.10.050

Sodaee, T., Ghasemi, A., & Razavi, R. S. (2016). Controlled growth of large-area arrays of gadolinium-substituted cobalt ferrite nanorods by hydrothermal processing without use of any template. *Ceramics International, 42*(15), 17420–17428. https://doi.org/10.1016/j.ceramint.2016.08.042

Sodaee, T., Ghasemi, A., & Razavi, R. S. (2017). Shape factors dependence of magnetic features of $CoFe_{2-x}Gd_xO_4$ nanocrystals. *Journal of Alloys and Compounds, 693*, 1231–1242. https://doi.org/10.1016/j.jallcom.2016.10.102

Sohn, B., Cohen, R., & Papaefthymiou, G. (1998). Magnetic properties of iron oxide nanoclusters within microdomains of block copolymers. *Journal of Magnetism and Magnetic Materials, 182*(1–2), 216–224. https://doi.org/10.1016/S0304-8853(97)00675-6

Speliotis, D. (1987). Distinctive characteristics of barium ferrite media. *IEEE Transactions on Magnetics, 23*(5), 3143–3145. https://doi.org/10.1109/TMAG.1987.1065475

Stergiou, C., & Litsardakis, G. (2011). Preparation and magnetic characterization of Co_2-W strontium hexaferrites doped with Ni and La. *Journal of Magnetism and Magnetic Materials, 323*(18–19), 2362–2368. https://doi.org/10.1016/j.jmmm.2011.04.021

Stergiou, C. A., & Litsardakis, G. (2011). Electromagnetic properties of Ni and La doped strontium hexaferrites in the microwave region. *Journal of Alloys and Compounds, 509*(23), 6609–6615. https://doi.org/10.1016/j.jallcom.2011.03.107

Stoner, E., & Wohlfarth, E. (1991). A mechanism of magnetic hysteresis in heterogeneous alloys. *IEEE Transactions on Magnetics, 27*(4), 3475–3518. https://doi.org/10.1109/TMAG.1991.1183750

Stoner, E. C., & Wohlfarth, E. (1948). A mechanism of magnetic hysteresis in heterogeneous alloys. *Philosophical Transactions of the Royal Society of London - Series A: Mathematical and Physical Sciences, 240*(826), 599–642. https://doi.org/10.1098/rsta.1948.0007

Sui, X., Kryder, M. H., Wong, B. Y., & Laughlin, D. E. (1993). Microstructural origin of the perpendicular anisotropy in M-type barium hexaferrite thin films deposited by RF magnetron sputtering. *IEEE Transactions on Magnetics, 29*(6), 3751–3753. https://doi.org/10.1109/20.281291

Suzuki, Y., Baba, M., Taki, M., Fukunaga, K., & Watanabe, S. (2006). Imaging the 3D temperature distributions caused by exposure of dielectric phantoms to high-frequency electromagnetic fields. *IEEE Transactions on Dielectrics and Electrical Insulation, 13*(4), 744–750. https://doi.org/10.1109/TDEI.2006.1667731

Suzuki, Y., Makanae, H., Kudo, H., Miyanaga, T., Nanke, T., & Kobayashi, T. (2004). Anomalous infrared and visible light absorption by spherical gold nanoparticles dispersed in a comb copolymer. *Applied Physics A, 78*(3), 335–338. https://doi.org/10.1007/s00339-003-2361-8

Tang, R., Jiang, C., Zhou, H., & Yang, H. (2016). Effects of composition and temperature on the magnetic properties of (Ba, Sr)$_3$Co$_2$Fe$_{24}$O$_{41}$Z type hexaferrites. *Journal of Alloys and Compounds, 658*, 132–138. https://doi.org/10.1016/j.jallcom.2015.10.207

Tebble, R. S., & Craik, D. J. (1969). *Magnetic materials*. John Wiley & Sons.

Tomitaka, A., Kobayashi, H., Yamada, T., Jeun, M., Bae, S., & Takemura, Y. (2010). Magnetic characterization and self-heating of various magnetic nanoparticles for medical applications. In *2010 3rd international nanoelectronics conference (INEC)* (pp. 896–897). IEEE.

Torkian, S., Ghasemi, A., & Razavi, R. S. (2017). Cation distribution and magnetic analysis of wideband microwave absorptive Co$_x$Ni$_{1-x}$Fe$_2$O$_4$ ferrites. *Ceramics International, 43*(9), 6987–6995. https://doi.org/10.1016/j.ceramint.2017.02.124

Torrejon, J., Adenot-Engelvin, A.-L., Bertin, F., Dubuget, V., Acher, O., & Vazquez, M. (2009). Sum rules and figures-of-merit on the microwave permeability of nanocrystalline microwires. *Journal of Magnetism and Magnetic Materials, 321*(9), 1227–1230. https://doi.org/10.1016/j.jmmm.2008.11.002

Van der Woude, F., & Sawatzky, G. (1971). Hyperfine magnetic fields at Fe^{57} nuclei in ferrimagnetic spinels. *Physical Review B, 4*(9), 3159–3165. https://doi.org/10.1103/PhysRevB.4.3159

Verma, S., Joy, P., Khollam, Y., Potdar, H., & Deshpande, S. (2004). Synthesis of nanosized MgFe$_2$O$_4$ powders by microwave hydrothermal method. *Materials Letters, 58*(6), 1092–1095. https://doi.org/10.1016/j.matlet.2003.08.025

Walser, R., Win, W., & Valanju, P. (1998). Shape-optimized ferromagnetic particles with maximum theoretical microwave susceptibility. *IEEE Transactions on Magnetics, 34*(4), 1390–1392. https://doi.org/10.1109/20.706558

Wang, X., Li, L., Su, S., Gui, Z., Yue, Z., & Zhou, J. (2003). Low-temperature sintering and high frequency properties of Cu-modified Co$_2$Z hexaferrite. *Journal of the European Ceramic Society, 23*(5), 715–720. https://doi.org/10.1016/S0955-2219(02)00157-7

Wang, X., Li, L., Su, S., & Yue, Z. (2004). Electromagnetic properties of low-temperature-sintered Ba$_3$Co$_{2-x}$Zn$_x$Fe$_{24}$O$_{41}$ ferrites prepared by solid state reaction method. *Journal of Magnetism and Magnetic Materials, 280*(1), 10–13. https://doi.org/10.1016/j.jmmm.2004.02.016

Weller, D., & Moser, A. (1999). Thermal effect limits in ultrahigh-density magnetic recording. *IEEE Transactions on Magnetics, 35*(6), 4423–4439. https://doi.org/10.1109/20.809134

Wohlfarth, E. (1980). The magnetic field dependence of the susceptibility peak of some spin glass materials. *Journal of Physics F: Metal Physics, 10*(9), 241–246. https://doi.org/10.1088/0305-4608/10/9/006

Wu, T., Su, H., Ding, Q., Zhang, H., Jing, Y., & Tang, X. (2013). Aluminum substituted low loss Z-type hexaferrites for antenna applications. *Physica B: Condensed Matter, 429*, 85–89. https://doi.org/10.1016/j.physb.2013.08.012

Xiao, G., Liou, S., Levy, A., Taylor, J., & Chien, C. (1986). Magnetic relaxation in Fe-(SiO$_2$) granular films. *Physical Review B, 34*(11), 7573–7578. https://doi.org/10.1103/PhysRevB.34.7573

Xu, H., Yang, H., Xu, W., & Feng, S. (2008). Magnetic properties of Ce, Gd-substituted yttrium iron garnet ferrite powders fabricated using a sol–gel method. *Journal of Materials Processing Technology, 197*(1–3), 296–300. https://doi.org/10.1016/j.jmatprotec.2007.06.061

Xu, J.-J., Yang, C.-M., Zou, H.-F., Song, Y.-H., Gao, G.-M., An, B.-C., & Gan, S.-C. (2009). Electromagnetic and microwave absorbing properties of Co_2Z-type hexaferrites doped with La^{3+}. *Journal of Magnetism and Magnetic Materials, 321*(19), 3231–3235. https://doi.org/10.1016/j.jmmm.2009.05.039

Yousaf, M., Akhtar, M. N., Wang, B., & Noor, A. (2020). Preparations, optical, structural, conductive and magnetic evaluations of RE's (Pr, Y, Gd, Ho, Yb) doped spinel nanoferrites. *Ceramics International, 46*(4), 4280–4288. https://doi.org/10.1016/j.ceramint.2019.10.149

Yasemian, A. R., Kashi, M. A., & Ramazani, A. (2019). Surfactant-free synthesis and magnetic hyperthermia investigation of iron oxide (Fe_3O_4) nanoparticles at different reaction temperatures. *Materials Chemistry and Physics, 230*, 9–16. https://doi.org/10.1016/j.matchemphys.2019.03.032

You, J.-H., & Yoo, S.-I. (2018). Improved magnetic properties of Zn-substituted strontium W-type hexaferrites. *Journal of Alloys and Compounds, 763*, 459–465. https://doi.org/10.1016/j.jallcom.2018.05.296

Youssef, J. B., & Brosseau, C. (2006). Magnetization damping in two-component metal oxide micropowder and nanopowder compacts by broadband ferromagnetic resonance measurements. *Physical Review B, 74*(21), 214413–214425. https://doi.org/10.1103/PhysRevB.74.214413

Zayed, M., Ahmed, M., Imam, N., & El Sherbiny, D. H. (2016). Preparation and structure characterization of hematite/magnetite ferro-fluid nanocomposites for hyperthermia purposes. *Journal of Molecular Liquids, 222*, 895–905. https://doi.org/10.1016/j.molliq.2016.07.082

Zhang, L., Papaefthymiou, G. C., & Ying, J. Y. (1997). Size quantization and interfacial effects on a novel γ-Fe_2O_3/SiO_2 magnetic nanocomposite via sol-gel matrix-mediated synthesis. *Journal of Applied Physics, 81*(10), 6892–6900. https://doi.org/10.1063/1.365233

Zhang, L., Papaefthymiou, G. C., & Ying, J. Y. (2001). Synthesis and properties of γ-Fe_2O_3 nanoclusters within mesoporous aluminosilicate matrices. *The Journal of Physical Chemistry B, 105*(31), 7414–7423. https://doi.org/10.1021/jp010174i

Zhang, M., Yin, L., Liu, Q., Kong, X., Zi, Z., Dai, J., & Sun, Y. (2017). Magnetic properties and magnetodielectric effect in Y-type hexaferrite $Ba_{0.5}Sr_{1.5}Zn_{2-x}Mg_xFe_{11}AlO_{22}$. *Journal of Alloys and Compounds, 725*, 1252–1258. https://doi.org/10.1016/j.jallcom.2017.07.286

Zhang, M., Yin, L., Liu, Q., Zi, Z., Dai, J., & Sun, Y. (2018). Indium doping effect on the magnetic properties of Y-type hexaferrite $Ba_{0.5}Sr_{1.5}Zn_2(Fe_{1-x}In_x)_{12}O_{22}$. *Current Applied Physics, 18*(9), 1001–1005. https://doi.org/10.1016/j.cap.2018.05.017

Zhang, H., Zhou, J., Wang, Y., Li, L., Yue, Z., & Gui, Z. (2002). The effect of Zn ion substitution on electromagnetic properties of low-temperature fired Z-type hexaferrite. *Ceramics International, 28*(8), 917–923. https://doi.org/10.1016/S0272-8842(02)00074-3

Zhang, X., & Zhang, J. (2020). Synthesis and magnetic properties of $Ba_2Fe_{14}O_{22}$ Y-type hexaferrite. *Materials Letters, 269*, 127642–127645. https://doi.org/10.1016/j.matlet.2020.127642

Zhao, D.-L., Teng, P., Xu, Y., Xia, Q.-S., & Tang, J.-T. (2010). Magnetic and inductive heating properties of Fe_3O_4/polyethylene glycol composite nanoparticles with core–shell structure. *Journal of Alloys and Compounds, 502*(2), 392–395. https://doi.org/10.1016/j.jallcom.2010.04.177

CHAPTER

Magnetoelectric ferrite nanocomposites

4

4.1 Introduction

The convergence of memory and logic functions may be realized thanks to passive magnetoelectric (ME) components in future spintronic devices, providing information technology and nanoscale electronics with new and promising standards. One way to optimally achieve this aim is to use novel multifunctional materials with superior performance and manipulate their physical properties by changing the degrees of freedom of the order parameters. Multiferroic (MF) materials are considered an advanced class of materials with the ME effect and have drawn considerable attention in recent years. For decades, materials science researchers have sought to couple magnetic and ferroelectric ordering within one material. Inspired by pioneering work in Russia, the coupling between magnetic and electrical order parameters was initially investigated by the development of MF and ME materials in the late 1950s and continued further in the 1960s. MF materials are extremely scarce in nature, and a few of them have been synthesized experimentally in laboratories. Nickel iodine boracite, with the chemical formula of $Ni_3B_2O_{13}I$, was the first MF material discovered. In continuance, several MFs comprising boracite compounds with complex structures were synthesized. The synthesis and characterization of different oxides with MF properties, including Bi-based compounds ($BiMnO_3$ and $BiFeO_3$), have been carried out since the 1990s. Recently, the driving force for the development of ferrite/ferroelectric nanocomposites has dramatically changed due to advanced fabrication approaches that have provided low-dimensional structures. In this chapter, ferrite/ferroelectric nanocomposites comprising ME components are comprehensively investigated, and the role of effective parameters in their magnetic and ferroelectric properties is studied.

4.2 Magnetoelectric effect

The ME effect may arise from mechanical interactions between the ferrite and piezoelectric phases of samples. Due to magnetostriction, mechanical stresses are induced by applying a magnetic field, leading to a direct ME effect. In turn, the stresses passed to the piezoelectric phase cause a polarization change because of

Magnetic Ferrites and Related Nanocomposites. https://doi.org/10.1016/B978-0-12-824014-4.00005-6

the piezoelectric effect. Conversely, mechanical stresses may be induced in the piezoelectric phase after applying an electric field to the sample and subsequently passed to the ferrite phase in an inverse ME effect, thus changing magnetization by magnetostriction.

The phase transition theory proposed by Landau can macroscopically describe the interactions between the electric field (*E*), magnetic field (*H*), and stress tensor (*s*) in the sample (Dzunuzovic et al., 2018). Typically, the ME response of a sample is measured in terms of the voltage (*V*) induced due to an external alternating current (AC) magnetic field (H_{ac}) (Vopson et al., 2017). The corresponding ME voltage coefficient (α_{ij}^E) is then calculated by the following equation:

$$\alpha_{ij}^E = \frac{dE_i}{dH_j} = \frac{1}{t}\left(\frac{dV}{dH}\right) \tag{4.1}$$

where α_{ij} is the second-rank tensor of ME susceptibility (in terms of V/cmOe) and t is the ferroelectric layer thickness.

In the case of an inverse ME effect, the coefficient α_{ji} represents the magnetic induction variation of the sample as a function of E, according to Eq. (4.2):

$$\alpha_{ji} = b/E = U_{\text{ind}}/(ENS2\pi f) \tag{4.2}$$

The parameter b results from a change in magnetic induction because of the ME interaction, as given below:

$$b = -U_{\text{ind}}/(NS2\pi f) \tag{4.3}$$

where U_{ind} is the induced voltage amplitude of the measuring coil, N is the coil turn number, S is the sample cross-sectional area, and f is the frequency of the varying E.

Physically speaking, the ME parameters can be better understood and visualized when the α_{ij} coefficients are expressed in terms of V/cmOe, as well as Gcm/V (in the case of inverse ME transformation). The first observation of the ME effect was reported for materials with single-phase crystals (e.g., Cr_2O_3 (Ma et al., 2011)), giving rise to $\alpha_{ij} \approx 20$ mV/cmOe. However, the magnitude of α_{ij} was so small that it could not be employed for practical device applications. The small magnitude of α_{ij} can be justified by the following relation:

$$\alpha_{ij}^2 \leq \varepsilon_0 \mu_0 \varepsilon_{ij} \mu_{ij} \tag{4.4}$$

where ε_0 and μ_0 are the respective vacuum dielectric permittivity and magnetic permeability. Moreover, ε_{ij} and μ_{ij} denote the relative dielectric permittivity and permeability of the medium, respectively.

As the magnetic nature of single-phase materials is often antiferromagnetic, the exchange interactions occurring between the magnetic and polar sublattices will be weak. Single-phase materials with leakage current and reduced coupling properties need to operate at extremely low operating temperatures, which limits them for further ME applications (Catalan & Scott, 2009; Gupta & Chatterjee, 2010; Hu et al., 2017; Khomskii, 2009; Nan et al., 2008; Ramana & Sankaram, 2010). As a solution, ferrite/ferroelectric nanocomposites have recently been proposed to enhance the ME coefficient. The nanocomposites can be electrically and magnetically poled to improve the corresponding ME signals. Before measuring ME signals, samples need to be exposed to alternating demagnetizing fields with slightly varying amplitude, thus eliminating long-term memory effects. Accordingly, the demagnetized samples may ensure appropriate repeatability of results. To pole the nanocomposites electrically, they are initially heated to above their ferroelectric Curie temperature and then slowly cooled in the presence of an electric field. The polarized nanocomposites are anisotropic, meaning that their physical parameters, such as the permittivity, piezoelectric constant, and Young's modulus, will depend on the polarization direction.

From a magnetic perspective, it may be initially assumed that the magnetic characteristics of nanocomposites will remain isotropic. However, considering that polarization causes nanocomposites to be deformed (thus inducing interactions between the phases and magnetoelastic effects), anisotropic magnetic properties may emerge from electric polarization. Moreover, a magnetic field can be applied to nanocomposites in the direction of E, thereby influencing ME characteristics by magnetic anisotropy.

4.3 Boomgaard's requirements

The main principles and important characteristics of the ME effect in composites were stipulated by Boomgaard in 1978 and are summarized as follows: (1) an equilibrium state should be established between two individual phases with minimal chemical interaction, (2) intergrain mismatch should not exist, (3) the respective magnetostriction and piezoelectric coefficients of the magnetostrictive and piezoelectric phases need to be large enough to result in a high ME coupling coefficient in the composites, (4) accumulated charges should not be able to leak via the magnetostrictive phase, (5) electric polarization of the composites needs to be carried out based on a deterministic strategy, and (6) the constituent phases of the composites should be sufficiently high in terms of electrical resistivity, thus preventing the charge carriers evolved during the ferroelectric phase from leaking (Boomgaard & Born, 1978).

In practice, the contribution of some factors may be responsible for inducing different properties of composites. Notably, any modification of the composite microstructure can significantly influence electrical and magnetic properties in addition to ME coupling coefficients. In this case, the formation of impurity phases and cracks at the interface between magnetic and piezoelectric phases can decrease the corresponding ME coefficient. To enhance ME coupling, the two aforementioned phases must be well dispersed while having minimal chemical interaction. It is also possible to increase the density of composites by sintering them at high temperatures over long periods. In this way, while one can prevent the individual phases of chemical interactions, the interdiffusion of constituent elements at the interface between the phases appears unavoidable. Depending on the sintering time and temperature, as well as element diffusivity, the extent of interdiffusion may vary as well.

Alternatively, by appropriately dispersing the ferrite phase in the piezoelectric matrix, it is feasible to suppress any percolation and facilitate the poling of the composites. Nevertheless, the different sintering kinetics of the magnetic and piezoelectric phases may cause pore formation in the composites, giving rise to small coupling coefficients. In this respect, some techniques such as spark plasma sintering, microwave sintering, and hot isostatic pressing can sufficiently increase the bulk density and reduce the chemical interactions between phases (DeVreugd et al., 2004; Jiang et al., 2007).

Based on the abovementioned factors and different applications, ferrimagnetic and ferroelectric materials selected to prepare MF composites can be varied. The following section introduces the most important ferroelectrics and magnetic ferrites frequently employed in ME components.

4.4 Ferroelectrics in magnetoelectric components

From an applied standpoint, ferroelectrics show a broad range of physical properties, including piezoelectricity, pyroelectricity, switchable polarizer, and nonlinear optical and dielectric behaviors. The most crucial and desired properties of ferroelectric material constituting an MF composite are a high dielectric constant, low piezoelectric and dielectric losses, and high piezoelectric coefficients. Other substantial parameters such as the Curie temperature (T_C), electric coercive field (E_C), and remanent polarization (P_R) should also be considered. Ferroelectric materials can be categorized into two groups, relaxor and normal. The most famous relaxor and normal ferroelectric materials are disordered perovskite crystal, with the chemical formula $PbMg_{1/3}Nb_{2/3}O_3$, and barium titanate ($BaTiO_3$), respectively. The dielectric response of a relaxor is generally explained in terms of separate polar microregions originating from the composition fluctuations possessing a spontaneous polarization vector. In general, materials exhibit ferroelectricity only below T_C. Spontaneous polarization disappears at temperatures above T_C, transforming the ferroelectric state into one that is paraelectric. With several exceptions, the phase transition between the ferroelectric and paraelectric states is considered a typical behavior of most normal ferroelectrics. In this case, while the relative permittivity decreases above T_C, it can increase as a function of temperature, thereby reaching a peak at T_C.

The phase transition at T_C can either be continuous (second-order transition) or discontinuous (first-order transition). For both continuous and discontinuous transitions, permittivity (ε) at temperatures above T_C can be given as $\varepsilon = C/(T - T_0)$, in which C is the Curie–Weiss constant, and T_0 is the Curie–Weiss temperature. In the case of a continuous phase transition, the parameter T_0 is only equal to T_C (Lines & Glass, 1977).

Due to their relatively high piezoelectric voltage coefficient (g_{33}) and charge coefficient (d_{33}), piezoceramic materials are advantageous in fabricating ME composites. The g_{33} and d_{33} coefficients depend on the dielectric constant and open-circuit voltage, according to Eqs. (4.5) and (4.6):

$$g_{33} = d_{33}/\varepsilon_0 K_3^T \tag{4.5}$$

$$V = g_{33}Tt \tag{4.6}$$

where K_3^T is the dielectric constant, ε_0 is the vacuum permittivity, V is the output voltage, T is the applied stress, and t is the piezoceramic thickness. Accordingly, one can tune d_{33} and K_3^T to enhance g_{33}. In other words, sensor output voltage and sensitivity can be improved by enhancing g_{33}. The output voltage of an ME composite (being another form of Eq. (4.1)) is given by

$$\frac{dE}{dH} = \frac{V}{H_{ac} \cdot t} = \frac{g_{33} \cdot T}{H_{ac}} \tag{4.7}$$

where dE/dH and H_{ac} are the ME coefficient and applied AC magnetic field, respectively. Based on Eq. (4.7), it is inferred that higher values of voltage sensitivity are expected to be obtained in piezoelectric and ME sensors with high g_{33}.

Ferroelectric materials consist of two main groups, inorganic and organic compounds. Metal oxides are inorganic. Among them, ferroelectrics (having the general chemical formula ABO_3) with a perovskite crystal structure are considered the most technologically important group. From an application point of view, the most interesting perovskites have the following structures: $BaTiO_3$ (BTO), $PbTiO_3$, and $Pb(Zr_{0.53}Ti_{0.47})O_3$ (PZT). At room temperature, investigations performed on Pb-based ME composites (e.g., PZT/Terfenol-D) (Chen et al., 2014) and PZT/$CoFe_2O_4$ (CFO) (Zhou et al., 2006) have indicated the occurrence of large ME coupling coefficients. Of late, research investigations have focused on Pb-free ME composites because of the environmental toxicity of Pb.

Apart from PZT, BTO has attracted considerable attention for its excellent piezoelectricity. Nevertheless, considering the low T_C of BTO ($\sim$120°C), its potential use in electronic devices is limited. Thus, it is necessary to enhance the T_C of BTO composites. In this regard, the $Na_{0.5}Bi_{0.5}TiO_3$ (NBT) compound has a relatively high T_C ($\sim$320°C) and large P_R ($\sim$38 μC/cm^2), thus rendering improved electrical properties and T_C compared with those of BTO. Consequently, NBT can be proposed as an appropriate Pb-free BTO-based material with excellent electrical properties (Suchanicz et al., 2003; Yang et al., 2015).

Pb-free relaxors, including $Ba(Zr_xTi_{1-x})O_3$ (BZT)-based materials, have potential use in tunable dielectric devices and dynamic random access memory. The diffuse phase transition in BZT can be advantageous in achieving high piezoelectric and electrocaloric properties. It has been observed that partial occupancy of Zr^{4+} at the Ti^{4+} site can reduce the corresponding T_C while decreasing the conductivity arising from electron hopping between the Ti^{4+} and Ti^{3+} ions (Cheng et al., 2006; Kumar et al., 2008).

The $Pb(Fe_{1/2}Nb_{1/2})O_3$ (PFN) compound is a ferroelectric relaxor with a high dielectric constant. In this case, a diffuse phase transition is expected to occur at about 385K. Moreover, two antiferromagnetic transitions may occur in the PFN involving Néel temperatures ranging between 140 and 155K and 9 and 19K (Bhat et al., 2004).

4.5 Ferrites in magnetoelectric components

Theoretical and experimental investigations have indicated low ME output for ferrite phases with low resistivity. This is because their low resistivity limits the electric field used during the poling of composites. Therefore, poor piezoelectric coupling along with a leakage current would contribute to loss in the induced voltage of composites.

Spinel ferrites, including nickel ferrite (NFO; $NiFe_2O_4$), cobalt ferrite (CFO), and nickel-zinc ferrite, are mostly employed as a magnetic component of ME composites. Among magnetic oxides, CFO is known to possess the highest magnetostriction constants. The randomly orientated magnetostriction constant of CFO range from $\lambda_S^R = -110 \times 10^{-6}$ to -225×10^{-6}, whereas the magnetostriction constants of single-crystal CFO along [100] and [111] directions are $\lambda_S^{100} = -110 \times 10^{-6}$ to -590×10^{-6}, and $\lambda_S^{111} = \sim -1/5\lambda_S^{100}$, respectively. The stoichiometry, thermal history, and Co/Fe ratio can determine the magnetostriction constant of CFO (Jiles & Lo, 2003). It must be noted that NFO shows larger resistivity (thus leading to a reduction in the leakage current of ME composites) and lower magnetostriction and coercive field values than CFO.

4.6 Heterostructural configuration of ferrite/ferroelectric materials in magnetoelectric components

As noted above, the development of composites or multiphase ME materials has been inspired to overcome the limitations of single-phase MFs. Controlled by the connectivity between the two phases in composite architectures, the heterostructure configuration of MEs can be readily established using various methods such as thin film-based and solid-state approaches (see Fig. 4.1) (Andrew et al., 2014; Newnham et al., 1978). Thereby, the most conventional configurations of ME composites are particulate-based (0–3), pillar (1–3), and layered (2–2) heterostructures (Fig. 4.1A). Notably, the 0–3 configuration represents the dispersion of zero-dimensional particles in a three-dimensional matrix, thus forming a particulate-based composite. For nanostructures, the ME components can be formed by incorporating ferrite and ferroelectric nanoparticles (0–0), as well as ferrite and ferroelectric nanowires or nanofibers in different configurations (1–1) (see Fig. 4.1B). It has been predicted that ideal nanofibers without substrate-based constraints will show a higher magnitude of ME signals than thin films.

Bilayer and trilayer structures are expected to show a high degree of polarization and enhanced responses in ME composites. They exploit the 2–2 connectivity, thus separating the conducting ferrimagnetic and insulating ferroelectric phases (Li et al.,

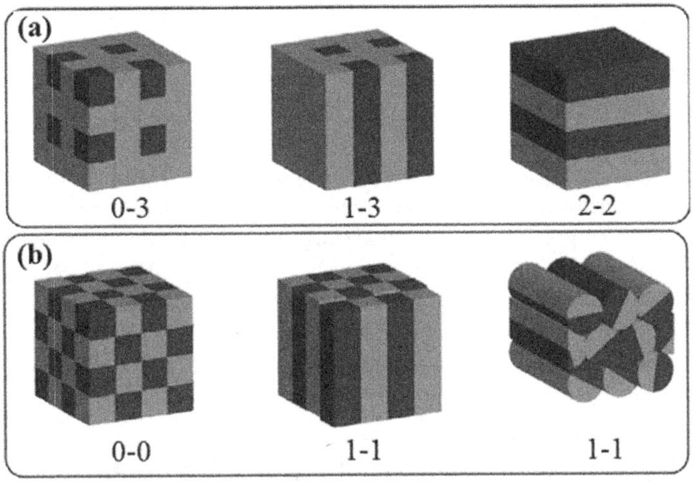

FIGURE 4.1

(A) Conventional configurations of magnetoelectric (ME) composites and (B) novel configurations of ME composites prepared by nanostructures.

From Andrew, J. S., Starr, J. D., & Budi, M. A. (2014). Prospects for nanostructured multiferroic composite materials. Scripta Materialia, *74, 38–43. https://doi.org/10.1016/j.scriptamat.2013.09.023.*

2018). Coupling between high magnetostriction ferrites and high piezoelectricity ceramics can also be facilitated by the 2–2 configuration. By realizing such laminated ME composites (which are impossible to form in the 0–3 configuration), they can easily give rise to practical applications in many different devices such as energy harvesters, magnetic field sensors, and electric transmission systems (Nan et al., 2008). Notwithstanding, these approaches may face challenges, as the ME response is limited by the clamping of the mechanical coupling from the substrates. Moreover, percolation of the ferrite phase may decrease for ferrite volume fractions ranging between 0.2 and 0.5 (Curecheriu et al., 2010; Lakhtakia & Mackay, 2005; Vaz et al., 2010).

Some studies (Srinivasan et al., 2001, 2003) have suggested cosintering ME composites comprising many thin layers of PZT/ferrite, thereby achieving high ME responses. The magnetostrictive/piezoelectric/magnetostrictive trilayer composite is preferred over other composites with a layered structure in terms of both fabrication ease and performance. It has been found that since the piezoelectric layer in the trilayer ME composite is stressed by two ferrite layers, enhanced mechanical coupling is achieved over that in a bilayer ME composite in which only one ferrite layer stresses the piezoelectric layer. In addition, the symmetric geometry of a trilayer composite can prevent the formation of any flexural strain, improving the ME response. As an influential parameter, shape demagnetization should also be considered when evaluating the ME response. Therefore, a trilayer composite with a piezoelectric layer sandwiched between two ferrite layers can show a higher

ME response than a single ferrite layer of the same thickness. This enhancement can be attributed to improved magnetic field penetration in the trilayer composite.

To electrically isolate the ferrite and ferroelectric phases from each other, novel core–shell ME particles consisting of a ferrite core surrounded by a ferroelectric shell have been proposed, leading to the maintenance of a high ferrite volume fraction. The MF composites can also be fabricated as magnetostrictive nanopillars inside a piezoelectric matrix, with ferrite and ferroelectric nanowires embedded in a template. In this respect, Nan et al. (2005) have investigated the influence of residual strain and ME interactions on polarization by modeling static ME effects in BTO-CFO nanobilayers and nanopillars according to the Landau-Ginzburg-Devonshire theory. They found that significantly stronger ferrimagnetic-ferroelectric coupling is involved in nanopillars than in nanobilayers and bulk counterparts.

4.7 Applications of magnetoelectric components

Since ME/MF composites allow for switching the magnetization by an electric field, they provide potential applications in information storage and the miniaturization of magnetostrictive random access memory (MRAM), which requires magnetic fields or large electric currents during the write operation. It is also possible to develop multiple-stable state memory bits and combined memory and logic functions (Bibes & Barthélémy, 2008; Chu et al., 2007; Gajek et al., 2007). Other potential applications of ME/MF composites can be found in magnetic and smart sensors and signal-processing systems (e.g., electric field tunable filters and phase shifters) (Scott, 2007; Spaldin & Fiebig, 2005).

4.8 Theoretical aspects of magnetoelectric effect

In the present chapter, the Maxwell–Wagner (MW) effect, Koop's theory, and Jahn–Teller distortions are employed to interpret dielectric and magnetic properties. Accordingly, brief explanations of them are included in the next sections.

4.8.1 Maxwell–Wagner effect

The MW effect considers the charge accumulation at the interface between two materials, depending on the difference between the relaxation times of their charge carrier. Essentially, the dielectric constant (ε) and conductivity (σ) can macroscopically determine the electrical properties of materials. The relaxation time (τ), describing the spreading time of excess free charge carriers, is obtained by the ratio of these two physical parameters, i.e., $\tau = \varepsilon/\sigma$. By flowing the current across the interface between the two materials with different values, charge accumulation occurs at the interface, resulting in the MW effect. This appears in various interfaces, including metal–insulator, insulator–semiconductor, metal–semiconductor, and ferrite-ferroelectric ME components.

The electric current flowing through materials is expressed by $J = \sigma E$, in which E denotes the electric field. In the case of a steady-state current, the relation $\nabla \cdot J = 0$ is satisfied across the two-material interface. Alternatively, E is obtained by $E = D/\varepsilon$, in which D is the electric flux density. Thus, the following relation describes D at the interface:

$$\nabla \cdot D = \nabla \cdot \varepsilon E = \nabla \cdot \left(\frac{\varepsilon}{\sigma}\right) J = q_s(\neq 0) \tag{4.8}$$

where $q_s(\neq 0)$ is the charge density accumulated at the interface. This indicates that $q_s(\neq 0)$ is accumulated with different τ values, thus satisfying the relation $\nabla(\varepsilon/\sigma) = \nabla\tau \neq 0$. Moreover, $q_s(\neq 0)$ can be expressed by the inner product of the spatial gradient of $\tau(\nabla\tau)$ and the steady-state J, making it credible for charge accumulation at various two-material interfaces (Maxwell, 1973).

4.8.2 Koop's theory

In Koop's theory, the dielectric structure of ferrite materials is formed by respective high-conducting and low-conducting layers of grains and grain boundaries so that charge motion under an electric field in the grains is suppressed at grain boundaries. This leads to the accumulation of localized charges at the interface while forming interfacial polarization. Increasing the frequency of the electric field can decrease the polarization because the electronic exchange between ions cannot follow the alternating field beyond a certain frequency, thereby reducing the real part of the dielectric constant (ε') (Koops, 1951).

4.8.3 Jahn–Teller distortions

A geometric distortion induced in a nonlinear molecular system can lead to the occurrence of the Jahn–Teller effect, reducing its symmetry and energy. Octahedral complexes and tetrahedral compounds have typically shown the presence of Jahn–Teller distortions. Two axial bonds of octahedral complexes are shorter or longer than equatorial bonds. The Jahn–Teller effect can vary depending on the electronic state of the nonlinear molecular system. Hermann Jahn and Edward Teller proposed in 1937 that it is impossible to simultaneously have stability and degeneracy in the electronic state of a nonlinear molecular system (Jahn & Teller, 1937). Therefore, a break in degeneracy can stabilize the molecular system by reducing its symmetry, as commonly reported in transition-metal octahedral complexes as well as tetrahedral compounds.

4.9 Ferrite/ferroelectric nanocomposites for magnetoelectric components

After introducing the requirement of ME nanocomposites and the related theories, the magnetic, dielectric, and ME coefficients of ferrite-ferroelectric nanocomposites should be considered. As noted earlier, nickel-based ferrite and cobalt ferrite are normally chosen along with the ferroelectric phase, acting as magnetic parts of the ME components. These are discussed in the following sections.

4.9.1 Nickel-based ferrite/ferroelectric components

As mentioned, to achieve ME responses close to the values predicted by theory for ceramic composites, some difficulties must be overcome. For example, the densification of composites is reduced due to the high thermal expansion mismatch between the ferrite and piezoelectric phases, giving rise to the formation of microcracks. Moreover, conventional sintering processes employing high sintering temperatures cause undesired phases to form during the chemical reaction. The local eutectic point of the boundary region is also lowered due to the interphase diffusion of constituent atoms, facilitating the formation of liquid phases and high-density defects. The formation of undesired phases and high-density defects results in low resistivity of composites, which can lead to a large leakage current and loss of charges induced by the piezoelectric phase.

Using a spark plasma sintering technique, ME bulk composites consisting of NFO and PZT with different NFO contents have been prepared at a temperature of 900°C for 3 min under a pressure of 50 MPa (Jiang et al., 2007). Fig. 4.2 shows the X-ray diffraction patterns of compacted (x)NFO/$(1-x)$PZT composites ($x = 0.3$,

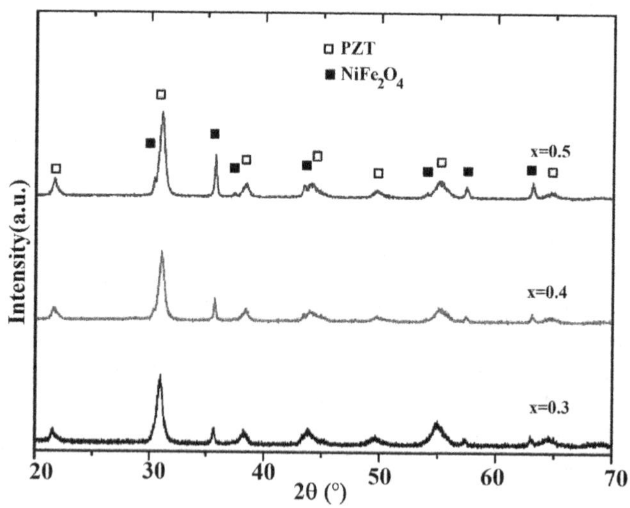

FIGURE 4.2

X-ray diffraction patterns of (x)NFO/$(1-x)$PZT composites ($x = 0.3$, 0.4, and 0.5).

From Jiang, Q., Shen, Z., Zhou, J., Shi, Z., & Nan, C.-W. (2007). Magnetoelectric composites of nickel ferrite and lead zirconnate titanate prepared by spark plasma sintering. Journal of the European Ceramic Society, *27(1), 279–284. https://doi.org/10.1016/j.jeurceramsoc.2006.02.041.*

0.4, and 0.5). As observed, NFO and PZT crystalline structures are formed without impurities for all NFO contents.

On the other hand, magnetostriction measurements of the NFO/PZT composites are shown in Fig. 4.3. Longitudinal (λ_{11}) and transverse (λ_{13}) magnetostriction directions are defined when the strain foil is parallel and perpendicular to the magnetic field, respectively. Regardless of the NFO content, the absolute λ values initially increase with increasing the magnetic field applied to the composites and then approach their saturation values for large magnetic fields. In addition, increasing the NFO content increases the saturation magnetostriction of the resulting composites.

In another research study, ME composites consisting of $Ni_{0.9}Zn_{0.1}Fe_2O_4$ (NZFO) (acting as the ferrite phase) and $PbZr_{0.52}Ti_{0.48}O_3$ (as the ferroelectric phase) with the general chemical formula of (x)NZFO+$(1-x)PbZr_{0.52}Ti_{0.48}O_3$ [(x)NZFO/$(1-x)$ PZT; $x = 0$, 0.15, 0.30, 0.45, and 1 mol%] have been prepared using a double-sintering ceramic technique (Chougule & Chougule, 2008). Fig. 4.4 shows the variation of the dielectric constant of ME composites as a function of temperature using $x = 0.45$ at different frequencies (1 kHz–1 MHz). In this case, two prominent peaks, one near T_C of the ferroelectric phase and another near T_C of the ferrite phase, are observed. Regardless of the frequency, the peak at 380°C can be assigned to a transition between ferroelectric and paraelectric phases. The dielectric polarization in

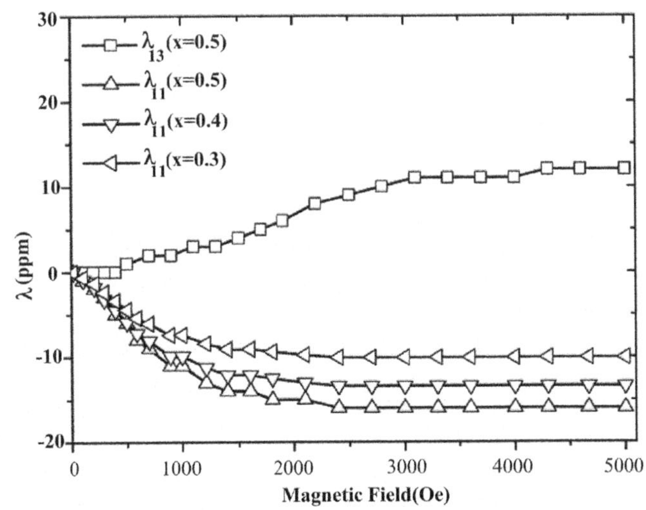

FIGURE 4.3

Magnetostriction measurements carried out for (x)NFO/$(1-x)$PZT composites ($x = 0.3$, 0.4, and 0.5).

From Jiang, Q., Shen, Z., Zhou, J., Shi, Z., & Nan, C.-W. (2007). Magnetoelectric composites of nickel ferrite and lead zirconnate titanate prepared by spark plasma sintering. Journal of the European Ceramic Society, *27(1), 279–284. https://doi.org/10.1016/j.jeurceramsoc.2006.02.041.*

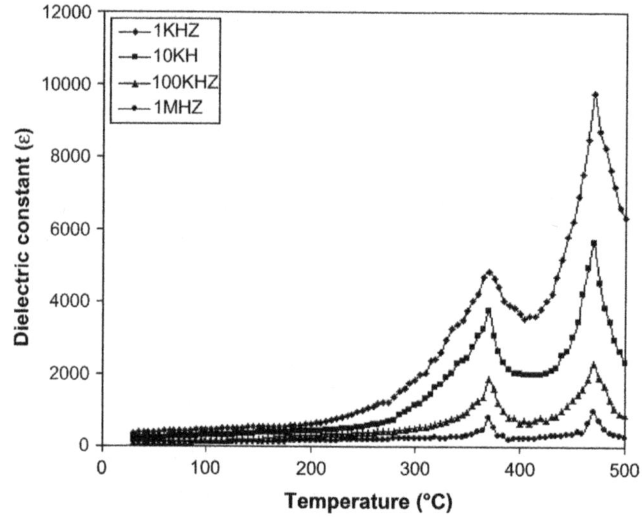

FIGURE 4.4

Variation of the dielectric constant as a function of temperature using 45% $Ni_{0.9}Zn_{0.1}Fe_2O_4$ + 55% $PbZr_{0.52}Ti_{0.48}O_3$ magnetoelectric composites at various frequencies.

From Chougule, S., & Chougule, B. (2008). Response of dielectric behavior and magnetoelectric effect in ferroelectric rich (x)$Ni_{0.9}Zn_{0.1}Fe_2O_4$ + (1–x) PZT ME composites. Journal of Alloys and Compounds, *456(1–2), 441–446. https://doi.org/10.1016/j.jallcom.2007.02.089.*

ferrites arises from the electron hopping phenomenon between the Fe^{2+} and Fe^{3+} ions. The electron hopping phenomenon is activated thermally. Therefore, increasing the temperature increases the electrical conductivity of the composite due to an enhancement in the corresponding drift mobility. In this way, the enhanced dielectric polarization increases the dielectric constant and loss tangent (tanδ). Above T_C, the random vibrational motion of electrons and ions causes the dielectric constant to decrease (Agrawal, 1997).

In a separate study, particulate composites of (x)NZFO/$(1-x)PbZr_{0.52}Ti_{0.48}O_3$ with $x = 0.15$, 0.3, 0.45, 0.6, 0.75, and 0.9 have been synthesized (Pandey et al., 2014). The variation of permittivity as a function of temperature for the (x)NZFO/$(1-x)$PZT composite ($x = 0.60$) at various frequencies (5–100 kHz) is shown in Fig. 4.5. In this case, significant frequency dispersion is seen, and the peak observed in the permittivity curve shifts to higher temperatures with increased frequency, indicating the relaxor properties of PZT due to the ferroelectric phase transition. Note that a pure PZT material is ferroelectric without showing relaxor properties. Consequently, the emergence of relaxor properties can be attributed to the diffusion of ions from NZFO to the PZT phase when forming the composite. In the case of perovskite solid solutions, local ionic charge disorder may contribute to the relaxor features of the ferroelectric phase transition (Bokov & Ye, 2006). Increasing the charge disorder

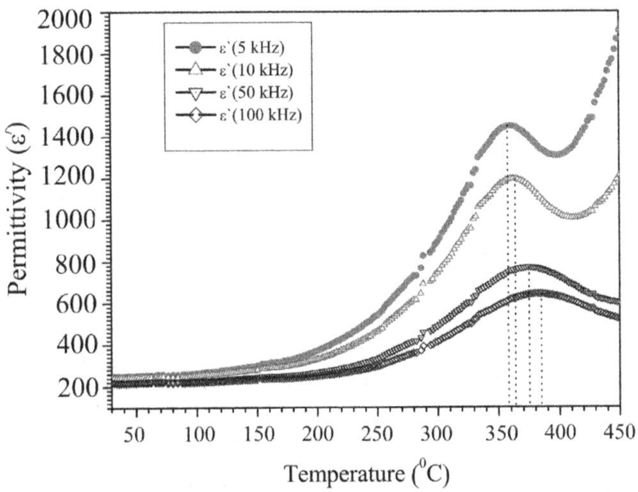

FIGURE 4.5

Variation of permittivity as a function of temperature for (*x*)NZFO/(1−*x*)PZT composite (*x* = 0.60) at various frequencies (5–100 kHZ).

From Pandey, R., Meena, B. R., & Singh, A. K. (2014). Structural and dielectric characterization on multiferroic *x*$Ni_{0.9}Zn_{0.1}Fe_2O_4/(1-x)\ PbZr_{0.52}Ti_{0.48}O_3$ *particulate composite.* Journal of Alloys and Compounds, 593, *224–229. https://doi.org/10.1016/j.jallcom.2014.01.007.*

beyond a certain level prevents the occurrence of long-range ferroelectric order, resulting in cubic crystal symmetry on average.

Variation in permittivity as a function of temperature for the (x)NZFO/(1−x)PZT composites with $x = 0$, 0.15, 0.30, 0.45, and 0.60 at a frequency of 10 kHz is shown in Fig. 4.6. In the case of pure PZT (i.e., $x = 0$), the permittivity curve has a peak at 380°C in accordance with the T_C value of $Pb(Zr_{0.52}Ti_{0.48})O_3$ (PZT) reported in the literature (Jaffe et al., 1971). Increasing the NZFO content continuously shifts the permittivity peak to lower temperatures, indicating the diffusive nature of the phase transition. In other words, stronger relaxor properties of the ferroelectric phase transition are pronounced when increasing the NZFO content in the composite, which can be assigned to the diffusion of higher amounts of Ni^{2+}, Zn^{2+}, and Fe^{3+} ions to the Zr^{4+} and Ti^{4+} sites in the PZT.

To elaborate on the strength of the relaxor properties, the full width at half maximum (FWHM) of the peak of the permittivity curves has been calculated as a function of NZFO content at a frequency of 10 kHz, and the results are shown in Fig. 4.7. As can be seen, increasing the NZFO content in the NZFO/PZT composite continuously increases the corresponding FWHM, indicative of the enhanced diffusive nature of the phase transition.

The variation of T_C of the ferroelectric phase as a function of NZFO content at 1 kHz is shown in Fig. 4.8. After introducing NZFO ($x = 0.15$) into the PZT structure, T_C drastically decreases from 380°C to about 350°C. Further increasing

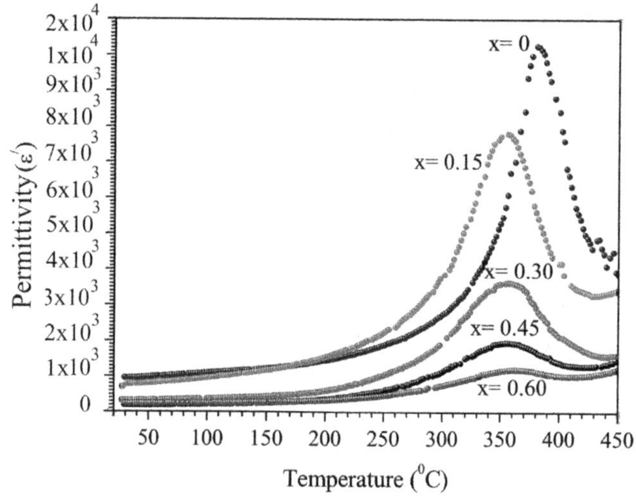

FIGURE 4.6

Variation of permittivity as a function of temperature for (x)NZFO/(1−x)PZT composites at a frequency of 10 kHz.

From Pandey, R., Meena, B. R., & Singh, A. K. (2014). Structural and dielectric characterization on multiferroic x$Ni_{0.9}Zn_{0.1}Fe_2O_4$/(1−x) $PbZr_{0.52}Ti_{0.48}O_3$ *particulate composite.* Journal of Alloys and Compounds, 593, *224–229. https://doi.org/10.1016/j.jallcom.2014.01.007.*

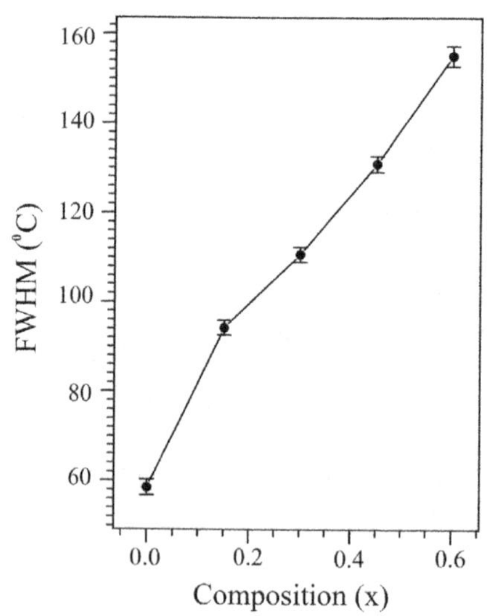

FIGURE 4.7

Variation of full width at half maximum of the peak of permittivity curves as a function of NZFO content for NZFO/PZT composites.

From Pandey, R., Meena, B. R., & Singh, A. K. (2014). Structural and dielectric characterization on multiferroic x$Ni_{0.9}Zn_{0.1}Fe_2O_4$/(1−x) $PbZr_{0.52}Ti_{0.48}O_3$ *particulate composite.* Journal of Alloys and Compounds, 593, *224–229. https://doi.org/10.1016/j.jallcom.2014.01.007.*

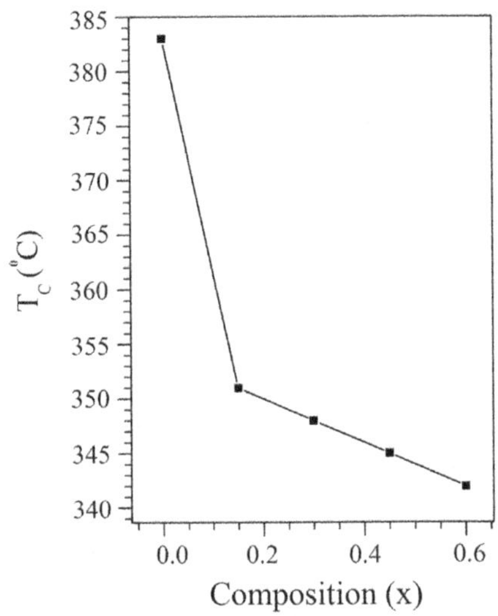

FIGURE 4.8

Variation of T_C as a function of NZFO content for (x)NZFO/(1−x)PZT composites at a frequency of 1 kHz.

From Pandey, R., Meena, B. R., & Singh, A. K. (2014). Structural and dielectric characterization on multiferroic $xNi_{0.9}Zn_{0.1}Fe_2O_4/(1-x)\ PbZr_{0.52}Ti_{0.48}O_3$ *particulate composite.* Journal of Alloys and Compounds, 593, *224–229. https://doi.org/10.1016/j.jallcom.2014.01.007.*

the NZFO content linearly decreases T_C. Overall, the formation of NZFO/PZT composites significantly changes T_C, indicating the reaction between the ferroelectric and ferrite phases.

In a separate research study, ME composites comprising (x)$NiFe_{1.9}Mn_{0.1}O_4$ ferrite and (1−x)$BaZr_{0.08}Ti_{0.92}O_3$ (BZTO) ferroelectric phases (x = 0.10, 0.20, and 0.30) have been synthesized using a conventional ceramic method (Kambale, Shaikh, Kolekar, et al., 2010). The $NiFe_{1.9}Mn_{0.1}O_4$ ferrite phase was selected for the capability of doping NFO with metal and magnetic ions (e.g., Cu, Mn, and Co), thereby enhancing its electrical and magnetic properties, including permeability, resistivity, magnetization, and coercive field (Islam et al., 2008; Newnham, 1989; Ryu et al., 2001). Notably, Van Uitert (1956) has reported an increase of up to 10^{10} or 10^{11} Ω-cm in the resistivity of NFO after doping it with a small amount of Mn (10%). Thus, high ME responses are obtained in ME composites since the ferrite phase has a large resistivity, according to Boomgaard's requirements. A linear decrease in DC resistivity as a function of temperature would reflect the semiconductor behavior of ME composites, which can be justified by the hopping conduction mechanism involving an increase in the charge carrier drift mobility. In addition, increasing the ferrite content can significantly decrease the resistivity of ME composites, arising from the formation of ferrite particle chains and the parallel connectivity created between grains of the ferrite and ferroelectric phases.

To determine electrical conduction, the activation energy (ΔE) of composites can be calculated using the following equation:

$$\rho = \rho_0 \exp\frac{\Delta E}{k_B T} \quad (4.9)$$

where ρ_0 is a temperature-independent constant, ρ is the room-temperature resistivity, and k_B is the Boltzmann constant. The occurrence of electron and hole hopping between Fe^{2+}/Fe^{3+}, Ni^{2+}/Ni^{3+}, Mn^{2+}/Mn^{3+}, Ba^{2+}/Ba^{3+}, Ti^{3+}/Ti^{4+}, and Zr^{3+}/Zr^{4+} ions with $\Delta E < 0.2$eV induces electrical conduction in the resultant $NiFe_{1.9}Mn_{0.1}O_4$/BZTO composites. The charge mobility is found to depend on the temperature and composition after obtaining values of 0.38, 0.33, 0.26, 0.22, and 0.20 eV for the composites with $x = 0$, 0.1, 0.2, 0.3, and 1, respectively.

Fig. 4.9 depicts the dependence of the ME response [defined as $(dE/dH)_H$] on DC magnetic field strength for $(x)NiFe_{1.9}Mn_{0.1}O_4/(1-x)$BZTO composites ($x = 0.10$, 0.20, and 0.30). As inferred, increasing the magnetic field initially increases $(dE/dH)_H$ up to a maximum value and then reduces it with further increases in the magnetic field. This variable behavior may be caused by a saturation state in the magnetostrictive coefficient of the spinel ferrite at a certain magnetic field. The decrease in $(dE/dH)_H$ is then related to the generation of a constant electric field in the piezoelectric phase due to the strain induced beyond the saturation state of magnetostriction when increasing the magnetic field.

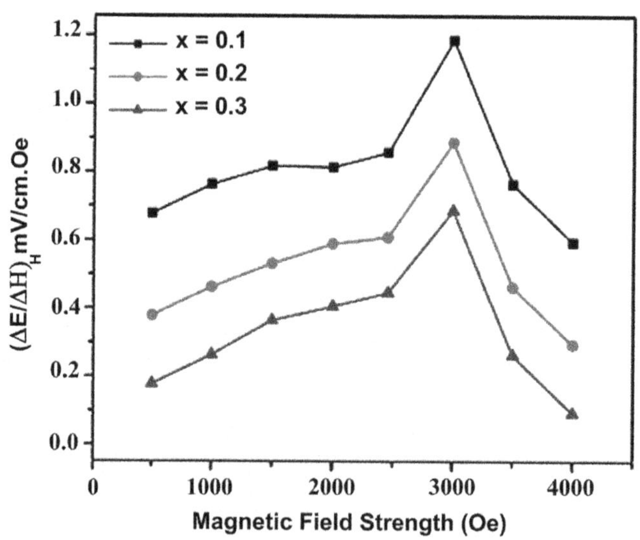

FIGURE 4.9

Dependence of magnetoelectric (ME) response on DC magnetic field strength for (x) $NiFe_{1.9}Mn_{0.1}O_4/(1-x)BaZr_{0.08}Ti_{0.92}O_3$ ME composites ($x = 0.10$, 0.20, and 0.30).

From Kambale, R., Shaikh, P., Bhosale, C., Rajpure, K., & Kolekar, Y. (2010). Studies on magnetic, dielectric and magnetoelectric behavior of (x) $NiFe_{1.9}Mn_{0.1}O_4$ and (1−x) $BaZr_{0.08}Ti_{0.92}O_3$ magnetoelectric composites. Journal of Alloys and Compounds, 489*(1), 310–315. https://doi.org/10.1016/j.jallcom.2009.09.080.*

Elsewhere, Zr and Pb-doped BTO composites with a high figure of merit (Suryanarayana, 1994) and excellent dielectric and piezoelectric properties (Prasad et al., 1996) have been employed as efficient ferroelectric phases. On the other hand, Cu-doped NFO composites have been prepared by the ceramic method to achieve large ME responses. In this case, the element Cu acted as the piezomagnetic phase, taking into account the emergence of mechanical coupling by Ni^{2+}, known as a Jahn–Teller ion. The electron hopping occurs between Fe^{2+} and Fe^{3+} and Cu^{2+} and Cu^{3+} ions in the Cu-doped NFO composites, giving rise to the appearance of conductivity and local displacement toward the external electric field, which can affect the dielectric polarization.

The variation of dielectric loss tangent as a function of frequency at room temperature for $(x)Ni_{0.8}Cu_{0.2}Fe_2O_4/(1-x)Ba_{0.9}Pb_{0.1}Ti_{0.9}Zr_{0.1}O_3$ composites ($x = 0$, 0.15, 0.30, and 0.45 and 1) is shown in Fig. 4.10 (Kanamadi et al., 2005). The dispersion behavior is similar to that of the dielectric constant. At low- and high-frequency regions, the dielectric constant sharply increases and decreases, respectively. The MW-type interfacial polarization is responsible for such a large dispersion of the

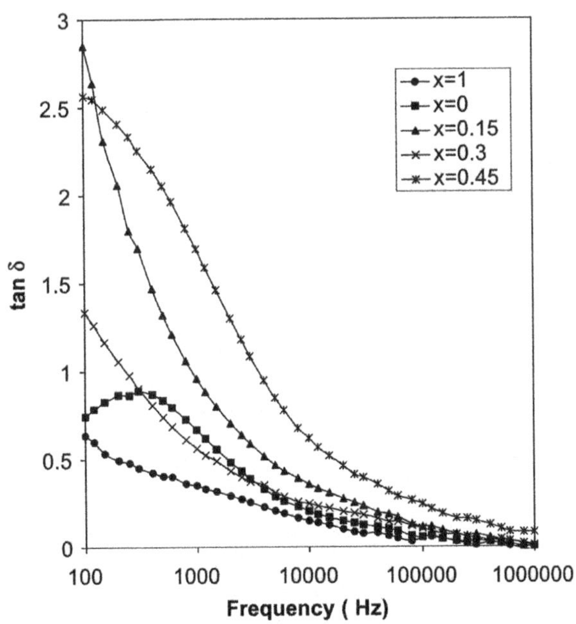

FIGURE 4.10

Variation of dielectric loss tangent (tanδ) as a function of frequency at room temperature for $(x)Ni_{0.8}Cu_{0.2}Fe_2O_4/(1-x)Ba_{0.9}Pb_{0.1}Ti_{0.9}Zr_{0.1}O_3$ composites ($x = 0$, 0.15, 0.30, 0.45, and 1).

From Kanamadi, C., Pujari, L., & Chougule, B. (2005). Dielectric behavior and magnetoelectric effect in (x) $Ni_{0.8}Cu_{0.2}Fe_2O_4/(1-x)Ba_{0.9}Pb_{0.1}Ti_{0.9}Zr_{0.1}O_3$ ME composites. Journal of Magnetism and Magnetic Materials, *295(2), 139–144. https://doi.org/10.1016/j.jmmm.2005.01.006.*

dielectric constant, in agreement with the phenomenological theory proposed by Koops.

The presence of space charge polarization leads to high dielectric constants at low-frequency regions for inhomogeneous dielectric structures involving impurity, porosity, and grains. Nevertheless, similar to the relaxor ferroelectric materials, ferroelectric regions surrounded by nonferroelectric ones are responsible for the high dielectric constants observed in composites.

In another study, $(x)Ba_{0.8}Pb_{0.2}TiO_3/(1-x)Ni_{0.5}Co_{0.5}Fe_2O_4$ ME composites ($x =$ 0, 0.55, 0.70, 0.85, and 1.0) have been synthesized using a double-sintering technique (Kadam et al., 2002). Fig. 4.11 shows the variation of dielectric constant versus temperature of 45% ferrite phase composites at different frequencies (1 kHz–1 MHz). Apart from the maximum dielectric constant corresponding to the ferroelectric T_C for all the frequencies, Fig. 4.11 demonstrates a second peak at frequencies of 1 and 10 kHz, evidencing ferrimagnetic T_C. It is found that the

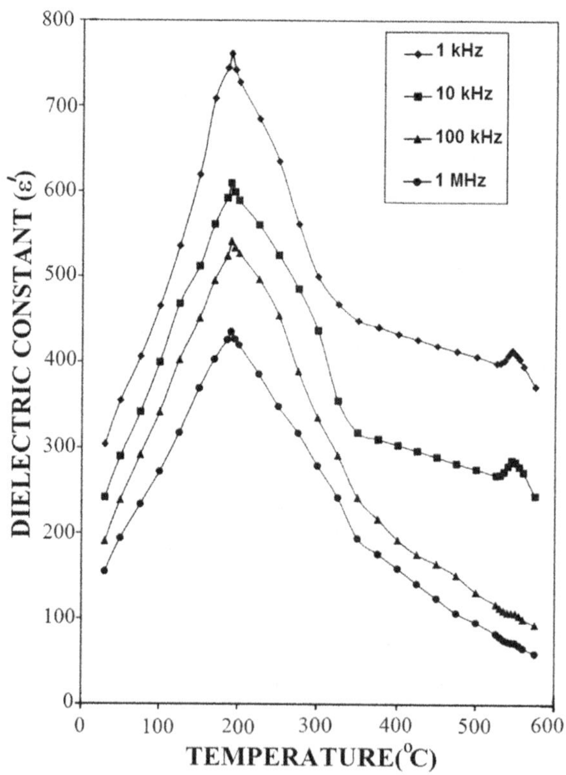

FIGURE 4.11

Variation of dielectric constant versus temperature of 45% $Ni_{0.50}Co_{0.50}Fe_2O_4$ + 55% $Ba_{0.8}Pb_{0.2}TiO_3$ composite at different frequencies (1 kHz–1 MHz).

From Kadam, S., Patankar, K., Mathe, V., Kothale, M., Kale, R., & Chougule, B. (2002). Dielectric Behavior and Magnetoelectric Effect in $Ni_{0.50}Co_{0.50}Fe_2O_4$ + $Ba_{0.8}Pb_{0.2}TiO$ ME Composites. Journal of Electroceramics, 9*(3), 193–198. https://doi.org/10.1016/j.matlet.2004.08.033.*

ferroelectric and ferrimagnetic T_C values of the composites do not depend on frequency.

The absence of the second peak at the frequencies of 100 kHz and 1 MHz occurs because the electron exchange mechanism and applied electric field are not efficiently linked above a certain frequency value (Ahmed et al., 1995; El Hiti, 1996).

Fig. 4.12 shows the variation of ME conversion factor (*dE/dH*) as a function of magnetic field (*H*) applied to the composites with $x = 0.85$, 0.70, and 0.55. As observed, increasing *H* continuously decreases *dE/dH*. As mentioned before, the magnetostrictive coefficient of spinel ferrites is saturated with magnetization at a certain *H*, leading to strain in the ferrite phase. In turn, a constant electric field is produced in the piezoelectric phase due to the strain, thus reducing *dE/dH* as a function of *H*. Moreover, increasing the ferrite content decreases *dE/dH*, which can be

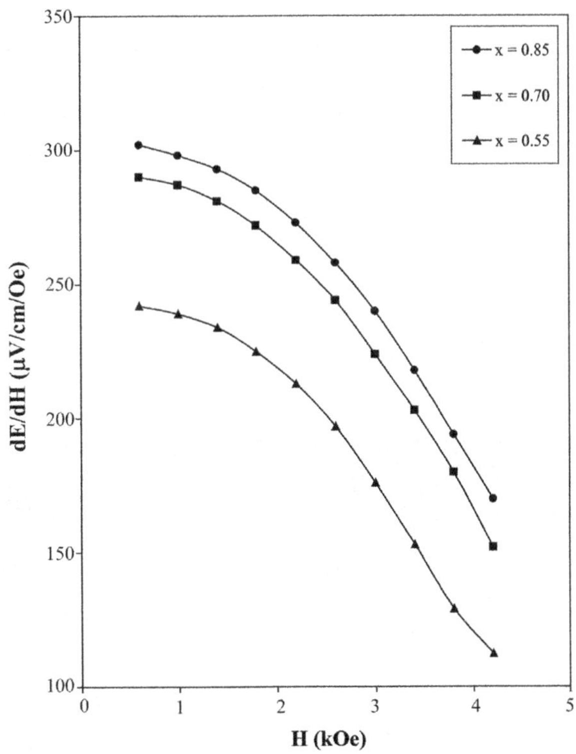

FIGURE 4.12

Variation of magnetoelectric conversion factor (*dE/dH*) as a function of magnetic field (*H*) applied to $(x)Ba_{0.8}Pb_{0.2}TiO_3/(1-x)Ni_{0.5}Co_{0.5}Fe_2O_4$ composites ($x = 0.85$, 0.70, and 0.55).

From Kadam, S., Patankar, K., Mathe, V., Kothale, M., Kale, R., & Chougule, B. (2002). Dielectric behavior and magnetoelectric effect in $Ni_{0.75}Co_{0.25}Fe_2O_4 + Ba_{0.8}Pb_{0.2}TiO_3$ ME composites. Journal of Electroceramics, *9(3), 193–198. https://doi.org/10.1016/j.matlet.2004.08.033.*

assigned to the lower resistivity of the ferrite phase than that of the ferroelectric phase. A low-resistivity path from the ferrite grains gives rise to charge leakage in the piezoelectric grains. Although the inelastic interaction is enhanced by increasing the ferrite content, the further addition of the ferrite phase is limited due to the reduction in the composite resistivity.

Using a tape-casting technique, cosintered ME composites consisting of piezoelectric ceramic multilayer [$Pb_{(1-x-3y/2)}Sr\ La_y(Zr_z,Ti_{(1-z)})O_3$, (PSLZT); $x = 0.06$, $y = 0.03$, and $z = 0.56$] and $Ni_{0.6}Zn_{0.4}Fe_2O_4$ (NZFO) have been synthesized, and their ferroelectric and piezoelectric behaviors have been investigated. To improve the contact area between the two phases, a Pt electrode was used (Premkumar et al., 2019). By inserting the Pt layer between the ferrimagnetic and ferroelectric layers, one can enhance the strain transfer efficiency, achieving a high ME coefficient.

Top-view and side-view photographs of the PSLZT/NZFO composite sample are depicted in Fig. 4.13A and B, respectively, indicating the absence of warpage

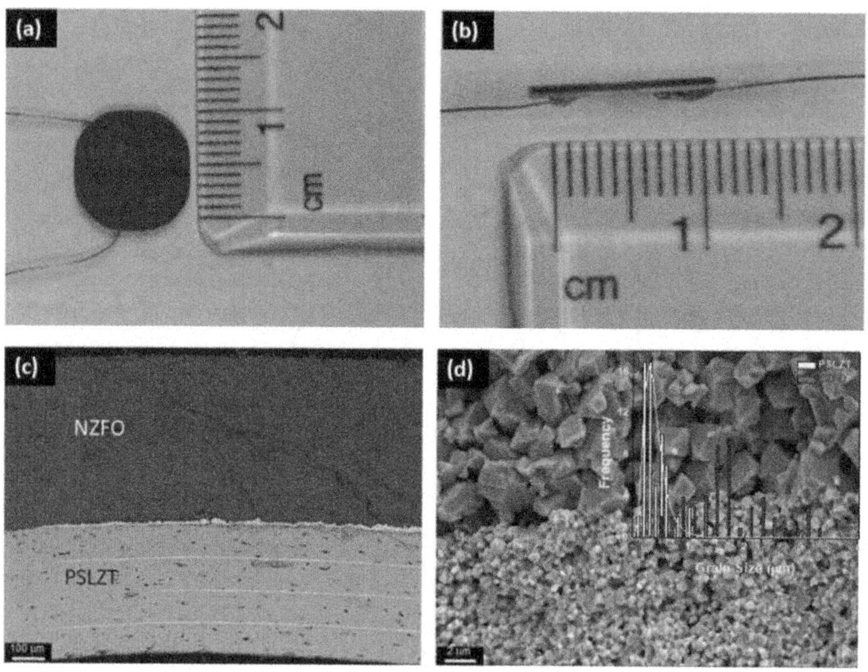

FIGURE 4.13

(A) Top-view and (B) side-view photographs of PSLZT/NZFO composite sample. (C) Cross-sectional scanning electron microscopy (SEM) image of PSLZT/NZFO composite. (D) A magnified cross-sectional SEM image of the composite. The inset in part (D) shows the corresponding grain size distributions of PSLZT and NZFO phases.

From Premkumar, S., Varadarajan, E., Rath, M., Rao, M. R., & Mathe, V. (2019). Ferroelectric and piezoelectric properties of PSLZT multilayer/NZFO cosintered magnetoelectric composites fabricated by tape casting. Journal of the European Ceramic Society, 39*(16), 5267–5276. https://doi.org/10.1016/j.jeurceramsoc.2019.08.010.*

in the composite and good adherence between the PSLZT and NZFO components. To study grain size distribution and the interface between the PSLZT multilayer and NZFO, morphological properties of the resulting ME composite have been investigated by scanning electron microscopy (SEM). The results obtained are shown in Fig. 4.13C and D. The cross-sectional SEM image depicted in Fig. 4.13C shows the formation of dispersed pores in the uniform and relatively dense PSLZT (light) and NZFO (dark) phases when the composite undergoes a sintering process at 1060°C for 2 h. The NZFO layer is about 1.2 times thicker than the PSLZT multilayer, indicating a net ferrite layer shrinkage of 15% compared with the piezoelectric multilayer when sintered under the same condition. It is important to mention that a clear adhesion between the PSLZT-Pt-NZFO is observed (Rosa & Venet, 2016). From Fig. 4.13D and the related inset, the grain size distributions of the PSLZT and NZFO phases range between 0.5 and 3 and 0.25 and 1.0 μm, respectively.

Fig. 4.14A shows polarization-voltage (P—V) hysteresis loops of the PSLZT/NZFO composite measured at 1 Hz. In this case, nearly square P—V hysteresis loops with P_R = 55.5 μC/cm^2, saturation polarization of 70.6 μC/cm^2, and a coercive voltage of 78.8 V are obtained for the composite, outperforming the polarization and P—V behavior of the PSLZT multilayer. In the case of the PSLZT/NZFO composite comprising three PSLZT layers (320 μm in total thickness) and approximately the same thickness as the NZFO layer, P_R increased up to about three times compared with a single-layer PSLZT/NZFO composite.

Fig. 4.14B shows current-voltage curves obtained from the PSLZT/NZFO composite involving one sharp peak in the switching current. This indicates a swift domain switching of the composite occurring within a narrow voltage range, similar to the switching behavior of bulk PSLZT ceramics. Therefore, one can infer that the PSLZT multilayer has a lower density than PSLZT loaded with NZFO in the layered composites, likely due to stresses induced during the sintering process.

The dynamic behavior of the PSLZT/NZFO composite was studied by investigating polarization curves at different applied frequencies, and the results are shown in Fig. 4.14C. The variations of polarization and coercive voltage as a function of frequency are also depicted in Fig. 4.14D. As can be seen, increasing the frequency reduces the polarization slightly. The variation behavior of P_R and E_C as a function of frequency in the PSLZT/NZFO composite is the same as that in the PSLZT multilayer (Bochenek & Niemiec, 2018; Qian et al., 2013), likely arising from the slow domain switching when increasing the applied field frequency, which leads to a decrease in the net polarization.

Fig. 4.14E shows capacitance-voltage and tanδ behaviors in the PSLZT/NZFO composite measured analogously to those in the PSLZT multilayer. A dielectric constant of 1450 is obtained after measuring the capacitance at a low bias field. Furthermore, despite the PSLZT multilayer, the asymmetric butterfly loop observed in the capacitance-voltage curve of the PSLZT/NZFO composite may be attributed to the accumulation of space charge in the dispersed pores, as well as at the interface between the PSLZT and NZFO layers. On the other hand, increasing the voltage

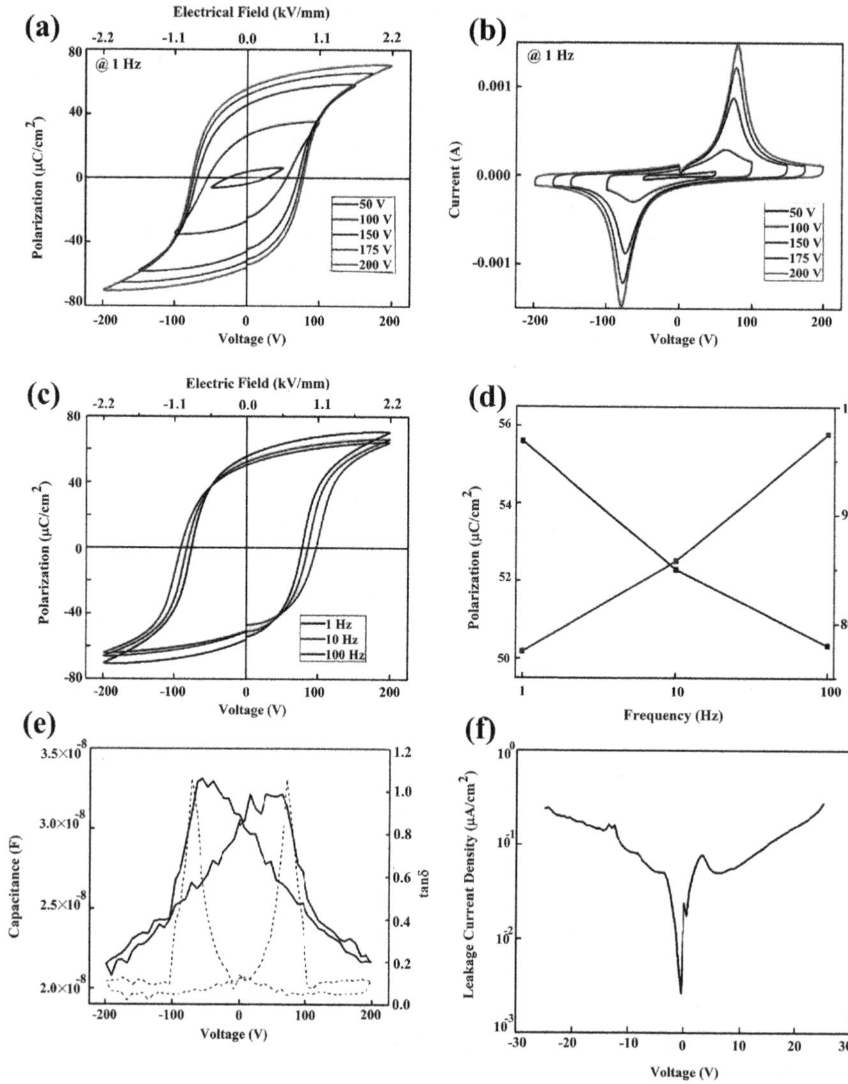

FIGURE 4.14

(A) Polarization-voltage hysteresis loops, (B) current-voltage curves, (C) polarization curves at different applied frequencies, (D) variations in polarization and coercive voltage as a function of frequency, (E) capacitance-voltage and tanδ behaviors, and (F) leakage current density-voltage curve of the PSLZT/NZFO composite.

From Premkumar, S., Varadarajan, E., Rath, M., Rao, M. R., & Mathe, V. (2019). Ferroelectric and piezoelectric properties of PSLZT multilayer/NZFO cosintered magnetoelectric composites fabricated by tape casting. Journal of the European Ceramic Society, 39*(16), 5267–5276. https://doi.org/10.1016/j.jeurceramsoc.2019.08.010.*

increases tanδ more sharply in the composite than in the PSLZT multilayer, indicating ion migration through the NZFO layer.

The leakage current density-voltage characteristics of the composite were recorded for applied voltages between −25 and +25 V, and the results are shown in Fig. 4.14F. By applying the positive bias, an asymmetric curve is also observed. Ion migration via the Pb^{2+} ion vacancy available in the PSLZT layer, as well as through the interface between the NZFO and PSLZT layers, may be responsible for the asymmetric behavior.

Nevertheless, the overall leakage current density level remains unchanged in the composite, similar to that in the PSLZT multilayer. The absence of interdiffusion of NZFO atoms into the PSLZT layer is also confirmed by the asymmetric behavior, as previously observed in the SEM images. As a result, the ferroelectric characteristics of the ferroelectric multilayer are not influenced by cosintering of the magnetostrictive NZFO and PSLZT layers mediated with the improved contact from the Pt electrode.

Composites with the composition of (x)NFO/$(1-x)$PZT ($x = 0, 0.1, 0.2, 0.3, 0.4, 0.5$, and 1) have been prepared by using a conventional solid-state reaction method in another study (Cheng et al., 2014). The dependence of dielectric behavior on magnetic field strength was studied by applying DC magnetic field (H_{dc}) values of 0 and 1 T to the NFO/PZT composites, thereby obtaining dielectric constant and complex impedance, respectively.

Variations in dielectric constant as a function of frequency together with Cole–Cole curves for 0.2NFO/0.8PZT and 0.5NFO/0.5PZT nanocomposites are depicted in Figs. 4.15 and 4.16, respectively. In this regard, the dielectric constant and complex impedance vary similarly in the two composites, so a resonance takes place at

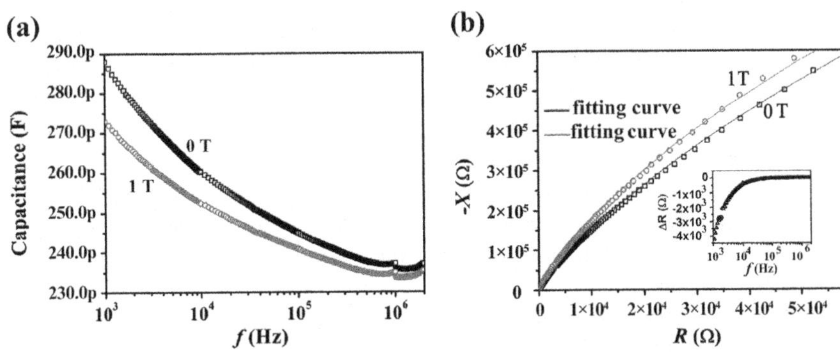

FIGURE 4.15

(A) Variation of dielectric constant as a function of frequency and (B) Cole–Cole curves for 0.2NFO/0.8PZT composite at H_{dc} of 0 and 1 T.

From Cheng, T., Xu, L. F., Qi, P. B., Yang, C. P., Wang, R. L., & Xiao, H. B. (2014). Tunable dielectric behaviors of magnetic field of PZT5/NiFe$_2$O$_4$ ceramic particle magnetoelectric composites at room temperature. Journal of Alloys and Compounds, 602, *269–274. https://doi.org/10.1016/j.jallcom.2014.03.032.*

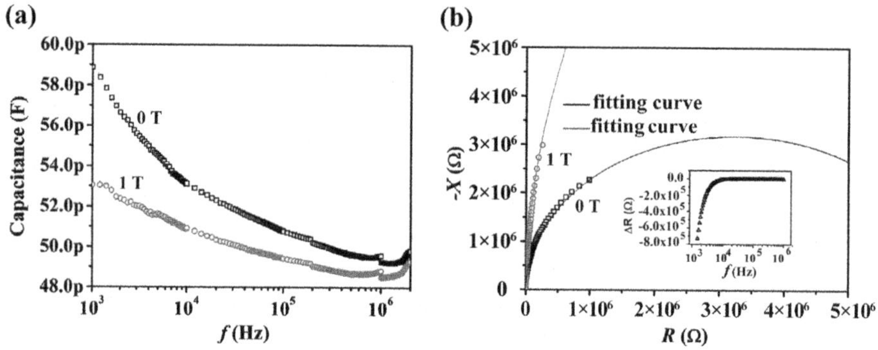

FIGURE 4.16

(A) Variation of dielectric constant as a function of frequency and (B) Cole–Cole curves for 0.5NFO/0.5PZT composite at H_{dc} of 0 and 1 T.

From Cheng, T., Xu, L. F., Qi, P. B., Yang, C. P., Wang, R. L., & Xiao, H. B. (2014). Tunable dielectric behaviors of magnetic field of PZT5/NiFe$_2$O$_4$ ceramic particle magnetoelectric composites at room temperature. Journal of Alloys and Compounds, 602, *269–274. https://doi.org/10.1016/j.jallcom.2014.03.032.*

the same frequency regardless of both the composition and the magnetic field. The results of $\Delta R(R(B)-R(0))$ representing the complex impedance real part as a function of frequency are shown in the inset of Figs. 4.15B and 4.16B. In the low-frequency region, only one semicircle is observed in the complex impedance. Changing the resistance as a function of H_{dc} depends on the frequency. Therefore, H_{dc} can affect the resistivity more at lower than at higher frequency. As well, for the entire frequency range, the ΔR values are negative.

As mentioned previously, unpoled ferroelectric composites with randomly oriented domains have negligible net polarization. In the case of poled composites, the ME effect is enhanced due to the domination of domain polarity (Ma et al., 2011). Thus, before performing the ME measurements, one can polarize the ferroelectric layer by applying an electric field perpendicular to the composite plane, as carried out for bilayer $BaCaZrTiO_3$ (BCZT)/NFO and trilayer BCZT/NFO/BCZT composites (Sowmya et al., 2018).

Fig. 4.17A shows the magnetization-magnetic field ($M-H$) loop obtained by applying H along the in-plane direction of NFO. The magnetic coercive field (H_c) and saturation magnetization (M_s) are 20 Oe and 40 emu/g, respectively, confirming that NFO is permeable.

For strain-mediated ME composites, it has been evidenced that the peak position of the ME value pertains to the maximum q value ($q = d\lambda/dH$). In the case of the NFO phase, λ and M values follow each other based on the relation given below (Palneedi et al., 2015):

$$\varphi \sim \frac{3\lambda\sigma}{K + 2\pi M^2} \quad \text{or} \quad \lambda \propto M^2 \tag{4.10}$$

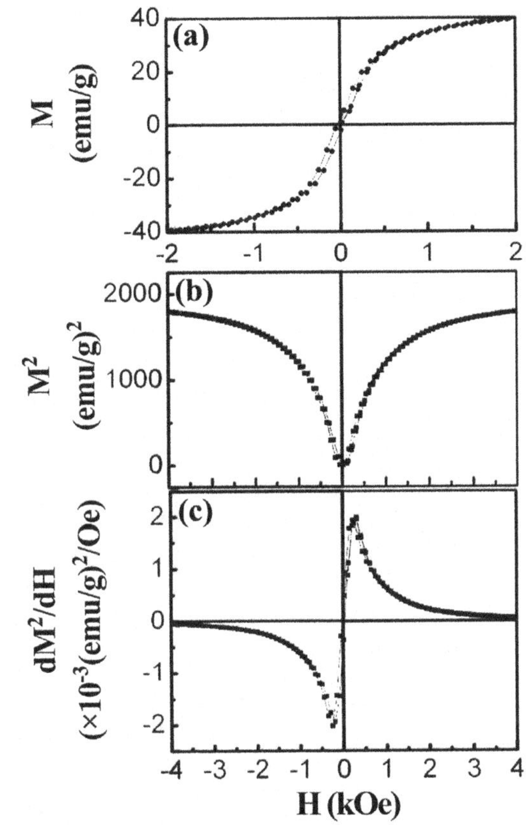

FIGURE 4.17

(A) Magnetization-magnetic field (M–H) loop obtained by applying H along the in-plane direction, (B) M^2–H curve, and (C) dM^2/dH–H curve of NFO.

From Sowmya, N. S., Srinivas, A., Saravanan, P., Reddy, K. V. G., Kamat, S., Praveen, J. P., Das, D., Murugesan, G., Kumar, S. D., & Subramanian, V. (2018). Studies on magnetoelectric coupling in lead-free [(0.5) BCT-(0.5) BZT]-NiFe$_2$O$_4$ laminated composites at low and EMR frequencies. Journal of Alloys and Compounds, 743, 240–248. *https://doi.org/10.1016/j.jallcom.2018.01.402.*

where ϕ represents the magnetic moment angle between Ni^{2+} and Fe^{3+} cations, K is the anisotropy constant, and σ denotes the stress. By considering the voltage coefficient $\alpha_E \propto q$, one can conclude that $\alpha_E \propto dM^2/dH$. Consequently, it is possible to evaluate the ME composite based on the variation behavior of M^2–H and dM^2/dH–H, as shown in Fig. 4.17B and C, respectively.

On the other hand, Fig. 4.18A schematically depicts the ME response in two directions [$\alpha_{E,31}(L–T)$] and [$\alpha_{E,33}(T–T)$]. Variation in the ME coefficient as a function of H_{dc} in the transverse and longitudinal directions for BCZT/NFO/BCZT and BCZT/NFO at a frequency of 1 kHz and H_{ac} of 1 Oe is also shown in Fig. 4.18B.

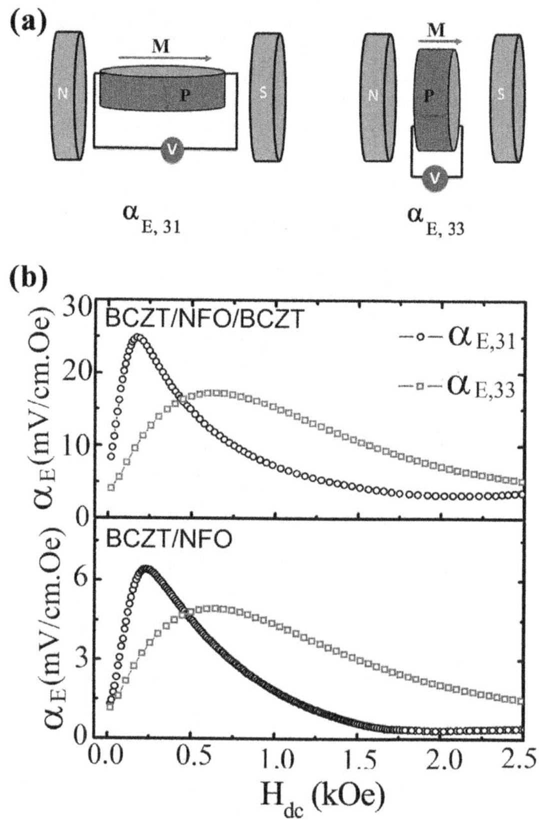

FIGURE 4.18

(A) Schematic representation of magnetoelectric (ME) response in the transverse and longitudinal directions, and (B) ME voltage coefficient (α_E) obtained from BCZT/NFO and BCZT/NFO/BCZT composites at a frequency of 1 kHz.

From Sowmya, N. S., Srinivas, A., Saravanan, P., Reddy, K. V. G., Kamat, S., Praveen, J. P., Das, D., Murugesan, G., Kumar, S. D., & Subramanian, V. (2018). Studies on magnetoelectric coupling in lead-free [(0.5) BCT-(0.5) BZT]-NiFe$_2$O$_4$ laminated composites at low and EMR frequencies. Journal of Alloys and Compounds, 743, *240–248. https://doi.org/10.1016/j.jallcom.2018.01.402.*

At $H_{dc} = 0$, the presence of the small H_{ac} and remanent magnetization (arising from the ferrite phase) can cause α_E to have a nonzero value. Increasing H_{dc} significantly increases α_E, leading to maximum $\alpha_{E,31}$ values of 6.5 and 25 mV/cmOe for the BCZT/NFO and BCZT/NFO/BCZT composites, respectively. Further increasing H_{dc} decreases α_E to a minimum value.

Although the variation behavior of $\alpha_{E,31}$ and $\alpha_{E,33}$ is similar, the following differences are present: (1) H_{dc} of the maximum $\alpha_{E,31}$ ($\sim$180 Oe) is smaller than that of the maximum $\alpha_{E,33}$ ($\sim$600 Oe), which arises from the demagnetizing field absence (Vlasko-Vlasov et al., 2000), and leads to *a* large *q* value in the transverse direction and (2) the saturation field of $\alpha_{E,33}$(>2000 Oe) is higher than that of $\alpha_{E,31}$ (~1500 Oe).

At higher H_{dc}, λ of the in-plane bias (λ_{11}) is saturated more quickly than that of the out-of-plane bias (λ_{13}). This results in an approximately constant electric field in the BCZT, which decreases $\alpha_{E,31}$ as a function of H_{dc}. In the case of the NFO phase, the maximum change in λ reflects the maximum ME coupling (response). When λ is saturated at H_{dc}, the corresponding ME coupling disappears accordingly. In the presence of H_{dc}, pseudo-piezomagnetic effects may be formed in the ferrimagnetic phase, leading to its coupling to the BCZT phase of the composite. Asymmetric and butterfly-like features of λ are evidenced, whereas the behavior of q_{11} follows that of dM^2/dH. In this way, one can infer that the NFO layer hysteretic response plays the main role in the corresponding ME effect.

It is possible to use the pulsed laser deposition technique to epitaxially grow NFO/PZT heterostructures on (001) $SrTiO_3$ substrates (Zheng et al., 2020). Depending on the required magnitude of the ME coupling coefficient, the growth mechanism involved in the formation of ferrimagnetic and ferroelectric layers and thin films can be changed. The effect of ferrimagnetic and ferroelectric layer thicknesses on ME response has also been studied (Li et al., 2018).

Polarization-electric field ($P-E$) hysteresis loops obtained from NFO/PZT heterostructures with different PZT thicknesses are shown in Fig. 4.19. For PZT thicknesses of 90, 130, 170, and 250 nm, the respective P_R values are 33, 40, 51, and 60 μC/cm^2. Due to the reduction in substrate clamping of PZT layers with higher

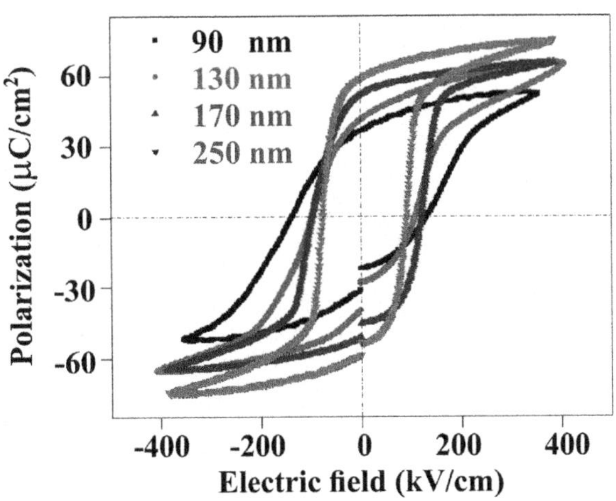

FIGURE 4.19

Polarization-electric field ($P-E$) hysteresis loops obtained from NFO/PZT heterostructures with PZT thicknesses ranging between 90–250 nm.

From Zheng, Z., Zhou, P., Liu, Y., Liang, K., Tanguturi, R., Chen, H., Srinivasan, G., Qi, Y., & Zhang, T. (2020). Strain effect on magnetoelectric coupling of epitaxial NFO/PZT heterostructure. Journal of Alloys and Compounds, *818, 152871. https://doi.org/10.1016/j.jallcom.2019.152871.*

thicknesses (thus improving the strain relaxation), increasing the PZT thickness continuously increases the corresponding P_R.

Strain-mediated ME coupling causes ME coupling in the NFO/PZT heterostructures. The mismatch between lattice or thermal expansion of the film and substrate is responsible for inducing the mechanical residual stress/strain, transferring it to the ferroelectric and ferrimagnetic layers.

According to the literature (Bichurin et al., 2003; Liang et al., 2017), the magnetoelectric voltage coefficient is calculated based on Eq. (4.11):

$$\alpha_{E31} = -\frac{dE_3}{dH_1} = \frac{-k^p d_{31}^m q f_m (1-f_m)}{\left[k(1-f_m)^m s + f_m^p s\right]\varepsilon_{33} + 2 f_m^p d_{31}^2} \tag{4.11}$$

where k is defined as the interface coupling coefficient $k = ({}^p s_i - {}^p s_{i0})/({}^m s_i - {}^p s_{i0})$, in which ${}^p s_{i0}$ is the frictionless strain, ${}^m s_i$ denotes the strain tensor component of the magnetic phase, and ${}^p s_i$ is the strain tensor component of the PZT phase (Liang et al., 2017). In the case of ideal coupling, $k = 1$, whereas the frictionless interface has $k = 0$. Moreover, ε_{33} is dielectric permittivity coefficient and ${}^p d_{31}$ denotes the piezoelectric coefficient of PZT, ${}^m s_{ij}$ is the compliance coefficient, ${}^m q_{kl}$ reflects the piezomagnetic coefficient, and parameter f_m is related to the ferrimagnetic (V_{FM}) and ferroelectric (V_{FE}) phase volume fractions as follows: $f_m = V_{FM}/(V_{FM} + V_{FE})$. Using Eq. (4.11), the ME coupling coefficients (α_{E31}) were calculated as a function of the parameter f_m for NFO/PZT composite films with different interface coupling coefficients (k) at an applied magnetic field of 3000 Oe, and the results are shown in Fig. 4.20.

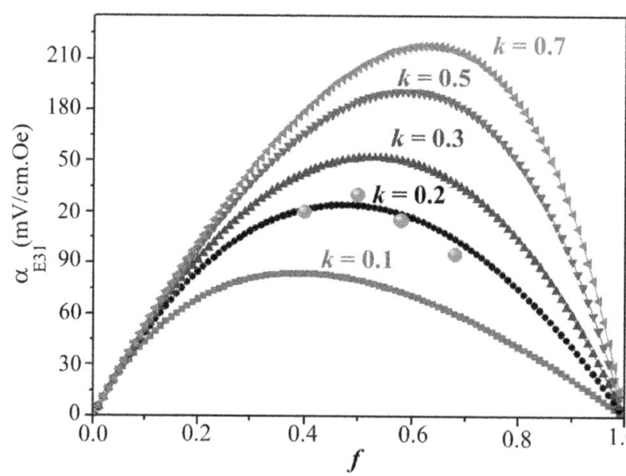

FIGURE 4.20

Variation of magnetoelectric coupling coefficient (αE31) as a function of NFO volume fraction calculated for NFO/PZT composite films with different interfacial factors ($0.1 \leq k \leq 0.7$). The solid orange points show the corresponding experimental α_{E31} values.

From Zheng, Z., Zhou, P., Liu, Y., Liang, K., Tanguturi, R., Chen, H., Srinivasan, G., Qi, Y., & Zhang, T. (2020). Strain effect on magnetoelectric coupling of epitaxial NFO/PZT heterostructure. Journal of Alloys and Compounds, 818, *152871. https://doi.org/10.1016/j.jallcom.2019.152871.*

It is found that α_{E31} depends on both V_{FM} and k, so it initially increases as a function of f and then decreases with increasing V_{FM} at a certain value of k. Alternatively, α_{E31} is gradually enhanced with increasing k at a certain V_{FM}. In the case of $k = 0.2$, the experimental α_{E31} values (the solid points) accord well with the calculated results, as shown in Fig. 4.20.

Interpenetration between layers of bilayer composite films can lead to the formation of coupled defects, thereby changing the lattice parameters and the dielectric constant. In this regard elsewhere, bilayer nanocomposite films consisting of a piezoelectric phase (0.65BT–0.35NBT) and ferrimagnetic phase (NFO) have been epitaxially grown on (001)-$SrRuO_3/SrTiO_3$ (SRO/STO) substrates using a pulsed laser deposition technique, resulting in the following structures: NFO/BT–NBT/SRO/STO and BT–NBT/NFO/SRO/STO (Dai et al., 2019). Note that the nanocomposite films were rich in the BT phase, and the SRO layer acted as the bottom electrode and buffer layer.

The ε_r of the NFO/BT–NBT/SRO/STO composite film is higher than that of the BT–NBT/NFO/SRO/STO one. This improvement in dielectric performance can be caused by the increased contact area of the NFO film with compact layers pulsed-laser deposited on the BT–NBT film.

P–E hysteresis loops and leakage properties of the composite films are shown in Fig. 4.21A and B, respectively (Dai et al., 2019). At an electric field of 300 kV/cm, the P_R and P_S of the NFO/BT–NBT/SRO/STO composite film are obtained as 31.8 and 85.6 μC/cm^2, respectively, outperforming those of BT–NBT/NFO/SRO/STO ($P_R = 27.3$ and $P_S = 78.3$ μC/cm^2) due to higher insulative qualities. However, under the same applied electric field, the NFO/BT–NBT/SRO/STO composite film

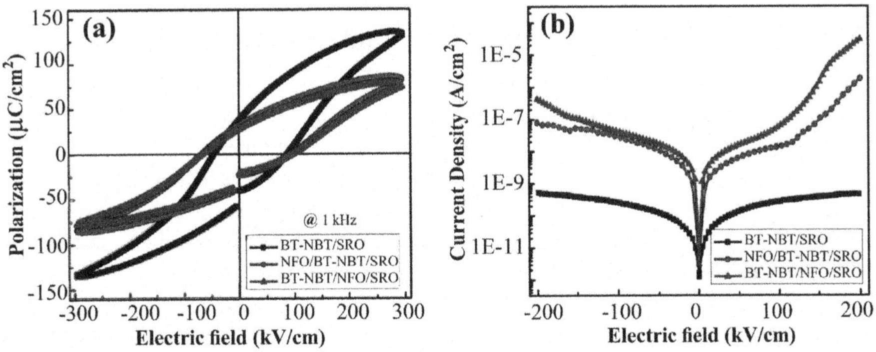

FIGURE 4.21

(A) P–E hysteresis loops and (B) leakage current density versus electric field curves of BT–NBT/SRO, NFO/BT–NBT/SRO, and BT–NBT/NFO/SRO composite films. NFO, BT, NBT, SRO denote $NiFe_2O_4$, $BaTiO_3$, $Na_{0.5}Bi_{0.5}TiO_3$ and $SrRuO_3$, respectively.

From Dai, Q., Guo, K., Zhang, M., Cui, R., & Deng, C. (2019). Magnetoelectric properties of $NiFe_2O_4$/0.65 $BaTiO_3$–0.35 $Na_{0.5}Bi_{0.5}TiO_3$ bilayer thin films deposited via pulsed laser deposition. Ceramics International, *45(7), 8448–8453. https://doi.org/10.1016/j.ceramint.2019.01.154.*

shows lower ferroelectric properties than the single-phase BT–NBT/SRO/STO composite film ($P_R = 38.1$ and $P_S = 135.1$ μC/cm^2).

It is evident that J_S of the NFO/BT–NBT/SRO/STO film is smaller than that of the BT–NBT/NFO/SRO/STO one, likely due to its higher insulating property. For small electric fields, leakage current density (J_S) and E have an almost linear relationship, according to Ohm's conduction mechanism expressed below (Cheng et al., 2004):

$$J_S = nev \cdot E \tag{4.12}$$

where n, e, and v are the carrier concentration, electron charge, and carrier mobility rate, respectively. According to Eq. (4.12), J_S is proportional to E. For composites, the proportionality of the relationship between J_S and E dramatically increases when $E > \pm$ 110 kV/cm, which can be justified by the Schottky emission theory. On the other hand, Eq. (4.13) expresses the relationship between J_S and E as follows (Pintilie et al., 2007):

$$J_S = \mathrm{A}^* \cdot T^2 \exp\left[\frac{-q\left(\varphi_B - \sqrt{\frac{qE}{4\pi\varepsilon_0\varepsilon_{0p}}}\right)}{\mathrm{k}_B T}\right] \tag{4.13}$$

where A* is the Richardson constant, q is the electron charge, T is the Kelvin temperature, ϕ_B is the Schottky barrier, ε_0 is the permittivity of vacuum, ε_{0p} is the dynamic (high frequency) dielectric constant, and k_B is the Boltzmann constant. Thus, for large values of E, Eq. (4.13) evidences a nonlinear relationship between J_S and E (Fig. 4.21B).

4.9.2 Cobalt ferrite-ferroelectric magnetoelectric components

Cobalt ferrites with the highest magnetostriction coefficient among the spinel ferrites are promising candidates for employment in ME components. By annealing metal-organic framework precursors, it is possible to induce a ferrimagnetic CFO phase with a porous structure, followed by in situ growth of a ferroelectric BTO phase in the pores using a hydrothermal technique (Wang et al., 2020). In this regard, CFO/BTO composite ceramics have been synthesized by sintering them at different temperatures (1000, 1100, and 1200°C).

Magnetic hysteresis loops of CFO/BTO composites measured at room temperature are shown in Fig. 4.22A. Obviously, the presence of hysteresis loops indicates the magnetic characteristics of the composites arising from the ferrimagnetic CFO phase. It is also inferred that the magnetic properties of the ferrimagnetic phase are not mainly influenced by the ferroelectric phase addition. Alternatively, H_c, M_s, and M_r of the resulting CFO/BTO composite ceramics are slightly changed by varying the sintering temperature, so the maximum M_s and minimum M_r are obtained at a sintering temperature of 1200°C. Furthermore, increasing the sintering temperature decreases H_c, resulting in a reduction in magnetocrystalline anisotropy,

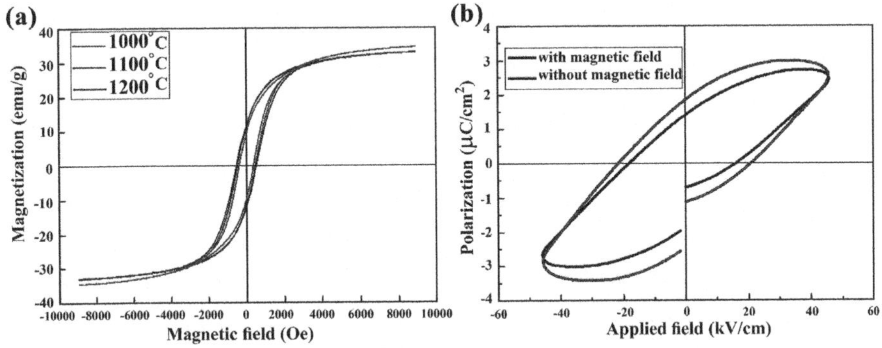

FIGURE 4.22

(A) Magnetic hysteresis loops of CFO/BTO composite ceramics with different sintering temperatures and (B) P–E hysteresis loops of the composite ceramics in the presence (with) and absence (without) of a magnetic field of 2.3 mT.

From Wang, Z., Gao, R., Chen, G., Deng, X., Cai, W., & Fu, C. (2020). Dielectric, ferroelectric and magneto-electric properties of in-situ synthesized $CoFe_2O_4/BaTiO_3$ composite ceramics. Ceramics International, *46(7), 9154–9160. https://doi.org/10.1016/j.ceramint.2019.12.165.*

as well as an improvement in the mechanical coupling effect. In other words, increasing the sintering temperature can enhance the ME coupling coefficient. In this respect, Fig. 4.22B shows *P–E* hysteresis loops in the presence (with) and absence (without) of a magnetic field obtained from the CFO/BTO composite with the sintering temperature of 1200°C. Applying the magnetic field causes P_R and E_C to be reduced at measuring conditions of 2.7 kV and 3 kHz, being indicative of the effective role of the magnetic field in the resultant ferroelectric properties. It is worth noting that the ferrimagnetic phase can play a role in composite polarization, although it is mainly determined by the ferroelectric phase.

Since the external magnetic field strength is not enough to induce strong coupling between the ferrimagnetic and ferroelectric phases, it would most influence the latter. The charge localization arising from the ferrimagnetic phase is reduced by the magnetic field, thus weakening polarization of the composite. For this reason, CFO/BTO composite ceramics showed a negative ME effect. The negative ME effect has already been reported elsewhere (Chen et al., 2014; Li et al., 2019).

Elsewhere, polycrystalline BZT and CFO phases have been separately synthesized using a solid-state reaction technique at a high temperature ($\sim$1000°C) (Sahoo et al., 2016). The polycrystalline phases have then been mixed in different ratios of BZT to CFO, resulting in BZT_{1-x}/CFO_x composites with $x = 0.10$, 0.15, 0.20, and 0.25. Fig. 4.23 shows the variations of relative permittivity (ε_r) and tanδ as a function of temperature for BZT/CFO composites at different frequencies ranging from 1 kHz to 1 MHz.

Relaxors are known for some of their typical features, including the broad response of permittivity versus temperature, the maximum permittivity $\left(\varepsilon_r^{max}\right)$ shift toward

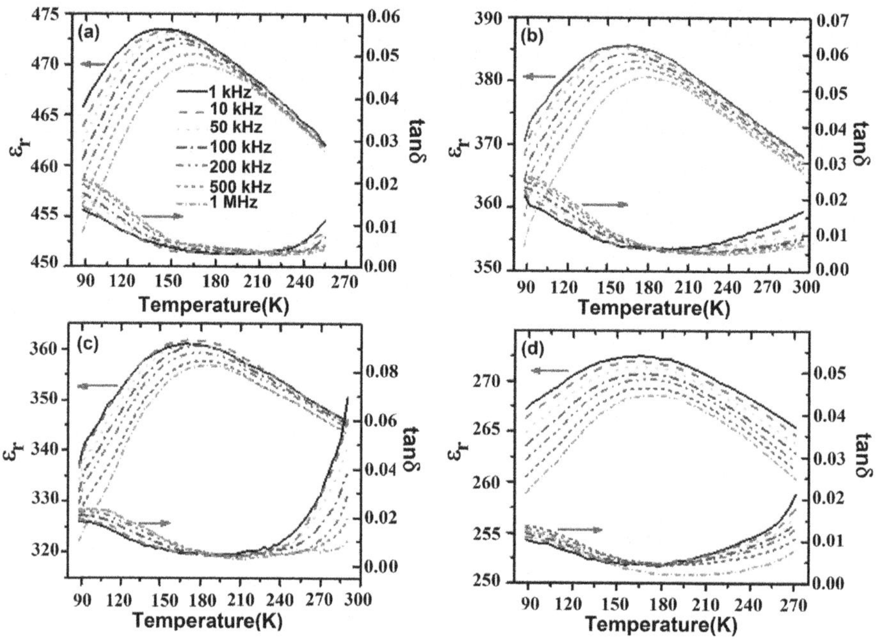

FIGURE 4.23

Variations of relative permittivity (ε_r) and tanδ as a function of temperature at different frequencies for BZT_{1-x}/CFO_x composites with: (A) $x = 0.10$, (B) $x = 0.15$, (C) $x = 0.20$, and (D) $x = 0.25$.

From Sahoo, M., Yajun, Z., Wang, J., & Choudhary, R. (2016). Composition control of magnetoelectric relaxor behavior in multiferroic $BaZr_{0.4}Ti_{0.6}O_3/CoFe_2O_4$ composites. Journal of Alloys and Compounds, 657, *12–20. https://doi.org/10.1016/j.jallcom.2015.10.040.*

higher temperatures with increasing frequency, Burns temperature (i.e., the merging of permittivity vs. temperature curves for some higher temperatures), and freezing temperature (i.e., the polar cluster to glassy phase transformation for some characteristic temperatures) (Pirc & Kutnjak, 2014; Sahoo & Choudhary, 2010).

It is interesting that all the features above can be achieved for the BZT_{1-x}/CFO_x composite when $x = 0.10$. Notably, the Burns temperature occurs at 210 K. While the $\left(\varepsilon_r^{max}\right)$ shifts toward higher temperatures when $x = 0.15$ and 0.20, the merging of the permittivity versus temperature curves does not occur.

For $0.10 \leq x \leq 0.20$, the number of polar clusters formed increases with decreasing temperature. Meanwhile, the number of BTO dipoles increases, whereas that of $BaZrO_3$ dipoles may be continuously reduced. These dipoles may initially orient along one of the following pseudocubic directions: [110], $[1\overline{1}0]$, $[\overline{1}10]$, $[\overline{1}\overline{1}0]$ (Prosandeev et al., 2015). Further reducing the temperature leads to an increase in the size of polar clusters while merging them. The polarization orientation in the resultant giant clusters would be along [111] (Akbarzadeh et al., 2012). In the case of $x = 0.25$, one can infer that the $\left(\varepsilon_r^{max}\right)$ does not shift toward higher temperatures and the permittivity curves do not merge, confirming the diffuse phase transition in the BZT/CFO composite. When the relaxor is transformed into the diffuse phase transition behavior, the mesoscopic heterogeneity may increase due to an increased CFO ratio (Li et al., 2018). Moreover, the maximum temperature (T_m) of permittivity occurs at 143K (1 kHz), followed by a slight shift to 166 K when $x = 0.15$. In other words, increasing the CFO content does not mainly influence the corresponding T_m.

Quantitatively speaking, the $\left(\varepsilon_r^{max}\right)$ value is obtained as 473 when $x = 0.10$ at 1 kHz, followed by a gradual decrease to 272 by increasing x to 0.25. As well, it is evident that ε_r does not considerably change over a wide range of temperatures regardless of the BZT to CFO ratio.

For the BZT_{1-x}/CFO_x composite with $x = 0.25$, an almost flattened response of permittivity versus temperature is observed, giving rise to a slight variation in $\varepsilon_r(\Delta\varepsilon_r = 2.5\%)$ over a wide range of temperatures ($\Delta T = 182°C$). The parameter $\Delta T_m = [T_m(1\text{MHz}) - T_m(1\text{kHz})]$ has also been obtained from Fig. 4.23, and used to quantify the frequency dispersion of T_m. In this respect, Table 4.1 presents ferroelectric characteristics for the different compositions. It is found that ΔT_m decreases from 23 to 5K with an increasing x from 0.10 to 0.25. This indicates that the frequency dispersion is gradually reduced while transforming the relaxor into the diffuse phase transition behavior. Additionally, tanδ of all the BZT/CFO composites are extremely small at T_m (~0.005 for the frequency of 1 kHz), almost independent of the CFO content.

To provide better evidence of the diffuse phase transition behavior, a modified Curie–Weiss law has been used as follows (Uchino & Nomura, 1982):

$$\frac{1}{\varepsilon_r - \frac{1}{\varepsilon_r^{max}}} = \frac{(T - T_m)^{\gamma}}{C} \tag{4.14}$$

where ε_r is the relative permittivity, $\left(\varepsilon_r^{max}\right)$ is the maximum permittivity, γ is the diffusivity parameter, and C is the Curie–Weiss constant. The γ value can change from 1 (representing a classical ferroelectric phase transition) to 2 (describing the diffuse phase transition behavior). The linear relationship between $\ln\left(\frac{1}{\varepsilon_r - \frac{1}{\varepsilon_r^{max}}}\right)$ and $\ln(T - T_m)$ is indicated in Fig. 4.24 for BZT_{1-x}/CFO_x composites with $x =$ 0.10, 0.15, 0.20, and 0.25. It is also possible to extract values by the linear fitting depicted in Fig. 4.24. In this case, γ increases from 1.62 to 1.98 with an increasing x from 0.10 to 0.25, thus confirming the enhancement in the diffusivity as a function of the CFO content.

By cooling the relaxor, one can thermally activate the nucleation sites to induce the growth of polar nanoregions. A flipping of relaxation time may take place if the neighboring polar nanoregions merge abruptly. The dynamic polar cluster state can

Table 4.1 Ferroelectric, diffuse phase transition and frequency dispersion characteristics of $Ba(Zr_xTi_{1-x})O_3$.

Composition x	$T_m(K)$ (1 kHz)	ε_r^{max}(1 kHz)	$\tan\delta(T_m)$ (1 kHz)	$\Delta T_m(K)$	γ	$f_0(\times 10^{10})$ Hz	T_f (K)	E_a(eV)
0.10	143	473	0.005	23	1.62	3.2	98(±5)	0.015
0.15	166	385	0.006	14	1.64	2.14	96(±7)	0.025
0.20	166	361	0.005	12	1.72	4.10	91(±4)	0.023
0.25	166	272	0.004	5	1.98	–	–	–

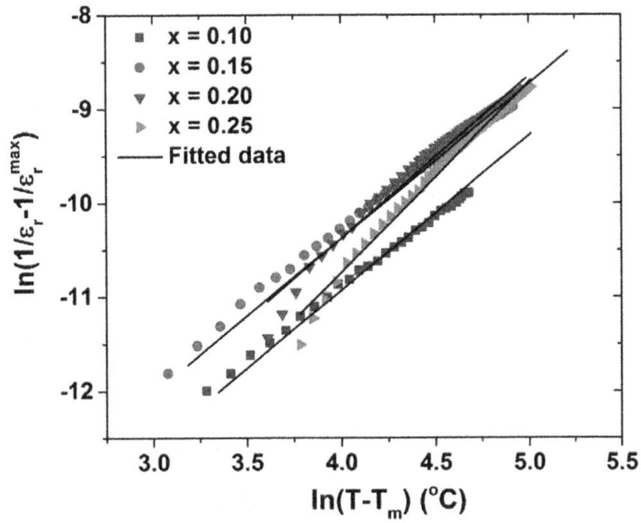

FIGURE 4.24

Variation of ln $\left(\frac{1}{\varepsilon_r - \frac{1}{\varepsilon_r^{max}}}\right)$ as a function of ln $(T - T_m)$ for BZT_{1-x}/CFO_x composites with different compositions ($0.1 \leq x \leq 0.25$).

From Sahoo, M., Yajun, Z., Wang, J., & Choudhary, R. (2016). Composition control of magnetoelectric relaxor behavior in multiferroic $BaZr_{0.4}Ti_{0.6}O_3/CoFe_2O_4$ composites. Journal of Alloys and Compounds, 657, *12–20. https://doi.org/10.1016/j.jallcom.2015.10.040.*

be progressively transformed into a static ferroglass one by cooling the relaxor system, leading to the freezing temperature.

To better understand the relaxation type while estimating the freezing temperature, the frequency responses measured for the composites have been fitted using the Vogel–Fulcher relation (Fulcher, 1925; Vogel, 1921) as given below:

$$f = f_0 \exp\left(\frac{-E_a}{k_B(T_m - T_f)}\right) \tag{4.15}$$

where f_0 is the Debye frequency, E_a is the polar cluster activation energy, k_B is the Boltzmann constant, and T_f is the freezing temperature. Fig. 4.25 shows the dependence of ln (f) on T_{max} for the BZT_{1-x}/CFO_x composites ($x = 0.10, 0.15$, and 0.20), along with the fitted data by the Vogel–Fulcher relation (the solid lines). Table 4.1 also presents the fit parameters, indicating that increasing the CFO content from 0.1 to 0.2 slightly decreases the corresponding T_f from 98 to 91K. Concurrently, E_a remains almost constant after changing the CFO content. Overall, one can conclude that growth in polar nanoregions is not considerably affected by the CFO content in the BZT/CFO composites.

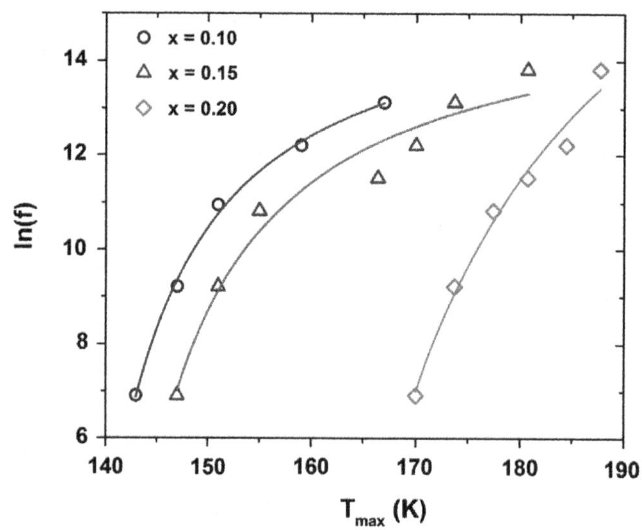

FIGURE 4.25

Dependence of ln(*f*) on T_{max}, along with the fitted data by the Vogel–Fulcher relation (*the solid lines*) for BZT_{1-x}/CFO_x composites with different compositions ($0.1 \leq x \leq 0.25$).

From Sahoo, M., Yajun, Z., Wang, J., & Choudhary, R. (2016). Composition control of magnetoelectric relaxor behavior in multiferroic $BaZr_{0.4}Ti_{0.6}O_3/CoFe_2O_4$ *composites.* Journal of Alloys and Compounds, 657, *12–20. https://doi.org/10.1016/j.jallcom.2015.10.040.*

The conduction mechanism of BZT/CFO composites with particulate structures can be explained based on a microscopic heterogeneous model (Panda & Behera, 2014) comprising grain interior/core and grain boundary formation, as illustrated in Fig. 4.26. In general, grain and grain boundary subcircuits play the main role in the conduction via grain interior and grain boundary. When the CFO content is low, grains with low resistance cause the grain interior to form conduction channels. Meanwhile, the conduction process is hindered by a high resistive barrier from the grain boundaries. In fact, since the grain boundary core of Ba-based ceramic materials is positively charged and surrounded by space charge or depletion layers, the grain boundary possesses resistive properties (Ricote et al., 2014). In the case of BZT/CFO composites, grains of both BZT and CFO phases play effective roles in the conduction process via the grain interior.

The simultaneous presence of different oxidation states (Fe^{2+}, Fe^{3+}, Co^{2+}, and Co^{3+}) arising from cations at tetrahedral or octahedral sites of the CFO phase with cubic spinel structure, together with their ligand field splitting, can induce a hopping-like conduction mechanism between Fe^{3+} and Fe^{2+} ions, and Co^{3+} and Co^{2+} ions (Jonker, 1959; Rahman et al., 2012). Moreover, a hopping-like mechanism with a combined variable range may be induced by grains of the BZT phase, leading to conduction through protons and Ti^{4+} and Ti^{3+} ions (Han et al., 2014; Iwahara et al., 1993).

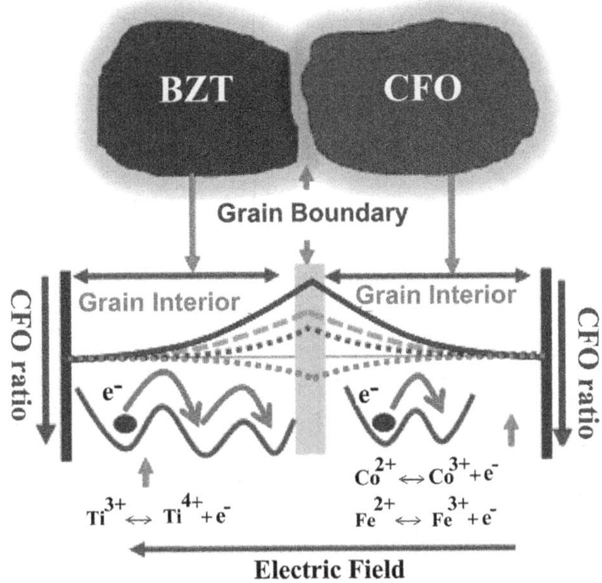

FIGURE 4.26

A schematic representation of the conduction mechanism in BZT/CFO composites via grain interior and grain boundary. Increasing the CFO ratio from $x = 0.10$ (the *solid blue line*) to $x = 0.25$ (the *pink dotted line*) reduces the resistive barrier progressively, allowing for the formation of the conduction channel by the grain boundary.

From Sahoo, M., Yajun, Z., Wang, J., & Choudhary, R. (2016). Composition control of magnetoelectric relaxor behavior in multiferroic $BaZr_{0.4}Ti_{0.6}O_3/CoFe_2O_4$ composites. Journal of Alloys and Compounds, 657, *12–20. https://doi.org/10.1016/j.jallcom.2015.10.040.*

On the other hand, the conduction via grain interior may occur due to the presence of oxygen vacancy inherently available during the synthesis process of perovskite oxides. Nevertheless, increasing the CFO content may reduce the resistivity of the barrier by decreasing the space charge or depletion layer width. This can be caused when negatively charged ions present in the space charge layers become segregated or when the crystallographic structure and the composition in the grain boundary vicinity change. Overall, one can expect the BZT/CFO composites to involve a mixed electronic-ionic conduction mechanism.

In another study, $Ba_{0.85}Ca_{0.15}Ti_{0.9}Zr_{0.1}O_3$ (BCTZ) and BCZT-CFO composites have been evaluated (Negi et al., 2018). *P–E* hysteresis loops of pure BCTZ and BCTZ–CFO composites measured at room temperature are shown in Fig. 4.27. As shown, the ferroelectric hysteresis loops obtained from the BCTZ and 0.9BCTZ–0.1CFO composite are narrow and not saturated. Increasing the CFO content from 10 to 50 wt.% significantly reduces the ferroelectric properties of the composites so that rounded features are observed in the hysteresis loops due to the leakage current and suppressed ferroelectric properties of the BCTZ phase.

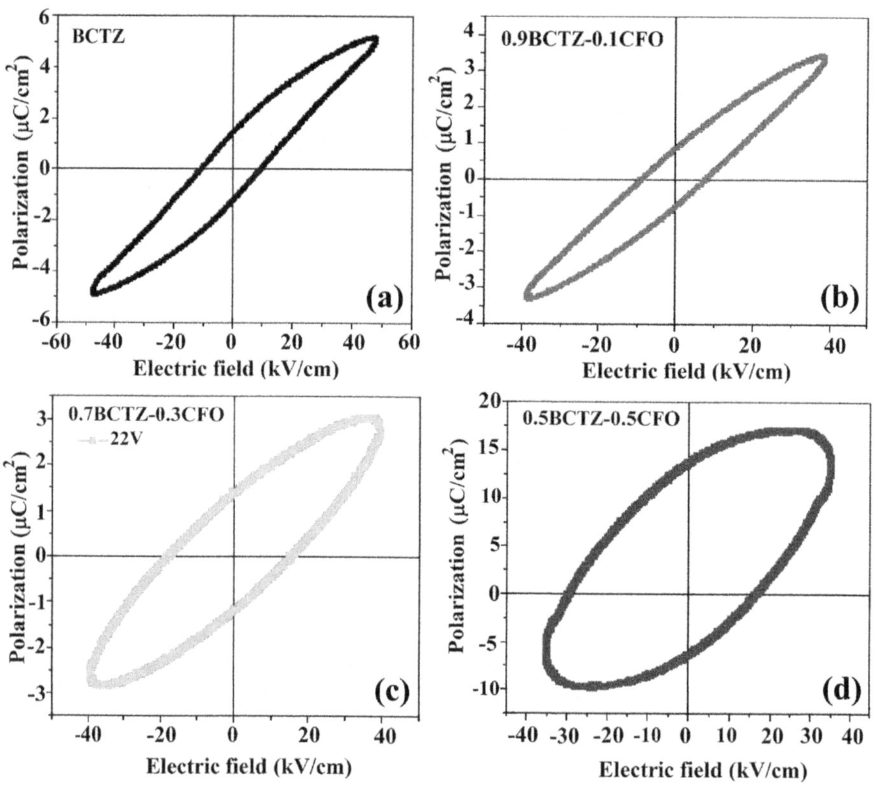

FIGURE 4.27

Room-temperature *P–E* hysteresis loops of pure BCTZ, and BCTZ–CFO composites.
From Negi, N., Kumar, R., Sharma, H., Shah, J., & Kotnala, R. (2018). Structural, multiferroic, dielectric and magnetoelectric properties of (1−x)$Ba_{0.85}Ca_{0.15}Ti_{0.90}Zr_{0.10}O_3$*–(x)*$CoFe_2O_4$ *lead-free composites.* Journal of Magnetism and Magnetic Materials, 456, *292–299. https://doi.org/10.1016/j.jmmm.2017.12.095.*

In other words, the addition of CFO content up to 30 wt.% maintains the ferroelectric behavior of the resultant BCTZ-CFO composites, and its further addition ($x > 0.3$) would lead to weakening of the ferroelectricity.

Apart from the significant influence of the sintering temperature, investigations have shown that the grain size and density of ceramic materials can also play a role in ferroelectric properties, including P_R, E_C, and the $P-E$ hysteresis loop shape (Jin et al., 2014). By decreasing the sintering temperature, grain size and density are usually reduced, which can increase the number of grain boundaries. Thus, the domain clamping effect for grain boundaries becomes evident in composites sintered at low temperatures, justifying the narrow, unsaturated, and highly coercive $P-E$ loop of pure BCTZ. It should be noted that interfacial reactions between the BCTZ and CFO phases were avoided by keeping the sintering temperature relatively low during composite formation.

Using a chemical solution deposition technique, Pb-free $Bi_4Ti_{2.9}Fe_{0.1}O_{12}$ (BTF) and CFO composite films have been fabricated on the $LaNiO_3$ (LNO)-coated Si substrate. The ME signal measurement setup of the resulting composite films is depicted in Fig. 4.28 (Duan et al., 2020).

Fig. 4.29A shows a cross-sectional transmission electron microscopy image of the BTF/CFO composite film, evidently indicating the formation of the interface between BTF and CFO layers. The respective thicknesses of LNO, CFO, and BTF layers are 270, 140, and 210 nm. A selected area electron diffraction pattern obtained from the BTF layer (the green circle in Fig. 4.29A) is shown in Fig. 4.29B, corresponding to (060), (151), and $(\bar{1}\bar{1}1)$ crystal planes with a crystal axis along the $(10\bar{1})$ direction. From the (060) crystal plane (perpendicular to the film surface direction), the BTF grains grow perpendicularly to the surface of the composite film along the *c*-axis.

The variation of α_E as a function of DC bias magnetic field (H_{bias}) for the BTF/CFO composite film at different AC magnetic field frequencies (10, 15, 20, and 25 kHz, with an applied AC magnetic field of 0.5 Oe) is shown in Fig. 4.30.

As observed, α_E initially increases with increasing H_{bias} at a frequency of 10 kHz and then decreases with further increases in H_{bias}. Considering the ME mechanism, α_E is given by Eq. (4.16) as follows (Tang et al., 2019):

$$\alpha_E = \frac{\partial E}{\partial H} = \frac{d_{ij}}{\varepsilon_0 \varepsilon_r} k_c d_{kl}^m \tag{4.16}$$

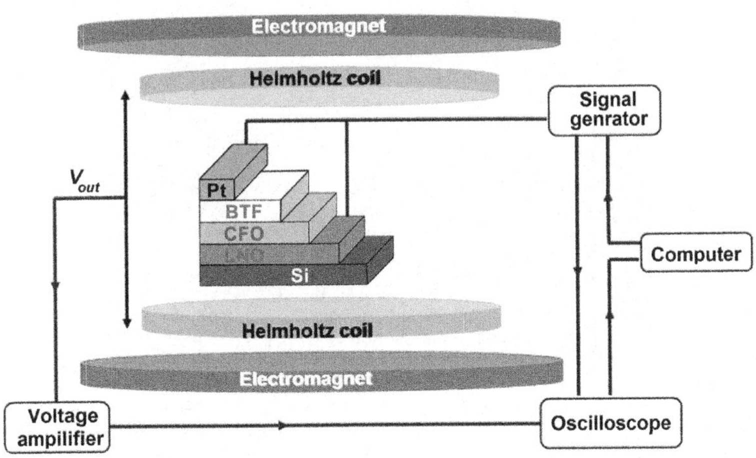

FIGURE 4.28

A schematic diagram of the magnetoelectric signal measurement setup of BTF/CFO composite film.

From Duan, Z., Fu, X., Mei, Y., Hu, Z., Mao, J., Ding, K., You, C., Wang, X., & Zhao, G. (2020). Annealing heating rate dependence of microstructure and multiferroic properties in $Bi_4Ti_{2.9}Fe_{0.1}O_{12}$/$CoFe_2O_4$ layered magnetoelectric composite films prepared by chemical solution deposition method. Ceramics International. *https://doi.org/10.1016/j.ceramint.2020.03.115.*

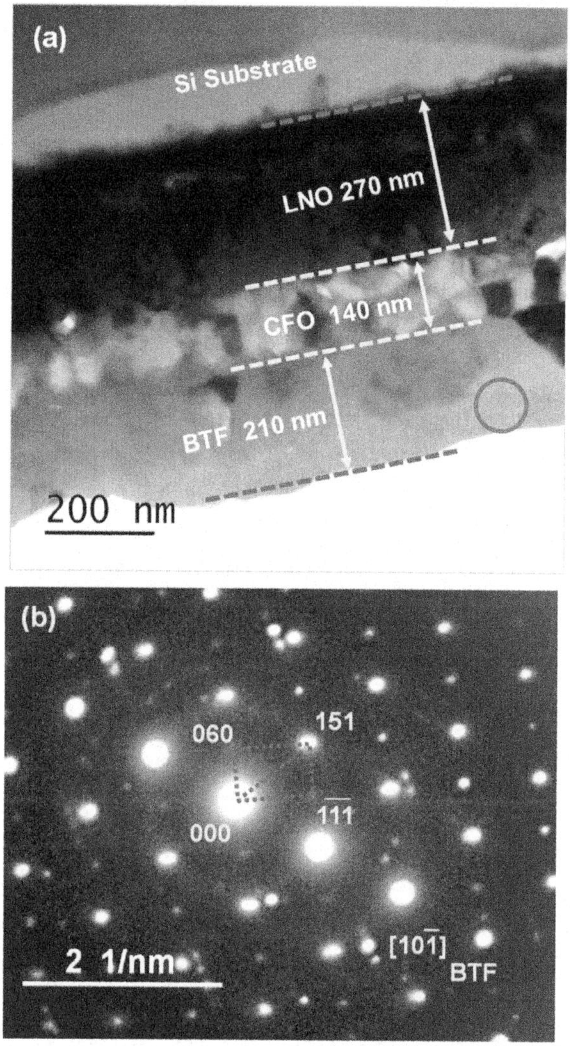

FIGURE 4.29

(A) Cross-sectional transmission electron microscopy image of BTF/CFO composite film fabricated on the LNO-coated Si substrate. (B) Corresponding selected area electron diffraction pattern of the BTF layer marked in part (A).

From Duan, Z., Fu, X., Mei, Y., Hu, Z., Mao, J., Ding, K., You, C., Wang, X., & Zhao, G. (2020). Annealing heating rate dependence of microstructure and multiferroic properties in $Bi_4Ti_{2.9}Fe_{0.1}O_{12}/CoFe_2O_4$ layered magnetoelectric composite films prepared by chemical solution deposition method. Ceramics International. *https://doi.org/10.1016/j.ceramint.2020.03.115.*

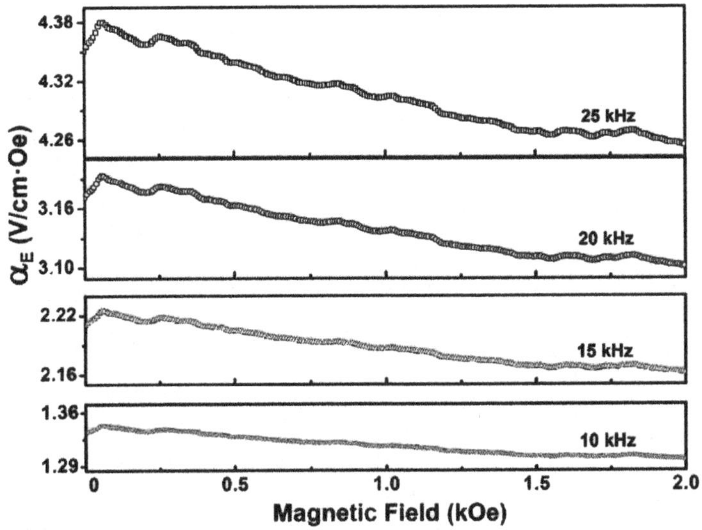

FIGURE 4.30

Variation of α_E as a function of DC bias magnetic field for the BTF/CFO composite film at different AC magnetic field frequencies (10–25 kHz).

From Duan, Z., Fu, X., Mei, Y., Hu, Z., Mao, J., Ding, K., You, C., Wang, X., & Zhao, G. (2020). Annealing heating rate dependence of microstructure and multiferroic properties in $Bi_4Ti_{2.9}Fe_{0.1}O_{12}/CoFe_2O_4$ layered magnetoelectric composite films prepared by chemical solution deposition method. Ceramics International. *https://doi.org/10.1016/j.ceramint.2020.03.115.*

where d_{ij}, ε_0, and k_c are the piezoelectric coefficient, vacuum dielectric constant, and coefficient representing coupling between the ferrite and piezoelectric phases, respectively. Furthermore, the parameter d^m_{kl} is the piezomagnetic coefficient, a derivative of the magnetic field to the magnetostrictive strain. According to Eq. (4.16), α_E is proportional to d^m_{kl}. The CFO phase magnetostriction increases up to a maximum value with increasing H_{bias} to 60 Oe, thereby enhancing α_E. Further increasing H_{bias} applied to the BTF/CFO film decreases α_E gradually. The variation behavior of α_E at 15, 20, and 25 kHz does not change compared with that at 10 kHz. The maximum α_E values are 1.34, 2.23, 3.19, and 4.38 V/cmOe at 10, 15, 20, and 25 kHz, respectively.

Elsewhere, (x)PZT/$(1-x)$CFO composites ($x = 0.0, 0.5, 0.6, 0.7, 0.8, 0.9$, and 1.0) have been prepared using a solid-state synthesis technique (Basu et al., 2012). Variations in ε_r and tanδ as a function of temperature for PZT, 0.9PZT/0.1CFO, 0.6PZT/0.4CFO, and CFO for a frequency range of 10 kHz to 1 MHz are shown in Fig. 4.31A–D, respectively. From Fig. 4.31A, a typical diffuse phase transition together with a dielectric anomaly is observed for the bulk PZT pellet. This dielectric anomaly can be assigned to a ferroelectric to paraelectric phase transition at $T_C \sim 377°C$. Slight dielectric constant frequency dispersion can also be seen around the T_C. Nevertheless, the dielectric constant frequency dispersion is not

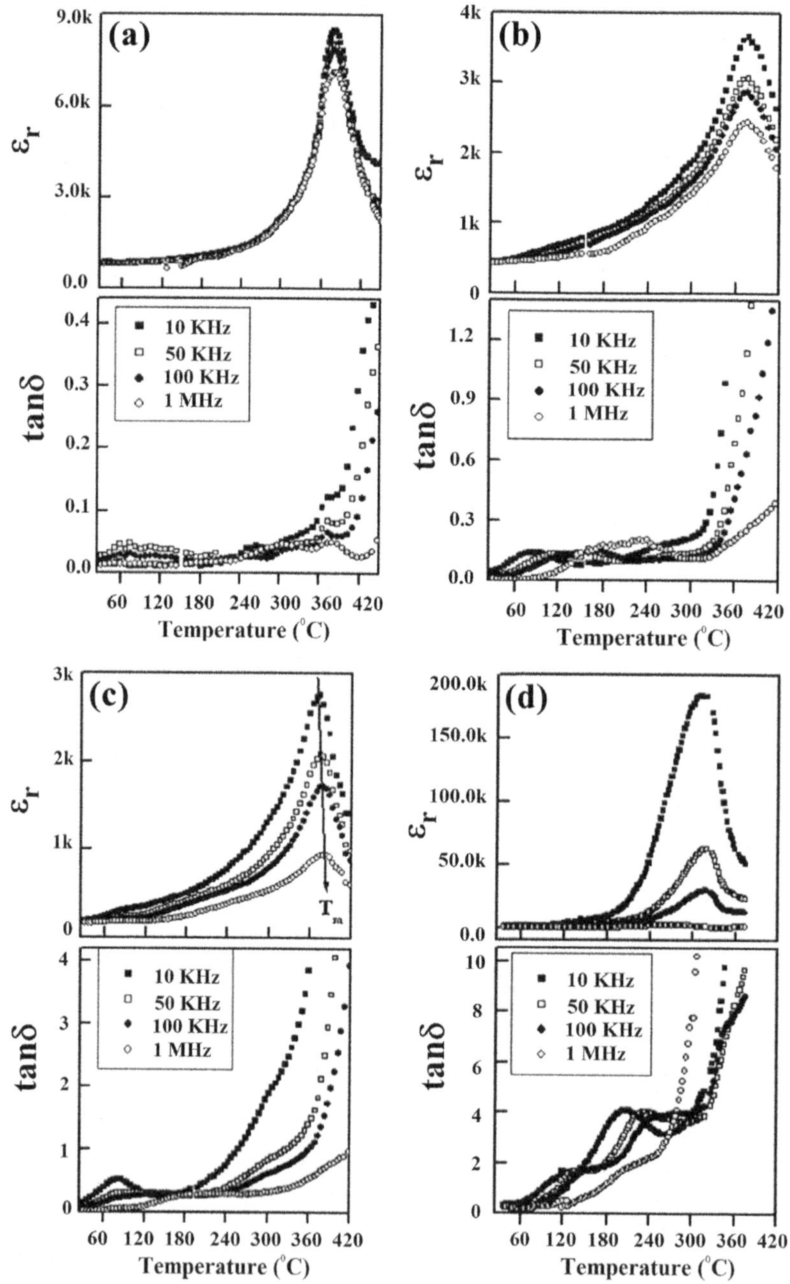

FIGURE 4.31

Variations of ε_r and tanδ as a function of temperature for (A) PZT, (B) 0.9PZT/0.1CFO, (C) 0.6PZT/0.4CFO, and (D) CFO pellets at different frequencies (10 kHz–1 MHz).

From Basu, S., Babu, K. R., & Choudhary, R. (2012). Studies on the piezoelectric and magnetostrictive phase distribution in lead zirconate titanate–cobalt iron oxide composites. Materials Chemistry and Physics, 132*(2–3), 570–580. https://doi.org/10.1016/j.matchemphys.2011.11.071.*

discernible over a wide temperature range in the bulk PZT. Alternatively, the tanδ of PZT suddenly increases near T_C, exhibiting frequency dispersion.

Increasing the CFO phase content in PZT/CFO composites systematically reduces the corresponding dielectric constants measured at room temperature. In the case of PZT/CFO composites, the dielectric constant frequency dispersion is significant and evident over a wide range of temperatures. Increased frequency shifts T_m to higher values, as marked in Fig. 4.31C. Such shifting behavior of T_m becomes more evident as the CFO content is increased.

By investigating the dielectric properties of the bulk CFO, high-frequency dispersion of ε_r and large values of tanδ are characterized (see Fig. 4.31D). As mentioned, tetrahedral or octahedral site cation distribution, Fe^{2+} to Fe^{3+} ion ratio, and the number of oxygen vacancies in the lattice can influence CFO dielectric properties. A perceptible increase in ε_r as a function of temperature may result from electron hopping (between Fe^{3+} and Fe^{2+} ions) activated thermally or from the transfer of holes (between Co^{3+} and Co^{2+} ions) (Mathe & Kamble, 2008). The electron hopping mechanism is relaxed at higher frequencies, giving rise to the frequency dispersion of ε_r.

In another study, $(1-x)$PZT/(x)CFO ME composites ($x = 0.20$, 0.35, and 0.50) have been synthesized by employing a hydrothermal technique along with a sintering process at various temperatures (Peng et al., 2015). The variation in direct current (DC) resistivity of the PZT−CFO20, PZT−CFO35, and PZT−CFO50 composites as a function of sintering temperature is shown in Fig. 4.32A. The optimal resistivity properties are obtained for the composites with a sintering temperature of 1100°C. However, increasing the CFO content decreases the DC resistivity at the sintering temperature of 1100°C. This is because the ferrite phase has a considerably lower resistivity than PZT, thereby hindering electric poling due to leakage current.

On the other hand, the dependence of the piezoelectric coefficient (d_{33}) on the ferrite content (mol%) for $(1-x)$PZT/(x)CFO composites ($x = 0.20$, 0.35, and 0.50) sintered at 1100°C is shown in Fig. 4.32B. As can be seen, the d_{33} of PZT, PZT−CFO20, PZT−CFO35, and PZT−CFO50 composites is found to be 36, 3.7, 1.0, and 0.6 pC/N, respectively, meaning that increasing the ferrite content sharply decreases the piezoelectric coefficient.

Fig. 4.33A shows the variation of α_E ($=dE/dH$) as a function of frequency for the PZT−CFO20 composite with a sintering temperature of 1100°C and $H_{\text{bias}} = 1000$ Oe. At 22.5 and 75.1 kHz, two α_E peaks are obtained for the composite. Several factors, including the phase components, composite form, geometrical dimensions, H_{bias}, and the electromechanical coupling coefficient, influence resonance frequency (Fang et al., 2011; Guo et al., 2010; Guzdek, 2014). The electromechanical resonance taking place in the PZT−CFO20 composite is responsible for the peaks observed in α_E at 25.5 and 75.1 kHz. However, previous reports have indicated that electromechanical resonance can result from magnetomechanical resonance. The maximum α_E of the PZT−CFO20 composite is 226 mV/cmOe at 75.1 kHz.

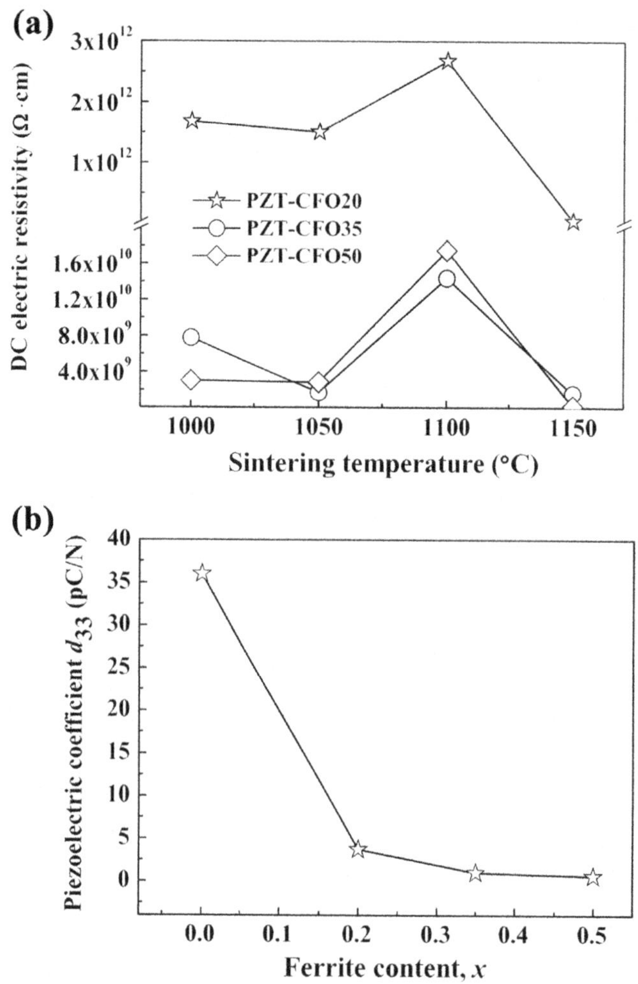

FIGURE 4.32

(A) Variation of direct current electric resistivity as a function of sintering temperature and (B) the dependence of piezoelectric coefficient (d_{33}) on the ferrite content (mol%) for (1-x)PZT/(x)CFO composites ($x = 0.20$, 0.35, and 0.50).

From Peng, J., Hojamberdiev, M., Li, H., Mao, D., Zhao, Y., Liu, P., Zhou, J., & Zhu, G. (2015). Electrical, magnetic, and direct and converse magnetoelectric properties of (1−x)Pb($Zr_{0.52}Ti_{0.48}$)O_3−(x)$CoFe_2O_4$ (PZT−CFO) magnetoelectric composites. Journal of Magnetism and Magnetic Materials, 378, *298–305. https://doi.org/10.1016/j.jmmm.2014.11.060.*

It is worth noting that frequency dispersion is not remarkable between 20 and 50 kHz, providing good frequency stability in the direct properties of ME composites.

Normally, $\alpha_E = dE/dH$ (V/cmOe) is a voltage induced by H_{ac} and employed as a unique characteristic of the ME coupling effect in composites with particulate

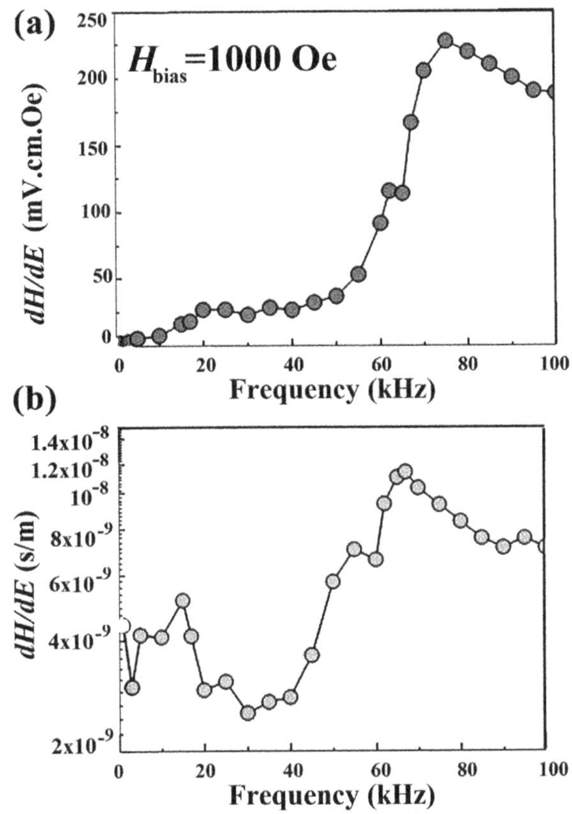

FIGURE 4.33

Variations of (A) *dE/dH* and (B) *dH/dE* as a function of frequency for PZT–CFO20 sintered at 1100°C.

From Peng, J., Hojamberdiev, M., Li, H., Mao, D., Zhao, Y., Liu, P., Zhou, J., & Zhu, G. (2015). Electrical, magnetic, and direct and converse magnetoelectric properties of (1–x)Pb($Zr_{0.52}Ti_{0.48}$)O_3–(x)$CoFe_2O_4$ (PZT–CFO) magnetoelectric composites. Journal of Magnetism and Magnetic Materials, 378, *298–305. https://doi.org/10.1016/j.jmmm.2014.11.060.*

structures. Notwithstanding this, the converse ME effect indicated by the susceptibility coefficient, i.e., $\alpha_e = dH/dE$ (s/m), can also be considered analogous to the dielectric constant and magnetic permeability and a fundamental ME coupling parameter (Zhai et al., 2007).

Fig. 4.33B shows the variation in $\alpha_e(dH/dE)$ as a function of frequency for PZT–CFO20 composites sintered at 1100°C. In the resonance region, the α_e value is enhanced, reaching a maximum of 1.15×10^8 s/m at 66.8 kHz. The maximum α_e of the PZT/CFO composite is higher (by about four orders of magnitude) than that of Cr_2O_3 MF material. Therefore, the PZT/CFO composite with a large converse ME effect can be employed as an efficient ME transducer in the resonance region.

Using a casting technique, three-phase ME composites consisting of PZT, CFO, and epoxy have been synthesized (Xu et al., 2015). Fig. 4.34A shows room-temperature $P-E$ hysteresis loops of composites with various PZT contents (0.4–0.7), confirming their ferroelectric nature. Increasing the PZT content from 0.4 to 0.51 increases the remnant polarization, P_R, of the composites. However, a comparison of the P_R values of composites with high PZT contents (0.51, 0.59,

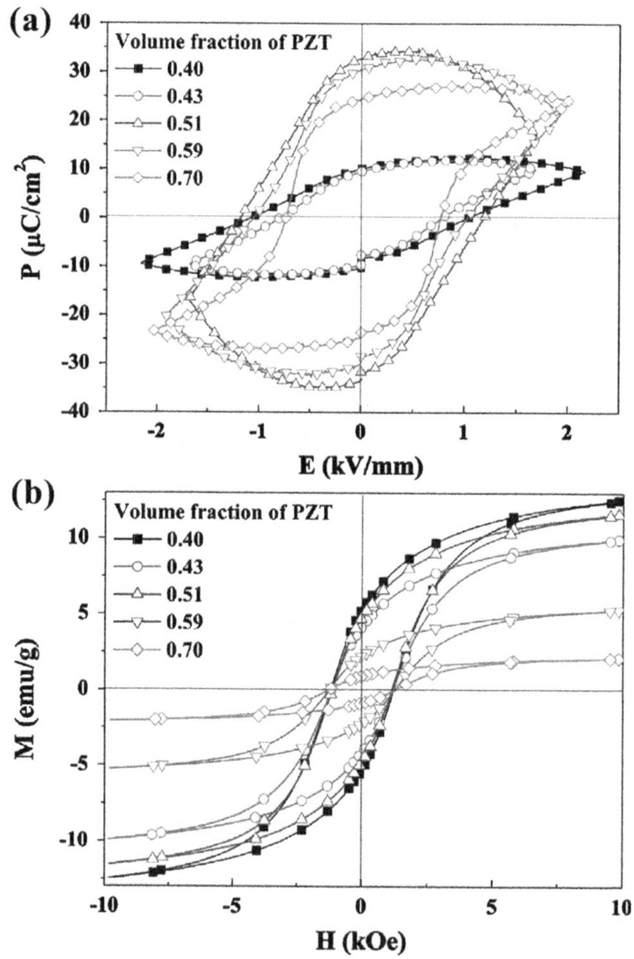

FIGURE 4.34

(A) P–E hysteresis loops and (B) M–H hysteresis loops obtained from PZT/CFO/epoxy composites with different PZT contents.

From Xu, T., Wang, C.-A., & Wang, C. (2015). Synthesis and magnetoelectric effect of composites with $CoFe_2O_4$-epoxy embedded in 3–1 type porous PZT ceramics. Ceramics International, *41(9), 11080–11085. https://doi.org/10.1016/j.ceramint.2015.05.054.*

and 0.7) shows that the corresponding P_R decreases with decreasing PZT content, which can be attributed to the significant amount of leakage current. Large leakage currents are expected in composites with relatively high CFO contents, making the saturation of $P-E$ hysteresis loops difficult under high applied voltages. Fig. 4.34B shows that the saturation magnetization and hysteresis loop area of nanocomposites decreases with increases in PZT content due to the nonmagnetic nature of PZT. However, changing the PZT content does not considerably affect the coercivity.

In another study, PZT/CFO ME composites with different compositions have been prepared using a conventional ceramic sintering process (Zhai et al., 2007). The variation of piezoelectric constant (d_{33}) as a function of PZT content is shown in Fig. 4.35. While the pure PZT ceramic has a d_{33} value of 302 pC/N, it is sharply reduced to 104 pC/N when mixed with 10% ferrite. Further increasing the ferrite content up to 50% decreases the d_{33} to zero. In other words, since the ferrite resistance is much lower than the PZT resistance, the piezoelectric constant is reduced to zero when the ferrite and PZT contents are the same. It is impossible to pole composites containing high ferrite contents with low resistivity at a high voltage, resulting in a weak piezoelectric effect.

Variations in the ME voltage coefficient as a function of PZT content measured under an H_{dc} of 1000 Oe and AC bias magnetic field frequency of 80 kHz are shown in Fig. 4.36. As observed, the PZT/CFO composite indicates no significant ME effect at the low PZT content (60%) due to the corresponding d_{33} (see Fig. 4.35). Increasing the PZT content leads to an increase in the dE/dH of the composites arising from the enhancement in the corresponding d_{33}. The composite reaches

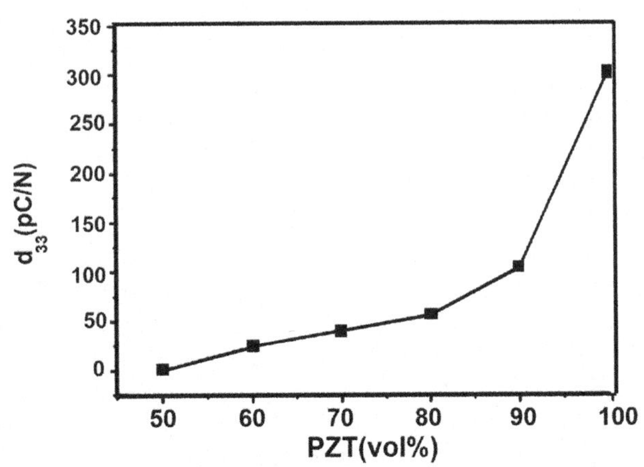

FIGURE 4.35

Variation of piezoelectric constant (d_{33}) as a function of PZT content for PZT/CFO magnetoelectric composites.

From Zhai, J. Y., Cai, N., Liu, L., Lin, Y. H., & Nan, C. W. (2003). Dielectric behavior and magnetoelectric properties of lead zirconate titanate/co-ferrite particulate composites. Materials Science and Engineering: B, *99(1–3), 329–331. https://doi.org/10.1016/S0921-5107(02)00565-2.*

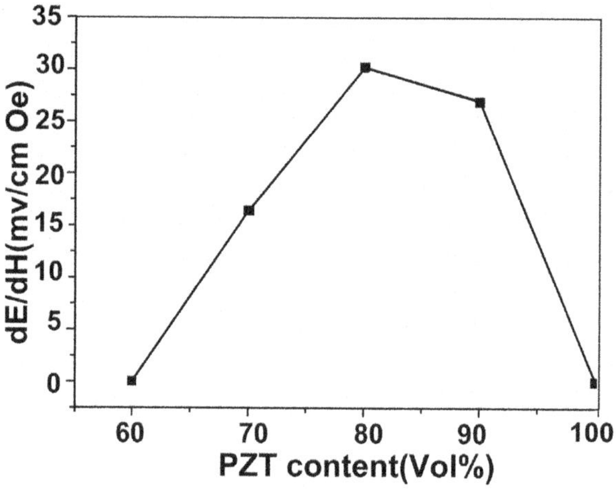

FIGURE 4.36

Variation of magnetoelectric voltage coefficient (*dE/dH*) as a function of PZT content for PZT/CFO composites.

From Zhai, J. Y., Cai, N., Liu, L., Lin, Y. H., & Nan, C. W. (2003). Dielectric behavior and magnetoelectric properties of lead zirconate titanate/co-ferrite particulate composites. Materials Science and Engineering: B, *99(1–3), 329–331. https://doi.org/10.1016/S0921-5107(02)00565-2.*

the highest ME voltage coefficient at a PZT content of 80%, as the ME effect is dominated by the piezoelectricity and magnetostriction of the piezoelectric and CFO phases, respectively. Further increasing the PZT content decreases *dE/dH*. This is because of the small magnetostriction of the composites according to the theoretical prediction.

In a separate research study, CFO/PFN composites (employed as bulk ceramics) have been compared with cofired composites (comprising alternate layers of ferrite and relaxor) for their ME and dielectric properties (Kulawik et al., 2012). A cross-sectional SEM image of the cofired CFO/PFN composite is depicted in Fig. 4.37. The relaxor-ferrite interfaces are free of cracks, delamination, and interlayers after performing the cofiring process. Moreover, the well-sintered CFO and PFN layers contain fine grains with respective sizes below 0.5 μm and from 0.5 to 2 μm.

The variation of dielectric permittivity as a function of temperature for the bulk CFO/PFN composite measured at different frequencies is shown in Fig. 4.38. At the frequency range between 10 Hz and 1 MHz, broad and high dielectric permittivity maxima with a magnitude of 10^3–10^4 are obtained for the composite. Increasing the frequency decreases the dielectric permittivity maxima while shifting them to higher temperatures.

The dielectric permittivity maximum positions of the composite do not correspond to the diffuse phase transition peaks of the relaxor phase. For example, at 1 kHz, the dielectric permittivity maximum of the composite is located at 563K, whereas the diffuse phase transition peak of the relaxor is found at 385K. Therefore,

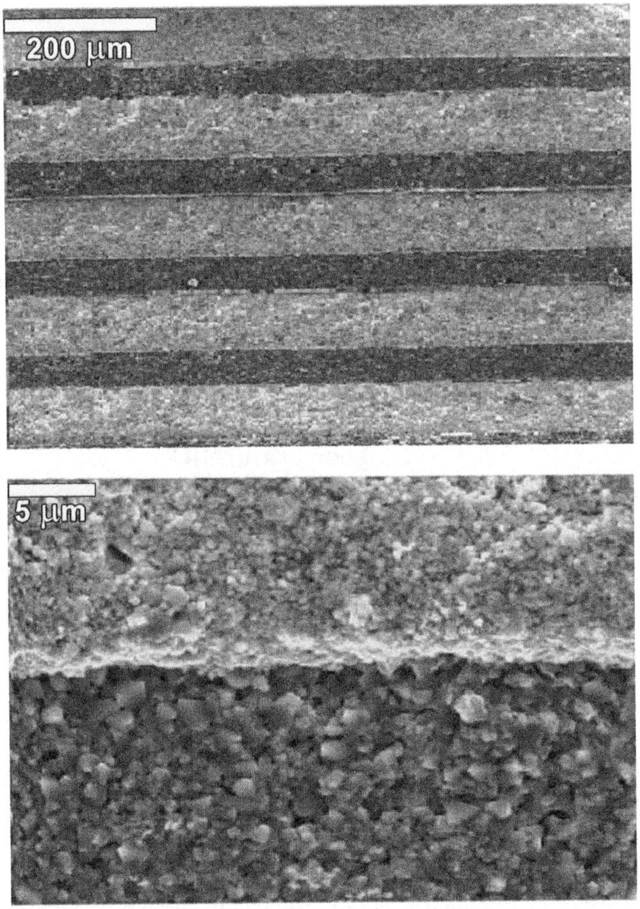

FIGURE 4.37

Cross-sectional scanning electron microscopy image of CFO/PFN composite comprising alternate layers of ferrite (*dark*) and relaxor (*light*).

From Kulawik, J., Szwagierczak, D., & Guzdek, P. (2012). Magnetic, magnetoelectric and dielectric behavior of $CoFe_2O_4$–$Pb(Fe_{1/2}Nb_{1/2})O_3$ particulate and layered composites. Journal of Magnetism and Magnetic Materials, *324(19), 3052–3057. https://doi.org/10.1016/j.jmmm.2012.04.056.*

the dielectric relaxation taking place in the ferrite and relaxor components is responsible for the emergence of the broad maxima in the bulk CFO/PFN composite. The alternate layers of ferrite and relaxor can form series capacitors, resulting in the lower dielectric permittivity of the layered composite compared with that of the particulate composite. Consequently, the layered composite has a considerably lower dielectric permittivity than the particulate composite due to the low permittivity of the ferrite layer.

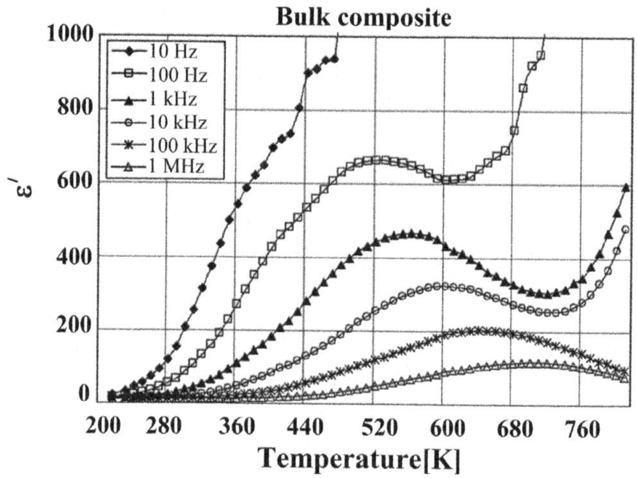

FIGURE 4.38

Variation of dielectric permittivity as a function of temperature for bulk CFO/PFN composite measured at different frequencies ranged between 10 Hz and 1 MHz.

From Kulawik, J., Szwagierczak, D., & Guzdek, P. (2012). Magnetic, magnetoelectric and dielectric behavior of $CoFe_2O_4$*–*$Pb(Fe_{1/2}Nb_{1/2})O_3$ *particulate and layered composites.* Journal of Magnetism and Magnetic Materials, *324(19), 3052–3057. https://doi.org/10.1016/j.jmmm.2012.04.056.*

By comparing Fig. 4.39A and B, one can infer that the maximum ME coefficient of the CFO/PFN layered composite (206 mV/cmOe) considerably outperforms the bulk composite (76 mV/cmOe). In fact, compared with layered composites, the ME coupling may be more restrained in bulk composites, arising from the elastic interaction between ferroelectric and ferrimagnetic domains. Under an external magnetic field, the domain wall movement is more facilitated in the alternate layers of ferrite and relaxor of the CFO/PFN composite than in bulk composites with mixed fine grains of the ferrite and relaxor. PFN grains with significantly lower magnetization in the bulk composite can surround the ferrite grains, hindering magnetic domain wall movement.

The ME properties of the composites $(x)Ba_{0.8}Pb_{0.2}TiO_3/(1-x)\ Cu_{0.6}Co_{0.4}Fe_2O_4$ ($x = 0.55$, 0.70, and 0.85) have been investigated in another study (Kothale et al., 2003). Fig. 4.40 shows the dependence of ME output (dE/dH) on the magnetic field (H) for the composites. It is found that the dE/dH remains almost constant for lower magnetic fields, and decreases at high H values, regardless of the composition. However, the ME output is maximized for the composite with $x = 0.85$. The presence of lattice distortion can induce strain in the composites, leading to the emergence of the ME signal. This can be explained by the occurrence of the Jahn–Teller ions (e.g.,

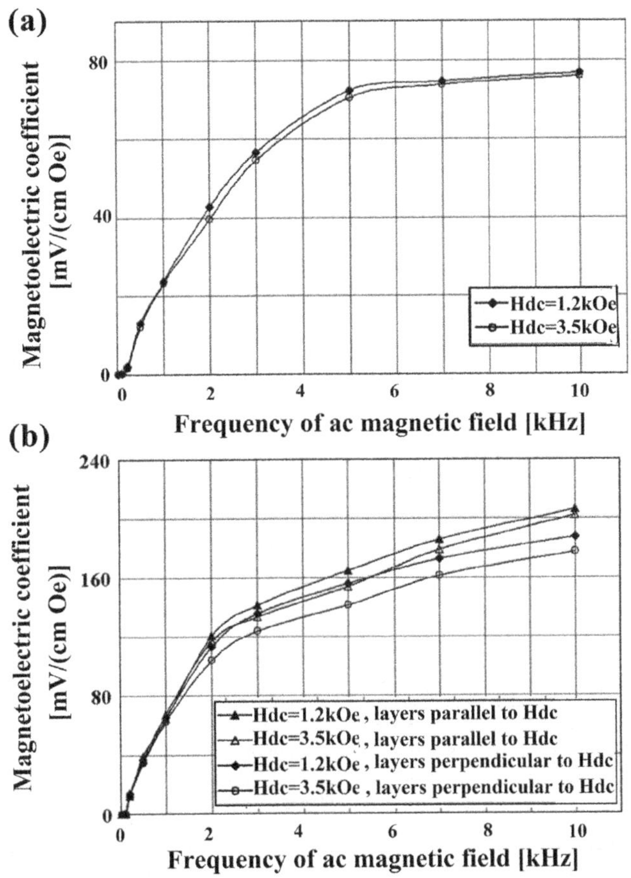

FIGURE 4.39

Variation of magnetoelectric coefficient as a function of frequency for (A) bulk and (B) layered CFO/PFN composites.

From Kulawik, J., Szwagierczak, D., & Guzdek, P. (2012). Magnetic, magnetoelectric and dielectric behavior of $CoFe_2O_4$–$Pb(Fe_{1/2}Nb_{1/2})O_3$ particulate and layered composites. Journal of Magnetism and Magnetic Materials, *324(19), 3052–3057. https://doi.org/10.1016/j.jmmm.2012.04.056.*

Cu) in the composites, resulting in the piezoelectric phase polarization. Above $H =$ 1.9 kOe, a constant electric field is generated in the piezoelectric phase due to the magnetostriction and strain induced, thereby decreasing the dE/dH. Furthermore, increasing the ferrite content causes the ME output to be reduced.

In a separate research study, ME composites consisting of ferrite [$Co_{1.2-y}Mn_yFe_{1.8}O_4$ (CMFO)] and ferroelectric BZT phases have been prepared (Kambale et al., 2010). The composite samples with $y = 0$, 0.1, 0.2, 0.3, and 0.4 have been labeled as $CMFO_0$+BZT, $CMFO_1$+BZT, $CMFO_2$+BZT, $CMFO_3$+BZT,

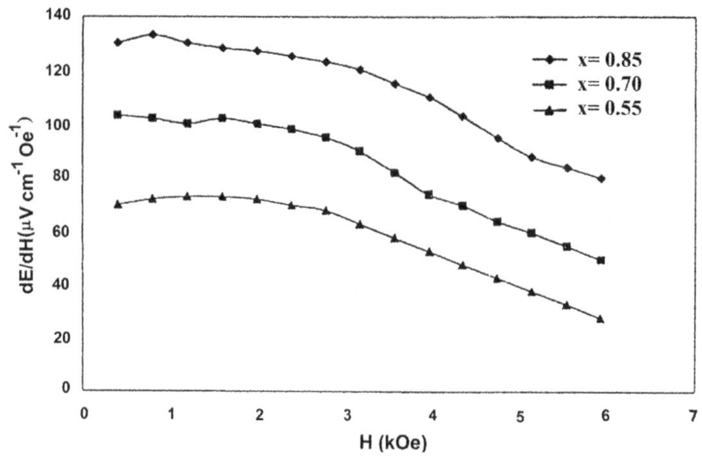

FIGURE 4.40

Dependence of magnetoelectric output (*dE/dH*) on the magnetic field (*H*) for (x) $Ba_{0.8}Pb_{0.2}TiO_3/(1-x)$ $Cu_{0.6}Co_{0.4}Fe_2O_4$ composites ($x = 0.55$, 0.70, and 0.85).

From Kothale, M., Patankar, K., Kadam, S., Mathe, V., Rao, A., & Chougule, B. (2003). Dielectric behavior and magnetoelectric effect in copper–cobalt ferrite + barium lead titanate composites. Materials Chemistry and Physics, *77(3), 691–696. https://doi.org/10.1016/S0254-0584(02)00135-9.*

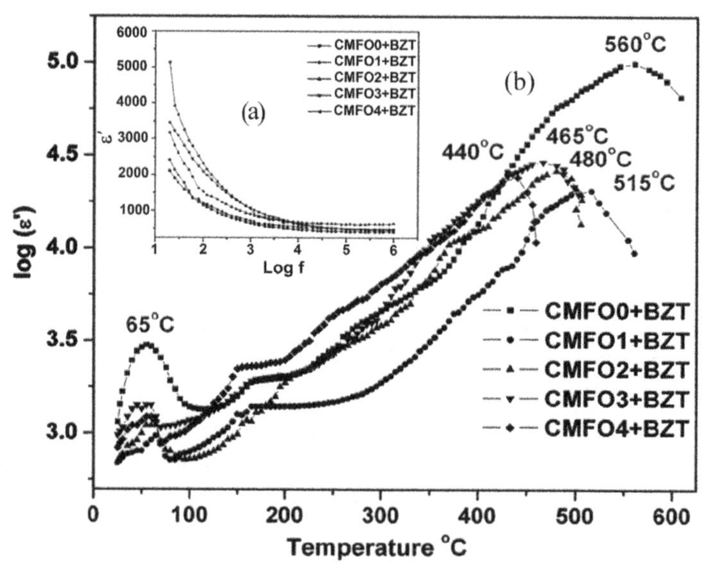

FIGURE 4.41

(A) Variation of dielectric constant as a function of frequency and (B) the dependence of the dielectric constant on temperature for CMFO/BZT composites.

From Kambale, R., Shaikh, P., Kolekar, Y., Bhosale, C., & Rajpure, K. (2010). Studies on dielectric and magnetoelectric behavior of 25% CMFO ferrite and 75% BZT ferroelectric multiferroic magnetoelectric composites. Materials Letters, *64(4), 520–523. https://doi.org/10.1016/j.matlet.2009.11.064.*

and $CMFO_4$+BZT, respectively. Fig. 4.41A (the inset) shows the variation of dielectric constant as a function of frequency for the CMFO/BZT composites. As can be seen, increasing the frequency decreases the dielectric constant. At the low frequency (1 kHz), the dielectric constant has a dispersion region, whereas it remains constant at high frequencies. The space charge polarization and the variation in the cation valence states can lead to the formation of dipoles, being responsible for the dielectric constant dispersion region at the low frequency. At high frequencies, the friction between diploes is enhanced since they cannot follow the H_{ac} with fast variation. For this reason, the dielectric constant behaves independently from the frequency at the high frequencies. Moreover, the enhanced friction between diploes can produce heat, and decrease the internal viscosity of the composite, thereby reducing the dielectric constant.

The polarization mechanism taking place in ferrites can determine the dielectric behavior of composites, providing conduction beyond the limits of the phase percolation. In other words, Fe^{3+} and Fe^{2+} ions available in the ferrites make them act as dipolar materials, so that orientational polarization is induced via the $Fe^{3+} \leftrightarrow Fe^{2+}$ dipole rotational displacement. In turn, this leads to alignment of the dipoles with the H_{ac} when the ions exchange electrons between themselves. In the CMFO composite system, p-type carriers arise from the Co^{2+} and Co^{3+}, and Mn^{2+} and Mn^{3+} ions. Apart from the n-type carriers, net polarization also results from the displacement of these ions toward the direction of the external electric field.

It should be noted that the polarization caused by the p-type carriers has an opposite sign, being smaller than that by the exchange of electrons between the Fe^{3+} and Fe^{2+} ions. Moreover, p-type carriers have lower mobility than that of n-type ones. Thus, even at low frequencies, the polarization contribution made by the p-type carriers decreases faster than that by the n-type ones. Accordingly, the net polarization initially increases as a function of frequency. However, further increasing the frequency reduces the corresponding net polarization. The dependence of the dielectric constant on temperature for composites measured at a frequency of 1 kHz is shown in Fig. 4.41B. The variation behavior of the dielectric constant versus temperature for the CMFO/BZT composites is the same as that for most composites containing ferrites and ferroelectrics (Devan & Chougule, 2007).

Elsewhere, using a chemical route, $[(GdMnO_3)_{0.7}(CoFe_2O_4)_{0.3}]_{0.5}[TiO_2]_{0.5}$ (GMO-CFO@TO) nanocomposites have been prepared (Mitra et al., 2020). Room-temperature $P-E$ hysteresis loops of GMO-CFO and GMO-CFO@TO nanocomposites were obtained at a frequency of 50 Hz under an applied voltage of 800 V, and the results are shown in Fig. 4.42A. As observed, the GMO-CFO@TO nanocomposite outperforms the GMO-CFO one in terms of room-temperature ferroelectric polarization. The respective maximum polarization, P_R, and E_C of the GMO-CFO@TO nanocomposite are ~0.2 μC/cm^2, 0.1 μC/cm^2, and 1.6 kV/cm. The extremely weak ferroelectric response of the GMO-CFO nanocomposite can be attributed to the high leakage current. Since the GMO and CFO are crystallized into centrosymmetric space groups, the TiO_2 crystal structure with the

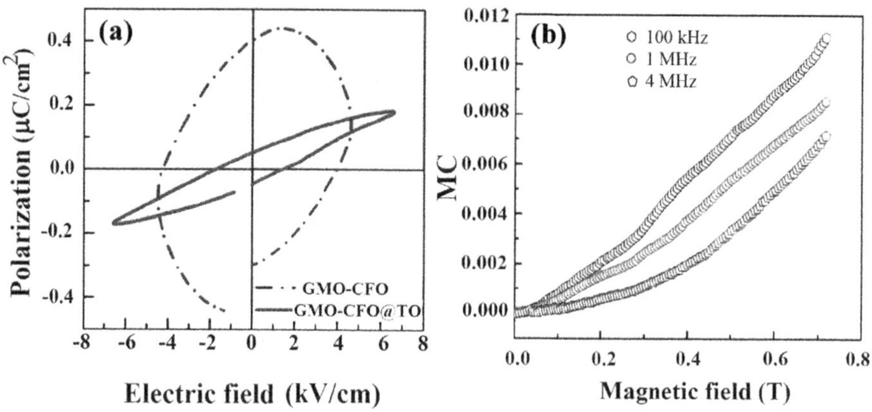

FIGURE 4.42

(A) Room-temperature *P–E* hysteresis loops of GMO-CFO and GMO-CFO@TO nanocomposites, (B) The dependence of magnetocapacitance on the applied magnetic field for the GMO-CFO@TO nanocomposite at different frequencies.

From Mitra, A., Shaw, A., & Chakrabarti, P. (2020). Microstructure, dielectric, ferroelectric and magnetoelectric coupling of a novel multiferroic of $[(GdMnO_3)_{0.7}(CoFe_2O_4)_{0.3}]_{0.5}[TiO_2]_{0.5}$ nanocomposite. Materials Chemistry and Physics, 240, *122242. https://doi.org/10.1016/j.matchemphys.2019.122242.*

noncentrosymmetric symmetry would induce the ferroelectric polarization in the GMO-CFO@TO nanocomposite.

A small distortion present in the TiO_2 crystal structure can cause it to deviate from the centrosymmetric position to the noncentrosymmetric one, thereby separating the negative and positive ion charge centers. This leads to the formation of a permanent electric dipole moment in the nanocomposite. The dipoles can thus be aligned along the external electric field direction after applying a voltage across the GMO-CFO@TO nanocomposite, resulting in a ferroelectric response induced by a local electric field.

In the presence and absence of the magnetic field, it is possible to measure the dielectric constant (ε') of the GMO-CFO@TO nanocomposite to confirm the coupling between magnetic and ferroelectric ordering. In this respect, the magnetocapacitance (MC) value is calculated as given by Eq. (4.17):

$$\mathrm{MC} = \frac{[\varepsilon'(H,T) - \varepsilon'(0,T)]}{\varepsilon'(0,T)} \tag{4.17}$$

where $\varepsilon'(H,T)$ denotes the dielectric constant measured in the presence of the magnetic field, and $\varepsilon'(0,T)$ is the dielectric constant in the absence of the magnetic field (Acharya et al., 2012; Mukhopadhyay et al., 2015).

Fig. 4.42B shows the dependence of MC on the applied magnetic field recorded at different frequencies (100 kHz, 1 and 4 MHz) for the GMO-CFO@TO nanocomposite. It is found that increasing the magnetic field increases the MC value nonlinearly, indicating positive ME coupling in the GMO-CFO@TO nanocomposite at room temperature. This kind of coupling does not occur for the GMO-CFO nanocomposite. Alternatively, increasing the frequency reduces the MC value of the nanocomposites, according to Fig. 4.42B. When the GMO-CFO@TO nanocomposite is exposed to the external magnetic field, the coupling between magnetic and electric ordering induces a strain that stresses the nanocomposites while producing an electric field. This can orient the ferroelectric domains of the nanocomposite, thus enhancing the polarization value. Moreover, increasing the magnetic field would induce more strain in the nanocomposite, thus changing the polarization value further.

It is well known that magnetostriction sensitivity ($d\lambda/dH$) of CFO increases after doping it with the element Mn (Bhame & Joy, 2007; Paulsen et al., 2005), making the Mn-doped CFO compound suitable for use in low magnetic fields. Using a pulsed laser deposition technique, PZT and CFO films have been deposited on the commercial Pt(111)-150 nm/Ti-10 nm/SiO_2-300 nm/Si(100) substrate. The CFO targets used were $CoFe_2O_4$, $Co_{0.8}Fe_{2.2}O_4$, and $Co_{0.6}Mn_{0.2}Fe_{2.2}O_4$, and the PZT target was $Pb(Zr_{0.52}Ti_{0.48})O_3$ (Khodaei et al., 2015).

Fig. 4.43A compares the $M-H$ hysteresis loops of single-layer CFO and bilayer PZT/CFO films measured in the in-plane direction. The resulting films show M_s values comparable to M_s reported for randomly oriented polycrystalline $CoFe_2O_4$ films fabricated on SiO_2/Si substrates using a pulsed laser deposition technique under various conditions (Raghunathan et al., 2010; Zhou et al., 2007). Nevertheless, M_s of CFO in the bulk form (425 emu/cm^3) is higher than in the CFO films (Cullity & Graham, 2011). This is because antiphase boundaries are created when merging islands with out-of-phase structures formed during the crystal growth of the ferrite films on different sites of the substrate (Margulies et al., 1997). Thus, spin frustration occurs at the boundaries, which reduces the resulting M_s in CFO films compared with bulk CFO.

The PZT/CFO bilayer film has a lower M_s than CFO single layers. This could be justified by the stress induced in CFO films with high magnetostriction, significantly changing their magnetic properties. Based on Le Chatelier's principle, the permeability, H_c, and M_s of magnetostrictive materials can be influenced by stress. Notably, for a material with a positive (negative) magnetostrictive coefficient exposed to tensile and compressive stresses, the corresponding M_s increases (decreases) as it is elongated (contracted) during the magnetizing process (Cullity & Graham, 2011).

Since CFO films are considered an anisotropic magnetostrictive material, they have different magnetostriction constants in the out-of-plane $\left(\lambda_S^{111} > 0\right)$ and in-plane $\left(\lambda_S^R < 0\right)$ directions. Thus, deposition of the PZT top layer causes M_s to be reduced in the in-plane direction due to the enhanced in-plane tensile stress of the CFO films. On the other hand, in the case of PZT/CFO bilayer films, the in-plane

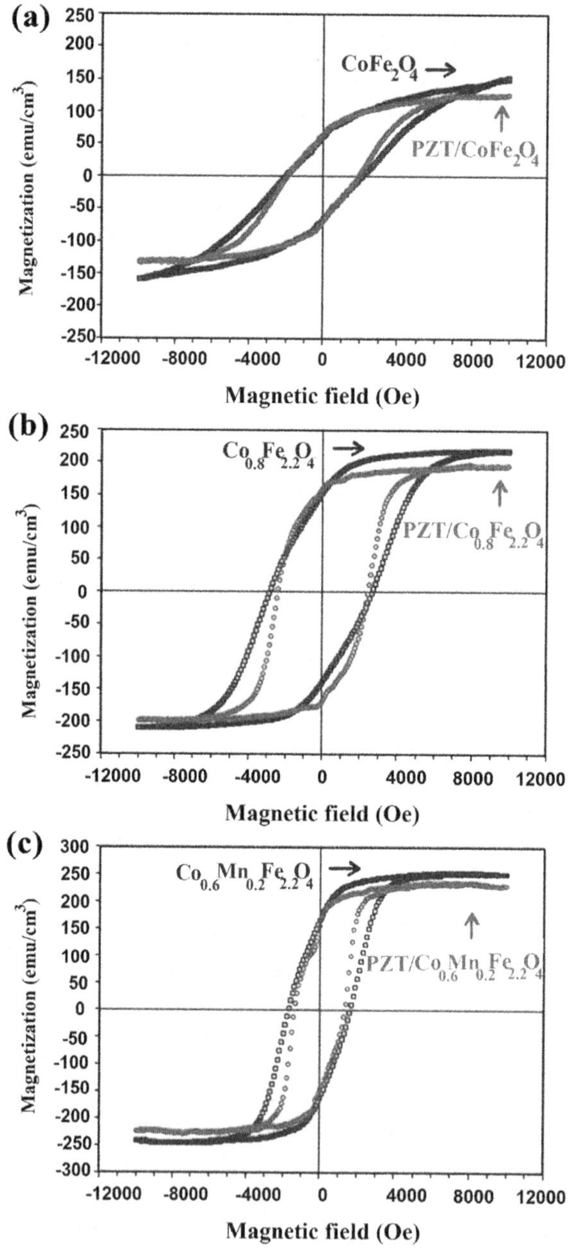

FIGURE 4.43

Comparison of room-temperature *M*–*H* hysteresis loops of (A) $CoFe_2O_4$ and PZT/$CoFe_2O_4$, (B) $Co_{0.8}Fe_{2.2}O_4$ and PZT/$Co_{0.8}Fe_{2.2}O_4$, and (C) $Co_{0.6}Mn_{0.2}Fe_{2.2}O_4$ and PZT/$Co_{0.6}Mn_{0.2}Fe_{2.2}O_4$ thin films, measured in the in-plane direction.

From Khodaei, M., Ebrahimi, S. S., Park, Y. J., Ok, J. M., Kim, J. S., Son, J., & Baik, S. (2014). Enhancement of in-plane magnetic anisotropy in (111)-oriented $Co_{0.8}Fe_{2.2}O_4$ thin film by deposition of PZT top layer. Applied Physics A, 117(3), 1153–1160. https://doi.org/10.1016/j.jmmm.2015.01.055.

H_c and squareness (M_r/M_s) decreases and increases, respectively, indicating the enhanced in-plane magnetic anisotropy of CFO films. CFO single layers oriented along the (111) direction show in-plane magnetic anisotropy due to the out-of-plane compression stress promoted by the enhanced residual stress of the CFO layer in the PZT/CFO bilayers. In this respect, the enhanced in-plane anisotropy of the PZT/$Co_{0.8}Fe_{2.2}O_4$ and PZT/$Co_{0.6}Mn_{0.2}Fe_{2.2}O_4$ bilayers is more noticeable than that of PZT/$CoFeO_4$, arising from the higher magnetostriction coefficient (Khodaei et al., 2014).

By applying a magnetic field to PZT/CFO bilayer films during $P-E$ measurements, the influence of the CFO layer composition on the resultant polarization was investigated. The $P-E$ hysteresis loops of PZT/CFO films with different CFO compositions measured at the maximum electric field (450 kV/cm) in the absence (0 Oe) and presence (2 kOe) of an in-plane magnetic field are shown in Fig. 4.44. The positive electric field polarization of all the bilayer films increases in the presence of the magnetic field, whereas the negative electric field polarization remains almost unchanged, likely arising from the leakage current and instability of the negative poled state.

In the presence of an out-of-plane magnetic field, the $P-E$ loop of PZT/CFO bilayer films does not considerably change, since CFO films with an out-of-plane hard magnetization axis oriented along the (111) direction show strong magnetic anisotropy. Nevertheless, applying the in-plane magnetic field to PZT/$CoFe_2O_4$, PZT/$Co_{0.8}Fe_{2.2}O_4$, and PZT/$Co_{0.6}Mn_{0.2}Fe_{2.2}O_4$ bilayer films enhances their corresponding positively poled P_R state by $\Delta P_R = 9\%$, 17%, and 21%, respectively. Note that CFO film dielectric properties are assumed to remain unchanged, ignoring their role in polarization variation. Due to their negative $\left(\lambda_S^R < 0\right)$ and positive $\left(\lambda_S^{111} > 0\right)$ magnetostriction constants, CFO films exposed to the in-plane magnetic field experience in-plane compressive and out-of-plane tensile strains, respectively. This causes the out-of-plane tensile strain to be transferred to the PZT top layer. In addition, the in-plane residual tensile strain of the PZT layer deposited on the Pt/Si substrate may be decreased by the strain transferred from the CFO film after applying the magnetic field (Zhou et al., 2006), thus changing the resultant polarization. As the composition and deposition conditions of the PZT top layer do not change in the PZT/CFO bilayers, the difference between the magnetostriction values of the CFO underlayers could be responsible for the different polarization enhancements observed in the bilayer films. Furthermore, the ΔP_R of the PZT/$Co_{0.8}Fe_{2.2}O_4$ is higher than that of PZT/$CoFe_2O_4$ due to the larger magnetostriction of the $Co_{0.8}Fe_{2.2}O_4$ layer. As well, the PZT/$Co_{0.6}Mn_{0.2}Fe_{2.2}O_4$ bilayer film outperforms PZT/$Co_{0.8}Fe_{2.2}O_4$ in terms of ΔP_R. Although the Mn-doped CFO bulk sample has a lower magnetostriction constant than that of pure CFO, doping the CFO with the element Mn increases its magnetostriction sensitivity in low magnetic fields (Bhame & Joy, 2007; Paulsen et al., 2005). Consequently, the PZT/$Co_{0.6}Mn_{0.2}Fe_{2.2}O_4$ bilayer exposed to a magnetic field of 2 kOe has higher magnetostriction than the PZT/$Co_{0.8}Fe_{2.2}O_4$ bilayer, giving rise to its larger ΔP.

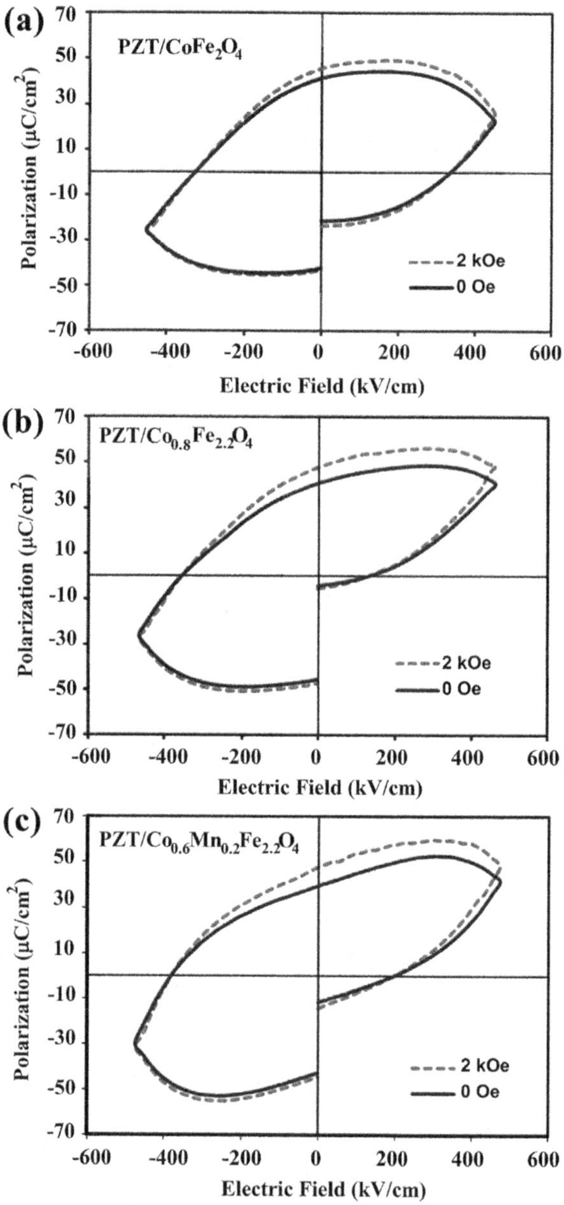

FIGURE 4.44

P–E hysteresis loops of (A) PZT/CoFe$_2$O$_4$, (B) PZT/Co$_{0.8}$Fe$_{2.2}$O$_4$, and (C) PZT/Co$_{0.6}$Mn$_{0.2}$Fe$_{2.2}$O$_4$ bilayer films measured in the absence (0 Oe) and presence (2 kOe) of an in-plane magnetic field.

From Khodaei, M., Ebrahimi, S. S., Park, Y. J., Ok, J. M., Kim, J. S., Son, J., & Baik, S. (2014). Enhancement of in-plane magnetic anisotropy in (111)-oriented Co$_{0.8}$Fe$_{2.2}$O$_4$ thin film by deposition of PZT top layer. Applied Physics A, 117*(3), 1153–1160. https://doi.org/10.1016/j.jmmm.2015.01.055.*

References

Acharya, S., Sutradhar, S., Mandal, J., Mukhopadhyay, K., Deb, A., & Chakrabarti, P. (2012). Sol−gel derived nanocrystalline multiferroic $BiFeO_3$ and R^{3+} (R= Er and Tm) doped therein: Magnetic phase transitions and enhancement of magnetic properties. *Journal of Magnetism and Magnetic Materials, 324*(24), 4209−4218. https://doi.org/10.1016/j.jmmm.2012.07.045

Agrawal, D. (1997). Relaxor ferroelectric materials. *Asian Journal of Physics, 6*, 108−114.

Ahmed, M., El Hiti, M. A., Mosaad, M. M., & Atoal, S. M. (1995). *Journal of Magnetism and Magnetic Materials, 146*, 85.

Akbarzadeh, A., Prosandeev, S., Walter, E. J., Al-Barakaty, A., & Bellaiche, L. (2012). Finite-temperature properties of Ba (Zr, Ti)O_3 relaxors from first principles. *Physical Review Letters, 108*(25), 257601−257605. https://doi.org/10.1103/PhysRevLett.108.257601

Andrew, J. S., Starr, J. D., & Budi, M. A. (2014). Prospects for nanostructured multiferroic composite materials. *Scripta Materialia, 74*, 38−43. https://doi.org/10.1016/j.scriptamat.2013.09.023

Basu, S., Babu, K. R., & Choudhary, R. (2012). Studies on the piezoelectric and magnetostrictive phase distribution in lead zirconate titanate−cobalt iron oxide composites. *Materials Chemistry and Physics, 132*(2−3), 570−580. https://doi.org/10.1016/j.matchemphys.2011.11.071

Bhame, S., & Joy, P. (2007). Magnetic and magnetostrictive properties of manganese substituted cobalt ferrite. *Journal of Physics D: Applied Physics, 40*(11), 3263−3267. https://doi.org/10.1088/0022-3727/40/11/001

Bhat, V., Ramanujachary, K., Lofland, S., & Umarji, A. (2004). Tuning the multiferroic properties of Pb $(Fe_{1/2}Nb_{1/2})O_3$ by cationic substitution. *Journal of Magnetism and Magnetic Materials, 280*(2−3), 221−226. https://doi.org/10.1016/j.jmmm.2004.03.016

Bibes, M., & Barthélémy, A. (2008). Towards a magnetoelectric memory. *Nature Materials, 7*(6), 425−426. https://doi.org/10.1038/nmat2189

Bichurin, M., Petrov, V., & Srinivasan, G. (2003). Theory of low-frequency magnetoelectric coupling in magnetostrictive-piezoelectric bilayers. *Physical Review B, 68*(5), 054402−054415. https://doi.org/10.1103/PhysRevB.68.054402

Bochenek, D., & Niemiec, P. (2018). Microstructure and physical properties of the multicomponent PZT-type ceramics doped by calcium, sodium, bismuth and cadmium. *Applied Physics A, 124*(11), 775−782. https://doi.org/10.1007/s00339-018-2203-3

Bokov, A., & Ye, Z.-G. (2006). Recent progress in relaxor ferroelectrics with perovskite structure. *Journal of Materials Science, 41*(1), 31−52. https://doi.org/10.1007/s10853-005-5915-7

Boomgaard, J., & Born, R. (1978). A sintered magnetoelectric composite material $BaTiO_3$-Ni(Co, Mn) Fe_2O_4. *Journal of Materials Science, 13*, 1538−1548. https://doi.org/10.1007/BF00553210

Chen, Z., Su, Y., & Meguid, S. (2014). The effect of field-orientation on the magnetoelectric coupling in Terfenol-D/PZT/Terfenol-D laminated structure. *Journal of Applied Physics, 116*(17), 173910−173919. https://doi.org/10.1063/1.4901069

Cheng, J., Lin, Y., & Tang, T. (2004). The effects of lanthanum's doping on PZT thin film leakage current characteristics. *Journal of Functional Materials, 6*(35), 692−695.

Catalan, G., & Scott, J. F. (2009). Physics and applications of bismuth ferrite. *Advanced Materials, 21*(24), 2463−2485. https://doi.org/10.1002/adma.200802849

Cheng, T., Xu, L. F., Qi, P. B., Yang, C. P., Wang, R. L., & Xiao, H. B. (2014). Tunable dielectric behaviors of magnetic field of PZT5/$NiFe_2O_4$ ceramic particle magnetoelectric composites at room temperature. *Journal of Alloys and Compounds, 602*, 269–274. https://doi.org/10.1016/j.jallcom.2014.03.032

Cheng, W., Ding, A., He, X., Zheng, X., & Qiu, P. (2006). Characterization of $Ba(Zr_{0.05}Ti_{0.95})O_3$ thin film prepared by sol-gel process. *Journal of Electroceramics, 16*(4), 523–526. https://doi.org/10.1007/s10832-006-9910-4

Chougule, S., & Chougule, B. (2008). Response of dielectric behavior and magnetoelectric effect in ferroelectric rich (x) $Ni_{0.9}Zn_{0.1}Fe_2O_{4+(1-x)}$ PZT ME composites. *Journal of Alloys and Compounds, 456*(1–2), 441–446. https://doi.org/10.1016/j.jallcom.2007.02.089

Chu, Y.-H., Martin, L. W., Holcomb, M. B., & Ramesh, R. (2007). Controlling magnetism with multiferroics. *Materials Today, 10*(10), 16–23. https://doi.org/10.1016/S1369-7021(07)70241-9

Cullity, B. D., & Graham, C. D. (2011). *Introduction to magnetic materials*. John Wiley & Sons.

Curecheriu, L. P., Buscaglia, M. T., Buscaglia, V., Mitoseriu, L., Postolache, P., Ianculescu, A., & Nanni, P. (2010). Functional properties of $BaTiO_3-Ni_{0.5}Zn_{0.5}Fe_2O_4$ magnetoelectric ceramics prepared from powders with core-shell structure. *Journal of Applied Physics, 107*(10), 104106–104117. https://doi.org/10.1063/1.3340844

Dai, Q., Guo, K., Zhang, M., Cui, R., & Deng, C. (2019). Magnetoelectric properties of $NiFe_2O_4$/0.65 $BaTiO_3$–0.35$Na_{0.5}Bi_{0.5}TiO_3$ bilayer thin films deposited via pulsed laser deposition. *Ceramics International, 45*(7), 8448–8453. https://doi.org/10.1016/j.ceramint.2019.01.154

Devan, R., & Chougule, B. (2007). Magnetic properties and dielectric behavior in ferrite/ferroelectric particulate composites. *Physica B: Condensed Matter, 393*(1–2), 161–166. https://doi.org/10.1016/j.physb.2007.01.001

DeVreugd, C., Flattery, C., & Srinivasan, G. (2004). Hot-pressed magnetoelectric composites. *Applied Physics Letters, 85*, 2550–2552.

Duan, Z., Fu, X., Mei, Y., Hu, Z., Mao, J., Ding, K., … Zhao, G. (2020). Annealing heating rate dependence of microstructure and multiferroic properties in $Bi_4Ti_{2.9}Fe_{0.1}O_{12}$/$CoFe_2O_4$ layered magnetoelectric composite films prepared by chemical solution deposition method. *Ceramics International, 46*(10), 15654–15664. https://doi.org/10.1016/j.ceramint.2020.03.115

Dzunuzovic, A., Petrovic, M. V., Bobic, J., Ilic, N., Ivanov, M., Grigalaitis, R., … Stojanovic, B. (2018). Magneto-electric properties of $xNi_{0.7}Zn_{0.3}Fe_2O_{4-(1-x)}BaTiO_3$ multiferroic composites. *Ceramics International, 44*(1), 683–694. https://doi.org/10.1016/j.ceramint.2017.09.229

El Hiti, M. (1996). Dielectric behavior and ac electrical conductivity of Zn-substituted Ni Mg ferrites. *Journal of Magnetism and Magnetic Materials, 164*(1–2), 187–196. https://doi.org/10.1016/S0304-8853(96)00368-X

Fang, F., Zhao, C., & Yang, W. (2011). Thickness effects on magnetoelectric coupling for Metglas/PZT/Metglas laminates. *Science China Physics, Mechanics & Astronomy, 54*(4), 581–585. https://doi.org/10.1007/s11433-011-4268-2

Fulcher, G. S. (1925). Analysis of recent measurements of the viscosity of glasses. *Journal of the American Ceramic Society, 8*(6), 339–355. https://doi.org/10.1111/j.1151-2916.1925.tb16731.x

Gajek, M., Bibes, M., Fusil, S., Bouzehouane, K., Fontcuberta, J., Barthelemy, A., & Fert, A. (2007). Tunnel junctions with multiferroic barriers. *Nature Materials, 6*(4), 296–302. https://doi.org/10.1038/nmat1860

Guo, Y.-Y., Zhou, J.-P., & Liu, P. (2010). Magnetoelectric characteristics around resonance frequency under magnetic field in Pb(Zr, Ti)O_3/Terfenol-D laminate composite. *Current Applied Physics, 10*(4), 1092–1095. https://doi.org/10.1016/j.cap.2010.01.003

Gupta, A., & Chatterjee, R. (2010). Study of dielectric and magnetic properties of $PbZr_{0.52}Ti_{0.48}O_3-Mn_{0.3}Co_{0.6}Zn_{0.4}Fe_{1.7}O_4$ composite. *Journal of Magnetism and Magnetic Materials, 322*(8), 1020–1025. https://doi.org/10.1016/j.jmmm.2009.12.007

Guzdek, P. (2014). The magnetostrictive and magnetoelectric characterization of $Ni_{0.3}Zn_{0.62}Cu_{0.08}Fe_2O_4-Pb(FeNb)_{0.5}O_3$ laminated composite. *Journal of Magnetism and Magnetic Materials, 349*, 219–223. https://doi.org/10.1016/j.jmmm.2013.08.064

Han, H., Davis, C., III, & Nino, J. C. (2014). Variable range hopping conduction in $BaTiO_3$ ceramics exhibiting colossal permittivity. *Journal of Physical Chemistry C, 118*(17), 9137–9142. https://doi.org/10.1021/jp502314r

Hu, J.-M., Duan, C.-G., Nan, C.-W., & Chen, L.-Q. (2017). Understanding and designing magnetoelectric heterostructures guided by computation: Progresses, remaining questions, and perspectives. *NPJ Computational Materials, 3*(1), 1–21. https://doi.org/10.1038/s41524-017-0020-4

Islam, R. A., Viehland, D., & Priya, S. (2008). Doping effect on magnetoelectric coefficient of $Pb(Zr_{0.52}Ti_{0.48})O_3-Ni_{(1-x)}Zn_xFe_2O_4$ particulate composite. *Journal of Materials Science, 43*(4), 1497–1500. https://doi.org/10.1007/s10853-007-2386-z

Iwahara, H., Yajima, T., Hibino, T., Ozaki, K., & Suzuki, H. (1993). Protonic conduction in calcium, strontium and barium zirconates. *Solid State Ionics, 61*(1–3), 65–69. https://doi.org/10.1016/0167-2738(93)90335-Z

Jaffe, B., Cook, W., & Jaffe, H. (1971). *Piezoelectric ceramics*. Academic Press.

Jahn, H. A., & Teller, E. (1937). Stability of polyatomic molecules in degenerate electronic states. I. Orbital Degeneracy. *Proceedings of the Royal Society A: Mathematical, Physical and Engineering Sciences*, 220–235. https://doi.org/10.1098/rspa.1937.0142

Jiang, Q., Shen, Z., Zhou, J., Shi, Z., & Nan, C.-W. (2007). Magnetoelectric composites of nickel ferrite and lead zirconnate titanate prepared by spark plasma sintering. *Journal of the European Ceramic Society, 27*(1), 279–284. https://doi.org/10.1016/j.jeurceramsoc.2006.02.041

Jiles, D., & Lo, C. (2003). The role of new materials in the development of magnetic sensors and actuators. *Sensors and Actuators A: Physical, 106*(1–3), 3–7. https://doi.org/10.1016/S0924-4247(03)00255-3

Jin, L., Li, F., & Zhang, S. (2014). Decoding the fingerprint of ferroelectric loops: Comprehension of the material properties and structures. *Journal of the American Ceramic Society, 97*(1), 1–27.

Jonker, G. (1959). Analysis of the semiconducting properties of cobalt ferrite. *Journal of Physics and Chemistry of Solids, 9*(2), 165–175. https://doi.org/10.1016/0022-3697(59)90206-9

Kadam, S., Patankar, K., Mathe, V., Kothale, M., Kale, R., & Chougule, B. (2002). Dielectric behavior and magnetoelectric effect in $Ni_{0.75}Co_{0.25}Fe_2O_4$ + $Ba_{0.8}Pb_{0.2}TiO_3$ ME composites. *Journal of Electroceramics, 9*(3), 193–198. https://doi.org/10.1016/j.matlet.2004.08.033

Kambale, R., Shaikh, P., Bhosale, C., Rajpure, K., & Kolekar, Y. (2010). Studies on magnetic, dielectric and magnetoelectric behavior of (x) $NiFe_{1.9}Mn_{0.1}O_4$ and (1−x)

$BaZr_{0.08}Ti_{0.92}O_3$ magnetoelectric composites. *Journal of Alloys and Compounds, 489*(1), 310–315. https://doi.org/10.1016/j.jallcom.2009.09.080

Kambale, R., Shaikh, P., Kolekar, Y., Bhosale, C., & Rajpure, K. (2010). Studies on dielectric and magnetoelectric behavior of 25% CMFO ferrite and 75% BZT ferroelectric multiferroic magnetoelectric composites. *Materials Letters, 64*(4), 520–523. https://doi.org/10.1016/j.matlet.2009.11.064

Kanamadi, C., Pujari, L., & Chougule, B. (2005). Dielectric behaviour and magnetoelectric effect in $(x)Ni_{0.8}Cu_{0.2}Fe_2O_4 + (1-x)Ba_{0.9}Pb_{0.1}Ti_{0.9}Zr_{0.1}O_3$ ME composites. *Journal of Magnetism and Magnetic Materials, 295*(2), 139–144. https://doi.org/10.1016/j.jmmm.2005.01.006

Khodaei, M., Ebrahimi, S. S., Park, Y. J., Ok, J. M., Kim, J. S., Son, J., & Baik, S. (2014). Enhancement of in-plane magnetic anisotropy in (111)-oriented $Co_{0.8}Fe_{2.2}O_4$ thin film by deposition of PZT top layer. *Applied Physics A, 117*(3), 1153–1160. https://doi.org/10.1007/s00339-014-8486-0

Khodaei, M., Ebrahimi, S. S., Park, Y. J., Son, J., & Baik, S. (2015). Magnetic field-dependent polarization of (111)-oriented PZT–Co ferrite nanobilayer: Effect of Co ferrite composition. *Journal of Magnetism and Magnetic Materials, 382*, 117–123. https://doi.org/10.1016/j.jmmm.2015.01.055

Khomskii, D. (2009). Trend: Classifying multiferroics: Mechanisms and effects. *Physics, 2*, 20–28. https://doi.org/10.1103/Physics.2.20

Koops, C. (1951). On the dispersion of resistivity and dielectric constant of some semiconductors at audiofrequencies. *Physical Review, 83*(1), 121–124. https://doi.org/10.1103/PhysRev.83.121

Kothale, M., Patankar, K., Kadam, S., Mathe, V., Rao, A., & Chougule, B. (2003). Dielectric behaviour and magnetoelectric effect in copper–cobalt ferrite+barium lead titanate composites. *Materials Chemistry and Physics, 77*(3), 691–696. https://doi.org/10.1016/S0254-0584(02)00135-9

Kulawik, J., Szwagierczak, D., & Guzdek, P. (2012). Magnetic, magnetoelectric and dielectric behavior of $CoFe_2O_4$–$Pb(Fe_{1/2}Nb_{1/2})O_3$ particulate and layered composites. *Journal of Magnetism and Magnetic Materials, 324*(19), 3052–3057. https://doi.org/10.1016/j.jmmm.2012.04.056

Kumar, M., Garg, A., Kumar, R., & Bhatnagar, M. (2008). Structural, dielectric and ferroelectric study of $Ba_{0.9}Sr_{0.1}Zr_xTi_{1-x}O_3$ ceramics prepared by the sol–gel method. *Physica B: Condensed Matter, 403*(10–11), 1819–1823. https://doi.org/10.1016/j.physb.2007.10.144

Li, S. B., Wang, C., Shen, Q., & Zhang, L. (2018). Multiferroic properties of (1-x)BCZT-xLCMO laminated composites. *Ceramics International, 44*(1), 231–235. https://doi.org/10.1016/j.ceramint.2017.09.163

Liang, K., Zhou, P., Ma, Z., Qi, Y., Mei, Z., & Zhang, T. (2017). Multiferroic magnetoelectric coupling effect of bilayer $La_{1.2}Sr_{1.8}Mn_2O_7/PbZr_{0.3}Ti_{0.7}O_3$ complex thin film. *Physics Letters A, 381*(17), 1504–1509. https://doi.org/10.1016/j.physleta.2017.02.029

Lines, M., & Glass, A. (1977). *Principles and applications of ferroelectrics and antiferroelectric materials*. Oxford: Clarendon.

Lakhtakia, A., & Mackay, T. (2005). Size-dependent Bruggeman approach for dielectric–magnetic composite materials. *AEU-International Journal of Electronics and Communications, 59*(6), 351–384. https://doi.org/10.1016/j.aeue.2004.10.004

Li, J., Pu, Y., Shi, Y., Shi, R., Wang, X., Yang, M., … Peng, X. (2019). Dielectric, multiferroic and magnetodielectric properties of $(1-x)BaTiO_{3-x}Sr_2CoMoO_6$ solid solution. *Ceramics International, 45*(13), 16353−16360. https://doi.org/10.1016/j.ceramint.2019.05.163

Ma, J., Hu, J., Li, Z., & Nan, C. (2011). Recent progress in multiferroic magnetoelectric composites: From bulk to thin films. *Advanced Materials, 23*(9), 1061−1087. https://doi.org/10.1002/adma.201003636

Margulies, D., Parker, F., Rudee, M., Spada, F., Chapman, J., Aitchison, P., & Berkowitz, A. (1997). Origin of the anomalous magnetic behavior in single crystal Fe_3O_4 films. *Physical Review Letters, 79*(25), 5162−5165. https://doi.org/10.1103/PhysRevLett.79.5162

Mathe, V., & Kamble, R. (2008). Anomalies in electrical and dielectric properties of nanocrystalline Ni−Co spinel ferrite. *Materials Research Bulletin, 43*(8−9), 2160−2165. https://doi.org/10.1016/j.materresbull.2007.09.001

Maxwell, J. (1973). *Electricity and magnetism* (1st ed.). Oxford: Clarendon Press.

Mitra, A., Shaw, A., & Chakrabarti, P. (2020). Microstructure, dielectric, ferroelectric and magnetoelectric coupling of a novel multiferroic of $[(GdMnO_3)_{0.7}(CoFe_2O_4)_{0.3}]_{0.5}[TiO_2]_{0.5}$ nanocomposite. *Materials Chemistry and Physics, 240*, 122242−122250. https://doi.org/10.1016/j.matchemphys.2019.122242

Mukhopadhyay, K., Mahapatra, A., & Chakrabarti, P. (2015). Modulated magneto-dielectric property and exchange bias effect of $BiFeO_3$ incorporated in $(BiFeO_3)_{0.50}(Li_{0.30}Zn_{0.35}Fe_{2.35}O_4)_{0.50}$ nanocomposite. *Journal of Magnetism and Magnetic Materials, 385*, 347−357. https://doi.org/10.1016/j.jmmm.2015.02.071

Nan, C.-W., Bichurin, M., Dong, S., Viehland, D., & Srinivasan, G. (2008). Multiferroic magnetoelectric composites: Historical perspective, status, and future directions. *Journal of Applied Physics, 103*(3), 31101−31136. https://doi.org/10.1063/1.2836410

Nan, C.-W., Liu, G., Lin, Y., & Chen, H. (2005). Magnetic-field-induced electric polarization in multiferroic nanostructures. *Physical Review Letters, 94*(19), 197203−197207. https://doi.org/10.1103/PhysRevLett.94.197203

Negi, N., Kumar, R., Sharma, H., Shah, J., & Kotnala, R. (2018). Structural, multiferroic, dielectric and magnetoelectric properties of $(1-x)Ba_{0.85}Ca_{0.15}Ti_{0.90}Zr_{0.10}O_3$-$(x)CoFe_2O_4$ lead-free composites. *Journal of Magnetism and Magnetic Materials, 456*, 292−299. https://doi.org/10.1016/j.jmmm.2017.12.095

Newnham, R. (1989). Electroceramics. *Reports on Progress in Physics, 52*(2), 123−156. https://doi.org/10.1088/0034-4885/52/2/001

Newnham, R., Skinner, D., & Cross, L. (1978). Connectivity and piezoelectric-pyroelectric composites. *Materials Research Bulletin, 13*(5), 525−536. https://doi.org/10.1016/0025-5408(78)90161-7

Palneedi, H., Maurya, D., Kim, G.-Y., Priya, S., Kang, S.-J. L., Kim, K.-H., … Ryu, J. (2015). Enhanced off-resonance magnetoelectric response in laser annealed PZT thick film grown on magnetostrictive amorphous metal substrate. *Applied Physics Letters, 107*(1), 12904−12909. https://doi.org/10.1063/1.4926568

Panda, R., & Behera, D. (2014). Investigation of electric transport behavior of bulk $CoFe_2O_4$ by complex impedance spectroscopy. *Journal of Alloys and Compounds, 587*, 481−486. https://doi.org/10.1016/j.jallcom.2013.10.195

Pandey, R., Meena, B. R., & Singh, A. K. (2014). Structural and dielectric characterization on multiferroic $xNi_{0.9}Zn_{0.1}Fe_2O_{4/(1-x)}PbZr_{0.52}Ti_{0.48}O_3$ particulate composite. *Journal of Alloys and Compounds, 593*, 224−229. https://doi.org/10.1016/j.jallcom.2014.01.007

Paulsen, J., Ring, A., Lo, C., Snyder, J. E., & Jiles, D. C. (2005). Manganese-substituted cobalt ferrite magnetostrictive materials for magnetic stress sensor applications. *Journal of Applied Physics, 97*(4), 44502–44505. https://doi.org/10.1063/1.1839633

Peng, J., Hojamberdiev, M., Li, H., Mao, D., Zhao, Y., Liu, P., ... Zhu, G. (2015). Electrical, magnetic, and direct and converse magnetoelectric properties of $(1-x)Pb(Zr_{0.52}Ti_{0.48})O_{3-(x)}CoFe_2O_4$(PZT–CFO) magnetoelectric composites. *Journal of Magnetism and Magnetic Materials, 378*, 298–305. https://doi.org/10.1016/j.jmmm.2014.11.060

Pintilie, L., Vrejoiu, I., Hesse, D., LeRhun, G., & Alexe, M. (2007). Ferroelectric polarization-leakage current relation in high quality epitaxial $Pb(Zr, Ti)O_3$ films. *Physical Review B, 75*(10), 104103–104116. https://doi.org/10.1103/PhysRevB.75.104103

Pirc, R., & Kutnjak, Z. (2014). Electric-field dependent freezing in relaxor ferroelectrics. *Physical Review B, 89*(18), 184110–184117. https://doi.org/10.1103/PhysRevB.89.184110

Prasad, K., Choudhary, R., Choudhary, S., & Sati, R. (1996). On the structural and electrical properties of modified $PbTiO_3$ ceramics. *Bulletin of Materials Science, 19*(3), 505–512. https://doi.org/10.1007/BF02744822

Premkumar, S., Varadarajan, E., Rath, M., Rao, M. R., & Mathe, V. (2019). Ferroelectric and piezoelectric properties of PSLZT multilayer/NZFO co-sintered magnetoelectric composites fabricated by tape casting. *Journal of the European Ceramic Society, 39*(16), 5267–5276. https://doi.org/10.1016/j.jeurceramsoc.2019.08.010

Prosandeev, S., Wang, D., Akbarzadeh, A., & Bellaiche, L. (2015). First-principles-based effective Hamiltonian simulations of bulks and films made of lead-free $Ba(Zr, Ti)O_3$ relaxor ferroelectrics. *Journal of Physics: Condensed Matter, 27*(22), 223202–223218. https://doi.org/10.1088/0953-8984/27/22/223202

Qian, K., Fang, B., Du, Q., Ding, J., Zhao, X., & Luo, H. (2013). Structural and electrical properties of $0.69Pb(Mg_{1/3}Nb_{2/3})O_{3-0.31}PbTiO_3$ ceramics prepared by partial oxalate route via different processings. *Ferroelectrics, 445*(1), 74–87. https://doi.org/10.1080/00150193.2013.814023

Raghunathan, A., Nlebedim, I., Jiles, D., & Snyder, J. E. (2010). Growth of crystalline cobalt ferrite thin films at lower temperatures using pulsed-laser deposition technique. *Journal of Applied Physics, 107*(9), 09A516–09A519. https://doi.org/10.1063/1.3357315

Rahman, A., Rafiq, M., Maaz, K., Karim, S., Oh Cho, S., & Hasan, M. (2012). Temperature induced delocalization of charge carriers and metallic phase in $Co_{0.6}Sn_{0.4}Fe_2O_4$ nanoparticles. *Journal of Applied Physics, 112*(6), 63718–63725. https://doi.org/10.1063/1.4754559

Ramana, E. V., & Sankaram, T. B. (2010). Magnetoelectric investigations in La-modified bismuth layered ferroelectromagnetics. *Materials Chemistry and Physics, 120*(2–3), 231–235. https://doi.org/10.1016/j.matchemphys.2009.12.003

Ricote, S., Bonanos, N., Manerbino, A., Sullivan, N. P., & Coors, W. (2014). Effects of the fabrication process on the grain-boundary resistance in $BaZr_{0.9}Y_{0.1}O_{3-\delta}$. *Journal of Materials Chemistry, 2*(38), 16107–16115. https://doi.org/10.1039/C4TA02848A

Rosa, W. S., & Venet, M. (2016). Exploring the processing conditions to optimize the interface in 2–2 composites based on $Pb(Zr, Ti)O_3$ and $NiFe_2O_4$. *Ceramics International, 42*(7), 7980–7986. https://doi.org/10.1016/j.ceramint.2016.01.196

Ryu, A., Carazo, A., Uchino, K., & Kim, H.-E. (2001). Piezoelectric and magnetoelectric properties of lead zirconate titanate/Ni–ferrite particulate composites. *Journal of Electroceramics, 7*, 17–24. https://doi.org/10.1023/A:1012210609895

Sahoo, M., & Choudhary, R. (2010). Evidence of relaxor freezing in $Pb(Mg_{1/3}Nb_{2/3})O_3$/ $BiFeO_3$ ceramics. *Solid State Communications, 150*(45–46), 2236–2238. https://doi.org/10.1016/j.ssc.2010.09.035

Sahoo, M., Yajun, Z., Wang, J., & Choudhary, R. (2016). Composition control of magneto-electric relaxor behavior in multiferroic $BaZr_{0.4}Ti_{0.6}O_3/CoFe_2O_4$ composites. *Journal of Alloys and Compounds, 657*, 12–20. https://doi.org/10.1016/j.jallcom.2015.10.040

Scott, J. (2007). Multiferroic memories. *Nature Materials, 6*(4), 256–257. https://doi.org/10.1038/nmat1868

Sowmya, N. S., Srinivas, A., Saravanan, P., Reddy, K. V. G., Kamat, S., Praveen, J. P., … Subramanian, V. (2018). Studies on magnetoelectric coupling in lead-free [(0.5)BCT-(0.5)BZT]-$NiFe_2O_4$ laminated composites at low and EMR frequencies. *Journal of Alloys and Compounds, 743*, 240–248. https://doi.org/10.1016/j.jallcom.2018.01.402

Spaldin, N. A., & Fiebig, M. (2005). The renaissance of magnetoelectric multiferroics. *Science, 309*(5733), 391–392. https://doi.org/10.1126/science.1113357

Srinivasan, G., Rasmussen, E., Gallegos, J., Srinivasan, R., Bokhan, Y. I., & Laletin, V. (2001). Magnetoelectric bilayer and multilayer structures of magnetostrictive and piezoelectric oxides. *Physical Review B, 64*(21), 214408–214414. https://doi.org/10.1103/PhysRevB.64.214408

Srinivasan, G., Rasmussen, E., & Hayes, R. (2003). Magnetoelectric effects in ferrite-lead zirconate titanate layered composites: The influence of zinc substitution in ferrites. *Physical Review B, 67*(1), 14418–14428. https://doi.org/10.1103/PhysRevB.67.014418

Suchanicz, J., Kusz, J., & Böhm, H. (2003). Structural and electric characteristics of $(Na_{0.5}Bi_{0.5})_{0.50}Ba_{0.50}TiO_3$ and $(Na_{0.5}Bi_{0.5})_{0.20}Ba_{0.80}TiO_3$ ceramics. *Materials Science and Engineering: B, 97*(2), 154–159. https://doi.org/10.1016/S0921-5107(02)00577-9

Suryanarayana, S. (1994). Magnetoelectric interaction phenomena in materials. *Bulletin of Materials Science, 17*(7), 1259–1270. https://doi.org/10.1007/BF02747225

Tang, Z., Yang, B., Chen, J., Lu, Q., & Zhao, S. (2019). Strong magnetoelectric coupling of Aurivillius phase multiferroic composite films with similar layered perovskite structure. *Journal of Alloys and Compounds, 772*, 298–305. https://doi.org/10.1016/j.jallcom.2018.09.101

Uchino, K., & Nomura, S. (1982). Critical exponents of the dielectric constants in diffused-phase-transition crystals. *Ferroelectrics, 44*(1), 55–61. https://doi.org/10.1080/00150198208260644

Van Uitert, L. (1956). High-resistivity nickel ferrites-the effect of minor additions of manganese or cobalt. *The Journal of Chemical Physics, 24*(2), 306–310. https://doi.org/10.1063/1.1742468.

Vaz, C.A., Hoffman, J., Ahn, C.H., & Ramesh, R. (2010). Magnetoelectric coupling effects in multiferroic complex oxide composite structures. *Advanced Materials, 22*(26-27), 2900–2918. https://doi.org/10.1002/adma.200904326.

Vlasko-Vlasov, V., Lin, Y., Miller, D., Welp, U., Crabtree, G., & Nikitenko, V. (2000). Direct magneto-optical observation of a structural phase transition in thin films of manganites. *Physical Review Letters, 84*(10), 2239. https://doi.org/10.1103/PhysRevLett.84.2239

Vogel, H. (1921). The law of the relation between the viscosity of liquids and the temperature. *Physikalische Zeitschrift, 22*, 645–646.

Vopson, M., Fetisov, Y., Caruntu, G., & Srinivasan, G. (2017). Measurement techniques of the magneto-electric coupling in multiferroics. *Materials, 10*(8), 963. https://doi.org/10.3390/ma10080963

Wang, Z., Gao, R., Chen, G., Deng, X., Cai, W., & Fu, C. (2020). Dielectric, ferroelectric and magnetoelectric properties of in-situ synthesized $CoFe_2O_4/BaTiO_3$ composite ceramics. *Ceramics International, 46*(7), 9154–9160. https://doi.org/10.1016/j.ceramint.2019.12.165

Xu, T., Wang, C.-A., & Wang, C. (2015). Synthesis and magnetoelectric effect of composites with $CoFe_2O_4$-epoxy embedded in 3–1 type porous PZT ceramics. *Ceramics International, 41*(9), 11080–11085. https://doi.org/10.1016/j.ceramint.2015.05.054

Zhai, J., Li, J., Viehland, D., & Bichurin, M. I. (2007). Large magnetoelectric susceptibility: The fundamental property of piezoelectric and magnetostrictive laminated composites. *Journal of Applied Physics, 101*(1), 14102–14105. https://doi.org/10.1063/1.2405015

Zhai, J. Y., Cai, N., Liu, L., Lin, Y. H., & Nan, C. W. (2003). Dielectric behavior and magnetoelectric properties of lead zirconate titanate/Co-ferrite particulate composites. *Materials Science and Engineering: B, 99*(1–3), 329–331. https://doi.org/10.1016/S0921-5107(02)00565-2

Yang, H., Zhang, G., Lin, Y., & Wang, F. (2015). Enhanced Curie temperature and magnetoelectric effects in the $BaTiO_3$-based piezoelectrics and $CoFe_2O_4$ laminate composites. *Materials Letters, 157*, 99–102. https://doi.org/10.1016/j.matlet.2015.05.072

Zheng, Z., Zhou, P., Liu, Y., Liang, K., Tanguturi, R., Chen, H., … Zhang, T. (2020). Strain effect on magnetoelectric coupling of epitaxial NFO/PZT heterostructure. *Journal of Alloys and Compounds, 818*, 152871–152880. https://doi.org/10.1016/j.jallcom.2019.152871

Zhou, J., He, H., & Nan, C.-W. (2007). Effects of substrate temperature and oxygen pressure on the magnetic properties and structures of $CoFe_2O_4$ thin films prepared by pulsed-laser deposition. *Applied Surface Science, 253*(18), 7456–7460. https://doi.org/10.1016/j.apsusc.2007.03.046

Zhou, J., He, H., Shi, Z., Liu, G., & Nan, C.-W. (2006). Dielectric, magnetic, and magnetoelectric properties of laminated $PbZr_{0.52}Ti_{0.48}O_3/CoFe_2O_4$ composite ceramics. *Journal of Applied Physics, 100*(9), 33503–33511. https://doi.org/10.1063/1.2220649

Further reading

Chen, K., Huang, C., Zhang, X., Yu, Y., Lau, K., Hu, W., … Liu, J. (2013). Negative magnetodielectric effect in $CaCu_3Ti_4O_{12}$. *Journal of Applied Physics, 114*(23), 234104–234110. https://doi.org/10.1063/1.4851815

Li, S., Eastman, J., Newnham, R., & Cross, L. (1997). Diffuse phase transition in ferroelectrics with mesoscopic heterogeneity: Mean-field theory. *Physical Review B, 55*(18), 12067–12079. https://doi.org/10.1103/PhysRevB.55.12067

Li, T., Li, K., & Hu, Z. (2014). Thickness and frequency dependence of magnetoelectric effect for epitaxial $La_{0.7}Sr_{0.3}MnO_3/BaTiO_3$ bilayer. *Journal of Alloys and Compounds, 592*, 266–270. https://doi.org/10.1016/j.jallcom.2014.01.021

Ma, F. D., Jin, Y. M., Wang, Y. U., Kampe, S., & Dong, S. (2014). Phase field modeling and simulation of particulate magnetoelectric composites: Effects of connectivity, conductivity, poling and bias field. *Acta Materialia, 70*, 45–55. https://doi.org/10.1016/j.actamat.2014.02.015

Palakkal, J. P., Cheriyedath, R. S., Lekshmi, P., Valant, M., Mihelj, M. V., & Varma, M. R. (2019). Large positive and negative magnetodielectric coupling in Fe half-doped

$LaMnO_3$. *Journal of Magnetism and Magnetic Materials, 474*, 183–186. https://doi.org/10.1016/j.jmmm.2018.10.121

Petrov, V., Srinivasan, G., Bichurin, M., & Gupta, A. (2007). Theory of magnetoelectric effects in ferrite piezoelectric nanocomposites. *Physical Review B, 75*(22), 224407–224412. https://doi.org/10.1103/PhysRevB.75.224407

Verma, K. C., Singh, D., Kumar, S., & Kotnala, R. (2017). Multiferroic effects in MFe_2O_4/$BaTiO_3$ (M = Mn, Co, Ni, Zn) nanocomposites. *Journal of Alloys and Compounds, 709*, 344–355. https://doi.org/10.1016/j.jallcom.2017.03.145

Wang, L. Y., Li, Q., Gong, Y. Y., Wang, D. H., Cao, Q. Q., & Du, Y. W. (2014). The positive and negative magnetodielectric effects in double perovskite Pr_2CoMnO_6. *Journal of the American Ceramic Society, 97*(7), 2024–2026. https://doi.org/10.1111/jace.13009

Zhou, Y., Yang, Z., & Zheng, X. (2003). Residual stress in PZT thin films prepared by pulsed laser deposition. *Surface and Coatings Technology, 162*(2–3), 202–211. https://doi.org/10.1016/S0257-8972(02)00581-9

CHAPTER

Exchange-spring ferrite nanocomposites

5

5.1 Introduction

Permanent magnets are used in high-density recording media, car gearboxes, electromotor components, microelectronic, sonar systems, microwave-absorbing composites, and high-energy production systems with superior corrosion and wear resistance. The most important parameter for practical evaluation of permanent magnet materials is the maximum energy product, or $(BH)_{max}$. To date, the demand for enhanced maximum energy products of magnetic materials, particularly rare-earth-free magnets, has increased considerably, marking the importance of magnetic hardening. In this regard, one efficient approach is to design nanostructured materials with a controlled mixture of hard and soft magnetic phases. Normally, incorporating a soft magnetic phase with a hard phase matrix has been considered harmful to the performance of permanent magnets, significantly deteriorating coercivity and consequently reducing the maximum energy product. Kneller and Hawig from the Institut für Werkstoffe der Elektrotechnik have proposed a theoretical model, indicating that specific configurations of the hard/soft magnetic composites not only prevent coercivity from deteriorating but also drastically enhance overall magnetic properties, resulting in innovative applications of energy-related machines that include hybrid vehicles and wind power generators. These new permanent magnetic materials are known as "exchange-spring magnets" (Kneller & Hawig, 1991). The term "exchange-spring" originated from fully reversible springlike magnetic interactions occurring between the spins of hard and soft phases when applying an external magnetic field. Since 1992, many experimental and theoretical studies have been devoted to investigating exchange-coupled nanocomposite systems, making exchange-spring magnets ideal candidates for fabricating the next generation of permanent magnets (Xiong et al., 2016).

5.2 Maximum energy product

The energy stored in or the strength of permanent magnets is measured by the maximum energy product. Using the $B-H$ loop measurement is the most popular technique for determining the $(BH)_{max}$ parameter. Based on the second-quadrant

Magnetic Ferrites and Related Nanocomposites. https://doi.org/10.1016/B978-0-12-824014-4.00004-4

branch of the $B-H$ loop with the largest area for an enclosed rectangle, it is possible to identify the $(BH)_{max}$ of a permanent magnet. The location of $(BH)_{max}$ is the point at which the magnet is most effectively used. The maximum energy product, remanence, and intrinsic coercivity measured by the second quadrant of the intrinsic hysteresis $(B-H)$ loop are shown in Fig. 5.1.

As a characteristic, the maximum energy product is twice the maximum magnetostatic energy accessible from a magnet. For magnets with a high coercive field $(H_c > 2\pi M_s)$, the theoretical limit of the $(BH)_{max}$ pertains only to the saturation polarization $(J_s = \mu_0 M_s)$ and is given by $(BH)_{max} \leq J_s^2/4\mu_0$. This value is only available for hard magnetic materials with large magnetocrystalline anisotropy, i.e., $K >> J_s^2/4\mu_0$, in which K is the magnetocrystalline anisotropy constant (Kneller & Hawig, 1991). For materials with $H_c < 2\pi M_s$, the theoretical limit of $(BH)_{max}$ is related only to the coercive field and is given by $(BH)_{max} \leq H_c^2$.

5.2.1 Soft phase thickness dependence

A schematic model of the structure of the exchange-coupled nanocomposite material is illustrated in Fig. 5.2. The magnetic behavior of the material is defined by $\kappa = K\ (J_s^2/4\mu_0) = 4\ K/\mu_0\ M_s^2$. In a hard magnetic material (k-material), magnetic properties are mainly controlled by the magnetocrystalline anisotropy involving $\kappa >> 1$. For a soft magnetic material (m-material), magnetostatic energy dominates

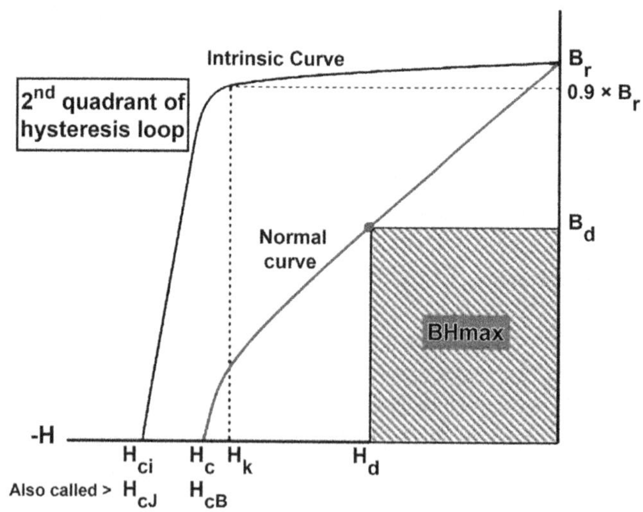

FIGURE 5.1

Second quadrant of the $B-H$ loop determining the maximum energy product.

From Lewis, L. H., & Jiménez-Villacorta, F. (2013). Perspectives on permanent magnetic materials for energy conversion and power generation. Metallurgical and Materials Transactions A, 44*(1), 2–20.*

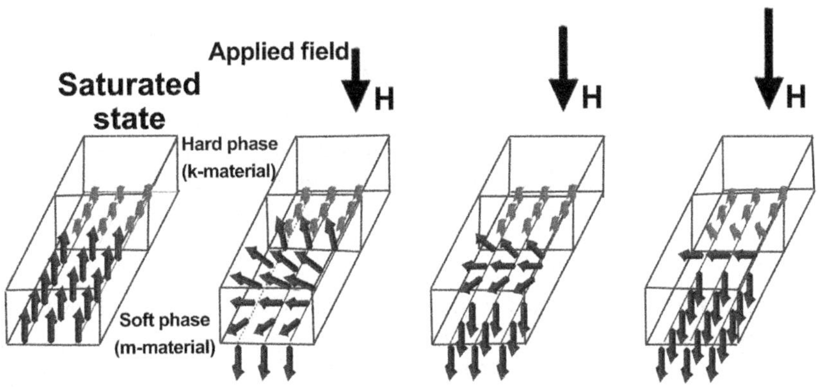

FIGURE 5.2

Schematic model of the structure of exchange-coupled magnets.

the magnetic characteristics involving $\kappa << 1$. It is well known that the saturation magnetization of the best attainable hard magnets is smaller than that of many soft magnetic materials (M_m). Consequently, it is possible to enhance the maximum energy product by making nanocomposites consisting of a hard phase with large H_c and high nucleation field (H_n) and a soft phase with high M_m. The nucleation field is given by

$$H_n \cong \frac{2K_k}{\mu_0 M_m} \tag{5.1}$$

where K_k and M_m are the anisotropy constant of the hard phase and saturation magnetization of the soft phase, respectively. For a microstructure with $b_m > b_{cm}$ (where b_m is the thickness of the soft phase and b_{cm} is the critical thickness of the soft phase), the coercivity depends on b_m as follows:

$$H_c = \frac{A_m \pi^2}{2\mu_0 M_m} \cdot \frac{1}{b_m^2} \tag{5.2}$$

where A_m is the exchange stiffness constant of the soft phase. As an example, the H_c of a sample with $b_m = 10$ nm is on the order of 3×10^5 A/m. Research investigations have indicated that H_n and H_c vary as a function of several factors, including the microstructure, synthetic route, defect type present in the structure, and working temperature (Fullerton et al., 1999; Kneller & Hawig, 1991; Lewis & Jiménez-Villacorta, 2013).

By covering a soft magnetic phase (having an optimal thickness) with a hard phase while also keeping a high specific surface area of the contact between the two phases, the well-established exchange-spring mechanism is realized. The optimal thickness of a soft magnet in the nanocomposite is twice the domain wall thickness (δ_k) of the hard phase. The domain wall thickness of the hard phase can be expressed as:

$$\delta_k = \pi\sqrt{\frac{A_k}{K_k}} \tag{5.3}$$

where A_k and K_k are the exchange stiffness and anisotropy constants of the hard phase, respectively. For practical purposes, the hard layer thickness is set equal to δ_k.

Alternatively, the critical thickness of a soft layer (b_{cm}) can be written as follows:

$$b_{cm} = \pi\sqrt{\frac{A_m}{2K_k}} \tag{5.4}$$

where A_m and K_k are the exchange stiffness and anisotropy constants of the soft and hard phases, respectively (Leineweber & Kronmüller, 1997a; Skomski & Coey, 1993). Notably, the optimal thicknesses of the domain wall, and consequently the soft layer thicknesses, do not exceed a few tens of nanometers. Below a critical soft layer thickness, the two magnetic phases are mutually reversed at similar nucleation fields, leading to a rectangular hysteresis loop with a large value for the maximum energy product. Above a critical thickness of the soft layer, the hard and soft magnetic phases are switched individually, giving rise to the appearance of "bee waist" type hysteresis loops (Fullerton et al., 1999; Kneller & Hawig, 1991). A schematic of the reversal mechanism occurring in the hard/soft exchange-coupled nanocomposite along with the corresponding hysteresis loops is shown in Fig. 5.3 (Lewis & Jiménez-Villacorta, 2013).

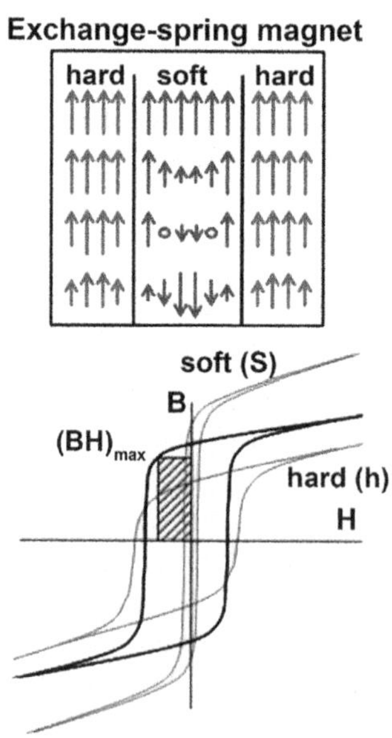

FIGURE 5.3

Magnetization reversal mechanism in exchange-coupled nanocomposite structures.

From Lewis, L. H., & Jiménez-Villacorta, F. (2013). Perspectives on permanent magnetic materials for energy conversion and power generation. Metallurgical and Materials Transactions A, *44(1), 2–20.*

Noticeably, below a critical soft layer thickness, the soft phase is strongly coupled to the hard phase so that the two phases act as a single phase, thus switching at similar nucleation fields. In this way, the nucleation field can be expressed as

$$H_n = \frac{2(t_m K_m + t_k K_k)}{t_m M_m + t_k M_k} \tag{5.5}$$

where t_m and t_k are the thicknesses, K_m and K_k are the anisotropy constants, and M_m and M_k are the magnetization values of the soft and hard layers, respectively. Therefore, a rectangular hysteresis loop is formed for $H_c = H_n$, and $(BH)_{\max}$ can be obtained according to Eq. (5.6) (Leineweber & Kronmüller, 1997b; Skomski & Coey, 1993):

$$(BH)_{\max} \approx (2\pi M_m)^2 \left[1 - \frac{2\pi(M_m - M_k)M_m}{K_k}\right] \tag{5.6}$$

It should be mentioned that above a critical soft layer thickness, the soft phase nucleates the magnetization reversal at considerably lower fields, thereby sharply decreasing the coercivity of the soft phase. Thus, the switching is defined based on inhomogeneous magnetization reversal. By considering a thin film of the soft phase deposited directly on a magnetically hard substrate while also applying a magnetic field higher than the exchange field, a continuous reversal occurs for the spins of the soft magnetic thin film. Furthermore, the angle of rotation increases with an increase in the thickness of the hard phase. If the reverse field is released, the soft phase will reverse back into alignment with the hard phase. This configuration demonstrates a reversible demagnetization curve comparable to the elastic motion of a mechanical spring (Fullerton et al., 1999).

The demagnetization curves of the hard/soft composite with various shapes and sizes are shown in Fig. 5.4. As mentioned for the exchange-coupled magnets of soft and hard phases, the reversible range (ΔM_{rev}) is related to various parameters, including the hard phase volume percentage (v_k), M_m/M_k ratio, and soft phase size (b_m). If v_k and M_m/M_k are constant, the reversible range will be smallest for $b_m \approx b_{cm}$ (Fig. 5.4B) and will increase when $b_m > b_{cm}$, which is attributed to the fixed H_n (Fig. 5.4C). In this case, the critical H_n is smaller than the anisotropy field of the hard phase. At high soft phase volume percentages (e.g., $v_m = 0.8$), the reversible range is larger than the remanence ($\Delta M_{\text{rev}} > M_r$). The demagnetization curves of the single ferromagnetic phase magnets show the distribution of the critical switching fields, indicating their irreversibility. It has been found that the composite with $b_m \approx b_{cm}$ has a single-phase demagnetization curve similar to that of a uniform permanent magnet (Kneller & Hawig, 1991). The composite with insufficient coupling (see Fig. 5.4D) reflects a shoulder in the graphs, reversing the hard and soft phases individually.

5.2.2 The role of exchange length

Exchange energy, dipolar interaction, and anisotropy energy play the main role in controlling the overall magnetic features of composites. Exchange interaction and

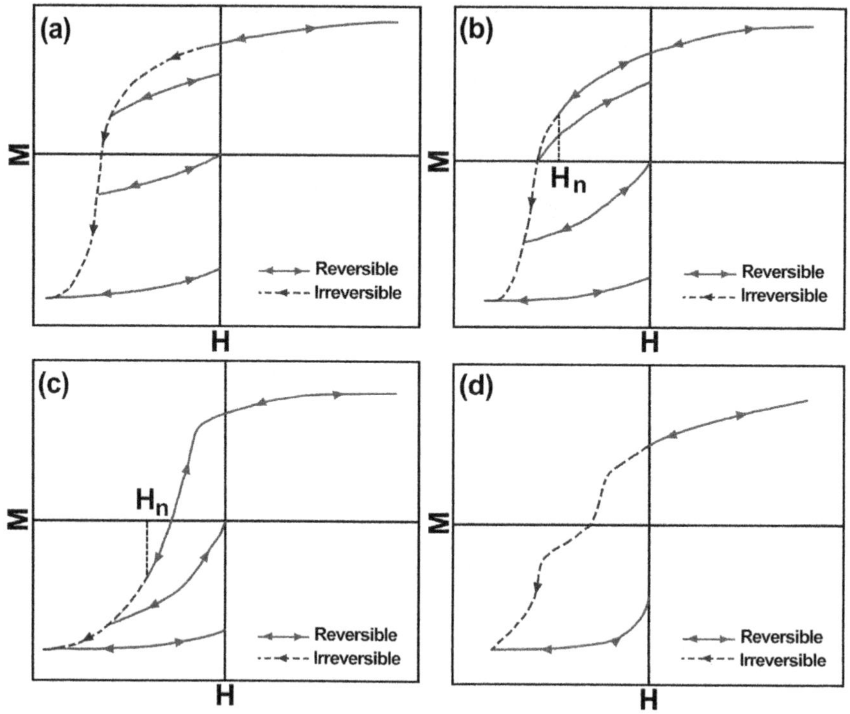

FIGURE 5.4

Demagnetization curve of hard/soft composite (A) single-phase magnet, (B) $b_m \approx b_{cm}$, (C) $b_m > b_{cm}$, and (D) composites of insufficiently coupled magnet.

magnetocrystalline anisotropy have an impact on remanent magnetization. To realize an effective exchange-coupled state, besides the magnetization alignment in the soft phase grains, collinear arrangement of moments at the interface between soft and hard phases must also be carried out, leading to a higher value of magnetization. There is competition between dipolar and exchange coupling interactions. The long-range nature of dipolar interactions can be suppressed by stronger exchange coupling interactions, increasing the remanent magnetization.

Schematic representations of hard/soft/hard thin layers with in-plane and perpendicular anisotropies are illustrated in Fig. 5.5.

The exchange length in the in-plane configuration can be expressed as

$$L_{ex} = \sqrt{\frac{A}{K_u}} \tag{5.7}$$

where A is the exchange stiffness constant and K_u is the anisotropy constant. It is then found that an increase in anisotropy decreases the exchange length, resulting

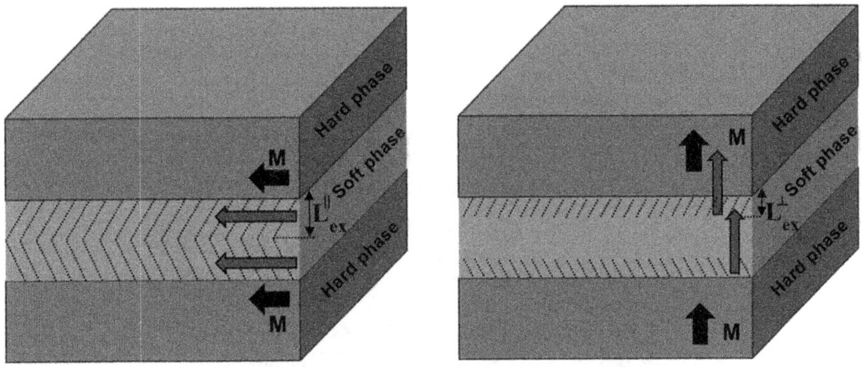

FIGURE 5.5

Trilayer hard/soft/hard configurations of in-plane and out-of-plane anisotropies.

in a decrease in the exchange coupling interaction. Coercivity is reduced as a result. On the other hand, the exchange length for uniaxial anisotropy can be expressed as

$$L_{ex}^{\perp} = \sqrt{\frac{A}{K_u + 2\pi\Delta M_s^2}} \tag{5.8}$$

where A is the exchange stiffness constant of the soft phase, ΔM_s is the difference between saturation magnetizations of soft and hard layers, and K_u is the uniaxial anisotropy constant.

A schematic representation of exchange interactions for hard and soft magnetic phases in a powder state is shown in Fig. 5.6. The contact area between the phases can be separated into an inner section without an exchange coupling interaction and an interfacial section with an exchange coupling interaction. Therefore, the exchange interaction is present only in the interface layer, similar to the exchange length (L_{ex}).

Normally, three magnetic interactions exist between the soft and hard phases, involving soft—hard, soft—soft, and hard—hard grain interactions of the nanocomposite. By applying an external magnetic field, the soft phase spins can be quickly switched in the direction of the field due to the lower magnetocrystalline anisotropy of the soft phase compared with that of the hard phase. DC demagnetization remanence curves of a trilayer structure with different soft phase thicknesses are shown in Fig. 5.7. Evidently, at an optimized thickness of the soft layer, sufficient exchange coupling is formed, leading to a hysteresis loop without the presence of a kink. Also, for soft phase thicknesses larger than the exchange length, "bee waist" type hysteresis loops are formed. By undergoing reversible switching of magnetization for the soft phase, the exchange-spring mechanism occurs when the applied magnetic fields are not large enough to reverse the hard phase magnetization.

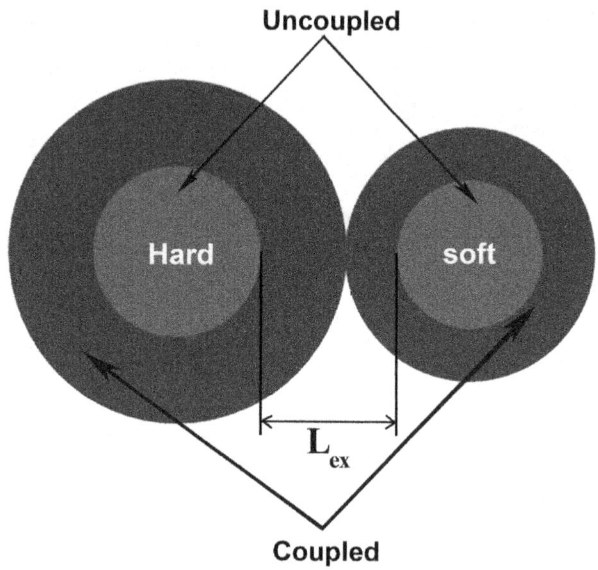

FIGURE 5.6

Schematic of exchange interactions for hard–soft particulate composite.

From Song, F., Shen, X., Liu, M., & Xiang, J. (2012). Microstructure, magnetic properties and exchange–coupling interactions for one-dimensional hard/soft ferrite nanofibers. Journal of Solid State Chemistry, 185, *31–36.*

5.3 Ferrite-based exchange-spring composites

While the highest values of the maximum energy product [$(BH)_{max}$ = 100–400 kJ/m^3] belong to rare-earth magnets and their composite structures, they have limited accessibility and pose unacceptable environmental hazards during the extraction process (Okabe, 2011). Currently used for industrial applications, exchange-spring rare-earth permanent magnets such as Nd–Fe–B/Fe and $SmCo_5$–FeCo suffer from low Curie temperature and poor oxidation and corrosion resistance. Notably, by heating metallic magnets in electric vehicle motors with high fundamental frequencies, i.e., high speed (e.g., a maximum rotational speed of 10,000 rpm) and a high pole number, partial irreversible demagnetization easily occurs in the nanocomposite with poor chemical stability and large electrical conductivity (Imaoka et al., 2019; Zhu & Howe, 2007). On the other hand, because of the crisis in the permanent magnets market in 2012 (thus causing a sharp increase in their price), big-name companies and academia began to consider replacements for rare-earth magnets for some specific applications based on hard/soft ferrite nanocomposites. Hard ferrites (e.g., strontium and barium ferrites in the hexagonal group and cobalt ferrites in the spinel group) with unique properties, including superior atmospheric corrosion resistance, high dielectric constant value, large magnetocrystalline anisotropy constant, high coercivity and nucleation field, moderate saturation magnetization, relatively high Curie

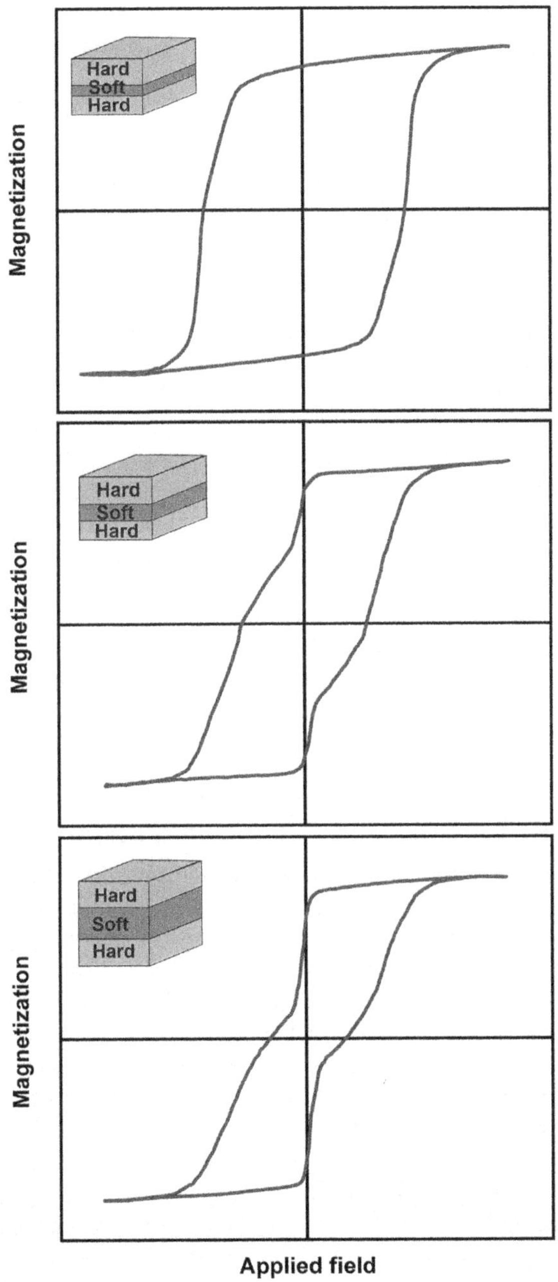

FIGURE 5.7

DC demagnetization remanence curves of hard/soft/hard trilayer configuration.

temperature (compared with NdFeB magnets), large remanence ratio, and low materials cost, have become suitable magnets employed the hard magnetic components of nanocomposites.

Alternatively, soft spinel ferrites, such as magnetite, nickel ferrite, Ni–Zn, and Mn–Zn ferrites with high saturation magnetization, low coercivity, and high electrical resistivity, could be employed as the soft components of bimagnetic materials. While achieving magnetization reversal at a low magnetic field is difficult for hard ferrites with large magnetocrystalline anisotropy, the magnetization can easily be switched to the direction of the applied field for soft ferrites with negligible magnetocrystalline anisotropy. For effectively tuning magnetic characteristics of bimagnetic nanoparticles, some key parameters should be sufficiently engineered to meet the requirements of specific applications. First, both hard and soft ferrite phases must have a large contact area in the form of multilayer, core–shell, particulate, and bulk composites. Second, the soft phase size should not exceed twice the hard phase domain wall thickness. Finally, since the single domain state results in the highest coercivity, the preference is for the dimension of the hard phase to be limited in the single domain size criteria. Thin film-related technologies such as pulsed laser deposition and molecular beam epitaxial growth have been used to make permanent magnet composites. For particulate nanocomposites, it is essential to reach a high level of homogeneous mixing of the hard and soft ferrite phases. In this case, wet-chemical synthetic routes have allowed us to control the microstructural parameters, including shape, composition, and size, thereby generating and developing exchange-coupled magnets with well-dispersed hard and soft phases.

5.4 Hard ferrite–soft iron oxide nanocomposites

Magnetite (Fe_3O_4) exhibits the strongest magnetism among the transition metal oxides, arising from the remarkable properties originating from four unpaired electrons in its three-dimensional orbital. Notably, magnetite has a ferrimagnetic state with $M_s = 75$ emu/g, H_c values ranging from 2.4 to 20 kA/m and a Curie temperature of 850 K (Cornell & Schwertmann, 2003; Teja & Koh, 2009). This makes the magnetite a suitable candidate for the soft phase in preparation of exchange-spring magnets. So far, $BaFe_{12}O_{19}/Fe_3O_4$ composites have been fabricated via a simple and effective approach (Remya et al., 2016), mixing ferrites in weight ratios of 3:1, 6:1, and 9:1. The hysteresis loops of the resulting nanocomposites sintered at 600°C are illustrated in Fig. 5.8. As inferred, magnetite and barium ferrite display soft and hard properties, respectively. In the absence of magnetite, the direct couplings among hard phase grains play the main role in magnetic properties, controlling the magnetization process. By adding the soft phase to barium ferrite, the exchange coupling becomes stronger in the absence of a reverse field so that the hard and soft phase magnetic moments interact parallelly. By applying a reverse field, the magnetic moments of magnetite rotate to align with the external field.

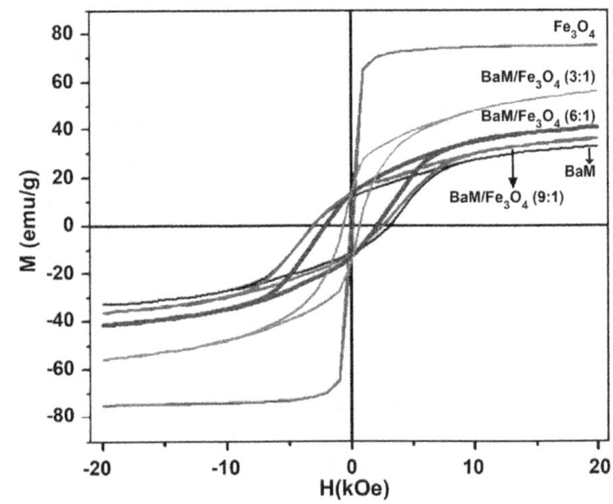

FIGURE 5.8

VSM graphs of composites with different content of soft magnetite sintered at 600°C.

From Remya, K., Prabhu, D., Amirthapandian, S., Viswanathan, C., & Ponpandian, N. (2016). Exchange spring magnetic behavior in $BaFe_{12}O_{19}/Fe_3O_4$ nanocomposites. Journal of Magnetism and Magnetic Materials, 406, *233–238.*

With an increase in the applied field, domain walls of soft magnetite move to the boundaries of the hard and soft phases. Consequently, at lower content of the soft phase, higher-exchange interactions are imposed by the hard magnet to the soft phase, thus increasing the corresponding coercivity.

An increase in the weight percentage of the soft phase leads to enhanced dipolar interactions between soft phase grains rather than between the soft and hard phases, which in turn results in decreased coercivity. The maximum energy product calculated for the whole series of the nanocomposite indicates a maximum value of 5.04 MGOe for the sample mixed with the 9:1 ratio. The simultaneous enhancement in $(BH)_{max}$ and remanence values confirms that the hard and soft phases are mutually reversed.

The specific contact area between hard and soft phases is relatively large in a rodlike configuration, involving stronger exchange coupling than other particulate composites. The composite structure of barium hexaferrite (*HF*) and soft maghemite (*S*) synthesized by a novel approach is illustrated in Fig. 5.9 (Primc & Makovec, 2015). The maghemite layer with homogeneous and uniform thickness has been covered topotactically on both basal-plane surfaces of the barium ferrite core. The presence of iron and oxygen is confirmed by EDS analysis. The close-packed (111) planes of maghemite are parallel to the (001) planes of hexaferrite, giving rise to remarkable coherency. Nevertheless, crystal matching between the side surfaces of the barium ferrite and soft magnet has not been established. Thus, the spinel cannot be situated at the side surfaces of barium ferrite.

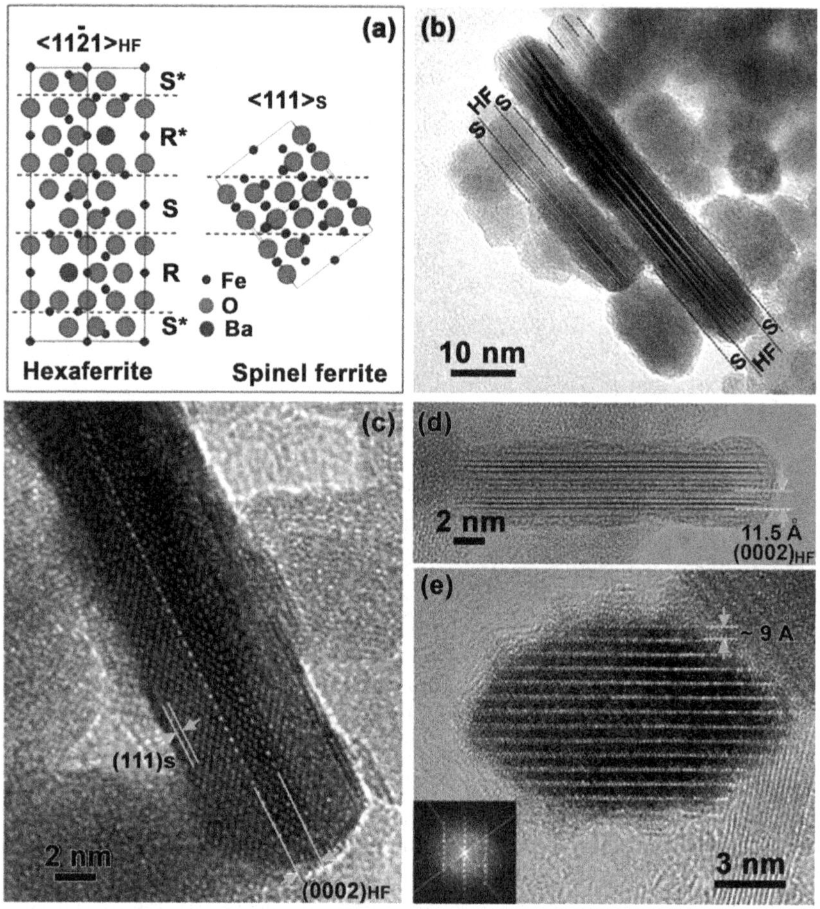

FIGURE 5.9

(A) Schematic presentation of barium ferrite (HF) and spinel ferrite (*S*) structures, (B) Transmission electron microscopy image of the platelet composite nanoparticles, (C) High-resolution transmission electron microscopy (HRTEM) micrograph of the platelet composite nanoparticle, (D) High-resolution electron microscopy image of the barium ferrite nanoparticles before and (E) after coating with the spinel (inset: fast Fourier transform (FTT) of the HRTEM image).

From Primc, D., & Makovec, D. (2015). Composite nanoplatelets combining soft-magnetic iron oxide with hard-magnetic barium hexaferrite. Nanoscale, *7(6), 2688–2697.*

Because of direct contacts between the barium ferrite and spinel ferrite in the composite, the hysteresis loop depicted in Fig. 5.10 confirms that the magnetization of barium ferrite and maghemite is mutually reversed. Meanwhile, the corresponding maximum energy product is enhanced more than two times, leading to $(BH)_{max} = 7.2$ and $15.2\ kJ/m^3$ for the barium ferrite and composite, respectively.

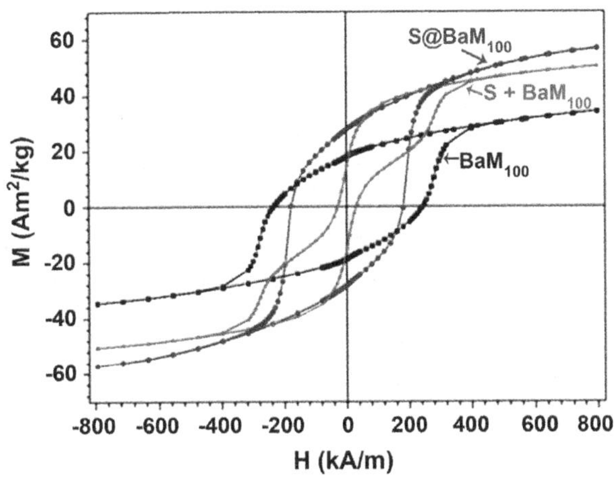

FIGURE 5.10

VSM curves of barium ferrite and composite prepared by physical mixing (S+BaM100) and novel chemical method (S@BaM100).

From Primc, D., & Makovec, D. (2015). Composite nanoplatelets combining soft-magnetic iron oxide with hard-magnetic barium hexaferrite. Nanoscale, *7(6), 2688–2697.*

The results demonstrate that the fabricated composite can be engineered to meet the requirements of specific applications such as high-frequency media for which metallic magnets cannot be used. For better comparison, the hysteresis loop of the composite prepared by a physical mixing method is also presented. As shown, the "bee waist" type loop appears because the physical mixing process results in a small contact area between the hard and soft phases.

In 2009, Roy and Kumar reported that an exchange-spring magnet consisting of Fe_3O_4 (the soft phase) and $BaCa_2Fe_{16}O_{27}$ (the hard phase) could be prepared by manipulating the volume percentage of the hard (H) and soft (S) phases, followed by heat treatment (Roy & Kumar, 2009). In this case, the volume percentage of the soft phase was varied, and the synthesized composites were divided into two batches. The first and second batches were heated at 400 and 800°C, respectively. The heat treatment time was held constant at 3 h. The X-ray diffraction (XRD) patterns (not shown here) confirmed the existence of both soft and hard ferrite peaks without impurity or a secondary phase.

The vibrating sample magnetometer (VSM) hysteresis loops of the nanocomposite are depicted in Fig. 5.11. The nanocomposite heated at 800°C shows single-phase magnetic behavior, revealing the presence of well-exchanged coupling between the S and H phases. However, the hysteresis loop of the nanocomposite heated at 400°C has a kink, indicating reduced exchange coupling between the two phases.

The Henkel plot of the nanocomposite heated at 800°C is shown in Fig. 5.12. The positive value of δM reflects the presence of exchange interactions between the Fe_3O_4 and $BaCa_2Fe_{16}O_{27}$ nanocomposite phases. The negative value of δM

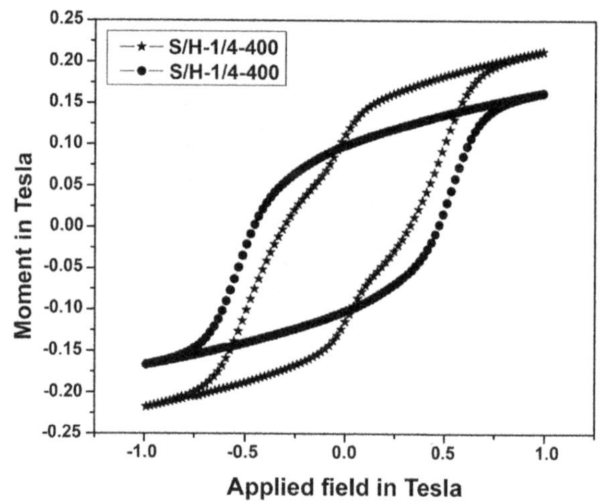

FIGURE 5.11

VSM graphs of the $Fe_3O_4/BaCa_2Fe_{16}O_{27}$ ferrite nanocomposite with the soft/hard ratio of 1/4 heated at 400 and 800°C.

From Roy, D., & Kumar, P. A. (2009). Enhancement of (BH)$_{max}$ in a hard-soft-ferrite nanocomposite using exchange spring mechanism. Journal of Applied Physics, 106*(7), 073902.*

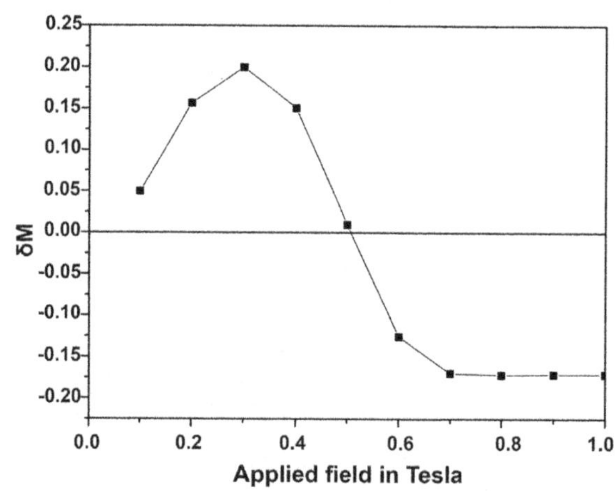

FIGURE 5.12

The Henkel plots of the $Fe_3O_4/BaCa_2Fe_{16}O_{27}$ ferrite nanocomposite with the soft/hard ratio of 1/4 heated at 800°C.

From Roy, D., & Kumar, P. A. (2009). Enhancement of (BH)$_{max}$ in a hard-soft-ferrite nanocomposite using exchange spring mechanism. Journal of Applied Physics, 106*(7), 073902.*

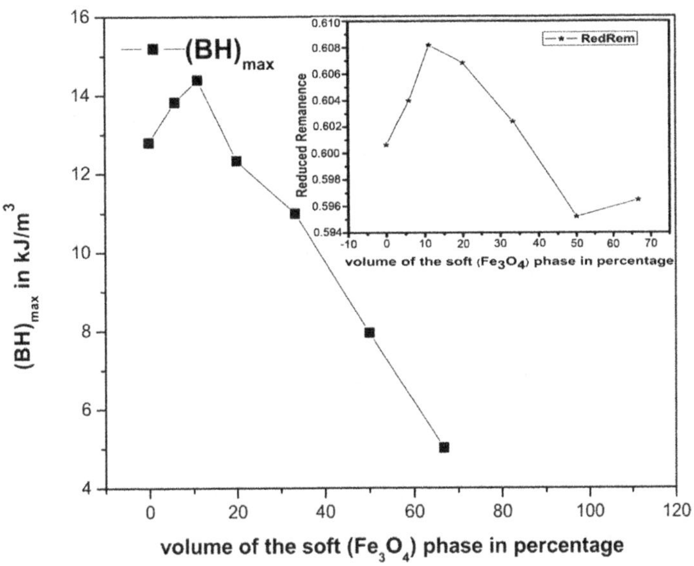

FIGURE 5.13

Variation of the maximum energy product with the soft phase content for the Fe_3O_4/$BaCa_2Fe_{16}O_{27}$ ferrite nanocomposite heated at 800°C.

From Roy, D., & Kumar, P. A. (2009). Enhancement of (BH)$_{max}$ in a hard-soft-ferrite nanocomposite using exchange spring mechanism. Journal of Applied Physics, 106*(7), 073902.*

correlates with dipolar interactions in the nanocomposite pertaining to the magnetization reversal.

From Fig. 5.13, the composite with the soft:hard ratio of 1:8 has the highest $(BH)_{max}$ = 14.31 kJ/m^3 (1.8 MGOe). The inset indicates an enhancement in the remanence when increasing the content of the soft phase. Further increasing the soft phase content sharply decreases the remanence, which might be attributed to dipolar interactions in the nanocomposite (Roy & Kumar, 2009).

Epitaxial magnetic heterostructures of Fe_3O_4/$CoFe_2O_4$ (001) have been grown using pulsed laser deposition (Lavorato et al., 2016). In this regard, soft ferrite Fe_3O_4 thicknesses were varied from 5 to 25 nm, whereas the cobalt ferrite thickness was kept constant at 25 nm. From magnetic data measured at 10 K and some supplementary evidence, magnetic interactions have been found to change from rigid-coupling to exchange-spring behavior. The type of coupling depends on the thickness of the soft phase. Remanence and coercivity have also increased for the thinnest Fe_3O_4 layer, indicating that the interface plays the dominant role in the overall magnetization process. For example, coercivity is increased from 13.1 to 16.5 kOe by changing the structure from cobalt ferrite to a bilayer with a soft magnetic thickness of 5 nm. Moreover, with an increase in the Fe_3O_4 thickness from 15 to 25 nm, the respective coercivity is decreased from 12.9 to 8.7 kOe, indicating that the switching field of the hard layer is lowered due to the exchange coupling with the soft layer. It must be stressed that some phenomena, including a nonstoichiometry increase in the antiphase boundaries (induced by a partial dislocation and symmetry mismatch), are activated by reducing the thickness (Eerenstein et al., 2002; Orna et al., 2010). Due to forming a high-quality interface between Fe_3O_4 and cobalt ferrite, the magnetization reversal process has been measured at a temperature of 10 K, and the results obtained are illustrated in Fig. 5.14. The maximum of the *M*(*H*) derivate confirms the existence of two switching events in the magnetization reversal process. The high-field maximum refers to the irreversible switching field of the cobalt ferrite, and the lowest reversal field is related to the soft phase nucleation field. Based on the exchange-spring mechanism, the mutual reversal of

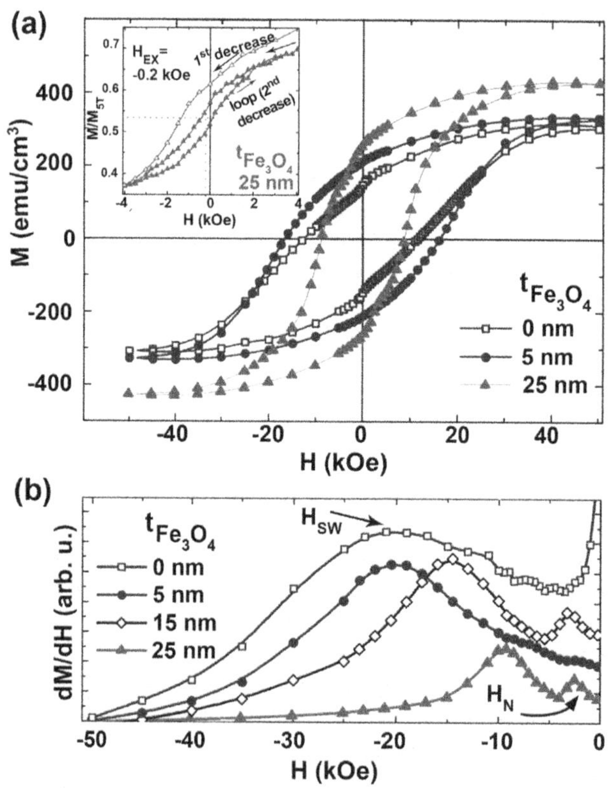

FIGURE 5.14

(A) In-plane hysteresis curves and (B) *dM*/*dH* curves of cobalt ferrite films and Fe_3O_4/cobalt ferrite bilayers for different Fe_3O_4 thicknesses measured at 10 K.

From Lavorato, G., Winkler, E., Rivas-Murias, B., & Rivadulla, F. (2016). Thickness dependence of exchange coupling in epitaxial $Fe_3O_4/CoFe_2O_4$ soft/hard magnetic bilayers. Physical Review B, *94(5), 054405.*

the hard and soft phases is controlled by the soft layer thickness and the exchange length. A bilayer with thicknesses smaller than 8 nm can act as a rigidly coupled magnet, and an exchange-spring magnet has been formed for thicker Fe_3O_4 layers.

5.5 Hexagonal ferrite–Ni-based spinel ferrite nanocomposites

Exchange-spring coupling between soft and hard phases depends on various parameters, including the distribution of phases in the composite and the individual phase grain sizes. Other parameters, such as the shape factor or relative orientations of crystallites, which are difficult to quantify and control, also play decisive roles.

A one-pot synthetic route has been employed for the preparation of $(NiFe_2O_4)_x$–$(BaFe_{12}O_{19})_{1-x}$ composites (Hazra et al., 2012), in which nickel ferrite and barium hexaferrite act as the soft and hard phases, respectively. For better comparison, a physical composite mixture has also been prepared. Fig. 5.15 shows the typical VSM loops of the composites prepared by two different methods. The composites prepared by the one-pot route exhibit a single-phase hysteresis loop, indicating that the hard and soft phases are well coupled with exchange interactions. Using the physical method, a "bee waist" type hysteresis loop appears, reflecting insufficient exchange coupling between the two phases.

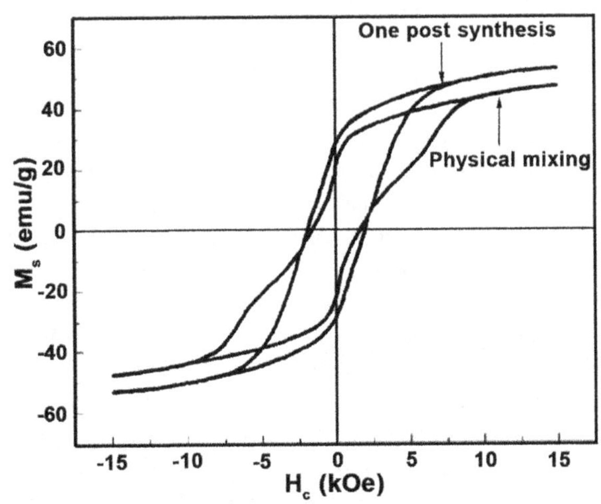

FIGURE 5.15

VSM curves for $(NiFe_2O_4)_{0.75}$–$(BaFe_{12}O_{19})_{0.25}$ composites.

From Hazra, S., Patra, M. K., Vadera, S. R., & Ghosh, N. N. (2012). A novel but simple "One-Pot" synthetic route for preparation of $(NiFe_2O_4)_x$–$(BaFe_{12}O_{19})_{1-x}$ composites. Journal of the American Ceramic Society, 95*(1), 60–63.*

FIGURE 5.16

Electron micrographs of $Ni_{0.5}Zn_{0.5}Fe_2O_4/BaFe_{12}O_{19}$ nanocomposite with ratio of 30/70 and sintering temperature of 1000°C.

From Xiong, R., Li, W., Fei, C., Liu, Y., & Shi, J. (2016). Exchange-spring behavior in $BaFe_{12}O_{19}$–$Ni_{0.5}Zn_{0.5}Fe_2O_4$ nanocomposites synthesized by a combustion method. Ceramics International, *42(10), 11913–11917.*

Elsewhere, an $Ni_{0.5}Zn_{0.5}Fe_2O_4$ (NZFO)/$BaFe_{12}O_{19}$ nanocomposite has been prepared by Xiong and his colleague (Xiong et al., 2016). The nanocomposites were sintered at various sintering temperatures. Fig. 5.16 shows typical electron micrographs of the nanocomposite with a weight ratio of 30:70 of NZFO to $BaFe_{12}O_{19}$ sintered at 1000°C. The average particle size has been determined to be about 100 nm. The presence of hard and soft magnetic phases has been confirmed by high-resolution transmission electron microscopy (HRTEM) micrographs. The phase boundary of soft and hard phases is evidenced as well.

From the VSM hysteresis loops depicted in Fig. 5.17, the whole series of nanocomposites annealed at 1000°C shows single magnetic phase behavior. The coercivity has an opposite trend as a function of the weight percentage of the soft phase associated with the contribution of dipolar interactions among the soft phase and the nucleation field of reverse domains of the soft phase.

Fig. 5.18 shows Henkel plots of the nanocomposites with the weight ratio of 30:70 sintered at 850 and 1000°C. Clearly, the sintering temperature strongly affects structural properties, such as particle size, particle shape, grain boundary type, and the distribution of magnetically hard and soft phases. In turn, such parameters have a significant role in the mechanism governing the magnetization reversal process. For samples sintered at 850°C, biphase magnetic behavior is observed in the hysteresis loops in conjunction with a negative δ_M value. On the contrary, at a low magnetic field, positive values of δ_M have been obtained for samples sintered at $T = 1000°C$.

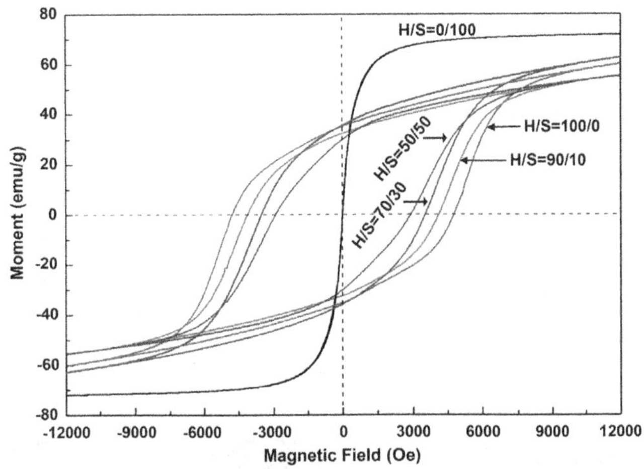

FIGURE 5.17

VSM graphs of $Ni_{0.5}Zn_{0.5}Fe_2O_4/BaFe_{12}O_{19}$ nanocomposite with different weight ratios sintered at temperature of 1000°C.

From Xiong, R., Li, W., Fei, C., Liu, Y., & Shi, J. (2016). Exchange-spring behavior in $BaFe_{12}O_{19}$–$Ni_{0.5}Zn_{0.5}Fe_2O_4$ nanocomposites synthesized by a combustion method. Ceramics International, *42(10), 11913–11917.*

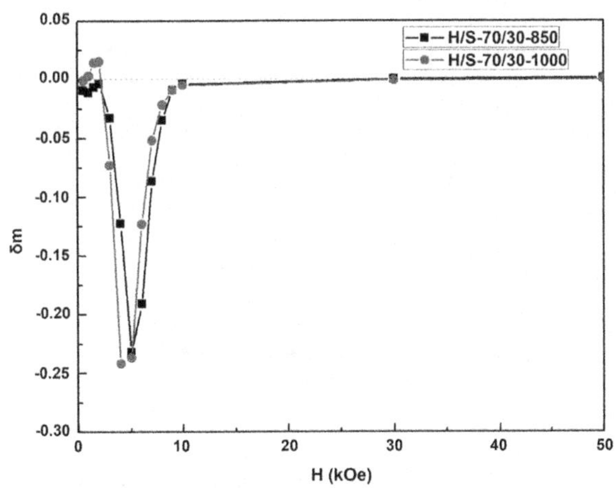

FIGURE 5.18

Henkel plots of $Ni_{0.5}Zn_{0.5}Fe_2O_4/BaFe_{12}O_{19}$ nanocomposite with the weight ratios of 30/70.

From Xiong, R., Li, W., Fei, C., Liu, Y., & Shi, J. (2016). Exchange-spring behavior in $BaFe_{12}O_{19}$–$Ni_{0.5}Zn_{0.5}Fe_2O_4$ nanocomposites synthesized by a combustion method. Ceramics International, *42(10), 11913–11917.*

In a separate work, microfibers of an NZFO/$SrFe_{12}O_{19}$ (SFO) composite have been prepared at various sintering temperature (Song et al., 2011). While microfibers calcined at $T < 850°C$ have shown the existence of α-Fe_2O_3 as an impurity phase, no secondary phases have been detected at $T \geq 900°C$. Fig. 5.19 shows scanning electron microscope (SEM) micrographs of the NZFO/SFO composite microfibers calcined at various temperatures. The surface morphology of the composite is composed of a combination of both hard and soft phases, as detected by EDS analysis. The diameters of the resulting microfibers are in the range of 1–2 μm. Noticeably, nanoparticles with a size of 100 nm have decorated the smooth outer surface of the NZFO microfibers. Moreover, the surface morphology of the strontium ferrite microfibers is rough, containing a uniform distribution of hexagonal platelike particles with submicron particle size.

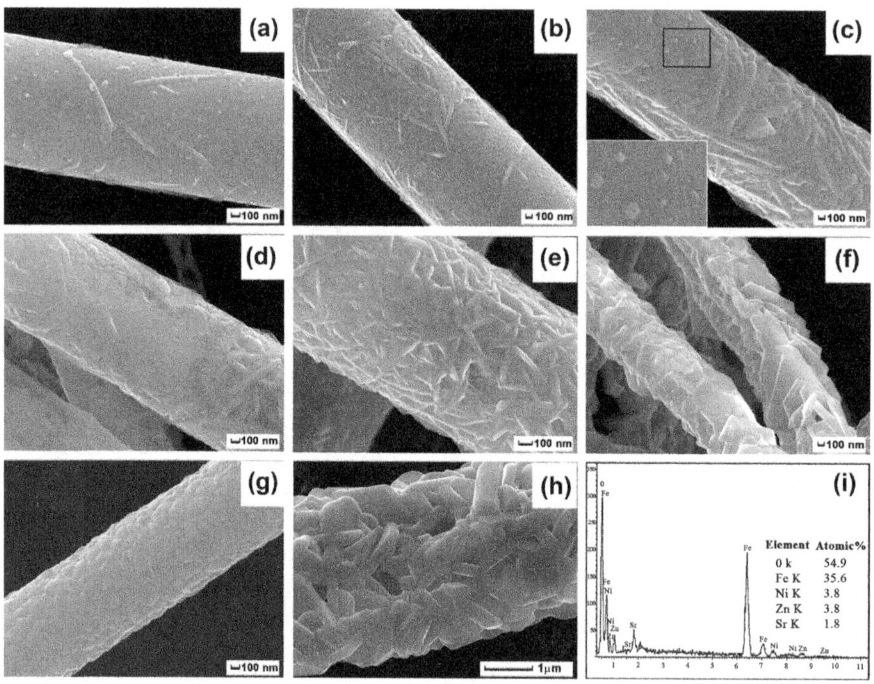

FIGURE 5.19

Scanning electron microscope micrographs of $Ni_{0.5}Zn_{0.5}Fe_2O_4/SrFe_{12}O_{19}$ composite microfibers calcined at (A) 750°C, (B) 850°C, (C) 900°C, (D) 950°C, (E) 1050°C, (F) 1150°C, and (G) $Ni_{0.5}Zn_{0.5}Fe_2O_4$, (H) $SrFe_{12}O_{19}$, (I) EDX spectrum of nanocomposite sintered at 900°C.

From Song, F., Shen, X., Liu, M., & Xiang, J. (2011). Preparation and magnetic properties of $SrFe_{12}O_{19}$/$Ni_{0.5}Zn_{0.5}Fe_2O_4$ nanocomposite ferrite microfibers via sol–gel process. Materials Chemistry and Physics, *126(3), 791–796.*

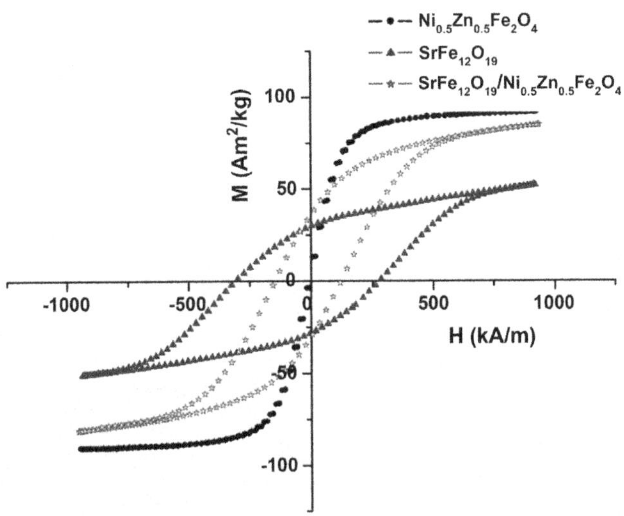

FIGURE 5.20

VSM graphs of $SrFe_{12}O_{19}$, $Ni_{0.5}Zn_{0.5}Fe_2O_4$ and $SrFe_{12}O_{19}/Ni_{0.5}Zn_{0.5}Fe_2O_4$ composite microfibers calcined at 900°C.

From Song, F., Shen, X., Liu, M., & Xiang, J. (2011). Preparation and magnetic properties of $SrFe_{12}O_{19}/Ni_{0.5}Zn_{0.5}Fe_2O_4$ nanocomposite ferrite microfibers via sol–gel process. Materials Chemistry and Physics, *126(3), 791–796.*

Evidently, the VSM hysteresis loops in Fig. 5.20 indicate single magnetic phase behavior for the composite microfibers calcined at 900°C. The coercivity and saturation magnetization of the soft phase and composites are 7.9 kA/m, 68.5 Am^2/kg, and 146.5 kA/m, 64.8 Am^2/kg, respectively. The remanent magnetization of composite microfibers is higher than that of the NZFO and SFO ferrites, which can be attributed to the exchange-spring behavior between the soft and hard ferrites.

Song and his colleagues have used HRTEM investigations to probe the phase analysis of $SrFe_{12}O_{19}$ (SFO)/$Ni_{0.5}Zn_{0.5}Fe_2O_4$ (NZFO) composite nanofibers (Song et al., 2012). They have shown that the particle sizes of soft and hard phases are about 18 and 42 nm, respectively. The nanoparticles are also distributed uniformly with coherent interface structures, reflecting the (224) plane of NZFO and (107) plane of SFO ferrite, according to the HRTEM micrographs in Fig. 5.21. The electron diffraction pattern demonstrates diffraction peaks reflected from the (022) and (224) planes of NZFO and the (107) and (406) planes of SFO.

Magnetic hysteresis loops of composite nanofibers with different weight ratios are shown in Fig. 5.22. Remanent magnetization and coercivity are enhanced by increasing the hard phase content and are maximized for a weight ratio of 20:80 of SFO to NZFO. Based on the theory discussed earlier, the respective hard and soft particle sizes should be higher than 40 nm and lower than 15 nm for effective

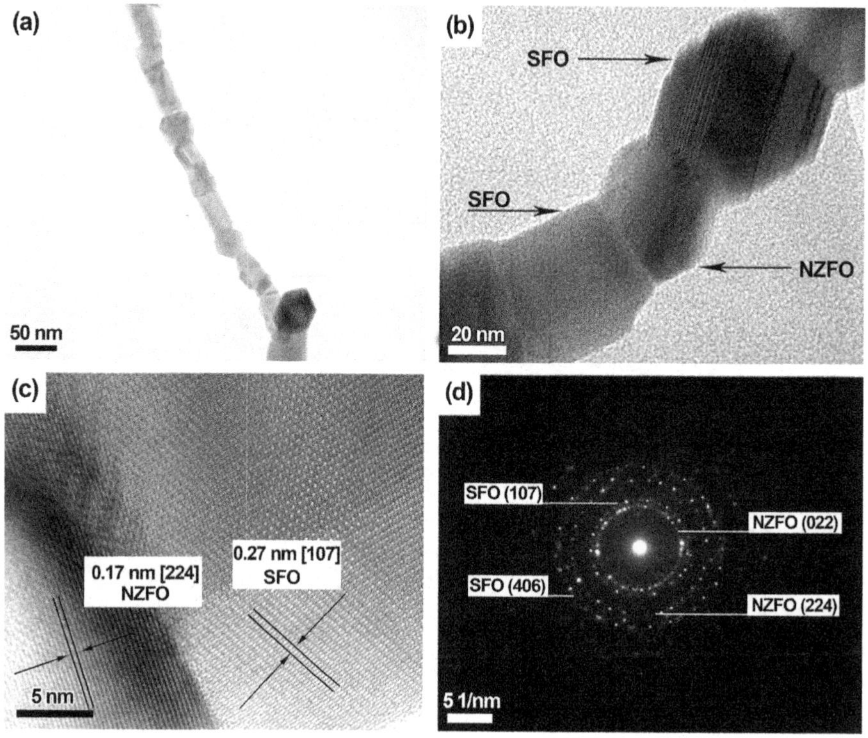

FIGURE 5.21

(A) Transmission electron microscopy, (B and C) high-resolution transmission electron microscopy micrographs, and (D) electron diffraction pattern of the $SrFe_{12}O_{19}$/$Ni_{0.5}Zn_{0.5}Fe_2O_4$ composite nanofibers with the weight ratios of 80/20 and sintering temperature of 900°C.

From Song, F., Shen, X., Liu, M., & Xiang, J. (2012). Microstructure, magnetic properties and exchange–coupling interactions for one-dimensional hard/soft ferrite nanofibers. Journal of Solid State Chemistry, 185, *31–36.*

exchange interactions. With an increase in the weight percentage of the hard phase, the exchange interaction increases, thereby improving remanent magnetization and coercivity.

$SrFe_{12}O_{19-x}$wt.%. $Ni_{0.6}Zn_{0.4}Fe_2O_4$ ($x = 10$, 20, and 30) composites have been prepared using an autocombustion method (Saeedi Afshar et al., 2018). SEM micrographs of the pure strontium hexaferrite and composites are shown in Fig. 5.23. As expected from the uniaxial alignment imposed in strontium ferrite, morphologies of pure SFO powders plate-shaped. Since Ni–Zn ferrite has a spinel structure, the composites have two distinct morphologies of spherical and platelet particles related

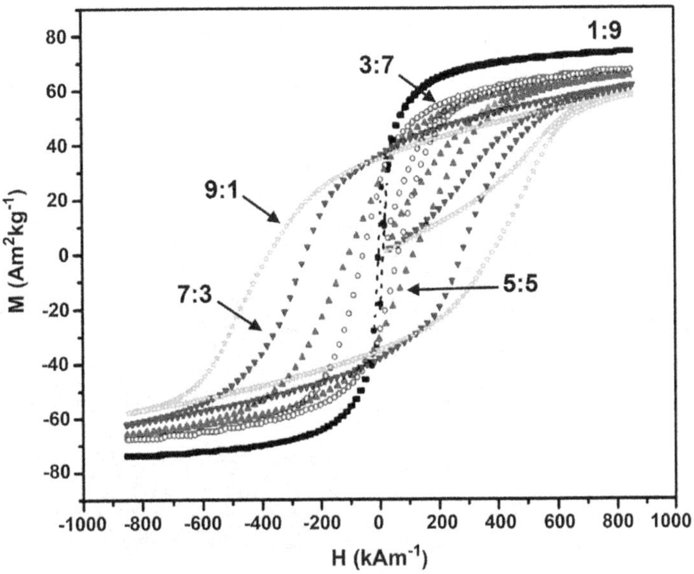

FIGURE 5.22

VSM graphs of the $SrFe_{12}O_{19}/Ni_{0.5}Zn_{0.5}Fe_2O_4$ composite nanofibers with the different weight ratios and sintered at temperature of 900°C.

From Song, F., Shen, X., Liu, M., & Xiang, J. (2012). Microstructure, magnetic properties and exchange–coupling interactions for one-dimensional hard/soft ferrite nanofibers. Journal of Solid State Chemistry, 185, *31–36.*

to spinel and hexagonal ferrites, respectively. The Ni–Zn ferrite phase heterogeneously nucleates on basal planes of hexagonal particles due to its similar crystal structure to S blocks of strontium hexaferrite and is absorbed on the surface.

According to the magnetization curves shown in Fig. 5.24, a hard phase with high coercivity and saturation magnetization is evidenced. The hard–soft nanocomposite with 10% soft phase is indicative of an 8600% enhancement in coercivity compared with the soft ferrite. In addition, it has higher saturation magnetization relative to the hard phase. The highest value for the M_r/M_s ratio has been obtained as 0.57 for a nanocomposite with 10 wt.% Ni–Zn ferrite. The competition between exchange and dipolar interactions also causes reduced coercivity. Single-phase hysteresis loops with smooth demagnetization curves prove the establishment of effective exchange coupling, resulting in successful magnetization switching of the available phases. If hard and soft phases are not strongly exchange-coupled with each other, the demagnetization curves will display superimposition of two loops corresponding to the hard and soft phases along with the presence of a noncollinear spin arrangement at the hard–soft interphase, leading to a reduction in saturation magnetization.

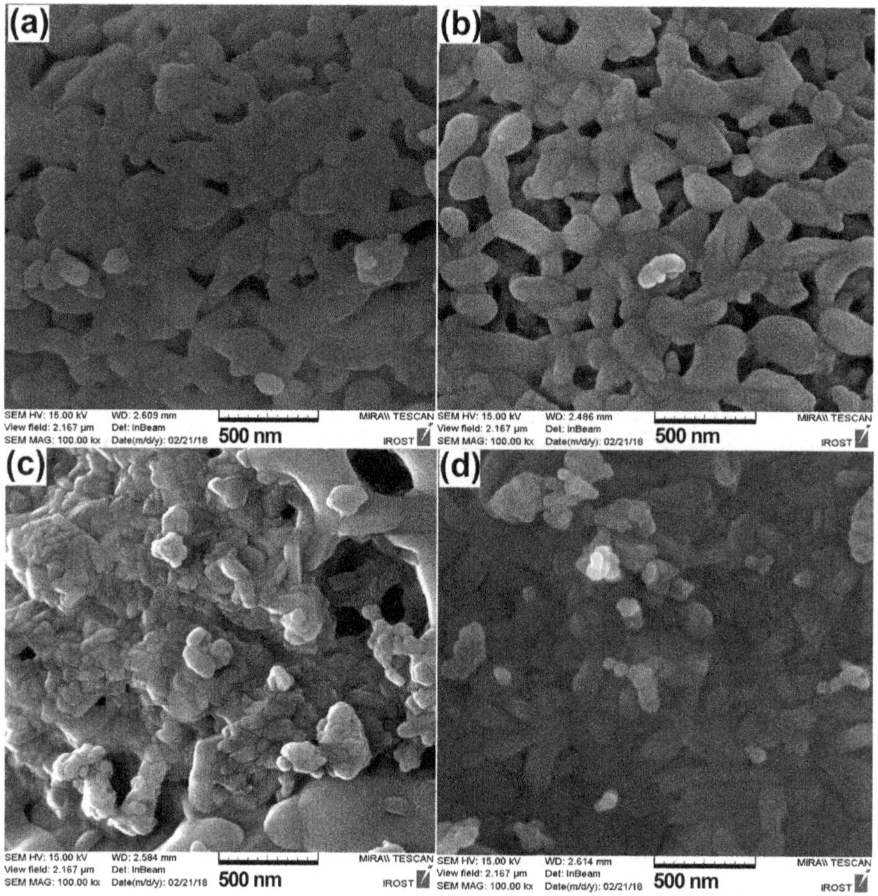

FIGURE 5.23

Scanning electron microscopy micrographs of the $SrFe_{12}O_{19-x}Ni_{0.6}Zn_{0.4}Fe_2O_4$ composites; (A) $SrFe_{12}O_{19}$, (B) $x = 10$ wt.%, (C) $x = 20$ wt.%, and (D) $x = 30$ wt.%.

From Saeedi Afshar, S. R., Hasheminiasari, M., & Masoudpanah, S. M. (2018). Structural, magnetic and microwave absorption properties of $SrFe_{12}O_{19}/Ni_{0.6}Zn_{0.4}Fe_2O_4$ composites prepared by one-pot solution combustion method. Journal of Magnetism and Magnetic Materials, 466, *1–6. https://doi.org/10.1016/j.jmmm.2018.06.061.*

5.6 Hexagonal ferrite–cobalt-based ferrite nanocomposites

It is well known that soft ferrites exhibit high saturation magnetization with low coercivity, whereas hard ferrites have high coercivity with moderate saturation magnetization. By providing an appropriate array of the soft phase in the hard magnetic matrix, one can achieve superior features with the ferrite composite

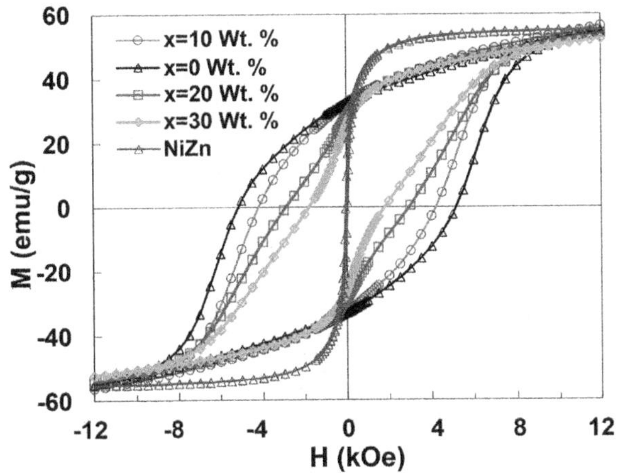

FIGURE 5.24

Magnetization curves of the $SrFe_{12}O_{19}$-x$Ni_{0.6}Zn_{0.4}Fe_2O_4$ composites.

From Saeedi Afshar, S. R., Hasheminiasari, M., & Masoudpanah, S. M. (2018). Structural, magnetic and microwave absorption properties of $SrFe_{12}O_{19}/Ni_{0.6}Zn_{0.4}Fe_2O_4$ composites prepared by one-pot solution combustion method. Journal of Magnetism and Magnetic Materials, 466, *1–6. https://doi.org/10.1016/j.jmmm.2018.06.061.*

compared with the individual hard and soft phases, thereby enhancing the magnetic energy product. In particular, the structural requirements associated with effective intergrain coupling, such as interfacial coherency and soft grain sizes on the order of a few nanometers, are often hard to meet in large-scale production methods. This is because maintaining the control of material structure in the nanoscale is difficult and challenging. To fabricate such a magnet, a high level of homogeneous hard and soft phase mixture is extremely essential. Various hard–soft composite systems have been studied thus far using various processing techniques, including rapid quench (for metallic bimagnets), mechanical alloying, one-pot chemical synthesis, spray pyrolysis, and thin film deposition. Nevertheless, the technical magnetic properties of the resultant products have not yet come close to their theoretical promise, arising from unresolved issues from uniformity, phase purity, and crystallographic alignment.

For example, the magnetic properties of an $SrFe_{10.5}O_{16.75}$ (SF)/$Co_{0.6}Zn_{0.4}Fe_2O_4$ (CZF) bimagnet have been studied (Poorbafrani et al., 2015). The hysteresis loops of the ferrite nanocomposites with different weight ratios of CZF:SF and sintering temperatures are shown in Figs. 5.25 and 5.26. The coercivity and saturation magnetization of pure strontium ferrite and cobalt zinc ferrite have been obtained as 5969 Oe and 66.2 emu/g and 75 Oe and 110 emu/g, respectively. Obviously, the magnetic characteristics of the nanocomposites abruptly change with an increase in the CZF concentration, corresponding to the alteration of exchange interactions between soft and hard magnetic phases. The CZF/SF nanocomposite with a weight

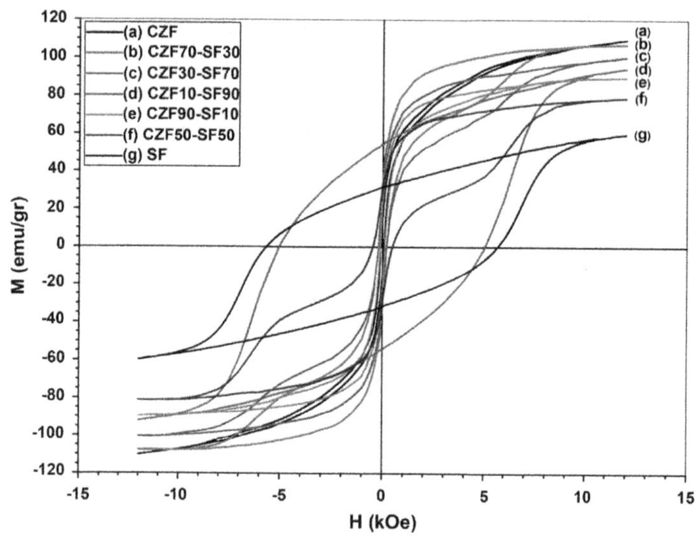

FIGURE 5.25

Hysteresis loops of $Co_{0.6}Zn_{0.4}Fe_2O_4/SrFe_{10.5}O_{16.75}$ nanocomposites with different weight ratios of CZF/SF and sintering temperature of 800°C.

From Poorbafrani, A., Salamati, H., & Kameli, P. (2015). Exchange spring behavior in $Co_{0.6}Zn_{0.4}Fe_2O_4$/ $SrFe_{10.5}O_{16.75}$ nanocomposites. Ceramics International, *41(1), 1603–1608.*

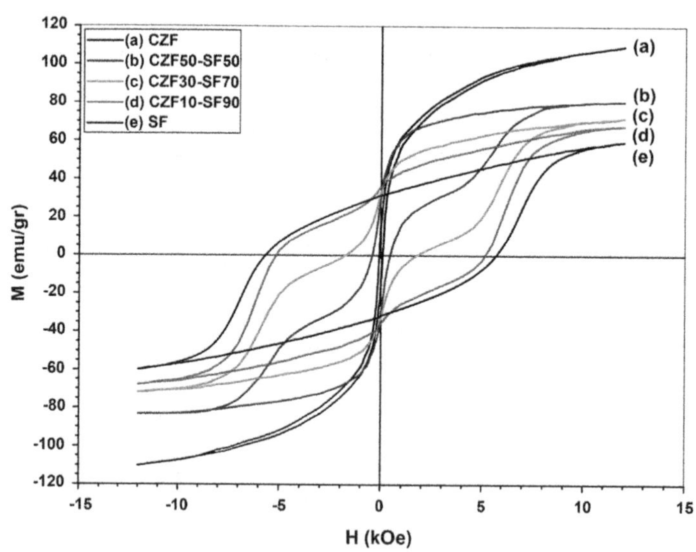

FIGURE 5.26

Hysteresis loops of $Co_{0.6}Zn_{0.4}Fe_2O_4/SrFe_{10.5}O_{16.75}$ nanocomposites with different weight ratios of CZF/SF and sintering temperature of 900°C.

From Poorbafrani, A., Salamati, H., & Kameli, P. (2015). Exchange spring behavior in $Co_{0.6}Zn_{0.4}Fe_2O_4$/ $SrFe_{10.5}O_{16.75}$ nanocomposites. Ceramics International, *41(1), 1603–1608.*

ratio of 10:90 and sintering temperature of 800°C contains a single magnetic phase, so the rest of the composite demonstrates an incomplete exchange mechanism, and kinks are observed in the magnetic hysteresis loops. The maximum energy products have not been measured in the aforementioned research.

The development of nanocomposites, including $SrAl_2Fe_{10}O_{19}$ ($SrAl_2M$) as the hard magnetic phase and $CoFe_2$ and Fe as the soft magnetic phase, have been carried out (Kahnes & Töpfer, 2019). In this respect, cobalt ferrite was used to synthesize FeCo-based alloys. $SrAl_2M$ particles and mixtures of $SrAl_2M/CoFe_2O_4$ and $SrAl_2M$/Fe were heat-treated in an Ar/5% H_2 gas atmosphere at 315°C. The processing parameters and developing trend of the composite preparation are illustrated in Fig. 5.27.

The results obtained from structural characteristics and the presence of kinks in hysteresis loops of hexaferrite/Fe($FeCo_2$) in Fig. 5.28 exhibit the absence of strong exchange coupling between the hard and soft magnetic phases. The authors have claimed that the whole outer surface of the hard phase has not been completely covered by the soft phase due to the agglomeration of particles. Moreover, impurities formed during the nucleation and growth steps could be an additional obstacle to obtaining completely efficient exchange-spring coupling. The hysteresis loops of $SrAl_2M$/Fe nanocomposites before and after heat treatment in Ar/5%H_2 atmosphere

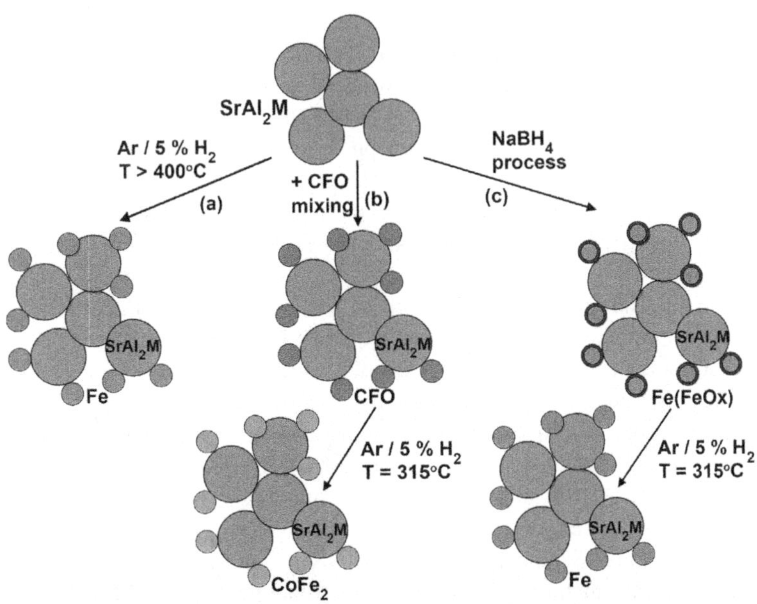

FIGURE 5.27

Scheme of approach for synthesizing composites.

From Kahnes, M., & Töpfer, J. (2019). Synthesis and magnetic properties of hard/soft $SrAl_2Fe_{10}O_{19}$/Fe ($FeCo_2$) nanocomposites. Journal of Magnetism and Magnetic Materials, 480, *40–46.*

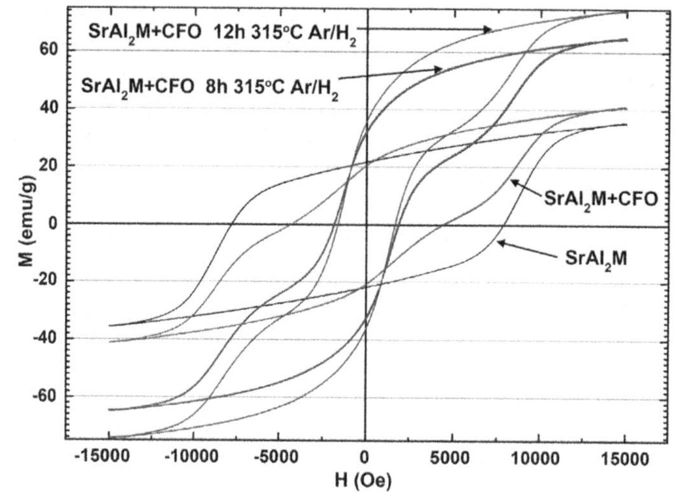

FIGURE 5.28

VSM graphs of $SrAl_2M$, $SrAl_2M$/CFO mixture, and composites sintered for 8 and 12 h at 315°C.

From Kahnes, M., & Töpfer, J. (2019). Synthesis and magnetic properties of hard/soft $SrAl_2Fe_{10}O_{19}$/Fe ($FeCo_2$) nanocomposites. Journal of Magnetism and Magnetic Materials, 480, *40–46.*

at 315°C are shown in Fig. 5.29. The VSM curve of $SrAl_2M$ was also included for comparison. The coercivity of nanocomposite is reduced while the saturation of magnetization increases compared to the pure $SrAl_2M$. It was found that the hysteresis loops of nanocomposites exhibit two-phase behavior, and no exchange coupling is completely established.

Controlling the effective processing parameters plays an important role in eliminating unexpected defects. Particles of hard and soft phases with narrow size distribution can satisfy the criteria for realizing a strong exchange mechanism. On the other hand, the quality enhancement of particle mixtures is another challenge to fully achieving exchange interactions between hard and soft phases.

The interface of hard and soft ferrites plays the dominant role in providing sufficient exchange coupling. Differences in lattice constant and crystal structure of phases may create partial bonding strength, leading to limited physical mixing (Fan et al., 2008; Palmstrom, 1995; Wang, 2000). My research group has focused on nanocomposites with suitable coherency and the contact area between hard and soft ferrites (Torkian, Ghasemi, ShojaRazavi, & Tavoosi, 2016; Torkian & Ghasemi, 2019). Notably, $SrFe_{10}Al_2O_{19}/Co_{0.8}Ni_{0.2}Fe_2O_4$ has been prepared by the sol–gel method and evaluated. Transmission electron microscopy (TEM) micrograph along with electron diffraction patterns of the nanocomposite (having a 15 wt.% soft phase) is depicted in Fig. 5.30. An almost uniform distribution of the soft ferrite with a mean particle size of 30 nm in the center of the hard particle with a rounded shape is observed. In fact, forming a uniform soft phase dot array in the heart of a

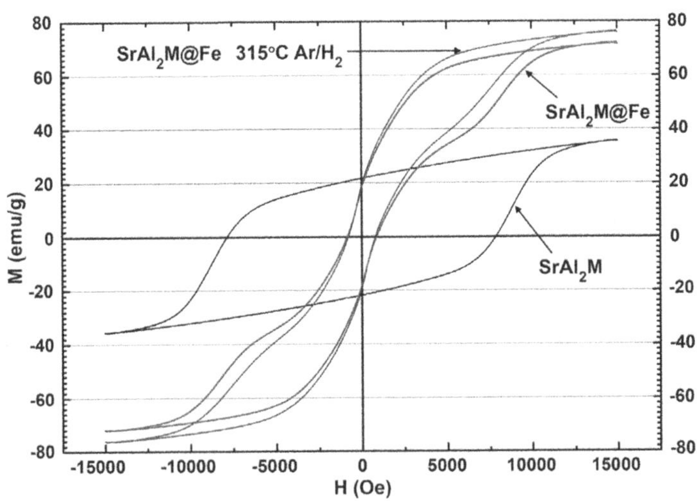

FIGURE 5.29

VSM graphs of $SrAl_2M$ and $SrAl_2M/Fe$ nanocomposite, after borohydride process, and treated for 12 h at 315°C.

From Kahnes, M., & Töpfer, J. (2019). Synthesis and magnetic properties of hard/soft $SrAl_2Fe_{10}O_{19}/Fe$ ($FeCo_2$) nanocomposites. Journal of Magnetism and Magnetic Materials, 480, *40–46.*

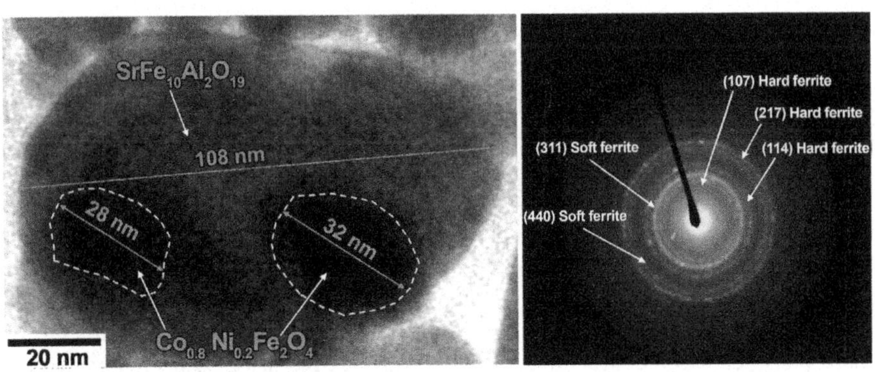

FIGURE 5.30

Transmission electron microscopy micrograph and electron diffraction pattern of (85 wt.%) $SrFe_{10}Al_2O_{19}$/(15 wt.%) $Co_{0.8}Ni_{0.2}Fe_2O_4$ nanocomposite.

From Torkian, S., Ghasemi, A., & Razavi, R. S. (2016). Magnetic properties of hard-soft $SrFe10Al_2O_{19}$/ $Co_{0.8}Ni_{0.2}Fe_2O_4$ ferrite synthesized by one-pot sol–gel auto-combustion. Journal of Magnetism and Magnetic Materials, 416, *408–416.*

hard magnet with controlled dot spacing is tough work. It has hardly been achieved in a particulate nanocomposite because of the random nature of the mixing process. The phase analyses of electron diffraction patterns have revealed the coexistence of strontium ferrite and cobalt nickel ferrite phases. Also, good crystallographic

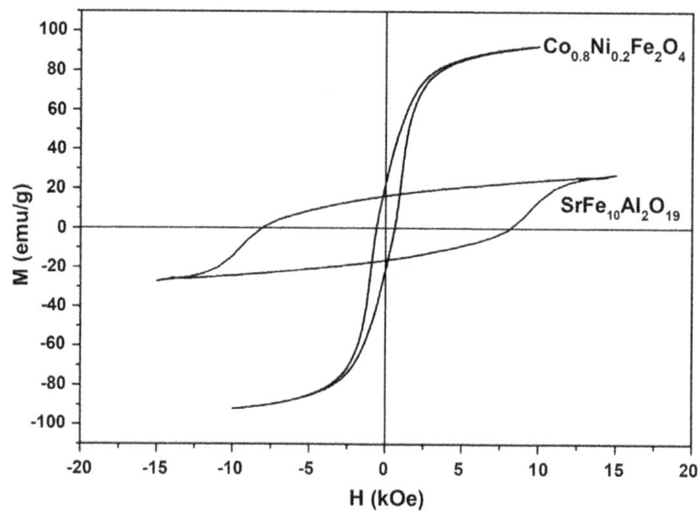

FIGURE 5.31

Hysteresis loops of hard and soft parts of nanocomposites.

From Torkian, S., Ghasemi, A., & Razavi, R. S. (2016). Magnetic properties of hard-soft $SrFe_{10}Al_2O_{19}$/$Co_{0.8}Ni_{0.2}Fe_2O_4$ ferrite synthesized by one-pot sol–gel auto-combustion. Journal of Magnetism and Magnetic Materials, 416, *408–416.*

matching has been found in the interface, essential for establishing a strong coupling mechanism (Coey, 2012).

Hysteresis loops of individual hard and soft ferrites are shown in Fig. 5.31. The saturation magnetization and coercivity of cobalt nickel ferrite are 92.7 emu/g and 600 Oe, respectively. The hard aluminum substituted strontium ferrite has respective coercivity and saturation magnetization of 8.1 kOe and 27.7 emu/g. It has been reported that the experimental coercivity of single-crystalline strontium ferrite is 6.7 kOe, whereas the theoretical coercivity value has been calculated to be 7.5 kOe (Pullar, 2012). The coercivity enhancement obtained in this report compared with other studies can be assigned to adding aluminum cations into the hard phase crystalline structure. The nanoparticle size has also decreased with an increase in the aluminum content, providing enhanced grain boundary density and limiting domain wall movement, which can cause the coercivity to increase (Torkian & Ghasemi, 2019; Torkian et al., 2016).

Nanocomposites with different weight percentages of the soft phase ranging from 5% to 90% have been synthesized by one-pot sol–gel and physical mixing methods (Torkian et al., 2016). The corresponding hysteresis loops of the resulting nanocomposites are shown in Fig. 5.32. The "bee waist" features formed in the hysteresis loops of the nanocomposites prepared by the physical mixing method confirm the presence of two phases rather than the formation of integrated nanostructures. This is because the interface between the soft and hard materials in the physical

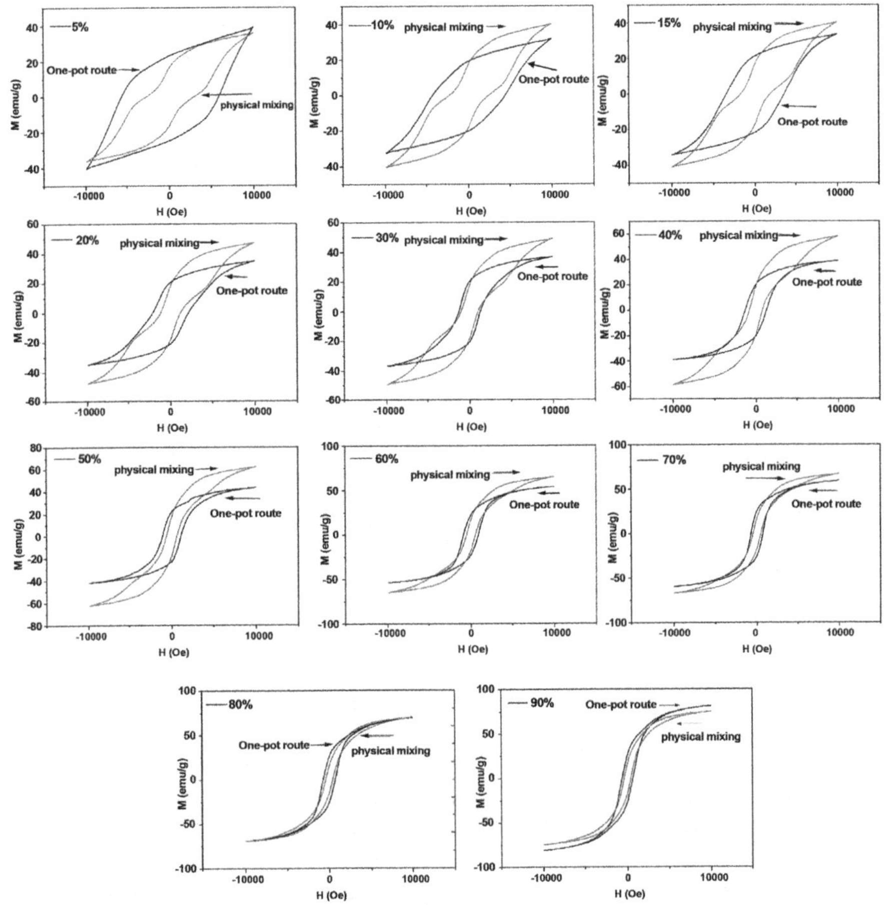

FIGURE 5.32

Hysteresis loops of $SrFe_{10}Al_2O_{19}/Co_{0.8}Ni_{0.2}Fe_2O_4$ nanocomposites prepared by one-pot sol–gel auto-combustion route and physical mixing method with various amount of soft phase.

From Torkian, S., Ghasemi, A., & Razavi, R. S. (2016). Magnetic properties of hard-soft $SrFe_{10}Al_2O_{19}/Co_{0.8}Ni_{0.2}Fe_2O_4$ ferrite synthesized by one-pot sol–gel auto-combustion. Journal of Magnetism and Magnetic Materials, 416, *408–416.*

mixing method is very irregular, avoiding the formation of strong-enough coupling. On the other hand, the sol–gel method can lead to the homogeneous mixing of hard and soft phases, causing the interdiffusion of cations in the interface to occur in the infinite regime as the mixing is carried out at the atomic scale. Consequently, the nanocomposites prepared by this method reflect single magnetic phase behavior.

The variation of δ_M versus the magnetic field depicted in Fig. 5.33 has been used to find the type of interactions in nanocomposites (Torkian & Ghasemi, 2019).

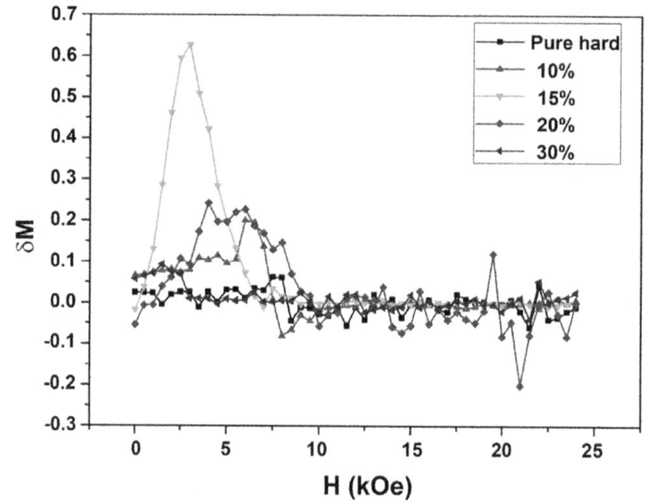

FIGURE 5.33

Variation of δ_M curve versus magnetic field of hard/soft nanocomposites.

From Torkian, S., & Ghasemi, A. (2019). Energy product enhancement in sufficiently exchange-coupled nanocomposite ferrites. Journal of Magnetism and Magnetic Materials, 469, *119–127.*

As mentioned previously, positive values of δM represent the existence of exchange interactions, whereas dipolar interactions are indicated by negative δM values (Shi et al., 2000; Wohlfarth, 1958). In this respect, positive δM values have been found for the whole series of nanocomposites when the magnetic fields were below 7.5 kOe. The nanocomposite with a 15% soft phase reflects the strongest positive peak, indicating fairly large interparticle coupling. Also, an increase of 20% in the soft phase gives rise to a mixture of dipolar and exchange interactions.

To find the optimum mixture content of the soft phase, it is required to measure the maximum energy products of the bulk sample by using a $B-H$ tracer. Accordingly, particles are cold-pressed to obtain a bulk ferrite. In fact, the optimum conditions for pressing and subsequent sintering processes make high-density nanocomposites with minimum porosity values.

Fig. 5.34 shows $B-H$ curves of the nanocomposite with various amounts of the soft phase. The $(BH)_{max}$ for hard strontium ferrite is obtained as 26.7 kJ/m^3. For the nanocomposite with a 15% soft phase, the maximum energy product increased to 29.5 kJ/m^3 (i.e., 10.5% enhancement compared with that of the hard phase), followed by a decrease with further increasing of the soft phase content. The results are consistent with the magnetic analysis obtained from the Henkel plot and magnetization curves. Importantly, controlling the grain size of hard and soft phases may allow optimal exchange coupling (López-Ortega et al., 2015). As indicated earlier, the critical size of the soft phase should be equal to or smaller than twice the domain wall thickness of the hard phase to achieve sufficient exchange coupling.

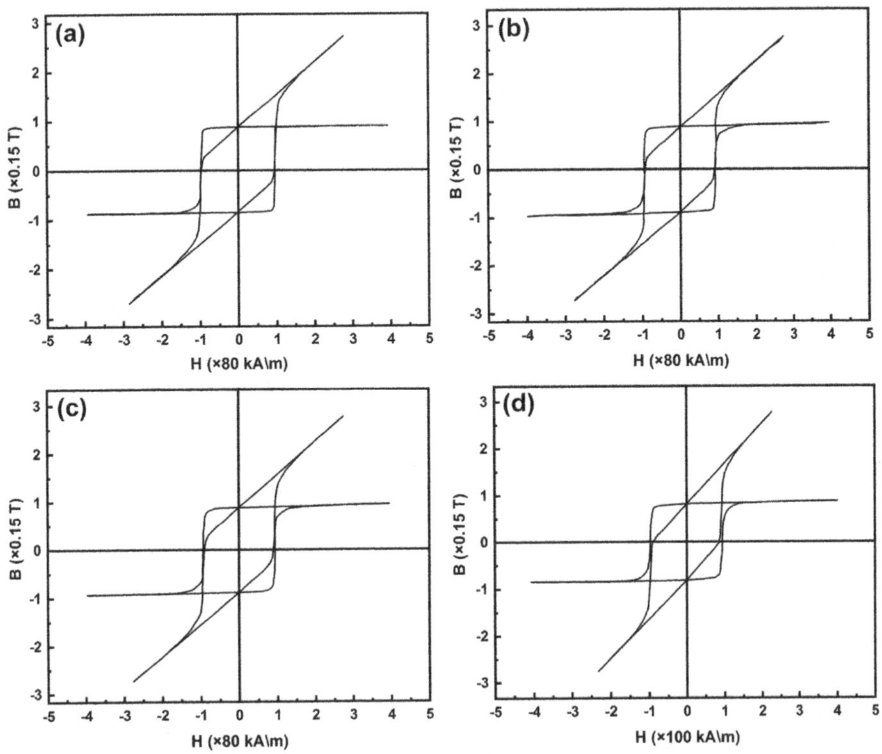

FIGURE 5.34

Hysteresis curves of hard/soft nanocomposites with soft phase content of (A) 10%, (B) 15%, (C) 20%, and (D) 30%.

From Torkian, S., & Ghasemi, A. (2019). Energy product enhancement in sufficiently exchange-coupled nanocomposite ferrites. Journal of Magnetism and Magnetic Materials, 469, *119–127.*

By considering $K_k \approx 2.38 \times 10^6$ erg/cm^3 and $A_m \approx 10^{-11}$ J/m in, the critical size of cobalt nickel ferrite is calculated to be 28 nm, consistent with the size obtained by electron microscopy micrographs. The enhancement in $(BH)_{max}$ of the nanocomposites confirms well-established exchange coupling. To the best of my knowledge, $(BH)_{max} = 29.5$ kJ/m^3 is the highest value of maximum energy product reported in the scientific literature for ferrite nanocomposites and is quite far from the values obtained from hard NdFeB magnets.

Hard-soft bulk magnets with a composition of $(100\%-x)SrFe_{12}O_{19}-_xCoFe_2O_4$, in which $x = 5$, 10, and 15 wt.%, have been prepared by spark plasma sintering (SPS) (Jenuš et al., 2016). It must be mentioned that cobalt ferrite has lower coercivity and higher saturation magnetization than strontium ferrite. Hysteresis loops of a 90% strontium ferrite–10% cobalt ferrite composite reflect a kink coupled with deteriorated magnetic characteristics before performing the SPS process. The processing parameters, including applied pressure, temperature, and sintering

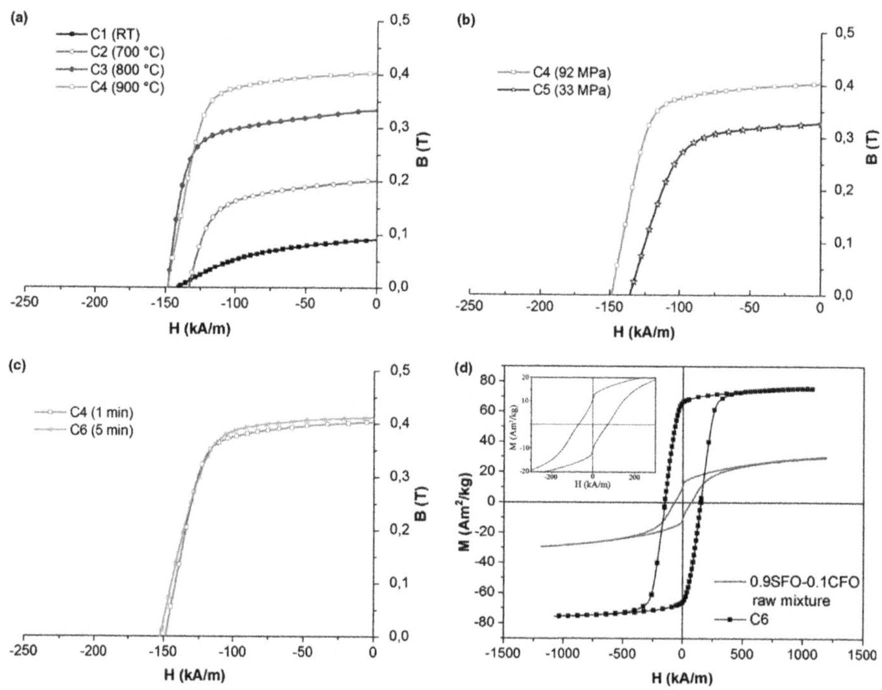

FIGURE 5.35

Demagnetizing curves of the strontium ferrite–cobalt ferrite composites, magnetic properties dependence on: (A) spark plasma sintering (SPS) temperature; (B) applied pressure; (C) time of sintering; (D) hysteresis loops of 90% strontium ferrite–10% cobalt ferrite composite before (raw mixture) and after SPS at 900°C for 5 min (C6).

From Jenuš, P., Topole, M., McGuiness, P., Granados-Miralles, C., Stingaciu, M., Christensen, M., Kobe, S., & Žužek Rožman, K. (2016). Ferrite-based exchange-coupled hard–soft magnets fabricated by spark plasma sintering. Journal of the American Ceramic Society, *99(6), 1927–1934.*

time, have been monitored to obtain a relatively high-density magnet (Fig. 5.35A–C). The highest relative density (93%) among the prepared composites is for the composite with a 10 wt.% cobalt ferrite phase sintered at 900°C for 5 min under an applied pressure of 92 MPa.

It is well established that the particle size of the hard phase must be maintained in the single domain state to reach maximum coercivity. In turn, this leads to an enhancement in the maximum energy product. In other words, while the particle size of strontium ferrite is allowed to be about 1 μm, the cobalt ferrite particle size should not exceed 28 nm, thereby archiving a high maximum energy product. Due to the low sintering temperatures and very short sintering times used in the SPS technique, it is possible to prevent the grains from growing in the structure. According to Fig. 5.35D, the coercivity remains high and is essential for keeping the successful exchange coupling between hard and soft magnets. The demagnetization curves reflect the achievement of a successful exchange coupling without forming a kink. As mentioned previously, the occurrence of a kink in the demagnetization curve indicates that the magnetization reversal is not realized for hard and soft phases under the same strength of an applied field. The maximum energy product of bulk magnets has improved from 21.9 to 26.1 kJ/m^3 for single-phase strontium ferrite and its composite, respectively.

5.7 Hard ferrite–soft iron cobalt core–shell nanocomposites

The core–shell structure can increase the mixture homogeneity and induce a strong interface contact. Core–shell composites could be used in many research areas such as biomedical applications, the mediators in magnetic hyperthermia (thus increasing the specific heating rate), next-generation magnetic recording media, sensing applications, permanent magnets, and microwave-absorbing media. The core–shell feature is also advantageous for controlling the nanoparticle structure during the processing. The core size is determined by controlling the effective processing parameters and chemical composition of core nanoparticles, whereas the shell thickness is monitored in the subsequent coating method or reduction–diffusion steps. Many literature studies have focused on developing bimagnetic nanoparticles containing a core–shell configuration (Song et al., 2011). One approach to achieving core–shell structures is a reduction–diffusion process. As an alternative approach, an oxidation–diffusion process can be considered, although its final products have no suitable properties compared with those of the reduction–diffusion mechanism. It is well known that cobalt ferrite is a moderately hard magnet, and iron cobalt has soft magnetic properties. The VSM hysteresis loops of both materials are shown in Fig. 5.36. The hysteresis loop of FeCo shows low coercivity, whereas the cobalt ferrite nanoparticles have rather high coercivity. By reducing cobalt ferrite in a

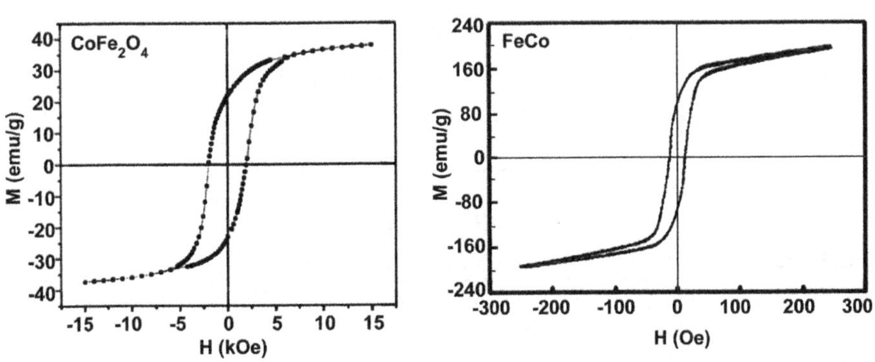

FIGURE 5.36

VSM graphs of hard cobalt ferrite and soft iron cobalt.

hydrogen atmosphere, one can induce the evolution of iron cobalt. On the other hand, under controlled oxidation of iron cobalt, it is feasible to achieve hard cobalt ferrite.

In separate research carried out by my group, exchange-spring magnets consisting of cobalt ferrite–iron cobalt ($CoFe_2O_4/CoFe_2$) were synthesized by a reduction–diffusion process (Safi et al., 2016, 2017). To this end, cobalt ferrite nanoparticles were first synthesized by a coprecipitation technique and then reduced in a pure hydrogen atmosphere using controlled heating and cooling programs. The reduction temperature also varied from 450 to 700°C in a step of 50°C. The heating time at the maximum temperature was kept constant for 10 min.

Fig. 5.37 depicts XRD patterns of the reduced nanocomposites. As can be seen, the nanocomposite sample annealed at 450 and 500°C contains no soft FeCo,

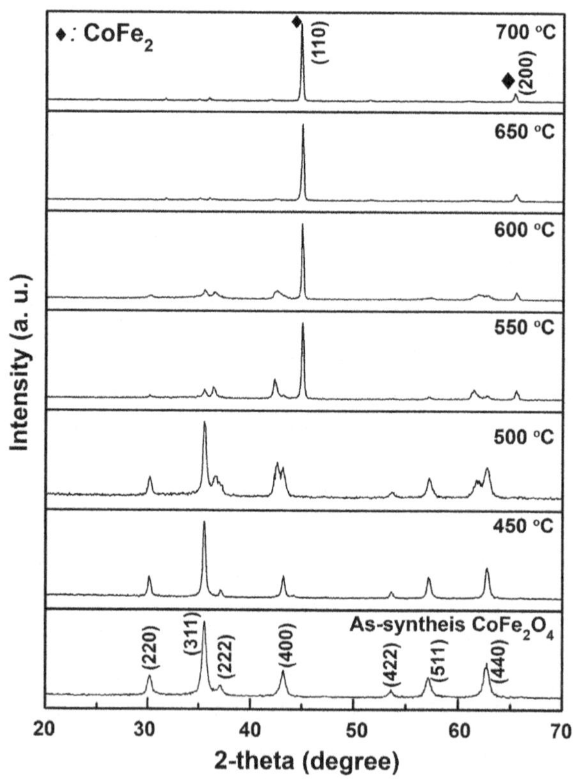

FIGURE 5.37

X-ray diffraction patterns of $CoFe_2O_4/CoFe_2$ nanocomposites annealed at reducing atmosphere at different temperature.

From Safi, R., Ghasemi, A., Shoja-Razavi, R., & Tavoosi, M. (2016). Development of novel exchange spring magnet by employing nanocomposites of $CoFe_2O_4$ and $CoFe_2$. Journal of Magnetism and Magnetic Materials, 419, *92–97.*

indicating that the reduction is not carried out in the cobalt ferrite. However, an intermediate FeO monoxide with an fcc structure is formed during the first step of the reduction process, transforming cobalt ferrite into metallic FeCo with a bcc structure. The evolution of the metallic FeCo phase is also indicated in the XRD pattern of the nanocomposite annealed at 550°C, confirming the coexistence of $CoFe_2O_4$–$CoFe_2$–FeO. Higher annealing temperatures (600, 650, and 700°C) result in the complete elimination of FeO, enhancing the weight percentage of the $CoFe_2$ phase in the nanocomposite.

Magnetic analysis of pure cobalt ferrite has demonstrated coercivity and saturation magnetization of 2.02 kOe and 40.3 emu/g, respectively. The fairly large value of the coercivity is attributed to the small particle size of ferrite originating from the coprecipitation method. By optimizing the processing parameter in this method, spherical nanoparticles with uniform size distribution have been achieved. The collective behavior shows that most nanoparticles are below the critical single domain size, preventing the domain wall from easy movement and consequently increasing the coercivity (Safi et al., 2016). Fig. 5.38 shows a smooth single-phase hysteresis loop, confirming that the two magnetic phases are well exchange-coupled to each other. The soft metallic phase content increases and saturation magnetization is enhanced with increasing reduction temperature. Nevertheless, the coercivity decreases, and the magnetic nanocomposite nature turns from a magnetically hard ferrite to a soft nanocomposite.

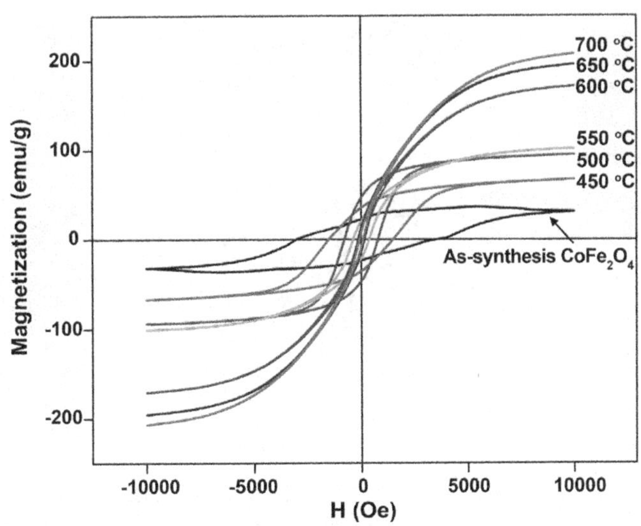

FIGURE 5.38

Hysteresis loops of $CoFe_2O_4/CoFe_2$ nanocomposites reduced at various temperatures.

From Safi, R., Ghasemi, A., Shoja-Razavi, R., & Tavoosi, M. (2016). Development of novel exchange spring magnet by employing nanocomposites of $CoFe_2O_4$ and $CoFe_2$. Journal of Magnetism and Magnetic Materials, 419, *92–97.*

FeCo can be formed on the outer surface of cobalt ferrite nanoparticles to fabricate a core–shell structure. To control the shell thickness, the reduction–diffusion parameters including hydrogen flux, temperature, and diffusion time must be precisely monitored. In this regard, $CoFe_2O_4$/CoFe core–shell structures with tunable shell thickness have been fabricated after being annealed at 550°C for different annealing times ranging from 10 to 30 min. TEM micrograph and related electron diffraction patterns of the core–shell sample annealed for 20 min are shown in Fig. 5.39. The core–shell configuration with a core diameter of 24 nm and a shell thickness of 8 nm is evidenced by the electron micrographs. The diffraction patterns show that the $CoFe_2O_4$ and FeCo coexist in the nanocomposite. The *d*-spacings of 0.48 and 0.28 nm in lattice fringes are related to $CoFe_2O_4$ (111) and FeCo (100) planes, respectively (Nlebedim et al., 2013). The results obtained are well consistent with the XRD patterns.

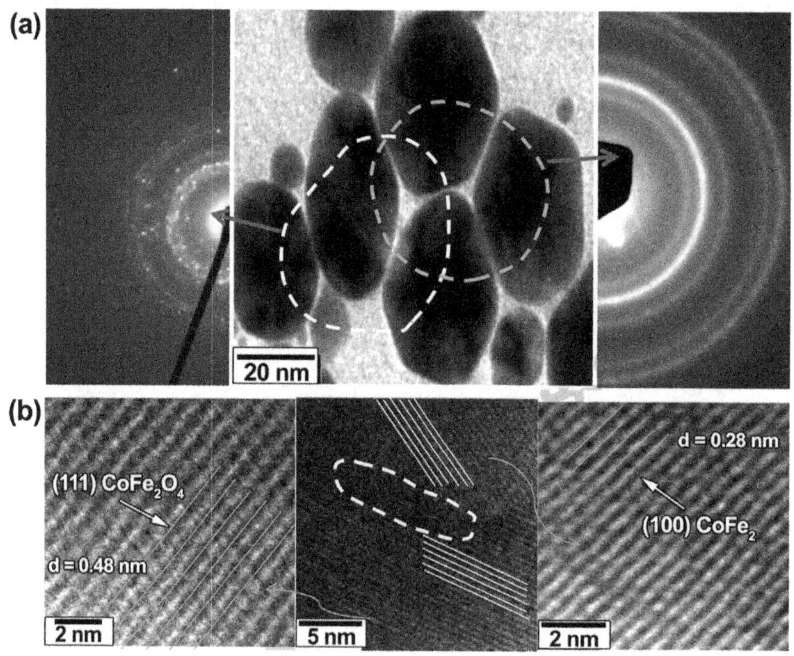

FIGURE 5.39

(A and B) High-resolution transmission electron microscopy micrograph and electron diffraction pattern of $CoFe_2O_4$/FeCo core–shell nanoparticles prepared at 550°C for 20 min along with the lattice fringes of cobalt ferrite, iron cobalt, and the interface of nanocomposite.

From Safi, R., Ghasemi, A., & Shoja-Razavi, R. (2017). The role of shell thickness on the exchange spring mechanism of cobalt ferrite/iron cobalt magnetic nanocomposites. Ceramics International, 43*(1), 617–624. https://doi.org/10.1016/j.ceramint.2016.09.203.*

The FeCo shell thickness measured from TEM micrographs of individual nanoparticles in Fig. 5.40 is found to vary from 1 to 12 nm when reduction time is increased.

By making bulk samples, $(BH)_{max}$ has been determined for the synthesized nanocomposites. In this case, high-density ferrites in cylindrical shape were prepared by cold pressing and sintering. Fig. 5.41 shows B–H curves of pure cobalt ferrite and $CoFe_2O_4$/FeCo core–shell nanocomposites fabricated at 550°C for annealing times of 15, 20, 25, and 30 min. The maximum energy product initially increases and then decreases with an increase in the annealing time (see Table 5.1).

While the maximum energy product of pure cobalt ferrite is found to be 19.5 kJ/m^3, the core–shell nanocomposite reduced for 20 min has $(BH)_{max} = 29.4$ kJ/m^3, exhibiting 50% enhancement. These results show that the synthesized nanocomposite samples have superior magnetic properties to those fabricated by Fukamachi et al. (25 kJ/m^3) (Fukamachi, 2010) and Cabral et al. (5 kJ/m^3) (Cabral et al., 2008). Strong exchange coupling between the hard and soft phases of the core–shell structure (resulting from the precise control of effective processing parameters in the reduction treatment) and the initial growth of cobalt ferrite have caused the remanence and coercivity to be higher than those reported by other researchers (Araújo et al., 2011; Cabral et al., 2008; Fukamachi, 2010).

Soares and his colleagues have grown iron cobalt on the surface of cobalt ferrite nanoparticles by a reduction–diffusion process at 280°C for annealing times of 30, 60, 180, 200, and 240 min (Soares et al., 2013). Careful investigation into the critical size of the shell thickness has been carried out, achieving a strong exchange mechanism between hard and soft phases. Fig. 5.42 shows typical micrographs of the sample annealed for 200 min. It displays the formation of core–shell composite structure. The optimum shell thickness has been calculated based on the theoretical model of the exchange-spring magnet and compared with other obtained results. It has been found that iron cobalt with a thickness of 8 nm can enhance the remanence and maximum energy product.

The reduction–diffusion process of cobalt ferrite has been carried out by employing activated charcoal in an inert atmosphere at elevated temperature based on the following chemical reaction, thereby developing $CoFe_2O_4$/FeCo core–shell nanocomposites (Leite et al., 2012):

$$CoFe_2O_4 + 2C \rightarrow CoFe_2 + 2CO_2 \tag{5.9}$$

The hysteresis loop analysis depicted in Fig. 5.43 indicates that the hard and soft magnetic phases can switch mutually, confirming the presence of an effective exchange-spring mechanism. The resulting maximum energy product is 115% higher than that of cobalt ferrite.

$CoFe_2O_4$/$CoFe_2$ nanofibers have also been obtained by reducing pure cobalt ferrite nanofibers prepared using a sol–gel-assisted electrospinning process (Xiang et al., 2013). The electron micrographs of the pure cobalt ferrite and core–shell nanofibers partially reduced at 300°C and fully reduced at 400°C are shown in Fig. 5.44. As observed, $CoFe_2$ nanoparticles are aligned along their long axes.

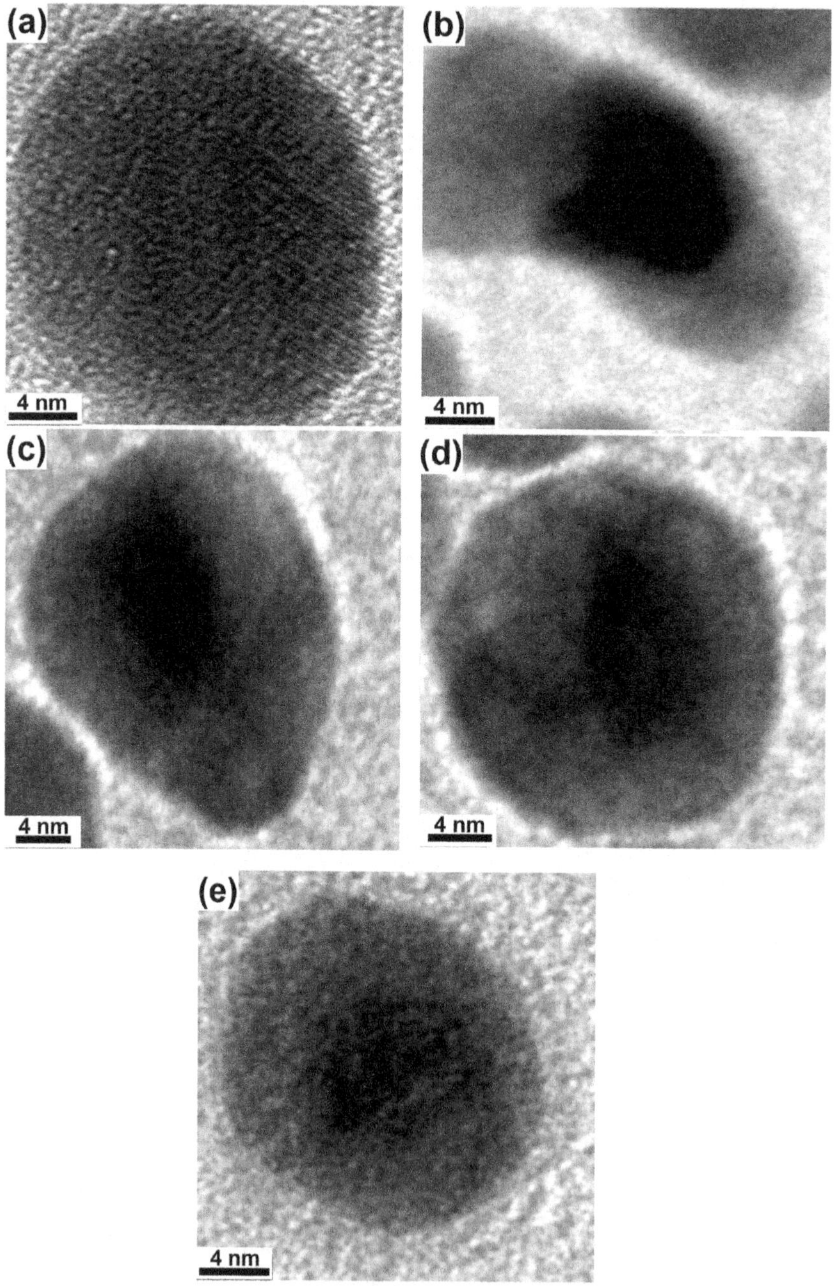

FIGURE 5.40

High-resolution transmission electron microscopy micrographs of single particle for nanocomposites reduced at 550°C for (A) 10 min, (B) 15 min, (C) 20 min, (D) 25 min, and (E) 30 min.

From Safi, R., Ghasemi, A., & Shoja-Razavi, R. (2017). The role of shell thickness on the exchange spring mechanism of cobalt ferrite/iron cobalt magnetic nanocomposites. Ceramics International, 43*(1), 617–624. https://doi.org/10.1016/j.ceramint.2016.09.203.*

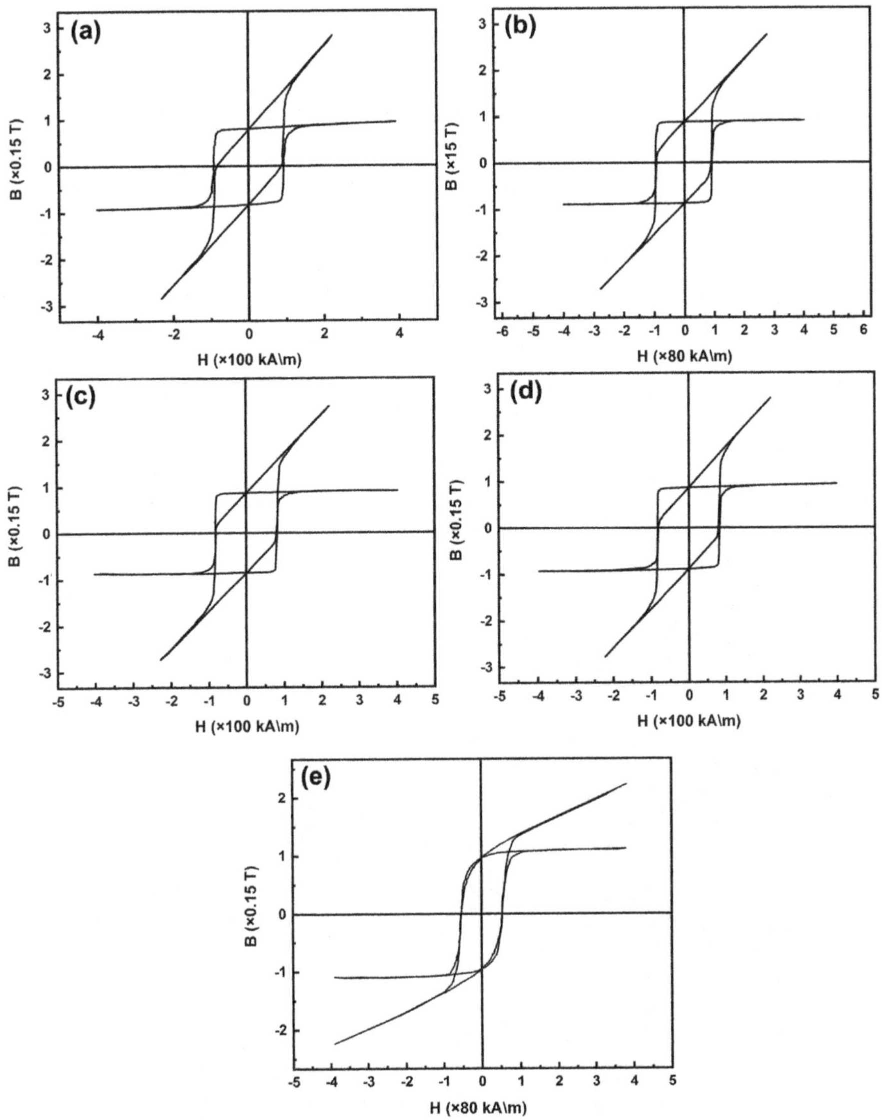

FIGURE 5.41

The *B–H* curve of (A) pure $CoFe_2O_4$ and nanocomposites prepared by reduction of cobalt ferrite at 550°C for various periods of time. (B) 15 min, (C) 20 min, (D) 25 min, and (E) 30 min.

From Safi, R., Ghasemi, A., & Shoja-Razavi, R. (2017). The role of shell thickness on the exchange spring mechanism of cobalt ferrite/iron cobalt magnetic nanocomposites. Ceramics International, 43*(1), 617–624. https://doi.org/10.1016/j.ceramint.2016.09.203.*

Table 5.1 Magnetic data of sample annealed at 550°C for different annealing time.

Reduction time (min)	M_s (emu/g)	M_r (emu/g)	M_r/M_s	H_c (Oe)	$(BH)_{max}$ (kJ/m^3)
0	52.7	26.5	0.5	2029	19.5
10	81.25	40.97	0.49	990	–
15	87.13	43.63	0.50	960	24.7
20	94.84	47.92	0.51	920	29.4
25	108.13	39.85	0.368	915	27.2
30	120.11	43.1	0.359	910	25.1

From Safi, R., Ghasemi, A., & Shoja-Razavi, R. (2017). The role of shell thickness on the exchange-spring mechanism of cobalt ferrite/iron cobalt magnetic nanocomposites. Ceramics International, *43(1), 617–624. https://doi.org/10.1016/j.ceramint.2016.09.203.*

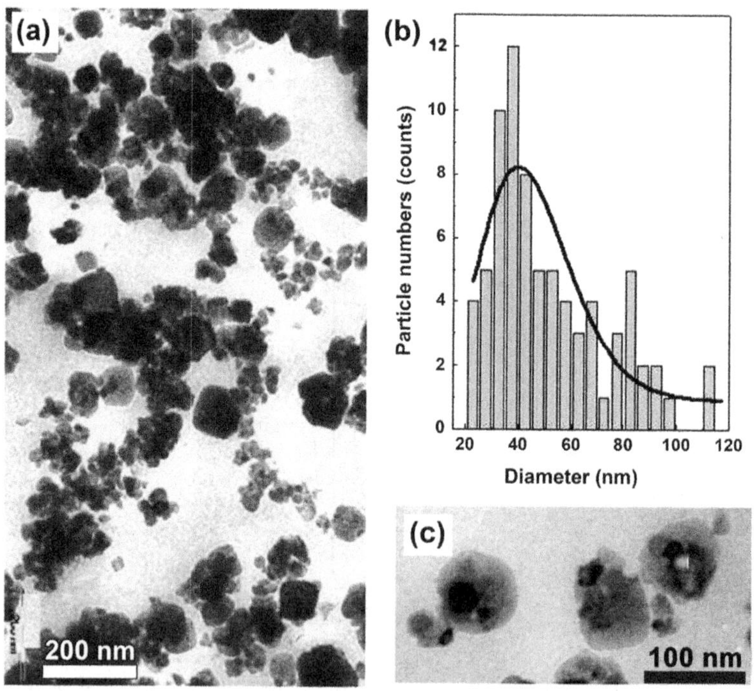

FIGURE 5.42

Transmission electron microscopy micrograph of typical composite annealed for 200 min along with particle size histogram and core–shell structure.

From Soares, J., Galdino, V., Conceição, O., Morales, M., De Araújo, J., & Machado, F. (2013). Critical dimension for magnetic exchange-spring coupled core/shell $CoFe_2O_4/CoFe_2$ nanoparticles. Journal of Magnetism and Magnetic Materials, 326, *81–84.*

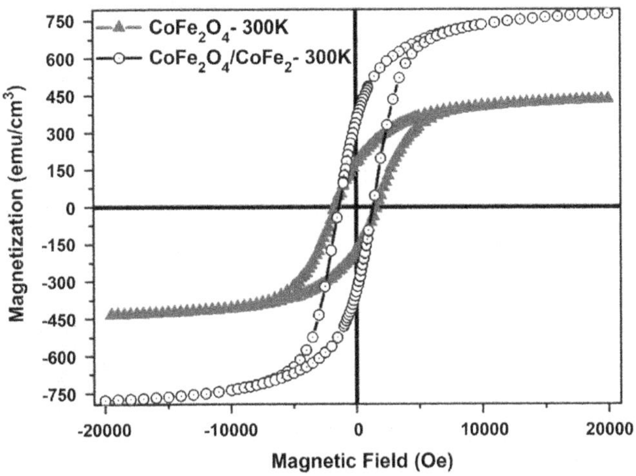

FIGURE 5.43

VSM curves of cobalt ferrite and core–shell structure of cobalt ferrite/FeCo.

From Leite, G. C., Chagas, E. F., Pereira, R., Prado, R. J., Terezo, A. J., Alzamora, M., & Baggio-Saitovitch, E. (2012). Exchange coupling behavior in bimagnetic $CoFe_2O_4/CoFe_2$ nanocomposite. Journal of Magnetism and Magnetic Materials, 324*(18), 2711–2716.*

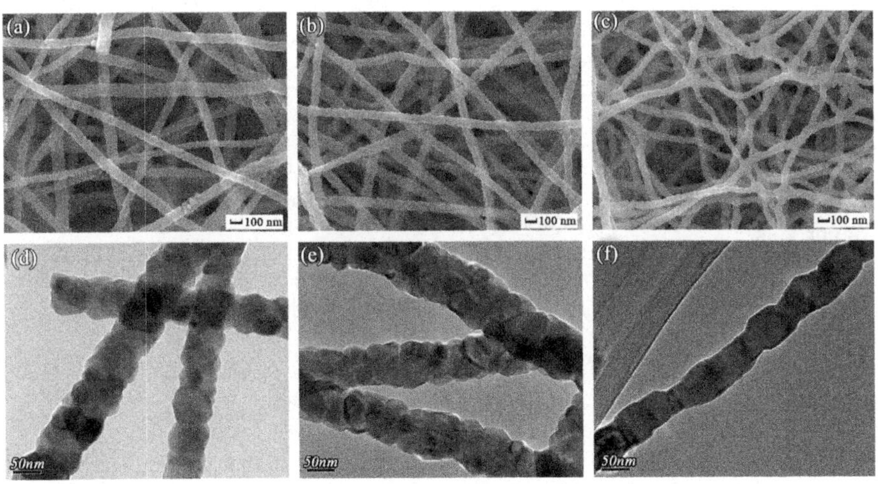

FIGURE 5.44

(A–C) Scanning electron micrographs and (D–F) transmission electron microscopy images of (A and D) cobalt ferrite nanofibers, (B and E) partially reduced sample at 300°C with an iron cobalt content of 54 wt.%, and (C and F) completely reduced sample at 400°C.

From Xiang, J., Zhang, X., Li, J., Chu, Y., & Shen, X. (2013). Fabrication, characterization, exchange coupling and magnetic behavior of $CoFe_2O_4/CoFe_2$ nanocomposite nanofibers. Chemical Physics Letters, 576, *39–43.*

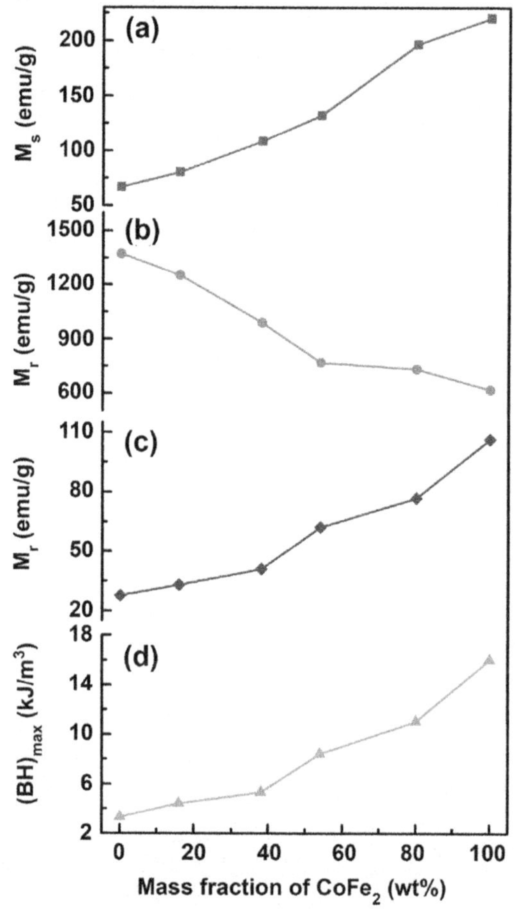

FIGURE 5.45

Magnetic parameters including (A) specific saturation magnetization, (B) coercivity, (C) remanent magnetization, and (D) maximum energy product as a function of mass fraction of iron cobalt.

From Xiang, J., Zhang, X., Li, J., Chu, Y., & Shen, X. (2013). Fabrication, characterization, exchange coupling and magnetic behavior of $CoFe_2O_4/CoFe_2$ nanocomposite nanofibers. Chemical Physics Letters, 576, *39–43.*

Magnetic data obtained from hysteresis loops indicate that the magnetization of both magnetic phases reverses mutually. From Fig. 5.45, the saturation magnetization increases when increasing the soft phase content, which is consistent with the expectation of high saturation magnetization of iron cobalt. Concurrently, coercivity has been found to decrease monotonously. Since the nanofiber sample has a large aspect ratio due to its small diameter, higher shape anisotropy and coercivity are obtained compared with cobalt ferrite nanoparticles. Remanence and maximum energy product also increase from 27.6 emu/g and 3.3 kJ/m^3 to 106.4 emu/g and

16.0 kJ/m^3, respectively. The agglomerated nanoparticles aligned along the long axis of nanofiber may result in enhanced exchange interactions between nanoparticles, thereby increasing the remanence.

A wet chemical method has been employed to decorate the surface of strontium ferrite particles with FeCo nanoparticles (Xu et al., 2015). As proved, the dimension of the soft phase needs to be smaller than twice the domain wall thickness of the hard phase (being less than 28 nm in the case of strontium ferrite) to develop an effective exchange-coupled magnet (Gregg et al., 2001). The schematic diagram shown in Fig. 5.46 explains the formation mechanism of FeCo on the strontium ferrite surface. In this way, FeCo nanoparticles can be magnetically attracted to SrM particles and form a core–shell structure according to the TEM micrograph. It should be noted that the surface of strontium ferrite has not been fully covered by FeCo, likely arising from an uneven magnetic field distribution.

Hysteresis loops of strontium ferrite and a core–shell structure with soft phase thicknesses of 15 and 7 nm are shown in Fig. 5.47. While the coercive field of the core–shell composite shows a decrease from 5.75 to 3.81 kOe compared with the strontium ferrite particles, the saturation magnetization indicates an increase from 65.3 to 80.5 emu/g. Based on the following equation (Kneller & Hawig, 1991; Park, 2016),

$$\sigma_{ex} = \sigma_s \chi + \sigma_h (1 - \chi) \tag{5.10}$$

where σ_{ex} is the magnetization of the exchange-coupled core–shell composite, σ_s and σ_h are the respective saturation magnetization of soft and hard phases, and χ

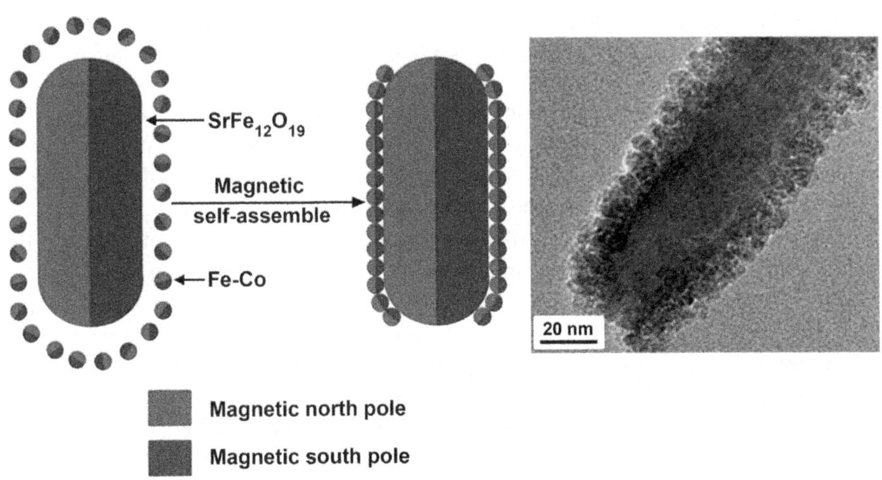

FIGURE 5.46

Schematic diagram and transmission electron microscopy micrograph of core–shell SrM/FeCo.

From Xu, X., Park, J., Hong, Y.-K., & Lane, A. M. (2015). Magnetically self-assembled SrFe$_{12}$O$_{19}$/Fe–Co core/shell particles. Materials Chemistry and Physics, 152, *9–12.*

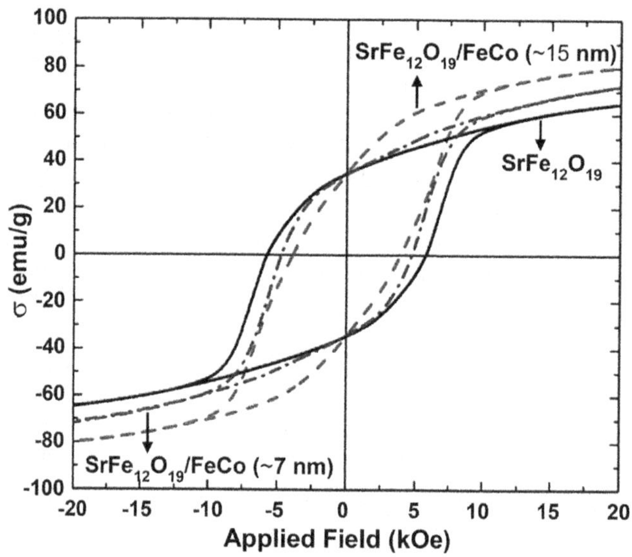

FIGURE 5.47

Magnetic hysteresis loops of strontium ferrite and SrM/FeCo composite.

From Xu, X., Park, J., Hong, Y.-K., & Lane, A. M. (2015). Magnetically self-assembled $SrFe_{12}O_{19}$/Fe–Co core/shell particles. Materials Chemistry and Physics, 152, *9–12.*

is the mass percentage of the soft phase, one can easily calculate the mass percentage of the soft phase. Accordingly, the soft phase with thicknesses of 15 and 7 nm corresponds to 6.1% and 13.0% mass percentages, respectively. Moreover, exchange coupling exists between the soft iron cobalt and hard strontium ferrite phases.

Elsewhere, SFO/$CoFe_2O_4$ nanofibers, using a homemade coaxial electrospinning setup followed by calcinating the as-spun nanofibers, have been developed in air and annealed in a reducing atmosphere (Dong et al., 2014). The TEM micrographs depicted in Fig. 5.48 confirm the formation of platelike nanoparticles along the length of strontium ferrite nanofibers. To prepare a core–shell configuration, the outer surface of strontium ferrite nanofibers has been decorated with cobalt ferrite. The resultant core and shell thicknesses have been obtained to be 25 and 10 nm, respectively. The interface between the two ferrites is rather smooth after the annealing process in air (see Fig. 5.48B). After reducing the nanofibers in a hydrogen ambient, the soft phase of iron cobalt appears, along with rough and nonuniform morphology of the soft and hard phase interface (Fig. 5.48C). The analysis demonstrates the elemental composition with an appropriate chemical alteration.

As expected from the FeCo formation, the saturation magnetization has increased, whereas the coercivity has decreased. The single-phase hysteresis loop of the reduced nanofibers without any shoulder or kink in Fig. 5.49 confirms the occurrence of a complete exchange-spring mechanism in the resulting composites.

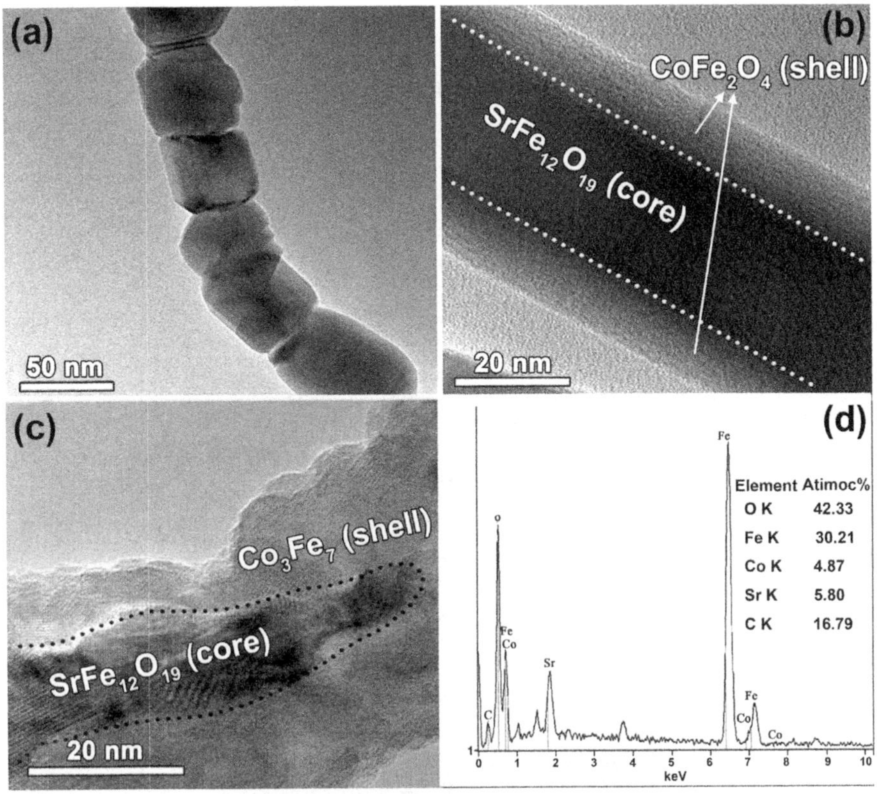

FIGURE 5.48

Transmission electron microscopy micrographs of (A) $SrFe_{12}O_{19}$, (B) $SrFe_{12}O_{19}/CoFe_2O_4$, (C) $SrFe_{12}O_{19}$/FeCo nanofibers, and (D) EDX analysis of samples (B).

From Dong, J., Zhang, Y., Zhang, X., Liu, Q., & Wang, J. (2014). Improved magnetic properties of $SrFe_{12}O_{19}$/ FeCo core–shell nanofibers by hard/soft magnetic exchange–coupling effect. Materials Letters, 120, *9–12.*

5.8 Ferrite thin-film bimagnets

Multilayers of iron cobalt/cobalt ferrite have been prepared by the deposition of a $CoFe_2$ layer, followed by partial oxidation in a mixture of argon and oxygen. In continuance, the deposition followed by the oxidation process has been repeated to fabricate multilayered stacks (Jurca et al., 2003). The HRTEM micrograph depicted in Fig. 5.50 reflects the morphology of the multilayered film. Despite the polycrystalline nature of the hard and soft layers, it is possible to develop perfect crystallographic matching and coherency between the bilayers. Although the authors have not measured the magnetic properties of the multilayers, the processing technique might be used to provide an exchange-spring permanent magnet by controlling the thickness of the hard and soft phases.

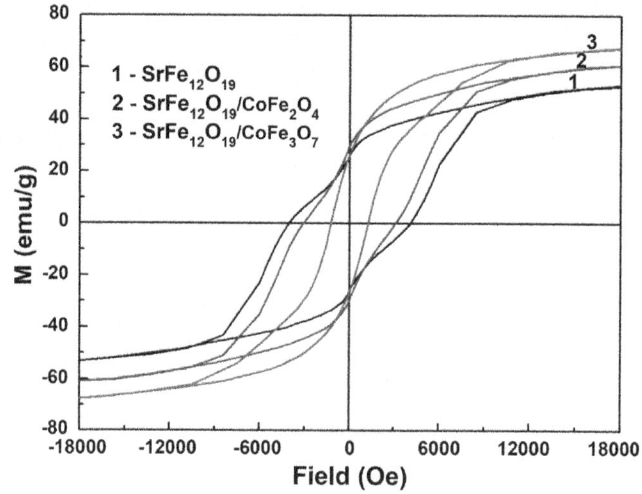

FIGURE 5.49

Hysteresis loops of $SrFe_{12}O_{19}$, $SrFe_{12}O_{19}/CoFe_2O_4$, $SrFe_{12}O_{19}$/FeCo nanofibers.

From Dong, J., Zhang, Y., Zhang, X., Liu, Q., & Wang, J. (2014). Improved magnetic properties of $SrFe_{12}O_{19}$/FeCo core–shell nanofibers by hard/soft magnetic exchange–coupling effect. Materials Letters, 120, *9–12.*

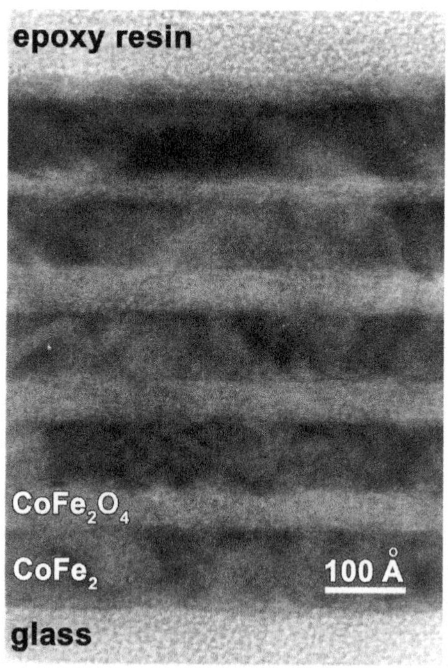

FIGURE 5.50

Transmission electron microscopy micrograph of multilayer iron cobalt/cobalt ferrite.

From Jurca, I., Meny, C., Viart, N., Ulhaq-Bouillet, C., Panissod, P., & Pourroy, G. (2003). Growth, structure and morphology of $CoFe_2/CoFe_2O_4$ multilayers. Thin Solid Films, 444*(1–2), 58–63.*

It is well known that cobalt ferrite can be grown on MgO (001) wafer. Depending on the thickness of cobalt ferrite film deposited on MgO (001) substrates, it has been feasible to tune the magnetic anisotropy. In this regard, perpendicular magnetic anisotropy has been obtained for films with thicknesses of <100 nm, owing to the large out-of-plane compressive strain applied to and the negative magnetostriction coefficient of the cobalt ferrite film (Chen et al., 2015). By increasing the film thickness, the shape anisotropy dominates the magnetic anisotropy, leading to an in-plane easy axis.

Elsewhere, metallic CoFe films have been grown by annealing as-deposited heteroepitaxial cobalt ferrite films in argon and hydrogen atmospheres. Fig. 5.51 shows the film microstructures before and after reduction. The cobalt ferrite film with a thickness of 100 nm has integrated morphology and a relatively sharp interface. While the resulting film obtained after reduction shows no obvious change in the film thickness, a porous structure is formed close to the substrate/film interface. Furthermore, the inset of Fig. 5.51B shows a mushroomlike structure consistent with the TEM results. The lattice matching between $CoFe_2$ (011) planes and the MgO substrate (001) is shown in Fig. 5.51C. The corresponding fast Fourier transform (FFT) image shows the misfit dislocations essential for the semicoherent structure formed between the two phases (Fig. 5.51D). As shown in the insets, the FFT diffraction patterns of the iron cobalt phase and substrate correspond to the regions above and below the dashed line, respectively. The $CoFe_2$ films show an out-of-plane texture that is induced by lattice matching between the film and substrate.

Kameli and his colleagues have used a pulsed laser deposition technique to fabricate cobalt ferrite (CFO) thin films with a thickness of 130 nm on MgO substrate. A lattice mismatch of 0.48% has been induced between the film and substrate wafer, confirming the imposed lattice strain and semicoherent interface. A reduction–diffusion process has also been carried out to form iron cobalt (Eskandari et al., 2019). Fig. 5.52 shows that the CFO (001) peaks become sharper with increased reduction time, indicating an increase in the structural factor of cobalt ferrite. On the other hand, the bcc iron cobalt peaks become clearer and sharper with increased hydrogen annealing time, indicating the formation of a soft phase in the nanocomposite.

The perpendicular and in-plane coercivities of cobalt ferrite extracted from the VSM hysteresis loops of Fig. 5.53 have been reported to be 0.549 and 0.300 T, respectively. In addition, the remanent magnetizations have been measured to be 139 and 200 kA/m, respectively. Therefore, the easy axis of the cobalt ferrite with a thickness of 130 nm is parallel to the film. After annealing cobalt ferrite films in a reducing atmosphere for 30 min, the iron cobalt has been developed in the film sample, decreasing the perpendicular and in-plane coercivities to 0.275 and 0.077 T. However, the saturation magnetization has been enhanced to 457 kA/m. Increasing the annealing time to 45 and 60 min has increased the respective saturation magnetization to 474 and 697 kA/m. The minimum perpendicular coercivity of

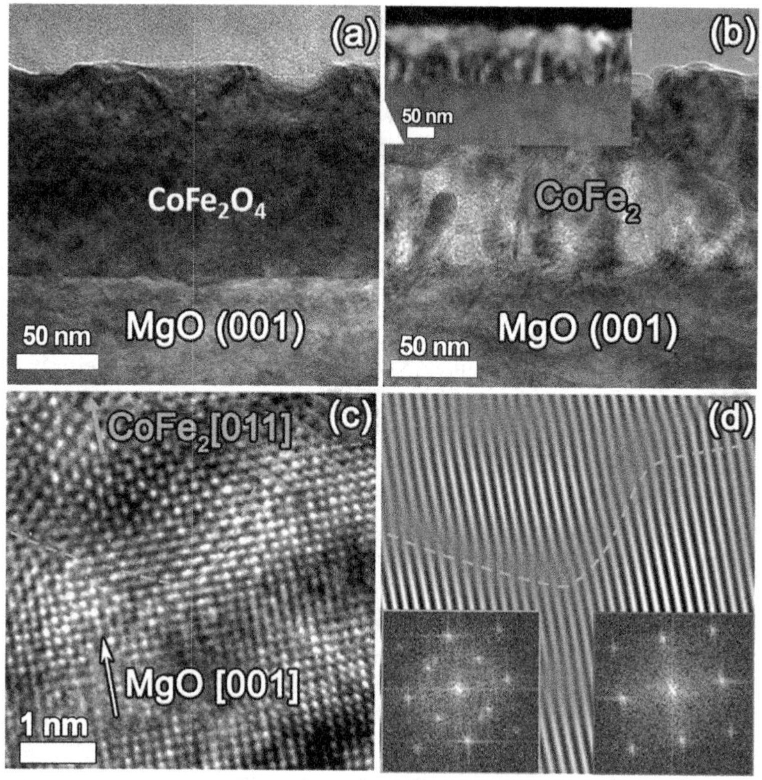

FIGURE 5.51

(A) Transmission electron microscopy (TEM) micrograph of CFO film with a thickness of 100 nm on the MgO (001) substrate. (B) The TEM micrograph of iron cobalt film after reduction–diffusion process. Inset of (B) is the scanning transmission electron microscopy micrograph of iron cobalt film. (C) A high-resolution transmission electron microscopy micrograph of interface of iron cobalt film and substrate. (D) The corresponding FFT image. Insets of (D) show the FFT diffraction patterns of iron cobalt and substrate. *Blue dashed lines* reflect the interface of iron cobalt film and substrate.
From Chen, A., Poudyal, N., Xiong, J., Liu, J. P., & Jia, Q. (2015). Modification of structure and magnetic anisotropy of epitaxial $CoFe_2O_4$ films by hydrogen reduction. Applied Physics Letters, 106*(11), 111907.*

the composite film is 0.118 T for an annealing time of 60 min (Fig. 5.53D). Although the in-plane coercivity decreases abruptly, a further increase in the in-plane coercivity from 0.077 to 0.101 T has been observed with increased annealing time. As noticed by the authors, the composite films fabricated at annealing times of 30 and 45 min reflect a perpendicular magnetic anisotropy. In contrast, the film sample prepared at an annealing time of 60 min shows comparable out-of-plane and in-plane coercivities, which means that the sample is almost isotropic. As mentioned earlier, cobalt ferrite contains the highest anisotropy among the spinel structures.

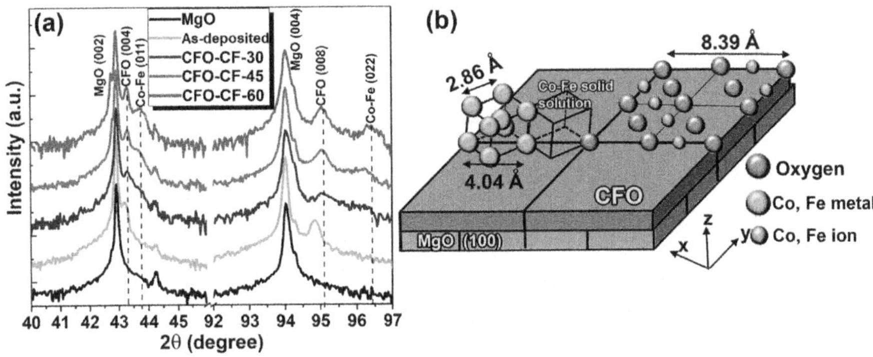

FIGURE 5.52

(A) X-ray diffraction patterns of MgO substrate, as deposited cobalt ferrite and CFO thin films annealed for 30 (CFO-CF-30), 45 (CFO-CF-45), and 60 (CFO-CF-60) minute, (B) schematic drawings of crystal matching between iron cobalt and cobalt ferrite on MgO substrate.

From Eskandari, F., Kameli, P., Salamati, H., & Esmaeily, A. S. (2019). Tuning the exchange coupling in pulse laser deposited cobalt ferrite thin films by hydrogen reduction. Journal of Magnetism and Magnetic Materials, *484, 188–195.*

It has been established that the existence of cobalt cations on the octahedral occupancies plays the main role in providing the strong cubic anisotropy of $K_1^c \approx 290\ \mathrm{kJ/m^3}$ with the <100> easy direction (Eskandari et al., 2019).

By increasing the annealing time, cobalt cations migrate from octahedral to tetrahedral sites. Since there is no orbital moment in the cubic $4\bar{3}m$ point symmetry, cobalt cations on the tetrahedral sites do not contribute to the cubic anisotropy, resulting in decreased coercivity. The formation of the soft iron cobalt in the hard cobalt ferrite matrix also reduces the magnetic anisotropy of the composite structure.

A single-phase hysteresis loop without a formed kink is indicated for cobalt ferrite films. This could also be proved by the variation of switching field distribution (SFD) curves, demonstrating the completion of magnetization reversal in a single step. The authors have claimed that despite the formation of a smooth single-phase hysteresis loop, the maximum energy product has been reduced with an increase in the annealing time, avoiding the applicability of the composites for permanent magnets. The reasoning behind this could be found in the phase distribution of the soft iron cobalt in the hard matrix. It is clear from the optical microscopy micrographs in Fig. 5.54 that there is a wide range of size distribution and scattering of the micron-size iron cobalt phase in the matrix, so the exchange coupling could not completely occur in the resulting composite films.

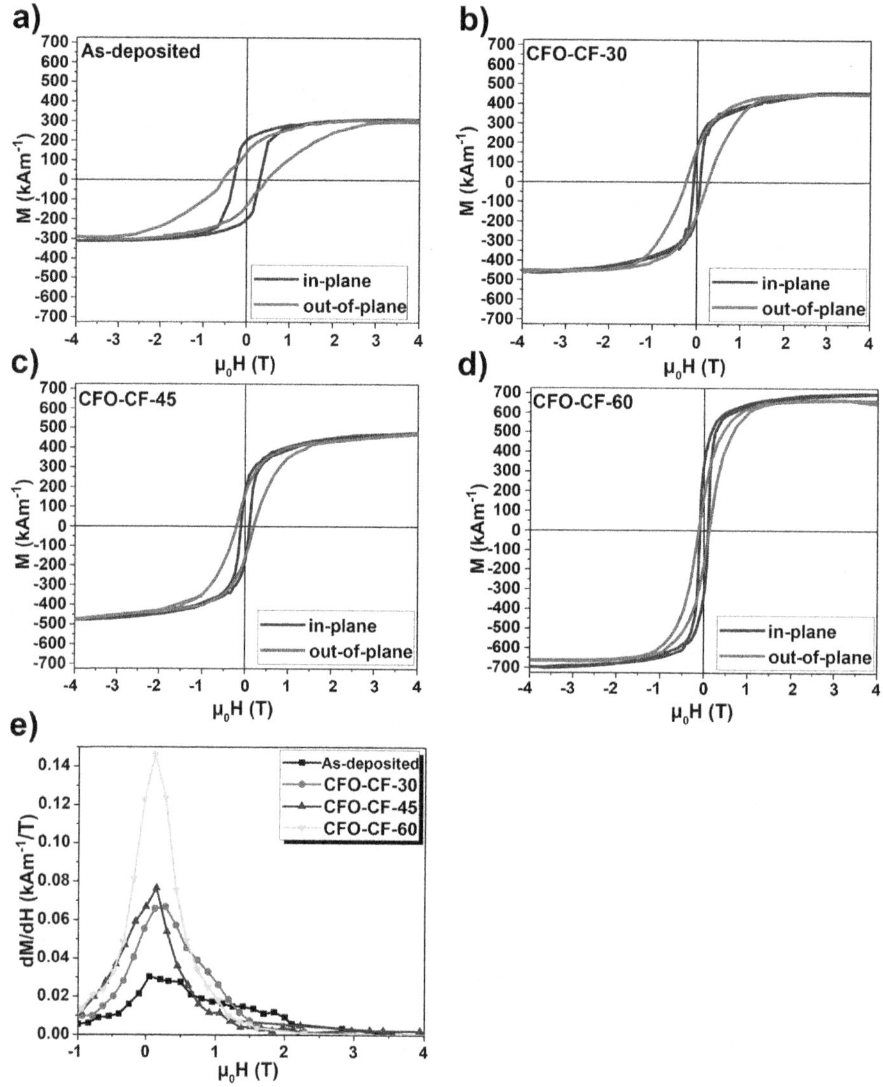

FIGURE 5.53

In-plane and perpendicular hysteresis loops of (A) cobalt ferrite, (B) CFO-CF-30, (C) CFO-CF-45, (D) CFO-CF-60 films, and (E) switching field distribution graphs of cobalt ferrite and hydrogen reduced cobalt ferrite films for perpendicular magnetization data.

From Eskandari, F., Kameli, P., Salamati, H., & Esmaeily, A. S. (2019). Tuning the exchange coupling in pulse laser deposited cobalt ferrite thin films by hydrogen reduction. Journal of Magnetism and Magnetic Materials, *484, 188–195.*

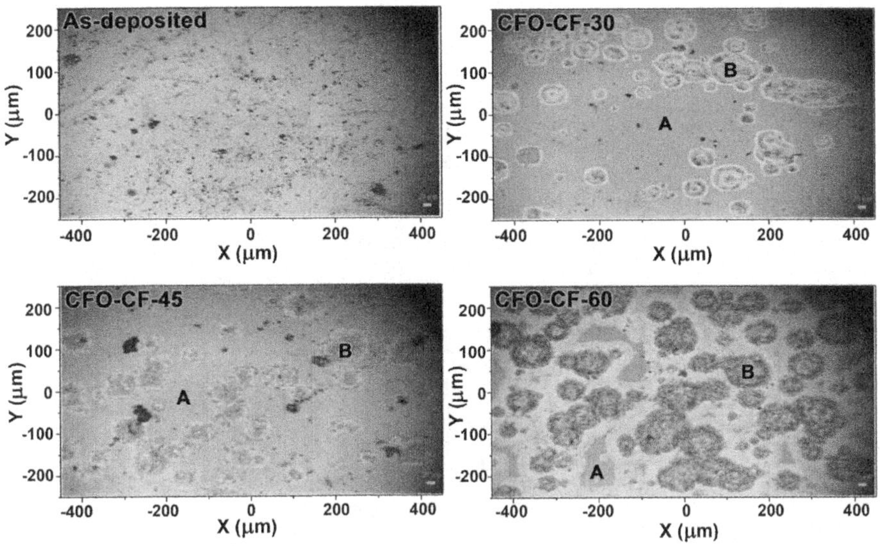

FIGURE 5.54

Optical microscopy micrographs of cobalt ferrite films annealed at different annealing times of 30, 45, and 60 min.

From Eskandari, F., Kameli, P., Salamati, H., & Esmaeily, A. S. (2019). Tuning the exchange coupling in pulse laser deposited cobalt ferrite thin films by hydrogen reduction. Journal of Magnetism and Magnetic Materials, *484, 188–195.*

5.9 Metallic magnet–ferrite and inverted nanocomposites

An excellent study has been carried out about the exchange-spring state in $Sm_2Fe_{17}N_3$/iron ferrite composites (Noriko Imaoka, 2013; Zhu & Howe, 2007). One of the challenges for using metallic permanent magnets in high-frequency applications including electric vehicles is eddy current loss. $Sm_2Fe_{17}N_3$ grains can be coated with soft ferrites (having relatively high dielectric constants) to overcome the Eddy current challenge, suppressing the electrical conductivity among the grains and increasing their resistivity while minimizing the deterioration of magnetic properties. This makes the resultant exchange-spring composite of $Sm_2Fe_{17}N_3$/ferrite a good candidate for use in electric vehicles. The conceptual design of the composite is illustrated in Fig. 5.55. The iron ferrite phase used in this research is intermediate between magnetite (Fe_3O_4) and maghemite (γ-Fe_2O_3) (Kitamoto et al., 2000; Nishimura et al., 2000; Tada et al., 2005).

TEM micrographs are also shown in Figs. 5.56 and 5.57. The grains of the hard phase can be seen to be coated with a thin layer of soft ferrite. The electron diffraction patterns demonstrate the diffracted rings from the soft phase without the formation of any secondary phase. Although the interface matching between ferrite and metallic magnets is still unknown, it appears that coherent matching occurs

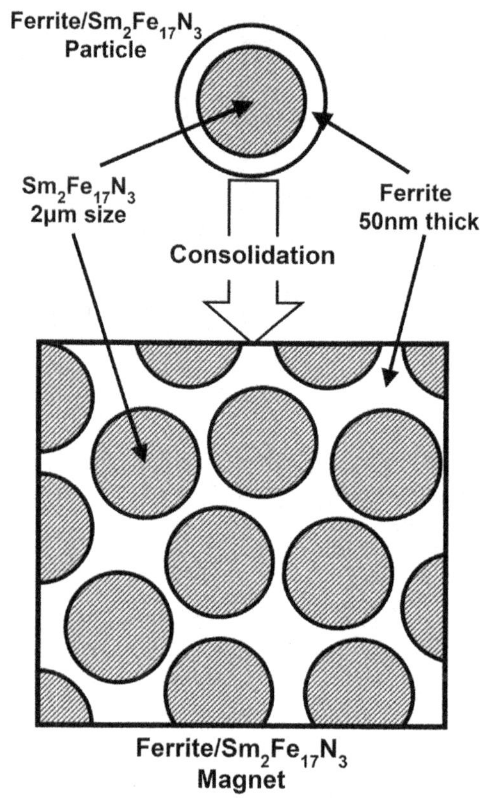

FIGURE 5.55

Conceptual design of $Sm_2Fe_{17}N_3$/ferrite exchange spring magnet.

From Imaoka, N., Kakimoto, E., Takagi, K., Ozaki, K., Tada, M., Nakagawa, T., & Abe, M. (2019). Progress with insulating nanocomposites based on ferrite plating of $Sm_2Fe_{17}N_3$ micropowders. Journal of Magnetism and Magnetic Materials, 476, *613–621.*

between the hard and soft grains (see Fig. 5.57). It should be mentioned that the exchange interactions between spins in $Sm_2Fe_{17}N_3$ and ferrite phases also remain unknown.

Fig. 5.58 shows the packing fraction dependence of resistivity in the hard phase and composites fabricated by powder compacting (PC) and explosive consolidation (EC) techniques. The resistivity of the composite is about 4000 μOhm cm, which is 10 times larger than the estimated resistivity of $Sm_2Fe_{17}N_3$ fully dense magnets (~400 μOhm cm). Since the exchange-spring mechanism is well established in the composite, resistivity increases, making the resultant magnet perfect for use in high-frequency applications of electric vehicles.

It must be noted that the authors have evidenced the existence of the exchange-spring mechanism by measuring the recoil permeability along with detailed

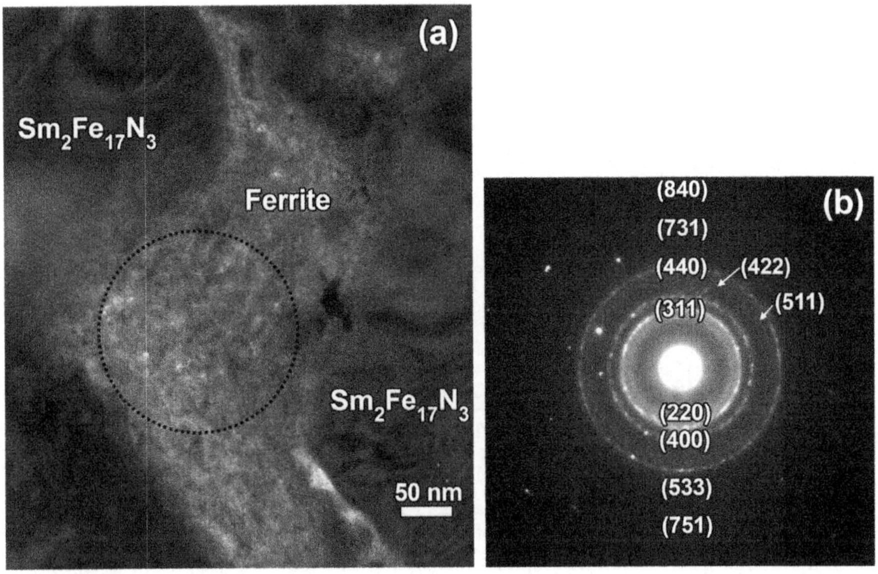

FIGURE 5.56

(A) Transmission electron microscopy micrograph of ferrite/$Sm_2Fe_{17}N_3$ composite and (B) corresponded electron diffraction pattern of circle area.

From Imaoka, N., Kakimoto, E., Takagi, K., Ozaki, K., Tada, M., Nakagawa, T., & Abe, M. (2019). Progress with insulating nanocomposites based on ferrite plating of $Sm_2Fe_{17}N_3$ micropowders. Journal of Magnetism and Magnetic Materials, 476, *613–621.*

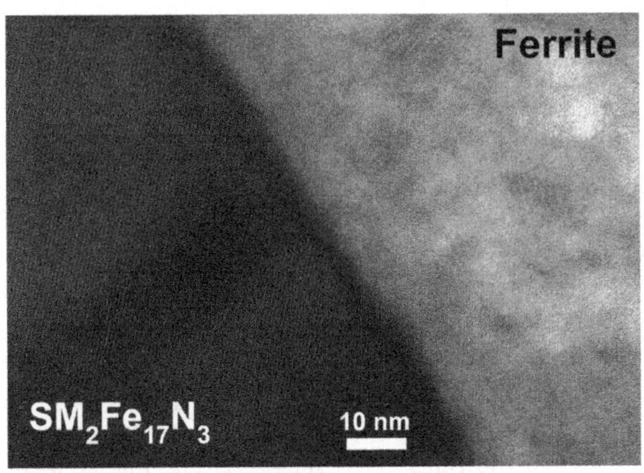

FIGURE 5.57

Transmission electron microscopy micrograph of $Sm_2Fe_{17}N_3$ and ferrite interface.

From Imaoka, N., Kakimoto, E., Takagi, K., Ozaki, K., Tada, M., Nakagawa, T., & Abe, M. (2019). Progress with insulating nanocomposites based on ferrite plating of $Sm_2Fe_{17}N_3$ micropowders. Journal of Magnetism and Magnetic Materials, 476, *613–621.*

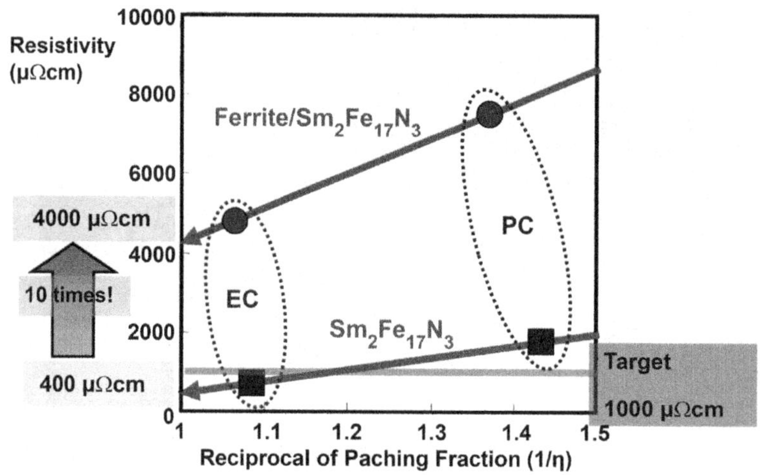

FIGURE 5.58

Variation of resistivity versus reciprocal of packing fractions in $Sm_2Fe_{17}N_3$ and ferrite/$Sm_2Fe_{17}N_3$ composite magnets fabricated by powder compacting (PC) and explosive consolidation (EC) techniques.

From Imaoka, N., Kakimoto, E., Takagi, K., Ozaki, K., Tada, M., Nakagawa, T., & Abe, M. (2019). Progress with insulating nanocomposites based on ferrite plating of $Sm_2Fe_{17}N_3$ micropowders. Journal of Magnetism and Magnetic Materials, 476, *613–621.*

microstructural observations. In this respect, the recoil permeability (μ_r) has been calculated by the following expression:

$$\mu_r = 1 + \frac{4\pi(M_0 - M_d)}{M_d} \tag{5.11}$$

where M_d is the magnetization at the reverse field with an absolute value of H_d, and M_0 is the magnetization at the zero-field after being reversed from H_d. As obtained from the graphs in Fig. 5.59, the recoil permeabilities for the composite are 1.31−1.35, which are slightly higher than those for $Sm_2Fe_{17}N_3$ (1.23−1.28). The results prove that exchange coupling has been well applied between the $Sm_2Fe_{17}N_3$ core and ferrite shell in the fabricated composites (Imaoka et al., 2010, 2016; Iriyama et al., 1992).

Fig. 5.60 compares the demagnetization and recoil curves of $Sm_2Fe_{17}N_3$ and ferrite/$Sm_2Fe_{17}N_3$ magnets. Evidently, while the composite shows fairly high coercivity and rectangularity, the remanence and coercivity decrease only slightly compared with those of the $Sm_2Fe_{17}N_3$ magnets. The magnetization curves show no kink, evidencing that the soft ferrite thin film is well exchange-coupled with the hard phase. In contrast, the composite prepared by the physical mixing method shows a kink, reflecting incomplete coupling between the $Sm_2Fe_{17}N_3$ and iron ferrite.

So far, less attention has been paid to inverse soft/hard core/shell nanocomposites, so only a few literature studies, such as FePt/$CoFe_2O_4$ (Zeng et al., 2004),

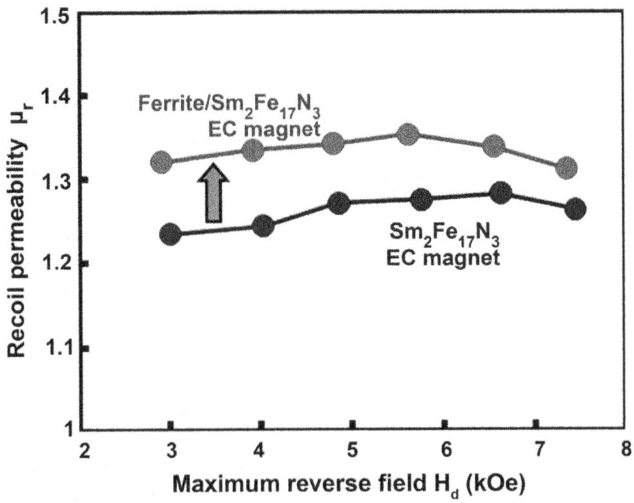

FIGURE 5.59

Recoil permeabilities of ferrite/$Sm_2Fe_{17}N_3$ and $Sm_2Fe_{17}N_3$.

From Imaoka, N., Kakimoto, E., Takagi, K., Ozaki, K., Tada, M., Nakagawa, T., & Abe, M. (2019). Progress with insulating nanocomposites based on ferrite plating of $Sm_2Fe_{17}N_3$ micropowders. Journal of Magnetism and Magnetic Materials, 476, *613–621.*

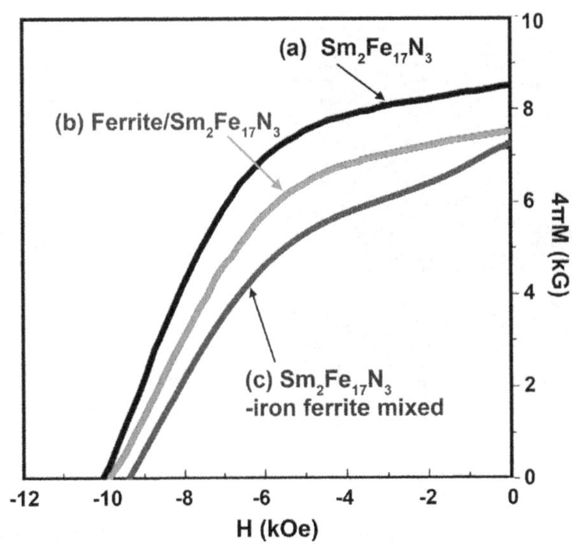

FIGURE 5.60

Demagnetization curves of (A) $Sm_2Fe_{17}N_3$, (B) ferrite/$Sm_2Fe_{17}N_3$, and (C) $Sm_2Fe_{17}N_3$-iron ferrite mixed magnets.

From Imaoka, N., Kakimoto, E., Takagi, K., Ozaki, K., Tada, M., Nakagawa, T., & Abe, M. (2019). Progress with insulating nanocomposites based on ferrite plating of $Sm_2Fe_{17}N_3$ micropowders. Journal of Magnetism and Magnetic Materials, 476, *613–621.*

$MFe_2O_4/CoFe_2O_4$ (M = Zn, Mn, and Fe) (Cheon et al., 2006; Lee et al., 2011; Masala et al., 2006), and γ-$Fe_2O_3/CoFe_2O_4$ (Salazar-Alvarez et al., 2007), could be found. In an interesting study of inverse soft/hard exchange-coupled magnets (in contrast to conventional hard/soft nanocomposites) carried out in 2012, Lopez-Ortega and his colleagues reported that $Mn_xFe_{3-x}O_4/Fe_xMn_{3-x}O_4$ soft/hard core/shell nanoparticles are capable of exhibiting strong exchange-spring behavior arising from the dominant role of the interfaces (López-Ortega et al., 2012). The core–shell configuration along with the schematic representation is shown in Fig. 5.61. The core diameters are 7.4 (CS1) and 11.5 nm (CS2), whereas the shell thickness is constant at 0.6 nm.

From X-ray absorption spectroscopy and X-ray magnetic circular dichroism (XMCD) depicted in Fig. 5.62, interdiffusion of iron and manganese ions between the interface of core and shell has caused the formation of mixed Fe–Mn spinel core–shell nanoparticles, providing an interface with graded composition. The XMCD spectra reflect the signal of the iron edge, which is consistent with that in the Fe_3O_4 and γ-Fe_2O_3 spectra. The existence of mixed $MnFe_2O_4$ has also been evidenced. The mixed XMCD signal also indicates the formation of tetragonal $FeMn_2O_4$ and cubic $MnFe_2O_4$ phases.

The magnetic analyses demonstrate that a strong exchange mechanism occurs between hard and soft phases, leading to smooth hysteresis loops. First-order reversal curve (FORC) analysis in Fig. 5.63 further investigates the presence of

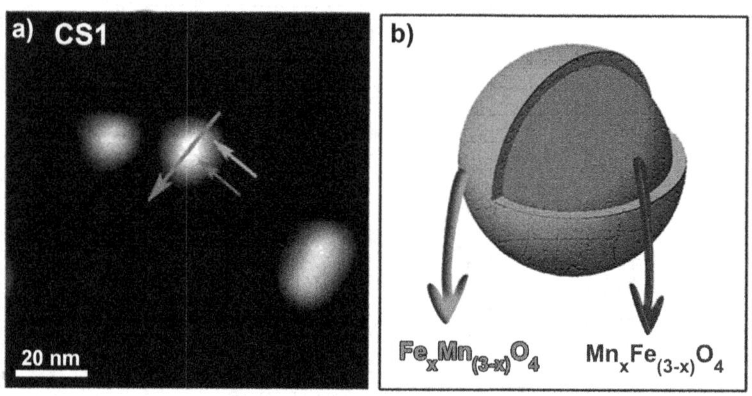

FIGURE 5.61

(A) High-angle annular dark field (HAADF) image of core–shell (CS1), (B) Schematic representation of the core–shell.

From López-Ortega, A., Estrader, M., Salazar-Alvarez, G., Estradé, S., Golosovsky, I. V., Dumas, R., Keavney, D., Vasilakaki, M., Trohidou, K., & Sort, J. (2012). Strongly exchange coupled inverse ferrimagnetic soft/hard, Mnx $Fe_{3-x}O_4/Fe_xMn_{3-x}O_4$, core/shell heterostructured nanoparticles. Nanoscale, *4(16), 5138–5147.*

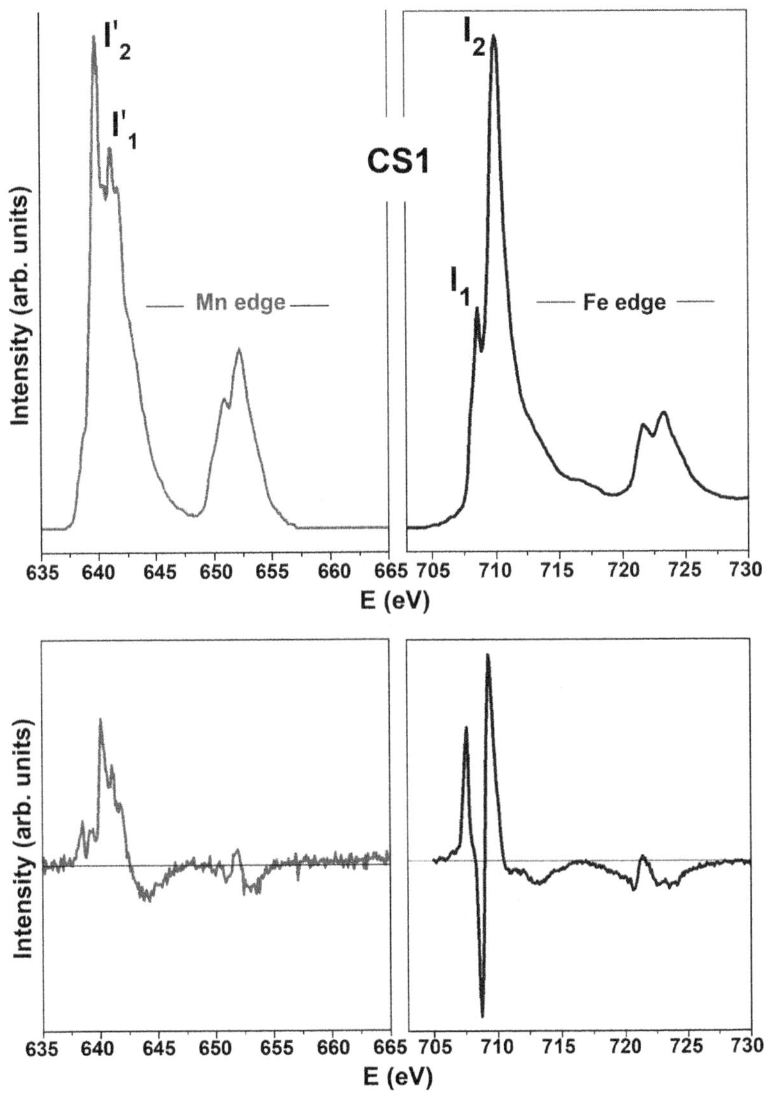

FIGURE 5.62

(Top) X-ray absorption spectroscopy and (bottom) X-ray magnetic circular dichroism spectra at the (left) Mn and (right) Fe edges of CS1 nanoparticles.

From López-Ortega, A., Estrader, M., Salazar-Alvarez, G., Estradé, S., Golosovsky, I. V., Dumas, R., Keavney, D., Vasilakaki, M., Trohidou, K., & Sort, J. (2012). Strongly exchange coupled inverse ferrimagnetic soft/hard, $Mn_xFe_{3-x}O_4/Fe_xMn_{3-x}O_4$, core/shell heterostructured nanoparticles. Nanoscale, *4(16), 5138–5147.*

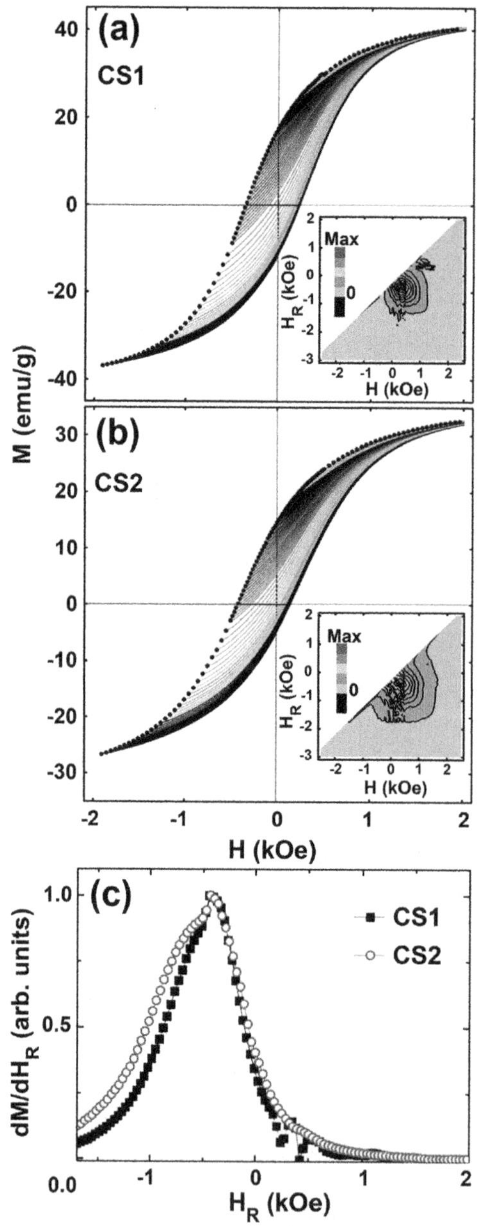

FIGURE 5.63

First-order reversal curves (FORCs) for nanocomposites (A) CS1 and (B) CS2 with the related FORC distributions plotted in (H, HR) coordinates shown as insets. FORC-switching force distributions are shown in (C).

From López-Ortega, A., Estrader, M., Salazar-Alvarez, G., Estradé, S., Golosovsky, I. V., Dumas, R., Keavney, D., Vasilakaki, M., Trohidou, K., & Sort, J. (2012). Strongly exchange coupled inverse ferrimagnetic soft/hard, $Mn_xFe_{3-x}O_4/Fe_xMn_{3-x}O_4$, core/shell heterostructured nanoparticles. Nanoscale, *4(16), 5138–5147.*

exchange coupling between core—shell structures. The FORC distributions are shown in Fig. 5.63A and B. The broad single peaks indicate the size distributions of nanoparticles, easy anisotropy axis direction and its strength, and the exchange coupling between the core and shell. The SFDs of FORC depicted in Fig. 5.63C confirm a clear asymmetry in both core—shell nanocomposites.

References

Araújo, J. H., de Cabral, F., Soares, J. M., & Machado, F. (2011). *Exchange-spring behavior in nanopowders of $CoFe_2O_4$—$CoFe_2$*. AIP Publishing.

Cabral, F., de, A. O., de Araujo Machado, F. L., de Araujo, J. H., Soares, J. M., Rodrigues, A. R., & Araujo, A. (2008). Preparation and magnetic study of the $CoFe_2O_4$—$CoFe_2$ nanocomposite powders. *IEEE Transactions on Magnetics, 44*(11), 4235—4238. https://doi.org/10.1109/TMAG.2008.2001545

Chen, A., Poudyal, N., Xiong, J., Liu, J. P., & Jia, Q. (2015). Modification of structure and magnetic anisotropy of epitaxial $CoFe_2O_4$ films by hydrogen reduction. *Applied Physics Letters, 106*(11), 111907—111910. https://doi.org/10.1063/1.4915504

Cheon, J., Park, J.-I., Choi, J., Jun, Y., Kim, S., Kim, M. G., & Kim, Y. J. (2006). Magnetic superlattices and their nanoscale phase transition effects. *Proceedings of the National Academy of Sciences, 103*(9), 3023—3027. https://doi.org/10.1073/pnas.0508877103

Coey, J. (2012). Permanent magnets: Plugging the gap. *Scripta Materialia, 67*(6), 524—529. https://doi.org/10.1016/j.scriptamat.2012.04.036

Cornell, R. M., & Schwertmann, U. (2003). *The iron oxides: Structure, properties, reactions, occurrences and uses*. John Wiley & Sons.

Dong, J., Zhang, Y., Zhang, X., Liu, Q., & Wang, J. (2014). Improved magnetic properties of $SrFe_{12}O_{19}$/FeCo core—shell nanofibers by hard/soft magnetic exchange—coupling effect. *Materials Letters, 120*, 9—12. https://doi.org/10.1016/j.matlet.2014.01.022

Eerenstein, W., Palstra, T., Hibma, T., & Celotto, S. (2002). Origin of the increased resistivity in epitaxial Fe_3O_4 films. *Physical Review B, 66*(20), 201101—201104. https://doi.org/10.1103/PhysRevB.66.201101

Eskandari, F., Kameli, P., Salamati, H., & Esmaeily, A. S. (2019). Tuning the exchange coupling in pulse laser deposited cobalt ferrite thin films by hydrogen reduction. *Journal of Magnetism and Magnetic Materials, 484*, 188—195. https://doi.org/10.1016/j.jmmm.2019.03.133

Fan, F.-R., Liu, D.-Y., Wu, Y.-F., Duan, S., Xie, Z.-X., Jiang, Z.-Y., & Tian, Z.-Q. (2008). Epitaxial growth of heterogeneous metal nanocrystals: From gold nano-octahedra to palladium and silver nanocubes. *Journal of the American Chemical Society, 130*(22), 6949—6951. https://doi.org/10.1021/ja801566d

Fukamachi, N., Tezuka, N., & Sugimoto, S. (2010). Preparation of Composite nanoparticles based on cobalte ferrite. *The Japan Institute of Metals, 74*(6), 345—350. https://doi.org/10.2320/jinstmet.74.345

Fullerton, E. E., Jiang, J., & Bader, S. (1999). Hard/soft magnetic heterostructures: Model exchange-spring magnets. *Journal of Magnetism and Magnetic Materials, 200*(1—3), 392—404. https://doi.org/10.1016/S0304-8853(99)00376-5

Gregg, J., Ziese, M., & Thornton, M. J. (2001). *Spin electronics*. Springer.

Hazra, S., Patra, M. K., Vadera, S. R., & Ghosh, N. N. (2012). A novel but simple "One-Pot" synthetic route for preparation of $(NiFe_2O_4)_x-(BaFe_{12}O_{19})_{1-x}$ composites. *Journal of the American Ceramic Society, 95*(1), 60–63. https://doi.org/10.1111/j.1551-2916.2011.04958.x

Imaoka, N. (2013). *Magnetic recording medium and method of manufacturing the same.* Google Patents.

Imaoka, N., Abe, M., Nakagawa, T., & Tada, S. (2010). *Magnetic material for high frequency wave, and method for production thereof.* Google Patents.

Imaoka, N., Kakimoto, E., Takagi, K., Ozaki, K., Tada, M., Nakagawa, T., & Abe, M. (2016). Exchange coupling between soft magnetic ferrite and hard ferromagnetic $Sm_2Fe_{17}N_3$ in ferrite/$Sm_2Fe_{17}N_3$ composites. *AIP Advances, 6*(5), 56022–56027. https://doi.org/10.1063/1.4944519

Imaoka, N., Kakimoto, E., Takagi, K., Ozaki, K., Tada, M., Nakagawa, T., & Abe, M. (2019). Progress with insulating nanocomposites based on ferrite plating of $Sm_2Fe_{17}N_3$ micropowders. *Journal of Magnetism and Magnetic Materials, 476*, 613–621. https://doi.org/10.1016/j.jmmm.2018.09.069

Iriyama, T., Kobayashi, K., Imaoka, N., Fukuda, T., Kato, H., & Nakagawa, Y. (1992). Effect of nitrogen content on magnetic properties of $Sm_2Fe_{17}N_x(0<x<6)$. *IEEE Transactions on Magnetics, 28*(5), 2326–2331. https://doi.org/10.1109/20.179482

Jenuš, P., Topole, M., McGuiness, P., Granados-Miralles, C., Stingaciu, M., Christensen, M., & Žužek Rožman, K. (2016). Ferrite-based exchange-coupled hard–soft magnets fabricated by spark plasma sintering. *Journal of the American Ceramic Society, 99*(6), 1927–1934. https://doi.org/10.1111/jace.14193

Jurca, I., Meny, C., Viart, N., Ulhaq-Bouillet, C., Panissod, P., & Pourroy, G. (2003). Growth, structure and morphology of $CoFe_2/CoFe_2O_4$ multilayers. *Thin Solid Films, 444*(1–2), 58–63. https://doi.org/10.1016/S0040-6090(03)01020-4

Kahnes, M., & Töpfer, J. (2019). Synthesis and magnetic properties of hard/soft $SrAl_2Fe_{10}O_{19}/Fe(FeCo_2)$ nanocomposites. *Journal of Magnetism and Magnetic Materials, 480*, 40–46. https://doi.org/10.1016/j.jmmm.2019.02.065

Kitamoto, Y., Nakayama, Y., & Abe, M. (2000). Spin-dependent intergranular transport in magnetite films deposited by ferrite plating. *Journal of Applied Physics, 87*(9), 7130–7132. https://doi.org/10.1063/1.372953

Kneller, E. F., & Hawig, R. (1991). The exchange-spring magnet: A new material principle for permanent magnets. *IEEE Transactions on Magnetics, 27*(4), 3558–3560. https://doi.org/10.1109/20.102931

Lavorato, G., Winkler, E., Rivas-Murias, B., & Rivadulla, F. (2016). Thickness dependence of exchange coupling in epitaxial $Fe_3O_4/CoFe_2O_4$ soft/hard magnetic bilayers. *Physical Review B, 94*(5), 54405–54410. https://doi.org/10.1103/PhysRevB.94.054405

Lee, J.-H., Jang, J., Choi, J., Moon, S. H., Noh, S., Kim, J., & Cheon, J. (2011). Exchange-coupled magnetic nanoparticles for efficient heat induction. *Nature Nanotechnology, 6*(7), 418–422. https://doi.org/10.1038/nnano.2011.95

Leineweber, T., & Kronmüller, H. (1997). Magnetisation reversal modes in inhomogeneous magnets. *Physica Status Solidi, 201*(1), 291–301.

Leineweber, T., & Kronmüller, H. (1997). Micromagnetic examination of exchange coupled ferromagnetic nanolayers. *Journal of Magnetism and Magnetic Materials, 176*(2–3), 145–154.

Leite, G. C., Chagas, E. F., Pereira, R., Prado, R. J., Terezo, A. J., Alzamora, M., & Baggio-Saitovitch, E. (2012). Exchange coupling behavior in bimagnetic $CoFe_2O_4/CoFe_2$ nanocomposite. *Journal of Magnetism and Magnetic Materials, 324*(18), 2711–2716. https://doi.org/10.1016/j.jmmm.2012.03.034

Lewis, L. H., & Jiménez-Villacorta, F. (2013). Perspectives on permanent magnetic materials for energy conversion and power generation. *Metallurgical and Materials Transactions A, 44*(1), 2–20. https://doi.org/10.1007/s11661-012-1278-2

López-Ortega, A., Estrader, M., Salazar-Alvarez, G., Estradé, S., Golosovsky, I. V., Dumas, R., & Sort, J. (2012). Strongly exchange coupled inverse ferrimagnetic soft/hard, $Mn_xFe_{3-x}O_4/Fe_xMn_{3-x}O_4$, core/shell heterostructured nanoparticles. *Nanoscale, 4*(16), 5138–5147. https://doi.org/10.1039/C2NR30986F

López-Ortega, A., Estrader, M., Salazar-Alvarez, G., Roca, A. G., & Nogués, J. (2015). Applications of exchange coupled bi-magnetic hard/soft and soft/hard magnetic core/shell nanoparticles. *Physics Reports, 553*, 1–32. https://doi.org/10.1016/j.physrep.2014.09.007

Masala, O., Hoffman, D., Sundaram, N., Page, K., Proffen, T., Lawes, G., & Seshadri, R. (2006). Preparation of magnetic spinel ferrite core/shell nanoparticles: Soft ferrites on hard ferrites and vice versa. *Solid State Sciences, 8*(9), 1015–1022. https://doi.org/10.1016/j.solidstatesciences.2006.04.014

Nishimura, K., Kohara, Y., Kitamoto, Y., & Abe, M. (2000). Magnetoresistance in magnetite films prepared from aqueous solution at room temperature. *Journal of Applied Physics, 87*(9), 7127–7129. https://doi.org/10.1063/1.372952

Nlebedim, I., Moses, A., & Jiles, D. (2013). Non-stoichiometric cobalt ferrite, $CoxFe_{3-x}O_4(x = 1.0$ to $2.0)$: Structural, magnetic and magnetoelastic properties. *Journal of Magnetism and Magnetic Materials, 343*, 49–54. https://doi.org/10.1016/j.jmmm.2013.04.063

Okabe, T. H. (2011). Current status and problems on rare earth metals. *Journal of Japanese Society of Tribologists, 56*(8), 460–465.

Orna, J., Algarabel, P., Morellón, L., Pardo, J., De Teresa, J., Antón, R. L., & Cezar, J. (2010). Origin of the giant magnetic moment in epitaxial Fe_3O_4 thin films. *Physical Review B, 81*(14), 144420–144426. https://doi.org/10.1103/PhysRevB.81.144420

Palmstrom, C. (1995). Epitaxy of dissimilar materials. *Annual Review of Materials Science, 25*(1), 389–415. https://doi.org/10.1146/annurev.ms.25.080195.002133

Park, J. (2016). *Rare-earth free permanent magnets and permanent magnet synchronous motors*. University of Alabama Libraries.

Poorbafrani, A., Salamati, H., & Kameli, P. (2015). Exchange spring behavior in $Co_{0.6}Zn_{0.4}Fe_2O_4/SrFe_{10.5}O_{16.75}$ nanocomposites. *Ceramics International, 41*(1), 1603–1608. https://doi.org/10.1016/j.ceramint.2014.09.097

Primc, D., & Makovec, D. (2015). Composite nanoplatelets combining soft-magnetic iron oxide with hard-magnetic barium hexaferrite. *Nanoscale, 7*(6), 2688–2697. https://doi.org/10.1039/C4NR05854B

Pullar, R. C. (2012). Hexagonal ferrites: A review of the synthesis, properties and applications of hexaferrite ceramics. *Progress in Materials Science, 57*(7), 1191–1334. https://doi.org/10.1016/j.pmatsci.2012.04.001

Remya, K., Prabhu, D., Amirthapandian, S., Viswanathan, C., & Ponpandian, N. (2016). Exchange spring magnetic behavior in $BaFe_{12}O_{19}/Fe_3O_4$ nanocomposites. *Journal of Magnetism and Magnetic Materials, 406*, 233–238. https://doi.org/10.1016/j.jmmm.2016.01.024

Roy, D., & Kumar, P. A. (2009). Enhancement of (BH) max in a hard-soft-ferrite nanocomposite using exchange spring mechanism. *Journal of Applied Physics, 106*(7), 73902–73905. https://doi.org/10.1063/1.3213341

Saeedi Afshar, S. R., Hasheminiasari, M., & Masoudpanah, S. M. (2018). Structural, magnetic and microwave absorption properties of $SrFe_{12}O_{19}/Ni_{0.6}Zn_{0.4}Fe_2O_4$ composites prepared by one-pot solution combustion method. *Journal of Magnetism and Magnetic Materials, 466*, 1–6. https://doi.org/10.1016/j.jmmm.2018.06.061

Safi, R., Ghasemi, A., & Shoja-Razavi, R. (2017). The role of shell thickness on the exchange spring mechanism of cobalt ferrite/iron cobalt magnetic nanocomposites. *Ceramics International, 43*(1), 617–624. https://doi.org/10.1016/j.ceramint.2016.09.203

Safi, R., Ghasemi, A., Shoja-Razavi, R., & Tavoosi, M. (2016). Development of novel exchange spring magnet by employing nanocomposites of $CoFe_2O_4$ and $CoFe_2$. *Journal of Magnetism and Magnetic Materials, 419*, 92–97. https://doi.org/10.1016/j.jmmm.2016.06.029

Salazar-Alvarez, G., Sort, J., Uheida, A., Muhammed, M., Surinach, S., Baró, M. D., & Nogués, J. (2007). Reversible post-synthesis tuning of the superparamagnetic blocking temperature of γ-Fe_2O_3 nanoparticles by adsorption and desorption of Co(II) ions. *Journal of Materials Chemistry, 17*(4), 322–328. https://doi.org/10.1039/B613026G

Shi, Y., Ding, J., & Yin, H. (2000). $CoFe_2O_4$ nanoparticles prepared by the mechanochemical method. *Journal of Alloys and Compounds, 308*(1–2), 290–295. https://doi.org/10.1016/S0925-8388(00)00921-X

Skomski, R., & Coey, J. (1993). Giant energy product in nanostructured two-phase magnets. *Physical Review B, 48*(21), 15812–15816. https://doi.org/10.1103/PhysRevB.48.15812

Soares, J., Galdino, V., Conceição, O., Morales, M., De Araújo, J., & Machado, F. (2013). Critical dimension for magnetic exchange-spring coupled core/shell $CoFe_2O_4/CoFe_2$ nanoparticles. *Journal of Magnetism and Magnetic Materials, 326*, 81–84. https://doi.org/10.1016/j.jmmm.2012.08.040

Song, F., Shen, X., Liu, M., & Xiang, J. (2011). Preparation and magnetic properties of $SrFe_{12}O_{19}/Ni_{0.5}Zn_{0.5}Fe_2O_4$ nanocomposite ferrite microfibers via sol–gel process. *Materials Chemistry and Physics, 126*(3), 791–796. https://doi.org/10.1016/j.matchemphys.2010.12.042

Song, F., Shen, X., Liu, M., & Xiang, J. (2012). Microstructure, magnetic properties and exchange–coupling interactions for one-dimensional hard/soft ferrite nanofibers. *Journal of Solid State Chemistry, 185*, 31–36. https://doi.org/10.1016/j.jssc.2011.10.009

Tada, M., Abe, T., Miyasaka, J., Matsushita, N., & Abe, M. (2005). Crystallite size dependence of natural resonance frequency for spin-sprayed Fe_3O_4/γ-Fe_2O_3 films highly permeable up to gigahertz range. *Journal of Applied Physics, 97*(10), 109–110. https://doi.org/10.1063/1.1862096

Teja, A. S., & Koh, P.-Y. (2009). Synthesis, properties, and applications of magnetic iron oxide nanoparticles. *Progress in Crystal Growth and Characterization of Materials, 55*(1–2), 22–45. https://doi.org/10.1016/j.pcrysgrow.2008.08.003

Torkian, S., & Ghasemi, A. (2019). Energy product enhancement in sufficiently exchange-coupled nanocomposite ferrites. *Journal of Magnetism and Magnetic Materials, 469*, 119–127. https://doi.org/10.1016/j.jmmm.2018.07.018

Torkian, S., Ghasemi, A., & Razavi, R. S. (2016). Magnetic properties of hard-soft $SrFe_{10}Al_2O_{19}/Co_{0.8}Ni_{0.2}Fe_2O_4$ ferrite synthesized by one-pot sol–gel auto-combustion. *Journal of Magnetism and Magnetic Materials, 416*, 408–416.

Torkian, S., Ghasemi, A., ShojaRazavi, R., & Tavoosi, M. (2016). Structural and magnetic properties of high coercive Al-substituted strontium hexaferrite nanoparticles. *Journal of Superconductivity and Novel Magnetism, 29*(6), 1627−1640.

Wang, Z. (2000). *Transmission electron microscopy of shape-controlled nanocrystals and their assemblies*. ACS Publications.

Wohlfarth, E. P. (1958). Relations between different modes of acquisition of the remanent magnetization of ferromagnetic particles. *Journal of Applied Physics, 29*(3), 595−596. https://doi.org/10.1063/1.1723232

Xiang, J., Zhang, X., Li, J., Chu, Y., & Shen, X. (2013). Fabrication, characterization, exchange coupling and magnetic behavior of $CoFe_2O_4/CoFe_2$ nanocomposite nanofibers. *Chemical Physics Letters, 576*, 39−43. https://doi.org/10.1016/j.cplett.2013.05.020

Xiong, R., Li, W., Fei, C., Liu, Y., & Shi, J. (2016). Exchange-spring behavior in $BaFe_{12}O_{19}$-$Ni_{0.5}Zn_{0.5}Fe_2O_4$ nanocomposites synthesized by a combustion method. *Ceramics International, 42*(10), 11913−11917.

Xu, X., Park, J., Hong, Y.-K., & Lane, A. M. (2015). Magnetically self-assembled $SrFe_{12}O_{19}$/Fe−Co core/shell particles. *Materials Chemistry and Physics, 152*, 9−12. https://doi.org/10.1016/j.matchemphys.2014.11.061

Zeng, H., Sun, S., Li, J., Wang, Z., & Liu, J. (2004). Tailoring magnetic properties of core/shell nanoparticles. *Applied Physics Letters, 85*(5), 792−794. https://doi.org/10.1063/1.1776632

Zhu, Z.-Q., & Howe, D. (2007). Electrical machines and drives for electric, hybrid, and fuel cell vehicles. *Proceedings of the IEEE, 95*(4), 746−765. https://doi.org/10.1109/JPROC.2006.892482

Further reading

Bigot, J.-Y., Kesserwan, H., Halté, V., Ersen, O., Moldovan, M. S., Kim, T. H., Jang, J., & Cheon, J. (2012). Magnetic properties of annealed core−shell CoPt nanoparticles. *Nano Letters, 12*(3), 1189−1197.

Eskandari, F., Porter, S., Venkatesan, M., Kameli, P., Rode, K., & Coey, J. (2017). Magnetization and anisotropy of cobalt ferrite thin films. *Physical Review Materials, 1*(7), 074413.

Jiang, J., Fullerton, E. E., Grimsditch, M., Sowers, C., & Bader, S. (1998). Exchange-spring behavior in epitaxial hard/soft magnetic bilayer films. *Journal of Applied Physics, 83*(11), 6238−6240.

Meffre, A., Mehdaoui, B., Kelsen, V., Fazzini, P. F., Carrey, J., Lachaize, S., Respaud, M., & Chaudret, B. (2012). A simple chemical route toward monodisperse iron carbide nanoparticles displaying tunable magnetic and unprecedented hyperthermia properties. *Nano Letters, 12*(9), 4722−4728.

Nandwana, V., Chaubey, G. S., Yano, K., Rong, C., & Liu, J. P. (2009). Bimagnetic nanoparticles with enhanced exchange coupling and energy products. *Journal of Applied Physics, 105*(1), 014303.

Sabiryanov, R., & Jaswal, S. (1998). Electronic structure and magnetic properties of hard/soft multilayers. *Journal of Magnetism and Magnetic Materials, 177*, 989−990.

Sabiryanov, R., & Jaswal, S. (1998). Magnetic properties of hard/soft composites: $SmCo_5$/$Co_{1-x}Fe_x$. *Physical Review B, 58*(18), 12071.

Song, Q., & Zhang, Z. J. (2012). Controlled synthesis and magnetic properties of bimagnetic spinel ferrite $CoFe_2O_4$ and $MnFe_2O_4$ nanocrystals with core–shell architecture. *Journal of the American Chemical Society, 134*(24), 10182–10190.

Wang, L., Wang, X., Luo, J., Wanjala, B. N., Wang, C., Chernova, N. A., Engelhard, M. H., Liu, Y., Bae, I.-T., & Zhong, C.-J. (2010). Core–shell-structured magnetic ternary nanocubes. *Journal of the American Chemical Society, 132*(50), 17686–17689.

Zeng, H., Li, J., Liu, J. P., Wang, Z. L., & Sun, S. (2002). Exchange-coupled nanocomposite magnets by nanoparticle self-assembly. *Nature, 420*(6914), 395–398.

CHAPTER

Microwave absorption of ferrite/carbon nanocomposites

6

6.1 Introduction

Society considers electromagnetic (EM) radiation the greatest source of environmental pollution after water, air, and noise sources. EM radiation is defined as EM energy waves propagated to distant points without reflecting back to the wave source. Since EM radiation can influence human health, it has become the focal point of recent research investigations (Hao et al., 2018). The EM wave radiation level in the atmosphere has increased because of the frequent use of high-frequency applications, resulting in serious damage to our ecology and health. In this regard, protracted exposure to microwave (MW) irradiation and EM interference (EMI) can severely affect human health and the environment, weakening the immune system while increasing heart rate and rearranging the protein structure, resulting in fatal cancers. Furthermore, localized heating can be induced in the human body by multiple resonances of the EM waves, which in turn causes dehydration around the eyes, brain, joints, heart, and abdomen. Serious diseases arising from EM localized heating, such as leukemia (Schüz & Ahlbom, 2008) and brain tumors (Baan et al., 2011), have been reported.

In the military sector, the absorption of EM waves in ultra-high-frequency regions, e.g., the L-(1−2 GHz), S-(2−4 GHz), C-(4−8 GHz), X-(8−12 GHz), K_u-(12−18 GHz), K-(18−26.5 GHz), and K_a-(26.5−40 GHz) bands, is used for high-resolution imaging and high-precision target identification. Alternatively, electrical and industrial sectors involve the X-band frequency range for various applications, including Doppler radar, weather radar, maritime radio navigation, satellite radiolocation, TV image transmission, and telephone MW relay systems. Chapter 3 discussed the role of hexagonal ferrites in absorbing media. The current chapter evaluates the MW absorption characteristics of various ferrites and carbon materials and interprets the mechanisms involved in absorption enhancement by incorporating carbon materials.

Magnetic Ferrites and Related Nanocomposites. https://doi.org/10.1016/B978-0-12-824014-4.00007-X

6.2 Shielding effectiveness

Partial differential equations express the EM wave equation as follows:

$$\nabla^2 E = \mu\varepsilon\left(\partial^2 E / \partial t^2\right) \text{ and } \nabla^2 H = \mu\varepsilon\left(\partial^2 H / \partial t^2\right) \quad (6.1)$$

in which μ and ε are permeability and permittivity, respectively, and the components of magnetic and electric fields sinusoidally vary as a function of time (t).

It is theoretically possible to find the average power per unit area of lossless dielectric media in the x-direction. In this case, the Poynting wave equation presented in Eq. (6.2) defines the instantaneous power density when the magnetic field (H) and electric field (E) are in phase:

$$P_x\left(W / m^2\right) = \left(\frac{1}{2}\right) E_m^2/\eta \quad (6.2)$$

where η is the characteristic impedance of the transmission medium. When H_z and E_y are not in time phase (e.g., in lossy dielectric media), the power propagation equation is written as follows:

$$P_x\left(W / m^2\right) = \left(\frac{1}{2}\right) E_m^2/\eta e^{-2\alpha x} \cos\theta_n \quad (6.3)$$

where α is the attenuation constant. Moreover, θ denotes the phase difference between magnetic (H_z) and electric (E_y) wave components. Thus, the power density of the EM wave moving in the medium decreases exponentially.

To obtain the shielding effectiveness (SE) of a material, one can use the ratio of incident (input) EM energy to transmitted (output) EM energy, according to the following equation:

$$\mathrm{SE}_{\mathrm{total}}(dB) = 10\log\left(\frac{P_{\mathrm{input}}}{P_{\mathrm{output}}}\right) = 20\log\left(\frac{E_{\mathrm{input}}}{E_{\mathrm{output}}}\right) = 20\log\left(\frac{H_{\mathrm{input}}}{H_{\mathrm{output}}}\right) \quad (6.4)$$

where P_{input} and P_{output} are the respective powers of the incident and transmitted waves. The total SE of EMI ($\mathrm{SE}_{\mathrm{total}}$) results from the sum of shielding due to reflection (SE_R), absorption (SE_A), and multiple reflection (SE_M), as given below:

$$\mathrm{SE}_{\mathrm{total}}(dB) = \mathrm{SE}_R + \mathrm{SE}_A + \mathrm{SE}_M \quad (6.5)$$

Commonly, for $\mathrm{SE}_A > 10$ dB, one can ignore the multiple reflections, since the reflected waves propagating in the material are absorbed (Jaroszewski et al., 2018).

The terms SE_A and SE_R can also be expressed by Eqs. (6.6) and (6.7), respectively:

$$\mathrm{SE}_A = 8.87\gamma d \quad (6.6)$$

$$\mathrm{SE}_R = 10\log(1 - R) \quad (6.7)$$

where d is the shielding material thickness, γ is the wave propagation constant in the conducting medium, and R is the reflection coefficient. The parameters γ and R can be expressed by

$$\gamma = \omega\sqrt{\left(\frac{\varepsilon'\mu}{2}\right)}\left[\sqrt{\left(1+\frac{(\sigma+\omega\varepsilon')}{\omega\varepsilon'}\right)^2}+1\right]^{\frac{1}{2}}, \omega = 2\pi f \tag{6.8}$$

$$R = \left(\frac{Z_0 - Z_{in}}{Z_0 + Z_{in}}\right)^2 \tag{6.9}$$

where ω, f, ε', μ, and σ are the angular frequency, frequency, real part of permittivity, permeability, and electrical conductivity, respectively. Moreover, Z_0 and Z_{in} are the characteristic impedance and input impedance, respectively.

The discontinuity between air and metal impedance causes EM waves to be reflected when they reach the shield surface. In this way, the absorbing process can attenuate the unreflected energy from the surface when entering the material. At the air–metal impedance interface, the unattenuated energy remaining is reflected, returning to the shielding body when passing to the side surface of the shielding material. This may cause multiple reflections of the EM waves at the interfaces of the two metals. The multiple reflections, defined as the reflected waves from various surfaces and interfaces, are normally neglected because they are absorbed in the material (Chung, 2001).

The presence of mobile charge carriers such as electrons or holes in the shielding materials is necessary to reflect EM waves as they interact with H and E fields. Thus, the shielding materials must be conductive to reach high reflectivity. The conductive material impedance is smaller than the air impedance. Consequently, the larger mismatch of the waves propagating in the air causes the EM waves to be reflected. The reflection loss (RL) depends on the ratio of conductivity to permeability; the reflection coefficient reaches about 1.0 for completely mismatched media at the interface. In other words, for completely matched media at the interface, the reflection coefficient will equal zero, thereby penetrating the EM waves into the material. Accordingly, the trade-off between intrinsic physical properties, including conductivity, permittivity, and permeability, is critical in obtaining the desired shielding.

6.3 Absorbing media

For EMI shielding applications, metal sheets are considered effective materials. Nevertheless, they suffer from processing difficulty, low flexibility, heavy weight, and high cost while being prone to oxidation, limiting the use of metal sheets for practical applications (Chung, 2001). Alternatively, while metal coatings resulting from electroplating and spraying processes, along with vacuum coatings and conductive paints, have been developed for EMI shielding applications, their poor environmental and wear resistance have made them undesirable for practical purposes (Geetha et al., 2009).

MW-absorbing materials have been suggested to enhance EM shielding properties. These materials have attracted considerable technical and scientific interest because of their increasing importance for specific applications such as silent rooms and radar systems. Furthermore, MW-absorbing materials can help in the long-term protection of army personnel and equipment from detection by radar devices, stealth aircraft and warships, airings from reduced EMI, and radar signatures (Matzui et al., 2007).

Magnetic or dielectric materials are required to realize SE through absorption. Internal *H* and *E* fields can be created when such materials are exposed to EM waves, leading or lagging the applied field. Thereby, the resultant SE caused by the absorption is induced by magnetic and dielectric losses (Mishra et al., 2014).

Ferrite materials with high magnetic loss can act as magnetic loss absorbers owing to their tunable coercivity and anisotropy field (controlled by adding substitutional cations in the crystal structure), high environmental stability, suitable permeability, and excellent EM wave absorption capacity. However, they suffer from high density and a narrow MW absorption band, limiting their practical applications (Zhu et al., 2017; Hapishah et al., 2018). Alternatively, carbon-based materials and their composites have been suggested as the best candidate materials for EMI shielding purposes, owing to their light weight, high flexibility, chemical inertness, low cost, simple processability, and environmental friendliness. As an intrinsic dielectric, carbon-based materials, including ferrite/carbon nanocomposites with intrinsic dielectric properties and a high degree of carbonation, can be employed for enhancing MW absorption (Jang et al., 2014). The power loss of EM waves generated is enhanced by the interactions of *E*- and *H*-components with electric and magnetic dipoles existing in the ferrite/carbon nanocomposites. The combined effects of magnetic and the dielectric losses enable us to enhance the corresponding absorption.

6.4 Types of carbon materials

Various carbon materials, including carbon nanotubes (CNTs), carbon black (CB), carbon nanofibers (CNFs), carbon nano-onions, graphene, carbon sheets, and porous carbons, are increasingly in demand as absorbing materials (Sattler, 2016; Sun et al., 2017). Discovered in 1991 by Iijima, CNTs are one of the most popular carbon nanomaterials (Iijima, 1991). CNTs are made of several graphite sheets rolled up to form a tube. Depending on the number of graphite walls, CNTs are classified into three types—single-walled CNTs (SWCNTs), double-walled CNTs (DWCNTs), and multiwalled CNTs (MWCNTs)—as shown schematically (from left to right) in Fig. 6.1. The diameter of SWCNTs is 1–2 nm, whereas DWCNTs and MWCNTs have respective outer diameters of 2 to 25 nm and larger than 100 nm. CNTs with nanometric dimensions and high aspect ratios have shown excellent chemical stability, high Young's modulus, and enhanced electrical properties such as large permittivity. Notably, the current-carrying capacity of CNTs is 1000 times greater than that of copper wire (Collins & Avouris, 2000). The strength of SWCNTs is 10–52 GPa. Furthermore, the respective Young's modulus and room-temperature thermal

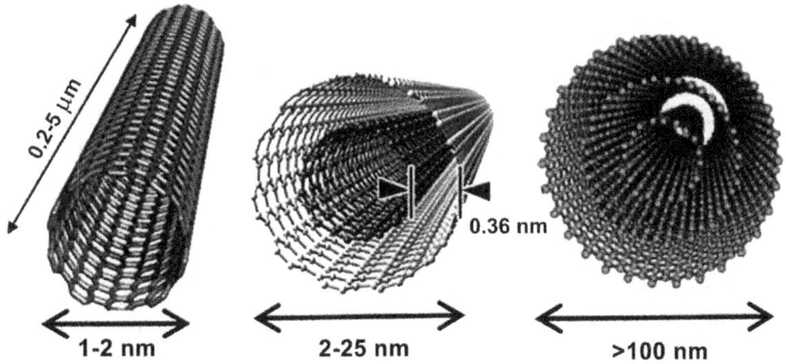

FIGURE 6.1

A schematic representation of single-walled, double-walled, and multiwalled carbon nanotubes.

Reproduced from Moghadam, A. D., Omrani, E., Menezes, P. L., & Rohatgi, P. K. (2015). Mechanical and tribological properties of self-lubricating metal matrix nanocomposites reinforced by carbon nanotubes (CNTs) and graphene–a review. Composites Part B: Engineering, 77, *402–420.*

conductivity of SWCNTs are 0.32–1.47 TPa and 1750–5800 W/mK (Moghadam et al., 2015).

Following electronic level splitting, the quantum size and macroscopic quantum tunneling effects in CNTs cause energy gaps to stay within the MW energy range. Concurrently, other CNTs characteristics, including high specific surface area and a large number of surface suspension parts, lead to multiple scattering and interfacial polarization, thereby enhancing the absorbing properties of CNTs (Liu, 2005). Essentially, MWCNTs networks are responsible for the MW absorption mechanism and involve more defects than SWCNTs.

As a result, the defect structure of MWCNTs with a high degree of functionalization and various bonding links leads to interfacial electric polarization, thus continuously dissipating energy in alternating EM fields. Under such EM fields, space charge polarization is formed by the defects and active groups on MWCNTs surfaces. Increasing the volume of functionalized MWCNTs in composites increases the formation of dense networks, thereby decreasing the distance between MWCNTs while facilitating electron flow through the network. Therefore, filler loading can increase electrical conductivity and RL. It should be noted that magnetic tangent loss is not considerably influenced by CNTs, as expected from their nonmagnetic behavior (Zhan et al., 2018).

Up to now, several techniques, including sol–gel, coprecipitation, chemical decoration, electrochemical deposition, electroless plating, solid-state reaction, chemical evaporation deposition, supercritical carbon dioxide deposition, layer-by-layer assembly, and polymer wrapping, have been developed to deposit nanoparticles onto CNTs. These depositions induce π interactions, forming bonding between nanoparticles and MWCNTs (Abdel Fattah et al., 2016).

When the EM wave is incident on magnetic nanocomposites containing ferrite nanoparticles and MWCNTs, eddy current is induced by the periodic changing of *H*. In turn, the eddy current flows in the MWCNTs due to their conductive properties, giving rise to eddy loss. By periodically reversing the applied field direction in each cycle, the magnetized ferrite nanoparticles encapsulated in the MWCNTs will not be capable of following the magnetic field without phase difference, which leads to magnetic hysteresis losses. An electromotive force is then induced, allowing for the flow of current through the conducting MWCNTs, which opposes the change in the flux density based on Lenz's law (Stavarache et al., 2011).

Graphite compounds with low density, large surface area, anti-high temperature, and low-cost preparation are known as expanded graphite (EG). They provide excellent electrical conductivity due to their large sheet spacing of up to 1 μm (Zhang et al., 2011).

As the thinnest material in the world, graphene with a monolayer of hexagonally arranged sp^2-bonded atoms has a honeycomb lattice packed into a two-dimensional (2-D) crystal of hybridized carbon. This unique carbon structure is considered the building block of all graphitic materials, which can be prepared by graphite nanolayers using chemical or physical approaches (MacDonald & MacDonald, 2007; Geim & Novoselov, 2010). Thus far, graphene sheets in the form of EG, EG oxide (EGO), and reduced EGO (REGO) have been employed as absorbers in various media.

It is well known that the absorption mechanism in graphene is influenced by its configuration. For example, graphene foam outperforms graphene film in terms of SE_{Total}. This arises from the porous foam structure, dominating the shielding process by the EM wave absorption and improving multiple internal reflections (Shen et al., 2016). Graphene has been doped with various elements, resulting in significantly enhanced electrical conductivity, space charge polarization, carrier density, EM wave trapping, and dielectric polarization while facilitating EMI shielding for absorption purposes (Umrao et al., 2015; Wan et al., 2017).

While graphene is intrinsically nonmagnetic, adding graphene sheets into nanocomposites might induce interfacial magnetic domains, slightly increasing magnetic permeability. Electronic transmission efficiency can be reduced by the C—C interfacial layers of reduced graphene oxide (RGO)/MWCNTs composites. In fact, the mismatched electronic structures of these composites further influence the transmission of EM waves (Zhou et al., 2017).

As a limitation, carbon materials, especially CNTs, show poor dispersion properties in ferrite solutions. Recent studies have evidenced that stable aqueous dispersion of graphene oxide (GO)/MWCNTs is achieved when the amphiphilic GO is directly added to the pristine MWCNTs, thus acting as a surfactant. This way, one can overcome poor MWCNTs dispersion properties and electronic structural damage.

CNFs are carbon materials that have received much interest due to their high chemical and thermal stability, large surface-to-volume ratio, high electrical and thermal conductivity, and extensive mechanical properties. Short conductive nanofibers have been employed as EM wave absorbers, acting as electric dipoles. On the other hand, the unique network structure of graphite nanosheets (GNs) allows for large storage and dissipation capacity of EM energy, which is advantageous for enhancing impedance matching levels in absorber materials. In this regard, the chemical and physical properties (e.g., thickness and surface area) of GNs are highly dependent on the oxidation and reduction processes occurring during the experiments.

Porous carbon materials could also provide multiple reflections in nanocomposites, thus enhancing MW attenuation. Several new approaches, including metal-organic frameworks, have been used for processing porous media, attracting tremendous attention due to their high surface area and tunable chemical structures (Lü et al., 2015; Mahmood et al., 2016). Interestingly, the size and structure of the resulting porous carbon materials can be tailored by arbitrarily combining nanostructures as building blocks.

6.5 Polymers as supporting media for absorption

EM wave absorption properties are expected to be enhanced by dispersing nanoscale composites of ferrite-carbon materials in a supporting matrix with suitable mechanical characteristics, good adhesion bond, and perfect conductivity. Such a supporting matrix can consist of conductive polymers such as polyaniline (PANI), polypyrrole (PPy), epoxy, polythiophene derivatives, polyacrylamide, and phenylalanine.

Known as "synthetic metals," intrinsically conducting polymers with shallow skin depths have been widely developed, providing reflection-dominant shielding (Wang & Jing, 2005). Among the intrinsically conducting polymers, PANI with tunable properties has attracted more attention from researchers owing to its unique nonredox doping, processability, variable configuration, special doping mechanism, and excellent thermal, chemical, and environmental stability (Bhadra et al., 2009; Qiu et al., 2010). Moreover, PANI conductivity can be easily controlled by changing the protonation and oxidation states. According to the literature (Yang et al., 2005), PANI shows the highest dielectric constant among conducting polymers, which is attributed to the disordered motion of its charge carriers along the conducting chains. In fact, arising from its intrinsic metallic nature, the dielectric properties of PANI can be tuned by conductivity and charge delocalization. Accordingly, high conductivity and dielectric properties are achieved by increasing the PANI content.

On the other hand, a p-type semiconducting polymer, PPy, has relatively low density, high conductivity, and excellent environmental stability. PPy-based composites can be easily prepared and used in various fields such as EMI shielding, rechargeable batteries, and MW-absorbing materials (Yang & Liu, 2009; Zhang et al., 2009).

Epoxy resin materials with thermoset properties and a low number of free charge carriers have also been employed as electrical insulator matrices, providing extraordinary applications in various industries. Depending on the filler shape and concentration, it is possible to incorporate conducting materials with the epoxy resin to manipulate its electrical conductivity (Yang & Liu, 2009; Zhang et al., 2009).

6.6 Permittivity and permeability of absorbing media

Different types of dielectric polarization, such as electron, atomic, ion, molecular, and space charge polarizations, are responsible for dielectric loss in ferrite/carbon nanocomposites. Ionic covalent bonds restrict positive/negative bond charges and avoid their directional migration, thereby forming dipole moments. In turn, these dipoles, along with the resultant relaxation polarization, can dominate the dielectric loss. The dipole interaction is responsible for the interaction between the material and MW, requiring efficient coupling between the target material components and the oscillating MW electric field. The polar ends of molecules are involved in the dipole interaction mechanism, tending to be aligned and oscillate in step with the rapidly oscillating MW electric field. In general, increasing the polarity of molecules enhances their coupling with the oscillating EM field.

Applying an alternating E field orients the electric dipoles when changing the field each time. In this respect, relaxation time is defined as the average time needed to orient dipoles with the E field direction. Moreover, the reciprocal of the relaxation time is known as the relaxation frequency. The polarization mechanism stops for applied field frequencies exceeding its relaxation frequency. In the MW region, orientation polarization is zero for dipoles failing to follow the field reversal, leading to a constant reduction in the dielectric properties.

Charge multipoles at the interface can interact with EM waves, facilitating the MW absorption by interfacial polarization. This kind of polarization is affected by structure and morphology. It has been found that increasing defects and dangling-bonded atoms in the material increase interface polarization. In the case of polycrystalline ferrite materials, the intrinsic electric dipole and interfacial polarization are responsible for the corresponding dielectric properties. The latter originates from the heterogeneous structure of ferrites with separated, highly conducting grains and poorly conducting layers, according to Koops theory (Koops, 1951).

In the case of composite materials, the dielectric loss increases due to the formation of large numbers of interfaces and voids, thereby facilitating the space-charge polarization.

On the other hand, the space-charge contribution is considerably affected by materials heterogeneity (or inhomogeneity). It is feasible to reach high heterogeneity using many MWCNTs, increasing the disordered motion of charge carriers. This results in higher dipole rotation, relaxation of space charge, and hopping of confined charges.

The dissipation of external electric energy in the bulk of materials is determined by the dielectric tangent loss ($\tan\delta_e$). In the case of magnetic materials, the energy attenuation ratio is determined by the magnetic tangent loss ($\tan\delta_m$), arising from the phase delay between the induced and applied H fields. In this way, based on

$\tan\delta_e$ and $\tan\delta_m$, the attenuation constant (α) indicates the material attenuation capability, as given below:

$$\alpha = \frac{\pi f}{c}\sqrt{\mu'\varepsilon'}\sqrt{2(\tan\delta_e \tan\delta_m - 1 + \sqrt{1 + \tan^2\delta_e + \tan^2\delta_m + \tan^2\delta_e\delta_m})} \tag{6.10}$$

where c, f, μ', and ε' are the velocity of light, frequency, the real part of permeability, and the real part of permittivity, respectively.

Increasing the ratio of imaginary permittivity to real permittivity ($\varepsilon''/\varepsilon'$) along with an increase in α will lead to an enhancement in the RL.

The dielectric loss mechanism involved in MW absorbers can be described by the Debye dipolar relaxation. In this respect, the permittivity can be given by Eq. (6.11):

$$\varepsilon_r = \varepsilon_\infty + \frac{\varepsilon_s - \varepsilon_\infty}{1 + j2\pi f\tau} = \varepsilon' - j\varepsilon'' \tag{6.11}$$

where ε_s is the static permittivity, ε_∞ is the relative dielectric permittivity at the high-frequency limit, τ is the polarization relaxation time, and f is the frequency. Therefore, ε' and ε'' are obtained by Eqs. (6.12) and (6.13) (He et al., 2013; Lin et al., 2015):

$$\varepsilon' = \varepsilon_\infty + \frac{\varepsilon_s - \varepsilon_\infty}{1 + (2\pi f)^2\tau^2} \tag{6.12}$$

$$\varepsilon'' = \frac{2\pi f\tau(\varepsilon_s - \varepsilon_\infty)}{1 + (2\pi f)^2\tau^2} \tag{6.13}$$

It is also possible to find the relationship between ε' and ε'' as follows (He et al., 2013; Lin et al., 2015):

$$\left(\varepsilon' - \frac{\varepsilon_s - \varepsilon_\infty}{2}\right)^2 + (\varepsilon'')^2 = \left(\frac{\varepsilon_s - \varepsilon_\infty}{2}\right)^2 \tag{6.14}$$

According to Eq. (6.14), single semicircles result from the ($\varepsilon'' - \varepsilon'$) curves, representing the Cole–Cole semicircles. A Debye relaxation process can correspond to each semicircle.

Apart from the permittivity, permeability can also play an important role in the resultant absorption properties. Based on the ferromagnetic theory, four frequency categories are considered for the permeability spectrum: low, intermediate, high, and ultra-high. Notably, the simultaneous appearance of domain wall resonance and natural resonance occurs in the high-frequency spectrum ($10^6 < f < 10^{10}$ Hz). For MW-absorbing materials, strong magnetic loss ability for high initial permeability (μ_i) is expected. Initial permeability is defined by the following formula (Ding et al., 2017):

$$\mu_i = \frac{M_s^2}{akH_cM_s + b\lambda\xi} \tag{6.15}$$

where a and b are two constants relating to the MW-absorbing material composition. Moreover, λ is a magnetostriction constant, ξ is the elastic strain parameter of the crystal, M_s is the saturation magnetization, k is a proportionality coefficient, and H_c is the coercivity.

In the following, the capability of carbon nanomaterials with various configurations in enhancing the absorption mechanism of ferrite nanocomposites will be discussed.

6.7 Hard ferrite/carbon nanotube nanocomposites

The outstanding physical and chemical properties of CNTs combined with the excellent magnetic properties of strontium ferrite (SrF) nanoparticles have made SrF/CNTs nanocomposites with high permeability and RL extremely attractive for practical applications in high-frequency MW-absorbing nanocomposites.

Several new MW absorbers have been proposed by my research group (Ghasemi, Liu, 2009; Ghasemi, Sepelak, et al., 2009; Ghasemi et al., 2010). In this regard, the first study was carried out on the formation of SrF thin films on functionalized MWCNTs in 2011 (Ghasemi, Sepelak, et al., 2011). Fig. 6.2A shows a typical high-resolution transmission electron microscopy (HRTEM) micrograph of substituted SrF nanoparticles with the composition of $SrFe_{12-x}(Cr_{0.5}Al_{0.5})_xO_{19}$. As observed, the particle size of ferrite nanoparticles ranges between 10 and 20 nm. Fig. 6.2B shows a scanning electron microscopy (SEM) micrograph of pristine MWCNTs, featuring the surface morphology. The diameter of the MWCNTs is in a range of approximately 50–100 nm. Fig. 6.2C and D show typical SEM micrographs of the substituted SrF thin film/MWCNTs ($x = 2.5$), indicating the formation of the thin film along the MWCNTs. Due to the bonding between the thin film and the MWCNTs, the resulting hard ferrite/CNTs nanocomposites would remain stable for subsequent applications.

The tendency of nanoparticles to decrease their surface energy is considered the main driving force for the aggregation of nanoparticles. The dipolar interactions between magnetic nanoparticles play a constructive role in their oriented aggregation.

The ferrite/CNTs composites suffer from two main problems: CNTs dispersion in the matrix and the interface between CNTs and the matrix. Essentially, CNTs are hardly dispersed in aqueous solutions. Moreover, weak interfacial strength is obtained due to the poor chemical compatibility between ferrite materials and CNTs. Therefore, the agglomeration of CNTs is rather unavoidable during the fabrication of ferrite/CNTs nanocomposites, requiring surface modification before the nanocomposite preparation. MWCNTs surface modifications involve physical or chemical treatments. Nevertheless, the physical treatment leads to poor miscibility of CNTs, followed by their reagglomeration over time. Thus, chemical treatment is usually considered to functionalize CNTs.

CNTs dispersion can be significantly improved by attaching aliphatic carbon chains and functional groups to the surface, chemically modifying CNTs through surface chemical bonds. Notably, in a separate study, Ghasemi and colleagues have functionalized CNTs using nitric acid, followed by their dispersion in

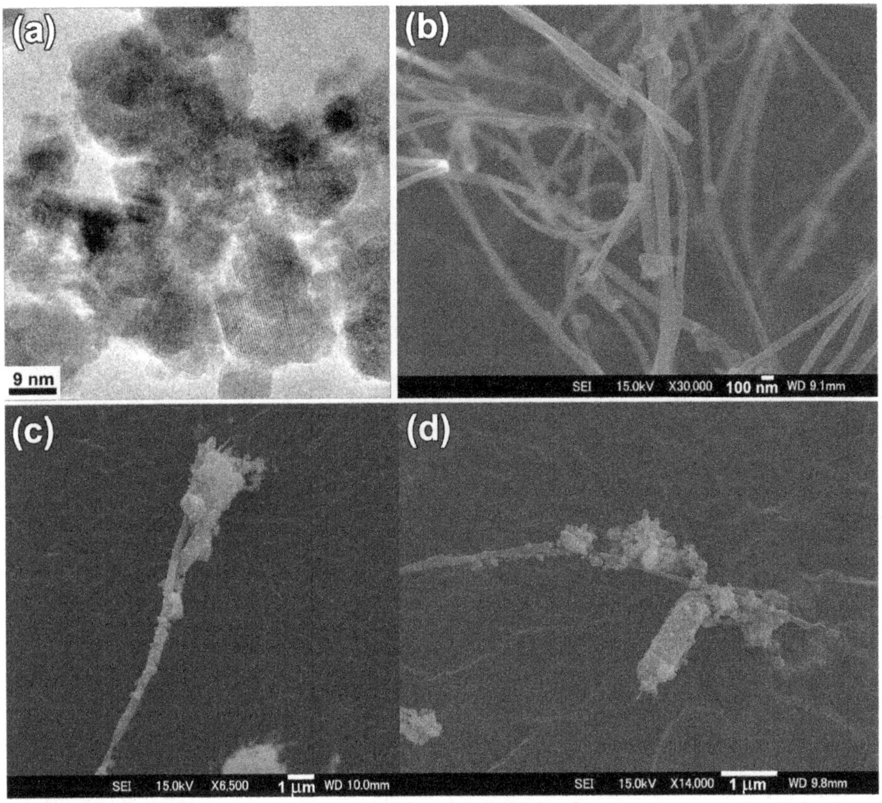

FIGURE 6.2

(A) High-resolution transmission electron microscopiy micrograph of ferrite nanoparticles with the composition of $SrFe_{12-x}(Cr_{0.5}Al_{0.5})_xO_{19}$. Scanning electron microscopy micrographs of (B) pristine multiwalled carbon nanotubes (MWCNTs) and (C) and (D) strontium ferrite thin film ($x = 2.5$)/MWCNTs nanocomposite.

poly(acrylic acid) (PAA) solution under high-power ultrasonication for 60 min. The solution containing CNTs was then dipped to $SrFe_{12-x}(Ni_{0.5}Co_{0.5}Sn)_{x/2}O_{19}$ solution under ultrasonication (Ghasemi et al., 2012).

Fourier-transform infrared (FTIR) spectroscopy provides very useful information, enabling us to investigate the presence of functional groups on the sidewalls of MWCNTs. Fig. 6.3 shows FTIR spectra obtained from acid-pretreated MWCNTs and acid-pretreated MWCNTs-PAA. In the 2800–3000 cm^{-1} region, the peaks observed are characteristics of C–H stretching. The peak at 1730 cm^{-1} in Fig. 6.3B is attributed to the carbonyl group stretching modes (C=O), indicating the presence of PAA chains in the MWCNTs-PAA. Additionally, the peaks at 3434 and 1630 cm^{-1} correspond to the O–H stretching modes. Due to the residual nitrate groups in the precursors, the peak appears at 1384 cm^{-1}, reflecting the

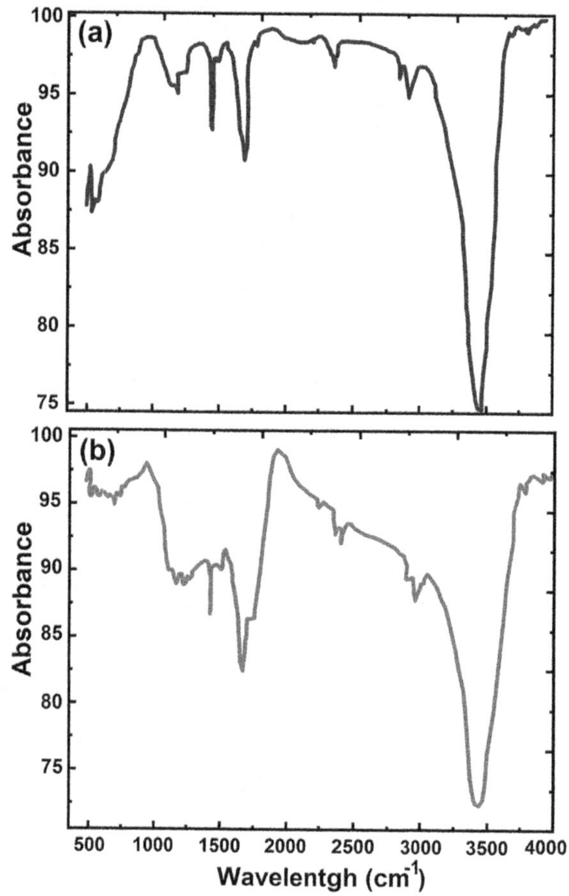

FIGURE 6.3

Fourier-transform infrared spectra of (A) acid-pretreated multiwalled carbon nanotubes (MWCNTs) and (B) acid-pretreated MWCNTs-poly(acrylic acid).

asymmetric NO_3^- stretching vibration. These results are in agreement with previous reports (Luo et al., 2001).

CNTs have represented some of the most promising matrix structures for ferrite composites. Especially, CNTs volume fractions can influence the MW properties of nanocomposites. In this case, MW properties of 8, 10, 12, and 14 vol% CNTs added to Ni–Co–Sn-substituted SrF solution have been investigated. Fig. 6.4 shows the variation in RL as a function of frequency for nanocomposite samples with a thickness of 1.5 mm. Here, the frequency width with RL of >−10 dB defines the bandwidth. As can be seen, SrF nanoparticles show RL values smaller than −10 dB. The bandwidth covered by Ni–Co–Sn-substituted SrF nanoparticles, i.e., $SrFe_{12-x}(Ni_{0.5}Co_{0.5}Sn)_{x/2}O_{19}$ ($x = 3$) is approximately 1 GHz.

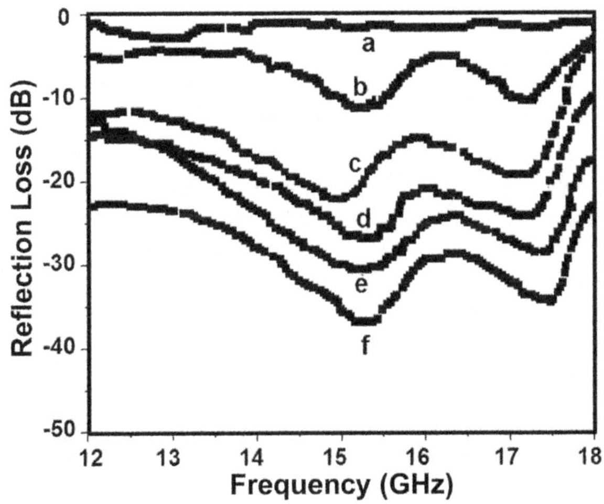

FIGURE 6.4

Reflection loss as a function of frequency for: (A) SrF, (B) $SrFe_{12-x}(Ni_{0.5}Co_{0.5}Sn)_{x/2}O_{19}$ ($x = 3$), and nanocomposites with carbon nanotube volume fractions of (C) 8 vol%, (D) 10 vol%, (E) 12 vol%, and (F) 14 vol%.

Reproduced from Ghasemi, A., Shirsath, S. E., Liu, X., & Morisako, A. (2012). A comparison of the magnetic and reflection loss characteristics of substituted strontium ferrite and nanocomposites of ferrite/carbon nanotubes. Journal of Applied Physics, 111*(7), 07B543.*

Pure SrF nanoparticles have a resonance frequency of about 48 GHz. It is well known that the respective natural resonance frequency (f_0) for easy c-plane and easy c-axis anisotropies can be given by Eqs. (6.16) and (6.17) (Smit, 1971):

$$f_0 = \frac{\gamma}{2\pi}(H_\theta H_\Phi)^{1/2} \tag{6.16}$$

$$f_0 = \frac{\gamma}{2\pi}(H_a) \tag{6.17}$$

where H_Φ and H_θ are the anisotropy fields in the c-plane and along the c-axis, respectively, and γ is the gyromagnetic ratio $\left(\frac{\gamma}{2\pi} = 2.8\,\text{MHz/Oe}\right)$. In general, the anisotropy field $(H_a)^2$ exceeds the product of H_θ and H_Φ due to the small value of H_Φ (a few hundred Oe). Increasing the substitutions decreases the anisotropy field, thereby rotating the magnetization direction from uniaxial to nearly planar. The corresponding resonance frequencies are thus reduced, revealing the resonant characteristics when $x = 3$ (Fig. 6.4B).

Fig. 6.4C–F display the frequency dependence of RL for substituted SrF($x = 3$)/CNTs nanocomposites at various CNTs volume fractions. The maximum RL of the Ku band is -39 dB at a frequency of 15.2 GHz, with another frequency at 17.5 GHz with an RL of -37 dB. Increasing the CNTs content in the nanocomposite increases the bandwidth associated with the RL value. By introducing 14 vol% CNTs into the ferrite nanoparticles, the corresponding maximum RL increases from −11 to −39 dB while enhancing the bandwidth from 1 to 6 GHz.

By matching the magnetic and dielectric contributions, MW absorption must increase to mathematically meet the requirement of the tan*h* function of input impedance (Formula 2.84). Dielectric loss only contributes to the EM wave energy loss of pure CNTs, whereas magnetic loss dominates the dielectric loss in the case of SrF particles. In other words, the magnetic and dielectric losses of both SrF and CNTs particles are out of balance, thus inducing poor EM wave absorption. Nevertheless, the MW absorption of SrF/CNTs nanocomposites is expected to improve due to the good matching between their dielectric and magnetic losses.

In principle, the magnetic spectrum of most magnetic materials comprises two resonances. Thus, the observation of a single resonance in some cases may indicate the coalescence of two resonances. In the case of nanoparticles, the detection of multiresonance behavior is a subject of controversy, according to the literature (Aharoni, 1991; Deng et al., 2007).

According to the exchange resonance mode, resonance frequencies can be obtained as follows:

$$\frac{\omega}{\gamma} = \frac{C\mu_{kn}^2}{R^2 M_s} + H_z \tag{6.18}$$

$$H_z = H_0 - NM_s + H_a \tag{6.19}$$

where γ is the gyromagnetic ratio, C is the exchange constant, R is the sphere particle radius, H_z is the static magnetic field, H_0 is the applied DC field, N is the demagnetization factor, M_s is the saturation magnetization, and H_a is the magnetocrystalline anisotropy field. Moreover, the eigenvalues of μ_{kn} are the roots of the equation, as expressed below:

$$\left[\frac{dj_n \mu_{kn} r/R}{dr}\right]_{r=R} = 0 \tag{6.20}$$

where j_n represents the spherical Bessel functions. By considering a nanoparticle as a nanograin (nanocrystallite), one can replace R with the crystallite size (D) and use H_c instead of H_z (Deng et al., 2007). Thus, Eq. (6.18) is rewritten as follows:

$$\frac{\omega}{\gamma} = \frac{C\mu_{kn}^2}{D^2 M_s} + H_c \tag{6.21}$$

Note that resonance frequency is strongly dependent on $(D^2 M_s)^{-1}$. In this regard, increasing D increases the exchange energy, so it is impossible to detect the exchange resonance peak for ferrite samples with a crystallite size of 50 nm under the given conditions. As a result, one can conclude that the modified Aharoni's method does not apply to this ferrites system. However, the results of Wen and colleagues were quite consistent with Aharoni's model, proposing nanoparticle size and geometry as important factors in the multiresonance phenomenon (Wen et al., 2008).

The natural and exchange resonances may overlap, which is advantageous for large-bandwidth MW absorbers. The magnetic spectra observed in Fig. 6.4 agree with the natural magnetic resonance mechanism, involving domain wall motion and spin rotation. The domain wall motion lags behind the magnetic field applied to the material, leading to magnetic loss. The relations corresponding to the resonance-relaxation phenomena near the domain wall motion or spin rotation characteristic frequency can be found elsewhere (Hur et al., 2004; Kaczmarek & Ninham, 1994).

Based on the RL curves depicted in Fig. 6.4, f_{dw} (i.e., the domain wall motion resonance frequency) and f_s (i.e., the spin rotation resonance frequency) are 15 and 17.1 GHz for ferrite nanoparticles, respectively. In addition, the respective f_{dw} and f_s of SrF/CNTs nanocomposites (14 vol%) are 15.1 and 17.5 GHz. Clearly, CNTs additions affect spin rotation resonance frequency much more significantly than domain wall motion resonance frequency.

The influence of CNTs volume fractions (1, 2, 3, and 4 vol%) on magnetic and MW properties of $SrFe_{10}(Sn_{0.5}Zn_{0.5})_2O_{19}$ nanocomposites has also been investigated (Ghasemi, 2011). The static repulsion occurring between active groups present on CNTs surfaces causes CNTs to be dispersed uniformly in the nanocomposites. Accordingly, ferrite nanoparticles have the opportunity to be arranged along the one-dimensional (1-D) radial structure of CNTs in addition to an increase in their orientational order. Moreover, the CNTs skeleton structure can act as a supporting matrix, reducing the agglomeration and magnetic interactions of ferrite particles with nanometer-scale size.

Fig. 6.5A shows a transmission electron microscopy (TEM) micrograph of a pristine MWCNT with a diameter of approximately 150–200 nm. A typical TEM micrograph of MWCNT/SrF thin film is also depicted in Fig. 6.5B, indicating the formation of uniform ferrite thin films with a thickness of 30 nm on the MWCNT. The nanocomposite was initially redispersed in water under sonication for 60 min and then prepared for TEM observation to examine the stability of the MWCNT-ferrite structure for possible subsequent applications. The nanocomposite structure remained intact after the sonication, indicating good stability of the ferrite film assemblies on the MWCNTs. The formation of the oriented thin films is mainly governed by the decreasing tendency of the high surface energy. The film thickness is relatively uniform even at the MWCNT tip, according to Fig. 6.5C. The active groups on MWCNTs surfaces may also adsorb ferrite nanoparticles, preventing their collision and reunion. Furthermore, the functionalized MWCNTs can be considered as nucleation sites for the directional growth of ferrite films along their walls.

The variation in RL versus frequency for a 1.5 mm thick SrF/MWCNTs nanocomposites with different MWCNTs volume fractions is shown in Fig. 6.6. The bandwidth covered by the nanocomposites is higher than 3 GHz. The corresponding RL of this band is obtained to be −25 dB at a frequency of 11.4 GHz. One can also find another frequency at 9.4 GHz, giving rise to an RL of −27 dB. The enhanced matching between the dielectric and magnetic losses improves the MW absorption of the nanocomposites.

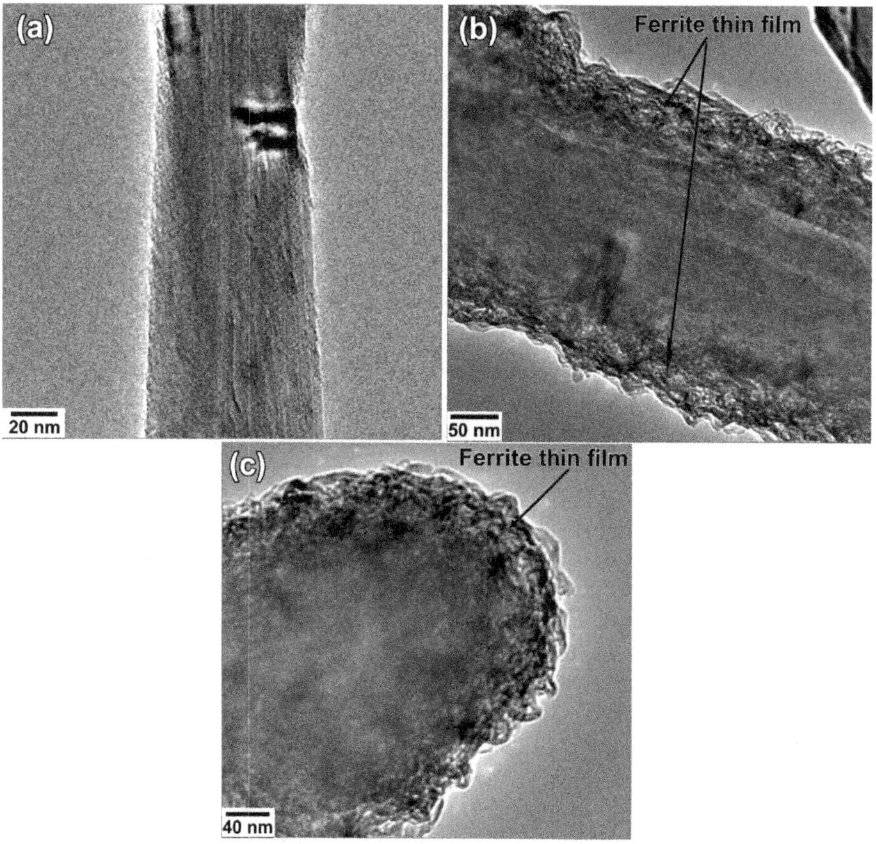

FIGURE 6.5

Transmission electron microscopic micrographs of (A) pristine multiwalled carbon nanotube (MWCNT), (B) MWCNT/strontium ferrite thin film, and (C) ferrite thin layer on the MWCNT tip.

Reproduced from Ghasemi, A. (2011). Remarkable influence of carbon nanotubes on microwave absorption characteristics of strontium ferrite/CNT nanocomposites. Journal of Magnetism and Magnetic Materials, *323(23), 3133–3137. https://doi.org/10.1016/j.jmmm.2011.06.070.*

The atoms existing on the crystalline grain surfaces comprise various coordination states for thin films. In this regard, decreasing the nanocrystalline grain size of the composite increases the number of unsaturated coordination atoms, so the multiple scatterings and interfacial polarizations govern the attenuation.

Dispersing ferrite-MWCNTs in PANI with superior mechanical characteristics and conductivity may hold several advantages for absorption properties. The PANI/SrF/MWCNTs nanocomposites have been synthesized using an in situ polymerization approach, followed by investigations of their EM properties (Li et al., 2012). From Fig. 6.7, an overall upward trend is observed in the frequency range

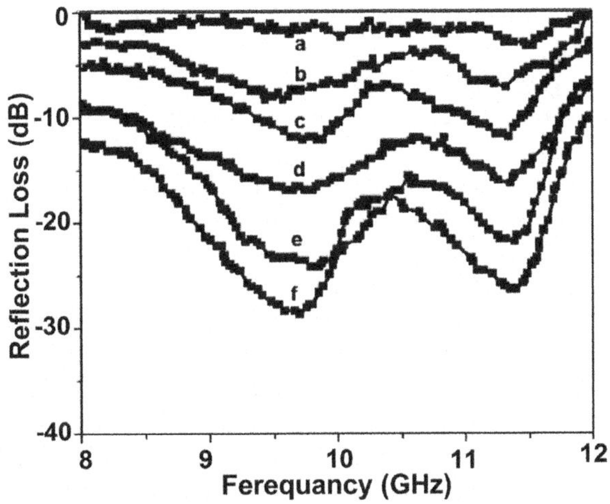

FIGURE 6.6

Reflection loss versus frequency for: (A) Multiwalled carbon nanotubes (MWCNTs) and SrF/MWCNTs nanocomposites with: (B) 0, (C) 1, (D) 2, (E) 3, and (F) 4 vol% MWCNTs.

Reproduced from Ghasemi, A. (2011). Remarkable influence of carbon nanotubes on microwave absorption characteristics of strontium ferrite/CNT nanocomposites. Journal of Magnetism and Magnetic Materials, *323(23), 3133–3137. https://doi.org/10.1016/j.jmmm.2011.06.070.*

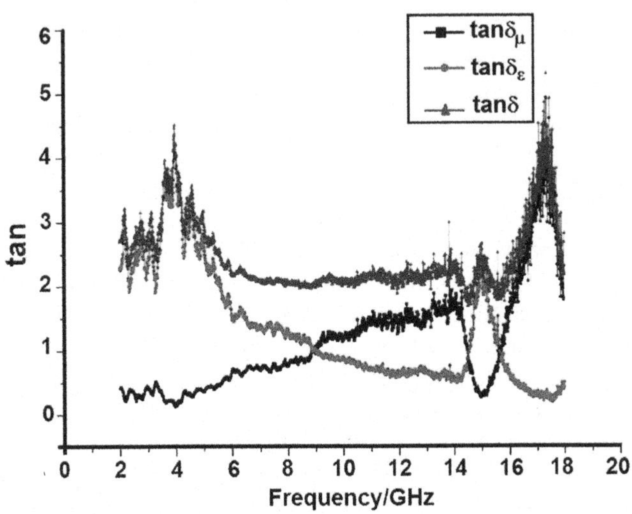

FIGURE 6.7

The electromagnetic parameter spectra of polyaniline/SrF/multiwalled carbon nanotube (MWCNTs) nanocomposites prepared with 2 g of MWCNTs.

Reproduced from Li, Y., Huang, Y., Qi, S., Niu, L., Zhang, Y., & Wu, Y. (2012). Preparation, magnetic and electromagnetic properties of polyaniline/strontium ferrite/multiwalled carbon nanotubes composite. Applied Surface Science, *258(8), 3659–3666. https://doi.org/10.1016/j.apsusc.2011.12.001.*

of 2–18 GHz for the magnetic loss angle tangent of the PANI/SrF/MWCNTs nanocomposites, reaching a minimum and maximum of 0.17 and 3.92 at the frequencies of 4 and 17 GHz, respectively.

The dielectric loss angle tangent outperforms the magnetic loss angle tangent in the frequency range of 2–9 GHz, reaching a maximum of 4.35 at 4 GHz. In addition, while the dielectric loss angle tangent slightly changes in the frequency range of 9–14 GHz, it changes significantly in the range of 14–18 GHz, reaching a maximum and minimum of 2.31 and 0.21 at 15 and 17.5 GHz, respectively. As well, the nanocomposite has minimum (1.7) and maximum (5.28) gross loss angle tangents at 14.5 and 17.3 GHz, respectively. Consequently, the frequency bands are optimized in the range of 2–6 GHz and 16–18 GHz, indicating that the MW absorption is mainly governed by the dielectric loss (2–9 GHz), and magnetic loss (9–18 GHz). The reasoning behind the higher dielectric loss in the range of 2–9 GHz could be attributed to the PANI coating and MWCNTs additions, thereby improving the nanocomposite conductivity while enhancing the dielectric loss because of the increased vibration of atoms and molecules under the alternating electric field.

As mentioned previously, combining ferrites with MWCNTs has attracted much interest as they pave the way for producing new materials with enhanced absorption. In this regard, structural and MW absorption characteristics of $SrFe_{9.5}(Ni_{0.5}Mn_{0.5}Zr)_{1.25}O_{19}$/MWCNTs nanocomposites have been investigated (Ghasemi, 2013). The X-ray diffraction (XRD) patterns obtained from ferrite nanoparticles and synthesized nanocomposites are shown in Fig. 6.8. The presence of MWCNTs is confirmed by comparing the XRD patterns of the nanocomposites and SrF nanoparticles. In this respect, the peak at $2\theta = 26.6$ degree is attributed to the (002) graphite plane, indicating the successful synthesis of the nanocomposite by the addition of MWCNTs to the SrF solution. Moreover, no secondary phase exists in the resulting ferrite nanoparticles and nanocomposite.

Essentially, the absorber performance changes as a function of the composite thickness. The relationship between matching thickness and frequency can be inferred from Fig. 6.9. Increasing the matching thickness from 1.5 to 1.8 mm linearly decreases the first reflection dip frequency from 9.5 to 8.3 GHz. Thus, the optimization of absorbing materials is realized by simultaneous matching between complex permittivity, complex permeability, frequency, and material thickness.

In another research, the role of MWCNTs in absorption characteristics of a wideband MW absorber has been investigated by fabricating $BaFe_9Mn_{0.75}Co_{0.75}Ti_{1.5}O_{19}$ (barium hexaferrite, or BaM)/MWCNT nanocomposites (Ghasemi, Sepelak, et al., 2011). In this case, the first anisotropy constant was determined using the law to approach saturation. The respective anisotropy constants calculated for BaM and substituted BaM ferrite were 5.38×10^5 and -1.2×10^5 erg/g. As a result, the easy axis changed from the c-axis to the basal plane due to the Mn–Co–Ti substitution.

The variation in RL as a function of frequency for substituted ferrite/MWCNTs with different MWCNTs volume fractions (0–4 vol%) is shown in Fig. 6.10. At a

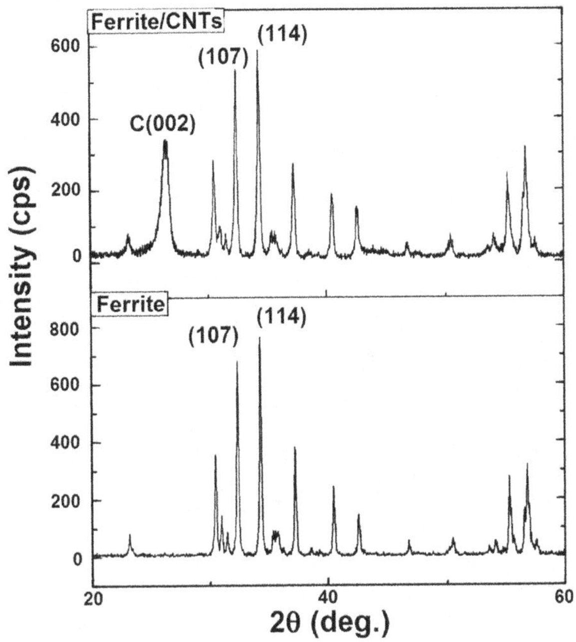

FIGURE 6.8

X-ray diffraction patterns obtained from $SrFe_{9.5}(Ni_{0.5}Mn_{0.5}Zr)_{1.25}O_{19}$ nanoparticles and $SrFe_{9.5}(Ni_{0.5}Mn_{0.5}Zr)_{1.25}O_{19}$/multiwalled carbon nanotube (8 vol%) nanocomposite.

Reproduced from Ghasemi, A. (2013). The role of multi-walled carbon nanotubes on the magnetic and reflection loss characteristics of substituted strontium ferrite nanoparticles. Journal of Magnetism and Magnetic Materials, 330, *163–168. https://doi.org/10.1016/j.jmmm.2012.10.033.*

frequency of 18.8 GHz, the maximum RL is −40 dB. Moreover, increasing the MWCNTs volume fractions in the nanocomposite enhances bandwidth.

Dispersions of ferrites with PPy (as a supporting matrix) and MWCNTs may also lead to enhanced absorption properties. Wang and colleagues have synthesized $BaFe_{12}O_{19}$–$Ni_{0.8}Zn_{0.2}Fe_2O_4$/PPy nanocomposites on MWCNTs using a deoxidation technique (Wang et al., 2014). Fig. 6.11A–C show TEM micrographs obtained from the treated MWCNTs and the nanocomposite with an MWCNTs content of 2 wt.%. Evidently, nanotube thickness increases after the formation of $BaFe_{12}O_{19}$–$Ni_{0.8}Zn_{0.2}Fe_2O_4$ and PPy on the MWCNTs. According to Fig. 6.11D–E, while $BaFe_{12}O_{19}$–$Ni_{0.8}Zn_{0.2}Fe_2O_4$ and PPy layers remain thin, some nanoparticles grow on the 5 wt.% MWCNTs surfaces. Fig. 6.11F shows an HRTEM micrograph obtained from the $BaFe_{12}O_{19}$–$Ni_{0.8}Zn_{0.2}Fe_2O_4$/PPy/MWCNTs nanocomposites, indicating the formation of highly crystalline ferrite nanoparticles while revealing well-defined lattice planes of $Ni_{0.8}Zn_{0.2}Fe_2O_4$ (333) and $BaFe_{12}O_{19}$ (107). The uniform phase distribution of nanoparticles consisting of $BaFe_{12}O_{19}$ and $Ni_{0.8}Zn_{0.2}$ Fe_2O_4 is also evidenced.

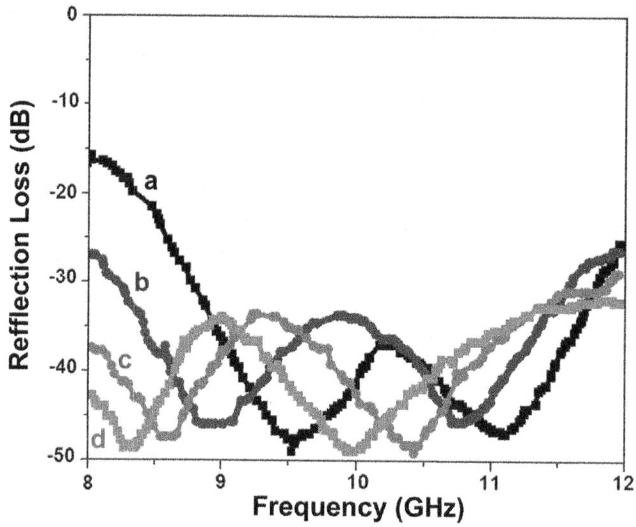

FIGURE 6.9

Reflection loss as a function of frequency for $SrFe_{9.5}(Ni_{0.5}Mn_{0.5}Zr)_{1.25}O_{19}$/multiwalled carbon nanotube (8 vol%) nanocomposite with thicknesses of (A) 1.5, (B) 1.6, (C) 1.7, and (D) 1.8 mm.

Reproduced from Ghasemi, A. (2013). The role of multi-walled carbon nanotubes on the magnetic and reflection loss characteristics of substituted strontium ferrite nanoparticles. Journal of Magnetism and Magnetic Materials, 330, *163–168. https://doi.org/10.1016/j.jmmm.2012.10.033.*

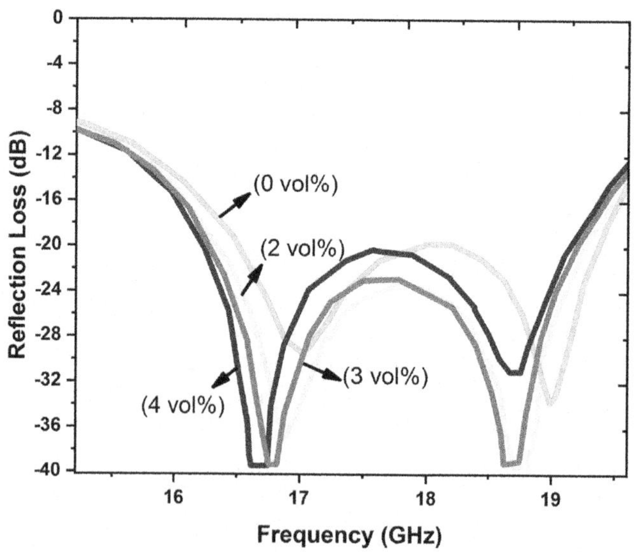

FIGURE 6.10

Reflection loss as a function of frequency for substituted ferrite-multiwalled carbon nanotube (MWCNTs) nanocomposites with different MWCNTs volume fractions of 0, 2, 3, and 4 vol%.

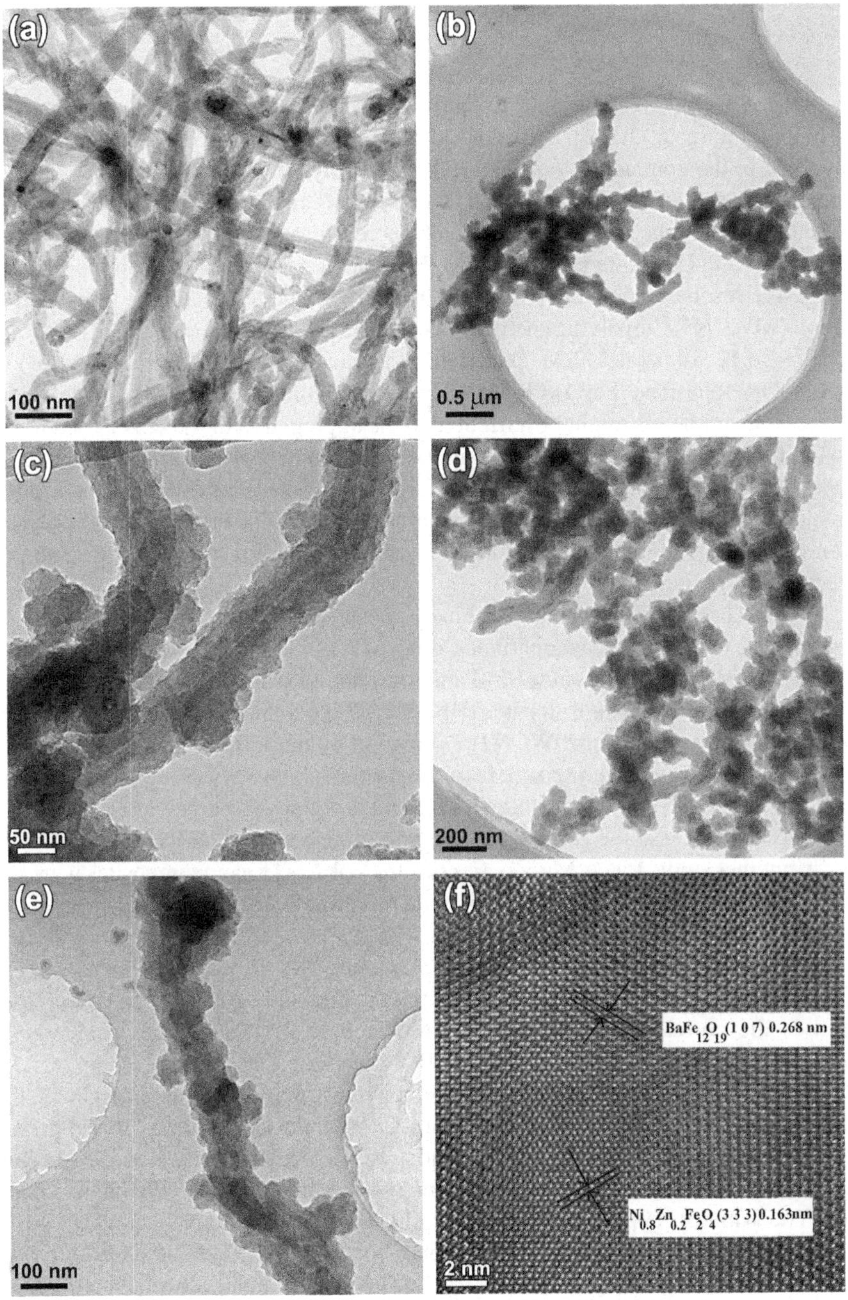

FIGURE 6.11

Transmission electron microscopic micrographs of (A) treated multiwalled carbon nanotubes (MWCNTs), (B and C) 2 wt.% MWCNTs, and (D and E) 5 wt.% MWCNTs, and (F) high-resolution transmission electron microscopy micrograph of $BaFe_{12}O_{19}$–$Ni_{0.8}Zn_{0.2}Fe_2O_4$/polypyrrole/MWCNTs nanocomposite.

Reproduced from Wang, Y., Huang, Y., & Ding, J. (2014). Synthesis and electromagnetic absorption properties of polypyrrole/$BaFe_{12}O_{19}$–$Ni_{0.8}Zn_{0.2}Fe_2O_4$/multi-walled carbon nanotube composites. Materials Science in Semiconductor Processing, 26, *632–641.*

Several factors, including the dopant type, doping level, molecular weight, crystallinity, conjugate chain length, and interactions between conjugated chains, are involved in the conductivity of materials. Basically, while ferrite nanoparticles are considered insulators, the MWCNT is a conducting material, so the nanocomposite conductivity is expected to increase with increasing the MWCNTs content. Since MWCNTs and PPy materials can act well as an electron acceptor and donor, respectively, the resultant conductivity may also improve due to the charge transfer from PPy to MWCNTs, involving the conducting bridge effect (Li et al., 2012).

Fig. 6.12 illustrates EM parameters as a function of frequency for the $BaFe_{12}O_{19}-Ni_{0.8}Zn_{0.2}Fe_2O_4$/PPy/MWCNTs nanocomposites. The wavy variation in the EM parameters might be attributed to the aggregation of MWCNTs when their content reaches a certain value, thus influencing the decentralization state. In addition, by propagating EM waves in the nanocomposites, an oscillatory current is created due the charge carrier's directional movement. The relaxation and polarization phenomena are also induced by the boundary charges, thereby converting the electric energy into heat energy.

Elsewhere, nanocomposites have been synthesized by assembling $BaFe_{10}Mg_{0.5}Co_{0.5}ZrO_{19}$ nanoparticles onto MWCNTs surfaces (Ghasemi, 2012). Fig. 6.13 shows field-emission SEM micrographs of pristine MWCNTs and Mg–Co–Zr-substituted barium ferrite (BF)/MWCNTs nanocomposites. As can be seen, the whole surface of MWCNTs is covered by the ferrite nanoparticles. Moreover, the resulting nanocomposite is in the form of dot arrays along the MWCNTs. The main driving force for forming the oriented dot arrays can be assigned to their tendency to decrease the surface energy, providing a wideband RL.

In another study, $BaCu_xMg_xZr_{2x}Fe_{12-4x}O_{19}$/MWCNTs nanocomposites were prepared using a hydrothermal route (Nikmanesh et al., 2016). RL curves of the aforementioned nanocomposites ($x = 0.5$) with various thicknesses are shown in Fig. 6.14. It is interesting to find that a maximum RL of −23 dB is achieved at a thickness of 2.1 mm. Increasing the matching thickness from 1.3 to 3.7 mm decreases the corresponding frequency from 13.3 to 4.9 GHz. The bandwidth of 2.1 mm thick nanocomposites with $x = 0.5$ is about 6 GHz.

It is highly expected that fabricating sandwich structures comprising CNTs, EG, BF, and conductive matrices provide various potential applications in the area of nanocomposites with superior absorption characteristics. Zhao and colleagues have been able to synthesize sandwich microstructured CNTs/EG/BF nanocomposite. The synthesis approach is schematically shown in Fig. 6.15 (Zhao et al., 2017).

The RL curves as a function of frequency for EG, BF, EG/BF, and CNTs/EG/BF samples are shown in Fig. 6.16. The CNTs/EG/BF nanocomposite with a thickness of 1.5 mm has a maximum RL of −49.5 dB at 11.2 GHz. When the nanocomposite thickness is 1 mm, the RL reaches −45.8 dB at 14.1 GHz.

The CNTs increased the RL due to their excellent electrical conductivity and three-dimensional (3-D) web structures. From Fig. 6.16, while the absorption properties of pure EG and BF are low, their combination greatly improves the RL. Moreover, the frequency dependence of RL of EG/BF and EG/CNTs/BF nanocomposites

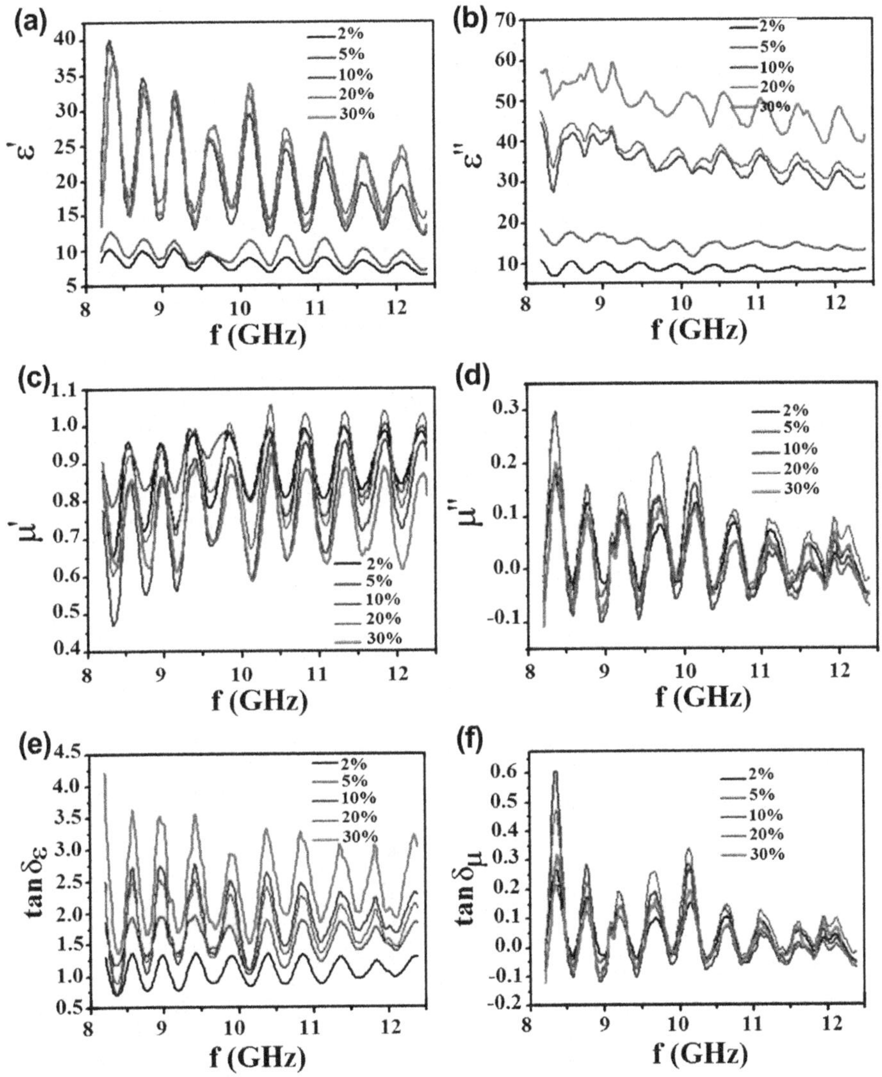

FIGURE 6.12

Electromagnetic parameters ($\varepsilon^* = \varepsilon' - j\varepsilon''$, $\mu^* = \mu' - j\mu''$, $\tan\delta_e$, and $\tan\delta_m$) as a function of frequency and multiwalled carbon nanotube (MWCNTs) content for the $BaFe_{12}O_{19}$–$Ni_{0.8}Zn_{0.2}Fe_2O_4$/polypyrrole/MWCNTs nanocomposites.

From Wang, Y., Huang, Y., & Ding, J. (2014). Synthesis and electromagnetic absorption properties of polypyrrole/$BaFe_{12}O_{19}$–$Ni_{0.8}Zn_{0.2}Fe_2O_4$/multi-walled carbon nanotube composites. Materials Science in Semiconductor Processing, 26, *632–641. https://doi.org/10.1016/j.mssp.2014.06.001.*

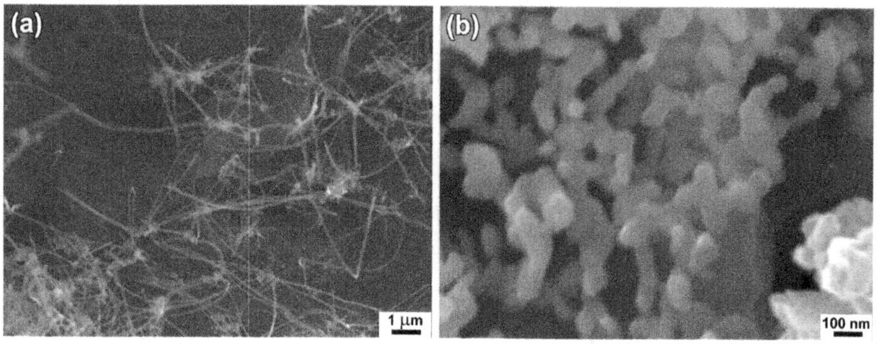

FIGURE 6.13

Field-emission scanning electron microscopy micrographs of (A) pristine multiwalled carbon nanotubes (MWCNTs) and (B) Mg–Co–Zr-substituted barium ferrite/MWCNT nanocomposites in the form of dot arrays.

From Ghasemi, A. (2012). Enhanced reflection loss and permittivity of self-assembled Mg–Co–Zr substituted barium ferrite dot array on carbon nanotube. Journal of Magnetism and Magnetic Materials, 324*(6), 1080–1083. https://doi.org/10.1016/j.jmmm.2011.10.027.*

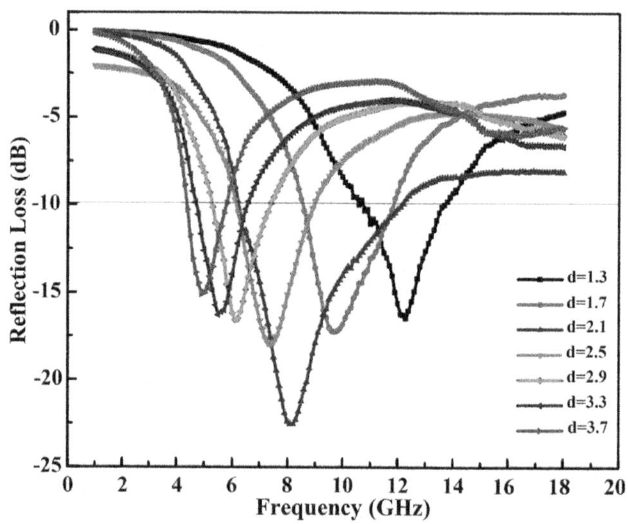

FIGURE 6.14

Reflection loss curves of $BaCu_xMg_xZr_{2x}Fe_{12-4x}O_{19}$/multiwalled carbon nanotube nanocomposites ($x = 0.5$) with various thicknesses.

From Nikmanesh, H., Moradi, M., Bordbar, G. H., & Alam, R. S. (2016). Synthesis of multi-walled carbon nanotube/doped barium hexaferrite nanocomposites: An investigation of structural, magnetic and microwave absorption properties. Ceramics International, 42*(13), 14342–14349. https://doi.org/10.1016/j.ceramint.2016.05.089.*

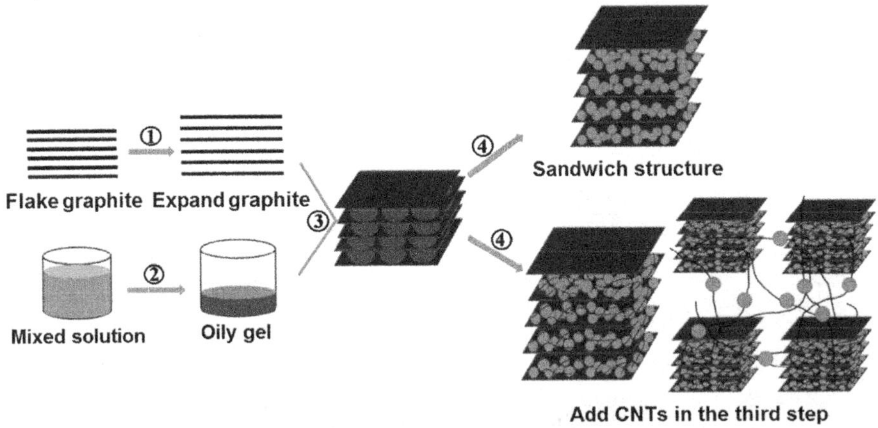

FIGURE 6.15

The schematic diagram of the synthesis approach of sandwich structure of expanding graphite (EG)/barium ferrite (BF) and carbon nanotube (CNT) nanocomposites.

Reproduced from Zhao, T., Jin, W., Ji, X., Yan, H., Jiang, Y., Dong, Y., Yang, Y., Dang, A., Li, H., & Li, T. (2017). Synthesis of sandwich microstructured expanded graphite/barium ferrite connected with carbon nanotube composite and its electromagnetic wave absorbing properties. Journal of Alloys and Compounds, 712, *59–68. https://doi.org/10.1016/j.jallcom.2017.04.070.*

is different, likely arising from the effect of the interconnected CNTs on the nanocomposite microstructure and BF grain size. Increasing the nanocomposite thickness causes the maximum of RL to shift toward low frequencies due to the synergistic effect of the MW absorption with various travel paths and time in the absorber.

6.8 Hard ferrite/single-walled carbon nanotube nanocomposites

SWCNTs have also been employed in the fabrication and study of nanocomposites. For example, the effects of SWCNTs volume percentage on magnetic and MW properties of $SrFe_8MgCoTi_2O_{19}$/SWCNTs nanocomposites have been studied (Gordani & Ghasemi, 2015). In this regard, SWCNTs were initially functionalized with active carboxylic groups. The SWCNTs were also packed with polymer molecules due to the presence of polyacrylic acid. In the following, ferrite nanoparticles were attached to the SWCNTs surfaces, leading to their rearrangement. Finally, the nanocomposite was mixed with epoxy resin to form a bulk sample suitable for measuring MW characteristics. Fig. 6.17 shows the schematic diagram of the synthesis of ferrite/SWCNTs nanocomposites.

The RL curves as a function of frequency for $SrFe_8MgCoTi_2O_{19}$/SWCNTs nanocomposites with different SWCNTs volume fractions are shown in Fig. 6.18. The

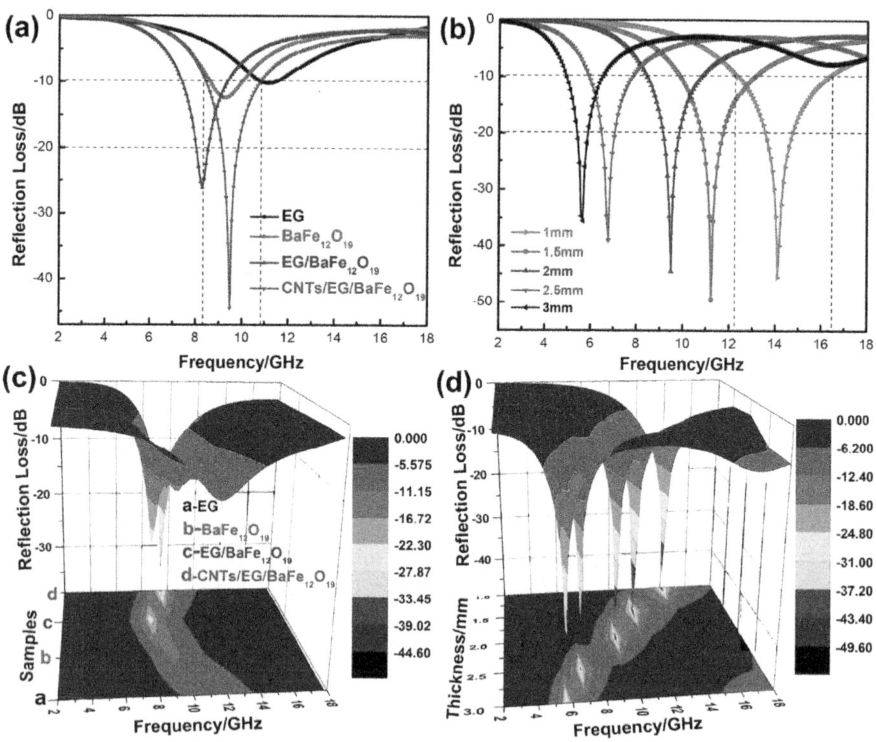

FIGURE 6.16

(A) and (C) Reflection loss curves as a function of frequency in the range of 2–18 GHz for expanded graphite (EG), barium ferrite (BF), EG/BF, and carbon nanotube (CNT)/EG/BF. (B) and (D) Reflection loss curves as a function of frequency in the range of 2–18 GHz for the CNTs/EG/BF nanocomposite with various thicknesses.

From Zhao, T., Jin, W., Ji, X., Yan, H., Jiang, Y., Dong, Y., Yang, Y., Dang, A., Li, H., & Li, T. (2017). Synthesis of sandwich microstructured expanded graphite/barium ferrite connected with carbon nanotube composite and its electromagnetic wave absorbing properties. Journal of Alloys and Compounds, 712, *59–68. https://doi.org/10.1016/j.jallcom.2017.04.070.*

TEM micrographs indicate a relatively uniform distribution of ferrites on the outer surface of SWCNTs. The obtained maximum RLs of $SrFe_8MgCoTi_2O_{19}$ nanoparticles are −21.64 and −22.18 dB at frequencies of 9.44 and 11.02 GHz, respectively. The $SrFe_8MgCoTi_2O_{19}$ nanoparticles have lower RLs than $SrFe_8MgCoTi_2O_{19}$/SWCNTs nanocomposites. The frequency bandwidth of the $SrFe_8MgCoTi_2O_{19}$ nanoparticles is also less than 2.6 GHz. Increasing the SWCNTs volume fraction to 4 vol% increases the corresponding bandwidth to a maximum value of 4 GHz. The maximum RL of the nanocomposite containing 4 vol% SWCNTs is −37 dB at a frequency of 9.31 GHz. Further increasing the SWCNTs volume fraction (>4 vol%) decreases the bandwidth of the nanocomposites. For instance, the absorption

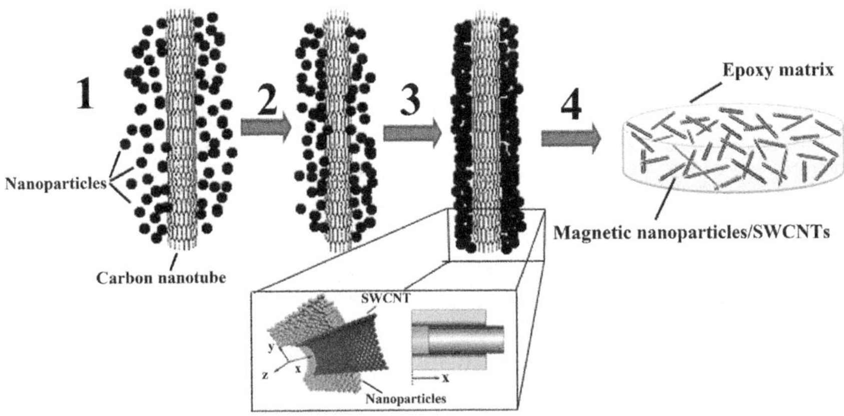

FIGURE 6.17

The schematic diagram of the synthesis of ferrite/single-walled carbon nanotube nanocomposites.

From Gordani, G. R., & Ghasemi, A. (2015). High frequency electromagnetic reflection loss performance of substituted Sr-hexaferrite nanoparticles/SWCNTs/epoxy nanocomposite. Journal of Magnetism and Magnetic Materials, 391, *184–190. https://doi.org/10.1016/j.jmmm.2015.05.002.*

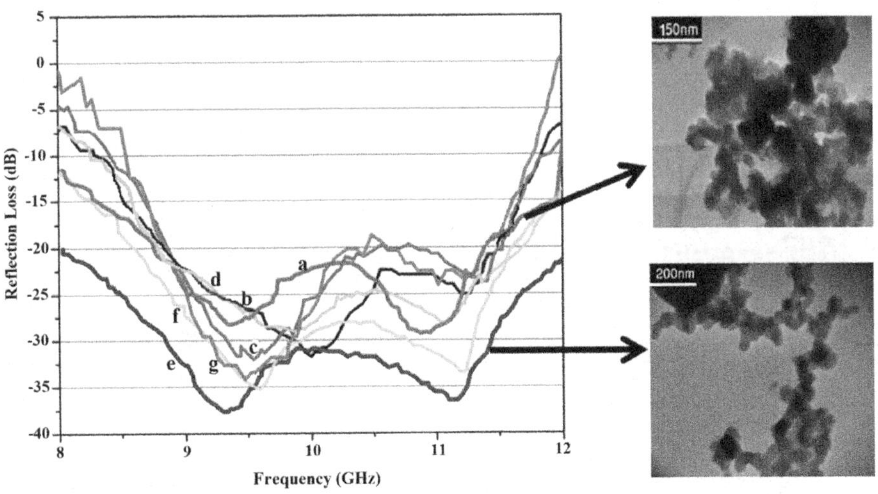

FIGURE 6.18

Reflection loss curves as a function of frequency for: (A) $SrFe_8MgCoTi_2O_{19}$ nanoparticles and nanocomposites containing: (B) 1 vol%, (C) 2 vol%, (D) 3 vol%, (E) 4 vol%, (F) 5 vol %, and (G) 6 vol% single-walled carbon nanotubes.

From Gordani, G. R., & Ghasemi, A. (2015). High frequency electromagnetic reflection loss performance of substituted Sr-hexaferrite nanoparticles/SWCNTs/epoxy nanocomposite. Journal of Magnetism and Magnetic Materials, 391, *184–190. https://doi.org/10.1016/j.jmmm.2015.05.002.*

bandwidth of the nanocomposite containing 6 vol% SWCNTs is 2.53 GHz. In this case, the maximum RL is obtained to be −34.02 dB at a frequency of 9.47 GHz.

Manipulating the loadings of nanocomposites may provide highly effective MW absorbers with lightweight. Wang and colleagues have synthesized a complex nanocomposite consisting of a high-quality structure of $CoFe_2$ coated SWCNTs ($CoFe_2$/SWCNTs) and BF. The $CoFe_2$ alloy is formed during SWCNTs preparation (Wang & Jing, 2005). The nanocomposite samples were prepared with various loadings of as-synthesized $CoFe_2$/SWCNTs (1, 3, 5, and 7 wt.%) and labeled as S1, S3, S5, and S7, respectively. The variation in permittivity and permeability as a function of frequency is shown in Fig. 6.19. Various factors may explain the phenomena observed in the low and high-frequency regions. In a range of 2–7 GHz, leakage conductance and polarization lag are probably involved in the EM parameters. Alternatively,

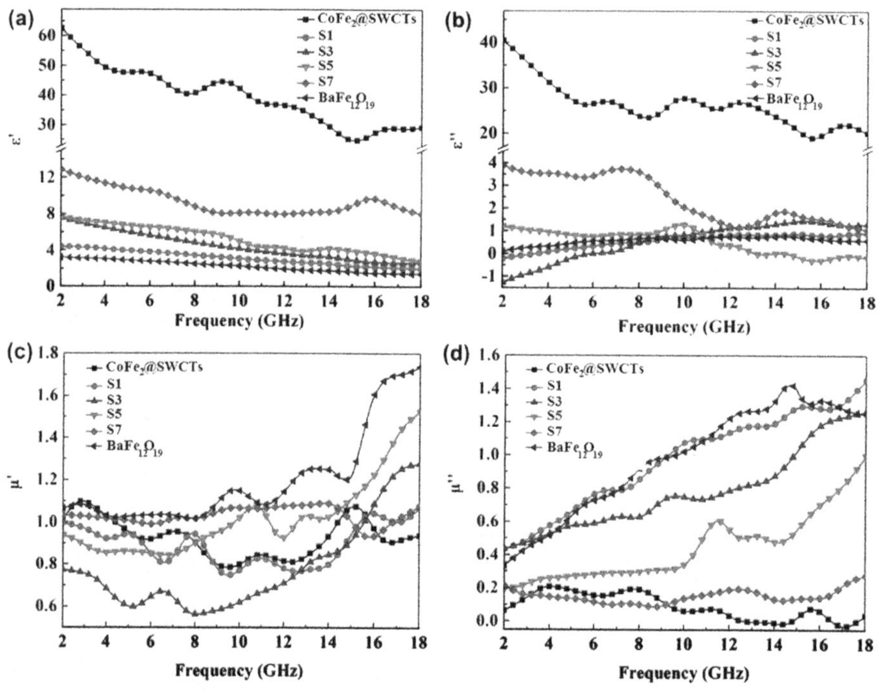

FIGURE 6.19

Electromagnetic (EM) parameters as a function of frequency in the range of 2–18 GHz for $CoFe_2$/single-walled carbon nanotube (SWCNTs)/barium ferrite nanocomposite samples: (A) real and (B) imaginary parts of the complex permittivity, and (C) real and (D) imaginary parts of complex permeability. For the sake of comparison, EM parameters of the $CoFe_2$/SWCNT nanocomposite and barium ferrite are also included.

Reproduced from Wang, M., An, K., Fang, Y., Wei, G., Yang, J., Sheng, L., Yu, L., & Zhao, X. (2017). The microwave absorbing properties of $CoFe_2$ attached single-walled carbon nanotube/$BaFe_{12}O_{19}$ nanocomposites. Journal of Materials Science: Materials in Electronics, 28*(17), 12475–12483. https://doi.org/10.1007/s10854-017-7069-1.*

interfacial polarization and the relaxation between $CoFe_2$/SWCNTs and impurities may contribute to the variation behavior of the EM parameters in the higher frequency range. However, $CoFe_2$/SWCNTs loadings <5 wt.% are not very effective in the absorption properties. When the loading amount reaches 5 wt.% or higher (thus achieving a possible threshold value), $CoFe_2$/SWCNTs nanocomposites play an important role in the EM parameters. This, in turn, implies that the nanocomposite dielectric loss is mainly governed by the $CoFe_2$@SWCNTs amount. Increasing the loading in samples $S1$ and $S3$ decreases ε'', being smaller than that in BF at the lower frequency. The electric dipolar polarization may be responsible for the variation in ε'' observed in Fig. 6.19B. Moreover, resonance peaks can be detected in μ'' values of the nanocomposite samples, arising from exchange resonance loss, local confinement, and natural resonance (Che et al., 2004; Wen et al., 2011).

Due to the high electric resistivity of BF, eddy current loss may be ignored. Overall, domain wall resonance of the magnetic BF nanoparticles and magnetic natural resonance and exchange resonance of the resulting nanocomposites make the most important contributions to the magnetic loss of the $CoFe_2$/SWCNTs/BF. The enhancement in the absorption properties of the nanocomposites is justified by the following: (1) SWCNTs bundles improve the electrical conductivity by forming a conductive network, increasing the dielectric loss in the nanocomposites, and (2) the interfaces between $CoFe_2$/SWCNTs and BF nanocomposites attenuate the EM waves due to interfacial polarization and interface scattering.

6.9 Spinel ferrite/carbon nanotube nanocomposites

Among spinel ferrites, cobalt ferrite ($CoFe_2O_4$) is a hard magnetic material with relatively high magnetocrystalline anisotropy, magnetostriction coefficient, coercivity, and electrical resistivity. $CoFe_2O_4$ structures also have moderate saturation magnetization, excellent thermal and chemical stability, and high mechanical hardness.

Magnetic and dielectric properties of $CoFe_2O_4$/CNTs nanocomposites have been studied by Routray et al. (2017). Magnetic hysteresis loops of $CoFe_2O_4$ nanoparticles and the $CoFe_2O_4$/CNTs nanocomposite, obtained using a vibrating sample magnetometer (VSM) at room temperature, are shown in Fig. 6.20. The VSM results indicate the ferrimagnetic behavior of the nanoparticles and nanocomposite. The coercivity and saturation magnetization of the $CoFe_2O_4$ nanoparticles and $CoFe_2O_4$/CNTs nanocomposite are 2053 Oe and 73 emu/g, and 1619 Oe and 46.5 emu/g, respectively. The coercivity and saturation magnetization reductions in the resulting nanocomposite are related to the presence of nonmagnetic CNTs. Magnetic properties of the $CoFe_2O_4$/CNTs nanocomposites depend not only on the ferrite nanoparticle volume fractions but also on the demagnetizing field induced by the nonmagnetic CNTs. This can be inferred by the equation $M_s = \varphi m_s$, in which M_s

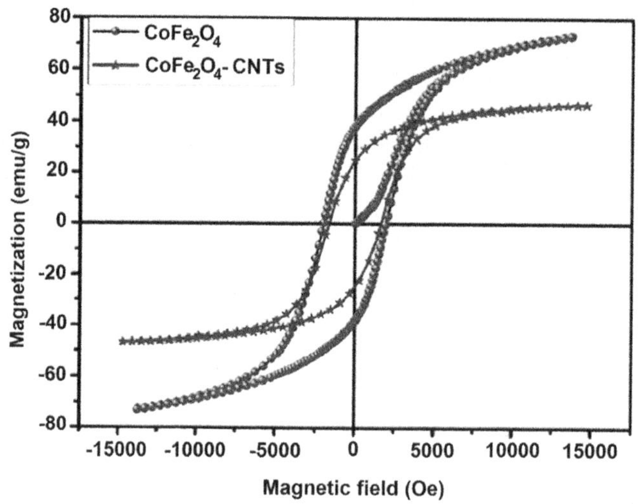

FIGURE 6.20

Vibrating sample magnetometer results of $CoFe_2O_4$ nanoparticles and $CoFe_2O_4$/carbon nanotube nanocomposite.

From Routray, K. L., Saha, S., & Behera, D. (2017). Effect of CNTs blending on the structural, dielectric and magnetic properties of nanosized cobalt ferrite. Materials Science and Engineering: B, 226, *199–205. https://doi.org/10.1016/j.mseb.2017.09.021.*

is the saturation magnetization, φ is the nanoparticle volume fraction, and m_s is the saturation moment of a single nanoparticle (Li et al., 2009).

The variation in dielectric constant versus frequency for the $CoFe_2O_4$ nanoparticles and $CoFe_2O_4$/CNTs nanocomposite at various temperatures are shown in Fig. 6.21A and B, respectively. As observed, the dielectric constant initially decreases with increasing frequency and then reaches a constant value at the high-frequency region. This can be explained by the Maxwell−Wagner theory. In fact, the main reason for obtaining high dielectric constants at low frequencies is the hopping of electrons between Fe^{2+} and Fe^{3+} ions in the grain boundaries. At high frequencies, the hopping of electrons does not occur, thus making the dielectric constant a frequency-independent parameter. In addition, crystalline grains are effective at high frequencies, whereas the grain boundaries dominate the absorption properties at low frequencies. Therefore, dipoles cannot rotate fast enough to respond to the applied electric field when increasing the frequency, thereby decreasing the probability of electrons to reach the grain boundary region while reducing the polarization (Koops, 1951; Narang et al., 2015).

By adding CNTs to the cobalt ferrite nanoparticles, the dielectric constant increases from 8.3×10^6 to 7×10^8. Interestingly, the dielectric constant also increases with increasing the temperature. In this respect, the mobility of charge carriers is enhanced at high temperatures due to the extra thermal energy produced. In turn, this increases the polarization, followed by an increase in the dielectric

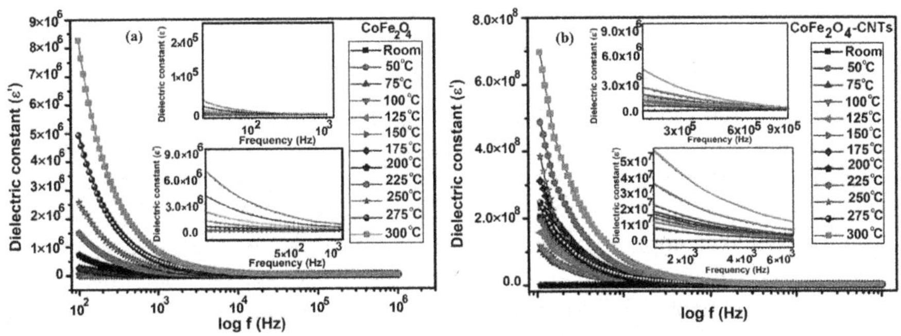

FIGURE 6.21

Variation in dielectric constant versus frequency for: (A) $CoFe_2O_4$ nanoparticles and (B) $CoFe_2O_4$/carbon nanotube nanocomposite at various temperatures.

From Routray, K. L., Saha, S., & Behera, D. (2017). Effect of CNTs blending on the structural, dielectric and magnetic properties of nanosized cobalt ferrite. Materials Science and Engineering: B, 226, *199–205. https://doi.org/10.1016/j.mseb.2017.09.021.*

constant. The dielectric loss curves versus frequency at various temperatures are shown in Fig. 6.22, indicating that the dielectric loss increases with increasing the temperature. The variation behavior of the dielectric loss versus frequency is the same as that of the dielectric constant. In addition, the dielectric loss is enhanced after adding CNTs to the nanoparticles.

Novel magnetic carbon-based MW absorbing materials with favorable characteristics may render the balance between dielectric loss and magnetic loss. For instance, nitrogen-doped MWCNTs (NMWCNTs) hybrid composites can be decorated by cobalt-zinc ferrite ($CoZnFe_2O_4$) microspheres using a solvothermal route

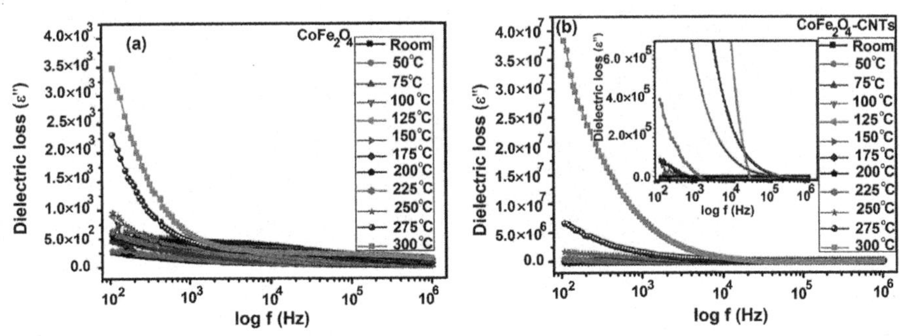

FIGURE 6.22

Variation in dielectric loss versus frequency for: (A) $CoFe_2O_4$ nanoparticles and (B) $CoFe_2O_4$/carbon nanotube nanocomposite at various temperatures.

From Routray, K. L., Saha, S., & Behera, D. (2017). Effect of CNTs blending on the structural, dielectric and magnetic properties of nanosized cobalt ferrite. Materials Science and Engineering: B, 226, *199–205. https://doi.org/10.1016/j.mseb.2017.09.021.*

(Shu et al., 2019). They synthesized $Co_{0.5}Zn_{0.5}Fe_2O_4$/NMWCNTs nanocomposites in the paraffin matrix with various NMWCNTs additive amounts of 0, 10, and 20 mg. Characterized by TEM, micromorphology and structure of the nanocomposite (20 mg NMWCNTs) are displayed in Fig. 6.23.

Evidently, $Co_{0.5}Zn_{0.5}Fe_2O_4$ nanoparticles have spherical morphology, completely covering the NMWCNTs surfaces with slight aggregation (Fig. 6.23A–C). Furthermore, under powerful ultrasonication, almost none of the $Co_{0.5}Zn_{0.5}Fe_2O_4$ nanoparticles is detached from the NMWCNTs surface. In other words, tightly anchored $Co_{0.5}Zn_{0.5}Fe_2O_4$ microspheres have decorated the surface of NMWCNTs. The HRTEM image depicted in Fig. 6.23D reveals interplanar

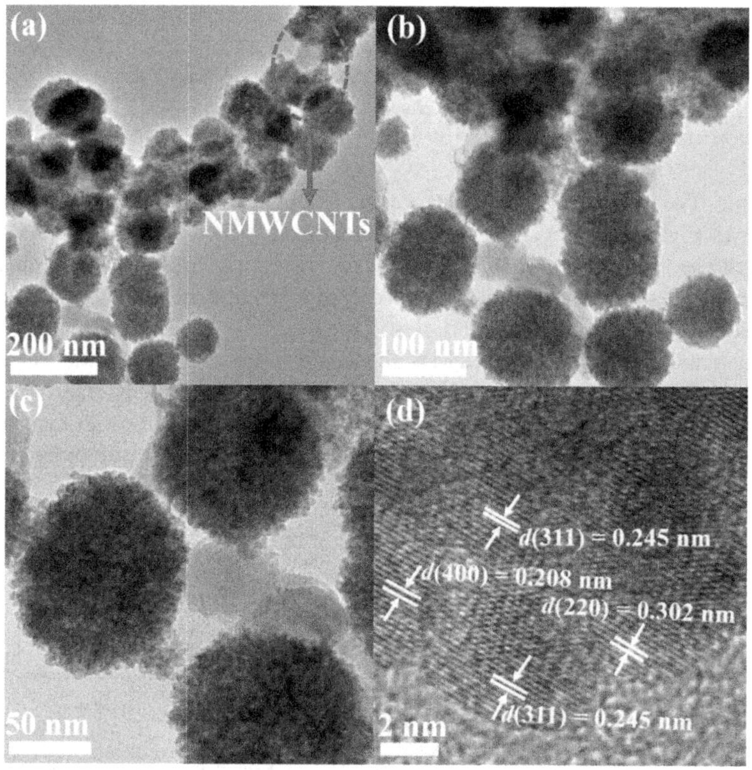

FIGURE 6.23

(A)–(C) Transmission electron microscopic micrograph with various magnifications and (D) high-resolution transmission electron microscopic micrograph obtained from $Co_{0.5}Zn_{0.5}Fe_2O_4$/nitrogen-doped multiwalled carbon nanotube nanocomposite.

From Shu, R., Wu, Y., Li, Z., Zhang, J., Wan, Z., Liu, Y., & Zheng, M. (2019). Facile synthesis of cobalt-zinc ferrite microspheres decorated nitrogen-doped multi-walled carbon nanotubes hybrid composites with excellent microwave absorption in the X-band. Composites Science and Technology, 184, *107839. https://doi.org/10.1016/j.compscitech.2019.107839.*

distances of 0.302, 0.245, and 0.208 nm, corresponding to (220), (311), and (400) planes of $Co_{0.5}Zn_{0.5}Fe_2O_4$ nanoparticles, respectively.

The authors have summarized the possible MW absorption mechanisms of the $Co_{0.5}Zn_{0.5}Fe_2O_4$/NMWCNTs hybrid nanocomposites, according to Fig. 6.24. At the first stage, charge asymmetric distributions are induced through residual oxygen-containing groups (e.g., COOH and OH) and NMWCNTs surface defects, leading to the formation of dipoles. Due to the relaxation loss, the EM energy is converted into thermal energy when the dipoles rotate toward the alternating EM field (Cao et al., 2018; Wu, Shu, Zhang, et al., 2019). In the second stage, under

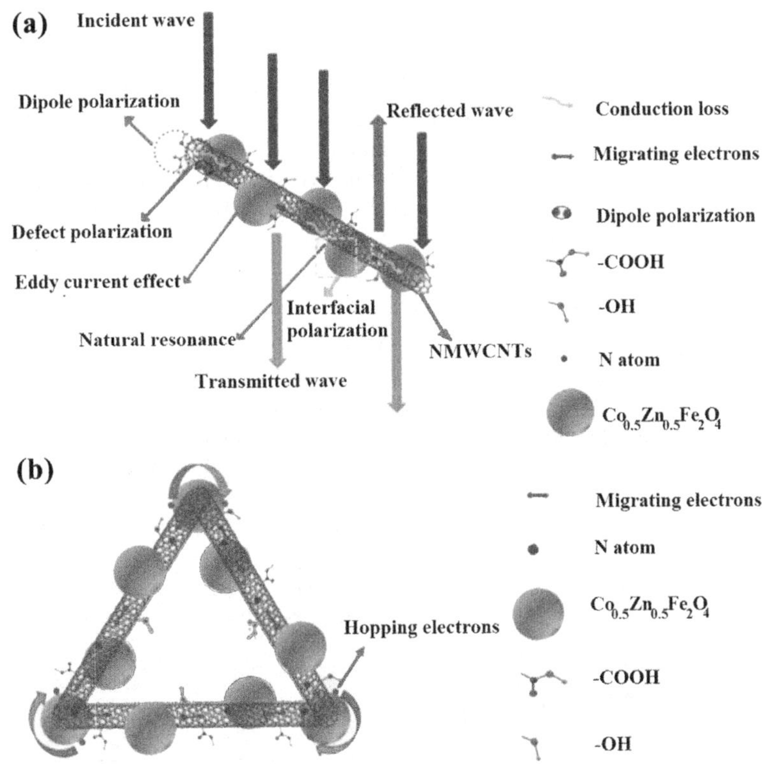

FIGURE 6.24

(A) Schematic illustration of the possible mechanisms involved in the microwave absorption of $Co_{0.5}Zn_{0.5}Fe_2O_4$/nitrogen-doped multiwalled carbon nanotube hybrid nanocomposites and (B) hopping conductivity in the three-dimensional conductive network.

Reproduced from Shu, R., Wu, Y., Li, Z., Zhang, J., Wan, Z., Liu, Y., & Zheng, M. (2019). Facile synthesis of cobalt-zinc ferrite microspheres decorated nitrogen-doped multi-walled carbon nanotubes hybrid composites with excellent microwave absorption in the X-band. Composites Science and Technology, 184, *107839. https://doi.org/10.1016/j.compscitech.2019.107839.*

alternating EM fields, heterogeneous interfaces formed between paraffin matrix, NMWCNTs, and $Co_{0.5}Zn_{0.5}Fe_2O_4$ nanoparticles can act as a capacitor-like structure, attenuating the incident MW power by aligning polar bonds or charges, according to the model proposed by Cao and colleagues (Cao et al., 2018; Wu, Shu, Zhang, et al., 2019). In the third stage, performing the nitrogen doping in the MWCNTs increases the number of nanocomposite defects and the capacity of the dipole polarization loss while optimizing the impedance matching (Li et al., 2017; Wang, Zhao, et al., 2018; Wang et al., 2019). In the fourth stage, the electrons initially absorb the EM energy needed to migrate on the NMWCNTs surfaces and then convert it into thermal energy after colliding with the lattice, according to Cao's electron hopping model (Cao et al., 2010).

On the other hand, at the contact areas between adjacent NMWCNTs and $Co_{0.5}Zn_{0.5}Fe_2O_4$ microspheres, the electrons may absorb more EM energy, thus jumping across the potential barriers. The extra hopped electrons enhance the conductivity of the 3-D NMWCNTs conductive networks, converting more EM energy into thermal energy. Finally, the MW attenuation capacity is significantly improved. The impedance matching is optimized due to the combinatory effects of the following phenomena: dielectric loss (originating from the interfacial polarization), magnetic loss of ferrimagnetic $Co_{0.5}Zn_{0.5}Fe_2O_4$ nanoparticles, conduction loss (derived from the migrated and hopped electrons in the 3-D NMWCNTs conductive networks), and dipole and defect polarizations.

Elsewhere, using a co-precipitation method, $Co_{0.5}Zn_{0.5}Fe_2O_4$/MWCNTs nanocomposites have been prepared by Mallick et al. (2016). Fig. 6.25 shows the variation in real and imaginary parts of permittivity and permeability of $Co_{0.5}Zn_{0.5}Fe_2O_4$/MWCNTs nanocomposites in the X- and Ku-band frequencies. Generally, the real

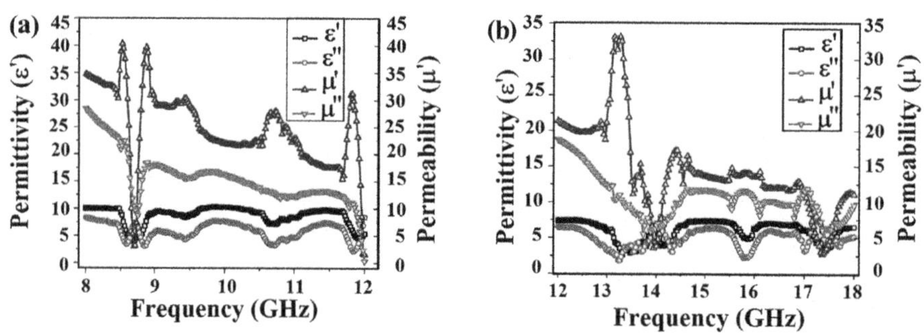

FIGURE 6.25

The variation in real and imaginary parts of permittivity and permeability of $Co_{0.5}Zn_{0.5}Fe_2O_4$/multiwalled carbon nanotube nanocomposite in (A) X- and (B) Ku-band frequencies.

Reproduced from Mallick, A., Mahapatra, A., Mitra, A., & Chakrabarti, P. (2016). Soft magnetic property and enhanced microwave absorption of nanoparticles of $Co_{0.5}Zn_{0.5}Fe_2O_4$ incorporated in MWCNT. Journal of Magnetism and Magnetic Materials, 416, *181–187. https://doi.org/10.1016/j.jmmm.2016.04.073.*

part of permittivity can be changed by varying the polarization properties of nanocomposites, whereas the imaginary part of permittivity is governed by energy dissipation. The permittivity (real and imaginary parts) is almost constant in both X and Ku bands.

Alternatively, the permeability real part is defined as the capacity to store magnetic energy, whereas the imaginary part of permeability is a function of magnetic loss. In this regard, the real and imaginary parts of permeability decrease with increasing frequency in both bands. In fact, the magnetic losses occurring in the MW region are induced by the natural and exchange resonances. The former dominates the latter in low frequencies. Note that the existence of several multiple exchange resonances arises from the narrow distribution of nanoparticle size (Deng et al., 2007; Yang et al., 2009).

The RL curves of $Co_{0.5}Zn_{0.5}Fe_2O_4$/MWCNTs nanocomposites with various thicknesses (1 and 2 mm) in the X- and Ku-band frequencies are shown in Fig. 6.26. For the thickness of 2 mm, the maximum RL of $Co_{0.5}Zn_{0.5}Fe_2O_4$ nanoparticles is −9 dB at 15.5 GHz. The respective maximum RLs of $Co_{0.5}Zn_{0.5}Fe_2O_4$/MWCNTs nanocomposites with thicknesses of 1 and 2 mm are −33.3 and −38.2 dB at 16.9 GHz. Therefore, the addition of MWCNTs to the $Co_{0.5}Zn_{0.5}Fe_2O_4$ nanoparticles increases the RL, resulting from the matching between magnetic and dielectric characteristics. The absorption bandwidth of the nanocomposite with a thickness of 2 mm at the Ku band is 1 GHz.

$NiZnCuFe_2O_4$ (NZCF) is magnetically soft spinel ferrite with high saturation magnetization, which can be used for high-frequency applications. In this regard, nanocomposites consisting of $Ni_{0.40}Zn_{0.40}Cu_{0.20}Fe_2O_4$ nanoparticles embedded in CNTs have been prepared (Sutradhar et al., 2013). Mössbauer spectra of the NZCF/CNTs nanocomposites annealed at 100 and 300°C are shown in Fig. 6.27.

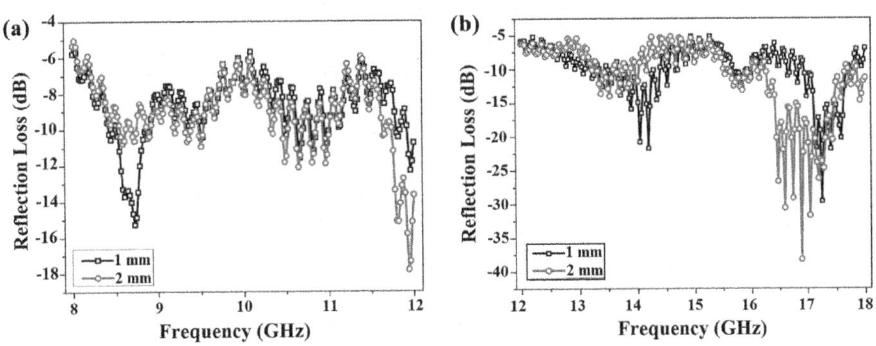

FIGURE 6.26

Reflection loss curves of $Co_{0.5}Zn_{0.5}Fe_2O_4$/multiwalled carbon nanotube nanocomposites with various thicknesses (1 and 2 mm) in: (A) X- and (B) Ku band frequencies.

Reproduced from Mallick, A., Mahapatra, A., Mitra, A., & Chakrabarti, P. (2016). Soft magnetic property and enhanced microwave absorption of nanoparticles of $Co_{0.5}Zn_{0.5}Fe_2O_4$ incorporated in MWCNT. Journal of Magnetism and Magnetic Materials, 416, *181–187. https://doi.org/10.1016/j.jmmm.2016.04.073.*

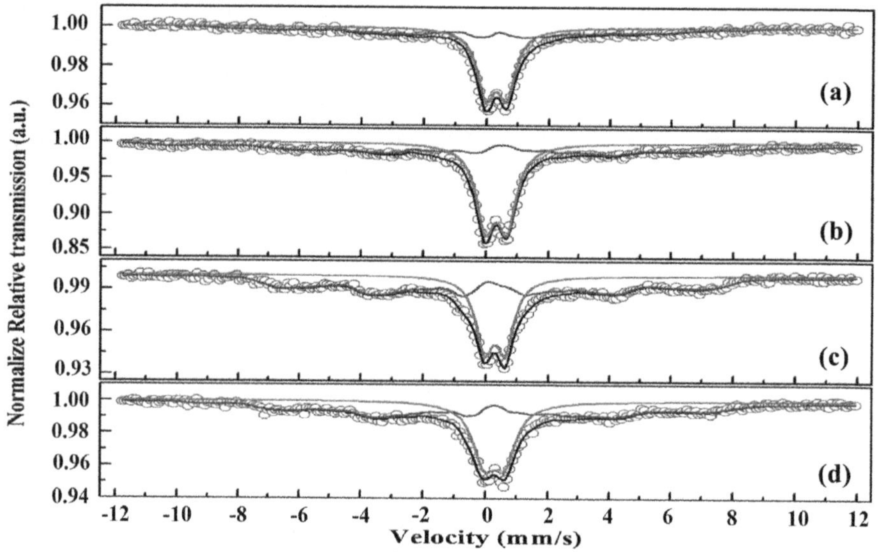

FIGURE 6.27

Mössbauer spectra of (A) $NiZnCuFe_2O_4$ (NZCF) nanoparticles annealed at 100°C, (B) NZCF/carbon nanotube (CNTs) nanocomposite annealed at 100°C, (C) NZCF nanoparticles annealed at 300°C, and (D) NZCF/CNT nanocomposite annealed at 300°C.

From Sutradhar, S., Mukhopadhyay, K., Pati, S., Das, S., Das, D., & Chakrabarti, P. (2013). Modulated magnetic property, enhanced microwave absorption and Mössbauer spectroscopy of $Ni_{0.40}Zn_{0.40}Cu_{0.20}Fe_2O_4$ nanoparticles embedded in carbon nanotubes. Journal of Alloys and Compounds, 576, *126–133. https://doi.org/10.1016/j.jallcom.2013.04.059.*

In this respect, the NZCF nanoparticles have a notable doublet absorption component, indicating that they are superparamagnetic. A mixture of doublet and weak sextet components is observed in the Mössbauer spectra of the NZCF/CNTs nanocomposites. The weak sextet pattern confirms that a small fraction of nanoparticles existing in the nanocomposites show ferrimagnetic behavior. In addition, the superior doublet pattern in the nanocomposites indicates that most of the NZCF nanoparticles are ultrafine, having superparamagnetic behavior. The isomer shift values show the existence of Fe^{3+} ions in the ferrites for all the nanocomposite samples.

Fig. 6.28 shows RL curves as a function of frequency in X and Ku bands for the NZCF/CNTs nanocomposites with various thicknesses. Obviously, the MW absorption properties of the NZCF nanoparticles are improved after adding CNTs. A maximum RL of −54 dB is obtained for the 2 mm thick nanocomposite annealed at 300°C. As well, the bandwidth of NZCF/CNTs nanocomposites is 6 GHz.

In a separate study, $Cu_{0.25}Ni_{0.25}Zn_{0.5}Fe_2O_4$/MWCNTs nanocomposites were prepared using co-precipitation and solution-mixing methods, and their absorption properties were investigated. Thermal analyses, including thermogravimetric analysis (TGA) and differential thermal analysis (DTA) of MWCNTs, are shown in

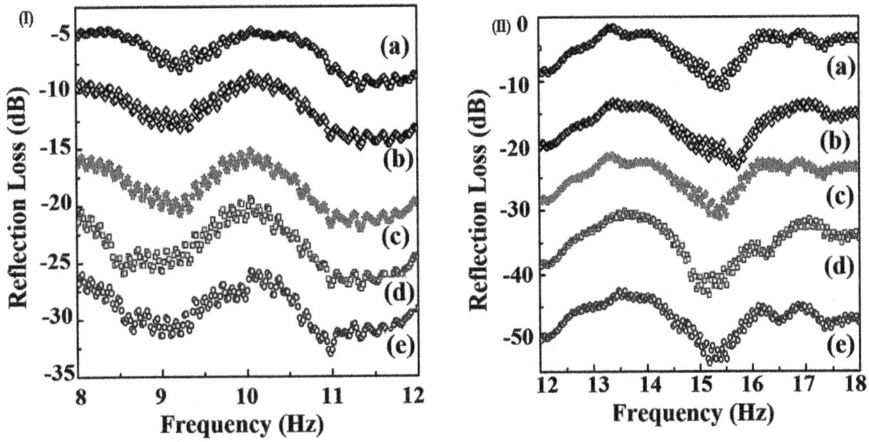

FIGURE 6.28

Reflection loss curves as a function of frequency in X and Ku bands for (A) $NiZnCuFe_2O_4$ (NZCF) nanoparticles annealed at 100°C, (B) 1 mm thick NZCF/carbon nanotube (CNTs) nanocomposite annealed at 100°C, (C) 2 mm thick NZCF/CNTs nanocomposite annealed at 100°C, (D) 1 mm thick NZCF/CNTs nanocomposite annealed at 300°C, and (E) 2 mm thick NZCF/CNTs nanocomposite annealed at 300°C.

From Sutradhar, S., Mukhopadhyay, K., Pati, S., Das, S., Das, D., & Chakrabarti, P. (2013). Modulated magnetic property, enhanced microwave absorption and Mössbauer spectroscopy of $Ni_{0.40}Zn_{0.40}Cu_{0.20}Fe_2O_4$ nanoparticles embedded in carbon nanotubes. Journal of Alloys and Compounds, 576, *126–133. https://doi.org/10.1016/j.jallcom.2013.04.059.*

Fig. 6.29. The TGA curve of MWCNTs shows an initial weight loss of about 1.35% at 500°C when they are exposed to the flowing air. On the other hand, a sharp endothermic peak is observed at 665°C, according to the DTA curve. The reaction involved in the strong endothermic process may be assigned to the oxygen functional groups (existing on the MWCNTs surface), graphitic decomposition, and carbon volatilization, giving rise to a total weight loss of 37% at 1000°C (Bibi et al., 2017).

MW absorption properties were measured after mixing the $Cu_{0.25}Ni_{0.25}Zn_{0.5}Fe_2O_4$/MWCNTs nanocomposite samples with epoxy resin and determined by the permittivity and permeability spectra. The results (not shown here) indicated that the complex permittivity increases and decreases with increasing the MWCNTs content and frequency, respectively.

Broadband adjustable RL may be provided by the incorporation of ferrite nanoparticles with the CNTs matrix. In this regard, CNTs ensheathed Ni–Zn–Co-ferrite (NZCoF) nanoparticles, synthesized by a sol-gel method, have been studied (Sarkar et al., 2014). The variation in RL versus frequency in the X and Ku bands is shown in Fig. 6.30. In the given frequency range, the magnetic losses observed may be caused by the magnetic spin resonances and/or hysteresis effects occurring in the

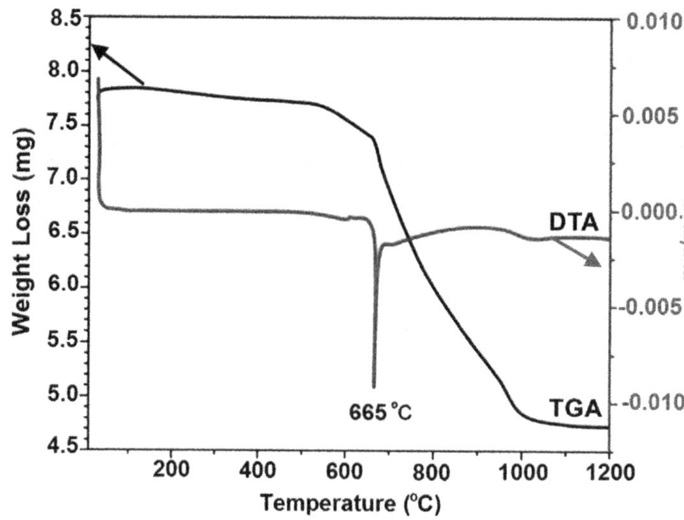

FIGURE 6.29

Thermogravimetric analysis and differential thermal analysis curves obtained from multiwalled carbon nanotubes.

From Bibi, M., Abbas, S. M., Ahmad, N., Muhammad, B., Iqbal, Z., Rana, U. A., & Khan, S. U.-D. (2017). Microwaves absorbing characteristics of metal ferrite/multiwall carbon nanotubes nanocomposites in X-band. Composites Part B: Engineering, 114, *139–148. https://doi.org/10.1016/j.compositesb.2017.01.034.*

nanocomposites, which in turn result from magnetic damping (e.g., eddy current and intrinsic damping) induced by domain wall motion, and grain boundary and dislocation movements in polycrystalline materials. Since the average size of NZCoF nanoparticles is 16 nm, the single domain effect will dominate the magnetic and MW absorption properties of the nanocomposites. As a matter of fact, nanoscale ferrite materials with extremely small sizes are nonconductive, thus showing a trivial micro-eddy current effect (Cullity & Graham, 2011). Note that the effect of intergrain exchange coupling interactions on the RL of the resulting nanocomposites is negligible.

$Ni_{0.3}Zn_{0.4}Co_{0.3}Fe_2O_4$ (N3Z4C3) nanoparticles have also been incorporated into the MWCNTs matrix to enhance MW absorption properties (Dalal et al., 2018). For comparison purposes, MW absorption of $Ni_{0.4}Zn_{0.4}Co_{0.2}Fe_2O_4$ (N4Z4C2) nanoparticles encapsulated in MWCNTs has been evaluated as well. The N3Z4C3 and N4Z4C2 nanoparticles can be considered an assembly of superparamagnets and ferrimagnets at the nanoscale.

Variations in loss tangent as a function of frequency in the Ku-band are depicted in Fig. 6.31, demonstrating the role of dielectric and magnetic losses in the maximum RL. As observed, the dielectric loss outperforms the magnetic loss in most of the Ku-band frequency range (12–18 GHz) for both N3Z4C3 and N4Z4C2 nanocomposites. Therefore, the MW absorption properties of the

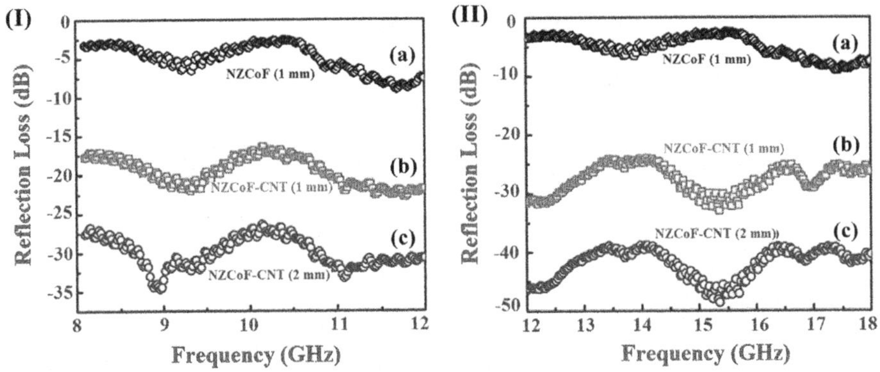

FIGURE 6.30

The variation in reflection loss versus frequency in the X and Ku bands for Ni–Zn–Co-ferrite (NZCoF) nanoparticles and NZCoF/carbon nanotube (CNTs) nanocomposites with various thicknesses: (I) X band (8–12 GHz): (A) NZCoF (1 mm), (B) NZCoF/CNTs (1 mm), and (C) NZCoF/CNTs (2 mm); and (II) Ku band (12–18 GHz): (A) NZCoF (1 mm), (B) NZCoF/CNTs (1 mm), and (C) NZCoF/CNTs (2 mm).

From Sarkar, D., Bhattacharya, A., Nandy, P., & Das, S. (2014). Enhanced broadband microwave reflection loss of carbon nanotube ensheathed Ni–Zn–Co-ferrite magnetic nanoparticles. Materials Letters, 120, *259–262. https://doi.org/10.1016/j.matlet.2014.01.089.*

nanocomposites are mainly attributed to the dielectric loss in accord with the results of Liu (2005). However, the MW absorption is optimized when both dielectric and magnetic loss tangents are equal to each other. In this case, the frequencies with maximum RLs in N3Z4C3/MWCNTs and N4Z4C2/MWCNTs nanocomposites are marked in Fig. 6.31. It is interesting to find that both loss tangents are almost equal at the aforementioned frequencies.

The addition of MWCNTs to Zn ferrite nanoparticles has also enhanced the corresponding EM wave absorption performance (Liu et al., 2018). In this regard, Zn ferrite/MWCNTs nanocomposites with various MWCNTs amounts (30, 40, and 50 mg) have been labeled as S_1, S_2, and S_3, respectively. Importantly, MWCNTs are capable of forming conductivity nets in the resulting Zn ferrite/MWCNTs nanocomposites, thus enhancing the dielectric loss.

The quarter-wave thickness criterion is given below:

$$t_m = \frac{n\lambda}{4} = \frac{nc}{4f_m\sqrt{|\mu_r||\varepsilon_r|}} (n = 1, 3, 5.....) \tag{6.22}$$

where t_m is the nanocomposite thickness, λ is the EM wavelength, f_m is the peak frequency of RL, and c is the EM wave velocity. In this way, if the relationship between the coating thicknesses matches well with t_m, a portion of the incident wave would be canceled by the EM wave reflected from the absorber. According to Fig. 6.32, the simulated values of t_m at $n = 1$ match well with the coating thicknesses at f_m. This suggests that the RL dip frequency mostly depends on the phase cancellation theory.

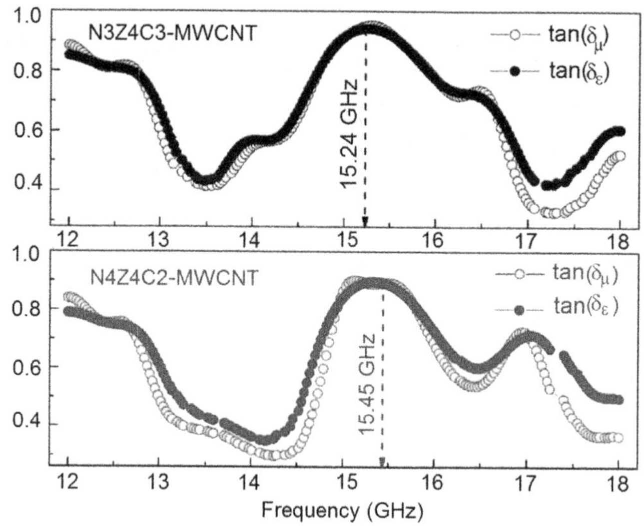

FIGURE 6.31

Variations in loss tangents as a function of frequency in the Ku band for $Ni_{0.3}Zn_{0.4}Co_{0.3}Fe_2O_4$/multiwalled carbon nanotubes (MWCNTs) and $Ni_{0.4}Zn_{0.4}Co_{0.2}Fe_2O_4$/MWCNTs nanocomposites.

From Dalal, M., Das, A., Das, D., Ningthoujam, R. S., & Chakrabarti, P. K. (2018). Studies of magnetic, Mössbauer spectroscopy, microwave absorption and hyperthermia behavior of Ni–Zn–Co-ferrite nanoparticles encapsulated in multi-walled carbon nanotubes. Journal of Magnetism and Magnetic Materials, 460, *12–27. https://doi.org/10.1016/j.jmmm.2018.03.048.*

Using a co-precipitation method, Dalal and colleagues have synthesized $Mn_{0.5}Zn_{0.5}Fe_2O_4$/MWCNTs nanocomposites while exploring the influence of MWCNTs on MW absorption properties (Dalal et al., 2018). From Fig. 6.33A,

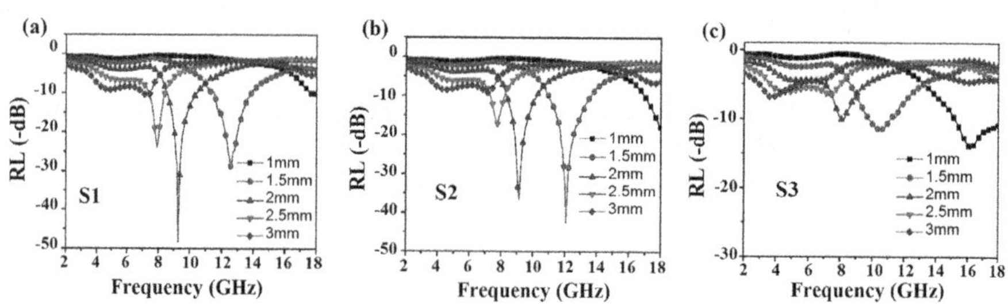

FIGURE 6.32

Reflection loss as a function of frequency for samples: (A) S_1, (B) S_2, and (C) S_3 at various thicknesses.

From Liu, Z., Xing, H., Liu, Y., Wang, H., Jia, H., & Ji, X. (2018). Hydrothermally synthesized Zn ferrite/multi-walled carbon nanotubes composite with enhanced electromagnetic-wave absorption performance. Journal of Alloys and Compounds, 731, *745–752. https://doi.org/10.1016/j.jallcom.2017.09.317.*

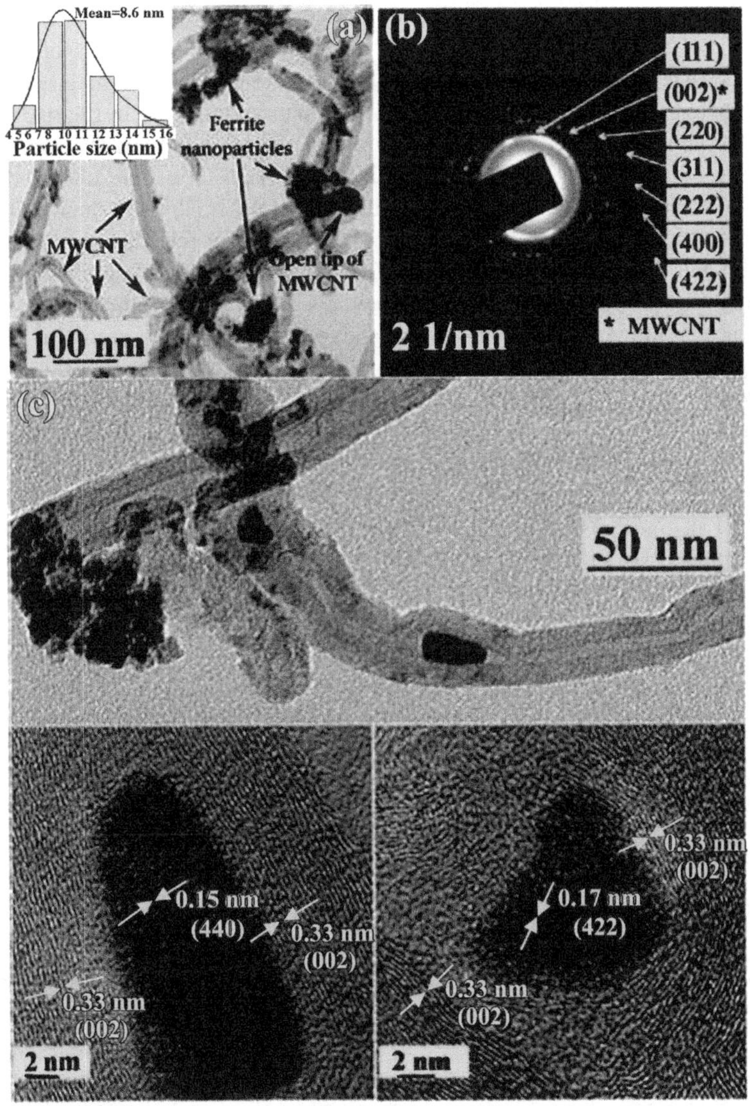

FIGURE 6.33

(A) Transmission electron microscopic micrograph, (B) selected area electron diffraction pattern, and (C) high-resolution transmission electron microscopy micrographs of $Mn_{0.5}Zn_{0.5}Fe_2O_4$/multiwalled carbon nanotube nanocomposite.

From Dalal, M., Mallick, A., & Chakrabarti, P. (2019). Microwave absorption of $Mn_{0.5}Zn_{0.5}Fe_2O_4$ nanoparticles integrated in multi-walled carbon nanotubes. Journal of the Australian Ceramic Society, 55*(1), 157–167. https://doi.org/10.1007/s41779-018-0221-4.*

the TEM micrograph shows that $Mn_{0.5}Zn_{0.5}Fe_2O_4$ nanoparticle clusters are attached to the MWCNTs outer walls. The size of nanoparticles is 4–16 nm. Moreover, it is observed that the nanoparticles tend to agglomerate due to their magnetic nature. The selected area electron diffraction (SAED) pattern depicted in Fig. 6.33B indicates the presence of different crystallographic planes of the ferrite nanoparticles. The ring corresponding to the (002) plane of MWCNTs is also observed in the SAED pattern, evidencing the incorporation of MWCNTs in the nanocomposite. HRTEM micrographs depicted in Fig. 6.33C reveal interplanar distances of 0.15 nm and 0.17 nm, corresponding to (440) and (422) crystallographic planes of the $Mn_{0.5}Zn_{0.5}Fe_2O_4$ nanoparticles, respectively. Furthermore, the interplanar distance of 0.33 nm pertains to the (002) plane of MWCNTs. Therefore, the HRTEM micrographs confirm the formation of appropriate crystallographic phases in the nanocomposite and suitable encapsulation of the nanoparticles by the MWCNTs.

The RL curves as a function of frequency in X and Ku bands for $Mn_{0.5}Zn_{0.5}Fe_2O_4$/MWCNTs nanocomposites with various thicknesses annealed at 200 and 400°C are presented in Fig. 6.34. While no considerable absorption is observed in the X band, several notable peaks are found in the Ku band. The maximum RL is obtained to be −20.3 dB at 15.45 GHz for the nanocomposite with a thickness of 1 mm annealed at 400°C (MZ400@CNTs). In this case, the RL decreases to −18.1 dB with an increase in the thickness to 2 mm. Moreover, the respective bandwidths of the nanocomposite samples with thicknesses of 1 and 2 mm are 4.32 and 2.28 GHz. Alternatively, the maximum RL of the nanocomposite annealed at 200°C (MZ200@CNT) increases from −17.8 to −18.5 dB with increasing the thickness from 1 to 2 mm.

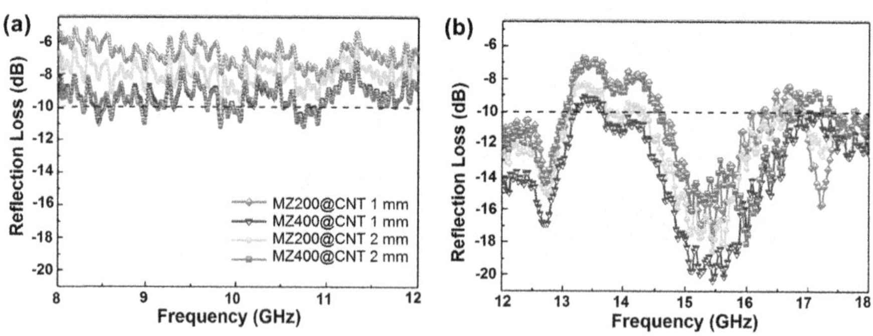

FIGURE 6.34

Reflection loss curves as a function of frequency in: (A) X and (B) Ku bands for $Mn_{0.5}Zn_{0.5}Fe_2O_4$/multiwalled carbon nanotube nanocomposites with various thicknesses annealed at 200 and 400°C.

From Dalal, M, Mallick, A., & Chakrabarti, P. (2019). Microwave absorption of $Mn_{0.5}Zn_{0.5}Fe_2O_4$ nanoparticles integrated in multi-walled carbon nanotubes. Journal of the Australian Ceramic Society, 55*(1), 157–167. https://doi.org/10.1007/s41779-018-0221-4.*

Recent investigations have opened up a new class of materials known as hybrid nanocomposites. These materials are made of a combination of electric nanoparticles and magnetic microparticles, having special properties (e.g., high mechanical stability and EM performance together with lightweight) and providing advantages over conventional composite materials (Ma et al., 2018, 2019).

In fact, adding electric nanoparticles to the magnetic microparticles forms conductive networks in the nanocomposite material, according to the EM wave percolation theory. In this regard, the roles of adding MWCNTs to the Mn–Zn ferrite hybrid and carbonyl iron (CI) into the silicone rubber matrix have been investigated (Lopes et al., 2020). Fig. 6.35A and B show schematic illustrations of Mn–Zn ferrite and CI magnetic nanocomposites. The corresponding hybrid magnetic nanocomposites made by the addition of CNTs are also shown in Fig. 6.35C and D.

The variation in RL as a function of frequency and thickness for the magnetic hybrid nanocomposites is shown in Fig. 6.36. The intensity of MW absorption is significantly enhanced for thicknesses larger than 2.0 mm. Concurrently, the peak frequency is reduced due to the thickness increase.

The formation of conductive networks by the CNTs with a large surface area may be responsible for the increase in the MW absorption of the hybrid nanocomposites, thus allowing for storing a higher amount of EM energy and reducing the reflected signal. In other words, the addition of CNTs leads to an increase in the surface of conductive networks between the magnetic microparticles, enhancing electric interactions occurring in the 3-D structure of the hybrid nanocomposites.

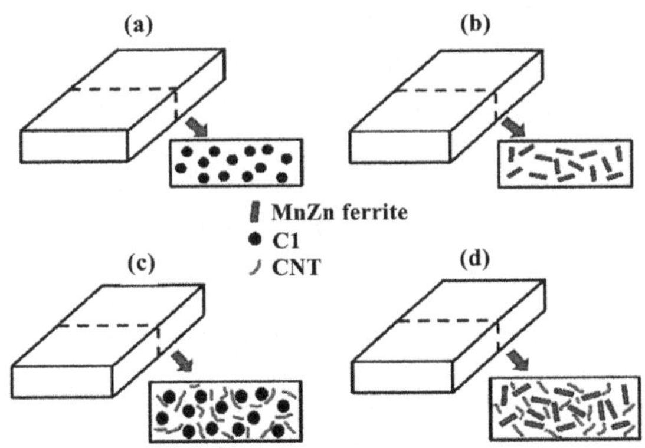

FIGURE 6.35

Schematic illustration of (A) and (B) Mn–Zn ferrite and carbonyl iron, and (C) and (D) corresponding hybrid magnetic nanocomposites made by the addition of carbon nanotubes.

From Lopes, B., Portes, R., do Amaral Junior, M., Florez-Vergara, D., Gama, A., Silva, V., Quirino, S., & Baldan, M. (2020). X Band electromagnetic property influence of multi-walled carbon nanotube in hybrid MnZn ferrite and carbonyl iron composites. Journal of Materials Research and Technology. *https://doi.org/10.1016/j.jmrt.2019.12.068.*

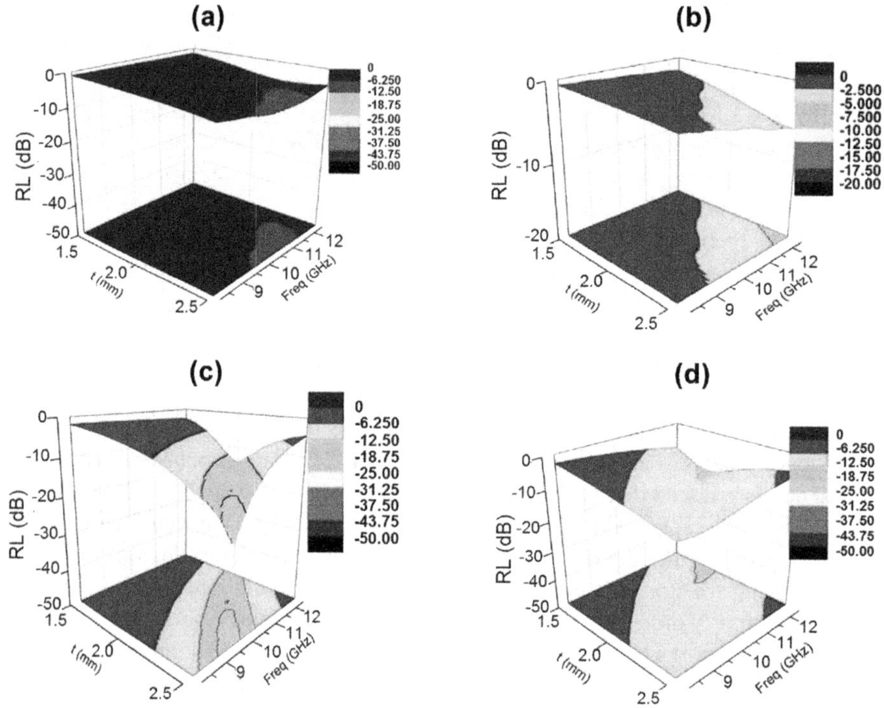

FIGURE 6.36

The reflection loss as a function of frequency and thickness for the magnetic hybrid nanocomposites: (A) 30 wt.% MnZn ferrite +1 wt.% carbon nanotubes (CNTs), (B) 50 wt.% MnZn ferrite +1 wt.% CNTs, (C) 30 wt.% carbonyl iron +1 wt.% CNTs, and (D) 50 wt.% carbonyl iron +1 wt.% CNTs.

From Lopes, B., Portes, R., do Amaral Junior, M., Florez-Vergara, D., Gama, A., Silva, V., Quirino, S., & Baldan, M. (2020). X Band electromagnetic property influence of multi-walled carbon nanotube in hybrid MnZn ferrite and carbonyl iron composites. Journal of Materials Research and Technology. *https://doi.org/10.1016/j.jmrt.2019.12.068.*

New kinds of candidates for MW absorbing materials have emerged from the incorporation of carbon and ferrites in porous matrices. Xiong and colleagues have prepared nanocomposites consisting of porous rod-like silica coated nickel ferrite and MWCNTs ($NiFe_2O_4$@SiO_2/MWCNTs) and investigated the role of filler loading in the MW absorption properties. The Cole−Cole semicircles obtained from $NiFe_2O_4$@SiO_2/MWCNTs nanocomposites with various filler loadings (30 wt.% and 40 wt.%) are shown in Fig. 6.37. As can be seen, the filler loadings of 30 wt.% and 40 wt.% have three and four Cole−Cole semicircles, indicating ternary and quaternary Debye dipolar relaxations, respectively. Under alternating EM fields, charge accumulations at the interfaces between $NiFe_2O_4$ and SiO_2 and SiO_2 and MWCNTs result in the interfacial polarization in the $NiFe_2O_4$@SiO_2/MWCNTs

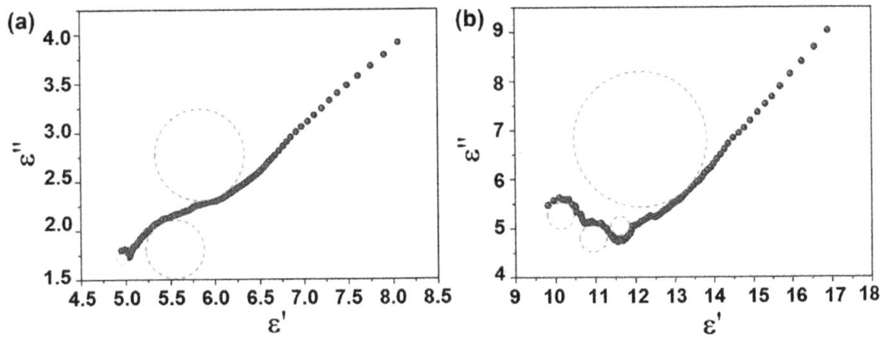

FIGURE 6.37

The Cole–Cole semicircles obtained from $NiFe_2O_4$@SiO_2/multiwalled carbon nanotube nanocomposites with different filler loadings: (A) 30 wt.% and (B) 40 wt.%.

From Xiong, Y., Luo, H., Nie, Y., Chen, F., Dai, W., Wang, X., Cheng, Y., & Gong, R. (2019). Synergistic effect of silica coated porous rodlike nickel ferrite and multiwalled carbon nanotube with improved electromagnetic wave absorption performance. Journal of Alloys and Compounds, 802, *364–372. https://doi.org/10.1016/j.jallcom.2019.06.174.*

nanocomposites. Furthermore, scattering and polarization centers may be created by the defects present in the MWCNTs, thereby inducing polarization relaxation under alternating EM fields. The C=O and C–O chemical bonds of MWCNTs can also give rise to dipole polarization by trapping electrons between carbon and oxygen atoms. Overall, the dielectric loss of $NiFe_2O_4$@SiO_2/MWCNT nanocomposites increases due to the dipole orientation and interfacial polarizations, facilitating the attenuation of EM waves (Yang & Liu, 2009; Zhang et al., 2009).

In general, the magnetic loss occurring in magnetic nanocomposite arises from the following contributions: magnetic hysteresis loss, eddy current loss, and domain wall, natural and exchange resonances. Under a weak EM field, one can ignore the magnetic hysteresis loss contribution. The domain wall resonance contribution has been mostly observed in the frequency range of 1–100 MHz, occurring only in multidomain magnetic structures. On the other hand, the exchange resonance contribution is induced for nanoparticle sizes smaller than 10 nm. Consequently, the magnetic hysteresis loss, domain wall resonance, and exchange resonance ($NiFe_2O_4$ nanoparticle sizes >50 nm) contributions are excluded in the $NiFe_2O_4$@SiO_2/MWCNTs nanocomposites. In other words, the natural resonance effect within the frequency range of 2–18 GHz plays the main role in the magnetic loss.

To evaluate the eddy current loss, one can consider the following equations:

$$\mu'' \approx 2\pi\mu_0(\mu')^2\sigma d^2 f/3 \tag{6.23}$$

$$C_0 = \mu''(\mu')^{-2}f^{-1} \tag{6.24}$$

where C_0 is the eddy current coefficient, σ is the electrical conductivity, μ_0 is the vacuum permeability, and d is the absorber thickness. If the magnetic loss is induced mainly by the eddy current effect, C_0 will remain constant as a function of frequency (Shu, Wu, et al., 2019).

Fig. 6.38 shows the frequency dependence of C_0 for $NiFe_2O_4$ nanoparticles and $NiFe_2O_4$@SiO_2/MWCNTs nanocomposites. In the case of $NiFe_2O_4$ nanoparticles, C_0 values range from 0.04 to 0.005 with increasing frequency from 2 to 18 GHz. Accordingly, the natural resonance contribution is mainly responsible for the magnetic loss, and the eddy current effect can be ignored. For $NiFe_2O_4$@SiO_2/MWCNTs nanocomposites, C_0 values decrease with increasing frequency in the range of 2–10 GHz and then remain almost constant in the frequency range of 10–18 GHz, corresponding to magnetic losses by natural resonance and eddy current effect, respectively.

Fig. 6.39 shows the variation in attenuation constant as a function of frequency for $NiFe_2O_4$ nanoparticles and $NiFe_2O_4$@SiO_2/MWCNTs nanocomposites. The nanocomposites outperform the nanoparticles in terms of attenuation constant in the whole frequency range. Therefore, it is expected that the EM wave absorption performance and attenuation capacity of the $NiFe_2O_4$@SiO_2/MWCNTs nanocomposites are higher than those of pure $NiFe_2O_4$ nanoparticles. It is also worth noting that increasing the filler loading enhances the corresponding attenuation capacity of the nanocomposite.

Variations in RL and impedance (Z) as a function of frequency for the 40 wt.% $NiFe_2O_4$@SiO_2/MWCNTs nanocomposite with various thicknesses are shown in Fig. 6.40A and B, respectively. The dashed lines feature the relationship between

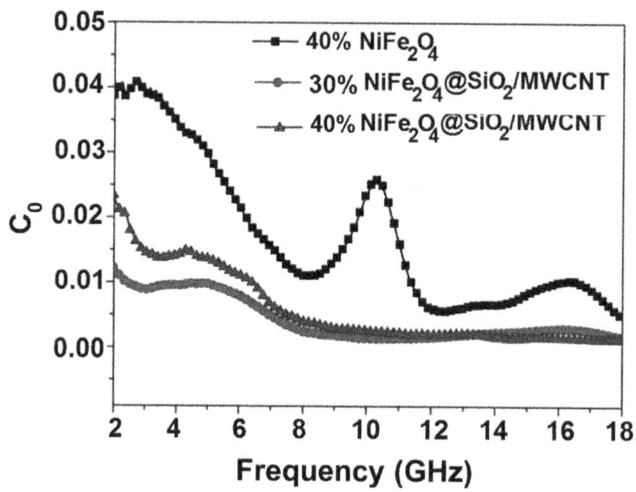

FIGURE 6.38

Frequency dependence of C_0 for $NiFe_2O_4$ nanoparticles and $NiFe_2O_4$@SiO_2/multiwalled carbon nanotube nanocomposites.

From Xiong, Y., Luo, H., Nie, Y., Chen, F., Dai, W., Wang, X., Cheng, Y., & Gong, R. (2019). Synergistic effect of silica coated porous rodlike nickel ferrite and multiwalled carbon nanotube with improved electromagnetic wave absorption performance. Journal of Alloys and Compounds, 802, *364–372. https://doi.org/10.1016/j.jallcom.2019.06.174.*

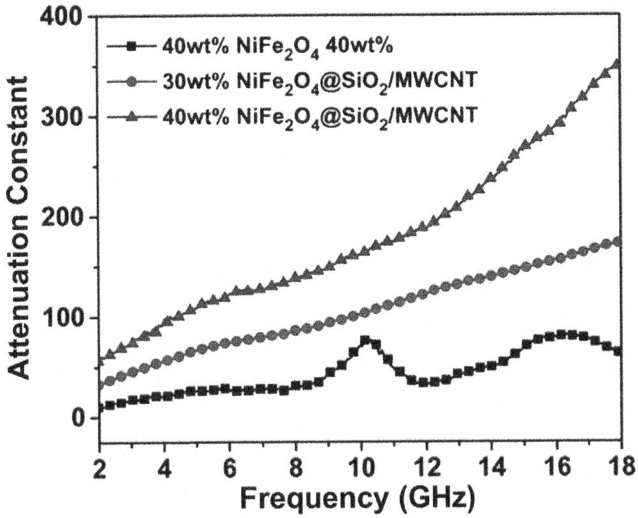

FIGURE 6.39

The variation in attenuation constant as a function of frequency for $NiFe_2O_4$ nanoparticles and $NiFe_2O_4@SiO_2$/multiwalled carbon nanotube nanocomposites.

From Xiong, Y., Luo, H., Nie, Y., Chen, F., Dai, W., Wang, X., Cheng, Y., & Gong, R. (2019). Synergistic effect of silica coated porous rodlike nickel ferrite and multiwalled carbon nanotube with improved electromagnetic wave absorption performance. Journal of Alloys and Compounds, 802, *364–372. https://doi.org/10.1016/j.jallcom.2019.06.174.*

the RL peak frequency and Z for the various matching thicknesses. In this respect, the RL is maximized when the corresponding impedance is close to unity at the matching frequency of 11.9 GHz.

The possible mechanism involved in the EM wave absorption of $NiFe_2O_4@SiO_2$/MWCNTs nanocomposites may be considered as follows: (1) MWCNTs surface defects can act as scattering and polarization centers, thus inducing dipole polarization (Shu, Li, et al., 2019); (2) the presence of multiple heterogeneous interfaces between the $NiFe_2O_4$ nanoparticles, SiO_2 coating, and MWCNTs leads to interfacial polarization, enhancing the attenuation capacity of EM waves; (3) natural resonance and eddy current are responsible for the magnetic loss and EM wave attenuation occurring in nanocomposites; and (4) the combination of magnetic and dielectric properties causes improvement in the impedance matching, facilitating the entrance of EM waves into the absorber nanocomposites.

6.10 Ferrite/graphene nanocomposites

Novel carbon nanomaterials including RGO have numerous potential applications in MW absorption due to their high specific surface area, structural defects, oxygen-

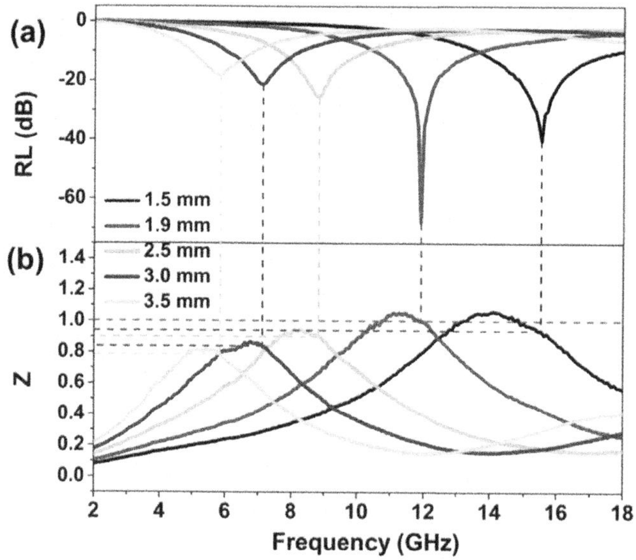

FIGURE 6.40

Variations in (A) reflection loss (RL) and (B) impedance (*Z*) as a function of frequency for the 40 wt.% $NiFe_2O_4@SiO_2$/multiwalled carbon nanotube nanocomposite with various thicknesses.

From Xiong, Y., Luo, H., Nie, Y., Chen, F., Dai, W., Wang, X., Cheng, Y., & Gong, R. (2019). Synergistic effect of silica coated porous rodlike nickel ferrite and multiwalled carbon nanotube with improved electromagnetic wave absorption performance. Journal of Alloys and Compounds, 802, *364–372. https://doi.org/10.1016/j.jallcom.2019.06.174.*

containing groups, and low density, in addition to considerable dielectric loss (Lv et al., 2017; Meng et al., 2018; Wang et al., 2011). Notably, BaM nanoparticles have been synthesized in a reaction medium comprising various graphene nanosheets, such as EG, EGO, and REGO, providing novel hexaferrite/graphene nanocomposites for use in MW absorbers (Durmus et al., 2015). The RL curves as a function of frequency are shown in Fig. 6.41. As observed, BaM-EG and BaM-REGO nanocomposites have a broad RL peak maximized at −58 dB for the frequency of 11.42 GHz. In the case of BaM-EGO nanocomposite, the RL curve changes slightly within the frequency range. The RL peaks of BaM-EG and BaM-REGO increase to more than −20 dB at 10.14 GHz, indicating a reflection of <1% up to 12.4 GHz with a frequency width of 2.26 GHz. Overall, BaM-EG and BaM-REGO nanocomposites with RL values higher than −10 dB show promise for MW absorption applications, absorbing 90% of incidence waves.

The RL of BaM-REGO nanocomposite with various thicknesses (ranging between 0.5 and 5.0 mm) has also been calculated, thereby investigating the nanocomposite thickness effect on the corresponding MW absorption performance. Fig. 6.42 shows the RL curve and thickness color plot as a function of frequency for the BaM-

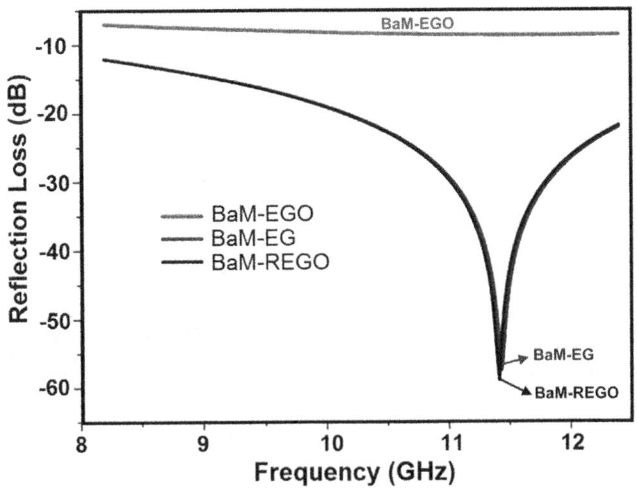

FIGURE 6.41

The reflection loss curves as a function of frequency for barium hexaferrite/graphene nanocomposites.

From Durmus, Z., Durmus, A., & Kavas, H. (2015). Synthesis and characterization of structural and magnetic properties of graphene/hard ferrite nanocomposites as microwave-absorbing material. Journal of Materials Science, 50*(3), 1201–1213. https://doi.org/10.1007/s10853-014-8676-3.*

REGO nanocomposite. As inferred, increasing the nanocomposite thickness up to 2.7 mm changes the absorption band to the X-band frequency range while maintaining the bandwidth almost constant when the thickness and RL are up to 5 mm, and < -20 dB, respectively. Bhattacharya and colleagues have also observed the same trend for graphene decorated with the hexagonal-shaped M-type ferrite and the PANI wrapper (Bhattacharya et al., 2014).

Recently, a new approach to tailoring chemical and electrical properties of ferrite/graphene nanocomposites has been introduced, in which RGO sheets were doped with heteroatoms. Notably, by doping the crystal lattice of RGO with nitrogen atoms [resulting in nitrogen-doped RGO (NRGO)-based nanocomposites], the number of defects, the dipole polarization, and the conduction loss are enhanced while optimizing the impedance matching condition (Feng et al., 2016; Shu, Zhang, et al., 2019).

The demand for hybrid nanocomposites with lightweight, broadband, and tunable EM properties is increasing day by day. Using a facile one-pot hydrothermal route, NRGO/$ZnFe_2O_4$ hybrid nanocomposites have been synthesized by Shu and colleagues (Shu, Zhang, et al., 2019). In this regard, they prepared pure $ZnFe_2O_4$ nanoparticles (labeled as S_1) and NRGO/$ZnFe_2O_4$ nanocomposites with various GO loadings [labeled as S_2 (60 mg) and S_3 (90 mg)]. Based on the RL curves depicted in Fig. 6.43A, RL of -2.2 dB is only obtained for S_1 with a thickness of 5.0 mm, indicating poor MW absorption performance. From Fig. 6.43B, the

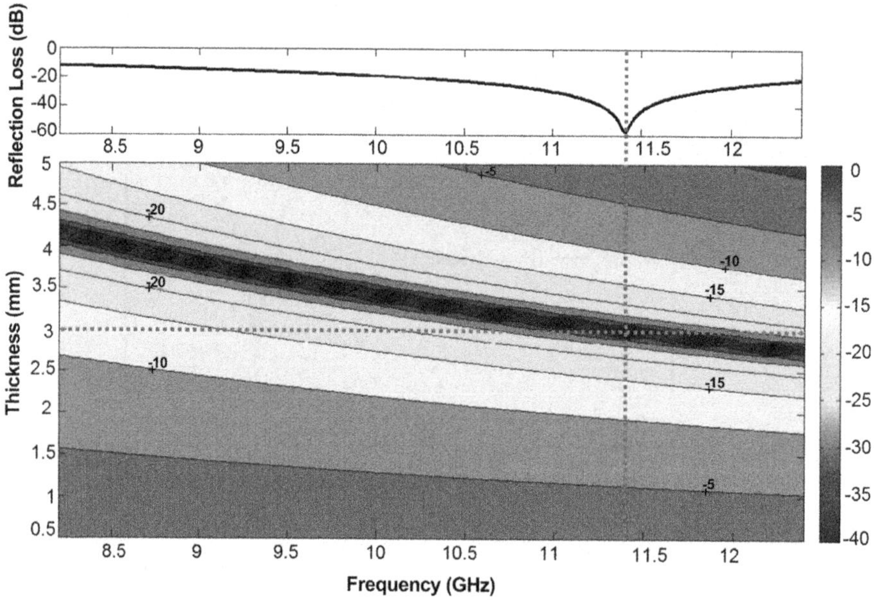

FIGURE 6.42

The reflection loss curve and thickness color plot as a function of frequency for the $BaFe_9Mn_{0.75}Co_{0.75}Ti_{1.5}O_{19}$-reduced expanded graphite oxide nanocomposite. Note: the reflection loss curve has been measured for a nanocomposite thickness of 3 mm.

From Durmus, Z., Durmus, A., & Kavas, H. (2015). Synthesis and characterization of structural and magnetic properties of graphene/hard ferrite nanocomposites as microwave-absorbing material. Journal of Materials Science, 50*(3), 1201–1213. https://doi.org/10.1007/s10853-014-8676-3.*

nanocomposite S_2 with a thickness of 3.6 mm exhibits an RL of −54.6 dB at 10.0 GHz (the X-band frequency). Moreover, an effective absorption bandwidth (EAB) of 3.6 GHz (12.7−16.3 GHz) is found for S_2 with a thickness of 2.5 mm.

According to Fig. 6.43C, the nanocomposite S_3 with a thickness of 5.0 mm presents an RL of −54.5 dB at 5.0 GHz. At a thickness of 2.0 mm, S_3 has an EAB of 4.2 GHz (11.8−16.0 GHz). Thus, the NRGO/$ZnFe_2O_4$ hybrid nanocomposites show perceptibly higher MW absorption properties than the pure $ZnFe_2O_4$ nanoparticles. By modulating the absorber thickness, it is possible to locate the RL at different frequencies, as depicted in the RL-frequency-thickness 3-D plot of the nanocomposite S_2 (see Fig. 6.43D).

In recent years, many researchers have focused on MW absorption features of zinc ferrite/graphene nanocomposites. For example, Yang and colleagues have demonstrated that the MW absorption of $ZnFe_2O_4$/RGO nanohybrids outperforms that of RGO, giving rise to an RL_{max} of −29.3 dB at 16.7 GHz along with an EAB of 2.6 GHz (15.4−18.0 GHz) at a thickness of 1.6 mm (Yang et al., 2015). Alternatively, Zhang and colleagues have successfully synthesized $ZnFe_2O_4$@SiO_2 core−shell hollow microsphere/RGO nanocomposites. They found an RL_{max}

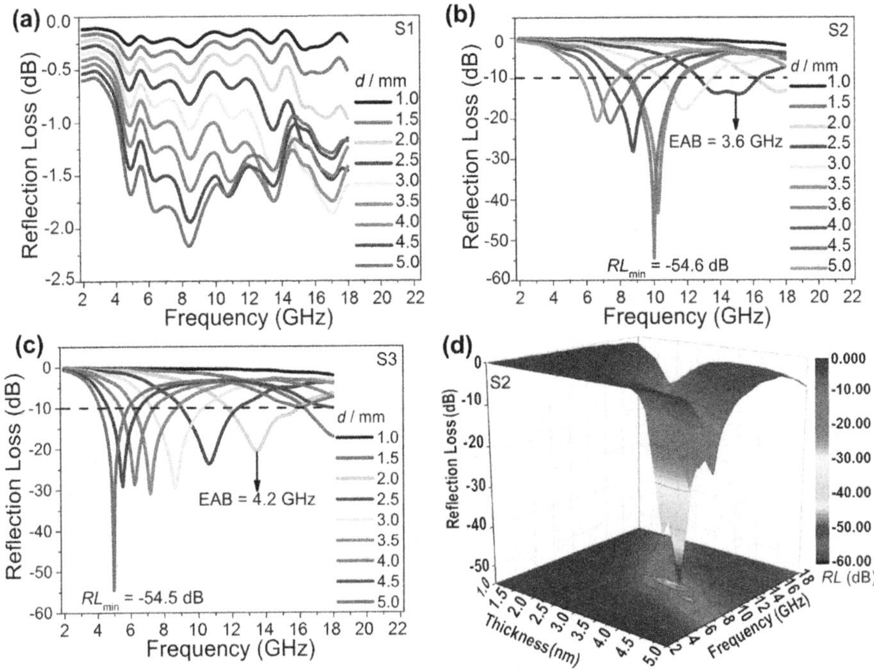

FIGURE 6.43

Reflection loss curves as a function of frequency for: (A) $ZnFe_2O_4$ nanoparticles (S_1); and nitrogen-doped reduced graphene oxide/$ZnFe_2O_4$ hybrid nanocomposites with GO loadings of (B) 60 mg (S_2), and (C) 90 mg (S_3) at various thicknesses. (D) Reflection loss-frequency-thickness three-dimensional plot of the nanocomposite S_2.

From Shu, R., Zhang, J., Wu, Y., Wan, Z., & Zheng, M. (2019). Facile design of nitrogen-doped reduced graphene oxide/zinc ferrite hybrid nanocomposites with excellent microwave absorption in the X-band. Materials Letters, *255, 126549. https://doi.org/10.1016/j.matlet.2019.126549.*

of −45.8 dB at 7.6 GHz for the $ZnFe_2O_4$@SiO_2/RGO nanocomposite with a thickness of 3.7 mm. The bandwidth also reached 4.0 GHz (7.7−11.7 GHz) at a thickness of 3.0 mm (Zhang et al., 2016). Feng and colleagues have combined a layer-by-layer coating technique with a chemical process to synthesize $ZnFe_2O_4$@RGO@TiO_2 yolk-shell nanocomposites with a thickness of 2.6 mm, achieving an RL_{max} of −44.3 dB at 15.92 GHz, together with an EAB of 4.1 GHz (Feng et al., 2016; Shu, Zhang, et al., 2019).

Elsewhere, RGO/MWCNTs/$ZnFe_2O_4$ nanocomposites were prepared using the following approach schematically illustrated in Fig. 6.44 (Shu, Zhang, et al., 2018).

Fig. 6.45A depicts the RL-frequency-thickness 3-D plot of the RGO/MWCNTs/$ZnFe_2O_4$ ternary nanocomposite. Evidently, controlling the absorber thickness enables us to reach the RL_{max} at different frequencies gradually, thus achieving the maximum absorption. The corresponding 2-D contour plot of the RL is also shown

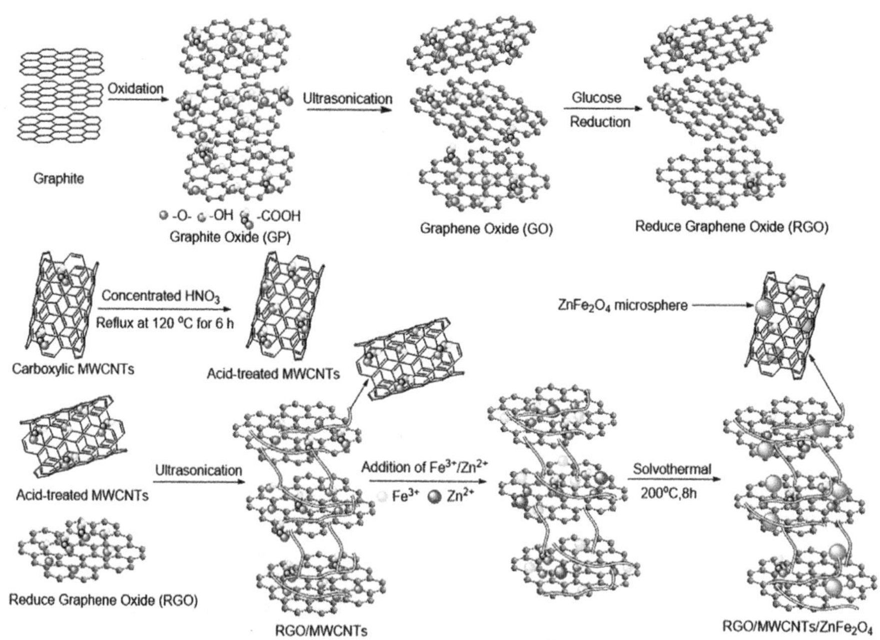

FIGURE 6.44

Schematic illustration of the synthesis process of reduced graphene oxide/multiwalled carbon nanotube/$ZnFe_2O_4$ nanocomposites.

From Shu, R., Zhang, G., Zhang, J., Wang, X., Wang, M., Gan, Y., Shi, J., & He, J. (2018). Fabrication of reduced graphene oxide/multi-walled carbon nanotubes/zinc ferrite hybrid composites as high-performance microwave absorbers. Journal of Alloys and Compounds, 736, *1–11. https://doi.org/10.1016/j.jallcom.2017.11.084.*

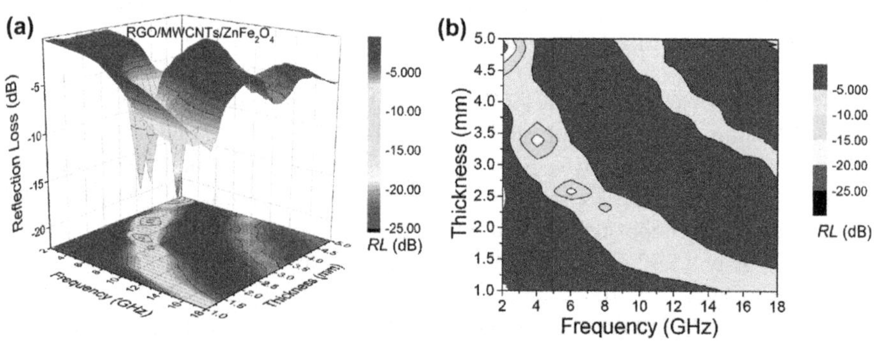

FIGURE 6.45

(A) Reflection loss-frequency-thickness three-dimensional plot and (B) corresponding two-dimensional contour plot of the reflection loss for reduced graphene oxide/ multiwalled carbon nanotube/$ZnFe_2O_4$ ternary nanocomposite.

From Shu, R., Zhang, G., Zhang, J., Wang, X., Wang, M., Gan, Y., Shi, J., & He, J. (2018). Fabrication of reduced graphene oxide/multi-walled carbon nanotubes/zinc ferrite hybrid composites as high-performance microwave absorbers. Journal of Alloys and Compounds, 736, *1–11. https://doi.org/10.1016/j.jallcom.2017.11.084.*

in Fig. 6.45B. As inferred, for an optimized absorber thickness, the absorption band is higher than −5.0 dB, spanning the entire frequency range (2.0−18.0 GHz).

The MW absorption properties of nickel ferrite/graphene nanocomposites have also been studied in the literature. For instance, Zhang and colleagues have synthesized $NiFe_2O_4$/RGO nanohybrids by a confined growth process, achieving an RL and EAB (RL ≤ -10 dB) of −58 dB and 4.08 GHz at a thickness of 2.7 mm, respectively (Zhang et al., 2018).

It is possible to prepare RGO/MWCNTs/$NiFe_2O_4$ ternary nanocomposites by the self-assembling of GO and MWCNTs in aqueous dispersion, followed by Ni^{2+}/Fe^{3+} adsorption via electrostatic interactions and employing a hydrothermal process (Wu et al., 2019). In this case, when GO is partially reduced into RGO, $NiFe_2O_4$ nanoparticles can nucleate and grow in situ on the resultant RGO surface. The RGO surface loaded with $NiFe_2O_4$ nanoparticles is then twined with MWCNTs based on the microcurrent network mechanism proposed by Cao and colleagues, thus allowing for the $NiFe_2O_4$ nanoparticles to bridge between neighboring RGO and MWCNTs due to the electron-hopping process. This, in turn, leads to the microcurrent network formation, enhancing the conduction loss of the RGO/MWCNTs/$NiFe_2O_4$ nanocomposites compared with that of pure $NiFe_2O_4$ nanoparticles (Wen et al., 2008). Accordingly, for the sake of comparison, RGO/$NiFe_2O_4$ binary nanocomposite (S_1) and RGO/MWCNTs/$NiFe_2O_4$ ternary nanocomposites with thicknesses of 1.4 mm (S_2) and 1.6 mm (S_3) were also synthesized. The EM wave absorption performance of the ternary nanocomposites (S_2 and S_3) was higher than that of the binary nanocomposite (S_1) and $NiFe_2O_4$ nanoparticles. It is worth mentioning that the MWCNTs additive amount needs to be carefully controlled to achieve excellent EM wave absorption properties. Since $NiFe_2O_4$ nanoparticles have a mean size ($\sim$12.41 nm) close to their exchange length ($\sim$10 nm), the exchange resonance contribution would play an important role in determining the magnetic loss.

Fig. 6.46 shows the variation in C_0 as a function of frequency for $NiFe_2O_4$, S_1, S_2, and S_3. Fluctuations are clearly observed in the high-frequency range of 8−18 GHz for all the nanocomposite samples, suggesting that the eddy current loss is not mainly involved in the magnetic loss mechanism. In fact, prominent C_0 peaks are observed in the low-frequency range, indicating the significant involvement of the natural resonance for all the samples. The absorption mechanism of the ternary nanocomposites is similar to that of hybrid nanocomposites explained previously.

In a separate study, the mechanism involved in the MW absorption of $Ni_{0.5}Zn_{0.5}Fe_2O_4$/NRGO nanocomposites has been investigated by Shu and colleagues (Shu et al., 2020). The XRD patterns (Fig. 6.47A) of as-prepared $Ni_{0.5}Zn_{0.5}Fe_2O_4$ nanoparticles (S_1) and $Ni_{0.5}Zn_{0.5}Fe_2O_4$/NRGO nanocomposite (S_2) show the formation of spinel ferrite structure without the presence of secondary phases. Raman spectroscopy has also been used to evaluate the degree of graphitization, and the results for S_1, S_2, and GO are compared in Fig. 6.47B. In this case, two prominent scattering peaks are observed at 1590 cm^{-1} (the G band related to sp^2 hybridization) and 1347 cm^{-1} (the D band related to sp^3 defects) for both S_2 and GO. The ratio of D band intensity to G band intensity (I_D/I_G) can be employed to evaluate

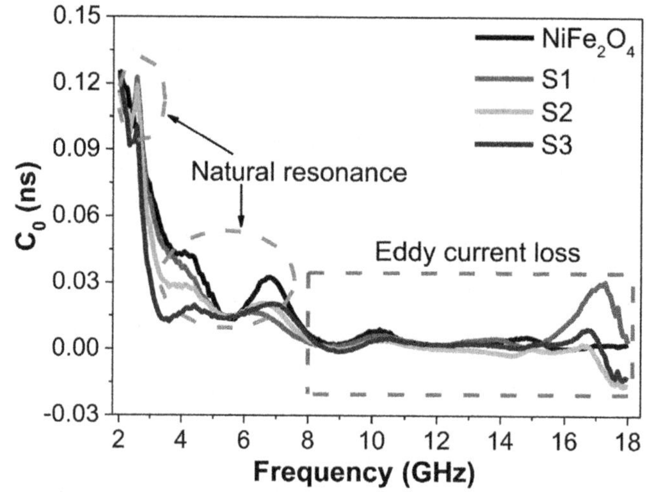

FIGURE 6.46

The variation in C_0 as a function of frequency for $NiFe_2O_4$ nanoparticles, reduced graphene oxide (RGO)/$NiFe_2O_4$ binary nanocomposite (S_1), and RGO/multiwalled carbon nanotube/$NiFe_2O_4$ ternary nanocomposites with thicknesses of 1.4 mm (S_2) and 1.6 mm (S_3).

From Wu, Y., Shu, R., Li, Z., Guo, C., Zhang, G., Zhang, J., & Li, W. (2019). Design and electromagnetic wave absorption properties of reduced graphene oxide/multi-walled carbon nanotubes/nickel ferrite ternary nanocomposites. Journal of Alloys and Compounds, 784, *887–896. https://doi.org/10.1016/j.jallcom.2019.01.139.*

the degree of defects (Shu, Zhang, et al., 2018, Shu, Zhang, et al., 2019). In this regard, I_D/I_G of S_2 and GO are 1.20 and 0.89, respectively, indicating a higher degree of defects in the as-prepared binary nanocomposite compared with that in the GO. This may arise from the RGO surface defects induced by nitrogen doping when performing the hydrothermal process. The Raman scattering peaks of S_1 and S_2 in the low wavenumber range (300–800 cm^{-1}) are also attributed to the $Ni_{0.5}Zn_{0.5}Fe_2O_4$ characteristic peaks.

TEM and HRTEM micrographs obtained from S_1 and S_2 are depicted in Fig. 6.48. Compared with the morphological properties of S_1, the formation of heterogeneous interfaces between $Ni_{0.5}Zn_{0.5}Fe_2O_4$ nanoparticles and NRGO may significantly enhance dielectric loss and interfacial polarization of S_2 under alternating EM fields. In fact, the $Ni_{0.5}Zn_{0.5}Fe_2O_4$ nanoparticles are firmly attached to the NRGO surfaces after performing ultrasound treatment (Shu, Zhang, et al., 2018). According to Fig. 6.48C and F, HRTEM micrographs reveal the interplanar distances of 0.305 and 0.245 nm corresponding to (220) and (311) planes of $Ni_{0.5}Zn_{0.5}Fe_2O_4$ nanoparticles, respectively.

Fig. 6.49 shows the variation in C_0 as a function of frequency for S_2 with various filler loadings. In this case, the increasing frequency decreases the corresponding C_0,

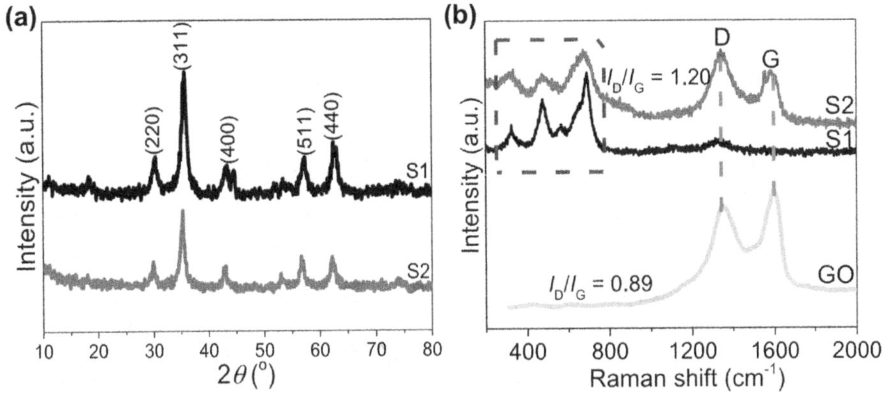

FIGURE 6.47

(A) The X-ray diffraction patterns of $Ni_{0.5}Zn_{0.5}Fe_2O_4$ nanoparticles (S_1) and $Ni_{0.5}Zn_{0.5}Fe_2O_4$/nitrogen-doped reduced graphene oxide nanocomposite (S_2). (B) Raman spectra of S_1, S_2, and GO.

From Shu, R., Zhang, J., Guo, C., Wu, Y., Wan, Z., Shi, J., Liu, Y., & Zheng, M. (2020). Facile synthesis of nitrogen-doped reduced graphene oxide/nickel-zinc ferrite composites as high-performance microwave absorbers in the X-band. Chemical Engineering Journal, 384, *123266. https://doi.org/10.1016/j.cej.2019.123266.*

indicating the relatively weak contribution of the eddy current loss in the absorption properties. A filler loading of 60 wt.% results in the highest ε′ and ε″, manifesting enhanced electric energy storage capability and dielectric loss induced by the magnetic resonance contribution.

In another research, the MW absorption behavior of graphene/$NiZnFe_2O_4$ nanocomposites with various compositions of Ni:Zn has been investigated (Zhou et al., 2017). According to Fig. 6.50, while pure $Ni_{0.5}Zn_{0.5}Fe_2O_4$ nanofibers show small attenuation constant α values in the frequency range of 2–18 GHz, the EM attenuation capability of the hybrid nanocomposites (prepared by combining $NiZnFe_2O_4$ nanofibers with RGO) is considerably improved, so the RGO/$Ni_{0.7}Zn_{0.3}Fe_2O_4$ nanocomposite has the maximum α value. Thereby, one can expect RGO/NiZn ferrite hybrid nanocomposites to show good MW absorption properties compared with pure nickel-zinc ferrites.

Fig. 6.51 shows the variation in relative input impedance (Z) as a function of frequency for pure $Ni_{0.5}Zn_{0.5}Fe_2O_4$ nanofibers and RGO/$Ni_xZn_{(1-x)}Fe_2O_4$ ($x = 0.7$, 0.5 and 0.3) nanocomposites. RL_{max} probably occurs when the matching frequency of the RGO/$Ni_{0.7}Zn_{0.3}Fe_2O_4$ nanocomposite is close to 12 GHz at a matching thickness of 2.0 mm, corresponding to a Z value of near one. In other words, the incident EM wave probably enters into the 2.0 mm thick RGO/$Ni_{0.7}Zn_{0.3}Fe_2O_4$ nanocomposite at a frequency of ~12 GHz.

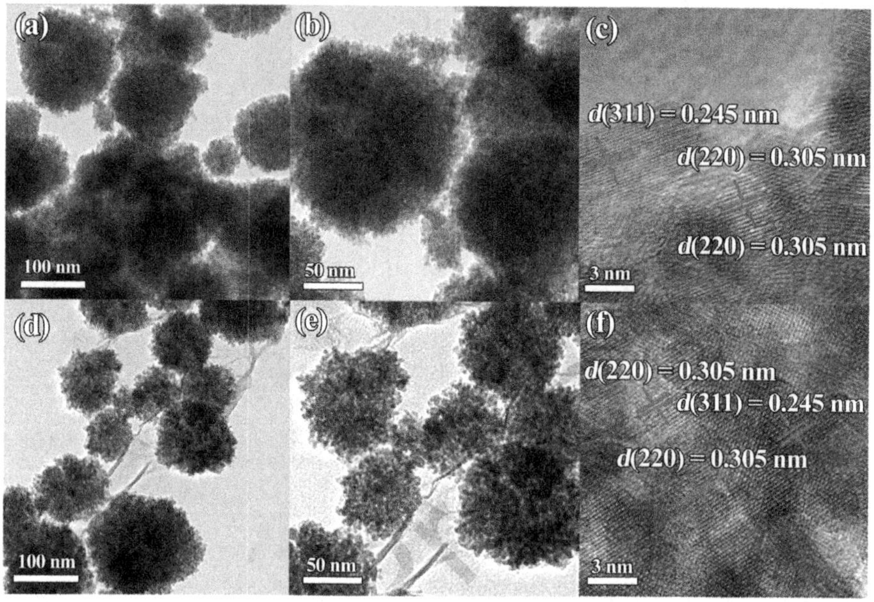

FIGURE 6.48

Transmission electron microscopic micrographs with different magnifications obtained from (A) and (B) $Ni_{0.5}Zn_{0.5}Fe_2O_4$ nanoparticles (S_1), and (D) and (E) $Ni_{0.5}Zn_{0.5}Fe_2O_4$/nitrogen-doped reduced graphene oxide nanocomposite (S_2). High-resolution transmission electron microscopy micrographs of (C) S_1 and (F) S_2.

From Shu, R., Zhang, J., Guo, C., Wu, Y., Wan, Z., Shi, J., Liu, Y., & Zheng, M. (2020). Facile synthesis of nitrogen-doped reduced graphene oxide/nickel-zinc ferrite composites as high-performance microwave absorbers in the X-band. Chemical Engineering Journal, 384, *123266. https://doi.org/10.1016/j.cej.2019.123266.*

With the help of structural and morphological optimization, one may address the challenge of ameliorating the EM properties of conventional absorbing materials. In this regard, cobalt ferrite/graphene nanocomposites have extensively been studied. Notably, $CoFe_2O_4$/graphene nanocomposites with flower-like morphology have been prepared, followed by investigating MW absorption properties of $CoFe_2O_4$/graphene/paraffin wax composites (Li et al., 2017; Wang, Zhao, et al., 2018; Wang et al., 2019). The resulting composites with a thickness of 2.0 mm and filler loading of 45 wt.% showed an RL_{max} of −42.0 dB and EAB of 4.59 GHz.

Alternatively, using a one-step hydrothermal process, Yin and colleagues have synthesized $Co_{0.5}Ni_{0.5}Fe_2O_4$/graphene composites, achieving an RL_{max} of −30.92 dB at a thickness of 4.0 mm and filler loading of 50 wt.% (Yin et al., 2018). Moreover, $Co_{0.8}Fe_{2.2}O_4$/RGO nanocomposites (having a filler loading of 50 wt.%) prepared by a facile one-pot MW hydrothermal process resulted in an RL_{max} of −51.2 dB and EAB of 7.2 GHz at thicknesses of 2.1 and 2.0 mm, respectively (Shen et al., 2016).

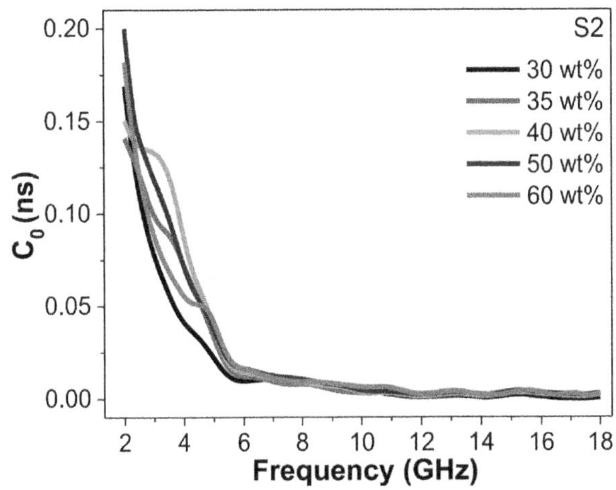

FIGURE 6.49

The variation in C_0 as a function of frequency for $Ni_{0.5}Zn_{0.5}Fe_2O_4$/nitrogen-doped reduced graphene oxide nanocomposite (S_2) with various filler loadings.

From Shu, R., Zhang, J., Guo, C., Wu, Y., Wan, Z., Shi, J., Liu, Y., & Zheng, M. (2020). Facile synthesis of nitrogen-doped reduced graphene oxide/nickel-zinc ferrite composites as high-performance microwave absorbers in the X-band. Chemical Engineering Journal, 384, *123266. https://doi.org/10.1016/j.cej.2019.123266.*

6.11 Ferrite/graphene oxide nanocomposites

Liang and colleagues have investigated MW absorption properties of GO/cobalt ferrite composites (Liang et al., 2017). Fig. 6.52 shows the variation in RL as a function of frequency for GO/cobalt ferrite composites with various GO filler loadings. From Fig. 6.52A, RL of the composite with a thickness of 2 mm is obtained to be −9.19 dB at 9.0 GHz for GO = 0.1 wt.%, having a bandwidth of 5.4 GHz (6.7−12.1 GHz) for RL higher than −5 dB. Increasing the composite thickness up to 5 mm shifts the corresponding maximum absorption frequencies to lower frequency ranges, resulting in an RL_{max} of −25.60 dB at 2.18 GHz.

In the case of GO = 0.5 wt.%, the composite with a thickness of 2 mm reaches the RL_{max} of −9.27 dB at 8.48 GHz, having a bandwidth of 4.6 GHz (6.4−11.0 GHz). Increasing the thickness from 3 to 3.5 mm changes the RL_{max} from −12.78 to −14.25 dB at 3.6 and 3.0 GHz, respectively. The maximum RL is obtained to be −18.80 dB at 1.7 GHz for the GO/cobalt ferrite composite with a thickness of 6 mm. Compared with the GO = 0.1 wt.%, RL is slightly higher for GO = 0.5 wt.%. For the 1 mm thick composite with GO = 0.5 wt.%, the RL higher than −5 dB is ranged between 10.9 and 17.7 GHz, having a bandwidth of about 6.8 GHz, which indicates the excellent absorption properties of the GO/cobalt ferrite composites. These may be assigned to the combined effects of electric and magnetic losses at GO = 0.5 wt.%. In the case of GO = 1.0 wt.% and a thickness

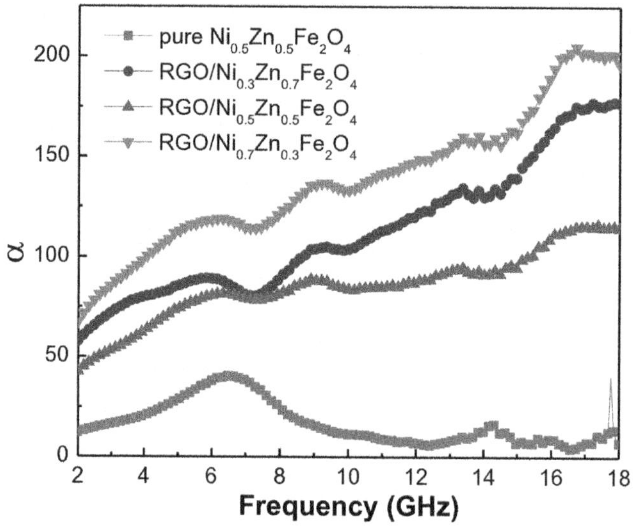

FIGURE 6.50

The variation in attenuation constant (α) as a function of frequency for pure $Ni_{0.5}Zn_{0.5}Fe_2O_4$ nanofibers and reduced graphene oxide (RGO)/$Ni_{0.3}Zn_{0.7}Fe_2O_4$, RGO/$Ni_{0.5}Zn_{0.5}Fe_2O_4$, and RGO/$Ni_{0.7}Zn_{0.3}Fe_2O_4$ hybrid nanocomposites.

From Zhou, X., Chuai, D., & Zhu, D. (2019). Electrospun synthesis of reduced graphene oxide (RGO)/NiZn ferrite nanocomposites for excellent microwave absorption properties. Journal of Superconductivity and Novel Magnetism, *32(8), 2687–2697. https://doi.org/10.1007/s10948-019-5039-y.*

of 1.5 mm, RL_{max} is −13.27 dB at 11.6 GHz, having a bandwidth of about 2.6 GHz (10.3–12.9 GHz). Moreover, RL_{max} is obtained to be about −13.29 dB at 8.3 GHz when the thickness is 2 mm, having a bandwidth of 2.1 GHz (7.3–9.4 GHz). The maximum absorption frequencies shift to the lower frequency range with increasing the thickness, thereby improving RL_{max}. For the 3.5 mm thick composite, twin peaks appear at 4.3 and 15.6 GHz. The RL_{max} is −16.09 dB at 2.3 GHz when the thickness increases up to 6 mm.

The combined effects of electric and magnetic losses lead to a significant increase in the absorption bandwidth while increasing the RL when enhancing the GO filler loading. Based on Fig. 6.52D, the presence of three RL peaks is indicated at S and Ku bands when the thickness is 6 mm. Increasing the thickness to 10 mm causes the peaks to appear at S, X, and Ku bands. In this case, two peaks are below −10 dB at 9.0 and 15.4 GHz. Compared with single ferrites prepared elsewhere, the GOs resulting from the roasting process can efficiently improve permittivity and conductivity of the composites, leading to enhanced EM parameters of MW absorption, particularly at low-frequency ranges.

Considering the relation $\varepsilon'' = \frac{1}{(2\pi\rho f \varepsilon_0)}$, in which ρ, f, and ε_0 are the composite resistivity, EM wave frequency, and the vacuum permittivity, respectively, GOs with

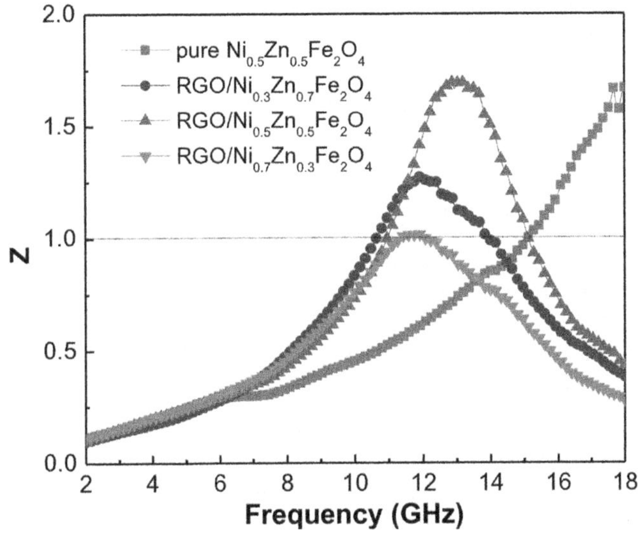

FIGURE 6.51

The variation in relative input impedance (Z) as a function of frequency for pure $Ni_{0.5}Zn_{0.5}Fe_2O_4$ nanofibers and reduced graphene oxide/$Ni_xZn_{(1-x)}Fe_2O_4$ ($x = 0.7$, 0.5, and 0.3) nanocomposites.

From Zhou, X., Chuai, D., & Zhu, D. (2019). Electrospun synthesis of reduced graphene oxide (RGO)/NiZn ferrite nanocomposites for excellent microwave absorption properties. Journal of Superconductivity and Novel Magnetism, 32*(8), 2687–2697. https://doi.org/10.1007/s10948-019-5039-y.*

high conductivity can decrease the composite resistivity, giving rise to an increase in ε''.

In the case of GO = 0.1 and 0.5 wt.%, the ferrite coverage ratio on the GO surface exceeds that on other surfaces. The electron transition from the ferrite surface to the GO causes an enhancement in its conductivity. Meanwhile, the relaxation effect arising from electron movement and dielectric dispersion enhances permittivity. Nevertheless, a rough interface only forms between the ferrite and GO when increasing the GO filler loading (e.g., GO = 1.0 or 2.0 wt.%) (Giovannetti et al., 2008; Pi et al., 2009).

For years, MW absorbing materials have suffered from low-frequency EM absorption. This challenge may be solved by the preparation of fibrous composites with conductive networks incorporated with natural cotton fabrics and ferrite nanoparticles. Cotton fabrics with enhanced SE values have various military applications. Fig. 6.53 shows the EM shielding performance of cotton fabrics coated with copper-cobalt-nickel ferrite/GO/PANI ternary nanocomposite (Sun et al., 2017). Increasing the absorbing agent content from 10% to 50% increases the corresponding coated cotton fabric shielding performance from −10 to −47 dB, followed by another increase to −45 dB at 2,250,000 kHz. For an absorbing agent content of approximately

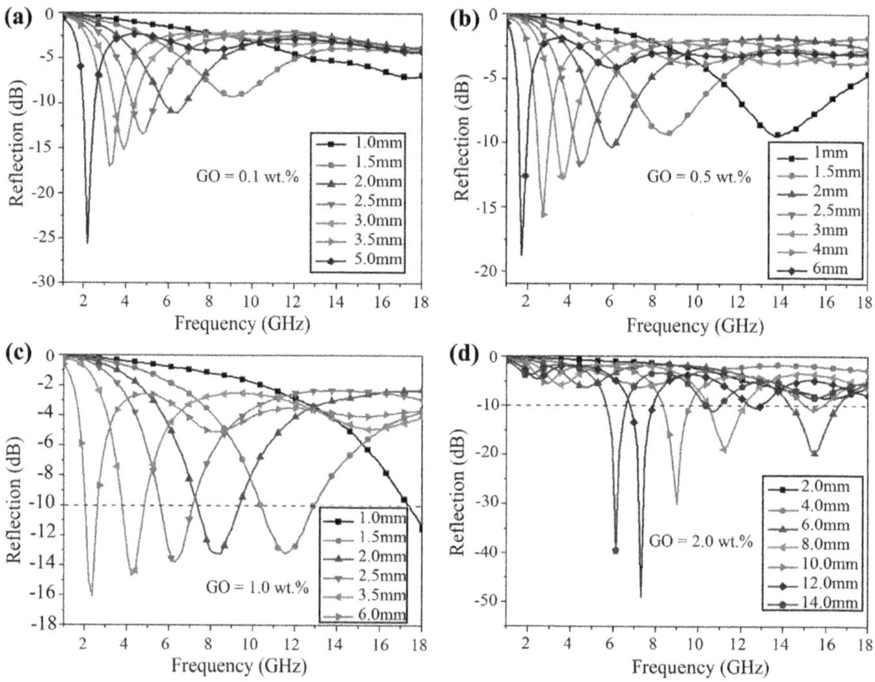

FIGURE 6.52

The variation in reflection loss as a function of frequency for GO/cobalt ferrite composites at various GO filler loadings: (A) 0.1 wt.%, (B) 0.5 wt.%, (C) 1.0 wt.%, and (D) 2.0 wt.%.

From Liang, K., Qiao, X.-J., Sun, Z.-G., Guo, X.-D., Wei, L., & Qu, Y. (2017). Preparation and microwave absorbing properties of graphene oxides/ferrite composites. Applied Physics A, *123(6), 445. https://doi.org/10.1007/s00339-017-1059-2.*

40%, SE is maximized at −47 dB due to the excessive loading of the ternary nanocomposite. Accordingly, the charge transfer occurring between newly emerged interfaces is facilitated by the covalent bonds formed between GO and copper-cobalt-nickel ferrite/PANI, weakening the EM energy. The occurrence of multiple reflections also enhances EM wave loss on the copper-cobalt-nickel ferrite/GO/PANI ternary nanocomposite surface.

6.12 Ferrite/carbon black nanocomposites

To further improve the MW absorption properties of ferrites, one can add a small amount of CB to the coating formulation, providing the absorber material with permittivity in addition to permeability. For synthetic and natural rubbers, CB is potentially used as a reinforcing filler to improve dielectric properties and tune the MW absorption bandwidth. In this regard, strontium ferrite (SrF)—CB—nitrile

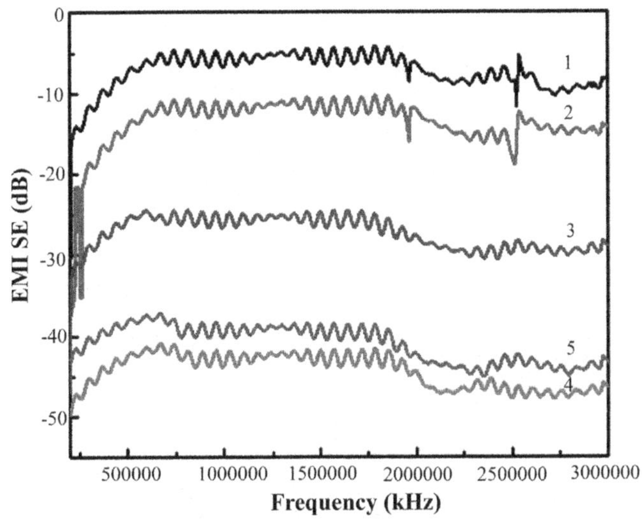

FIGURE 6.53

Electromagnetic shielding performance of copper-cobalt-nickel ferrite/GO/polyaniline ternary nanocomposite coated cotton fabrics with different coating agent contents of 10%, 20%, 30%, 40%, and 50% (1–5).

From Sun, J., Wang, L., Yang, Q., Shen, Y., & Zhang, X. (2020). Preparation of copper-cobalt-nickel ferrite/graphene oxide/polyaniline composite and its applications in microwave absorption coating. Progress in Organic Coatings, 141, *105552. https://doi.org/10.1016/j.porgcoat.2020.105552.*

rubbers have been synthesized and designed for MW absorption bands (Vinayasree et al., 2013). From Fig. 6.54, increasing the CB concentration shifts the maximum RL frequency toward lower frequencies. Conversely, decreasing the absorber thickness shifts the maximum RL frequency toward high frequencies. The sharp increase observed in the real part of dielectric permittivity arises from space charge accumulation due to the heterogeneity of the SrF-CB-nitrile rubbers and the individual component contributions to the effective medium theory. A dipole moment is induced by the accumulation of charges in the boundaries between the CB aggregate interface and the rubber. Within the MW frequency range, dielectric loss results from relaxation polarization and electric conductance.

Metal surfaces and fabrics can be coated with MW absorption materials for radar absorption applications. However, significantly less attention has been paid to preparing thin MW shielding fabrics, conducting polymer composites (e.g., PANI/PPy comprising dielectric/magnetic fillers), and sandwich structures (e.g., textile/resin/foam). These composites and structures perform in a wide frequency band and may suitably conceal defense vehicles and equipment from radar detection and safely protect humans from harmful EM waves.

While effective EMI shielding performance can be realized by conducting polymer-coated fabrics, their adhesion to the textile substrate makes commercial

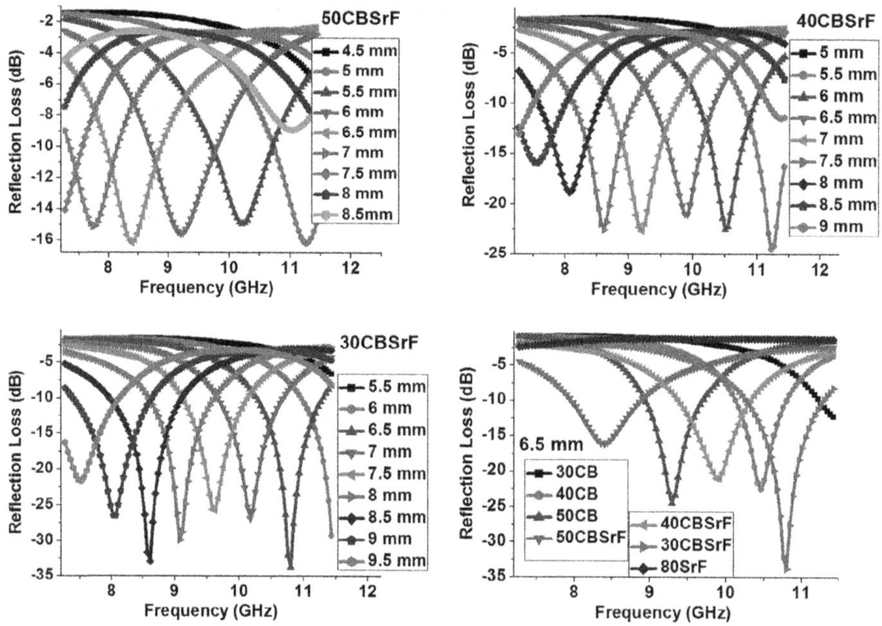

FIGURE 6.54

The variation in reflection loss as a function of frequency in the X band for strontium ferrite composites with various thicknesses and carbon black content.

From Vinayasree, S., Soloman, M., Sunny, V., Mohanan, P., Kurian, P., & Anantharaman, M. (2013). A microwave absorber based on strontium ferrite–carbon black–nitrile rubber for S and X-band applications. Composites Science and Technology, *82, 69–75. https://doi.org/10.1016/j.compscitech.2013.04.010.*

applications of polymer-coated fabrics doubtful. Moreover, the performance of low thickness coated fabrics decreases after several washings while suffering from a weight penalty after increasing the coating thickness.

Multilayered composites or foam absorbers filled with conducting fillers have recently been proposed to properly address these limitations. Nevertheless, increasing composite or absorber thickness is not considered useful for decreasing the radar cross-section detection of covering fabrics and defense equipment (Folgueras et al., 2007; Saini et al., 2012).

On the other hand, cotton fabrics coated with PANI (containing BF powders as fillers during polymerization) have provided 48% absorption in a frequency range of 6–14 GHz. Sheath–core bicomponent fibers comprising ferrite/bronze fillers have also exhibited good radar absorption performance. Additionally, fabrics coated with ferrite fibers have provided tuned responses to EM radiation at MW frequencies (Abdel Fattah et al., 2016; Goldberg & Marikar, 1988; Yu et al., 2007).

Attempts have also been made to coat cotton fabrics with thin layers of nanomaterials made of CB and ferrite (e.g., Ni–Zn) nanoparticle fillers in polyurethane (Gupta et al., 2014). The polyurethane-based CB and Ni–Zn ferrite nanofiller coatings possess intrinsic conductivity, permeability, and permittivity, allowing for MW interactions with the *E*- and *H*-components of EM waves. Experimentally speaking, Ni–Zn ferrite coatings at various concentrations (30, 40, 50, 60, and 70 g labeled as F1–F5), together with 5 g of CB, have been prepared in a 100 mL polyurethane solution. The RL curves as a function of frequency (8–18 GHz) for F1–F5 are shown in Fig. 6.55. Single-layer coatings on cotton fabrics with thicknesses of 1.6–1.8 mm have shown absorption, reflection, and transmission of 40%, 40%, and 20%, respectively, in conjunction with a maximum RL of 22.5 dB at 13.5 GHz. The RL is higher than 7.5 dB over nearly the entire Ku band.

Increasing the ferrite concentration from 30 to 70 g enhances RL in a range of 8–18 GHz. In this regard, sample F5 has the maximum RL, arising from the coupling of absorption, destructive interference, and internal reflections. In fact, by matching the optical thickness with the quarter wavelength ($\frac{\lambda}{4}$) of the wave, a path difference of $\frac{\lambda}{2}$ is created in the waves reflected from the front and back surfaces of the ferrite coated fabrics, thereby maximizing RL due to destructive interference.

In turn, destructive interferences are induced at various frequencies for ferrite-coated fabrics with different thicknesses, leading to specified peak losses. Notably, sample F5 shows RLs higher than 7.5 dB in a frequency range of 11–18 GHz, resulting in a maximum RL of 22.5 dB at 13.5 GHz. In this case, the refractive index: $[\sqrt{(\mu\varepsilon)} = \sqrt{(6.15 \times 1.388)}]$ is obtained as 2.92 at 13.5 GHz for $\lambda = 5.5$ mm.

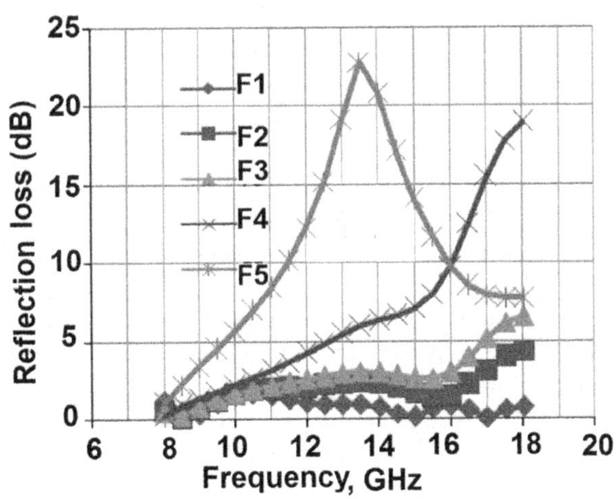

FIGURE 6.55

The reflection loss curves as a function of frequency for polyurethane-based carbon black and Ni–Zn ferrite coatings with various concentrations (30, 40, 50, 60, and 70 g labeled as F1–F5).

From Gupta, K., Abbas, S., Goswami, T., & Abhyankar, A. (2014). Microwave absorption in X and Ku band frequency of cotton fabric coated with Ni–Zn ferrite and carbon formulation in polyurethane matrix. Journal of Magnetism and Magnetic Materials, 362, *216–225. https://doi.org/10.1016/j.jmmm.2014.03.040.*

The thickness of sample F5 (1.68 mm) is very close to the corresponding optical thickness (1.88 mm), giving rise to the maximum MW absorption at 13.5 GHz due to destructive interference.

6.13 Ferrite/carbon fiber nanocomposites

As mentioned earlier, carbon fibers can be used as conductive networks to tailor the electrical properties of nanocomposites. In this respect, using an electrospinning method combined with stabilization and carbonization treatments, Zhang and colleagues (Zhang et al., 2013) have synthesized magnetic CNFs loaded with Fe_3O_4 nanoparticles, achieving an RL_{max} of −45 dB at 8 GHz. The absorption frequency bandwidth was also 13 GHz below −5 dB for an absorber layer (consisting of 5 wt.% CNFs/Fe_3O_4 nanoparticle) with a thickness of 5 mm.

The MW absorbers introduced to this point have consisted of a single-layer structure on perfect conductors. To realize the importance of double-layer and multilayer absorbers with thicknesses up to the millimeter range, a brief explanation is provided in the following.

Absorbing materials can be divided into single-layer, double-layer, and multilayer structures (Peng et al., 2013; Shen et al., 2007). While it is difficult to obtain high absorption and a wide absorption frequency range in a single-layer structure with a shallow layer thickness, absorber materials with double-layer and multilayer structures consisting of two or more absorbents with different thicknesses can provide enhanced EM properties after designing and optimizing their layered structures.

Eq. (6.22) gives the input impedance (Z_{in}) of a double-layer absorber as follows (Melvin et al., 2015):

$$Z_{in} = \frac{\sqrt{\frac{\mu_2}{\varepsilon_2}}\left(\sqrt{\frac{\mu_1}{\varepsilon_1}}\tanh\left[j\left(\frac{2\pi f d_1}{c}\right)\sqrt{\mu_1\varepsilon_1}\right] + \sqrt{\frac{\mu_2}{\varepsilon_2}}\tanh\left[j\left(\frac{2\pi f d_2}{c}\right)\sqrt{\mu_2\varepsilon_2}\right]\right)}{\sqrt{\frac{\mu_2}{\varepsilon_2}} + \sqrt{\frac{\mu_1}{\varepsilon_1}}\tanh\left[j\left(\frac{2\pi f d_1}{c}\right)\sqrt{\mu_1\varepsilon_1}\right]\tanh\left[j\left(\frac{2\pi f d_2}{c}\right)\sqrt{\mu_2\varepsilon_2}\right]} \quad (6.25)$$

where ε_1, ε_2, and μ_1, μ_2 are the relative complex permittivity and complex permeability of each layer, respectively. Moreover, d_1 and d_2 are the layer thicknesses. For the double-layer structure, the absorption and matching layer thicknesses considerably affect MW absorption properties. Fig. 6.56 shows RL curves as a function of frequency for 3 mm thick Ni−Zn ferrite NFs and CNFs with a double-layer structure (Hou et al., 2018).

Absorbers with a double-layer structure outperform the corresponding single-layer structure with the same thickness regardless of the MW effective bandwidth and absorption intensity. This indicates effective coupling between the absorption and matching layers, improving MW absorption properties and the broad absorption frequency range.

Compared with single-layer absorbers, Z_{in}/Z_0 for the double-layer structure is closer to 1 in a wider frequency range, facilitating the entrance of incident EM waves

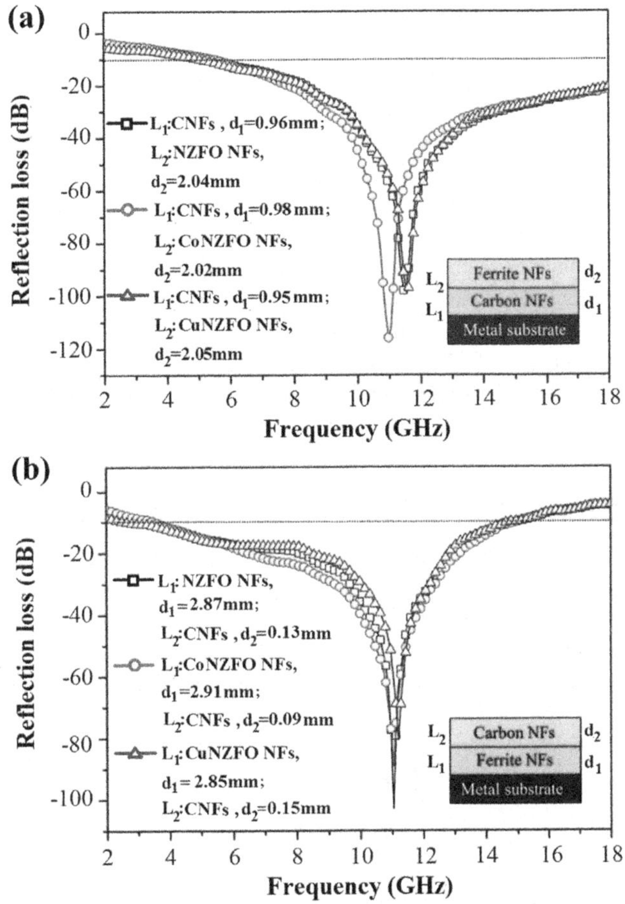

FIGURE 6.56

The reflection loss curves as a function of frequency for 3 mm thick double-layer absorbers with optimized structure: (A) 5 wt.% carbon nanofiber (CNF) filled absorbing layer together with 70 wt.% ferrite NF filled matching layer, and (B) 70 wt.% ferrite NF filled absorbing layer together with 5 wt.% CNF filled matching layer.

From Hou, Z., Xiang, J., Zhang, X., Gong, L., Mi, J., Shen, X., & Zhang, K. (2018). Microwave absorption properties of single-and double-layer absorbers based on electrospun nickel–zinc spinel ferrite and carbon nanofibers. Journal of Materials Science: Materials in Electronics, *29(14), 12258–12268. https://doi.org/10.1007/s10854-018-9338-z.*

into the double-layer absorbers while leading to higher MW absorption performance.

Elsewhere, the exploration of high-performance MW absorbers has been boosted by novel double-layer structures of doped ferrite/carbon composites. In this respect, W-type La-doped BF and carbon fiber composites with double-layer structures have been studied (Shen et al., 2006). The absorption curves as a function of frequency for the composites with single-layer and double-layer structures are shown in Figs. 6.57 and 6.58, respectively. As inferred, the La-doped ferrite composites outperform the undoped ferrite composite for absorption properties. The undoped ferrite composite has a larger magnetic loss, whereas the doped ferrite composite has a higher dielectric loss.

Fig. 6.57 shows that certain MW absorption properties can be specified for the short carbon fiber composite arising from its electric dipole polarization. The short conductive carbon fiber resonates with the incident EM wave, as it can act as an electric dipole. Particularly, absorption significantly improves when the La-doped ferrite composite is employed as the outer layer. In this case, an RL higher than −10 dB starts from a frequency of 12.8 GHz, giving rise to a bandwidth of 5.2 GHz. Moreover, RL_{max} is −18.6 dB at a frequency of approximately 16.0 GHz (see Fig. 6.58).

6.14 Soft ferrite/amorphous carbon nanocomposites

Gupta and colleagues have prepared a nanocomposite consisting of amorphous carbon (obtained from fresh cucumber by using a pyrolysis process under Ar atmosphere at 800°C), ZnO nanorods, and NiZn ferrite (Gupta et al., 2014). The resulting flat-sheet-like amorphous carbon is several micrometers long and wide, significantly facilitating electron migration in nanocomposite networks.

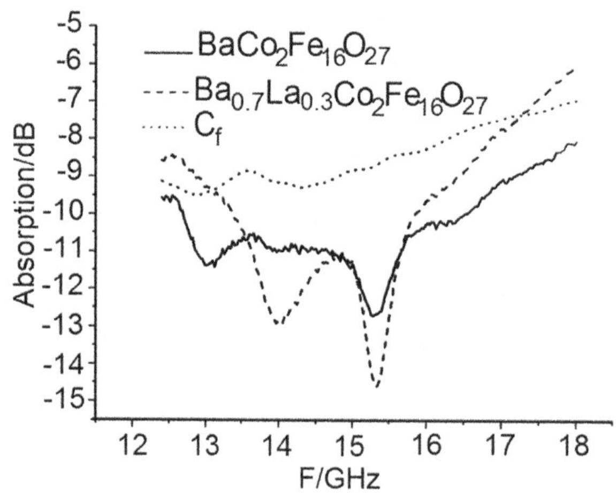

FIGURE 6.57

The absorption curves as a function of frequency for W-type La-doped barium ferrite and carbon fiber composites with single-layer structure.

From Shen, G., Xu, Z., & Li, Y. (2006). Absorbing properties and structural design of microwave absorbers based on W-type La-doped ferrite and carbon fiber composites. Journal of Magnetism and Magnetic Materials, 301 *(2), 325–330. https://doi.org/10.1016/j.jmmm.2005.07.007.*

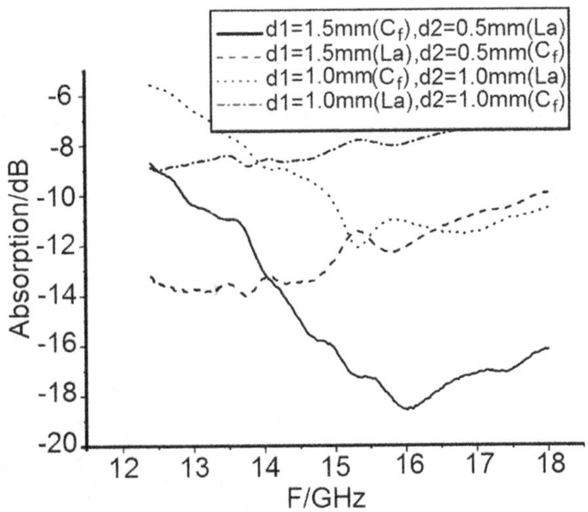

FIGURE 6.58

The absorption curves as a function of frequency for W-type La-doped barium ferrite and carbon fiber composites with double-layer structure.

Reproduced from Shen, G., Xu, Z., & Li, Y. (2006). Absorbing properties and structural design of microwave absorbers based on W-type La-doped ferrite and carbon fiber composites. Journal of Magnetism and Magnetic Materials, 301*(2), 325–330.*

Microstructural and morphological analyses of the amorphous carbon and the ferrite powder have been carried out by TEM.

Fig. 6.59A shows a TEM micrograph of amorphous carbon with a flake morphology and a length and width of several microns. Moreover, Fig. 6.59B shows a phase-contrast HRTEM micrograph of the carbon flakes associated with the corresponding SAED pattern at the inset. Accordingly, the amorphous structure of the flakes is confirmed due to the absence of lattice fringes and spots.

A TEM micrograph of the ferrite powder is depicted in Fig. 6.59C, simultaneously revealing nanorod and nanosheet morphologies of the ZnO and NiZn, respectively. The ZnO nanorod is approximately 3 μm and 100 nm in length and diameter, respectively, preferably growing along the [011] direction (see Fig. 6.59D) with a single-crystal structure due to its symmetrical system. By magnifying the micrograph of the NiZn ferrite nanosheets (see Fig. 6.59E), their length and width range from 150 to 200 nm. The SAED pattern depicted in Fig. 6.59F indicates that the NiZn nanosheets are highly crystalline.

Composites containing ZnO nanorods and NiZn nanosheets can be tuned for their electrical, magnetic, and EMI shielding properties. For example, EMI shielding properties considerably decrease when the powder percentage in the nanocomposite increases. Overall, as-prepared amorphous carbon nanocomposites with 20 wt.% powder loading (containing ZnO nanorods and NiZn ferrite) outperform other

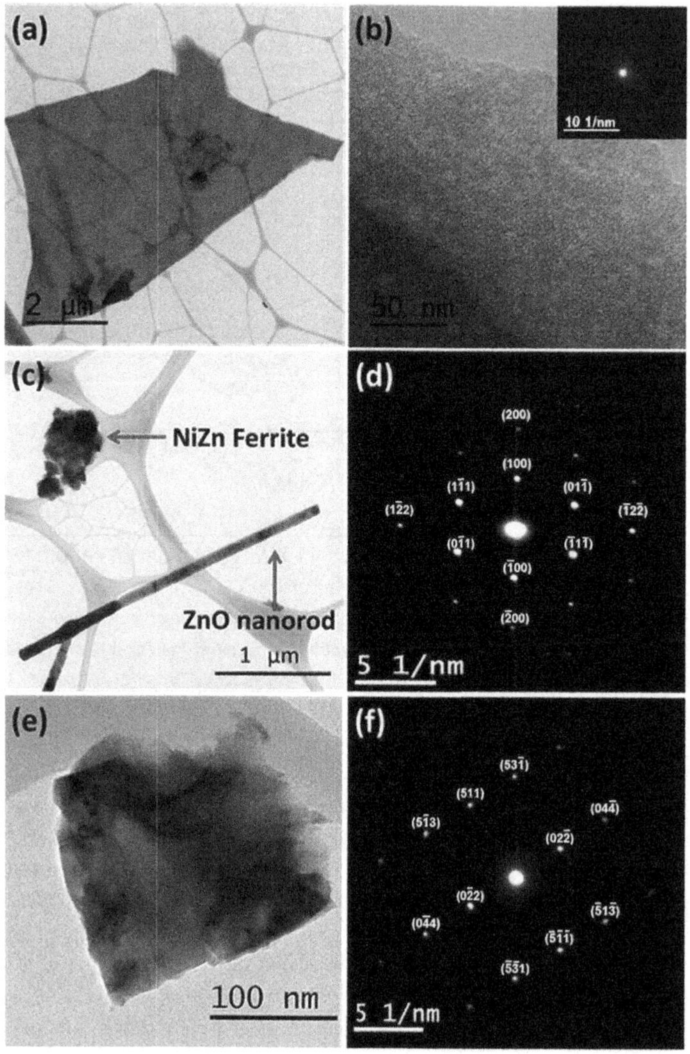

FIGURE 6.59

(A) Transmission electron microscopy (TEM) micrograph obtained from amorphous carbon flakes. (B) High-resolution transmission electron microscopy micrograph of the amorphous carbon. The inset in part (B) shows the corresponding selected area electron diffraction (SAED) pattern. (C) and (E) TEM micrographs of the powder containing ZnO and NiZn ferrite. (D) and (F) SAED patterns obtained from the ZnO nanorod and NiZn nanosheet, respectively.

From Gupta, S., Chang, C., Lai, C.-H., & Tai, N.-H. (2019). Hybrid composite mats composed of amorphous carbon, zinc oxide nanorods and nickel zinc ferrite for tunable electromagnetic interference shielding. Composites Part B: Engineering, 164, *447–457. https://doi.org/10.1016/j.compositesb.2019.01.060.*

conventional materials in terms of EMI shielding, holding great promise for MW absorption applications.

6.15 Ferrite/porous carbon nanocomposites

Instead of focusing on the synthesis of 1-D or 2-D MW-absorbing materials, the suitable tailoring of nanoporous structures with light weights and wide absorption ranges has been proposed as a new avenue for achieving efficient MW absorption properties. In this way, nanoporous absorbing materials with micro-nano multilevel structures have drastically enhanced MW absorption, leading to multiple internal reflections and scattering involving multiple loss mechanisms.

Zhang and colleagues have synthesized biocarbon/ferrite porous absorbers in the S-band (Zhang et al., 2013). Bamboo-based biocarbon with tubular and porous microstructure has attracted considerable attention as a novel environmentally compatible material due to its abundance in availability and improved scattering effect for EM wave absorption. In this case, the pore size and porous structure of the carbon are determined by the bamboo's physiological characteristics.

A schematic illustration of the synthesis of bamboo-based biocarbon/ferrite composite is depicted in Fig. 6.60. Porous carbon can be easily prepared using a pyrolysis process of bamboo materials under an N atmosphere, followed by dipping them in a saturated solution of $Fe(NO_3)_3$. The several-micron-diameter polyhedral particles firmly attach to the biocarbon matrix surface so that the carbon and iron atom diffusion occurs mutually between the particles and the matrix. In turn, this may lead to the formation of metallurgic bonding during the pyrolysis process.

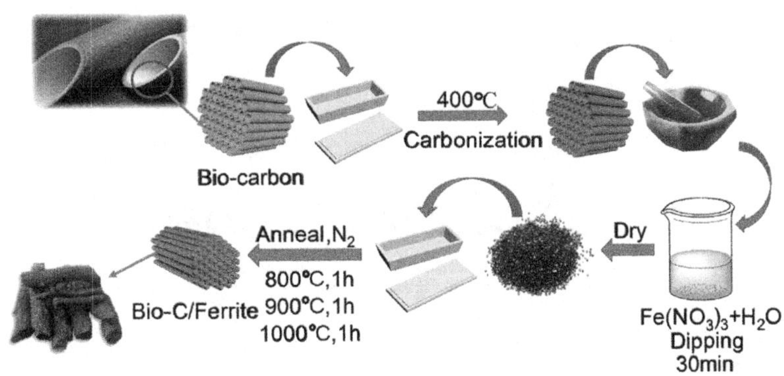

FIGURE 6.60

Schematic illustration of the synthesis of bamboo-based biocarbon/ferrite composite.

From Zhang, T., Zhao, D., Wang, L., Meng, R., Zhao, H., Zhou, P., Xia, L., Zhong, B., Wang, H., & Wen, G. (2020). A facile precursor pyrolysis route to bio-carbon/ferrite porous architecture with enhanced electromagnetic wave absorption in S-band. Journal of Alloys and Compounds, 819, *153269. https://doi.org/10.1016/j.jallcom.2019.153269.*

Raman spectra of the biocarbon/ferrite composites annealed at various temperatures (800, 900, and 1000°C) are shown in Fig. 6.61. As mentioned, the *D* peak mainly results from defects or lattice distortion in the graphite phase structure, representing the sp^3 hybrid bond stretching vibration. The *G* peak also indicates the *C* atom sp^2 hybrid bond vibration in the in-plane bonding of the 2-D graphite phase structure (Fang et al., 2016).

According to Fig. 6.61, all the Raman spectra of the composites annealed at various temperatures show broad *D* and *G* peaks at 1363 cm^{-1} and 1580 cm^{-1}, respectively. Increasing the annealing temperature increases the corresponding I_D/I_G ratio, indicating the formation of more structural defects in the bamboo-based biocarbon/ferrite composites due to the enhanced heat treatment temperature.

The absorption properties measured in mid-frequency and high-frequency regions are significantly improved after adding ferrimagnetic particles to the pure carbon materials. The resulting biocarbon/ferrite composites with tubular structures have shown high EM wave absorption performance. The synergistic effects between the biocarbon with dielectric loss property and the magnetic ferrite with magnetic loss property may regulate EM parameters, giving rise to enhanced impedance matching and strong EM wave attenuation inside the composites. Furthermore, combining the biocarbon with the ferrite induces many reflective interfaces, thereby consuming the EM energy effectively. The absorbers are made of the bamboo-based biocarbon/ferrite composites with promising characteristics such as environmental

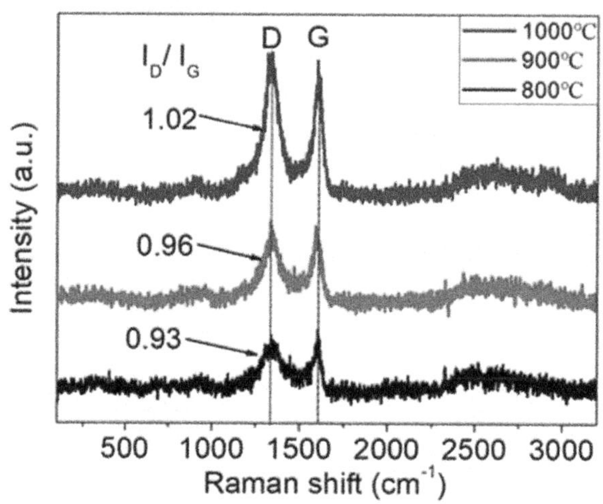

FIGURE 6.61

Raman spectra of biocarbon/ferrite composites annealed at various temperatures (800, 900, and 1000°C).

From Zhang, T., Zhao, D., Wang, L., Meng, R., Zhao, H., Zhou, P., Xia, L., Zhong, B., Wang, H., & Wen, G. (2020). A facile precursor pyrolysis route to bio-carbon/ferrite porous architecture with enhanced electromagnetic wave absorption in S-band. Journal of Alloys and Compounds, 819, *153269. https://doi.org/10.1016/j.jallcom.2019.153269.*

protection, cost-effectiveness, and renewability, providing an RL of −43.2 dB at a thickness of 2.0 mm, together with an EAB of 4.72 GHz.

6.16 Ferrite/graphite nanosheet nanocomposites

Highly conductive GNs with a low density and a high aspect ratio have shown excellent chemical stability and mechanical properties, making them attractive for MW absorbers. GNs have strong EM wave attenuation capacity due to their high dielectric loss and conductive networks. As discussed, it is possible to reduce the density and obtain enhanced absorption in absorbers with a double-layer structure using lightweight carbon-based materials with high dielectric loss (Quan et al., 2017). It should be noted that the matching layer needs to have good impedance matching performance so MWs can enter into the absorbers in the most efficient way. Owing to the dielectric or magnetic loss of the absorption layers, MWs entering the absorbers are then dissipated.

$BaFe_{12}O_{19}$ ferrite (BFO)/GN composites with a double-layer structure have been synthesized, and their MW absorption properties have been studied (Min, 2020). BFO composites with high permeability and low permittivity at low frequencies can act as suitable matching layers for designing double-layer absorbing materials.

RL curves as a function of frequency (2.6–18 GHz) have been measured for double-layer BFO/GN composites with various thicknesses in two configurations: a BFO matching layer with a GN absorption layer (Fig. 6.62A) and a GN matching

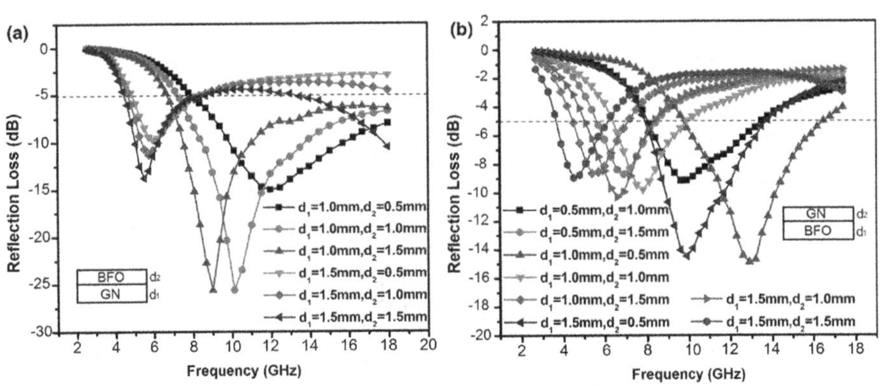

FIGURE 6.62

The reflection loss curves as a function of frequency (2.6–18 GHz) for double-layer $BaFe_{12}O_{19}$ (BFO)/graphite nanosheet (GN) composites with various thicknesses in two different configurations: (A) a BFO matching layer with a GN absorption layer and (B) a GN matching layer with a BFO absorption layer.

From Min, D. (2020). Enhanced microwave absorption performance of double-layer absorbers containing $BaFe_{12}O_{19}$ ferrite and graphite nanosheet composites. Journal of Electronic Materials, 49*(1), 819–825. https://doi.org/10.1007/s11664-019-07730-5.*

layer with a BFO absorption layer (Fig. 6.62B). A comparison of Fig. 6.62A and B indicates that double-layer absorbers in the first configuration enhance MW absorption performance, including a wider absorption band and a higher absorption capacity.

It is well known that absorber absorption performance is drastically affected by impedance matching levels. For absorbers with high impedance matching levels, a small reflection is expected to occur after the entrance of MWs, followed by their dissipation due to the dielectric and magnetic losses of the absorbers. Accordingly, the absorber impedance needs to be close to the free space impedance, i.e., $\sqrt{\mu_r/\varepsilon_r} \sim 1$.

BFO composites outperform GN composites in impedance matching performance. The good impedance matching properties and improved absorption performance of absorbers can be obtained by adjusting the permittivity or permeability while controlling the combinatory effects of permittivity and permeability. In this regard, Fig. 6.63 shows the input impedance as a function of frequency for BFO and GN composites with various thicknesses. At higher frequencies, the BFO composite input impedance matches well with the air input impedance (377 Ω) for thicknesses of 1.2 and 1.4 mm, which is favorable for enhancing the impedance matching performance between the BFO composite absorbers and the air. Meanwhile, the GN composites show good absorption properties, effectively dissipating the MWs

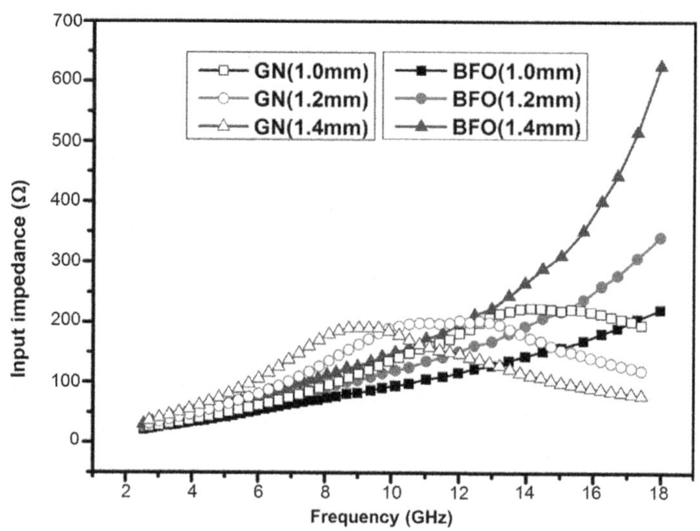

FIGURE 6.63

The input impedance as a function of frequency for $BaFe_{12}O_{19}$ ferrite and graphite nanosheet composites with various thicknesses.

From Min, D. (2020). Enhanced microwave absorption performance of double-layer absorbers containing $BaFe_{12}O_{19}$ ferrite and graphite nanosheet composites. Journal of Electronic Materials, *49(1), 819–825. https://doi.org/10.1007/s11664-019-07730-5.*

entered into the absorbers. In other words, when the BFO and GN composites are employed as the matching layer and the absorption layer, respectively, MWs can efficiently enter into the absorbers and be highly attenuated.

6.17 Ferrite-MoS_2@nitrogen-doped carbon hybrid structure

A one-pot green approach has been employed for synthesizing an MoS_2@nitrogen-doped carbon (n-C) hybrid structure decorated with $CoFe_2O_4$ ultrasmall nanoparticles at annealing temperatures of 500–800°C (Zhu et al., 2017; Hapishah et al., 2018). A schematic representation of the synthesis approach is depicted in Fig. 6.64.

Variations in RL and thickness versus frequency for MoS_2@n-C@$CoFe_2O_4$ hybrids are shown in Fig. 6.65. At an annealing temperature of 600°C, the highest EM wave absorption capacity is obtained, mainly arising from optimum EM matching. The resonance effect occurring in the frequency range of 11–15 GHz for the MoS_2@n-C@$CoFe_2O_4$ annealed at 600°C is mainly induced by the interfaces between MoS_2, n-C, and $CoFe_2O_4$, which involve exchange resonance loss, natural resonance, and local confinement.

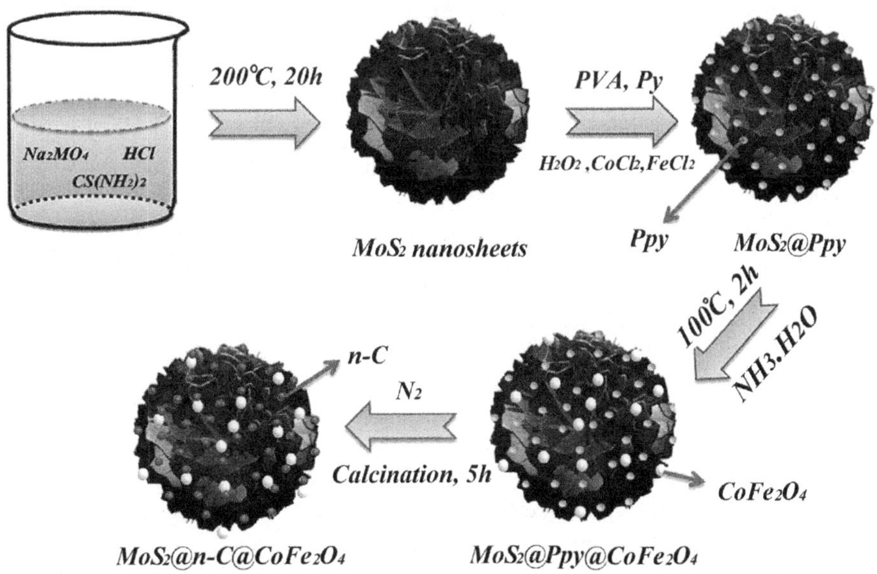

FIGURE 6.64

A schematic representation of the synthesis of MoS_2@n-C@$CoFe_2O_4$ hybrid structure.

From Chen, X., Wang, W., Shi, T., Wu, G., & Lu, Y. (2020). One pot green synthesis and EM wave absorption performance of MoS2@ nitrogen doped carbon hybrid decorated with ultrasmall cobalt ferrite nanoparticles. Carbon. *https://doi.org/10.1016/j.carbon.2020.03.005.*

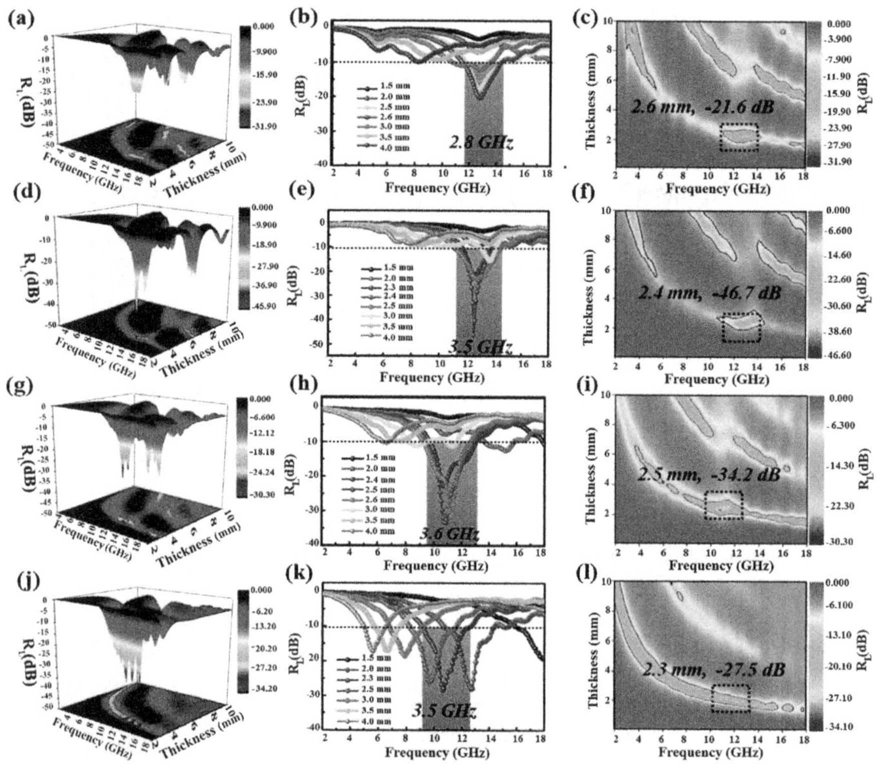

FIGURE 6.65

Variations in reflection loss and thickness versus frequency for MoS_2@n-C@$CoFe_2O_4$ hybrids annealed at various temperatures: (A–C), 500°C, (D–F), 600°C, (G–I) 700°C, and (J–L) 800°C.

From Chen, X., Wang, W., Shi, T., Wu, G., & Lu, Y. (2020). One pot green synthesis and EM wave absorption performance of MoS_2@ nitrogen doped carbon hybrid decorated with ultrasmall cobalt ferrite nanoparticles. Carbon. *https://doi.org/10.1016/j.carbon.2020.03.005.*

At an extremely high annealing temperature, EM wave absorption performance is reduced due to partial damage to the hybrid structure. Superior EM wave absorption process in the MoS_2@n-C@$CoFe_2O_4$ hybrid structure is proposed by the following mechanism: (1) the combination of MoS_2 (as a perfect dielectric material) with the n-C decorated with $CoFe_2O_4$ ultrasmall nanoparticles induces conductivity, leading to good EM matching and enhanced EM loss; (2) the presence of multiple interfaces (e.g., MoS_2-carbon, MoS_2–$CoFe_2O_4$ and carbon-$CoFe_2O_4$) favoring multiple interfacial polarization and relaxation, resulting in an improvement in the EM wave RL; (3) free electrons that generate microcurrents in the structure after entering the EM waves into the surface, followed by their effective consumption through the $CoFe_2O_4$ nanoparticles embedded in the carbon matrix; (4) EM wave interfacial reflection enhanced by the spaces formed by the MoS_2 nanosheet lamellar structure while increasing the Joule heat and consuming the EM wave energy due to promoted free-electron movements; and (5) polarization centers induced by the doped nitrogen element to enhance molecular and dipole polarizations, thereby improving the efficiency of the EM wave absorption (Fang et al., 2016).

References

Abdel Fattah, A. R., Majdi, T., Abdalla, A. M., Ghosh, S., & Puri, I. K. (2016). Nickel nanoparticles entangled in carbon nanotubes: Novel ink for nanotube printing. *ACS Applied Materials and Interfaces, 8*(3), 1589–1593. https://doi.org/10.1021/acsami.5b11700

Aharoni, A. (1991). Exchange resonance modes in a ferromagnetic sphere. *Journal of Applied Physics, 69*(11), 7762–7764. https://doi.org/10.1063/1.347502

Baan, R., Grosse, Y., Lauby-Secretan, B., El Ghissassi, F., Bouvard, V., Benbrahim-Tallaa, L., … Straif, K. (2011). Carcinogenicity of radiofrequency electromagnetic fields. *The Lancet Oncology, 12*(7), 624–626. https://doi.org/10.1016/S1470-2045(11)70147-4

Bhadra, S., Khastgir, D., Singha, N. K., & Lee, J. H. (2009). Progress in preparation, processing and applications of polyaniline. *Progress in Polymer Science, 34*(8), 783–810. https://doi.org/10.1016/j.progpolymsci.2009.04.003

Bhattacharya, P., Dhibar, S., Hatui, G., Mandal, A., Das, T., & Das, C. K. (2014). Graphene decorated with hexagonal shaped M-type ferrite and polyaniline wrapper: A potential candidate for electromagnetic wave absorbing and energy storage device applications. *RSC Advances, 4*(33), 17039–17053. https://doi.org/10.1039/C4RA00448E

Bibi, M., Abbas, S. M., Ahmad, N., Muhammad, B., Iqbal, Z., Rana, U. A., & Khan, S. U.-D. (2017). Microwaves absorbing characteristics of metal ferrite/multiwall carbon nanotubes nanocomposites in X-band. *Composites Part B: Engineering, 114*, 139–148. https://doi.org/10.1016/j.compositesb.2017.01.034

Cao, M., Wang, X., Cao, W., Fang, X., Wen, B., & Yuan, J. (2018). Thermally driven transport and relaxation switching self-powered electromagnetic energy conversion. *Small, 14*(29), 1800987–1800995. https://doi.org/10.1002/smll.201800987

Cao, M.-S., Song, W.-L., Hou, Z.-L., Wen, B., & Yuan, J. (2010). The effects of temperature and frequency on the dielectric properties, electromagnetic interference shielding and microwave-absorption of short carbon fiber/silica composites. *Carbon, 48*(3), 788–796. https://doi.org/10.1016/j.carbon.2009.10.028

Che, R. C., Peng, L., Duan, X. F., Chen, Q., & Liang, X. L. (2004). Microwave absorption enhancement and complex permittivity and permeability of Fe encapsulated within carbon nanotubes. *Advanced Materials, 16*(5), 401–405. https://doi.org/10.1002/adma.200306460

Chung, D. (2001). Electromagnetic interference shielding effectiveness of carbon materials. *Carbon, 39*(2), 279–285. https://doi.org/10.1016/S0008-6223(00)00184-6

Collins, P. G., & Avouris, P. (2000). Nanotubes for electronics. *Scientific American, 283*(6), 62–69. https://doi.org/10.2307/26058971

Cullity, B. D., & Graham, C. D. (2011). *Introduction to magnetic materials*. John Wiley & Sons.

Dalal, M., Das, A., Das, D., Ningthoujam, R. S., & Chakrabarti, P. K. (2018). Studies of magnetic, Mössbauer spectroscopy, microwave absorption and hyperthermia behavior of Ni-

Zn-Co-ferrite nanoparticles encapsulated in multi-walled carbon nanotubes. *Journal of Magnetism and Magnetic Materials, 460*, 12–27. https://doi.org/10.1016/j.jmmm.2018.03.048

Deng, L., Zhou, P., Xie, J., & Zhang, L. (2007). Characterization and microwave resonance in nanocrystalline FeCoNi flake composite. *Journal of Applied Physics, 101*(10), 103916–103921. https://doi.org/10.1063/1.2733610

Ding, D., Wang, Y., Li, X., Qiang, R., Xu, P., Chu, W., … Du, Y. (2017). Rational design of core-shell Co@C microspheres for high-performance microwave absorption. *Carbon, 111*, 722–732. https://doi.org/10.1016/j.carbon.2016.10.059

Durmus, Z., Durmus, A., & Kavas, H. (2015). Synthesis and characterization of structural and magnetic properties of graphene/hard ferrite nanocomposites as microwave-absorbing material. *Journal of Materials Science, 50*(3), 1201–1213. https://doi.org/10.1007/s10853-014-8676-3

Fang, J., Liu, T., Chen, Z., Wang, Y., Wei, W., Yue, X., & Jiang, Z. (2016). A wormhole-like porous carbon/magnetic particles composite as an efficient broadband electromagnetic wave absorber. *Nanoscale, 8*(16), 8899–8909. https://doi.org/10.1039/c6nr01863g

Feng, J., Pu, F., Li, Z., Li, X., Hu, X., & Bai, J. (2016). Interfacial interactions and synergistic effect of CoNi nanocrystals and nitrogen-doped graphene in a composite microwave absorber. *Carbon, 104*, 214–225. https://doi.org/10.1016/j.carbon.2016.04.006

Folgueras, L.d. C., Nohara, E. L., Faez, R., & Rezende, M. C. (2007). Dielectric microwave absorbing material processed by impregnation of carbon fiber fabric with polyaniline. *Materials Research, 10*(1), 95–99. https://doi.org/10.1590/S1516-14392007000100020

Geetha, S., Satheesh Kumar, K., Rao, C. R., Vijayan, M., & Trivedi, D. (2009). EMI shielding: Methods and materials—a review. *Journal of Applied Polymer Science, 112*(4), 2073–2086. https://doi.org/10.1002/app.29812

Geim, A. K., & Novoselov, K. S. (2010). The rise of graphene. In *Nanoscience and technology: A collection of reviews from nature journals* (pp. 11–19). World Scientific.

Ghasemi, A. (2011). Remarkable influence of carbon nanotubes on microwave absorption characteristics of strontium ferrite/CNT nanocomposites. *Journal of Magnetism and Magnetic Materials, 323*(23), 3133–3137. https://doi.org/10.1016/j.jmmm.2011.06.070

Ghasemi, A. (2012). Enhanced reflection loss and permittivity of self assembled Mg–Co–Zr substituted barium ferrite dot array on carbon nanotube. *Journal of Magnetism and Magnetic Materials, 324*(6), 1080–1083. https://doi.org/10.1016/j.jmmm.2011.10.027

Ghasemi, A. (2013). The role of multi-walled carbon nanotubes on the magnetic and reflection loss characteristics of substituted strontium ferrite nanoparticles. *Journal of Magnetism and Magnetic Materials, 330*, 163–168. https://doi.org/10.1016/j.jmmm.2012.10.033

Ghasemi, A., Liu, X., & Morisako, A. (2009). Effect of additional elements on the structural properties, magnetic characteristics and natural resonance frequency of strontium ferrite nanoparticles/polymer composite. *IEEE Transactions on Magnetics, 45*(10), 4420–4423.

Ghasemi, A., Sepelak, V., Liu, X., & Morisako, A. (2009). Microwave absorption properties of Mn–Co–Sn doped barium ferrite nanoparticles. *IEEE Transactions on Magnetics, 45*(6), 2456–2459.

Ghasemi, A., Šepelák, V., Liu, X., & Morisako, A. (2010). Mössbauer spectroscopy and magnetic susceptibility studies of Cr–Al substituted strontium ferrite particles. *Journal of Applied Physics, 107*(9), 09A743–09A745. https://doi.org/10.1063/1.3339004

Ghasemi, A., Sepelak, V., Liu, X., & Morisako, A. (2011). First study on the formation of strontium ferrite thin films on functionalized multi-walled carbon nanotube. *IEEE Transactions on Magnetics, 47*(10), 2800–2803. https://doi.org/10.1109/TMAG.2011.2143391

Ghasemi, A., Shirsath, S. E., Liu, X., & Morisako, A. (2012). A comparison between magnetic and reflection loss characteristics of substituted strontium ferrite and nanocomposites of ferrite/carbon nanotubes. *Journal of Applied Physics, 111*(7), 07B543–07B545. https://doi.org/10.1063/1.3683012

Giovannetti, G., Khomyakov, P. A., Brocks, G., Karpan, V., van den Brink, J., & Kelly, P. J. (2008). Doping graphene with metal contacts. *Physical Review Letters, 101*(2), 026803–026806. https://doi.org/10.1103/PhysRevLett.101.026803

Goldberg, H. A., & Marikar, Y. F. (1988). Fiber structure and method for obtaining tuned response to high frequency electromagnetic radiation. *Google Patents*.

Gordani, G. R., & Ghasemi, A. (2015). High frequency electromagnetic reflection loss performance of substituted Sr-hexaferrite nanoparticles/SWCNTs/epoxy nanocomposite. *Journal of Magnetism and Magnetic Materials, 391*, 184–190. https://doi.org/10.1016/j.jmmm.2015.05.002

Gupta, K., Abbas, S., Goswami, T., & Abhyankar, A. (2014). Microwave absorption in X and Ku band frequency of cotton fabric coated with Ni–Zn ferrite and carbon formulation in polyurethane matrix. *Journal of Magnetism and Magnetic Materials, 362*, 216–225. https://doi.org/10.1016/j.jmmm.2014.03.040

Hao, Y. H., Li, Z., & Peng, R. Y. (2018). Effects of electromagnetic radiation on autophagy and its regulation. *Biomedical and Environmental Sciences, 31*(1), 57–65. https://doi.org/10.3967/bes2018.006

Hapishah, A., Syazwan, M., & Hamidon, M. (2018). Synthesis and characterization of magnetic and microwave absorbing properties in polycrystalline cobalt zinc ferrite ($Co_{0.5}Zn_{0.5}Fe_2O_4$) composite. *Journal of Materials Science: Materials in Electronics, 29*(24), 20573–20579. https://doi.org/10.1007/s10854-018-0192-9

He, S., Wang, G.-S., Lu, C., Liu, J., Wen, B., Liu, H., … Cao, M.-S. (2013). Enhanced wave absorption of nanocomposites based on the synthesized complex symmetrical CuS nanostructure and poly (vinylidene fluoride). *Journal of Materials Chemistry, 1*(15), 4685–4692. https://doi.org/10.1039/c3ta00072a

Hou, Z., Xiang, J., Zhang, X., Gong, L., Mi, J., Shen, X., & Zhang, K. (2018). Microwave absorption properties of single-and double-layer absorbers based on electrospun nickel–zinc spinel ferrite and carbon nanofibers. *Journal of Materials Science: Materials in Electronics, 29*(14), 12258–12268. https://doi.org/10.1007/s10854-018-9338-z

Hur, N., Park, J., Dho, J., Kim, S., & Lee, E. (2004). Magnetic and microstructural properties on Mn-and Ru-doped hexagonal barium ferrites. *IEEE Transactions on Magnetics, 40*(4), 2790–2792. https://doi.org/10.1109/TMAG.2004.832277

Iijima, S. (1991). Helical microtubules of graphitic carbon. *Nature, 354*(6348), 56–58. https://doi.org/10.1038/354056a0

Jang, K., Lee, S., Yu, S., Salunkhe, R. R., Chung, I., Choi, S., & Ahn, H. (2014). Facile low-temperature chemical synthesis and characterization of a manganese oxide/multi-walled carbon nanotube composite for supercapacitor applications. *Bulletin of the Korean Chemical Society, 35*(10), 2975. https://doi.org/10.5012/bkcs.2014.35.10.2974

Jaroszewski, M., Thomas, S., & Rane, A. V. (2018). *Advanced materials for electromagnetic shielding: Fundamentals, properties, and applications*. John Wiley & Sons.

Kaczmarek, W., & Ninham, B. (1994). Preparation of high-coercivity fine barium ferrite powder. *Journal of Applied Physics, 76*(10), 6065–6067. https://doi.org/10.1063/1.358367

Koops, C. (1951). On the dispersion of resistivity and dielectric constant of some semiconductors at audiofrequencies. *Physical Review, 83*(1), 121–124. https://doi.org/10.1103/PhysRev.83.121

Liang, K., Qiao, X.-J., Sun, Z.-G., Guo, X.-D., Wei, L., & Qu, Y. (2017). Preparation and microwave absorbing properties of graphene oxides/ferrite composites. *Applied Physics A, 123*(6), 445–455. https://doi.org/10.1007/s00339-017-1059-2

Li, Y., Huang, Y., Qi, S., Niu, L., Zhang, Y., & Wu, Y. (2012). Preparation, magnetic and electromagnetic properties of polyaniline/strontium ferrite/multiwalled carbon nanotubes composite. *Applied Surface Science, 258*(8), 3659–3666. https://doi.org/10.1016/j.apsusc.2011.12.001

Li, Z., Li, X., Zong, Y., Tan, G., Sun, Y., Lan, Y., ... Zheng, X. (2017). Solvothermal synthesis of nitrogen-doped graphene decorated by superparamagnetic Fe_3O_4 nanoparticles and their applications as enhanced synergistic microwave absorbers. *Carbon, 115*, 493–502. https://doi.org/10.1016/j.carbon.2017.01.036

Lin, L., Xing, H., Shu, R., Wang, L., Ji, X., Tan, D., & Gan, Y. (2015). Preparation and microwave absorption properties of multi-walled carbon nanotubes decorated with Ni-doped SnO_2 nanocrystals. *RSC Advances, 5*(115), 94539–94550. https://doi.org/10.1039/C5RA17303E

Li, Li, Qin, Z.-Y., Liang, X., Fan, Q.-Q., Lu, Y.-Q., Wu, W.-H., & Zhu, M.-F. (2009). Facile fabrication of uniform core–shell structured carbon nanotube–polyaniline nanocomposites. *Journal of Physical Chemistry C, 113*(14), 5502–5507. https://doi.org/10.1021/jp808582f

Liu, P. (2005). Modifications of carbon nanotubes with polymers. *European Polymer Journal, 41*(11), 2693–2703. https://doi.org/10.1016/j.eurpolymj.2005.05.017

Liu, Z., Xing, H., Liu, Y., Wang, H., Jia, H., & Ji, X. (2018). Hydrothermally synthesized Zn ferrite/multi-walled carbon nanotubes composite with enhanced electromagnetic-wave absorption performance. *Journal of Alloys and Compounds, 731*, 745–752. https://doi.org/10.1016/j.jallcom.2017.09.317

Lopes, B., Portes, R., do Amaral Junior, M., Florez-Vergara, D., Gama, A., Silva, V., ... Baldan, M. (2020). X Band electromagnetic property influence of multi-walled carbon nanotube in hybrid MnZn ferrite and carbonyl iron composites. *Journal of Materials Research and Technology, 9*(2), 2369–2375. https://doi.org/10.1016/j.jmrt.2019.12.068

Lü, Y., Wang, Y., Li, H., Lin, Y., Jiang, Z., Xie, Z., ... Zheng, L. (2015). MOF-derived porous Co/C nanocomposites with excellent electromagnetic wave absorption properties. *ACS Applied Materials and Interfaces, 7*(24), 13604–13611. https://doi.org/10.1021/acsami.5b03177

Luo, H., Shi, Z., Li, N., Gu, Z., & Zhuang, Q. (2001). Investigation of the electrochemical and electrocatalytic behavior of single-wall carbon nanotube film on a glassy carbon electrode. *Analytical Chemistry, 73*(5), 915–920. https://doi.org/10.1021/ac000967l

Lv, H., Guo, Y., Yang, Z., Cheng, Y., Wang, L. P., Zhang, B., ... Ji, G. (2017). A brief introduction to the fabrication and synthesis of graphene based composites for the realization of electromagnetic absorbing materials. *Journal of Materials Chemistry C, 5*(3), 491–512. https://doi.org/10.1039/C6TC03026B

MacDonald, K. G., & MacDonald, A. H. (2007). Graphene: Exploring carbon flatland. *Physics Today, 60*(8), 35–41. https://doi.org/10.1063/1.2774096

Mahmood, A., Guo, W., Tabassum, H., & Zou, R. (2016). Metal-organic framework-based nanomaterials for electrocatalysis. *Advanced Energy Materials, 6*(17), 1600423–1600448. https://doi.org/10.1002/aenm.201600423

Ma, J., Shu, J., Cao, W., Zhang, M., Wang, X., Yuan, J., & Cao, M. (2019). A green fabrication and variable temperature electromagnetic properties for thermal stable microwave absorption towards flower-like Co_3O_4@rGO/SiO_2 composites. *Composites Part B: Engineering, 166*, 187–195. https://doi.org/10.1016/j.compositesb.2018.11.119

Mallick, A., Mahapatra, A., Mitra, A., & Chakrabarti, P. (2016). Soft magnetic property and enhanced microwave absorption of nanoparticles of $Co_{0.5}Zn_{0.5}Fe_2O_4$ incorporated in MWCNT. *Journal of Magnetism and Magnetic Materials, 416*, 181–187. https://doi.org/10.1016/j.jmmm.2016.04.073

Matzui, L., Vovchenko, L., Prylutskyy, Y., Korotash, I., Matzui, V., Eklund, P., ... Scharff, P. (2007). Electromagnetic losses in carbon–epoxy composites. *Materials Science and Engineering: C, 27*(5–8), 1007–1009. https://doi.org/10.1016/j.msec.2006.06.017

Ma, J., Wang, X., Cao, W., Han, C., Yang, H., Yuan, J., & Cao, M. (2018). A facile fabrication and highly tunable microwave absorption of 3D flower-like Co_3O_4-rGO hybrid-architectures. *Chemical Engineering Journal, 339*, 487–498. https://doi.org/10.1016/j.cej.2018.01.152

Melvin, G. J. H., Ni, Q.-Q., Natsuki, T., Wang, Z., Morimoto, S., Fujishige, M., ... Endo, M. (2015). Ag/CNT nanocomposites and their single-and double-layer electromagnetic wave absorption properties. *Synthetic Metals, 209*, 383–388. https://doi.org/10.1016/j.synthmet.2015.08.017

Meng, F., Wang, H., Huang, F., Guo, Y., Wang, Z., Hui, D., & Zhou, Z. (2018). Graphene-based microwave absorbing composites: A review and prospective. *Composites Part B: Engineering, 137*, 260–277. https://doi.org/10.1016/j.compositesb.2017.11.023

Min, D. (2020). Enhanced microwave absorption performance of double-layer absorbers containing $BaFe_{12}O_{19}$ ferrite and graphite nanosheet composites. *Journal of Electronic Materials, 49*(1), 819–825. https://doi.org/10.1007/s11664-019-07730-5

Mishra, M., Singh, A. P., Singh, B., Singh, V., & Dhawan, S. (2014). Conducting ferrofluid: A high-performance microwave shielding material. *Journal of Materials Chemistry, 2*(32), 13159–13168. https://doi.org/10.1039/C4TA01681E

Moghadam, A. D., Omrani, E., Menezes, P. L., & Rohatgi, P. K. (2015). Mechanical and tribological properties of self-lubricating metal matrix nanocomposites reinforced by carbon nanotubes (CNTs) and graphene–a review. *Composites Part B: Engineering, 77*, 402–420. https://doi.org/10.1016/j.compositesb.2015.03.014

Narang, S. B., Chawla, S., Mudsainiyan, R., & Pubby, K. (2015). Comparative dielectric analysis of Co-Zr doped M-type barium hexaferrites $BaCo_xZr_xFe_{(12\text{-}2x)}O_{19}$ prepared by different wet chemical routes. *Integrated Ferroelectrics, 167*(1), 98–106. https://doi.org/10.1080/10584587.2015.1106882

Nikmanesh, H., Moradi, M., Bordbar, G. H., & Alam, R. S. (2016). Synthesis of multi-walled carbon nanotube/doped barium hexaferrite nanocomposites: An investigation of structural, magnetic and microwave absorption properties. *Ceramics International, 42*(13), 14342–14349. https://doi.org/10.1016/j.ceramint.2016.05.089

Peng, Z., Hwang, J.-Y., & Andriese, M. (2013). Design of double-layer ceramic absorbers for microwave heating. *Ceramics International, 39*(6), 6721–6725. https://doi.org/10.1016/j.ceramint.2013.01.114

Pi, K., McCreary, K., Bao, W., Han, W., Chiang, Y., Li, Y., ... Kawakami, R. (2009). Electronic doping and scattering by transition metals on graphene. *Physical Review B, 80*(7), 75406–75410. https://doi.org/10.1103/PhysRevB.80.075406

Qiu, H., Qi, S., Wang, D., Wang, J., & Wu, X. (2010). Synthesis of polyaniline nanostructures via soft template of sucrose octaacetate. *Synthetic Metals, 160*(11–12), 1179–1183. https://doi.org/10.1016/j.synthmet.2010.03.005

Quan, L., Qin, F., Estevez, D., Wang, H., & Peng, H. (2017). Magnetic graphene for microwave absorbing application: Towards the lightest graphene-based absorber. *Carbon, 125*, 630–639. https://doi.org/10.1016/j.carbon.2017.09.101

Routray, K. L., Saha, S., & Behera, D. (2017). Effect of CNTs blending on the structural, dielectric and magnetic properties of nanosized cobalt ferrite. *Materials Science and Engineering: B, 226*, 199–205. https://doi.org/10.1016/j.mseb.2017.09.021

Saini, P., Choudhary, V., Vijayan, N., & Kotnala, R. (2012). Improved electromagnetic interference shielding response of poly (aniline)-coated fabrics containing dielectric and magnetic nanoparticles. *Journal of Physical Chemistry C, 116*(24), 13403–13412. https://doi.org/10.1021/jp302131w

Sarkar, D., Bhattacharya, A., Nandy, P., & Das, S. (2014). Enhanced broadband microwave reflection loss of carbon nanotube ensheathed Ni–Zn–Co-ferrite magnetic nanoparticles. *Materials Letters, 120*, 259–262. https://doi.org/10.1016/j.matlet.2014.01.089

Sattler, K. D. (2016). *Carbon nanomaterials sourcebook: Graphene, fullerenes, nanotubes, and nanodiamonds, 1*. CRC Press.

Schüz, J., & Ahlbom, A. (2008). Exposure to electromagnetic fields and the risk of childhood leukaemia: A review. *Radiation Protection Dosimetry, 132*(2), 202–211. https://doi.org/10.1093/rpd/ncn270

Shen, B., Li, Y., Yi, D., Zhai, W., Wei, X., & Zheng, W. (2016). Microcellular graphene foam for improved broadband electromagnetic interference shielding. *Carbon, 102*, 154–160. https://doi.org/10.1016/j.carbon.2016.02.040

Shen, G., Xu, Z., & Li, Y. (2006). Absorbing properties and structural design of microwave absorbers based on W-type La-doped ferrite and carbon fiber composites. *Journal of Magnetism and Magnetic Materials, 301*(2), 325–330. https://doi.org/10.1016/j.jmmm.2005.07.007

Shen, G., Xu, M., & Xu, Z. (2007). Double-layer microwave absorber based on ferrite and short carbon fiber composites. *Materials Chemistry and Physics, 105*(2–3), 268–272. https://doi.org/10.1016/j.matchemphys.2007.04.056

Shu, R., Li, W., Wu, Y., Zhang, J., & Zhang, G. (2019). Nitrogen-doped Co-C/MWCNTs nanocomposites derived from bimetallic metal–organic frameworks for electromagnetic wave absorption in the X-band. *Chemical Engineering Journal, 362*, 513–524.

Shu, R., Li, W., Zhou, X., Tian, D., Zhang, G., Gan, Y., ... He, J. (2018). Facile preparation and microwave absorption properties of RGO/MWCNTs/$ZnFe_2O_4$ hybrid nanocomposites. *Journal of Alloys and Compounds, 743*, 163–174. https://doi.org/10.1016/j.jallcom.2018.02.016

Shu, R., Wu, Y., Li, Z., Zhang, J., Wan, Z., Liu, Y., & Zheng, M. (2019). Facile synthesis of cobalt-zinc ferrite microspheres decorated nitrogen-doped multi-walled carbon nanotubes hybrid composites with excellent microwave absorption in the X-band. *Composites Science and Technology, 184*, 107839–107850. https://doi.org/10.1016/j.compscitech.2019.107839

Shu, R., Zhang, J., Guo, C., Wu, Y., Wan, Z., Shi, J., ... Zheng, M. (2020). Facile synthesis of nitrogen-doped reduced graphene oxide/nickel-zinc ferrite composites as high-performance microwave absorbers in the X-band. *Chemical Engineering Journal, 384*, 123266–123302. https://doi.org/10.1016/j.cej.2019.123266

Shu, R., Zhang, J., Wu, Y., Wan, Z., & Zheng, M. (2019). Facile design of nitrogen-doped reduced graphene oxide/zinc ferrite hybrid nanocomposites with excellent microwave absorption in the X-band. *Materials Letters, 255*, 126549–126551. https://doi.org/10.1016/j.matlet.2019.126549

Shu, R., Zhang, G., Zhang, J., Wang, X., Wang, M., Gan, Y., Shi, J., & He, J. (2018). Fabrication of reduced graphene oxide/multi-walled carbon nanotubes/zinc ferrite hybrid composites as high-performance microwave absorbers. *Journal of Alloys and Compounds, 736*, 1–11. https://doi.org/10.1016/j.jallcom.2017.11.084

Smit, J. (1971). *Magnetic properties of materials, 13*. McGraw-Hill.

Stavarache, I., Lepadatu, A.-M., Teodorescu, V. S., Ciurea, M. L., Iancu, V., Dragoman, M., … Buiculescu, R. (2011). Electrical behavior of multi-walled carbon nanotube network embedded in amorphous silicon nitride. *Nanoscale Research Letters, 6*(1), 1–6. https://doi.org/10.1186/1556-276X-6-88

Sun, S., Sun, X., Chen, Z., Liu, Y., Wilkinson, D. P., & Zhang, J. (2017). *Carbon nanomaterials for electrochemical energy technologies: Fundamentals and applications*. CRC Press.

Sutradhar, S., Mukhopadhyay, K., Pati, S., Das, S., Das, D., & Chakrabarti, P. (2013). Modulated magnetic property, enhanced microwave absorption and Mössbauer spectroscopy of $Ni_{0.40}Zn_{0.40}Cu_{0.20}Fe_2O_4$ nanoparticles embedded in carbon nanotubes. *Journal of Alloys and Compounds, 576*, 126–133. https://doi.org/10.1016/j.jallcom.2013.04.059

Umrao, S., Gupta, T. K., Kumar, S., Singh, V. K., Sultania, M. K., Jung, J. H., … Srivastava, A. (2015). Microwave-assisted synthesis of boron and nitrogen co-doped reduced graphene oxide for the protection of electromagnetic radiation in Ku-band. *ACS Applied Materials and Interfaces, 7*(35), 19831–19842. https://doi.org/10.1021/acsami.5b05890

Vinayasree, S., Soloman, M., Sunny, V., Mohanan, P., Kurian, P., & Anantharaman, M. (2013). A microwave absorber based on strontium ferrite–carbon black–nitrile rubber for S and X-band applications. *Composites Science and Technology, 82*, 69–75. https://doi.org/10.1016/j.compscitech.2013.04.010

Wang, Y., Gao, X., Wu, X., Zhang, W., Luo, C., & Liu, P. (2019). Facile design of 3D hierarchical $NiFe_2O_4$/N-GN/ZnO composite as a high performance electromagnetic wave absorber. *Chemical Engineering Journal, 375*, 121942–121951. https://doi.org/10.1016/j.cej.2019.121942

Wang, C., Han, X., Xu, P., Zhang, X., Du, Y., Hu, S., … Wang, X. (2011). The electromagnetic property of chemically reduced graphene oxide and its application as microwave absorbing material. *Applied Physics Letters, 98*(7), 72906–72909. https://doi.org/10.1063/1.3555436

Wang, Y., Huang, Y., & Ding, J. (2014). Synthesis and electromagnetic absorption properties of polypyrrole/$BaFe_{12}O_{19}$–$Ni_{0.8}Zn_{0.2}Fe_2O_4$/multi-walled carbon nanotube composites. *Materials Science in Semiconductor Processing, 26*, 632–641. https://doi.org/10.1016/j.mssp.2014.06.001

Wang, S., Zhao, Y., Xue, H., Xie, J., Feng, C., Li, H., … Jiao, Q. (2018). Preparation of flower-like $CoFe_2O_4$@graphene composites and their microwave absorbing properties. *Materials Letters, 223*, 186–189. https://doi.org/10.1016/j.matlet.2018.04.050

Wan, Y.-J., Zhu, P.-L., Yu, S.-H., Sun, R., Wong, C.-P., & Liao, W.-H. (2017). Graphene paper for exceptional EMI shielding performance using large-sized graphene oxide sheets and doping strategy. *Carbon, 122*, 74–81. https://doi.org/10.1016/j.carbon.2017.06.042

Wang, Y., & Jing, X. (2005). Intrinsically conducting polymers for electromagnetic interference shielding. *Polymers for Advanced Technologies, 16*(4), 344–351. https://doi.org/10.1002/pat.589

Wen, F., Yi, H., Qiao, L., Zheng, H., Zhou, D., & Li, F. (2008). Analyses on double resonance behavior in microwave magnetic permeability of multiwalled carbon nanotube composites containing Ni catalyst. *Applied Physics Letters, 92*(4), 42507–42509. https://doi.org/10.1063/1.2839382

Wen, F., Zhang, F., & Liu, Z. (2011). Investigation on microwave absorption properties for multiwalled carbon nanotubes/Fe/Co/Ni nanopowders as lightweight absorbers. *Journal of Physical Chemistry C, 115*(29), 14025–14030. https://doi.org/10.1021/jp202078p

Wu, Y., Shu, R., Li, Z., Guo, C., Zhang, G., Zhang, J., & Li, W. (2019). Design and electromagnetic wave absorption properties of reduced graphene oxide/multi-walled carbon nanotubes/nickel ferrite ternary nanocomposites. *Journal of Alloys and Compounds, 784*, 887–896. https://doi.org/10.1016/j.jallcom.2019.01.139

Wu, Y., Shu, R., Zhang, J., Sun, R., Chen, Y., & Yuan, J. (2019). Oxygen vacancy defects enhanced electromagnetic wave absorption properties of 3D net-like multi-walled carbon nanotubes/cerium oxide nanocomposites. *Journal of Alloys and Compounds, 785*, 616–626. https://doi.org/10.1016/j.jallcom.2019.01.227

Xiong, Y., Luo, H., Nie, Y., Chen, F., Dai, W., Wang, X., Cheng, Y., & Gong, R. (2019). Synergistic effect of silica coated porous rodlike nickel ferrite and multiwalled carbon nanotube with improved electromagnetic wave absorption performance. *Journal of Alloys and Compounds, 802*, 364–372. https://doi.org/10.1016/j.jallcom.2019.06.174

Yang, C., & Liu, P. (2009). Core-shell attapulgite@polypyrrole composite with well-defined corn cob-like morphology via self-assembling and in situ oxidative polymerization. *Synthetic Metals, 159*(19–20), 2056–2062. https://doi.org/10.1016/j.synthmet.2009.07.022

Yang, R., Hsu, S., & Lin, C. (2009). Frequency-dependent complex permittivity and permeability of iron-based powders in 2–18 GHz. *Journal of Applied Physics, 105*(7), 7A527–7A529. https://doi.org/10.1063/1.3068039

Yang, Z., Wan, Y., Xiong, G., Li, D., Li, Q., Ma, C., … Luo, H. (2015). Facile synthesis of $ZnFe_2O_4$/reduced graphene oxide nanohybrids for enhanced microwave absorption properties. *Materials Research Bulletin, 61*, 292–297. https://doi.org/10.1016/j.materresbull.2014.10.004

Yin, P., Deng, Y., Zhang, L., Wu, W., Wang, J., Feng, X., … Tao, Y. (2018). One-step hydrothermal synthesis and enhanced microwave absorption properties of $Ni_{0.5}Co_{0.5}Fe_2O_4$/graphene composites in low frequency band. *Ceramics International, 44*(17), 20896–20905. https://doi.org/10.1016/j.ceramint.2018.08.096

Yu, B., Qi, L., Sun, H., & Ye, J. (2007). Radar wave absorbing characterization of bicomponent fibers. *Journal of Materials Science, 42*(11), 3783–3788. https://doi.org/10.1007/s10853-006-0414-z

Zhang, C., Li, Q., & Ye, Y. (2009). Preparation and characterization of polypyrrole/nano-$SrFe_{12}O_{19}$ composites by in situ polymerization method. *Synthetic Metals, 159*(11), 1008–1013. https://doi.org/10.1016/j.synthmet.2009.01.013

Zhang, H., Hong, M., Chen, P., Xie, A., & Shen, Y. (2016). 3D and ternary rGO/MCNTs/Fe_3O_4 composite hydrogels: Synthesis, characterization and their electromagnetic wave absorption properties. *Journal of Alloys and Compounds, 665*, 381–387. https://doi.org/10.1016/j.jallcom.2016.01.091

Zhang, Q., Jiao, Q., Xuan, Z., Li, T., & Yu, W. (2011). Preparation of carbon nanotube/expanded graphite composite material and its 8 mm wave attenuation properties. *Acta Armamentarii, 32*, 1510–1513.

Zhang, T., Huang, D., Yang, Y., Kang, F., & Gu, J. (2013). Fe_3O_4/carbon composite nanofiber absorber with enhanced microwave absorption performance. *Materials Science and Engineering: B, 178*(1), 1–9. https://doi.org/10.1016/j.mseb.2012.06.005

Zhan, Y., Long, Z., Wan, X., Zhang, J., He, S., & He, Y. (2018). 3D carbon fiber mats/nano-Fe_3O_4 hybrid material with high electromagnetic shielding performance. *Applied Surface Science, 444*, 710–720. https://doi.org/10.1016/j.apsusc.2018.03.006

Zhang, Y., Wang, X., & Cao, M. (2018). Confinedly implanted $NiFe_2O_4$-rGO: Cluster tailoring and highly tunable electromagnetic properties for selective-frequency microwave absorption. *Nano Research, 11*(3), 1426–1436. https://doi.org/10.1007/s12274-017-1758-1

Zhao, T., Jin, W., Ji, X., Yan, H., Jiang, Y., Dong, Y., Yang, Y., Dang, A., Li, H., & Li, T. (2017). Synthesis of sandwich microstructured expanded graphite/barium ferrite connected with carbon nanotube composite and its electromagnetic wave absorbing properties. *Journal of Alloys and Compounds, 712*, 59–68. https://doi.org/10.1016/j.jallcom.2017.04.070

Zhou, N., An, Q., Xiao, Z., Zhai, S., & Shi, Z. (2017). Solvothermal synthesis of three-dimensional, Fe_2O_3 NPs-embedded CNT/N-doped graphene composites with excellent microwave absorption performance. *RSC Advances, 7*(71), 45156–45169. https://doi.org/10.1039/C7RA06751H

Zhu, X., Liu, Q., & Fu, M. (2017). Preparation of urchin-like strontium ferrites as microwave absorbing materials. *Materials Letters, 209*, 425–428. https://doi.org/10.1016/j.matlet.2017.08.075

Further reading

Cao, M.-S., Yang, J., Song, W.-L., Zhang, D.-Q., Wen, B., Jin, H.-B., … Yuan, J. (2012). Ferroferric oxide/multiwalled carbon nanotube vs polyaniline/ferroferric oxide/multiwalled carbon nanotube multiheterostructures for highly effective microwave absorption. *ACS Applied Materials and Interfaces, 4*(12), 6949–6956. https://doi.org/10.1021/am3021069

Chen, X., Wang, W., Shi, T., Wu, G., & Lu, Y. (2020). One pot green synthesis and EM wave absorption performance of MoS_2@nitrogen doped carbon hybrid decorated with ultrasmall cobalt ferrite nanoparticles. *Carbon, 163*, 202–212. https://doi.org/10.1016/j.carbon.2020.03.005

Chung, D. D. L. (2000). Materials for electromagnetic interference shielding. *Journal of Materials Engineering and Performance, 9*(3), 350–354. https://doi.org/10.1361/105994900770346042

Dalal, M., Mallick, A., & Chakrabarti, P. (2019). Microwave absorption of $Mn_{0.5}Zn_{0.5}Fe_2O_4$ nanoparticles integrated in multi-walled carbon nanotubes. *Journal of the Australian Ceramic Society, 55*(1), 157–167. https://doi.org/10.1007/s41779-018-0221-4

Fang, Y., Tang, X., Sun, X., Zhang, Y., Zhao, J., Yu, L., … Zhao, X. (2017). Preparation and enhanced microwave absorption properties of Ni-Co attached single-walled carbon nanotubes and $CoFe_2O_4$ nanocomposites. *Journal of Applied Physics, 121*(22), 224301–224308. https://doi.org/10.1063/1.4984937

Feng, J., Wang, Y., Hou, Y., & Li, L. (2017). Tunable design of yolk–shell $ZnFe_2O_4$@RGO@TiO_2 microspheres for enhanced high-frequency microwave absorption. *Inorganic Chemistry Frontiers, 4*(6), 935–945. https://doi.org/10.1039/C7QI00120G

Ghasemi, A., Shirsath, S. E., Liu, X., & Morisako, A. (2011). Enhanced reflection loss characteristics of substituted barium ferrite/functionalized multi-walled carbon nanotube nanocomposites. *Journal of Applied Physics, 109*(7), 7A507–7A509. https://doi.org/10.1063/1.3551727

Gupta, S., Chang, C., Lai, C.-H., & Tai, N.-H. (2019). Hybrid composite mats composed of amorphous carbon, zinc oxide nanorods and nickel zinc ferrite for tunable electromagnetic interference shielding. *Composites Part B: Engineering, 164*, 447–457. https://doi.org/10.1016/j.compositesb.2019.01.060

Jia, Z., Kou, K., Yin, S., Feng, A., Zhang, C., Liu, X., … Wu, G. (2020). Magnetic Fe nanoparticle to decorate N dotted C as an exceptionally absorption-dominate electromagnetic shielding material. *Composites Part B: Engineering, 189*, 107895–107902. https://doi.org/10.1016/j.compositesb.2020.107895

Lee, C., Song, H., Jang, K., Oh, E., Epstein, A., & Joo, J. (1999). Electromagnetic interference shielding efficiency of polyaniline mixtures and multilayer films. *Synthetic Metals, 102*(1–3), 1346–1349. https://doi.org/10.1016/S0379-6779(98)00234-3

Li, L., Xiang, C., Liang, X., & Hao, B. (2010). $Zn_{0.6}Cu_{0.4}Cr_{0.5}Fe_{1.46}Sm_{0.04}O_4$ ferrite and its nanocomposites with polyaniline and polypyrrole: Preparation and electromagnetic properties. *Synthetic Metals, 160*(1–2), 28–34. https://doi.org/10.1016/j.synthmet.2009.09.026

Liu, P., Yao, Z., & Zhou, J. (2015). Preparation of reduced graphene oxide/ $Ni_{0.4}Zn_{0.4}Co_{0.2}Fe_2O_4$ nanocomposites and their excellent microwave absorption properties. *Ceramics International, 41*(10), 13409–13416. https://doi.org/10.1016/j.ceramint.2015.07.129

Shen, W., Ren, B., Cai, K., Song, Y., & Wang, W. (2019). Synthesis of nonstoichiometric $Co_{0.8}Fe_{2.2}O_4$/reduced graphene oxide (rGO) nanocomposites and their excellent electromagnetic wave absorption property. *Journal of Alloys and Compounds, 774*, 997–1008. https://doi.org/10.1016/j.jallcom.2018.09.361

Sun, J., Wang, L., Yang, Q., Shen, Y., & Zhang, X. (2020). Preparation of copper-cobalt-nickel ferrite/graphene oxide/polyaniline composite and its applications in microwave absorption coating. *Progress in Organic Coatings, 141*, 105552–105559. https://doi.org/10.1016/j.porgcoat.2020.105552

Wang, M., An, K., Fang, Y., Wei, G., Yang, J., Sheng, L., … Zhao, X. (2017). The microwave absorbing properties of $CoFe_2$ attached single walled carbon nanotube/$BaFe_{12}O_{19}$ nanocomposites. *Journal of Materials Science: Materials in Electronics, 28*(17), 12475–12483. https://doi.org/10.1007/s10854-017-7069-1

Wang, X.-X., Ma, T., Shu, J.-C., & Cao, M.-S. (2018). Confinedly tailoring Fe_3O_4 clusters-NG to tune electromagnetic parameters and microwave absorption with broadened bandwidth. *Chemical Engineering Journal, 332*, 321–330. https://doi.org/10.1016/j.cej.2017.09.101

Wang, Y., Huang, Y., Wang, Q., He, Q., & Chen, L. (2012). Preparation and electromagnetic properties of Polyaniline (polypyrrole)-$BaFe_{12}O_{19}$/$Ni_{0.8}Zn_{0.2}Fe_2O_4$ ferrite nanocomposites. *Applied Surface Science, 259*, 486–493. https://doi.org/10.1016/j.apsusc.2012.07.072

Wen, B., Cao, M.-S., Hou, Z.-L., Song, W.-L., Zhang, L., Lu, M.-M., … Yuan, J. (2013). Temperature dependent microwave attenuation behavior for carbon-nanotube/silica composites. *Carbon, 65*, 124–139. https://doi.org/10.1016/j.carbon.2013.07.110

Yang, R.-B., Reddy, P. M., Chang, C.-J., Chen, P.-A., Chen, J.-K., & Chang, C.-C. (2016). Synthesis and characterization of Fe_3O_4/polypyrrole/carbon nanotube composites with tunable microwave absorption properties: Role of carbon nanotube and polypyrrole content. *Chemical Engineering Journal, 285*, 497–507. https://doi.org/10.1016/j.cej.2015.10.031

Zhang, N., Huang, Y., Zong, M., Ding, X., Li, S., & Wang, M. (2016). Synthesis of core-shell $ZnFe_2O_4@SiO_2$ hollow microspheres/reduced graphene oxides for a high-performance EM wave absorber. *Ceramics International, 42*(16), 18879–18886. https://doi.org/10.1016/j.ceramint.2016.09.035

Zhang, T., Zhao, D., Wang, L., Meng, R., Zhao, H., Zhou, P., … Wen, G. (2020). A facile precursor pyrolysis route to bio-carbon/ferrite porous architecture with enhanced electromagnetic wave absorption in S-band. *Journal of Alloys and Compounds, 819*, 153269–153296. https://doi.org/10.1016/j.jallcom.2019.153269

Zhou, X., Chuai, D., & Zhu, D. (2019). Electrospun synthesis of reduced graphene oxide (RGO)/NiZn ferrite nanocomposites for excellent microwave absorption properties. *Journal of Superconductivity and Novel Magnetism, 32*(8), 2687–2697. https://doi.org/10.1007/s10948-019-5039-y

CHAPTER

Nanoferrite photocatalysts

7

7.1 Introduction

Photocatalysts are outstanding materials that can easily alter solar energy for use in oxidation and reduction activities. Photocatalysts are used in several areas, such as eliminating pollutants from air and water, water splitting to generate H_2, odor control, cancer cell inactivation, and bacterial inactivation. Over the last decade, photocatalyst materials have attracted great interest because of their potential for eliminating toxic and dangerous compounds from the environment. Ferrites are photocatalyst materials that effectively use visible light for oxidation processes. Their size, shape, and overall structure can influence ferritic photocatalytic activity. The activity changes with heat treatment of the photocatalyst, as the crystal structure and surface defects can change with heat treatment. Photocatalysts cause the generation of reactive oxygen, which helps in the decomposition of contaminants. Incorporating two photocatalysts with various bandgaps leads to a greater separation of electron/hole pairs, where more species would exist for reactions to degrade contaminants.

7.2 Basic definition of photocatalysts

Photocatalyst materials employ light energy (hυ) to perform oxidation and reduction processes. With the irradiation of light energy, an electron (e^-) is excited from the valence band to the conduction band of the photocatalyst, where a photogenerated hole (h^+) is left (Dutta et al., 2005; Hoffmann et al., 1995; Shifu et al., 2009a, 2009b). Fig. 7.1 displays the photocatalytic light energy-absorbing process to generate e^-/h^+ pairs (Casbeer et al., 2012).

When photocatalytic reactions happen in aqueous solutions, hydroxyl radicals (•OH) are generated by the reaction of water and hydroxide ions with a photogenerated hole (h^+). The hydroxyl radicals (•OH) are the initial oxidant in the photocatalytic oxidation of organic compounds, which has higher oxidation than other oxidants such as ozone (O_3) and hydrogen peroxide (H_2O_2). The production of hydroxyl radicals (•OH) occurs in two steps. Initially, the O_2 of water is reduced to form $O_2^{\cdot-}$. Then, $O_2^{\cdot-}$ reacts with H^+ to create •OOH, wherein •OOH decomposes

Magnetic Ferrites and Related Nanocomposites. https://doi.org/10.1016/B978-0-12-824014-4.00006-8

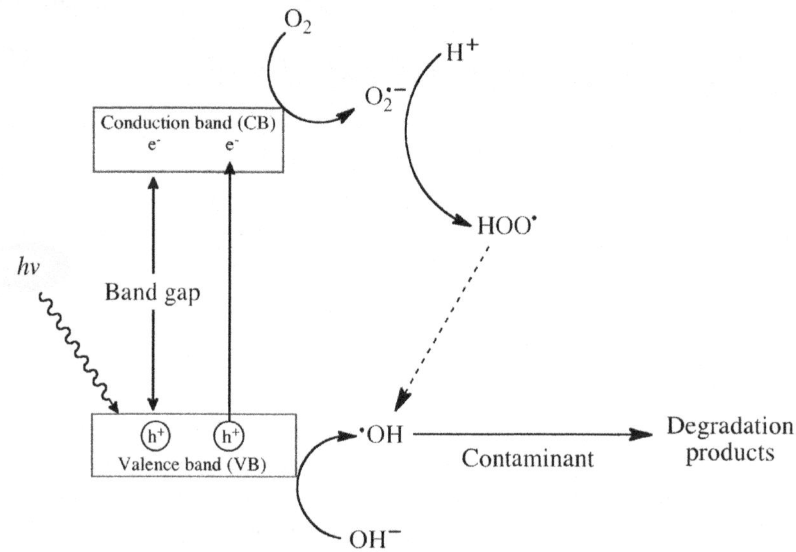

FIGURE 7.1

Schematic of reactions occurring by a photocatalyst to generate •OH to react with contaminants.

From Casbeer, E., Sharma, V. K., & Li, X.-Z. (2012). Synthesis and photocatalytic activity of ferrites under visible light: a review. Separation and Purification Technology, 87, *1–14. https://doi.org/10.1016/j.seppur.2011.11.034.*

to •OH rapidly (Casbeer et al., 2012). The bandgap of the photocatalyst reveals the wavelength of light that it can absorb. The wavelength of light depends on the difference between the valence band and the conduction band. Fig. 7.2 reveals the absorbance of the solar energy spectrum in terms of wavelength across the Earth's surface. It is important to employ visible solar energy because of the large amount that annually reaches the Earth's surface. TiO_2 is a suitable photocatalyst, but it has a wide bandgap (3.03 eV rutile and 3.18 eV anatase) which causes small absorption of sunlight. Applying dopants into TiO_2 such as nitrogen, carbon, sulfur, and combinations of nitrogen, carbon, sulfur, and silicon could enhance the photocatalytic properties of TiO_2, thereby absorbing visible light effectively. Heavy metals such as cobalt, silver, gold, platinum, and ruthenium have also been doped in TiO_2 to become visible light active (Sen, 2008; Zang & Farnood, 2008).

Spinel ferrites are chemically stable magnetic materials employed in many industries such as magnetic resonance imaging, electronic devices, information storage, and drug delivery, as well as in adsorbents for the removal of toxic substances. The bandgaps of various photocatalysts are seen in Fig. 7.3. It is clear that some photocatalysts such as TiO_2, ZnS, and $SrTiO_3$ are suitable for ultraviolet (UV) irradiation because of their wide bandgaps; on the other hand, $CaFe_2O_4$, $MgFe_2O_4$, $ZnFe_2O_4$, CdS, and WO_3 are effective for visible light because of smaller bandgaps (Casbeer et al., 2012).

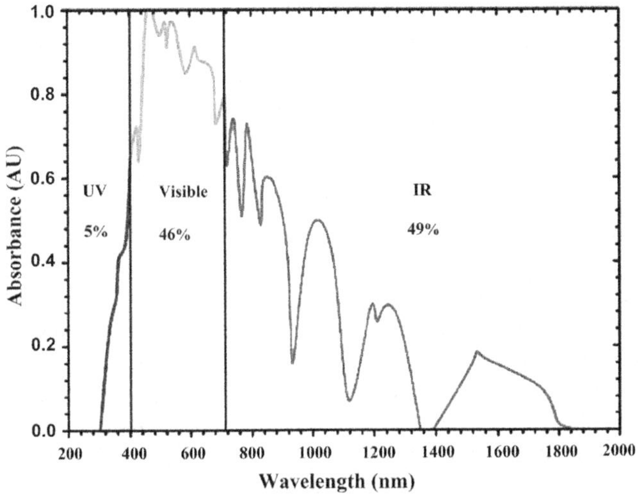

FIGURE 7.2

The absorbance of solar energy spectrum by wavelength across the Earth's surface.

From Casbeer, E., Sharma, V. K., & Li, X.-Z. (2012). Synthesis and photocatalytic activity of ferrites under visible light: a review. Separation and Purification Technology, 87, *1–14. https://doi.org/10.1016/j.seppur.2011.11.034.*

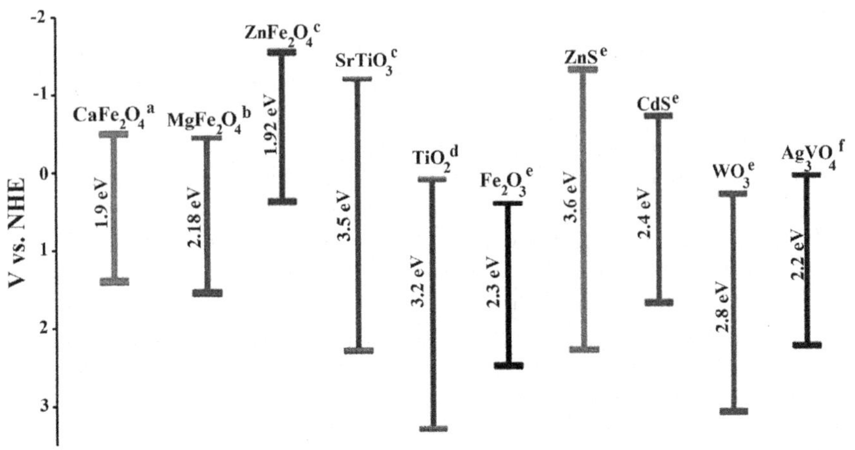

FIGURE 7.3

Bandgaps of various photocatalysts.

From Casbeer, E., Sharma, V. K., & Li, X.-Z. (2012). Synthesis and photocatalytic activity of ferrites under visible light: a review. Separation and Purification Technology, 87, *1–14. https://doi.org/10.1016/j.seppur.2011.11.034.*

The photocatalytic activity of ferrites involves removing inorganic contaminants from the environment, disinfection of Escherichia coli when combined with other materials, and the photocatalytic degradation of organic dyes in water treatment. The percentage of photocatalytic degradation can be calculated by the following equation:

$$\%\ \text{degradation} = ((A_0 - A_t)\,/\,A_0) \times 100 \tag{7.1}$$

where A_0 and A_t represent the initial concentration and dye concentration at time t, respectively. There are several important dyes. Methyl orange (MO) dye is a famous acid/base indicator that has been studied as a model of a series of general azo dyes. The structure of MO dye consists of two benzene rings attached to an azo group. One ring includes dimethylamine, while the other contains a sulfonic acid group. MO dye can be degraded by ferrites employing UV and visible light irradiation sources.

Methylene blue (MB) is another famous dye and is a model for organic contaminants. MB dye is usually employed in chemistry and biology as a redox indicator and stain, respectively. The structure of MB consists of phenothiazine with dimethylamine on each side. MB dye can have relatively high degradation potential with ferrites. The degradation activity of the dyes is related to the ratios of the photocatalysts and temperature treatments.

Rhodamine B (RhB) is a xanthene dye that is potentially toxic and carcinogenic and is employed in the textile area. Thus, it is not used in foods and cosmetics.

4-Nitrophenol (4-NP) is an organic compound consisting of a benzene ring bearing both a hydroxyl group and a nitro group on two different ring carbon atoms. 4-NP has been found as a solid, slightly soluble (in water), and a very weakly acidic compound. 4-NP is a potentially toxic compound.

Alizarin Red S (ARS) is an azo dye widely employed in wool and nylon dyeing. It is toxic and carcinogenic and can cause a serious hazard to aquatic ecosystems.

Safranin dye is a biological stain employed in histology and cytology. Safranin is an azonium compound of symmetrical 2,8-dimethyl-3,7-diaminophenazine. This dye is a counterstain for some staining protocols, coloring cell nuclei red. It can also be used to detect cartilage, mucin, and mast cell granules.

Eriochrome Black T (EBT) dye is a complexometric indicator employed in complexometric titrations, such as the water hardness determination process. It is an azo dye.

Crystal violet dye is a triarylmethane dye utilized as a histological stain and in Gram's method of classifying bacteria. Crystal violet has antibacterial, antifungal, and anthelmintic characteristics. Chloramine-T dye is an organic compound with the formula $CH_3C_6H_4SO_2NClNa$. Both the anhydrous salt and its trihydrate are known. Fig. 7.4 shows the structure of some dyes.

Ferrites that have reactivity with MO (Ao et al., 2008; Cheng et al., 2004; Jadhav et al., 2011; Rana et al., 2005; Xu et al., 2007; Zhang, Sun et al., 2009; Zhang, Hou et al., 2009), MB (Chen et al., 2010; Dom et al., 2011; Fu et al., 2010; Hou et al., 2010; Li et al., 2009; Shifu et al., 2009a; Wang et al., 2009), and RhB (Chen et al., 2010; Dom et al., 2011; Fu et al., 2010; Hou et al., 2010; Li et al., 2009; Shifu et al., 2009a; Wang et al., 2009) dyes are listed in Tables 7.1–7.3, respectively.

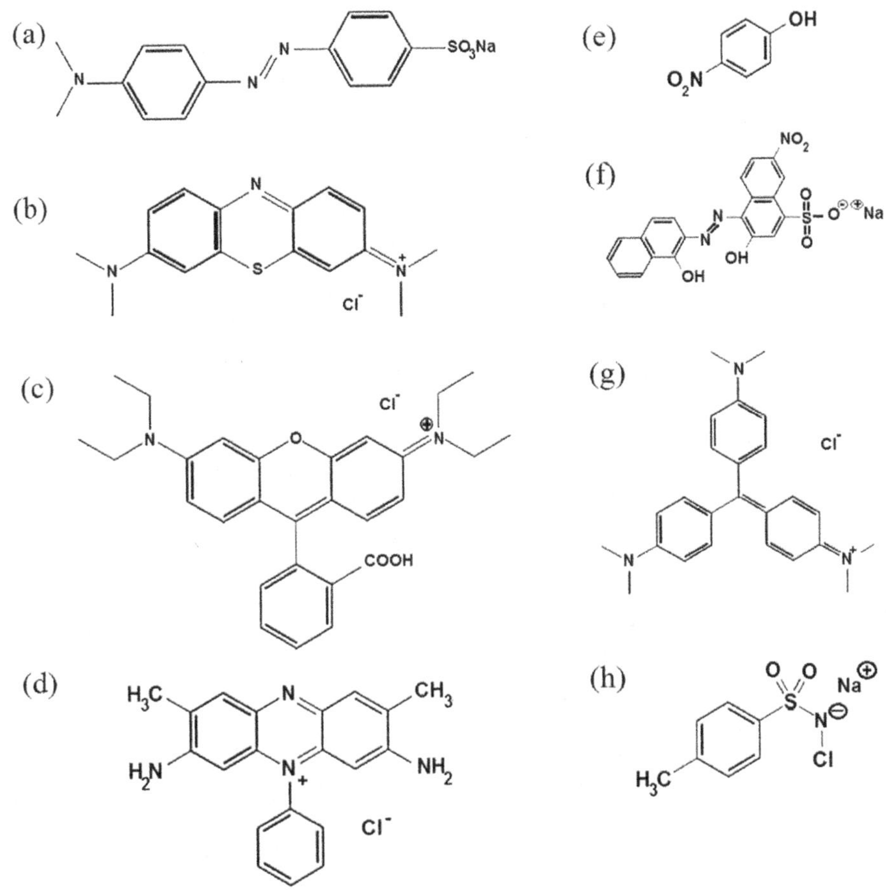

FIGURE 7.4

The structure of (A) methyl orange, (B) methylene blue, (C) rhodamine B, (D) safranin dye, (E) 4-nitrophenol, (F) Eriochrome Black T, (G) Crystal violet, and (H) Chloramine-T dyes.

From Casbeer, E., Sharma, V. K., & Li, X.-Z. (2012). Synthesis and photocatalytic activity of ferrites under visible light: a review. Separation and Purification Technology, 87, *1–14. https://doi.org/10.1016/j.seppur.2011.11.034.*

The following is a discussion of the photocatalytic features of several ferrites and related nanocomposites.

7.3 Nanocrystalline Co-ferrite photocatalyst

The relationship between the rate of the photocatalytic decomposition reaction of 4-NP and cobalt ferrite was evaluated (Frolova, 2020). It was seen that the activity of the photocatalytic decomposition reaction of 4-NP diminished with an increase in

Table 7.1 Summary of ferrites reacting with methyl orange dye.

Photocatalyst	Dye (mg/L)	Catalyst (g/L)	Irradiation source	Irradiation time (min)	Degradation (%)
$BaFe_2O_4$ prepared by citrate–nitrate method	20	5	Ultraviolet (UV)	120	65
$BaFe_2O_4$ prepared by thermal method	20	5	UV	120	90
$BaFe_2O_4$ prepared by thermal method	20	5	UV	120	35
TiO_2(43.6%)/$CoFe_2O_4$(8.4%)/SiO_2(28%)	6	n/a	UV	250	60
$CoFe_2O_4$/ZnO	50	30	UV	300	93.9
$SiO_2/NiFe_2O_4$	10	1	$\lambda > 400$ nm	60	5
$Bi_{12}TiO_{20}/SiO_2/NiFe_2O_4$	10	1.1	$\lambda > 400$ nm	30	85
$TiO_2/NiFe_2O_4$	8	0.8	UV	420	80
$TiO_2/ZnFe_2O_4$	10	5	$\lambda > 400$ nm	180	5
$ZnFe_2O_4$	10	4	$\lambda > 400$ nm	60	75

Table 7.2 Summary of ferrites reacting with methylene blue dye.

Photocatalyst	Dye (mg/L)	Catalyst (g/L)	Irradiation source	Irradiation time (min)	Degradation (%)
$Ag_3VO_4/CaFe_2O_4$ (0.5%)	32	2	$\lambda > 470$ nm	60	59.3
$ZnO/CaFe_2O_4$ (0.1%)	32	2	UV	20	57
$TiO_2/CoFe_2O_4$ (10%)	5	0.1	UV	60	56
TiO_2/Ni–Cu–Zn ferrite	20	1.33	UV	120	82
TiO_2/Ni–Cu–Zn ferrite	20	2.67	UV	120	99
$MgFe_2O_4$	10	0.6	400–700 nm	180	26
$CaFe_2O_4$	10	0.6	400–700 nm	180	28
$ZnFe_2O_4$	10	0.6	400–700 nm	180	32
$MnFe_2O_4$	7	0.3	Visible	1200	15.17
$ZnFe_2O_4$	10	1	$\lambda > 420$ nm	360	8
$ZnFe_2O_4 + H_2O_2$	10	1	$\lambda > 420$ nm	360	52
$ZnFe_2O_4 + H_2O_2$	10	1	Dark	360	45
$ZnFe_2O_4$/multiwalled carbon nanotubes (MWCNTs)	10	1	$\lambda > 420$ nm	360	25
$ZnFe_2O_4$/MWCNTs + H_2O_2	10	1	$\lambda > 420$ nm	360	99
$ZnFe_2O_4$/MWCNTs + H_2O_2	10	1	Dark	360	95
$Fe_3O_4 + H_2O_2$	10	1	$\lambda > 420$ nm	120	12
Fe_3O_4/FePc + H_2O_2	10	1	$\lambda > 420$ nm	120	78

Table 7.3 Summary of ferrites reacting with rhodamine B dye.

Photocatalyst	Dye (mg/L)	Catalyst (g/L)	Irradiation source	Irradiation time (min)	Degradation (%)
$ZnFe_2O_4$	10	2	200 −700 nm	150	60
$ZnFe_2O_4$/Ag (8.1%)	10	2	200 −700 nm	150	95
$TiO_2/ZnFe_2O_4$	9.6	1	254 nm	150	99.7
$ZnFe_2O_4$	20	0.8	200 −700 nm	300	38.4
$ZnFe_2O_4$	20	0.8	200 −700 nm	300	100
$Ag/Fe_3O_4/SiO_2$	25	0.5	$\lambda > 420$ nm	150	10
$BaFe_2O_4/SiO_2/TiO_2$	10	0.8	UV	80	59
$BaFe_2O_4/SiO_2/TiO_2$	10	0.8	UV	80	71
$ZnFe_2O_4$	20	0.5	254 nm	360	45
$ZnFe_2O_4$	20	0.5	254 nm	360	98

the amount of cobalt in the composition of $Co_xFe_{3-x}O_4$ ($x = 0.25$, 0.5, 0.75, and 1) nanoparticles. Fig. 7.5 depicts the ultraviolet−visible (UV−V) absorption spectra of 4-NP solution using $Co_xFe_{3-x}O_4$ ferrite. The absorption peak at 317 nm is a signature for characterizing 4-NP. It was found that subjecting the samples to UV irradiation is key for decomposing 4-NP. The destruction process was accelerated in the presence of a ferrite catalyst. During the decomposing process with time, the sequential smoothing of the main absorption peak and the appearance of an additional peak are observed. The additional peak corresponds to the formation of intermediation, such as the phenolic OH group. Fig. 7.6 shows the dependence of degradation of 4-NP on the amount of cobalt in the $Co_xFe_{3-x}O_4$ ferrite photocatalyst. It was observed that with an increase in cobalt content in cobalt ferrite nanoparticles from 0.25 to 1, the rate of photocatalytic degradation decreased.

It was found that the catalytic activity diminished with an increase in the amount of cobalt in the $Co_xFe_{3-x}O_4$ ferrite photocatalyst, which is attributed to the increased ion-ion octahedral distance (Frolova, 2020).

In research work, cobalt ferrite nanoparticles were prepared via the coprecipitation method, and the photocatalytic performance of ARS dye was evaluated (Ali, Ali, Said et al. 2020). It was illustrated that various parameters such as the irradiation time, pH, adsorbent dose, dye concentration, and solvent could affect the photocatalytic degradation of the ARS dye. The photodegradation of ARS dye by three types of cobalt ferrite nanoparticles [bare (B), annealed (A), and surface functionalized (F)] was 87%, 88%, and 90%, respectively. The highest photodegradation

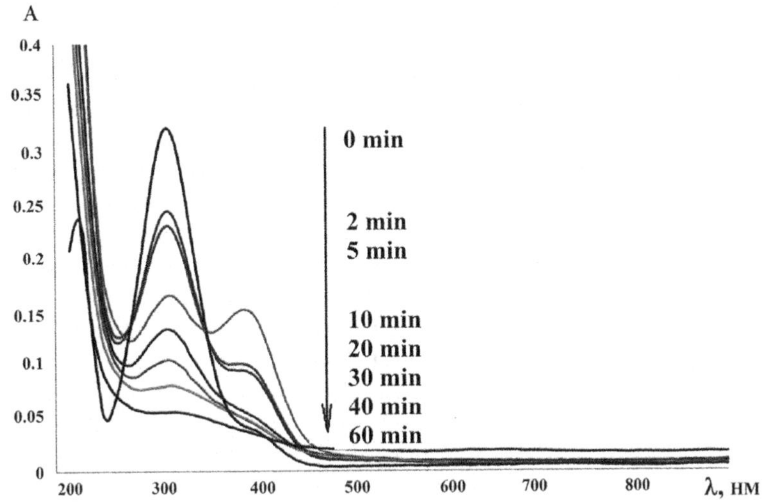

FIGURE 7.5

Ultraviolet–visible absorption spectra of 4-nitrophenol solution using $Co_xFe_{3-x}O_4$ ferrite.

From Frolova, L. (2020). Photocatalytic activity of spinel ferrites $Co_xFe_{3-x}O_4$ (0.25 < x < 1) obtained by treatment contact low-temperature nonequilibrium plasma. Applied Nanoscience, *1–6. https://doi.org/10.1007/s13204-020-01344-8.*

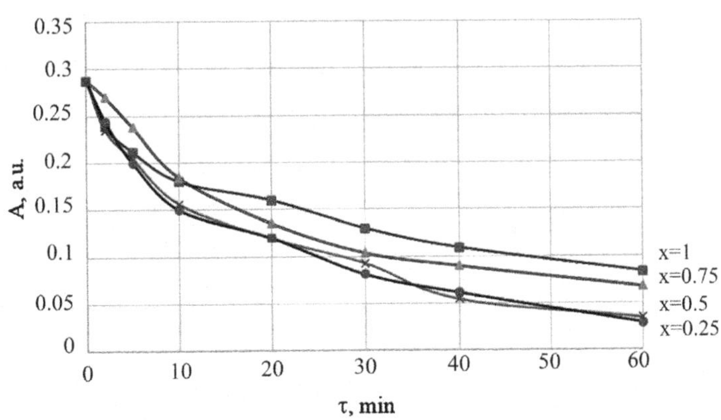

FIGURE 7.6

Kinetics of 4-nitrophenol decomposition in $Co_xFe_{3-x}O_4$ ferrite photocatalyst.

From Frolova, L. (2020). Photocatalytic activity of spinel ferrites $Co_xFe_{3-x}O_4$ (0.25 < x < 1) obtained by treatment contact low-temperature nonequilibrium plasma. Applied Nanoscience, *1–6. https://doi.org/10.1007/s13204-020-01344-8.*

performance was achieved for the surface functionalized with cobalt ferrite nanoparticles because of the larger surface area, thereby facilitating the adsorption of organic molecules. Fig. 7.7 presents the effect of adsorbent dose (from 10 to 60 mg) on the photodegradation performance of ARS dye by cobalt ferrite nanoparticles. It was observed that the degradation activity increased with a rise in the adsorbent dose from 10 to 30 mg. However, at a catalyst quantity higher than 30 mg, the degradation rate was not altered. Indeed, the higher amount of the catalyst leads to the aggregation and reduction of the irradiation field because of increased light scattering. Thus, a higher amount of the catalyst is not suitable. Interestingly, the optimum catalyst quantity is related to the initial solute concentration. The influence of initial dye concentration on the photodegradation performance of ARS dye by cobalt ferrite nanoparticles is shown in Fig. 7.8. It was observed that with an increase in the dye concentration, the degradation activity diminished. The highest photodegradation performance was achieved for the surface of functionalized cobalt ferrite nanoparticles, 91%. The values of degradation activity of bare and annealed cobalt ferrite nanoparticles were 82% and 87%, respectively. Fig. 7.9 depicts the photodegradation performance of ARS dye by cobalt ferrite nanoparticles as a function of pH. It was observed that photocatalytic degradation increased with a rise in pH from 1 to 8. With a further increase in pH (i.e., pH > 8), the photocatalytic degradation rate decreased. Indeed, the photodegradation performance was higher in an acidic

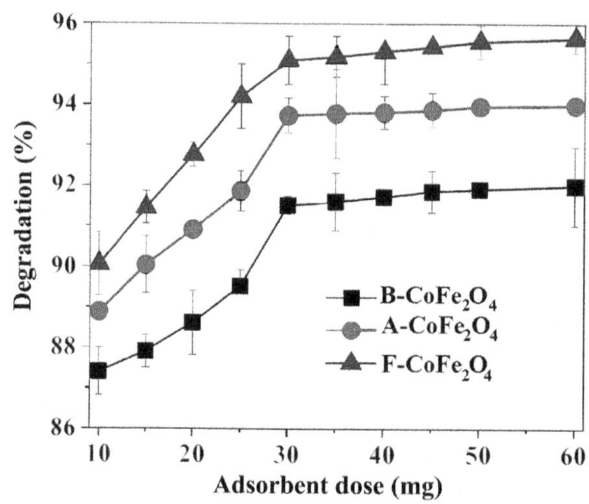

FIGURE 7.7

The effect of adsorbent dose on the photodegradation performance of Alizarin Red S dye by cobalt ferrite nanoparticles.

From Ali, N., Ali, F., Said, A., Begum, T., Bilal, M., Rab, A., Sheikh, Z. A., Iqbal, H. M., & Ahmad, I. (2020). Characterization and deployment of surface-engineered cobalt ferrite nanospheres as photocatalyst for highly efficient remediation of alizarin red S dye from aqueous solution. Journal of Inorganic and Organometallic Polymers and Materials, *1–11. https://doi.org/10.1007/s10904-020-01654-y.*

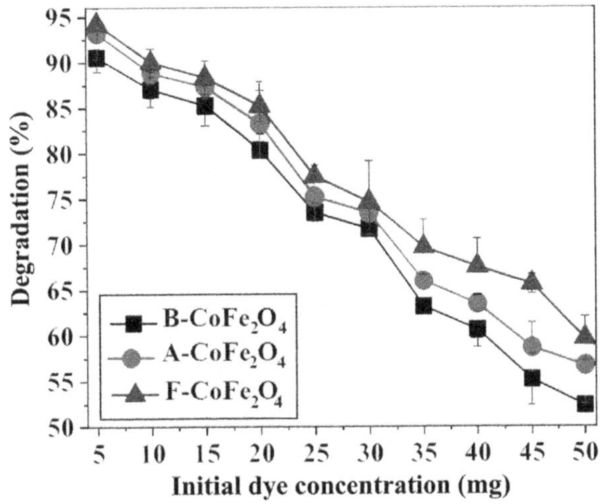

FIGURE 7.8

The effect of dye concentration on the photodegradation performance of Alizarin Red S dye by cobalt ferrite nanoparticles.

From Ali, N., Ali, F., Said, A., Begum, T., Bilal, M., Rab, A., Sheikh, Z. A., Iqbal, H. M., & Ahmad, I. (2020). Characterization and deployment of surface-engineered cobalt ferrite nanospheres as photocatalyst for highly efficient remediation of alizarin red S dye from aqueous solution. Journal of Inorganic and Organometallic Polymers and Materials, *1–11. https://doi.org/10.1007/s10904-020-01654-y.*

environment compared with an alkaline environment (Ali et al., 2020). The enhancement of photodegradation performance in acidic media is attributed to the higher quantities of oxygen anion radical formed through the reaction of electron of the photocatalyst ferrite and O_2 molecules (Badawy et al., 2020; Nezamzadeh-Ejhieh & Zabihi-Mobarakeh, 2014; To Loan et al., 2019).

It was observed that the acidic environment could influence the photocatalytic performance of the dye. Thus, photocatalytic degradation was investigated in the presence of various acids which were added to the reaction solution. Fig. 7.10 shows the effect of various acids on the photodegradation performance of ARS dye by cobalt ferrite nanoparticles. Lower photodegradation activity was achieved in the presence of H_2SO_4. Nitric acid was more effective at degrading ARS dye. The linear state reveals that photocatalytic degradation performance follows pseudo-first-order kinetics (Ali, Ali, Said et al., 2020).

It was reported that the catalytic activity of the ferrites grows with an increase in transition metal concentration. It was observed that with an increase in the concentration of metal ions, the saturation magnetization of the metal substituted cobalt ferrites diminished because of the replacement of Fe^{3+} with ions that have a lesser magnetic metal. The maximum saturation magnetization was achieved for $CoCr_{0.2}Fe_{1.8}O_4$, at 69.3 emu/g, while the minimum saturation magnetization was achieved for $CoZnFeO_4$, which was 5.8 emu/g. It was also observed that with an

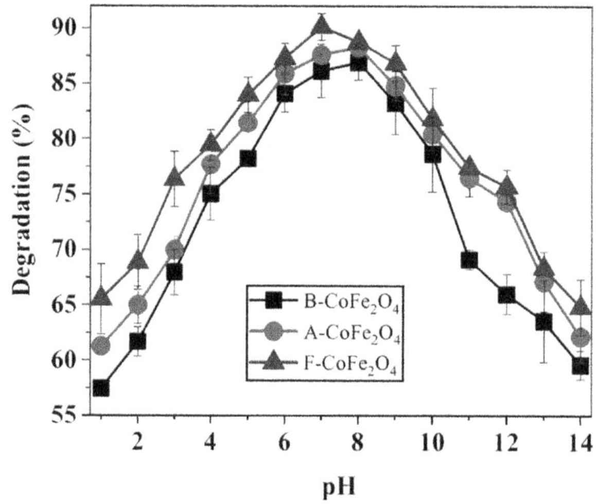

FIGURE 7.9

The effect of pH on the photodegradation performance of Alizarin Red S dye by cobalt ferrite nanoparticles.

From Ali, N., Ali, F., Said, A., Begum, T., Bilal, M., Rab, A., Sheikh, Z. A., Iqbal, H. M., & Ahmad, I. (2020). Characterization and deployment of surface-engineered cobalt ferrite nanospheres as photocatalyst for highly efficient remediation of alizarin red S dye from aqueous solution. Journal of Inorganic and Organometallic Polymers and Materials, *1–11. https://doi.org/10.1007/s10904-020-01654-y.*

increase in Cr^{3+} and Zn^{2+} contents, the coercive field decreased. However, in $CoNi_xFe_{2-x}O_4$ nanoparticles, with an increase in Ni^{2+} concentration, the coercive field first grew up to $x = 0.8$ and then dropped. For $CoCu_xFe_{2-x}O_4$ nanoparticles, the coercive field first diminished and then remained constant. Several parameters can influence the potency of ferrites to catalyze the decomposition of H_2O_2, such as the potential of transition metal ions with various oxidation states, the surface area of the catalyst, substitution ions, cation distribution, and the existence of oxygen vacancies on the surface of the catalyst. Note that the octahedral sites in spinel ferrites are the active positions of the catalyst, and the tetrahedral sites are catalytically inactive. The variations of concentration of H_2O_2 versus time for $CoM_xFe_{2-x}O_4$ (M = Cr^{2+}, Ni^{2+}, Cu^{2+}, and Zn^{2+}) annealed at 400°C are shown in Fig. 7.11 (Jauhar & Singhal, 2014).

The catalytic activity of cobalt ferrite is related to the existence of Co^{2+} ions in the octahedral sites and electron movement between Co^{2+}–Co^{3+} and Fe^{2+}–Fe^{3+} ion pairs. For $CoCr_xFe_{2-x}O_4$ nanoparticles, the maximum rate of catalysis was achieved for composition with $x = 1.0$. With an increase in the Cr^{3+} contents from 0.2 to 0.4, the rate of catalysis was reduced. However, with a further increase in Cr^{3+} concentration ($x > 0.4$), the rate of catalysis increased because of the enhancement of Co^{2+} ions in the octahedral B-sites. For $CoNi_xFe_{2-x}O_4$ nanoparticles, with a rise in the Ni^{2+} contents, the catalytic activity of the ferrites increased

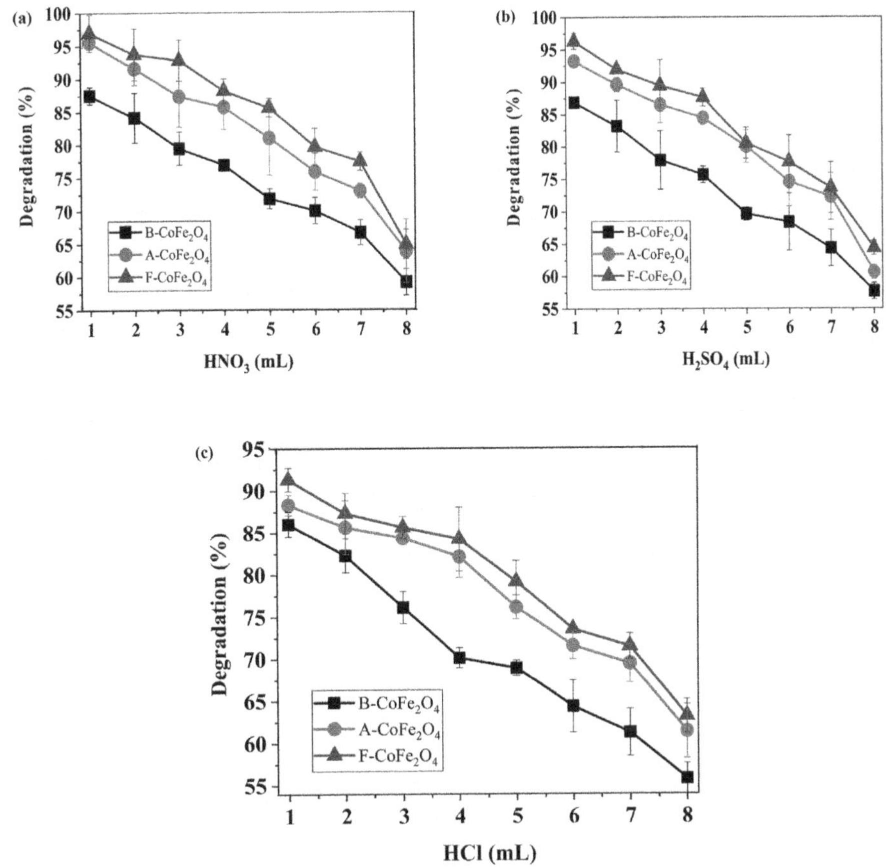

FIGURE 7.10

The effect of various acids (A) HNO_3, (B) H_2SO_4, and (C) HCl on the photodegradation performance of Alizarin Red S dye by cobalt ferrite nanoparticles.

From Ali, N., Ali, F., Said, A., Begum, T., Bilal, M., Rab, A., Sheikh, Z. A., Iqbal, H. M., & Ahmad, I. (2020). Characterization and deployment of surface-engineered cobalt ferrite nanospheres as photocatalyst for highly efficient remediation of alizarin red S dye from aqueous solution. Journal of Inorganic and Organometallic Polymers and Materials, *1–11. https://doi.org/10.1007/s10904-020-01654-y.*

because of the presence of oxygen vacancies on the surface of the ferrites. It was also seen that the catalytic activity rose with an increase in Cu^{2+} and Zn^{2+} contents, which is related to the presence of oxygen vacancies. Fig. 7.12 indicates the H_2O_2 decomposition of $CoCu_{0.4}Fe_{1.6}O_4$ annealed at different temperatures. It was found that the reaction obeys first-order kinetics. It was clear that with an increase in the annealing temperature, the decomposition rate of H_2O_2 in the presence of ferrite catalyst dropped. Indeed, with an increase in the annealing temperature, the particle

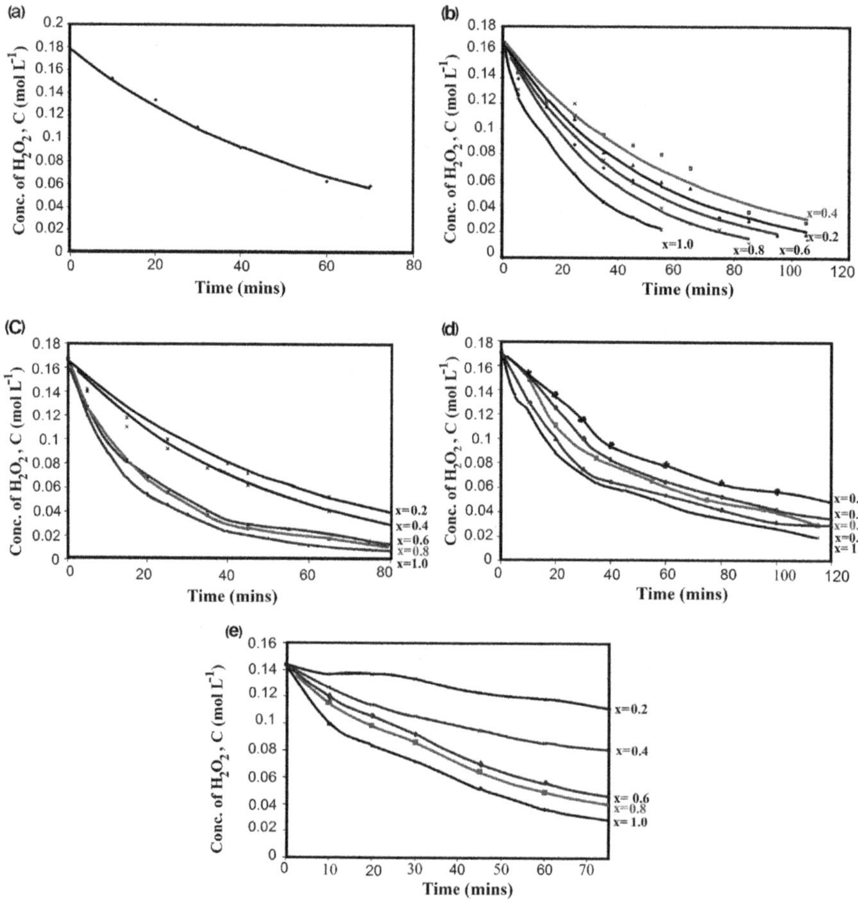

FIGURE 7.11

Variations of concentration of H_2O_2 versus time for (A) $CoFe_2O_4$, (B) $CoCr_xFe_{2-x}O_4$, (C) $CoNi_xFe_{2-x}O_4$, (D) $CoCu_xFe_{2-x}O_4$, and (E) $CoZn_xFe_{2-x}O_4$ annealed at 400°C.

From Jauhar, S., & Singhal, S. (2014). Substituted cobalt nanoferrites, $CoM_xFe_{2-x}O_4$ ($M = Cr^{3+}$, Ni^{2+}, Cu^{2+}, Zn^{2+}; $0.2 \le x \le 1.0$) as heterogeneous catalysts for modified Fenton's reaction. Ceramics International, *40(8), 11845–11855. https://doi.org/10.1016/j.ceramint.2014.04.019.*

size of the ferrite increased, and hence, the surface area of ferrites diminished. Thus, the catalytic activity of the ferrites decreased (Jauhar & Singhal, 2014).

7.4 Nanocrystalline Zn-ferrite photocatalyst

It was reported that the catalytic activity of ferrites depends on several factors, such as the surface area and morphology of ferrite nanoparticles. In a study, zinc ferrite

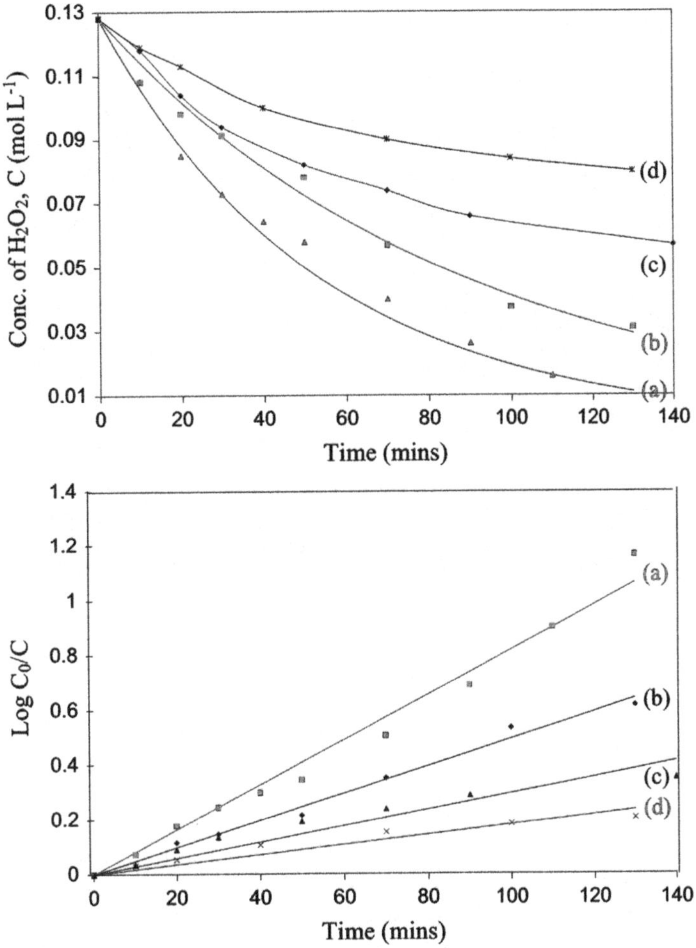

FIGURE 7.12

Hydrogen peroxide decomposition for $CoCu_{0.4}Fe_{1.6}O_4$ annealed at (A) 400°C, (B) 600°C, (C) 800°C, and (D) 1000°C.

From Jauhar, S., & Singhal, S. (2014). Substituted cobalt nanoferrites, $CoM_xFe_{2-x}O_4$ ($M = Cr^{3+}$, Ni^{2+}, Cu^{2+}, Zn^{2+}; $0.2 \leq x \leq 1.0$) as heterogeneous catalysts for modified Fenton's reaction. Ceramics International, 40*(8), 11845–11855. https://doi.org/10.1016/j.ceramint.2014.04.019.*

nanostructures were prepared via a hydrothermal method with different shapes (Dhiman et al., 2016). The relationship between the shape of ferrite and photocatalytic activity was defined as the following order of porous nanorods > nanoparticles > nanoflowers > hollow microspheres. The field emission scanning electron microscopy micrographs of zinc ferrite nanostructures with different shapes are shown in Fig. 7.13. The average diameter of nanorods was seen to be around 60 nm.

FIGURE 7.13

The field emission scanning electron microscopy micrographs of zinc ferrite nanostructures with different shapes (A) nanoparticles, (B) porous nanorods, (C) nanoflowers, and (D) hollow microspheres.

From Dhiman, M., Sharma, R., Kumar, V., & Singhal, S. (2016). Morphology controlled hydrothermal synthesis and photocatalytic properties of $ZnFe_2O_4$ nanostructures. Ceramics International, *42(11), 12594–12605. https://doi.org/10.1016/j.ceramint.2016.04.115.*

Many nanoparticles with an average size of 15–20 nm were complexed together to make a flower-like structure. The spherical characteristic of the ferrite with an average particle size of ~200 nm was seen. Interestingly, it was found from the vibrating sample magnetometer (VSM) loops that the coercivity and remanence are almost zero for all samples with different shapes. The samples were not saturated even at a magnetic field of 10 kOe, which exhibits superparamagnetism.

Fig. 7.14 displays the absorption spectra of zinc ferrite nanostructures with different shapes. Initially, the absorption increased sharply at ~400 nm, then absorption diminished from ~700 nm onwards. This confirms the wide visible light absorption range of the ferrite nanostructures. The bandgap values of nanoparticles, porous nanorods, nanoflowers, and hollow microspheres were around 1.82 eV, 1.97 eV, 1.81 eV, and 1.76 eV, respectively. The narrow bandgap values indicated the potential of all ferrite nanostructures to operate as visible light photocatalytic materials. The sample surface areas were calculated by the Brunauer–Emmett–Teller (BET) technique. The surface area values of nanoparticles, porous nanorods, nanoflowers, and hollow microspheres were 121.74, 138.43, 92.78, and 115.23 m^2/g, respectively. The large surface area values confirmed the potential of the prepared

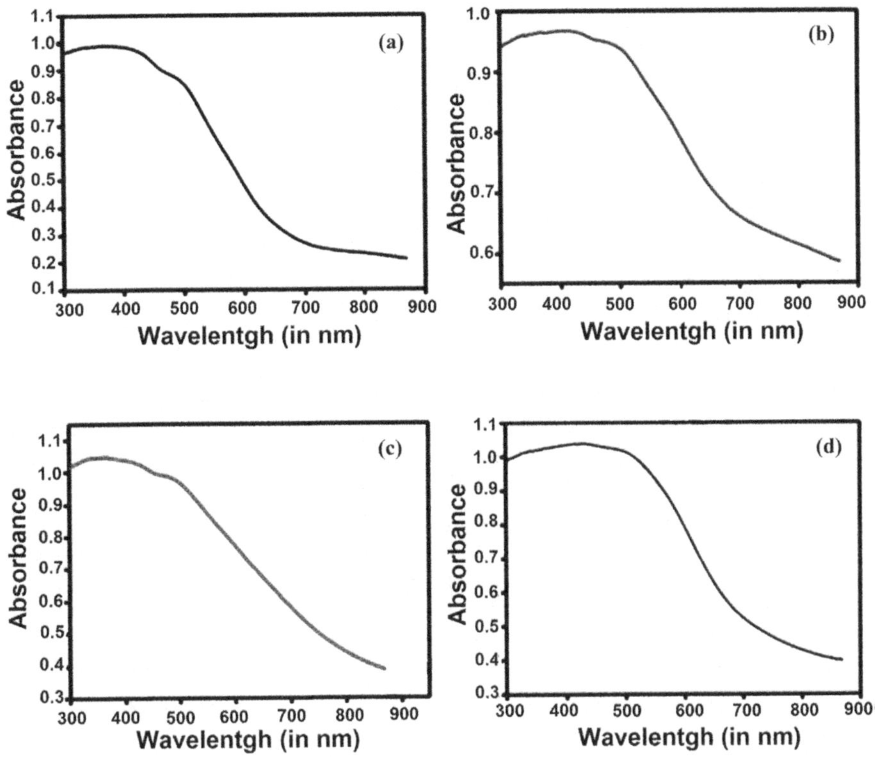

FIGURE 7.14

Absorption spectra of zinc ferrite nanostructures with different shapes (A) nanoparticles, (B) porous nanorods, (C) nanoflowers, and (D) hollow microspheres.

From Dhiman, M., Sharma, R., Kumar, V., & Singhal, S. (2016). Morphology controlled hydrothermal synthesis and photocatalytic properties of $ZnFe_2O_4$ nanostructures. Ceramics International, *42(11), 12594–12605. https://doi.org/10.1016/j.ceramint.2016.04.115.*

ferrite nanostructures as suitable catalysts. The absorption spectra of safranine-O (cationic) dye solution using $ZnFe_2O_4$ nanostructures with different shapes are shown in Fig. 7.15. The maximum absorption of (SO) and RBY became visible at 274 and 529 nm, respectively. The absorption diminished with prolongation of the irradiation time (Dhiman et al., 2016).

Fig. 7.16 reveals the variations of degradation percentage versus time for safranine-O and Remazol Brilliant Yellow of $ZnFe_2O_4$ nanostructures with different shapes. It was clear that the degradation of ferrite nanostructures was greater than 90% for both dyes. The catalytic activity followed the order: $ZnFe_2O_4$ porous nanorods > nanoparticles > nanoflowers > hollow microspheres. In the degradation process, •OH is the main active species. The generation of •OH is due to the production of an e^-/h^+ pair on the catalyst surface. The catalyst with a larger surface area has more active sites, producing a larger number of hydroxyl radicals. Thus, the rate of

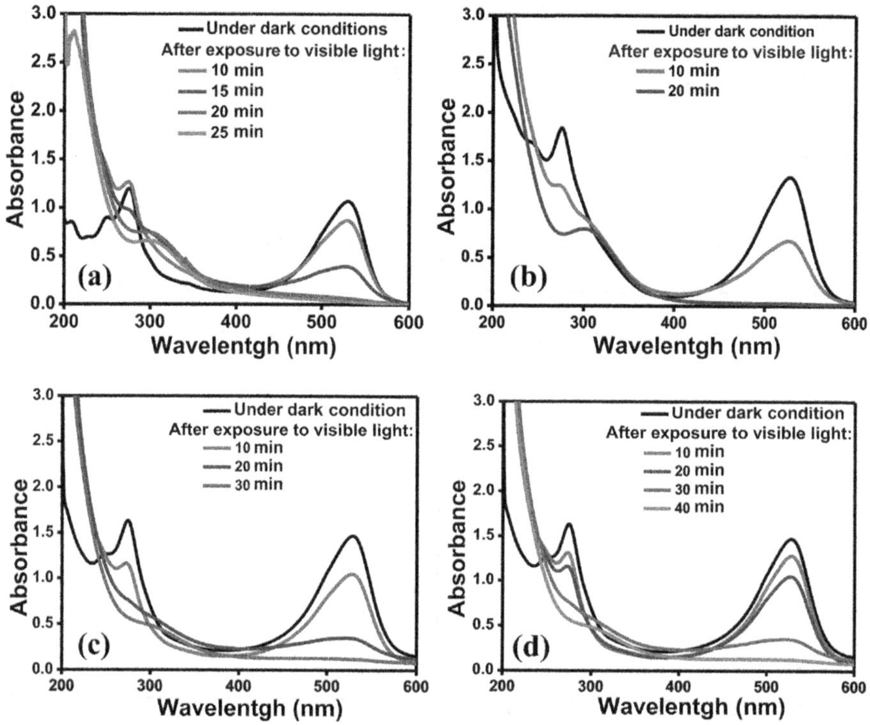

FIGURE 7.15

Ultraviolet–visible spectra of safranine-O dye solution using $ZnFe_2O_4$ nanostructures with different shapes (A) nanoparticles, (B) porous nanorods, (C) nanoflowers, and (D) hollow microspheres.

From Dhiman, M., Sharma, R., Kumar, V., & Singhal, S. (2016). Morphology controlled hydrothermal synthesis and photocatalytic properties of $ZnFe_2O_4$ nanostructures. Ceramics International, *42(11), 12594–12605. https://doi.org/10.1016/j.ceramint.2016.04.115.*

degradation depends on the rate of production of hydroxyl radicals and the surface area of the catalyst. The surface areas of the catalysts change with the alteration in the morphologies. The BET analyzer revealed that the highest surface area was obtained for the porous nanorods; hence, the maximum photocatalytic efficiency was achieved for the porous zinc ferrite nanorods. In addition, the photoactivity can decline by a smaller bandgap, which is attributed to the increased recombination rate between photogenerated e^-/h^+ pairs. Thus, the porous nanorods with a slow recombination rate of the excited electrons show higher photocatalytic activity. Hence, the maximum photocatalytic activity of nanorods could be related to the higher surface area, greater bandgap, and better delocalization of charged species along the rods (Dhiman et al., 2016; Gregor et al., 2010; Kim & Kang, 2012; Sharma et al., 2015; Singh & Pal, 2013).

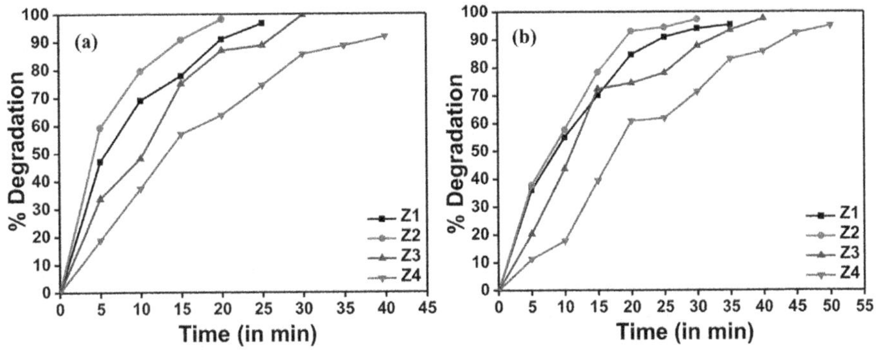

FIGURE 7.16

Variations of degradation percentage versus time for (A) safranine-O (B) Remazol Brilliant Yellow of $ZnFe_2O_4$ nanostructures with different shapes (Z1) nanoparticles, (Z2) porous nanorods, (Z3) nanoflowers, and (Z4) hollow microspheres.

From Dhiman, M., Sharma, R., Kumar, V., & Singhal, S. (2016). Morphology controlled hydrothermal synthesis and photocatalytic properties of $ZnFe_2O_4$ nanostructures. Ceramics International, *42(11), 12594–12605. https://doi.org/10.1016/j.ceramint.2016.04.115.*

Elsewhere, the photocatalytic performance of the zinc ferrite nanoparticles was evaluated for EBT dye under unfiltered sunlight (Ali, Ali, Said et al. 2020). It was observed that the photocatalytic rates of EBT dye for nanoparticles annealed at 400°C was 91%. Degradation kinetics are confirmed in pseudo-first-order models. The bandgaps in the acidic and basic media obtained from absorption spectra were 2.47 and 2.7 eV, respectively. Thus, the zinc ferrite nanostructures are suitable photocatalysts in a tolerable environment for degrading organic pollutants. It was noted that the degradation of toxic pollutants from wastewater occurs through photocatalytic degradation. Fig. 7.17 shows the influence of contact time on the photocatalytic performance of the zinc ferrite nanoparticles prepared via a co-precipitation method. It was clear that the degradation performance of ferrite nanoparticles was quick initially, as abundant molecules of the dye are accessible for photocatalytic degradation in the early stages. However, the concentration of the dye molecules decreased over time, and thus the rate of degradation became limited. It was observed that the annealed nanoparticles exhibited a higher degradation performance than blank ferrite, which is related to the smaller particle size and higher surface-to-volume ratio. For functionalization, the zinc ferrite nanoparticles were treated in two steps. First, the nanoparticles were immersed in a specific molar ratio of hydrochloric acid, followed by subsequent filtration and washing, and then 3-APTES was added dropwise. The final product was then ground mechanically to obtain the functionalized nanoparticles. Functionalized nanoparticles have the highest degradation because they have the smallest relative size (Ali, Ali, Sheikh et al., 2020; El-Salamony et al., 2018; Kansal et al., 2013).

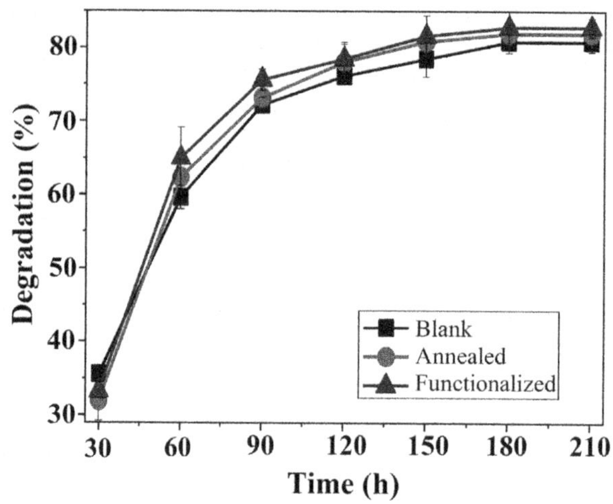

FIGURE 7.17

The influence of contact time on the photocatalytic performance of Eriochrome Black T of the zinc ferrite nanoparticles.

From Ali, N., Ali, F., Sheikh, Z. A., Bilal, M., & Ahmad, I. (2020). Photocatalytic performance of zinc ferrite magnetic nanostructures for efficient Eriochrome Black T degradation from the aqueous environment under unfiltered sunlight. Water, Air, & Soil Pollution, 231*(2), 59. https://doi.org/10.1007/s11270-020-4437-x.*

Likewise, photocatalytic dye degradation is controlled by pH. The pH has an important effect on the dissociation and adsorption of the molecules. Indeed, the variation of pH can change the dye structure. It was observed that with the shift of the medium from acidic to the basic regime, the rate of degradation rose, which is related to the higher concentration of OH^- ions in the basic regime (Bouanimba et al., 2011). Fig. 7.18 demonstrates the variations of photocatalytic degradation of the EBT with pH. It was seen that the rate of degradation increased from 37% at pH $=$ 5 to 92% at pH $=$ 14. Thus, the photocatalytic degradation rate increased as the nature of the medium changed from the acidic to the basic regime.

Fig. 7.19 exhibits the influence of the adsorbent dose on the photocatalytic degradation of the EBT of the zinc ferrite nanoparticles. The photocatalytic degradation rate of EBT increased with a rise in the content of catalyst or adsorbent dose. Indeed, with an increase in the quantity of catalyst, the surface area and thus photocatalytic degradation rate were enhanced. Interestingly, with a further increase in the adsorbent dose, the degradation rate did not grow significantly. Indeed, the higher amounts of adsorbent dose can increase the agglomeration; thus, the accessible surface areas of the adsorbent site are reduced (Sakthivel et al., 2003; Ullah et al., 2018). The maximum degradation was achieved for 25 mg of zinc ferrite.

In another study, zinc ferrite nanoparticles were prepared by an aqueous seed extract of *Piper nigrum* as a bio-reducing and stabilizing agent (Din et al., 2020). Fig. 7.20 presents the UV–visible absorption spectra of MB dye using *P. nigrum*

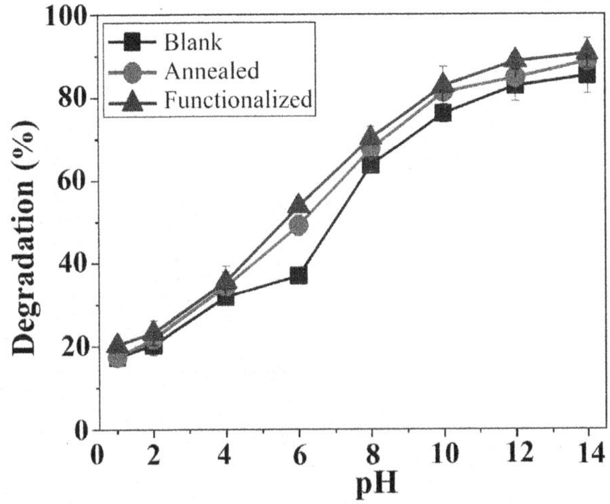

FIGURE 7.18

Variations of photocatalytic degradation of Eriochrome Black T.

From Ali, N., Ali, F., Sheikh, Z. A., Bilal, M., & Ahmad, I. (2020). Photocatalytic performance of zinc ferrite magnetic nanostructures for efficient Eriochrome Black T degradation from the aqueous environment under unfiltered sunlight. Water, Air, & Soil Pollution, 231*(2), 59. https://doi.org/10.1007/s11270-020-4437-x.*

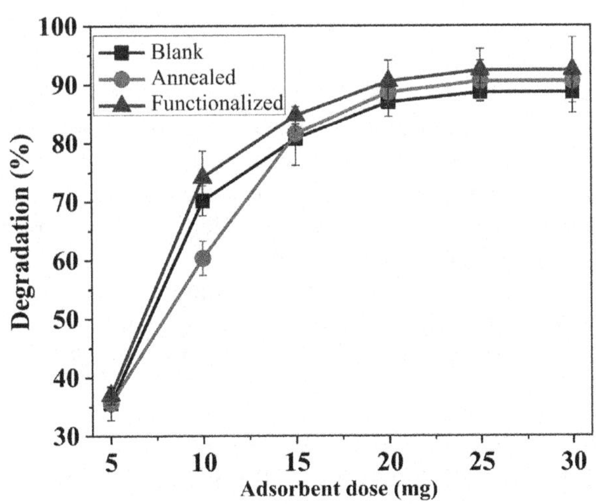

FIGURE 7.19

The influence of the adsorbent dose on the photocatalytic performance of Eriochrome Black T.

From Ali, N., Ali, F., Sheikh, Z. A., Bilal, M., & Ahmad, I. (2020). Photocatalytic performance of zinc ferrite magnetic nanostructures for efficient Eriochrome Black T degradation from the aqueous environment under unfiltered sunlight. Water, Air, & Soil Pollution, 231*(2), 59. https://doi.org/10.1007/s11270-020-4437-x.*

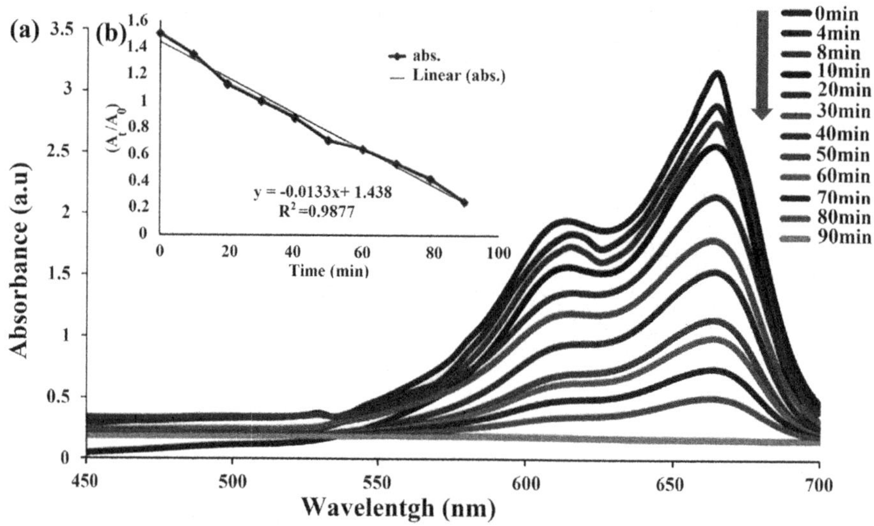

FIGURE 7.20

(A) Absorption spectra of methylene blue dye using *Piper nigrum* stabilized zinc ferrite nanoparticles (reaction conditions: [ZF-NPs] = 5 mg; [MB] = 15 ppm; sunlight; pH = neutral and temperature = 20 ± 2°C), and (B) linear plot between ln [MB] versus time.

From Din, M. I., Jabbar, S., Najeeb, J., Khalid, R., Ghaffar, T., Arshad, M., Khan, S. A., & Ali, S. (2020). Green synthesis of zinc ferrite nanoparticles for photocatalysis of methylene blue. International Journal of Phytoremediation, *22(13), 1440–1447. https://doi.org/10.1080/15226514.2020.1781783.*

stabilized zinc ferrite nanoparticles. It was observed that the surface plasmon resonance peak in the spectra did not change because of the presence of very small amounts of catalyst. The photocatalytic decolorization of MB using *P. nigrum* stabilized zinc ferrite nanoparticles was excellent as the photocatalyst creates a reaction that can cross the activation energy barrier. In addition, it was observed that the reaction rate could be affected by the initial concentration of dye. The slope of the logarithmic curve of reactant concentration versus time can indicate the reaction rate. Thus, *P. nigrum* stabilized zinc ferrite nanoparticles were efficient, cost-executable, and suitable for MB decolorization (Patil et al., 2018).

7.5 Nanocrystalline Ni–Zn-ferrite photocatalyst

The application of $Zn_xNi_{1-x}Fe_2O_4$ ferrite was also evaluated in the photodegradation of MB (Sharma & Singhal, 2013). X-ray diffraction patterns exhibited the formation of a cubic spinel structure without any secondary phases. It was reported that the saturation magnetization increased with a rise in the Zn^{2+} content up to $x = 0.4$, and then it diminished. On the other hand, the coercivity was found to drop with an increase in the Zn^{2+} concentration. The decline in coercivity is related to the reduction of the anisotropy constant. Fig. 7.21 displays the variations of drift mobility versus temperature for $Zn_xNi_{1-x}Fe_2O_4$ ferrite with different compositions. It is clear that with a rise in temperature, mobility increases because the energy for hopping conductivity of charge carriers grows from one site to another. However, the drift mobility declined with increasing Zn^{2+} concentration and was inversely related to the electrical resistivity.

Fig. 7.22 shows the absorption spectra of MB solution using $Zn_xNi_{1-x}Fe_2O_4$ ($x =$ 0) ferrite. It is observed that with increasing irradiation time, the absorption peak intensity decreases and reaches zero after 200 min, which indicates the complete degradation of MB after this time. In the spectrum of MB, four peaks were seen at 246, 292, 617, and 663 nm. In addition, the degradation of MB occurred within a shorter time with increasing the Zn^{2+} concentration. The dye degradation was accelerated because of the greater bandgap of $NiFe_2O_4$ (2.19 eV) compared with $ZnFe_2O_4$ (1.92 eV).

7.6 Nanocrystalline Co–Zn–ferrite photocatalyst

Evaluation of the photocatalytic activity of $Co_{0.6}Zn_{0.4}Mn_xFe_{2-x}O_4$ ferrites through MO dye degradation showed that MO dye degradation increased with a rise in Mn^{3+}

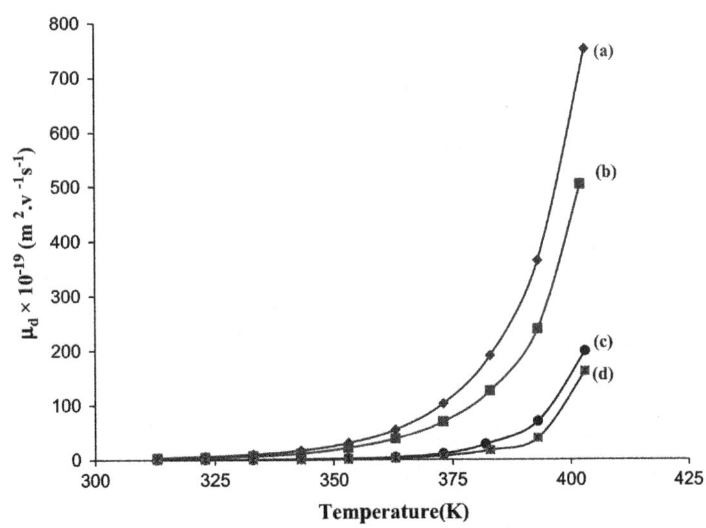

FIGURE 7.21

Variations of drift mobility versus temperature for $Zn_xNi_{1-x}Fe_2O_4$ ferrite with composition of (A) $x = 0.2$, (B) $x = 0.4$, (C) $x = 0.6$, and (D) $x = 0.8$ annealed at 1000°C.

From Sharma, R., & Singhal, S. (2013). Structural, magnetic and electrical properties of zinc doped nickel ferrite and their application in photo catalytic degradation of methylene blue. Physica B: Condensed Matter, *414, 83–90. https://doi.org/10.1016/j.physb.2013.01.015.*

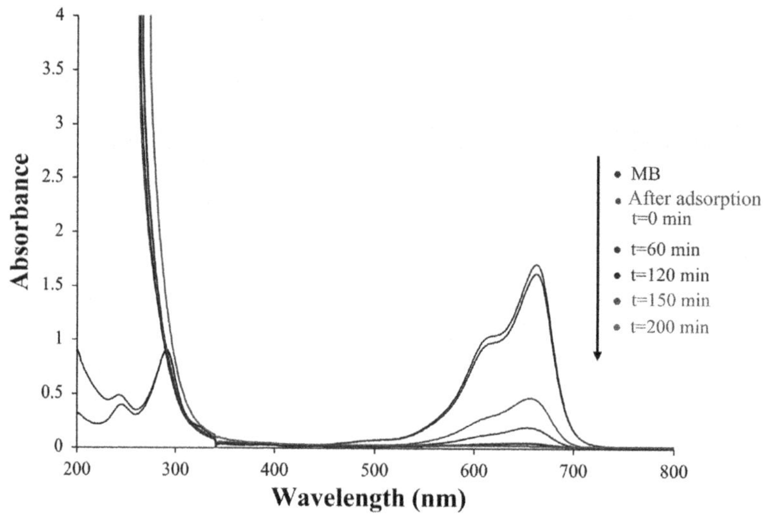

FIGURE 7.22

Absorption spectra of methylene blue solution using $Zn_xNi_{1-x}Fe_2O_4$ ($x = 0$) ferrite annealed at 1000°C.

From Sharma, R., & Singhal, S. (2013). Structural, magnetic and electrical properties of zinc doped nickel ferrite and their application in photo catalytic degradation of methylene blue. Physica B: Condensed Matter, 414, *83–90. https://doi.org/10.1016/j.physb.2013.01.015.*

ion content (Bhukal et al., 2014). The saturation magnetization decreased, which is related to the lower magnetic moment of Mn^{3+} ($4\mu_B$). In addition, the coercivity of $Co_{0.6}Zn_{0.4}Mn_xFe_{2-x}O_4$ nanoparticles decreased from 165 Oe for $x = 0.2$ to 100 Oe for $x = 0.8$, and then increased to 190 Oe for $x = 1.0$. The variations of drift mobility of $Co_{0.6}Zn_{0.4}Mn_xFe_{2-x}O_4$ nanoparticles are seen in Fig. 7.23. The drift mobility of all samples increased with elevated temperature due to the semiconductor nature of nanoferrites and the hopping conductivity of charge carriers from one site to another site. The plots of ln C_0/C_t versus irradiation time for photocatalytic degradation of MO are shown in Fig. 7.24. It was found that MO photodegradation obeys the first-order reaction kinetics, which can be described according to the Langmuir-Hinshelwood equation.

Fig. 7.25 reveals the absorption spectra of MO solution using $Co_{0.6}Zn_{0.4}Mn_xFe_{2-x}O_4$ nanoparticles. It was seen that the photocatalytic activity of ferrite nanoparticles was enhanced with an increase in Mn^{3+} content, which is attributed to the octahedral occupancy of manganese ions. Likewise, the reduction potential of Mn^{3+} (1.51 V) is higher than that of the Fe^{3+} (0.77 V).

In separate work, $Co_{0.6}Zn_{0.4}Cu_{0.2}M_xFe_{1.8-x}O_4$ (M = Mn^{3+}, Co^{2+}, Ni^{2+}, and Zn^{2+}) nanoferrites were synthesized via a sol–gel auto-combustion method (Bhukal et al., 2016). The XRD patterns of the samples displayed the formation of a single-phase cubic spinel structure. It was found that the saturation magnetization and

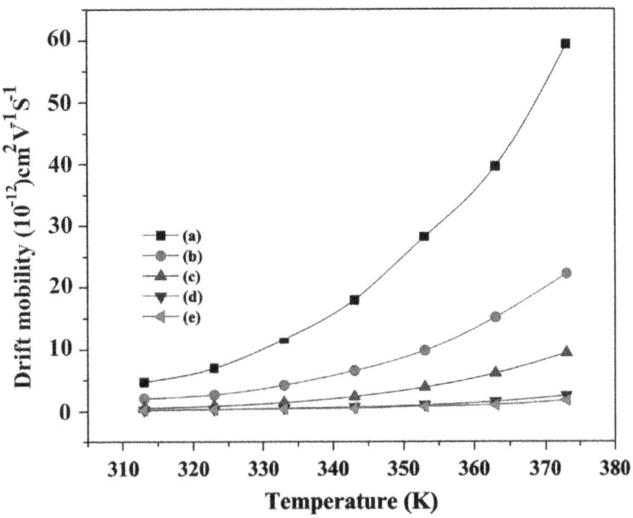

FIGURE 7.23

Variations of drift mobility with temperature for $Co_{0.6}Zn_{0.4}Mn_xFe_{2-x}O_4$ nanoparticles annealed at 1000°C with composition of (A) $x = 0.2$, (B) $x = 0.4$, (C) $x = 0.6$, (D) $x = 0.8$, and (E) $x = 1.0$.

From Bhukal, S., Bansal, S., & Singhal, S. (2014). Magnetic Mn substituted cobalt zinc ferrite systems: structural, electrical and magnetic properties and their role in photocatalytic degradation of methyl orange azo dye. Physica B: Condensed Matter, 445, *48–55. https://doi.org/10.1016/j.physb.2014.03.088.*

coercivity diminished with the substitution of all metal ions in Co–Zn–Cu ferrites. The ferrite with the composition of $Co_{0.6}Zn_{0.4}Cu_{0.2}Zn_xFe_{1.8-x}O_4$ had the lowest coercivity and anisotropy constant. Fig. 7.26 demonstrates the absorption spectra of MO solution using $Co_{0.6}Zn_{0.4}Cu_{0.2}Mn_xFe_{1.8-x}O_4$ ferrites. Interestingly, $Co_{0.6}Zn_{0.4}Cu_{0.2}Mn_xFe_{1.8-x}O_4$ ferrites exhibited the best catalytic activity, with the dye completely degraded in 40 min. The catalytic activity of $Co_{0.6}Zn_{0.4}Cu_{0.2}Mn_xFe_{1.8-x}O_4$ ferrites increased with a rise in manganese substitution due to the ability of manganese to take part in the electron movement process. The catalytic activity of $Co_{0.6}Zn_{0.4}Cu_{0.2}Zn_xFe_{1.8-x}O_4$ and $Co_{0.6}Zn_{0.4}Cu_{0.2}Co_xFe_{1.8-x}O_4$ ferrites rose with an increase in substitution ions concentration. Indeed, substituting trivalent iron ions with divalent zinc/cobalt ions can lead to the development of oxygen vacancies on the catalyst surface; thus, degradation capacity increases. For $Co_{0.6}Zn_{0.4}Cu_{0.2}Ni_xFe_{1.8-x}O_4$ ferrites, the best photocatalytic activity was achieved for $x = 0.2$, where Fe:Ni ratio was maximum. With a further increase in Ni^{2+} concentration, the photocatalytic activity of $Co_{0.6}Zn_{0.4}Cu_{0.2}Ni_xFe_{1.8-x}O_4$ decreased. Fig. 7.27 reveals the variations of rate constant with metal ion content in $Co_{0.6}Zn_{0.4}Cu_{0.2}M_xFe_{1.8-x}O_4$ ferrites for photocatalytic degradation of MO. It was clear that the rate constants increased with a rise in all-metal ion contents except for nickel. The best photocatalytic activity was obtained for

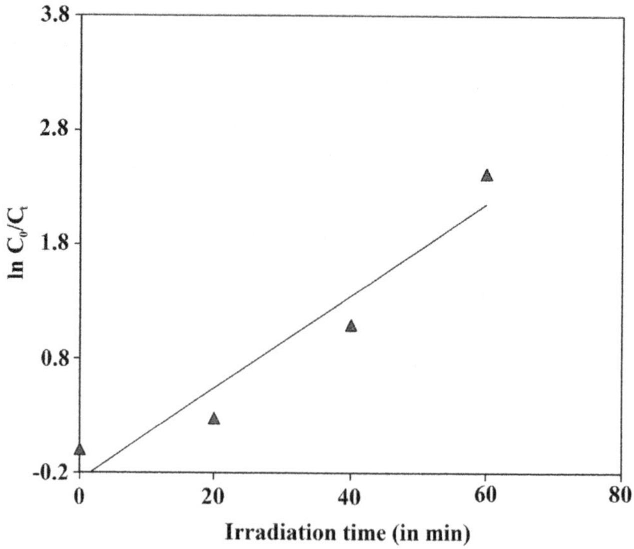

FIGURE 7.24

Plots of ln C_0/C_t versus irradiation time for photocatalytic degradation of methyl orange.

From Bhukal, S., Bansal, S., & Singhal, S. (2014). Magnetic Mn substituted cobalt zinc ferrite systems: structural, electrical and magnetic properties and their role in photocatalytic degradation of methyl orange azo dye. Physica B: Condensed Matter, 445, *48–55. https://doi.org/10.1016/j.physb.2014.03.088.*

$Co_{0.6}Zn_{0.4}Cu_{0.2}Mn_xFe_{1.8-x}O_4$ ferrites. The recyclability of $Co_{0.6}Zn_{0.4}Cu_{0.2}M_xFe_{1.8-x}O_4$ ferrites for the degradation of MO for four cycles is shown in Fig. 7.28. The recyclability experiment was performed to test the stability and long-term use of ferrites as heterogeneous photo-Fenton catalysts. It was observed that the activity of samples slightly diminished because of the loss occurring during the recovery process. For all ferrites, no reduction in the efficiency was seen after several consecutive rounds of reaction.

Elsewhere, the usage of $Co_{0.6}Zn_{0.4}Cu_xFe_{2-x}O_4$ ferrite was studied in the photodegradation of MO (Bhukal & Singhal, 2014). X-ray diffraction patterns demonstrated the formation of a cubic spinel structure without any impurities. It was observed that with a rise in the copper concentration, the saturation magnetization increased up to $x = 0.4$ due to the growth of B–B exchange interaction and then decreased because of the spin canting. The plots of ln C_t/C_0 versus irradiation time with and without H_2O_2 are shown in Fig. 7.29. It is clear that the dye was fully degraded after 90 min in the existence of H_2O_2; however, no degradation occurred when the reaction was performed without H_2O_2.

The absorption spectra of MO solution using $Co_{0.6}Zn_{0.4}Cu_xFe_{2-x}O_4$ ferrite annealed at 1000°C with different compositions are shown in Fig. 7.30. The catalytic performance of the samples has been defined within the visible light range, and it was revealed that the catalytic performance of ferrite was enhanced with an increase

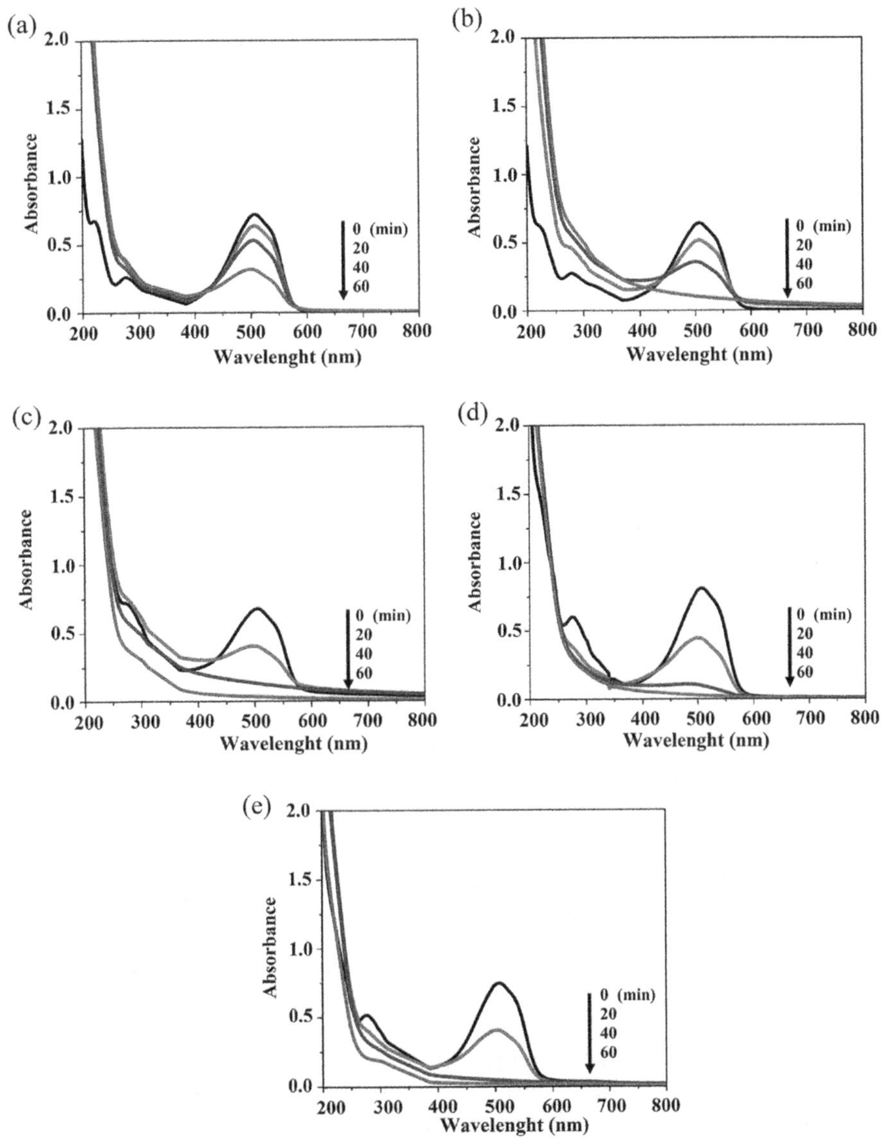

FIGURE 7.25

Absorption spectra of methyl orange solution using $Co_{0.6}Zn_{0.4}Mn_xFe_{2-x}O_4$ nanoparticles with composition of (A) $x = 0.2$, (B) $x = 0.4$, (C) $x = 0.6$, (D) $x = 0.8$, and (E) $x = 1.0$. *From Bhukal, S., Bansal, S., & Singhal, S. (2014). Magnetic Mn substituted cobalt zinc ferrite systems: structural, electrical and magnetic properties and their role in photocatalytic degradation of methyl orange azo dye.* Physica B: Condensed Matter, 445, *48–55. https://doi.org/10.1016/j.physb.2014.03.088.*

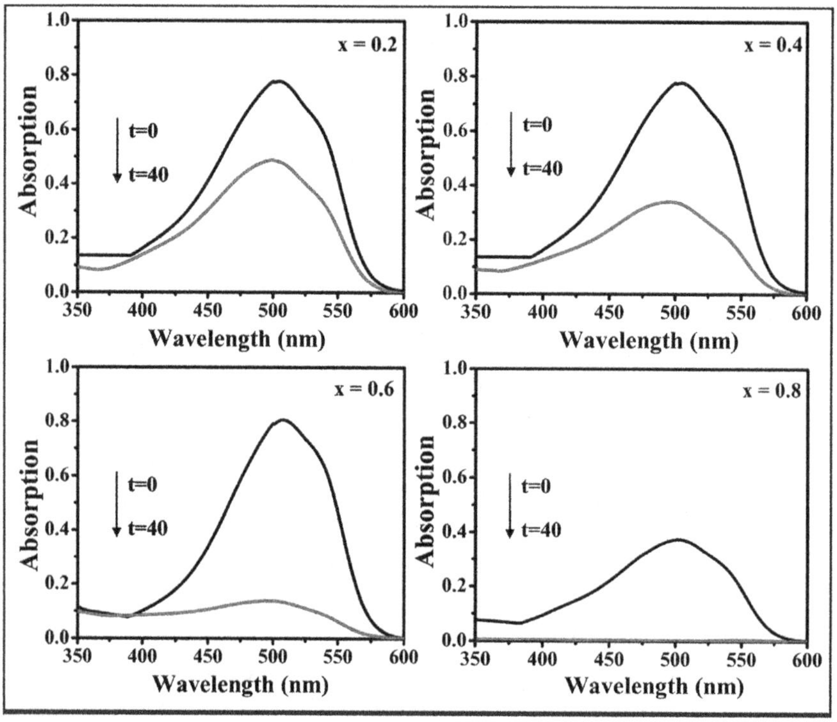

FIGURE 7.26

Absorption spectra of methyl orange solution using $Co_{0.6}Zn_{0.4}Cu_{0.2}Mn_xFe_{1.8-x}O_4$ annealed at 1000°C.

From Bhukal, S., Dhiman, M., Bansal, S., Tripathi, M. K., & Singhal, S. (2016). Substituted Co–Cu–Zn nanoferrites: synthesis, fundamental and redox catalytic properties for the degradation of methyl orange. RSC Advances, 6*(2), 1360–1375. https://doi.org/10.1039/C5RA22561B.*

in copper content. The variations of percentage degradation of $Co_{0.6}Zn_{0.4}Cu_xFe_{2-x}O_4$ ferrite annealed at 1000°C versus concentration of Cu^{2+} are shown in Fig. 7.31. It was clear that the degradation of MO increased with elevated copper concentration. The remarkable enhancement in MO photodegradation is related to the B-site occupancy of copper ions. By employing an external magnetic field, this catalyst is also well separable from water media; hence, it can be used as a suitable catalyst for the correction of textile wastewater.

7.7 Nanocrystalline Mn-rich ferrite photocatalyst

Lanthanum substituted Mn-ferrite nanoparticles ($La_xMnFe_{2-x}O_4$) were prepared by a reverse-micelle method (Baig et al., 2020). It was revealed from the XRD pattern that the substitution of La^{3+} ion in ferrites would lead to the formation of a

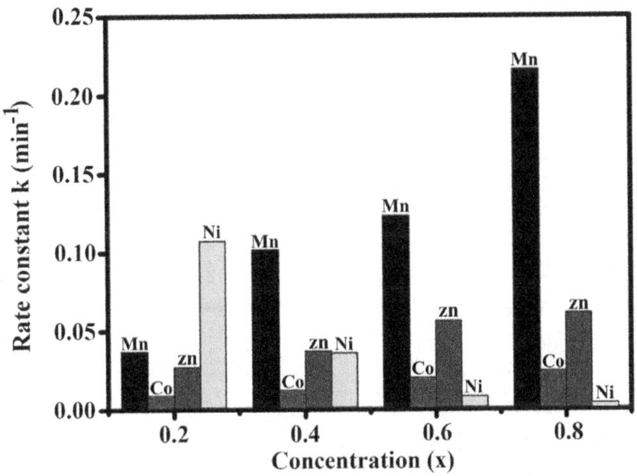

FIGURE 7.27

Plots of rate constant with metal ion content in $Co_{0.6}Zn_{0.4}Cu_{0.2}M_xFe_{1.8-x}O_4$ ferrites for photocatalytic degradation of methyl orange.

From Bhukal, S., Dhiman, M., Bansal, S., Tripathi, M. K., & Singhal, S. (2016). Substituted Co–Cu–Zn nanoferrites: synthesis, fundamental and redox catalytic properties for the degradation of methyl orange. RSC Advances, *6(2), 1360–1375. https://doi.org/10.1039/C5RA22561B.*

secondary phase. For determining the bandgap value, the variations of $(\alpha h\nu)^{1/n}$ versus photon energy ($h\nu$) were investigated for many n values ($n = 1/2, 3/2, 2$), known as a Tauc plot (Tauc, 2012). It was observed that the value of the redshift further increased with the substitution of lanthanum ions in manganese ferrite. The value of the energy bandgap for pure manganese ferrite was 2.64 eV. However, a large redshift of 2.40 eV was seen for the lanthanum substituted ferrite nanoparticles. The larger energy bandgap of pure manganese ferrite would cause lower photocatalytic performance. Indeed, the large energy bandgap of catalysts would lead to the absorption of shorter wavelengths and the production of electron-hole pairs. The substitution of lanthanum rare-earth ions with larger ionic radius in manganese ferrites generated the metastable La-4f energy levels, which are very close to the lower edge of the conduction energy bandgap, thereby increasing the absorption and degradation efficiency. The metastable La-4f energy states can decrease the bandgap between the conduction band and the valence band, which would lead to the enhancement of the photocatalytic activity (Kislov et al., 2008; Li et al., 2003). The reduced bandgap of the lanthanum-substituted ferrite nanoparticles can be attributed to the overlapping of wave functions of the electrons, which increased with a rise in the amount of rare-earth ions (Bahmanrokh et al., 2020; Jain et al., 1991). The mechanism of crystal violet dye degradation using $La_xMnFe_{2-x}O_4$ ferrite nanoparticles under visible light irradiation is shown in Fig. 7.32. The illumination of ferrites by light would result in electrons jumping from the valence bands to the conduction band generating photoinduced holes in valence bands. The photoinduced

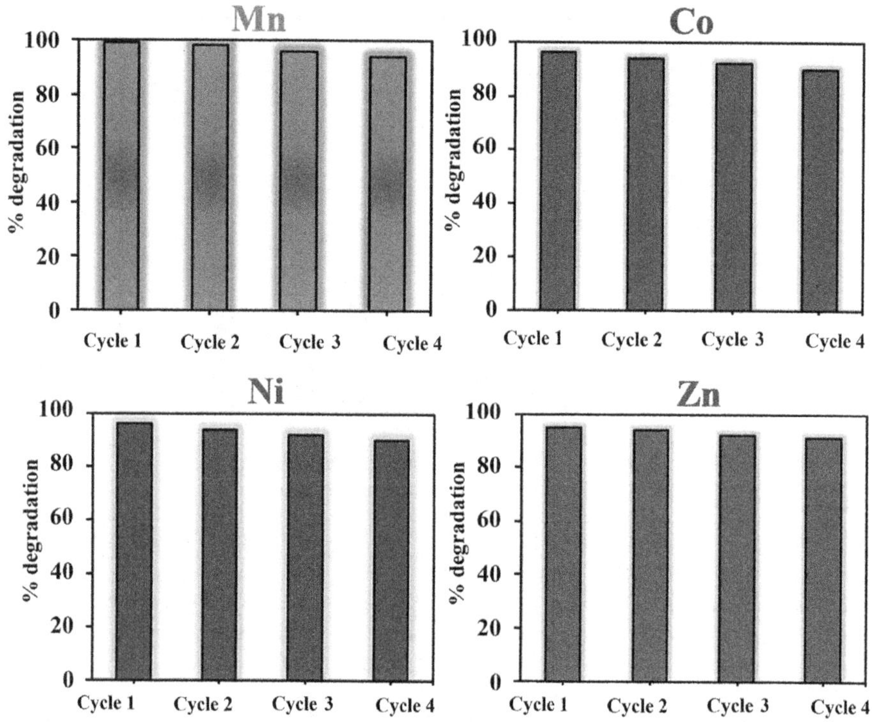

FIGURE 7.28

Recyclability of $Co_{0.6}Zn_{0.4}Cu_{0.2}M_xFe_{1.8-x}O_4$ ferrites for the degradation of methyl orange for four cycles.

From Bhukal, S., Dhiman, M., Bansal, S., Tripathi, M. K., & Singhal, S. (2016). Substituted Co–Cu–Zn nanoferrites: synthesis, fundamental and redox catalytic properties for the degradation of methyl orange. RSC Advances, 6*(2), 1360–1375. https://doi.org/10.1039/C5RA22561B.*

holes reacted with water and prepared highly reactive OH• radical species. The OH• radical species functions as a good oxidant in the dye degradation performance.

Fig. 7.33 indicates the absorption spectra of crystal violet dye using $MnFe_2O_4$ and $La_xMnFe_{2-x}O_4$ ferrite nanoparticles. The system was stirred in the absence of light to reach the absorption–desorption equilibrium. Then the reaction container was subjected to visible light irradiation. After irradiation at regular intervals, the supernatant without photocatalyst materials was achieved. The kinetic study and degradation rate of $MnFe_2O_4$ and $La_xMnFe_{2-x}O_4$ ferrite nanoparticles are also observed in Fig. 7.34. It was found that the pure manganese ferrite nanoparticles revealed 30% degradation. The degradation efficiency increased to 95% with the substitution of lanthanum ions in manganese ferrite nanoparticles. In addition, the degradation rate constant of pure manganese ferrite is 0.002 min^{-1}, which increased to 0.019 min^{-1} with the substitution of lanthanum ions in ferrites. Thus, the

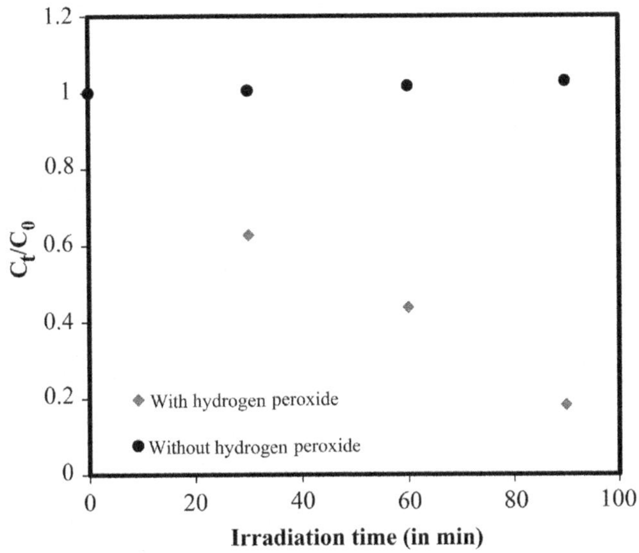

FIGURE 7.29

Plots of ln C_t/C_0 versus irradiation time with and without H_2O_2.

From Bhukal, S., & Singhal, S. (2014). Magnetically separable copper substituted cobalt–zinc nano-ferrite photocatalyst with enhanced photocatalytic activity. Materials Science in Semiconductor Processing, 26, *467–476. https://doi.org/10.1016/j.mssp.2014.05.023.*

lanthanum-substituted manganese ferrite nanoparticles can be used as a useful photocatalyst for environmental reparation under visible light.

In other research, the $Mn_{0.5}Zn_{0.5-x}Mg_xFe_2O_4$ ferrite nanoparticles were prepared via the sol–gel method (Abdel Maksoud et al., 2020). The absorption spectra of chloramine-T and RhB dyes of $Mn_{0.5}Zn_{0.125}Mg_{0.375}Fe_2O_4$ ferrite nanoparticles indicated that the maximum values of absorption of chloramine-T and RhB were achieved at wavelengths of 225 and 553 nm, respectively. It was found that with an increase in irradiation time, the maximum value of absorption of chloramine-T and RhB diminished. The photocatalytic degradation of Chloramine-T and RhB of $Mn_{0.5}Zn_{0.5-x}Mg_xFe_2O_4$ ferrite nanoparticles is shown in Fig. 7.35. The figures indicated that the photocatalytic activity of ferrite nanoparticles increased with the rise in the Mg content. The highest photocatalytic activity belonged to the $Mn_{0.5}Zn_{0.125}Mg_{0.375}Fe_2O_4$ ferrite nanoparticles. The photocatalytic degradation of chloramine-T and RhB of $Mn_{0.5}Zn_{0.125}Mg_{0.375}Fe_2O_4$ ferrite nanoparticles was 47.5% and 25.5%, respectively, within the irradiation time of 40 min. With an increase in the irradiation time to 80 min, the photocatalytic degradation of chloramine-T and RhB of $Mn_{0.5}Zn_{0.125}Mg_{0.375}Fe_2O_4$ ferrite nanoparticles increased to 61% and 39%, respectively. Thus, the photocatalytic activity of $Mn_{0.5}Zn_{0.125}Mg_{0.375}Fe_2O_4$ ferrite nanoparticles was superior to that of

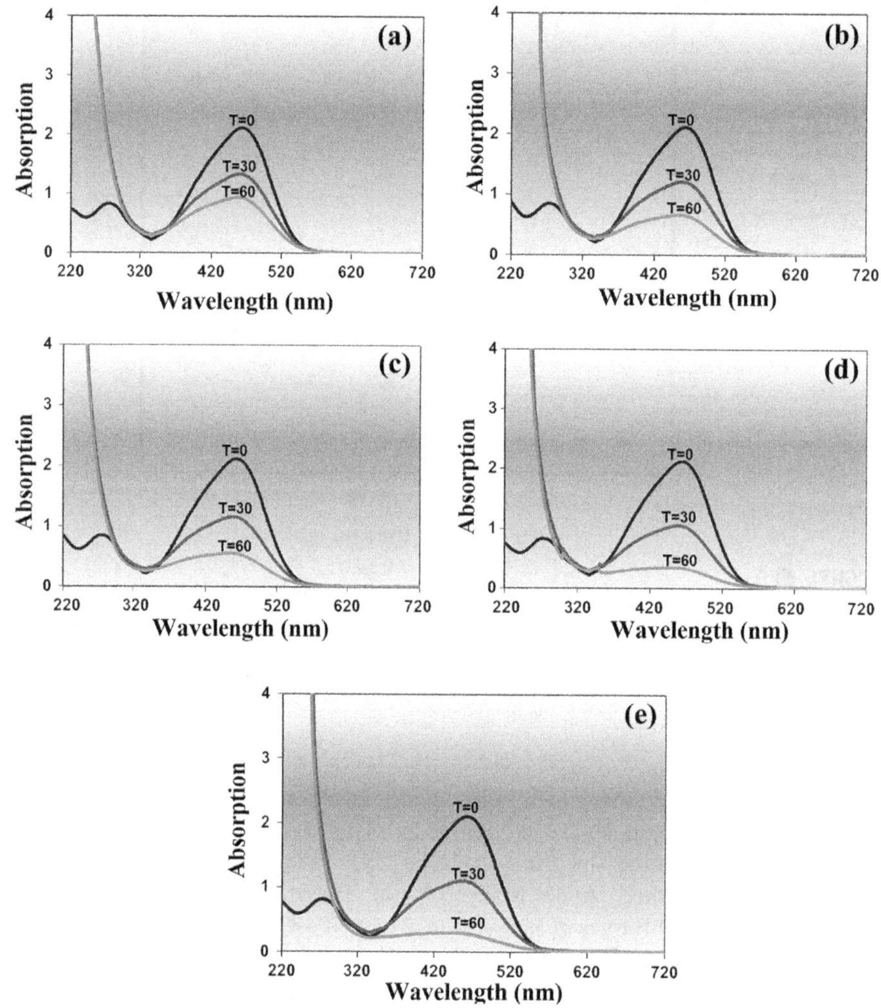

FIGURE 7.30

Absorption spectra of methyl orange solution using $Co_{0.6}Zn_{0.4}Cu_xFe_{2-x}O_4$ ferrite annealed at 1000°C with composition of (A) $x = 0.2$, (B) $x = 0.4$, (C) $x = 0.6$, (D) $x = 0.8$, and (E) $x = 1.0$.

From Bhukal, S., & Singhal, S. (2014). Magnetically separable copper substituted cobalt–zinc nano-ferrite photocatalyst with enhanced photocatalytic activity. Materials Science in Semiconductor Processing, 26, *467–476. https://doi.org/10.1016/j.mssp.2014.05.023.*

$Mn_{0.5}Zn_{0.5}Fe_2O_4$ and $Mn_{0.5}Mg_{0.5}Fe_2O_4$ photocatalysts in degradation of chloramine-T or RhB dyes under the same conditions.

Fig. 7.36 depicts the effect of catalyst dose on the degradation of chloramine-T and RhB of $Mn_{0.5}Zn_{0.125}Mg_{0.375}Fe_2O_4$ ferrite nanoparticles. For chloramine-T

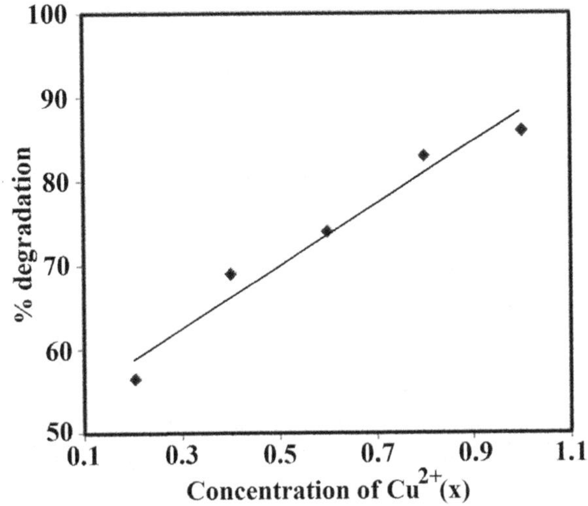

FIGURE 7.31

Variations of percentage degradation of $Co_{0.6}Zn_{0.4}Cu_xFe_{2-x}O_4$ ferrite annealed at 1000°C versus concentration of Cu^{2+}.

From Bhukal, S., & Singhal, S. (2014). Magnetically separable copper substituted cobalt–zinc nano-ferrite photocatalyst with enhanced photocatalytic activity. Materials Science in Semiconductor Processing, 26, *467–476. https://doi.org/10.1016/j.mssp.2014.05.023.*

$$La_xMnFe_{2-x}O_4 + h\nu \longrightarrow La_xMnFe_{2-x}O_4\ (e^-_{cb} + h^+_{vb})$$

$$La_xMnFe_{2-x}O_4 + h\nu \longrightarrow La_xMnFe_{2-x}O_4 + \text{Energy}$$

$$La_xMnFe_{2-x}O_4\ (h^+_{vb}) + H_2O \longrightarrow H^+ + OH^\bullet$$

$$La_xMnFe_{2-x}O_4\ (h^+_{vb}) + OH^- \longrightarrow OH^\bullet$$

$$La_xMnFe_{2-x}O_4\ (e^-_{cb}) + O_2 \longrightarrow O_2^{\bullet-}$$

$$O_2^{\bullet-} + H^+ \longrightarrow HO_2^\bullet$$

$$2e_{cb} + HO_2^\bullet + H^+ \longrightarrow OH^- + OH^\bullet$$

$$[\,OH^\bullet\ O_2^{\bullet-}\ HO_2^\bullet\ h^+_{vb}\,] + \text{Crystal Violet Dye} \longrightarrow \text{Degradation Product}$$

FIGURE 7.32

The mechanism of crystal violet dye degradation using $La_xMnFe_{2-x}O_4$ ferrite nanoparticles under visible light irradiation.

From Baig, M. M., Zulfiqar, S., Yousuf, M. A., Touqeer, M., Ullah, S., Agboola, P., Warsi, M. F., & Shakir, I. (2020). Structural and photocatalytic properties of new rare earth La3+ substituted $MnFe_2O_4$ ferrite nano-particles. Ceramics International. *https://doi.org/10.1016/j.ceramint.2020.06.103.*

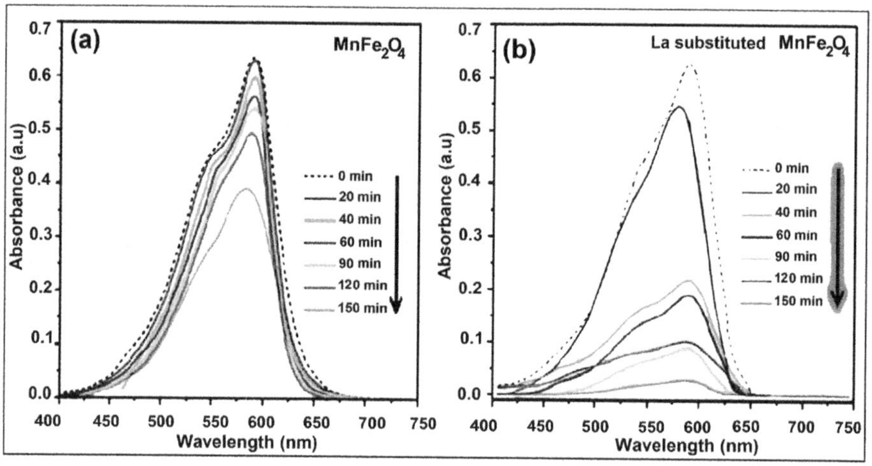

FIGURE 7.33

Absorption spectra of crystal violet dye using (A) $MnFe_2O_4$ and (B) $La_xMnFe_{2-x}O_4$ ferrite nanoparticles.

From Baig, M. M., Zulfiqar, S., Yousuf, M. A., Touqeer, M., Ullah, S., Agboola, P., Warsi, M. F., & Shakir, I. (2020). Structural and photocatalytic properties of new rare earth La^{3+} substituted $MnFe_2O_4$ ferrite nanoparticles. Ceramics International. *https://doi.org/10.1016/j.ceramint.2020.06.103.*

photodegradation, the value of $(C_t/C_0) \times 100$ decreased from 60% to 30%, with an increase in the content of the catalyst from 10 to 30 mg. Thus, the photodegradation of Chloramine-T of $Mn_{0.5}Zn_{0.125}Mg_{0.375}Fe_2O_4$ ferrite nanoparticles grew from 40% to 70% after 80 min. For RhB degradation, the value of $(C_t/C_0) \times 100$ also dropped from 60% to 50% with an increase in the amount of the catalyst from 10 to 30 mg, and the photodegradation rose from 40% to 50% after 80 min. Indeed, enhanced photocatalytic activity at higher catalyst dosages is attributed to the enhanced surface-active positions for photocatalytic reactions. In addition, the higher dosage of the catalyst would result in higher numbers of photons absorbed by the catalyst and hence, the enhancement of the photocatalytic activity.

Fig. 7.37 indicates the influence of initial pH on the degradation of Chloramine-T and RhB of $Mn_{0.5}Zn_{0.125}Mg_{0.375}Fe_2O_4$ ferrite nanoparticles. It was observed that with an increase in pH value, the photodegradation of the dyes increased. The photodegradation of Chloramine-T and RhB of $Mn_{0.5}Zn_{0.125}Mg_{0.375}Fe_2O_4$ ferrite nanoparticles for irradiation time of 80 min at pH $= 11$ was 90% and 95%, respectively. It was noted that surface charge properties and the degree of photocatalyst agglomeration can be influenced by the initial pH of the solution. The higher surface activity of the $Mn_{0.5}Zn_{0.125}Mg_{0.375}Fe_2O_4$ ferrite nanoparticles is suitable for absorption of the dye and transportation of the produced charge carriers to the surface of the photocatalyst.

The photodegradation mechanism at various pH values involves hydroxyl radical assault, oxidation by the holes, and reduction by the electrons at the conducting

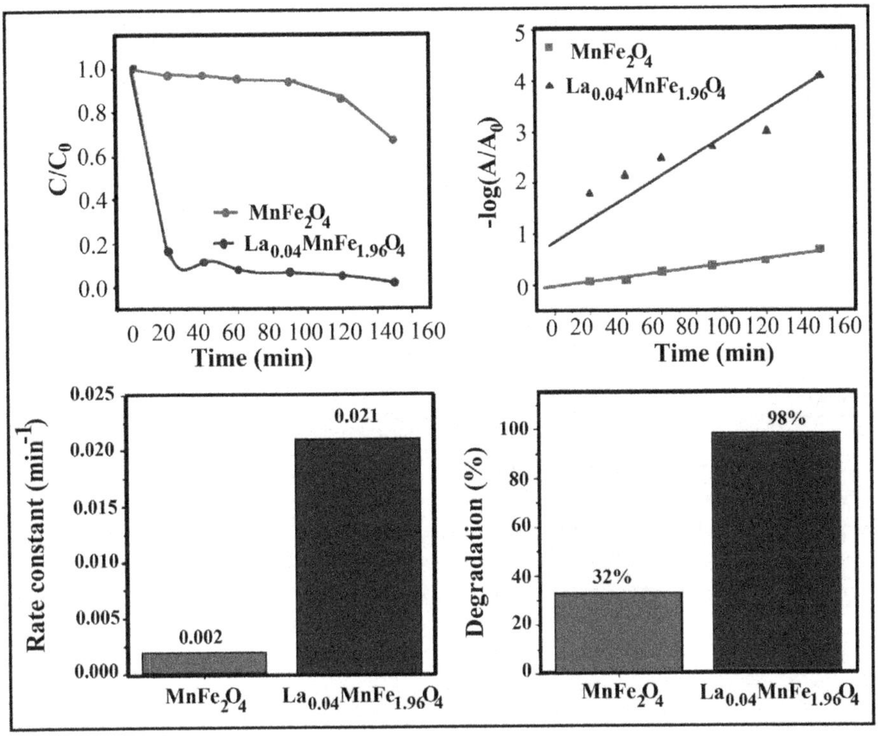

FIGURE 7.34

Kinetic study and degradation rate of $MnFe_2O_4$ and $La_xMnFe_{2-x}O_4$ ferrite nanoparticles.

From Baig, M. M., Zulfiqar, S., Yousuf, M. A., Touqeer, M., Ullah, S., Agboola, P., Warsi, M. F., & Shakir, I. (2020). Structural and photocatalytic properties of new rare earth La^{3+} substituted $MnFe_2O_4$ ferrite nanoparticles. Ceramics International. *https://doi.org/10.1016/j.ceramint.2020.06.103.*

band. Interestingly, the production of the electron–hole pairs on the surface of the photocatalyst by UV-irradiation can define the photodegradation process. The formation of hydroxyl radicals is due to the reaction of the holes with OH^- ions to generate the degradation product. Thus, the pH value is an important factor in the reaction mechanism and hence dye degradation (Abdel Maksoud et al., 2020).

Fig. 7.38 exhibits the variations of photodegradation efficiency versus the number of cycles for $Mn_{0.5}Zn_{0.125}Mg_{0.375}Fe_2O_4$ ferrite nanoparticles. The photodegradation efficiency of the Chloramine-T and RhB dyes is almost similar. A small reduction was observed with an increase in the number of cycles due to the absorption of uncleaned dyes and partially decolorized intermediates on the $Mn_{0.5}Zn_{0.125}Mg_{0.375}Fe_2O_4$ ferrite nanoparticles. Thus, the degradation of the dyes diminished with an increase in the number of cycles in the photocatalytic process. The photocatalyst activity in the fourth cycle was ~23% lower than in the first cycle. Likewise, photodegradation efficiency can be influenced by photocatalyst

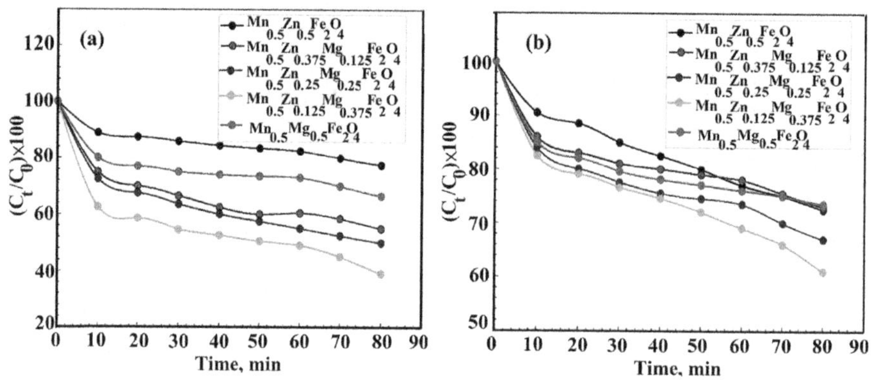

FIGURE 7.35

Photocatalytic degradation of (A) Chloramine-T and (B) Rhodamine-B using 10 mg of $Mn_{0.5}Zn_{0.5-x}Mg_xFe_2O_4$ ferrite nanoparticles in 50 mL dye solution.

From Abdel Maksoud, M. I. A., El-Sayyad, G. S., El-Khawaga, A. M., Abd Elkodous, M., Abokhadra, A., Elsayed, M. A., Gobara, M., Soliman, L. I., El-Bahnasawy, H. H., & Ashour, A. H. (2020). Nanostructured Mg substituted Mn–Zn ferrites: A magnetic recyclable catalyst for outstanding photocatalytic and antimicrobial potentials. Journal of Hazardous Materials, 399, *123000. https://doi.org/10.1016/j.jhazmat.2020.123000.*

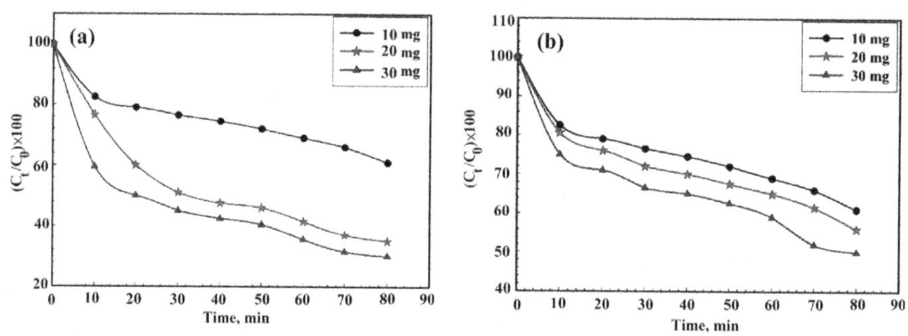

FIGURE 7.36

The effect of catalyst dose on degradation of (A) Chloramine-T and (B) Rhodamine-B of $Mn_{0.5}Zn_{0.125}Mg_{0.375}Fe_2O_4$ ferrite nanoparticles (50 mL solution and 20 mg/L dye concentration).

From Abdel Maksoud, M. I. A., El-Sayyad, G. S., El-Khawaga, A. M., Abd Elkodous, M., Abokhadra, A., Elsayed, M. A., Gobara, M., Soliman, L. I., El-Bahnasawy, H. H., & Ashour, A. H. (2020). Nanostructured Mg substituted Mn–Zn ferrites: a magnetic recyclable catalyst for outstanding photocatalytic and antimicrobial potentials. Journal of Hazardous Materials, 399, *123000. https://doi.org/10.1016/j.jhazmat.2020.123000.*

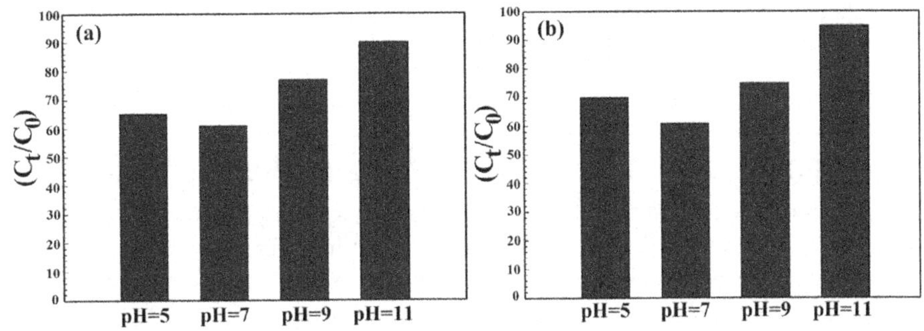

FIGURE 7.37

The influence of initial pH on degradation of (A) Chloramine-T and (B) Rhodamine-B of $Mn_{0.5}Zn_{0.125}Mg_{0.375}Fe_2O_4$ ferrite nanoparticles (50 mL solution, 10 mg catalyst, and 20 mg/L dye concentration).

From Abdel Maksoud, M. I. A., El-Sayyad, G. S., El-Khawaga, A. M., Abd Elkodous, M., Abokhadra, A., Elsayed, M. A., Gobara, M., Soliman, L. I., El-Bahnasawy, H. H., & Ashour, A. H. (2020). Nanostructured Mg substituted Mn–Zn ferrites: A magnetic recyclable catalyst for outstanding photocatalytic and antimicrobial potentials. Journal of Hazardous Materials, 399, *123000. https://doi.org/10.1016/j.jhazmat.2020.123000.*

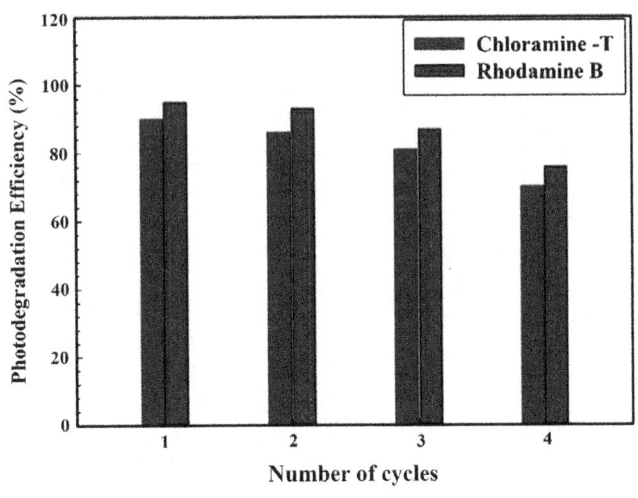

FIGURE 7.38

Variations of photodegradation efficiency versus number of cycles for $Mn_{0.5}Zn_{0.125}Mg_{0.375}Fe_2O_4$ ferrite nanoparticles.

From Abdel Maksoud, M. I. A., El-Sayyad, G. S., El-Khawaga, A. M., Abd Elkodous, M., Abokhadra, A., Elsayed, M. A., Gobara, M., Soliman, L. I., El-Bahnasawy, H. H., & Ashour, A. H. (2020). Nanostructured Mg substituted Mn–Zn ferrites: A magnetic recyclable catalyst for outstanding photocatalytic and antimicrobial potentials. Journal of Hazardous Materials, 399, *123000. https://doi.org/10.1016/j.jhazmat.2020.123000.*

aggregation as the annealing process after each cycle run can reduce the photocatalyst surface area.

7.8 Nanocrystalline Li-ferrite and Mg-ferrite photocatalysts

Another work examined the photocatalytic and magnetic properties of nickel and magnesium substituted lithium ferrite nanocrystalline prepared by sol–gel autocombustion process (Gandomi et al., 2019). The XRD patterns of nanoparticles calcined at 900°C revealed a good crystalline form and cubic spinel-phase lattice structure for samples with the compositions of $Li_{0.5}Fe_{2.5}O_4$, $LiMg_{0.5}Fe_2O_4$, and $LiNi_{0.5}Fe_2O_4$. It was revealed that the remanence magnetization and saturation magnetization of $Li_{0.5}Fe_{2.5}O_4$ nanoparticles decreased with the substitution of Ni or Mg in the crystal structure of ferrite. Fig. 7.39 displays the UV–V diffuse reflectance spectra of $Li_{0.5}Fe_{2.5}O_4$, $LiMg_{0.5}Fe_2O_4$, and $LiNi_{0.5}Fe_2O_4$ nanoparticles. The maximum absorption peak of $Li_{0.5}Fe_{2.5}O_4$, $LiMg_{0.5}Fe_2O_4$, and $LiNi_{0.5}Fe_2O_4$ nanoparticles was observed at around 370, 388, and 405 nm, respectively. Thus, the absorption increased with the substitution of Ni and Mg in nanocrystalline lithium ferrite. The UV–V absorption spectra of MO dye using $Li_{0.5}Fe_{2.5}O_4$, $LiMg_{0.5}Fe_2O_4$, and $LiNi_{0.5}Fe_2O_4$ nanoparticles are shown in Fig. 7.40. In the first stage, the system was kept dark until the adsorption–desorption equilibrium was

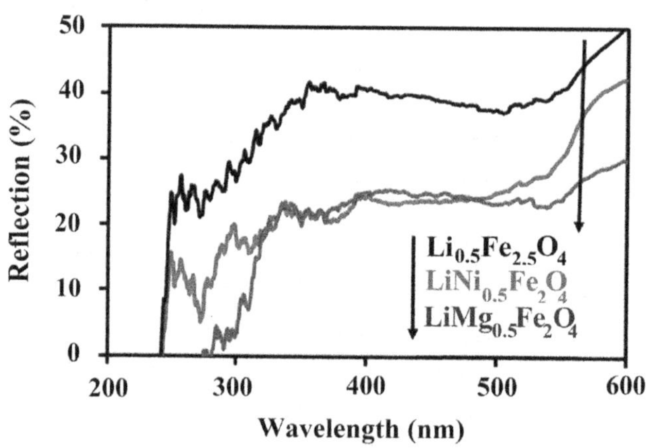

FIGURE 7.39

Ultraviolet–visible diffuse reflectance spectra of the $Li_{0.5}Fe_{2.5}O_4$, $LiMg_{0.5}Fe_2O_4$, and $LiNi_{0.5}Fe_2O_4$ nanoparticles.

From Gandomi, F., Peymani-Motlagh, S. M., Rostami, M., Sobhani-Nasab, A., Fasihi-Ramandi, M., Eghbali-Arani, M., Ahmadian, R., Gholipour, N., Rahimi-Nasrabadi, M., & Ganjali, M. R. (2019). Simple synthesis and characterization of $Li_{0.5}Fe_{2.5}O_4$, $LiMg_{0.5}Fe_2O_4$ and $LiNi_{0.5}Fe_2O_4$, and investigation of their photocatalytic and anticancer properties on hela cells line. Journal of Materials Science: Materials in Electronics, 30*(22), 19691–19702. https://doi.org/10.1007/s10854-019-02320-x.*

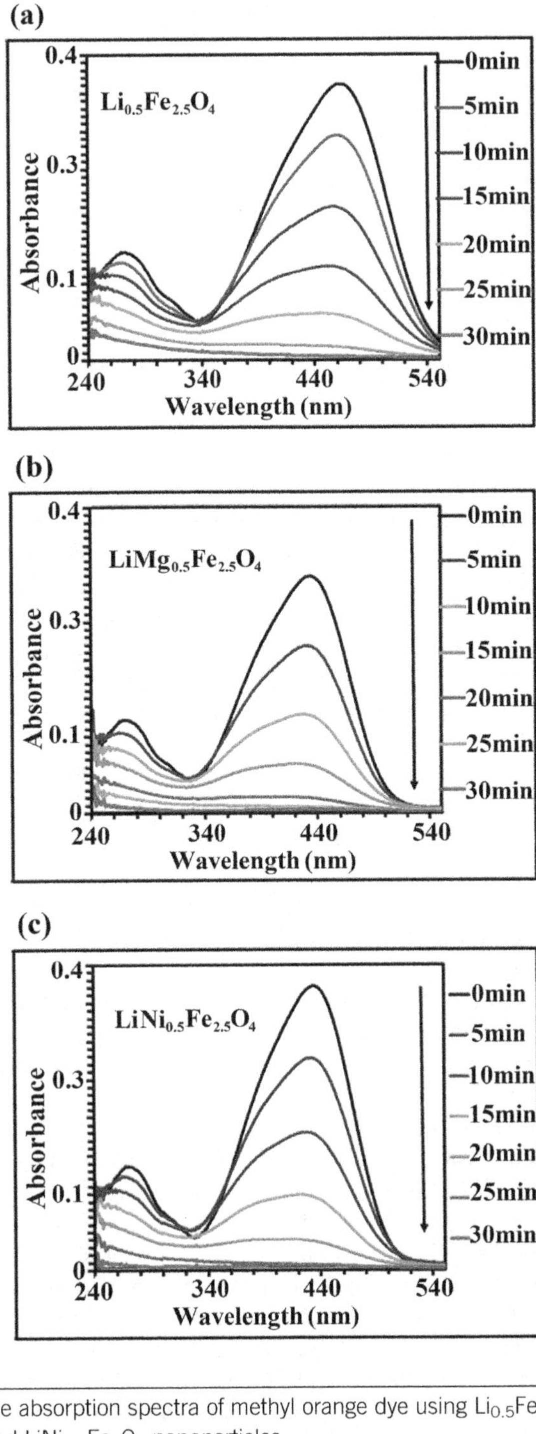

FIGURE 7.40

Ultraviolet–visible absorption spectra of methyl orange dye using $Li_{0.5}Fe_{2.5}O_4$, $LiMg_{0.5}Fe_2O_4$, and $LiNi_{0.5}Fe_2O_4$ nanoparticles.

From Gandomi, F., Peymani-Motlagh, S. M., Rostami, M., Sobhani-Nasab, A., Fasihi-Ramandi, M., Eghbali-Arani, M., Ahmadian, R., Gholipour, N., Rahimi-Nasrabadi, M., & Ganjali, M. R. (2019). Simple synthesis and characterization of $Li_{0.5}Fe_{2.5}O_4$, $LiMg_{0.5}Fe_2O_4$ and $LiNi_{0.5}Fe_2O_4$, and investigation of their photocatalytic and anticancer properties on hela cells line. Journal of Materials Science: Materials in Electronics, 30*(22), 19691–19702. https://doi.org/10.1007/s10854-019-02320-x.*

obtained. Then the UV irradiation with a constant interval time was applied to the suspension. The photodegradation of $Li_{0.5}Fe_{2.5}O_4$, $LiMg_{0.5}Fe_2O_4$, and $LiNi_{0.5}Fe_2O_4$ nanoparticles was 95%, 98.04%, and 99.09%, respectively. It was seen that with the substitution of Ni or Mg in lithium ferrite nanocrystalline, the removal of MO dye increased, which is related to the enhancement of the bandgap. The bandgap value can be influenced by different parameters such as strain, lattice distortion, and impurity. The highest photocatalytic activity was obtained for $LiNi_{0.5}Fe_2O_4$ nanoparticles. Thus, the photocatalytic performance of Ni or Mg substituted with lithium ferrite has been better than that of the pure nanocrystalline lithium ferrite. Indeed, the Mg^{2+} and Ni^{2+} ions can function as both electron–hole traps and increase the lifetimes of $e^- - h^+$ pairs (Jing et al., 2007).

The photocatalytic performance can be influenced by the crystallinity of the photocatalyst and surface area. The density functional theory (DFT) computations of spinel MFe_2O_4 structures illustrated that the density of states of the Fe–d orbitals near the Fermi level might be affected by the M-ion. Thus, the M-ion in spinel MFe_2O_4 structures can determine the band energetics and hence visible light photocatalytic performance. Fig. 7.41 exhibits the UV-diffuse reflectance spectra of spinel MFe_2O_4 (M: Mg, Ca, and Zn) nanoparticles prepared by the microwave sintering method (Dom et al., 2011). The inset of Fig. 7.41A reflects an almost linear variation of photodegradation percentage of MB versus time for MFe_2O_4 ferrites. In addition, the variations of photocatalytic degradation activity of MB with the irradiation time for $ZnFe_2O_4$ nanoparticles were observed. The bandgap of all the samples has been nearly ~1.9 eV. It was clear that the spinel ferrites had a high absorptivity within the range of wavelength of 400–700 nm, which is related to the hybridized Fe–d orbital. It was noted that microwave irradiation time has an important role in photocatalytic behavior. Sintering time has a significant effect on the morphology, crystallinity, and surface area of the sample. Perfect crystallinity, combined with a well-maintained and suitable surface area, results in greater photocatalytic activity. Fig. 7.41B clearly shows that a sample sintered for 30 min contains nanostructure morphology with a high surface area, which provides suitable photocatalytic activity.

It was demonstrated that the Fe-orbital could play an important role in the spinel MFe_2O_4 ferrite photocatalysts. The electronic structure of spinel MFe_2O_4 ferrite photocatalysts via the full-potential linearized augmented plane wave method was calculated within DFT (Gateshki et al., 2005; McCarty et al., 1981; Waerenborgh et al., 1994). Fig. 7.42 displays the density of states of spinel MFe_2O_4 (M: Mg, Ca, and Zn) nanoparticle prepared by the microwave sintering method. It was seen that the Fe–d orbital has a significant contribution to the states of the bandgap for all samples. The highest contribution of Fe–d orbital belonged to the $ZnFe_2O_4$ ferrite photocatalyst. However, the $MgFe_2O_4$ ferrite nanoparticles had the minimum affectability by Fe–d orbital. The DOS confirms the n-type electronic structure of Mg and Zn ferrites as well as the p-type conductivity of Ca ferrite semiconductors. Indeed, the Mg and Zn ferrites have an analogous cubic crystal structure. On the other hand, the p-type of Ca-ferrite semiconductor possessed an orthorhombic

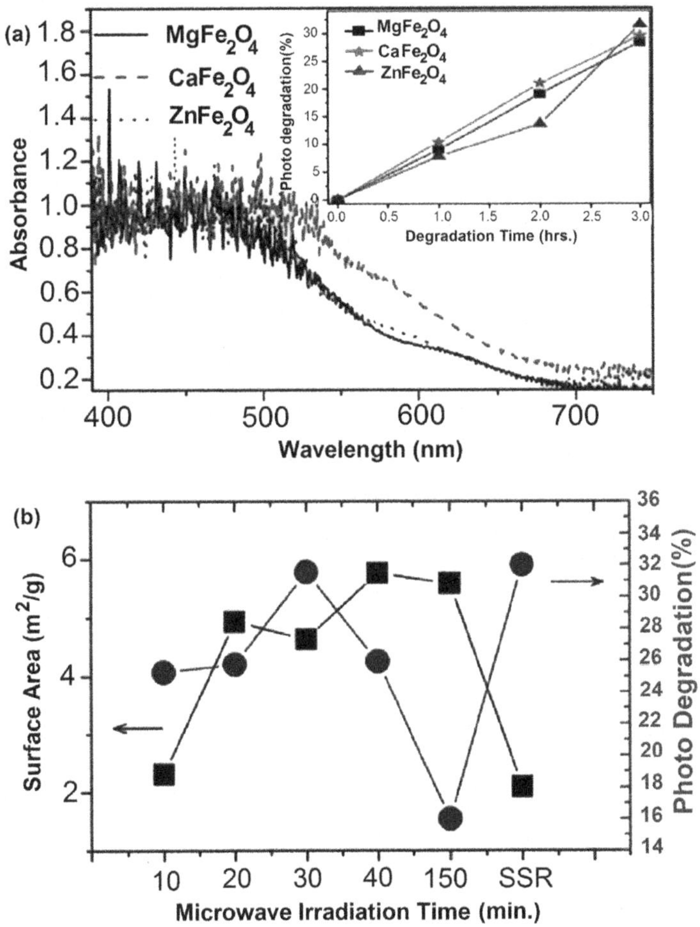

FIGURE 7.41

(A) Diffuse reflectance spectra of spinel MFe_2O_4 (M: Mg, Ca, and Zn) nanoparticles prepared by the microwave sintering method and (B) variations of photocatalytic degradation activity of methylene blue with the irradiation time for $ZnFe_2O_4$ nanoparticles.

From Dom, R., Subasri, R., Radha, K., & Borse, P. H. (2011). Synthesis of solar active nanocrystalline ferrite, MFe2O4 (M: Ca, Zn, Mg) photocatalyst by microwave irradiation. Solid State Communications, 151*(6), 470–473. https://doi.org/10.1016/j.ssc.2010.12.034.*

crystal structure. As such, it is expected that the Ca-ferrite would own a displaced DOS compared with the Mg and Zn ferrites. The photocatalytic reactions, such as photodegradation and photodecomposition of water, can be affected by the active radicals reacting with the electron/hole. The schema displays that all samples are suitable candidates for MB photodegradation. The *n*-type of Mg and Zn ferrites would be appropriate candidates for photooxidation of water. On the other hand, the p-type Ca-ferrite semiconductor would be best for the photoreduction of water.

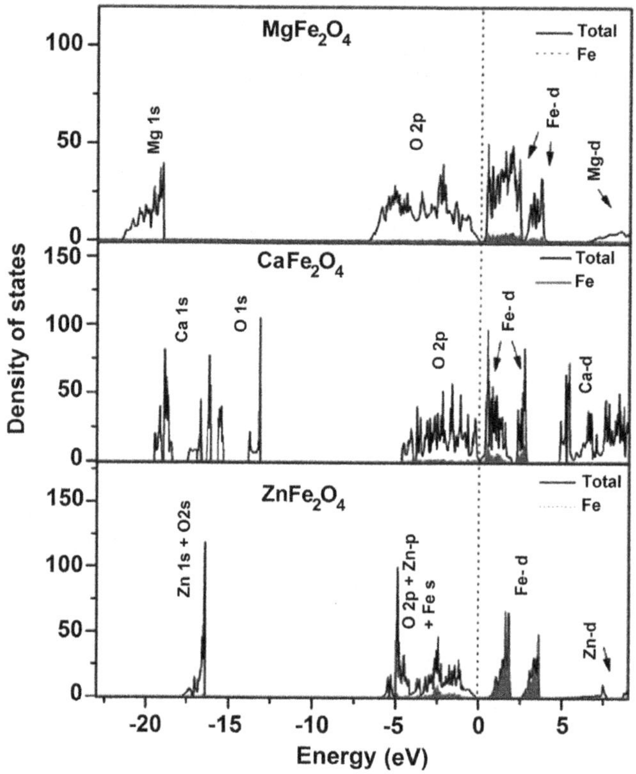

FIGURE 7.42

Density of states of spinel MFe_2O_4 (M: Mg, Ca, and Zn) nanoparticles prepared by the microwave sintering method.

From Dom, R., Subasri, R., Radha, K., & Borse, P. H. (2011). Synthesis of solar active nanocrystalline ferrite, MFe_2O_4 (M: Ca, Zn, Mg) photocatalyst by microwave irradiation. Solid State Communications, 151*(6), 470–473. https://doi.org/10.1016/j.ssc.2010.12.034.*

7.9 Nanocrystalline Bi-ferrite photocatalyst

The catalytic activity of bismuth ferrite for photodegradation of bromophenol blue using H_2O_2 was investigated (Krasnyakova et al., 2020). It was illustrated that the photocatalytic activity of $BiFeO_3$ increased with a rise in the mechanical preactivation in an argon atmosphere because of the enhancement of the specific surface area. In addition, it was seen that the photocatalytic activity increased when using a hydrogen atmosphere as the surface concentration of oxygen vacancies increased. Fig. 7.43 depicts the kinetics of the optical density of the reaction solution in the bismuth ferrite for photodegradation of bromophenol blue using H_2O_2. It was clear that the optical density of the reaction solution decreased over time. The rate constant was 1.2×10^{-3} min^{-1}. The variations of the rate constant of bromophenol blue consumption with the time of the preliminary activation for bismuth ferrite

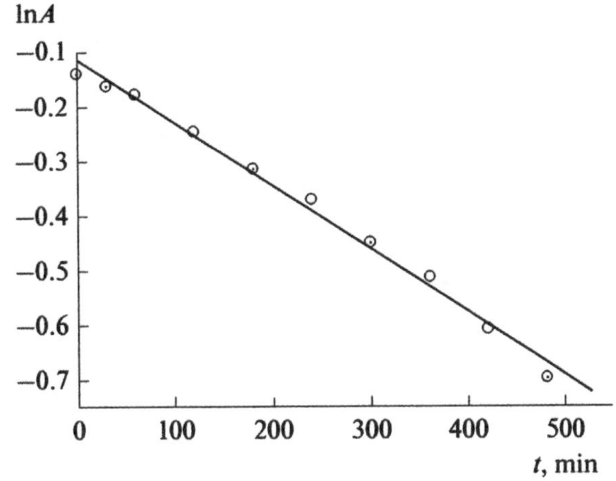

FIGURE 7.43

Kinetics of the optical density A of the reaction solution in the bismuth ferrite for photodegradation of bromophenol blue using hydrogen peroxide.

From Krasnyakova, T., Yurchilo, S., Morenko, V., Nosolev, I., Glazunova, E., Khasbulatov, S., Verbenko, I., & Mitchenko, S. (2020). Effect of mechanochemical pretreatment on the photocatalytic activity of bismuth ferrite. Kinetics and Catalysis, 61*(3), 384–389. https://doi.org/10.1134/S0023158420030155.*

nanoparticles in two atmospheres of argon and hydrogen are shown in Fig. 7.44. Initially, the samples were exposed to mechanical activation in an argon atmosphere for 30 min until an almost limited specific surface area was reached. From Fig. 7.44, it can be seen that the rate constant of reaction increased with the pretreatment time of the catalyst. The enhancement of the catalytic performance of bismuth ferrite nanoparticles under the mechanical activation is related to the increase in the specific surface area of the catalyst. Activation in a hydrogen atmosphere can increase photocatalytic activity by twice in an argon atmosphere. This is because the hydrogen atmosphere creates more oxygen vacancies on the surface of the specimen.

In another work, the photodegradation activity of bismuth ferrite embedded into macroporous liquid natural rubber (LNR) matrices was investigated (Mohd Ridzuan et al., 2020). Fig. 7.45 displays the effect of catalyst dosages and irradiation time on the photodegradation performance of MB dye by heterostructure-based $BiFeO_3$ embedded into LNR matrices. It was observed that the photodegradation efficiency increased with a rise in $BiFeO_3$ content because of the presence of more active positions on the surface of the catalyst, which can absorb more photons. The greater absorption of photons can result in the further formation of electron-hole pairs. In addition, the photodegradation efficiency increased with irradiation time. The effects of initial dye concentration values on the photodegradation performance of MB dye are shown in Fig. 7.46. It was seen that the photodegradation efficiency diminished with an increase in the initial dye concentration from 10 ppm to 100 ppm. Indeed, the light path length shrank with increasing the number of MB

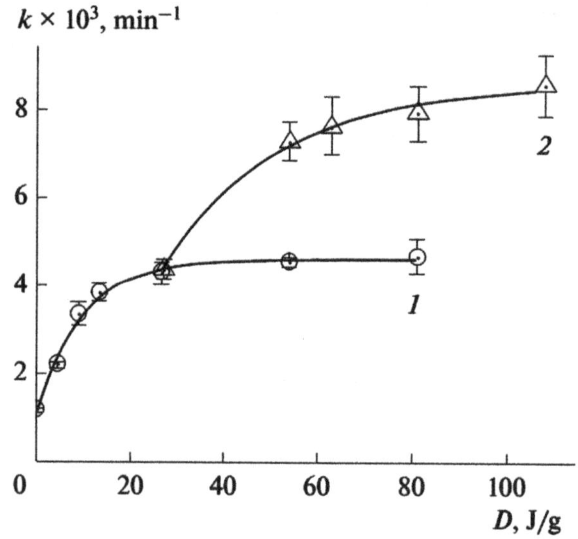

FIGURE 7.44

Variations of the rate constant of bromophenol blue consumption with the time of the preliminary activation for bismuth ferrite nanoparticles in atmospheres of (1) argon and (2) hydrogen.

From Krasnyakova, T., Yurchilo, S., Morenko, V., Nosolev, I., Glazunova, E., Khasbulatov, S., Verbenko, I., & Mitchenko, S. (2020). Effect of mechanochemical pretreatment on the photocatalytic activity of bismuth ferrite. Kinetics and Catalysis, 61*(3), 384–389. https://doi.org/10.1134/S0023158420030155.*

molecules. Fig. 7.47 shows the effect of pH values on the photodegradation performance of MB dye in sunlight irradiation and the dark. It was clear that at pH = 6.36, the charge of the surface was neutral. At pH < 6.36, the surface of $BiFeO_3$ found a positive charge because of the greater existence of H^+ ions sited on the surface. Thus, the anions can be more adsorbed on the surface. At pH > 6.36, the surface of ferrite had a negative charge because of the greater presence of OH^- ions which can result in more participation of H^+ in the solution with a high pH.

7.10 Ferrite nanocomposites photocatalysts

In the following, the photocatalyst activity of some typical ferrite nanocomposites is discussed.

7.10.1 Photocatalytic properties of $Ni_{1-x}Co_xFe_2O_4$/multiwalled carbon nanotube nanocomposites

It was reported that the photocatalytic activity of ferrite nanoparticles could enhance in the presence of multiwalled carbon nanotubes (MWCNTs) due to the synergetic

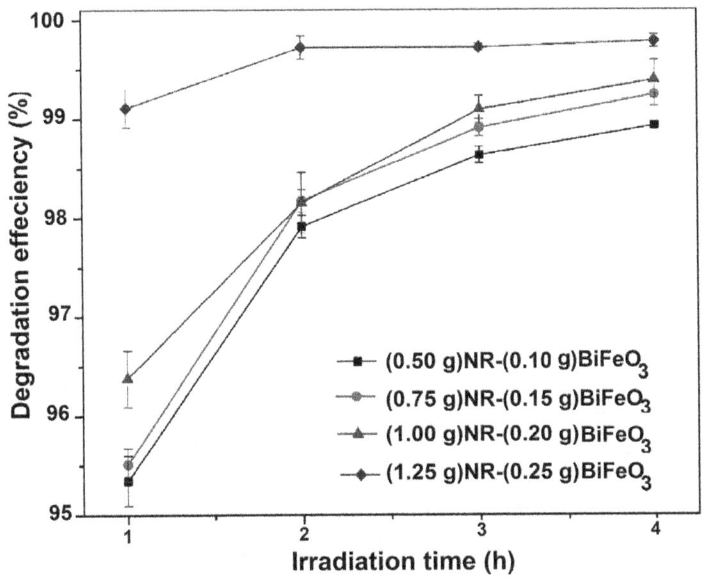

FIGURE 7.45

The effect of catalyst dosages and irradiation time on the photodegradation performance of methylene blue dye by bismuth ferrite.

From Mohd Ridzuan, N. D., Mohd Kaus, N. H., Md Lazim, M. A. S., Kobayashi, T., Adnan, R., Othman, M. Z., & Mohamad Kassim, M. H. (2020). Photocatalytic heterostructures-based $BiFeO_3$ embedded liquid natural rubber (LNR) for highly removal of cationic dye under direct sunlight. Journal of Environmental Chemical Engineering, *8(5), 104152. https://doi.org/10.1016/j.jece.2020.104152.*

effect. Accordingly, the magnetic, optical, and photocatalytic properties of $Ni_{1-x}Co_xFe_2O_4$ (NCFO)/MWCNT nanocomposites were evaluated (Dhiman et al., 2016; Gregor et al., 2010; Kim & Kang, 2012; Sharma et al., 2015; Singh & Pal, 2013). The photocatalytic activities of NCFO/MWCNT nanocomposites were studied to degrade RhB dye. The X-ray diffraction patterns of nanocomposites annealed at 400°C exhibited the presence of diffraction peaks of (220), (311), (400), (511), (440), and (002) planes, corresponding to the spinel structure of NCFO ferrite nanoparticles and MWCNTs, respectively. The TEM micrographs of nanocomposites demonstrated that the NCFO nanoparticles are well dispersed on the surface of MWCNTs. UV−V spectra of NCFO/MWCNT nanocomposites annealed at 400°C are shown in Fig. 7.48. UV−V diffuse reflectance spectroscopy technique was used to achieve the band edge absorption of nanocomposites. It was clear that with the substitution of cobalt ions in the nickel ferrite, the band edge absorption values diminished. The band edge absorption values of NCFO/MWCNT nanocomposites were 810−990 nm. The band edge absorption values of NCFO/MWCNT nanocomposites with $x = 0.0$, 0.2, 0.4, 0.6, and 0.8 were 810, 840, 900, 820, and 990 nm, respectively. These large variations in the band edge absorption values are quite important for photocatalytic activity.

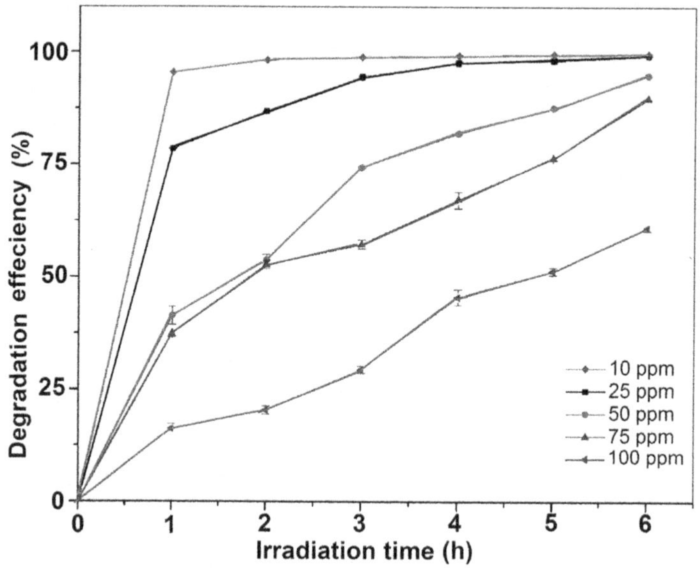

FIGURE 7.46

The effect of initial dye concentration values on the photodegradation performance of methylene blue dye by heterostructure-based $BiFeO_3$.

From Mohd Ridzuan, N. D., Mohd Kaus, N. H., Md Lazim, M. A. S., Kobayashi, T., Adnan, R., Othman, M. Z., & Mohamad Kassim, M. H. (2020). Photocatalytic heterostructures-based $BiFeO_3$ embedded liquid natural rubber (LNR) for highly removal of cationic dye under direct sunlight. Journal of Environmental Chemical Engineering, *8(5), 104152. https://doi.org/10.1016/j.jece.2020.104152.*

The bandgap of samples was calculated by the following equation:

$$\alpha h\upsilon = A\left(h\upsilon - E_g\right)^{\frac{1}{2}} \tag{7.2}$$

where α, h, υ, E_g, and A represent the absorption coefficient, Planck's constant, light frequency, bandgap, and the proportionality constant, respectively (Chavan et al., 2010). The bandgap energy value of $NiFe_2O_4$/CNTs nanocomposite was 1.53 eV. However, with the substitution of cobalt ions in the nickel ferrite, the bandgap decreased. The bandgap energy values of NCFO/MWCNT nanocomposites with $x = 0.2$, 0.4, 0.6, and 0.8 were 1.47, 1.37, 1.51, and 1.25 eV, respectively. It was also observed that the addition of MWCNTs to ferrite nanoparticles could decrease the bandgap of nanocomposites, which is attributed to the smaller bandgap of MWCNTs (around 1.1 eV) (Cao et al., 2007; Fan et al., 2009; Guo et al., 2011; Lee et al., 2006; Li et al., 2011; Zhang et al., 2010).

Interestingly, with the substitution of cobalt ions in the nickel ferrite/CNTs nanocomposites, the saturation magnetization increased from 12.5 emu/g for $x = 0.0$ to 15.0 emu/g for $x = 0.8$, which is due to the lower magnetic moment of Ni^{2+} ion ($2\mu_B$) compared with the Co^{2+} ion ($3\mu_B$). The lower values of magnetization for

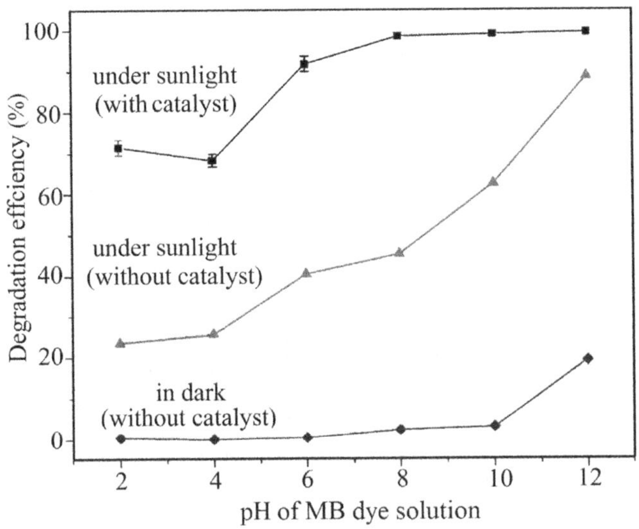

FIGURE 7.47

The effect of pH values on the photodegradation performance of methylene blue dye.

From Mohd Ridzuan, N. D., Mohd Kaus, N. H., Md Lazim, M. A. S., Kobayashi, T., Adnan, R., Othman, M. Z., & Mohamad Kassim, M. H. (2020). Photocatalytic heterostructures-based $BiFeO_3$ embedded liquid natural rubber (LNR) for highly removal of cationic dye under direct sunlight. Journal of Environmental Chemical Engineering, *8(5), 104152. https://doi.org/10.1016/j.jece.2020.104152.*

ferrite/CNTs nanocomposites are related to the presence of nonmagnetic MWCNTs and dead layer at the surface of nanoparticles or spin canting effect. Likewise, with the substitution of cobalt ions in the nickel ferrite/CNTs nanocomposites, the coercivity rose from 1.0 Oe for $x = 0.0$ to 110.0 Oe for $x = 0.8$. Fig. 7.49 depicts the photocatalytic degradation percentage of 50 μM of RhB dye (100 mL) versus time for 50 mg of NCFO/MWCNT nanocomposites and 100 μL of 30% H_2O_2 at a pH of 2.5. The complete degradation of the dye was achieved within 15–25 min. It was observed that the existence of CNTs increased the catalytic activity of nanocomposites. Indeed, in visible light, the electron jumps from the valence band to the conduction band of MWCNTs by making a photogenerated hole. This electron can be trapped by ferrite nanoparticles. Thus, the electron-hole recombination rate becomes limited, and the photochemical decomposition of H_2O_2 results in the formation of hydroxyl radical, which would decolorize the dye solution (Casbeer et al., 2012; Rashad et al., 2012).

7.10.2 Photocatalytic properties of $Cu_{1-x}Co_xFe_2O_4$/multiwalled carbon nanotube nanocomposites

Cobalt substituted copper ferrite nanoparticles and their nanocomposites with carbon nanotubes were prepared by a microemulsion technique (Singh et al., 2015).

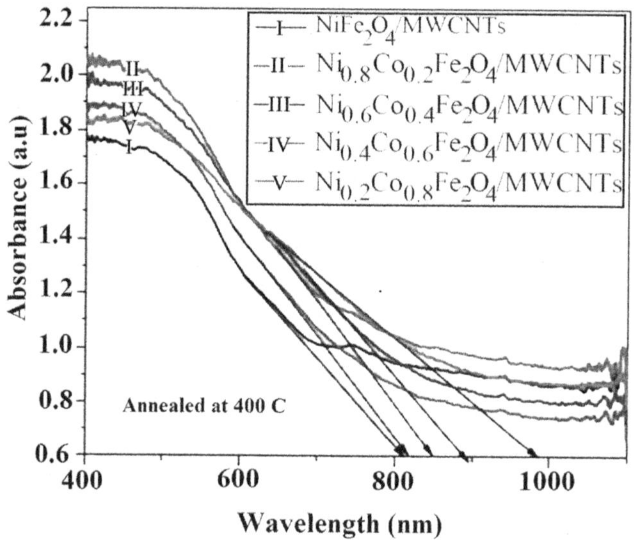

FIGURE 7.48

Diffused ultraviolet–visible spectra of $Ni_{1-x}Co_xFe_2O_4$/multiwalled carbon nanotube nanocomposites.

From Singh, C., Bansal, S., Kumar, V., & Singhal, S. (2015). Beading of cobalt substituted nickel ferrite nanoparticles on the surface of carbon nanotubes: a study of their synthesis mechanism, structure, magnetic, optical and their application as photocatalyst. Ceramics International, *41(3), 3595–3604. https://doi.org/10.1016/j.ceramint.2014.11.021.*

In the microemulsion method, for controlling the particle size and shape of nanoparticles, sodium dodecyl sulfate was employed as a soft templating agent. UV–V spectra of $Cu_{1-x}Co_xFe_2O_4$ nanoparticles and nanocomposites are shown in Fig. 7.50. The absorption edge values of $Cu_{1-x}Co_xFe_2O_4$ nanoparticles ranged from 825 to 1050 nm. It was observed that with the substitution of cobalt ions in the copper ferrite, band edge absorption values increased. The absorption edge values of $Cu_{1-x}Co_xFe_2O_4$ nanoparticles with $x = 0.2$, 0.6, 0.8, and 1.0 were 856, 902, 960, and 1050 nm, respectively. The absorption edge values of $Cu_{1-x}Co_xFe_2O_4$/MWCNT nanocomposites ranged from 680 to 995 nm. It was clear that the presence of CNTs can decrease the absorption edge of ferrites. The bandgap energy values were calculated by a plot of $(\alpha h\upsilon)^2$ versus photon energy $(h\upsilon)$. The bandgap values of $Cu_{1-x}Co_xFe_2O_4$ nanoparticles with $x = 0.0$, 0.2, 0.6, 0.8, and 1.0 were 1.40, 1.44, 1.37, 1.29, and 1.2 eV, respectively. Likewise, the bandgap values of $Cu_{1-x}Co_xFe_2O_4$/MWCNT nanocomposites with $x = 0.2$, 0.6, 0.8, and 1.0 were 1.76, 1.70, 1.42, and 1.25 eV, respectively. Interestingly, it was found that the addition of CNTs to the ferrite can increase the bandgap values because of the reduction of the electron-hole recombination rate. It was found that with the substitution of cobalt ions in the copper ferrite composition, the saturation

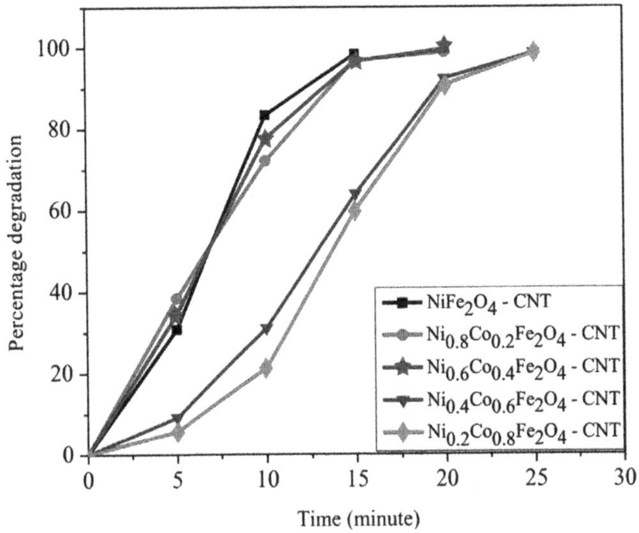

FIGURE 7.49

Degradation percentage of 50 μM Rhodamine B dye versus the time for 50 mg of $Ni_{1-x}Co_xFe_2O_4$/multiwalled carbon nanotube nanocomposites and 100 μL of 30% H_2O_2 at a pH of 2.5.

From Singh, C., Bansal, S., Kumar, V., & Singhal, S. (2015). Beading of cobalt substituted nickel ferrite nanoparticles on the surface of carbon nanotubes: a study of their synthesis mechanism, structure, magnetic, optical and their application as photocatalyst. Ceramics International, 41*(3), 3595–3604. https://doi.org/10.1016/j.ceramint.2014.11.021.*

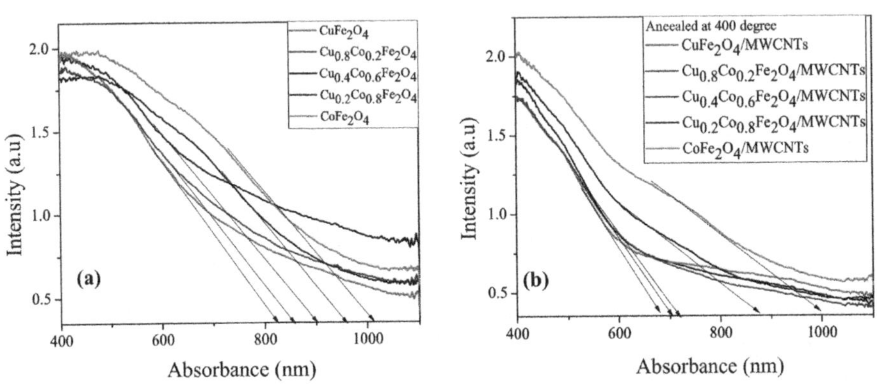

FIGURE 7.50

Ultraviolet–visible diffuse spectra of (A) $Cu_{1-x}Co_xFe_2O_4$ nanoparticles and (B) $Cu_{1-x}Co_xFe_2O_4$/multiwalled carbon nanotube nanocomposites.

From Singh, C., Bansal, S., Kumar, V., Tikoo, K., & Singhal, S. (2015). Encrustation of cobalt doped copper ferrite nanoparticles on solid scaffold CNTs and their comparison with corresponding ferrite nanoparticles: a study of structural, optical, magnetic and photo catalytic properties. RSC Advances, 5*(49), 39052–39061. https://doi.org/10.1039/C5RA03330F.*

magnetization increased, which is related to the higher magnetic moment of Co^{2+} ion ($3\mu_B$) than with the Cu^{2+} ion ($1\mu_B$). Likewise, the coercivity increased. The photocatalytic degradation percentage of 50 μM RhB dye versus the time for 50 mg of $Cu_{1-x}Co_xFe_2O_4$ nanoparticles and $Cu_{1-x}Co_xFe_2O_4$/MWCNT nanocomposites and 100 μL of 30% H_2O_2 at a pH of 2.5 are shown in Fig. 7.51. The copper ferrite completely degraded the RhB dye in 120 min. It was observed that with the substitution of cobalt ions in copper ferrite composition, the degradation rate increased. For pure cobalt ferrite, first, the reaction rate increased sharply after 40 min. The sample with the composition of $Cu_{0.2}Co_{0.8}Fe_2O_4$ had the highest rate of reaction and degraded the dye in 100 min. In addition, the catalytic activity for $Cu_{1-x}Co_xFe_2O_4$/MWCNT nanocomposites was better than for $Cu_{1-x}Co_xFe_2O_4$ nanoparticles. The $Cu_{0.2}Co_{0.8}Fe_2O_4$/MWCNT nanocomposites degrade the RhB dye in 60 min. Fig. 7.52 displays the energy level diagram exhibiting the mechanism of degradation of RhB dye. Indeed, with the substitution of cobalt ions in copper ferrite composition, the bandgap diminishes, and the electron jumps easily from the higher conduction band of copper ferrite to the lower conduction band of cobalt-doped copper ferrite. Thus, the electron-hole recombination rate becomes limited, and the photochemical decomposition of H_2O_2 leads to the formation of hydroxyl radical, which would decolorize the dye solution (Albuquerque et al., 2012; Fenton, 1894).

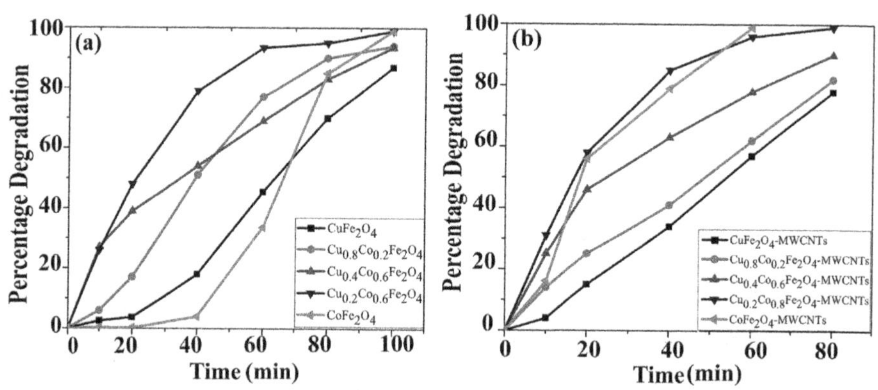

FIGURE 7.51

Degradation percentage of 50 mM Rhodamine B dye versus the time for 50 mg of (A) $Cu_{1-x}Co_xFe_2O_4$ nanoparticles and (B) $Cu_{1-x}Co_xFe_2O_4$/multiwalled carbon nanotube nanocomposites.

From Singh, C., Bansal, S., Kumar, V., Tikoo, K., & Singhal, S. (2015). Encrustation of cobalt doped copper ferrite nanoparticles on solid scaffold CNTs and their comparison with corresponding ferrite nanoparticles: a study of structural, optical, magnetic and photo catalytic properties. RSC Advances, 5*(49), 39052–39061. https://doi.org/10.1039/C5RA03330F.*

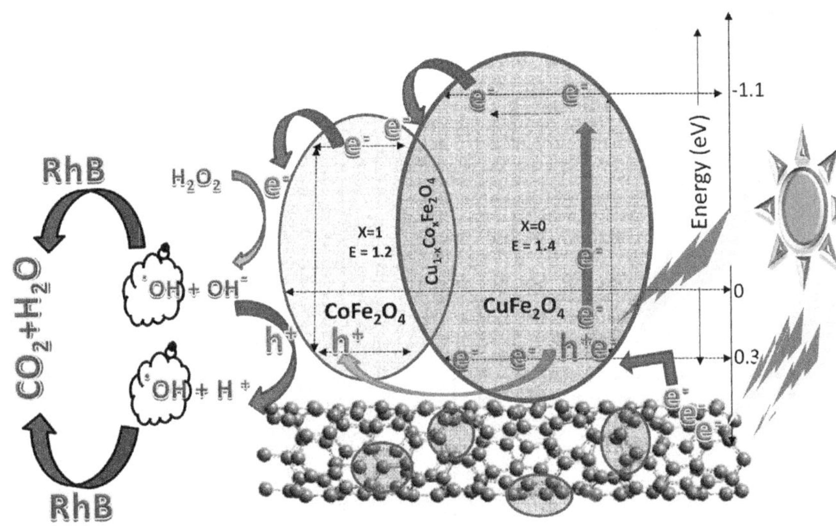

FIGURE 7.52

Energy level diagram exhibiting the mechanism of degradation of Rhodamine B dye.

From Singh, C., Bansal, S., Kumar, V., Tikoo, K., & Singhal, S. (2015). Encrustation of cobalt doped copper ferrite nanoparticles on solid scaffold CNTs and their comparison with corresponding ferrite nanoparticles: a study of structural, optical, magnetic and photo catalytic properties. RSC Advances, 5*(49), 39052–39061. https://doi.org/10.1039/C5RA03330F.*

7.10.3 Photocatalytic properties of reduced-graphene oxide-$Ni_{0.65}Zn_{0.35}Fe_2O_4$ ferrite nanohybrids

The photocatalytic activity of reduced graphene oxide (rGO) -$Ni_{0.65}Zn_{0.35}Fe_2O_4$ ferrite nanohybrids was also evaluated by photochemical decomposition of MB under visible light irradiation (Javed et al., 2019). The rGO-$Ni_{0.65}Zn_{0.35}Fe_2O_4$ nanohybrids were produced via ultrasonication technique. It was observed that the shape of particles was almost spherical. The interaction between the magnetic particles would lead to agglomeration. The particle size was within the range of 0.25–0.5 μm. In addition, the $Ni_{0.65}Zn_{0.35}Fe_2O_4$ nanoparticles were well attached to the graphene sheets. The bandgap of ferrite nanoparticles was 1.91 eV, which decreased to 1.84 eV when they were blended with rGO to make the nanohybrids.

The photocatalytic activity of rGO-$Ni_{0.65}Zn_{0.35}Fe_2O_4$ nanohybrids was better than that of the pure ferrite nanoparticles, which is attributed to the suitable separation of photoproduced electron-hole pairs. The surface area, good adsorption capacity, low bandgap, nontoxic nature, minimum recombination of electron–hole pairs, recyclability, and visible light range efficiency are important parameters for the photocatalytic activity of semiconductors materials. UV–V absorption spectra and bandgap of the $Ni_{0.65}Zn_{0.35}Fe_2O_4$ nanoparticles and rGO-$Ni_{0.65}Zn_{0.35}Nd_{0.03}Fe_{1.97}O_4$ nanohybrids are shown in Fig. 7.53. The absorption intensity of $Ni_{0.65}Zn_{0.35}Fe_2O_4$ nanoparticles changed in the UV region and increased in the visible region with an increase in

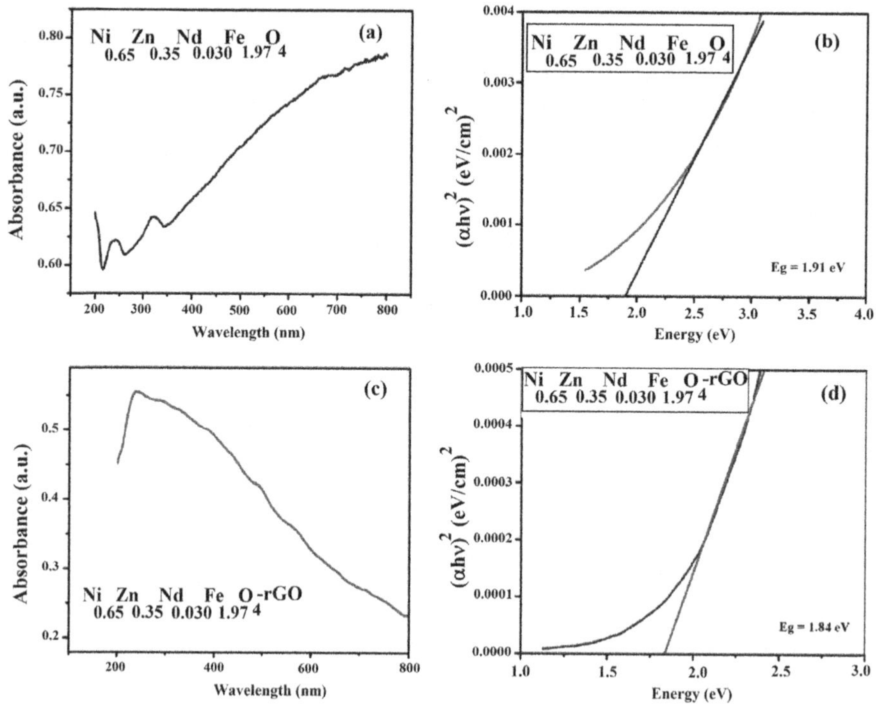

FIGURE 7.53

Ultraviolet–visible absorption spectra and bandgap of the (A–B) $Ni_{0.65}Zn_{0.35}Fe_2O_4$ nanoparticles and (C–D) rGO-$Ni_{0.65}Zn_{0.35}Nd_{0.03}Fe_{1.97}O_4$ nanohybrids.

From Javed, H., Rehman, A., Mussadiq, S., Shahid, M., Khan, M. A., Shakir, I., Agboola, P. O., Aboud, M. F. A., & Warsi, M. F. (2019). Reduced graphene oxide-spinel ferrite nano-hybrids as magnetically separable and recyclable visible light driven photocatalyst. Synthetic Metals, 254, *1–9. https://doi.org/10.1016/j.synthmet.2019.05.013.*

the wavelength. For rGO-$Ni_{0.65}Zn_{0.35}Nd_{0.03}Fe_{1.97}O_4$ nanohybrids, the absorption intensity diminished with an increase in wavelength. The bandgaps of $Ni_{0.65}Zn_{0.35}Fe_2O_4$ nanoparticles and rGO-$Ni_{0.65}Zn_{0.35}Nd_{0.03}Fe_{1.97}O_4$ nanohybrids were 1.91 and 1.84 eV, respectively. The bandgaps of pure $ZnFe_2O_4$ and $NiFe_2O_4$ are 2.43 and 1.50 eV, respectively. The narrow bandgap in the visible region is related to the nature of two-dimensional graphene sheets in rGO-$Ni_{0.65}Zn_{0.35}Nd_{0.03}Fe_{1.97}O_4$ nanohybrids. Fig. 7.54 indicates the UV–V spectra of MB dye solution using nanoparticles and nanohybrids. Two peaks were observed at 615 and 665 nm. Indeed, when rGO-$Ni_{0.65}Zn_{0.35}Fe_2O_4$ nanohybrids were subjected to visible light, an electron transferred from the valence band to the conduction band, and hence an electron-hole pair was created.

It is notable that the graphene sheets are strongly conductive. Thus, graphene takes the electron from the conduction band and creates a simple way for the transition of the electron as expressed in the following:

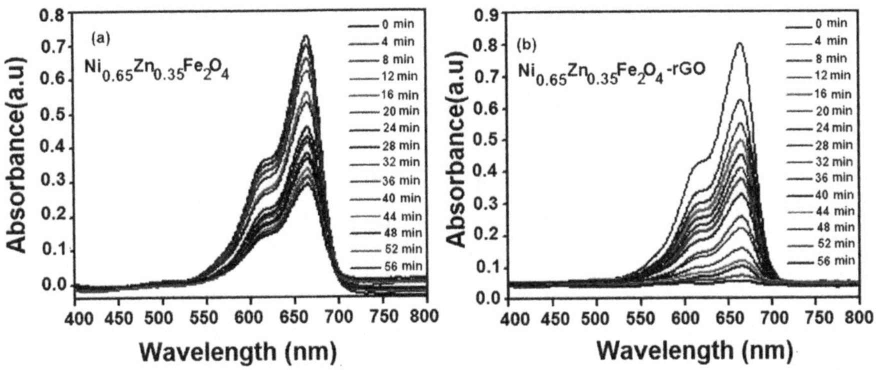

FIGURE 7.54

Ultraviolet–visible absorption spectra of methylene blue dye solution using (A) $Ni_{0.65}Zn_{0.35}Fe_2O_4$ nanoparticles and (B) rGO-$Ni_{0.65}Zn_{0.35}Fe_2O_4$ nanohybrids.

From Javed, H., Rehman, A., Mussadiq, S., Shahid, M., Khan, M. A., Shakir, I., Agboola, P. O., Aboud, M. F. A., & Warsi, M. F. (2019). Reduced graphene oxide-spinel ferrite nano-hybrids as magnetically separable and recyclable visible light driven photocatalyst. Synthetic Metals, 254, *1–9. https://doi.org/10.1016/j.synthmet.2019.05.013.*

$$\text{Graphene}(e) + O_2 \rightarrow O_2^{\cdot} + \text{Graphene} \tag{7.3}$$

$$Ni_{0.65}Zn_{0.35}Fe_2O_4(h) + H_2O \rightarrow Ni_{0.65}Zn_{0.35}Fe_2O_4 + H^+ + OH^{\cdot} \tag{7.4}$$

$$Ni_{0.65}Zn_{0.35}Fe_2O_4(h) + OH^{\cdot} + O_2^{\cdot} + MB \rightarrow CO_2 + H_2O \text{ etc} \tag{7.5}$$

Fig. 7.55 compares the rate of degradation of MB dye solution using TiO_2, $Ni_{0.65}Zn_{0.35}Fe_2O_4$ nanoparticles, and rGO-$Ni_{0.65}Zn_{0.35}Fe_2O_4$ nanohybrids. It is clear that the degradation of MB without any catalyst was 2%, and it was stable in visible light. The degradation percentages for nanoparticles and nanohybrids were about 55% and $>$ 90%, respectively. Thus, the photocatalytic activity of nanohybrids was better than that of the pure ferrite nanoparticles. The recycling stability of $Ni_{0.65}Zn_{0.35}Fe_2O_4$-rGO up to 5 cycles confirmed a decreasing trend from 100% to 92%. This can be attributed to the degradation of graphene during exposure to light and consequently reducing its activity (Park et al., 2019).

7.10.4 Photocatalytic properties of $MnFe_2O_4$ ferrite-graphene nanocomposites

In another study, the high photocatalytic activity of magnetically separable $MnFe_2O_4$ ferrite-graphene nanocomposites was examined, and exceptional photocatalytic activity was found in the composite (Chen et al., 2010; Dom et al., 2011; Fu et al., 2010; Hou et al., 2010; Li et al., 2009; Shifu et al., 2009a; Wang et al., 2009). It was found that the graphene sheets are homogeneously decorated with $MnFe_2O_4$ nanoparticles. The photodegradation of $MnFe_2O_4$-graphene nanocomposites and

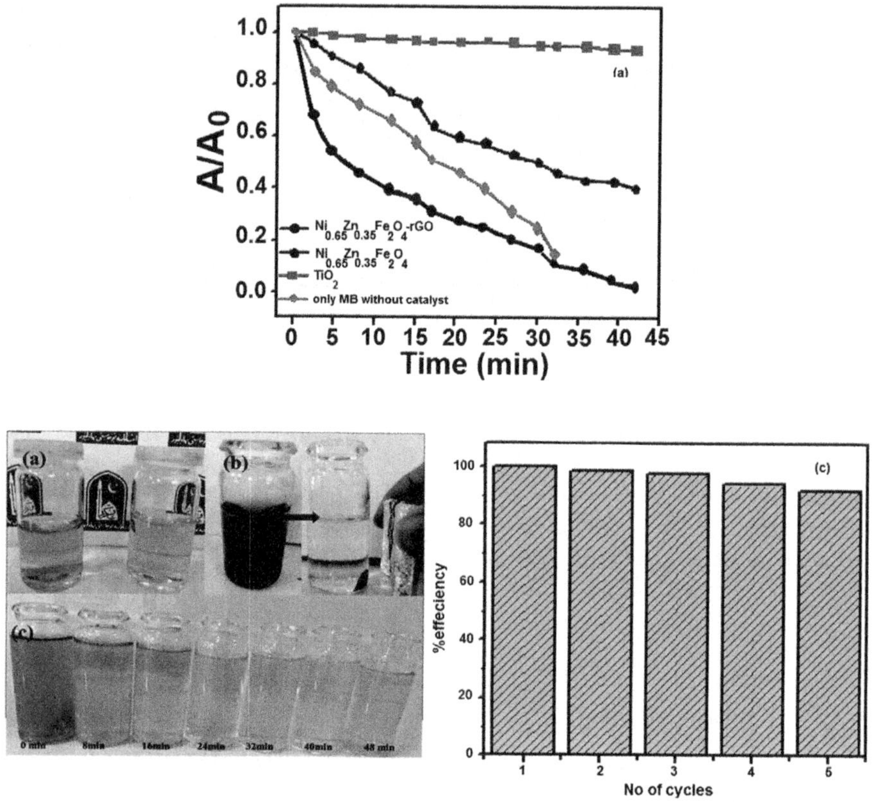

FIGURE 7.55

(A) Variations of degradation percentage versus time for methylene blue (MB) dye solution using TiO_2, $Ni_{0.65}Zn_{0.35}Fe_2O_4$ nanoparticles, and rGO-$Ni_{0.65}Zn_{0.35}Fe_2O_4$ nanohybrids, (B) complete degradation of MB in 48 min, and (C) cyclic stability of rGO-$Ni_{0.65}Zn_{0.35}Fe_2O_4$ nanohybrids.

From Javed, H., Rehman, A., Mussadiq, S., Shahid, M., Khan, M. A., Shakir, I., Agboola, P. O., Aboud, M. F. A., & Warsi, M. F. (2019). Reduced graphene oxide-spinel ferrite nano-hybrids as magnetically separable and recyclable visible light driven photocatalyst. Synthetic Metals, 254, *1–9. https://doi.org/10.1016/j.synthmet.2019.05.013.*

pure $MnFe_2O_4$ ferrite for MB and phenol under UV light irradiation is shown in Fig. 7.56. The photodegradation rates for MB and phenol in pure $MnFe_2O_4$ ferrite after irradiation for 300 min were 80% and 68%, respectively. Thus, the pure $MnFe_2O_4$ ferrite can be employed as a photocatalyst activity under UV irradiation. The photodegradation rates for MB and phenol in $MnFe_2O_4$-graphene (0.3) after irradiation for 300 min were 100% and 85%, respectively. Hence, the catalytic activity of $MnFe_2O_4$-graphene nanocomposites is better than that of the pure $MnFe_2O_4$ ferrite (Fu et al., 2012).

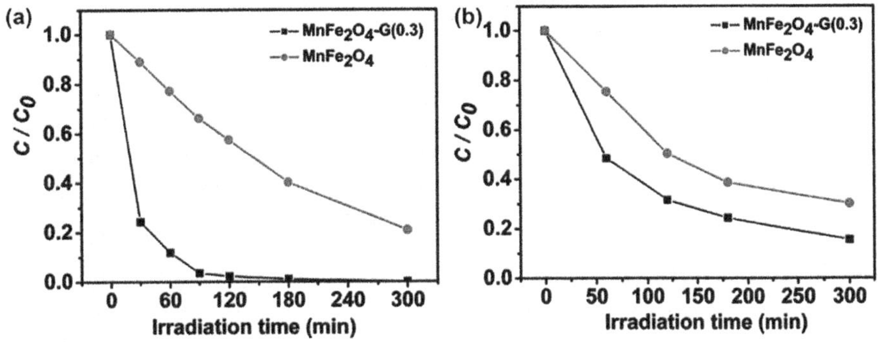

FIGURE 7.56

Plots of ln C_t/C_0 versus irradiation time of $MnFe_2O_4$-G (0.3) and pure $MnFe_2O_4$ ferrites for (A) methylene blue and (B) phenol under ultraviolet light irradiation.

From Fu, Y., Xiong, P., Chen, H., Sun, X., & Wang, X. (2012). High photocatalytic activity of magnetically separable manganese ferrite–graphene heteroarchitectures. Industrial & Engineering Chemistry Research, *51(2), 725–731. https://doi.org/10.1021/ie2026212.*

The visible-light photocatalytic activities of $MnFe_2O_4$-graphene nanocomposites with differing graphene contents in the composite for the degradation of MB are displayed in Fig. 7.57. It can be observed that the pure $MnFe_2O_4$ ferrite under visible light irradiation was photocatalytically inactive. However, the visible light photocatalytic performance of $MnFe_2O_4$-graphene nanocomposites was good, with the MB degradation occurring faster with an increase in graphene content. The best performance of photocatalytic activity was obtained for $MnFe_2O_4$-graphene(0.3) nanocomposites. The pseudo-first-order rate constant (k) for the photodecomposition of MB on $MnFe_2O_4$-graphene nanocomposites with different graphene contents is also seen in Fig. 7.57. The $MnFe_2O_4$-graphene (0.3) nanocomposites had the highest rate constant of 0.00974 min^{-1}. The $MnFe_2O_4$-graphene nanocomposites with a lower graphene concentration ($< 20\%$) exhibited a lower rate constant. It was found from the inset of Fig. 7.57B that 96.2% of total organic carbon (TOC) was removed in 360 min. VSM graphs of pure $MnFe_2O_4$ and $MnFe_2O_4$-graphene (0.3) are also displayed in Fig. 7.57. The saturation magnetization and coercivity of pure $MnFe_2O_4$ ferrite and $MnFe_2O_4$-graphene (0.3) nanocomposites were 55.58 emu/g, 150.32 Oe, and 3.34 emu/g, 142.83 Oe, respectively. It is notable that the $MnFe_2O_4$-graphene (0.3) nanocomposites can be simply separated from the solution phase by applying an external magnet. The photodegradation rate of MB for three cycles employing $MnFe_2O_4$-graphene (0.3) nanocomposites under visible light irradiation is also shown in Fig. 7.57. The recyclability experiment was performed in order to test the stability and long-term use of $MnFe_2O_4$-graphene nanocomposites as heterogeneous photo-Fenton catalysts. No significant reduction in the efficiency of photocatalytic activity of the recycled catalyst was seen after several consecutive rounds of reaction. Thus, the magnetically separable photocatalyst $MnFe_2O_4$-

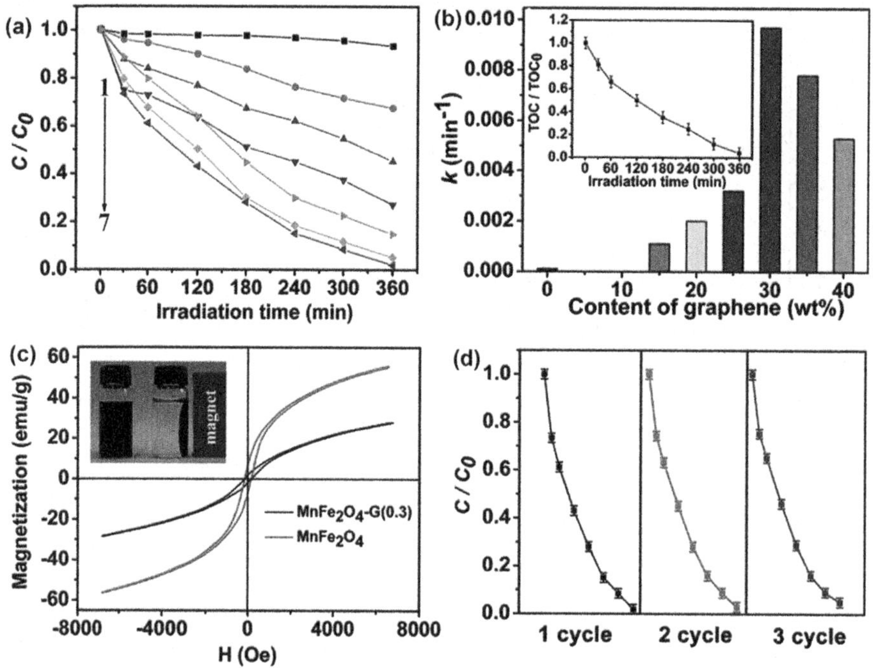

FIGURE 7.57

(A) Photocatalytic degradation of methylene blue (MB) over (1) pure $MnFe_2O_4$, (2) $MnFe_2O_4$-graphene (0.15), (3) $MnFe_2O_4$-graphene (0.20), (4) $MnFe_2O_4$-graphene (0.25), (5) $MnFe_2O_4$-graphene (0.40), (6) $MnFe_2O_4$-graphene (0.35), and (7) $MnFe_2O_4$-graphene (0.30). (B) Rate constant for the photodecomposition of MB on $MnFe_2O_4$-graphene photocatalysts with differing graphene content. (C) Vibrating sample magnetometer graphs of pure $MnFe_2O_4$ and $MnFe_2O_4$-graphene (0.3). (D) Photodegradation rate of MB in the solution for three cycles using $MnFe_2O_4$-graphene (0.3) photocatalyst.

From Fu, Y., Xiong, P., Chen, H., Sun, X., & Wang, X. (2012). High photocatalytic activity of magnetically separable manganese ferrite–graphene heteroarchitectures. Industrial & Engineering Chemistry Research, *51(2), 725–731. https://doi.org/10.1021/ie2026212.*

graphene was stable. The electrochemical impedance spectra of $MnFe_2O_4$-graphene (0.3), pure $MnFe_2O_4$, and GO are revealed in Fig. 7.58. The radius of $MnFe_2O_4$-graphene (0.3) in the impedance plot is smaller than that of GO and pure $MnFe_2O_4$ electrodes. The smaller radius leads to lower resistance of the interfacial charge transfer, which is suitable for enhancing photocatalytic activity. The transient photocurrent responses of $MnFe_2O_4$-graphene (0.3), $ZnFe_2O_4$-graphene (0.2), and pure $MnFe_2O_4$ electrodes under visible light irradiation are also observed in Fig. 7.58. The photocurrent of the $MnFe_2O_4$-graphene (0.3) electrode (0.098 μA) was higher than that of the pure $MnFe_2O_4$ electrode (0.02 μA). This is related to the electronic interaction

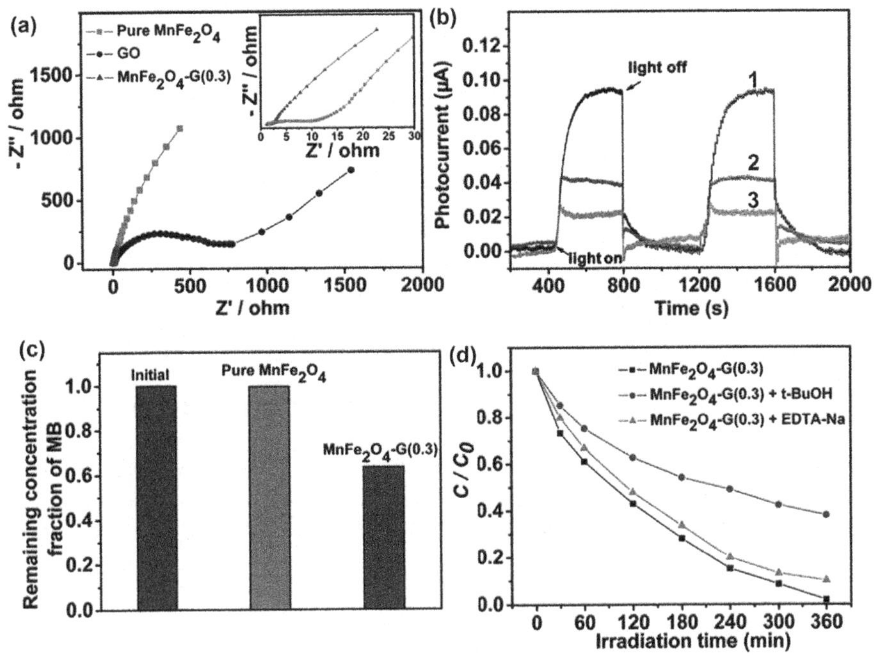

FIGURE 7.58

(A) Electrochemical impedance spectra of $MnFe_2O_4$-graphene (0.3), pure $MnFe_2O_4$, and GO. (B) Photocurrent transient responses of (1) $MnFe_2O_4$-graphene (0.3), (2) $ZnFe_2O_4$-graphene (0.2), and (3) pure $MnFe_2O_4$ electrodes under visible light irradiation. (C) The methylene blue (MB) remains in the solution after reaching the adsorption–desorption equilibrium in the dark for 60 min with stirring. (D) Photocatalytic degradation of MB under visible light irradiation at 25°C with $MnFe_2O_4$-graphene (0.3) + t-BuOH, $MnFe_2O_4$-graphene (0.3) + EDTA-2Na, and $MnFe_2O_4$-graphene (0.3).

From Fu, Y., Xiong, P., Chen, H., Sun, X., & Wang, X. (2012). High photocatalytic activity of magnetically separable manganese ferrite–graphene heteroarchitectures. Industrial & Engineering Chemistry Research, *51(2), 725–731. https://doi.org/10.1021/ie2026212.*

between $MnFe_2O_4$ nanoparticles and graphene sheets, causing enhanced separation efficiency of photoinduced electrons and holes. The photocurrent of the $ZnFe_2O_4$-graphene (0.2) electrode (0.05 μA) was far lower than that of the $MnFe_2O_4$-graphene (0.3) electrode. This is because the zinc ferrite has high photocatalytic activity only in the presence of H_2O_2.

Fig. 7.58 depicts the remaining MB after reaching the adsorption–desorption equilibrium in the dark for $MnFe_2O_4$-graphene (0.3) and pure $MnFe_2O_4$ ferrites. The pure $MnFe_2O_4$ does not show remarkable adsorption of MB dye. However, a large amount of MB molecules (37%) was adsorbed on the surface of $MnFe_2O_4$-graphene (0.3) because of the electrostatic attraction between MB and Π-conjugation regions of the graphene sheets. The photocatalytic degradation of MB under visible

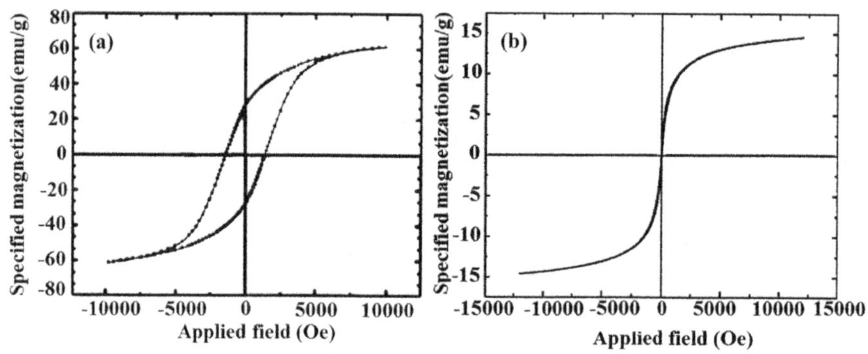

FIGURE 7.59

The magnetic hysteresis loops of (A) $CoFe_2O_4$ nanoparticles and (B) $TiO_2/CoFe_2O_4/Ag$ nanocomposites.

From Hashemi, F., Rizi, H. B., Gheisari, K., Motamedi, H., Pereira, M., & Shirinbayan, M. (2020). Investigation of magnetic composites using as photocatalyst and antibacterial application. Inorganic Chemistry Communications, 119, *108031. https://doi.org/10.1016/j.inoche.2020.108031.*

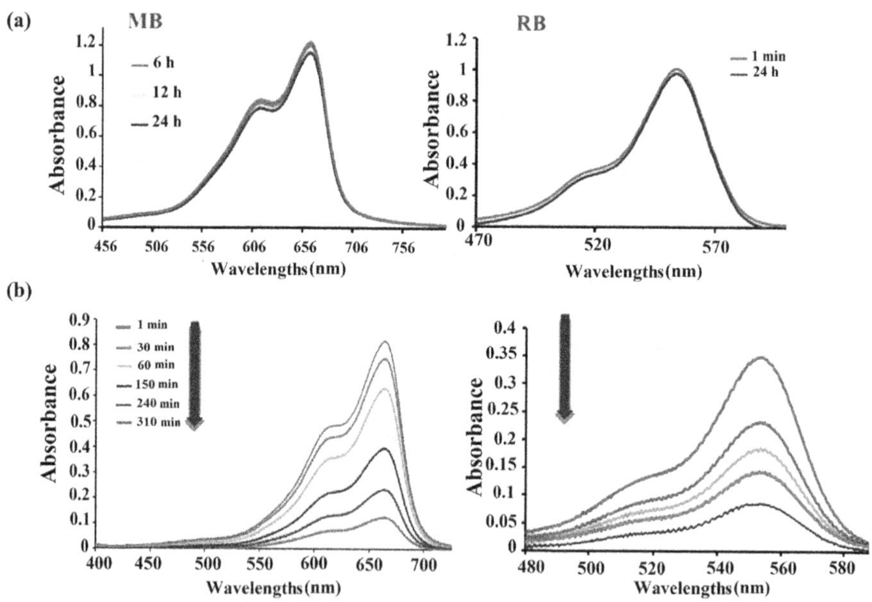

FIGURE 7.60

The (A) absorption and (B) photodegradation performance of $TiO_2/CoFe_2O_4/Ag$ nanocomposite for MB and RB dyes.

From Hashemi, F., Rizi, H. B., Gheisari, K., Motamedi, H., Pereira, M., & Shirinbayan, M. (2020). Investigation of magnetic composites using as photocatalyst and antibacterial application. Inorganic Chemistry Communications, 119, *108031. https://doi.org/10.1016/j.inoche.2020.108031.*

light irradiation of $MnFe_2O_4$-graphene (0.3) + t-BuOH, $MnFe_2O_4$-graphene (0.3) + EDTA-2Na, and $MnFe_2O_4$-graphene (0.3) is also shown in Fig. 7.58. It was observed that the photodegradation rate of MB in the $MnFe_2O_4$-graphene (0.3) suspension strongly decreases with the addition of t-BuOH. However, the addition of EDTA-2Na has no significant influence on the photodegradation rate of MB. Thus, most of the holes can react with the adsorbed water (or hydroxide anions) to create hydroxyl radicals, and only a few of them can oxidize MB molecules.

7.10.5 Photocatalytic properties of TiO_2/ferrite nanocomposites

Other research has examined the dye removal in an aqueous solution of silver deposited titanium dioxide/cobalt ferrite nanocomposites (Hashemi et al., 2020). The magnetic hysteresis loops of $CoFe_2O_4$ nanoparticles and $TiO_2/CoFe_2O_4/Ag$ (TCA) nanocomposites are shown in Fig. 7.59. The saturation magnetization of $TiO_2/CoFe_2O_4/Ag$ nanocomposites was lower than that of the $CoFe_2O_4$ nanoparticles, which is related to the deposition of Ag nanoparticles on the surfaces of $TiO_2/CoFe_2O_4$ nanocomposites. The absorption and photodegradation performance of

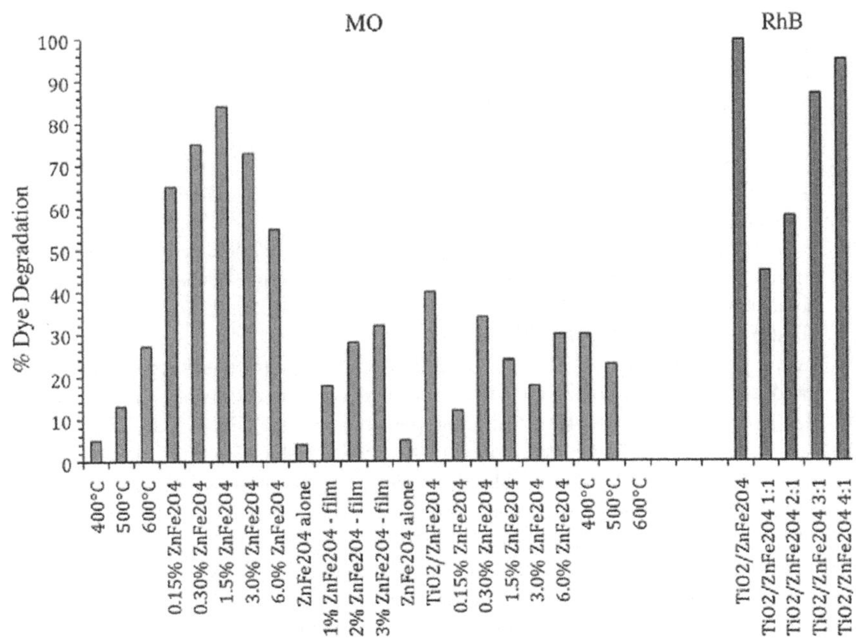

FIGURE 7.61

Degradation of methyl orange and rhodamine B dyes by $TiO_2/ZnFe_2O_4$ composite photocatalysts.

From Casbeer, E., Sharma, V. K., & Li, X.-Z. (2012). Synthesis and photocatalytic activity of ferrites under visible light: a review. Separation and Purification Technology, 87, *1–14. https://doi.org/10.1016/j.seppur.2011.11.034.*

TCA nanocomposites for MB and RB dyes is indicated in Fig. 7.60. The absorption of MB and RB dyes of TCA nanocomposites decreased over time under UV illumination. In addition, the photodegradation of the MB dye was higher than that of the RB dye, as RB dye molecules are stable.

Fig. 7.61 presents MO and RhB dye degradation by various $TiO_2/ZnFe_2O_4$ composite photocatalysts (Casbeer et al., 2012). Visible light irradiated the MO solutions, while UV light irradiated the RhB solutions. It can be seen that both dyes were successfully degraded, though RhB dye had greater degradation efficiency and was degraded more by around 100% (Jia et al., 2020).

References

Abdel Maksoud, M. I. A., El-Sayyad, G. S., El-Khawaga, A. M., Abd Elkodous, M., Abokhadra, A., Elsayed, M. A., … Ashour, A. H. (2020). Nanostructured Mg substituted Mn-Zn ferrites: A magnetic recyclable catalyst for outstanding photocatalytic and antimicrobial potentials. *Journal of Hazardous Materials, 399*, 123000–123018. https://doi.org/10.1016/j.jhazmat.2020.123000

Albuquerque, A. S., Tolentino, M. V., Ardisson, J. D., Moura, F. C., de Mendonça, R., & Macedo, W. A. (2012). Nanostructured ferrites: Structural analysis and catalytic activity. *Ceramics International, 38*(3), 2225–2231. https://doi.org/10.1016/j.ceramint.2011.10.071

Ali, N., Ali, F., Said, A., Begum, T., Bilal, M., Rab, A., Sheikh, Z. A., Iqbal, H. M., & Ahmad, I. (2020). Characterization and deployment of surface-engineered cobalt ferrite nanospheres as photocatalyst for highly efficient remediation of alizarin red S dye from aqueous solution. *Journal of Inorganic and Organometallic Polymers and Materials*, 1–11. https://doi.org/10.1007/s10904-020-01654-y

Ali, N., Ali, F., Sheikh, Z. A., Bilal, M., & Ahmad, I. (2020). Photocatalytic performance of zinc ferrite magnetic nanostructures for efficient Eriochrome black-T degradation from the aqueous environment under unfiltered sunlight. *Water, Air, & Soil Pollution, 231*(2), 59. https://doi.org/10.1007/s11270-020-4437-x

Ao, Y., Xu, J., Fu, D., Shen, X., & Yuan, C. (2008). A novel magnetically separable composite photocatalyst: titania-coated magnetic activated carbon. *Separation and Purification Technology, 61*(3), 436–441. https://doi.org/10.1016/j.seppur.2007.12.007

Badawy, A. A., Ibrahim, S. M., & Essawy, H. A. (2020). Enhancing the textile dye removal from aqueous solution using cobalt ferrite nanoparticles prepared in presence of fulvic acid. *Journal of Inorganic and Organometallic Polymers and Materials, 30*, 1798–1813. https://doi.org/10.1007/s10904-019-01355-1

Bahmanrokh, G., Cazorla, C., Mofarah, S. S., Shahmiri, R., Yao, Y., Ismail, I., … Sorrell, C. C. (2020). Band gap engineering of Ce-doped anatase TiO_2 through solid solubility mechanisms and new defect equilibria formalism. *Nanoscale, 12*(8), 4916–4934. https://doi.org/10.1039/C9NR08604H

Baig, M. M., Zulfiqar, S., Yousuf, M. A., Touqeer, M., Ullah, S., Agboola, P., … Shakir, I. (2020). Structural and photocatalytic properties of new rare earth La^{3+} substituted $MnFe_2O_4$ ferrite nanoparticles. *Ceramics International, 46*(14), 23208–23217. https://doi.org/10.1016/j.ceramint.2020.06.103

Bhukal, S., Bansal, S., & Singhal, S. (2014). Magnetic Mn substituted cobalt zinc ferrite systems: structural, electrical and magnetic properties and their role in photo-catalytic degradation of methyl orange azo dye. *Physica B: Condensed Matter, 445*, 48–55. https://doi.org/10.1016/j.physb.2014.03.088

Bhukal, S., Dhiman, M., Bansal, S., Tripathi, M. K., & Singhal, S. (2016). Substituted Co–Cu–Zn nanoferrites: synthesis, fundamental and redox catalytic properties for the degradation of methyl orange. *RSC Advances, 6*(2), 1360–1375. https://doi.org/10.1039/C5RA22561B

Bhukal, S., & Singhal, S. (2014). Magnetically separable copper substituted cobalt–zinc nano-ferrite photocatalyst with enhanced photocatalytic activity. *Materials Science in Semiconductor Processing, 26*, 467–476. https://doi.org/10.1016/j.mssp.2014.05.023

Bouanimba, N., Zouaghi, R., Laid, N., & Sehili, T. (2011). Factors influencing the photocatalytic decolorization of Bromophenol blue in aqueous solution with different types of TiO_2 as photocatalysts. *Desalination, 275*(1–3), 224–230. https://doi.org/10.1016/j.desal.2011.03.005

Cao, X., Gu, L., Lan, X., Zhao, C., Yao, D., & Sheng, W. (2007). Spinel $ZnFe_2O_4$ nanoplates embedded with Ag clusters: Preparation, characterization, and photocatalytic application. *Materials Chemistry and Physics, 106*(2–3), 175–180. https://doi.org/10.1016/j.matchemphys.2007.05.033

Casbeer, E., Sharma, V. K., & Li, X.-Z. (2012). Synthesis and photocatalytic activity of ferrites under visible light: a review. *Separation and Purification Technology, 87*, 1–14. https://doi.org/10.1016/j.seppur.2011.11.034

Chen, C.-H., Liang, Y.-H., & Zhang, W.-D. (2010). $ZnFe_2O_4$/MWCNTs composite with enhanced photocatalytic activity under visible-light irradiation. *Journal of Alloys and Compounds, 501*(1), 168–172. https://doi.org/10.1016/j.jallcom.2010.04.072

Chavan, S., Babrekar, M., More, S., & Jadhav, K. (2010). Structural and optical properties of nanocrystalline Ni–Zn ferrite thin films. *Journal of Alloys and Compounds, 507*(1), 21–25. https://doi.org/10.1016/j.jallcom.2010.07.171

Cheng, P., Li, W., Zhou, T., Jin, Y., & Gu, M. (2004). Physical and photocatalytic properties of zinc ferrite doped titania under visible light irradiation. *Journal of Photochemistry and Photobiology A: Chemistry, 168*(1–2), 97–101. https://doi.org/10.1016/j.jphotochem.2004.05.018

Dhiman, M., Sharma, R., Kumar, V., & Singhal, S. (2016). Morphology controlled hydrothermal synthesis and photocatalytic properties of $ZnFe_2O_4$ nanostructures. *Ceramics International, 42*(11), 12594–12605. https://doi.org/10.1016/j.ceramint.2016.04.115

Din, M. I., Jabbar, S., Najeeb, J., Khalid, R., Ghaffar, T., Arshad, M., Khan, S. A., & Ali, S. (2020). Green synthesis of zinc ferrite nanoparticles for photocatalysis of methylene blue. *International Journal of Phytoremediation, 22*(13), 1440–1447. https://doi.org/10.1080/15226514.2020.1781783

Dom, R., Subasri, R., Radha, K., & Borse, P. H. (2011). Synthesis of solar active nanocrystalline ferrite, MFe_2O_4 (M: Ca, Zn, Mg) photocatalyst by microwave irradiation. *Solid State Communications, 151*(6), 470–473. https://doi.org/10.1016/j.ssc.2010.12.034

Dutta, P. K., Pehkonen, S., Sharma, V. K., & Ray, A. K. (2005). Photocatalytic oxidation of arsenic (III): Evidence of hydroxyl radicals. *Environmental Science & Technology, 39*(6), 1827–1834. https://doi.org/10.1021/es0489238

El-Salamony, R. A., Amdeha, E., Badawy, N. A., Ghoneim, S. A., & Al-Sabagh, A. M. (2018). Visible light sensitive activated carbon-metal oxide (TiO_2, WO_3, NiO, and SnO) nano-catalysts for photo-degradation of methylene blue: a comparative study.

Toxicological and Environmental Chemistry, 100(2), 143–156. https://doi.org/10.1080/02772248.2018.1497634

Fan, G., Gu, Z., Yang, L., & Li, F. (2009). Nanocrystalline zinc ferrite photocatalysts formed using the colloid mill and hydrothermal technique. *Chemical Engineering Journal, 155*(1–2), 534–541. https://doi.org/10.1016/j.cej.2009.08.008

Fenton, H. (1894). LXXIII—oxidation of tartaric acid in presence of iron. *Journal of the Chemical Society Transactions, 65*, 899–910. https://doi.org/10.1039/CT8946500899

Frolova, L. (2020). Photocatalytic activity of spinel ferrites $Co_xFe_{3-x}O_4$ (0.25< x < 1) obtained by treatment contact low-temperature non-equilibrium plasma. *Applied Nanoscience, 10*(12), 4585–4590. https://doi.org/10.1007/s13204-020-01344-8

Fu, Y.-P., Chang, W.-K., Wang, H.-C., Liu-Wen, C., & Lin, C.-H. (2010). Synthesis and characterization of anatase TiO_2 nanolayer coating on Ni-Cu-Zn ferrite powders for magnetic photocatalyst. *Journal of Materials Research, 25*(1), 134–140. https://doi.org/10.1557/JMR.2010.0015

Fu, Y., Xiong, P., Chen, H., Sun, X., & Wang, X. (2012). High photocatalytic activity of magnetically separable manganese ferrite–graphene heteroarchitectures. *Industrial & Engineering Chemistry Research, 51*(2), 725–731. https://doi.org/10.1021/ie2026212

Gandomi, F., Peymani-Motlagh, S. M., Rostami, M., Sobhani-Nasab, A., Fasihi-Ramandi, M., Eghbali-Arani, M., … Ganjali, M. R. (2019). Simple synthesis and characterization of $Li_{0.5}Fe_{2.5}O_4$, $LiMg_{0.5}Fe_2O_4$ and $LiNi_{0.5}Fe_2O_4$, and investigation of their photocatalytic and anticancer properties on hela cells line. *Journal of Materials Science: Materials in Electronics, 30*(22), 19691–19702. https://doi.org/10.1007/s10854-019-02320-x

Gateshki, M., Petkov, V., Pradhan, S. K., & Vogt, T. (2005). Structure of nanocrystalline $MgFe_2O_4$ from X-ray diffraction, Rietveld and atomic pair distribution function analysis. *Journal of Applied Crystallography, 38*(5), 772–779. https://doi.org/10.1107/S0021889805024477

Gregor, C., Hermanek, M., Jancik, D., Pechousek, J., Filip, J., Hrbac, J., & Zboril, R. (2010). The effect of surface area and crystal structure on the catalytic efficiency of iron (III) oxide nanoparticles in hydrogen peroxide decomposition. *European Journal of Inorganic Chemistry, 2010*(16), 2343–2351. https://doi.org/10.1002/ejic.200901066

Guo, J.-F., Ma, B., Yin, A., Fan, K., & Dai, W.-L. (2011). Photodegradation of rhodamine B and 4-chlorophenol using plasmonic photocatalyst of Ag–AgI/Fe_3O_4@SiO_2 magnetic nanoparticle under visible light irradiation. *Applied Catalysis B: Environmental, 101*(3–4), 580–586. https://doi.org/10.1016/j.apcatb.2010.10.032

Hashemi, F., Rizi, H. B., Gheisari, K., Motamedi, H., Pereira, M., & Shirinbayan, M. (2020). Investigation of magnetic composites using as photocatalyst and antibacterial application. *Inorganic Chemistry Communications, 119*, 108031–108065. https://doi.org/10.1016/j.inoche.2020.108031

Hoffmann, M. R., Martin, S. T., Choi, W., & Bahnemann, D. W. (1995). Environmental applications of semiconductor photocatalysis. *Chemical Reviews, 95*(1), 69–96. https://doi.org/10.1021/cr00033a004

Hou, X., Feng, J., Xu, X., & Zhang, M. (2010). Synthesis and characterizations of spinel $MnFe_2O_4$ nanorod by seed–hydrothermal route. *Journal of Alloys and Compounds, 491*(1–2), 258–263. https://doi.org/10.1016/j.jallcom.2009.10.029

Jadhav, S., Hankare, P., Patil, R., & Sasikala, R. (2011). Effect of sintering on photocatalytic degradation of methyl orange using zinc ferrite. *Materials Letters, 65*(2), 371–373. https://doi.org/10.1016/j.matlet.2010.10.004

Jain, S. C., Mertens, R., & Van Overstraeten, R. (1991). Bandgap narrowing and its effects on the properties of moderately and heavily doped germanium and silicon. In *Advances in electronics and electron physics, 82*Elsevier.

Jauhar, S., & Singhal, S. (2014). Substituted cobalt nano-ferrites, $CoM_xFe_{2-x}O_4$ ($M = Cr^{3+}$, Ni^{2+}, Cu^{2+}, Zn^{2+}; $0.2 \leq x \leq 1.0$) as heterogeneous catalysts for modified Fenton's reaction. *Ceramics International, 40*(8), 11845–11855. https://doi.org/10.1016/j.ceramint.2014.04.019

Javed, H., Rehman, A., Mussadiq, S., Shahid, M., Khan, M. A., Shakir, I., … Warsi, M. F. (2019). Reduced graphene oxide-spinel ferrite nano-hybrids as magnetically separable and recyclable visible light driven photocatalyst. *Synthetic Metals, 254*, 1–9. https://doi.org/10.1016/j.synthmet.2019.05.013

Jia, Y., Zhang, W., Do, J. Y., Kang, M., & Liu, C. (2020). Z-scheme $SnFe_2O_4/\alpha\text{-}Fe_2O_3$ micro-octahedron with intimated interface for photocatalytic CO_2 reduction. *Chemical Engineering Journal, 402*, 126193–126202. https://doi.org/10.1016/j.cej.2020.126193

Jing, J., Liangchao, L., & Feng, X. (2007). Structural analysis and magnetic properties of Gd-doped Li-Ni ferrites prepared using rheological phase reaction method. *Journal of Rare Earths, 25*(1), 79–83. https://doi.org/10.1016/S1002-0721(07)60049-0

Kansal, S. K., Sood, S., Umar, A., & Mehta, S. K. (2013). Photocatalytic degradation of Eriochrome Black T dye using well-crystalline anatase TiO_2 nanoparticles. *Journal of Alloys and Compounds, 581*, 392–397. https://doi.org/10.1016/j.jallcom.2013.07.069

Kim, J., & Kang, M. (2012). High photocatalytic hydrogen production over the band gap-tuned urchin-like Bi_2S_3-loaded TiO_2 composites system. *International Journal of Hydrogen Energy, 37*(10), 8249–8256. https://doi.org/10.1016/j.ijhydene.2012.02.057

Kislov, N., Srinivasan, S., Emirov, Y., & Stefanakos, E. (2008). Optical absorption red and blue shifts in $ZnFe_2O_4$ nanoparticles. *Materials Science and Engineering: B, 153*(1–3), 70–77. https://doi.org/10.1016/j.mseb.2008.10.032

Krasnyakova, T., Yurchilo, S., Morenko, V., Nosolev, I., Glazunova, E., Khasbulatov, S., Verbenko, I., & Mitchenko, S. (2020). Effect of mechanochemical pretreatment on the photocatalytic activity of bismuth ferrite. *Kinetics and Catalysis, 61*(3), 384–389. https://doi.org/10.1134/S0023158420030155

Li, H., Zhang, Y., Wang, S., Wu, Q., & Liu, C. (2009). Study on nanomagnets supported TiO_2 photocatalysts prepared by a sol–gel process in reverse microemulsion combining with solvent-thermal technique. *Journal of Hazardous Materials, 169*(1–3), 1045–1053. https://doi.org/10.1016/j.jhazmat.2009.04.040

Li, W., Wang, Y., Lin, H., Ismat Shah, S., Huang, C., Doren, D., … Barteau, M. (2003). Band gap tailoring of Nd^{3+}-doped TiO_2 nanoparticles. *Applied Physics Letters, 83*(20), 4143–4145. https://doi.org/10.1063/1.1627962

Lee, S., Drwiega, J., Mazyck, D., Wu, C.-Y., & Sigmund, W. M. (2006). Synthesis and characterization of hard magnetic composite photocatalyst—barium ferrite/silica/titania. *Materials Chemistry and Physics, 96*(2–3), 483–488. https://doi.org/10.1016/j.matchemphys.2005.07.039

Li, X., Hou, Y., Zhao, Q., & Wang, L. (2011). A general, one-step and template-free synthesis of sphere-like zinc ferrite nanostructures with enhanced photocatalytic activity for dye degradation. *Journal of Colloid and Interface Science, 358*(1), 102–108. https://doi.org/10.1016/j.jcis.2011.02.052

McCarty, R. D., Hord, J., & Roder, H. M. (1981). *Selected properties of hydrogen (engineering design data), 168*. US Department of Commerce, National Bureau of Standards.

Mohd Ridzuan, N. D., Mohd Kaus, N. H., Md Lazim, M. A. S., Kobayashi, T., Adnan, R., Othman, M. Z., & Mohamad Kassim, M. H. (2020). Photocatalytic heterostructures-based $BiFeO_3$ embedded liquid natural rubber (LNR) for highly removal of cationic dye under direct sunlight. *Journal of Environmental Chemical Engineering, 8*(5), 104152. https://doi.org/10.1016/j.jece.2020.104152

Nezamzadeh-Ejhieh, A., & Zabihi-Mobarakeh, H. (2014). Heterogeneous photodecolorization of mixture of methylene blue and bromophenol blue using CuO-nano-clinoptilolite. *Journal of Industrial and Engineering Chemistry, 20*(4), 1421–1431. https://doi.org/10.1016/j.jiec.2013.07.027

Park, C. M., Kim, Y. M., Kim, K.-H., Wang, D., Su, C., & Yoon, Y. (2019). Potential utility of graphene-based nano spinel ferrites as adsorbent and photocatalyst for removing organic/inorganic contaminants from aqueous solutions: a mini review. *Chemosphere, 221*, 392–402. https://doi.org/10.1016/j.chemosphere.2019.01.063

Patil, S., Naik, H. B., Nagaraju, G., Viswanath, R., & Rashmi, S. (2018). Sugarcane juice mediated eco-friendly synthesis of visible light active zinc ferrite nanoparticles: Application to degradation of mixed dyes and antibacterial activities. *Materials Chemistry and Physics, 212*, 351–362. https://doi.org/10.1016/j.matchemphys.2018.03.038

Rana, S., Srivastava, R., Sorensson, M., & Misra, R. (2005). Synthesis and characterization of nanoparticles with magnetic core and photocatalytic shell: Anatase $TiO_2-NiFe_2O_4$ system. *Materials Science and Engineering: B, 119*(2), 144–151. https://doi.org/10.1016/j.mseb.2005.02.043

Rashad, M., Mohamed, R., Ibrahim, M., Ismail, L., & Abdel-Aal, E. (2012). Magnetic and catalytic properties of cubic copper ferrite nanopowders synthesized from secondary resources. *Advanced Powder Technology, 23*(3), 315–323. https://doi.org/10.1016/j.apt.2011.04.005

Sakthivel, S., Neppolian, B., Shankar, M., Arabindoo, B., Palanichamy, M., & Murugesan, V. (2003). Solar photocatalytic degradation of azo dye: comparison of photocatalytic efficiency of ZnO and TiO_2. *Solar Energy Materials and Solar Cells, 77*(1), 65–82. https://doi.org/10.1016/S0927-0248(02)00255-6

Sen, Z. (2008). *Solar energy fundamentals and modeling techniques: Atmosphere, environment, climate change and renewable energy*. Springer Science & Business Media.

Sharma, R., Bansal, S., & Singhal, S. (2015). Tailoring the photo-Fenton activity of spinel ferrites (MFe_2O_4) by incorporating different cations (M= Cu, Zn, Ni and Co) in the structure. *RSC Advances, 5*(8), 6006–6018. https://doi.org/10.1039/c4ra13692f

Sharma, R., & Singhal, S. (2013). Structural, magnetic and electrical properties of zinc doped nickel ferrite and their application in photo catalytic degradation of methylene blue. *Physica B: Condensed Matter, 414*, 83–90. https://doi.org/10.1016/j.physb.2013.01.015

Shifu, C., Wei, Z., Wei, L., Huaye, Z., & Xiaoling, Y. (2009a). Preparation, characterization and activity evaluation of p–n junction photocatalyst p-$CaFe_2O_4$/n-ZnO. *Chemical Engineering Journal, 155*(1–2), 466–473.

Shifu, C., Wei, Z., Wei, L., Huaye, Z., Xiaoling, Y., & Yinghao, C. (2009b). Preparation, characterization and activity evaluation of p-n junction photocatalyst p-$CaFe_2O_4$/n-Ag_3VO_4 under visible light irradiation. *Journal of Hazardous Materials, 172*(2–3), 1415–1423. https://doi.org/10.1016/j.jhazmat.2009.08.007

Singh, C., Bansal, S., Kumar, V., & Singhal, S. (2015). Beading of cobalt substituted nickel ferrite nanoparticles on the surface of carbon nanotubes: A study of their synthesis mechanism, structure, magnetic, optical and their application as photocatalyst. *Ceramics International, 41*(3), 3595–3604. https://doi.org/10.1016/j.ceramint.2014.11.021

Singh, C., Bansal, S., Kumar, V., Tikoo, K., & Singhal, S. (2015). Encrustation of cobalt doped copper ferrite nanoparticles on solid scaffold CNTs and their comparison with corresponding ferrite nanoparticles: A study of structural, optical, magnetic and photo catalytic properties. *RSC Advances, 5*(49), 39052–39061. https://doi.org/10.1039/C5RA03330F

Singh, R., & Pal, B. (2013). Study of excited charge carrier's lifetime for the observed photoluminescence and photocatalytic activity of CdS nanostructures of different shapes. *Journal of Molecular Catalysis A: Chemical, 371*, 77–85. https://doi.org/10.1016/j.molcata.2013.01.024

Tauc, J. (2012). *Amorphous and liquid semiconductors*. Springer Science & Business Media.

To Loan, N. T., Hien Lan, N. T., Thuy Hang, N. T., Quang Hai, N., Tu Anh, D. T., Thi Hau, V., … Van Tran, T. (2019). $CoFe_2O_4$ nanomaterials: Effect of annealing temperature on characterization, magnetic, photocatalytic, and photo-fenton properties. *Processes, 7*(12), 885. https://doi.org/10.3390/pr7120885

Ullah, I., Ali, F., Ali, Z., & Humayun, M. (2018). Glycol stabilized magnetic nanoparticles for photocatalytic degradation of xylenol orange. *Materials Research Express, 5*(5), 055509. https://doi.org/10.1088/2053-1591/aac3b8

Waerenborgh, J., Figueiredo, M., Cabral, J., & Pereira, L. (1994). Temperature and composition dependence of the cation distribution in synthetic $ZnFe_yAl_{2-y}O_4$ ($0 \leq y \leq 1$) spinels. *Journal of Solid State Chemistry, 111*(2), 300–309. https://doi.org/10.1006/jssc.1994.1231

Wang, K., Yu, L., Yin, S., Li, H., & Li, H. (2009). Photocatalytic degradation of methylene blue on magnetically separable $FePc/Fe_3O_4$ nanocomposite under visible irradiation. *Pure and Applied Chemistry, 81*(12), 2327–2335. https://doi.org/10.1351/PAC-CON-08-11-23

Xu, S., Shangguan, W., Yuan, J., Shi, J., & Chen, M. (2007). Preparations and photocatalytic degradation of methyl orange in water on magnetically separable $Bi_{12}TiO_{20}$ supported on nickel ferrite. *Science and Technology of Advanced Materials, 8*(1–2), 40–46. https://doi.org/10.1016/j.stam.2006.09.009

Zhang, G.-Y., Sun, Y.-Q., Gao, D.-Z., & Xu, Y.-Y. (2010). Quasi-cube $ZnFe_2O_4$ nanocrystals: Hydrothermal synthesis and photocatalytic activity with TiO_2 (Degussa P25) as nanocomposite. *Materials Research Bulletin, 45*(7), 755–760. https://doi.org/10.1016/j.materresbull.2010.03.025

Zhang, H., Hou, R., Lu, Z.-L., & Duan, X. (2009). A novel magnetic nanocomposite involving anatase titania coating on silica-modified cobalt ferrite via lower temperature hydrolysis of a water-soluble titania precursor. *Materials Research Bulletin, 44*(10), 2000–2008.

Zang, Y., & Farnood, R. (2008). Photocatalytic activity of $AgBr/TiO_2$ in water under simulated sunlight irradiation. *Applied Catalysis B: Environmental, 79*(4), 334–340. https://doi.org/10.1016/j.apcatb.2007.10.019

Zhang, G., Xu, W., Li, Z., Hu, W., & Wang, Y. (2009). Preparation and characterization of multi-functional $CoFe_2O_4$–ZnO nanocomposites. *Journal of Magnetism and Magnetic Materials, 321*(10), 1424–1427.

CHAPTER

Ferrite synthesis methods 8

8.1 Introduction

Despite considerable advances in the study of ferrites, more research is needed to discover the best techniques for synthesizing ferrites with high purity, narrow size distribution, and uniform morphology. It is essential to correlate such properties with the relevant chemical and structural features. More suitable approaches for synthesizing ferrites can be developed by controlling these characteristics. In synthesized ferrite, gross inhomogeneity can be quite harmful; however, by applying final heat treatment in a proper atmosphere, one can correct chemical inhomogeneity. In the last few years, many synthesis techniques have been developed for ferrites, each with advantages and drawbacks. In the following, the most common techniques for ferrite synthesis are presented and compared.

8.2 Synthesis techniques for ferrite nanoparticles

Ferrite nanoparticles are promising candidates to be employed in microwave-absorbing media, hyperthermia, exchange-spring magnets, and many other practical components. Several processing approaches, including sol−gel, coprecipitation, reverse and normal micelles, hydrothermal and solvothermal, and mechanical milling, are used to fabricate ferrite particles. Subsequent heat treatment is typically required to generate ferrite with a high crystalline feature.

8.2.1 Sol−gel technique

The sol−gel technique is a multifaceted solution-based method used in fabricating ceramics and glassy materials. This technique generally includes a suspension as a sol [the sol is a metal alkoxide solution, $M(OR)_x$, where R = an alkyl group] that undergoes hydrolysis and condensation polymerization reactions to form a viscous (3-D) network as a gel at room temperature. A graphical representation of the sol−gel technique is shown in Fig. 8.1. The properties of the resultant oxide material change with the nature of the metal and alkyl groups. In fact, the molecular weight of the oxide material is continuously enhanced to produce a viscous gel by

Magnetic Ferrites and Related Nanocomposites. https://doi.org/10.1016/B978-0-12-824014-4.00008-1

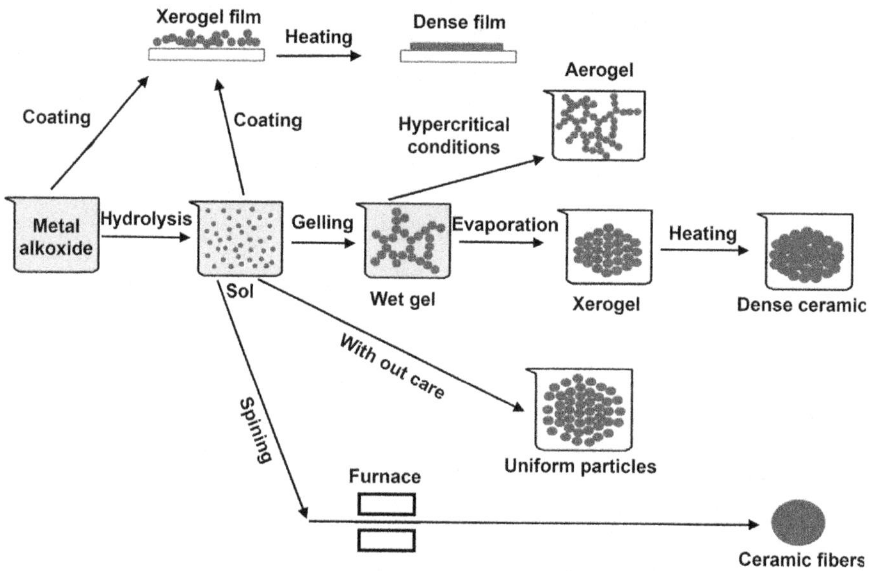

FIGURE 8.1

A graphical representation of the sol–gel technique.

From Kotnala, R., & Shah, J. (2015). Ferrite materials: nano to spintronics regime. In Handbook of magnetic materials *(Vol. 23, pp. 291–379). Elsevier.*

step-growth polymerization, known as a condensation process. The citrate-gel process or modified sol–gel technique involves the formation of complexes of metals and nonmetals with organic agents such as citric acid. In this process, a polyalcohol (e.g., ethylene glycol) creates linkages between the chelates through a polyesterification reaction, giving rise to gelation of the reaction mixture. The schematic representation of the modified sol–gel technique is shown in Fig. 8.2. As an advantage, ferrites with considerable complexity can be easily synthesized using the citrate-gel technique (Kotnala & Shah, 2015).

It should be noted that subsequent heat treatments are required to eliminate any vaporizable compounds and obtain a crystalline condition. The sol–gel technique is considered a significant synthesis technique, as it has a low cost and is carried out without special instruments (Kefeni et al., 2017).

The reaction temperature in the sol–gel technique alters between 25 and 200°C, making it feasible to synthesize ferrite nanoparticles with narrow size distributions. In addition, the sol–gel method is a simple route by which the composition, microstructure, purity, and morphology of ferrite nanoparticles can be improved after adjusting several key factors, including sol concentration, pH, stirring rate, aging time, and annealing temperature. The sol–gel technique has been used to synthesize ferrite nanoparticles, films and coatings, as shown in Fig. 8.1. The principal

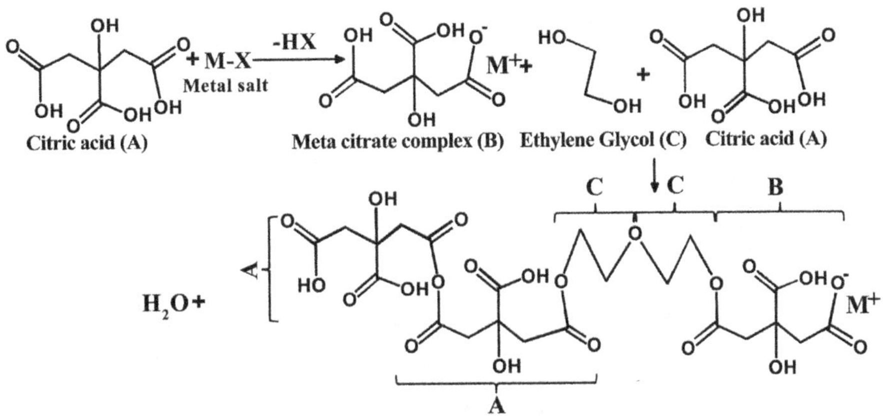

FIGURE 8.2

A schematic representation of the modified sol–gel technique.

From Kotnala, R., & Shah, J. (2015). Ferrite materials: nano to spintronics regime. In Handbook of magnetic materials *(Vol. 23, pp. 291–379). Elsevier.*

disadvantage of this technique is the formation of secondary phases such as iron oxides along with the resultant ferrite materials. Normally, further heat treatment is required to obtain pure ferrite nanoparticles (Avazpour et al., 2015; Larumbe et al., 2012).

8.2.2 Coprecipitation technique

Thermal decomposition of soluble compounds in aqueous solutions leads to oxide formation through a coprecipitation technique. The basic mechanisms of this technique have not been completely realized up to this time because of difficulties in isolating each process. The particle size and morphology of nanoparticles obtained from the coprecipitation technique can be influenced by nucleation and growth. At the beginning of precipitation reactions, nucleation involves the formation of many small crystals. However, since large-size crystals are thermodynamically more stable, small crystals are quickly assembled to create larger ones. This is known as the growth process or coarsening phenomenon controlled by diffusion.

The concentration of reactants, temperature, and pH may influence reaction and transport rates. It is worth noting that reaction rates and impurities can affect particle structure and crystallinity. In addition, the particle size and morphology of the precipitate can be controlled by several parameters such as colloidal stability, supersaturation, recrystallization, nucleation and growth rates, and aging phenomena. Notably, particles are small and compact at low supersaturation. Conversely, they become large and dendritic at high supersaturation. Thus, the morphology of the precipitate depends on the supersaturation, surface energy, and diffusion coefficients of cations and anions (Buschow, 2015).

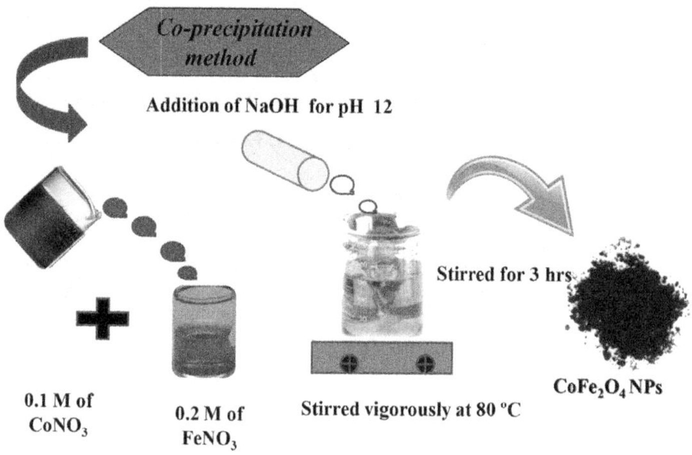

FIGURE 8.3

Schematic representation of the synthesis of $CoFe_2O_4$ nanoparticles by coprecipitation method.

From Annie Vinosha, P., Ansel Mely, L., Emima Jeronsia, J., Raja, K., Queen Sahaya Tamilarsi, D., Fernandez, A., Krishnan, C. S., & Jerome Das, S. (2016). Investigation of optical, electrical and magnetic properties of cobalt ferrite nanoparticles by naive co-precipitation technique. Optik, *127(20), 9917–9925. https://doi.org/10.1016/j.ijleo.2016.07.063.*

To produce ferrite nanoparticles using a coprecipitation technique, aqueous solutions comprising a combination of divalent and trivalent transition metals are blended together. Soluble salts including Fe(III) generally act as the origin of trivalent metal ions, and the corresponding experiment is usually carried out in an alkaline medium. The synthesis technique needs precise adjustment and monitoring of pH to prepare high-quality ferrite nanoparticles. Ammonium or sodium hydroxide solution is commonly used to adjust the pH. In the absence or presence of heat, the solution is stirred vigorously. A schematic representation of $CoFe_2O_4$ nanoparticle synthesis by the coprecipitation technique is shown in Fig. 8.3. While this technique is significant for preparing uniform nanoparticles, the crystallinity of ferrite nanoparticles synthesized in aqueous solutions is typically low. Consequently, subsequent heat treatment is needed to enhance the degree of crystallinity of the resulting ferrite nanoparticles (Cruz et al., 2018; Nabiyouni et al., 2015; Annie Vinosha et al., 2016; Xing et al., 2015).

8.2.3 Microemulsion technique

A microemulsion includes at least three components: water as a polar phase, oil as a nonpolar phase, and a surfactant. The stabilization of nanoparticles against aggregation in solution is realized by surfactants. Surfactants can be adsorbed onto or fully cover the surface of particles, thus preventing their aggregation. In the microemulsion technique, surfactant molecules thermodynamically stabilize the dispersion of two relatively immiscible liquids. The polar head groups of the surfactant molecules are attracted to water, whereas the nonpolar head groups are adsorbed on the oil surface. The particles enclosed within a membrane are prevented from aggregation with other particles because the membrane acts as a barrier. This technique can be categorized into two forms, namely, reverse micelle (water-in-oil) and normal micelle (oil-in-water). Reverse micelles are water pools in a hydrocarbon solvent. The size of the water pools depends on the water/surfactant ratio. A schematic representation of a reverse microemulsion synthesis process is shown in Fig. 8.4.

The La Mer diagram illustrates variations in particle size as a function of precursor concentration and aqueous droplet size. The nucleation and growth of nanoparticles in the La Mer model are observed in Fig. 8.5, revealing the variation of concentration as a function of time during the precipitation reaction. Initially, the concentration increases with increases in time, followed by the occurrence of the nucleation process when the concentration reaches the critical supersaturation value between C^*_{max} and C^*_{min}. In continuance, the concentration is reduced as a function of time during the growth of particles until reaching the solubility value (Buschow, 2015).

The surfactant, cosurfactant, oil:water ratio, and reaction conditions can be changed in the microemulsion technique, allowing for highly controlled particle sizes. Nevertheless, the resulting nanoparticles may not be highly crystalline, as this synthesis technique is carried out at low temperatures (Rashad et al., 2015). Alternatively, the structure of nanoparticles formed by microemulsion depends on the geometric characteristics of the surfactant molecule. The packing parameter of the surfactant molecule is estimated as follows:

$$\text{Packing parameter} = V/(a_0 l_c) \tag{8.1}$$

where V, a_0, and l_c are the volume, optimal area, and length of the hydrocarbon chain of the surfactant molecule, respectively. Fig. 8.6 shows the packing shapes of the surfactant and the structure of their aggregates. Accordingly, ferrite nanoparticles with various morphologies can be prepared using the microemulsion technique (Calero-Díaz del Castillo, 2005).

8.2.4 Hydrothermal and solvothermal techniques

The hydrothermal technique is an inorganic process involving chemical reactions in an aqueous solution above the water boiling point. This technique can be used to prepare ferrite nanoparticles for industrial and biomedical applications. Meanwhile, the solvothermal technique is carried out at relatively high temperatures. Hydrothermal and solvothermal reactions occur in a high-pressure autoclave or completely sealed receptacle. A reaction chamber or autoclave is a suitable medium for any hydrothermal or solvothermal system. Compared with organic solvents, water is an inexpensive and available material without any toxicity. Therefore, it has been widely used as a solvent in hydrothermal and solvothermal processes. It has been found that

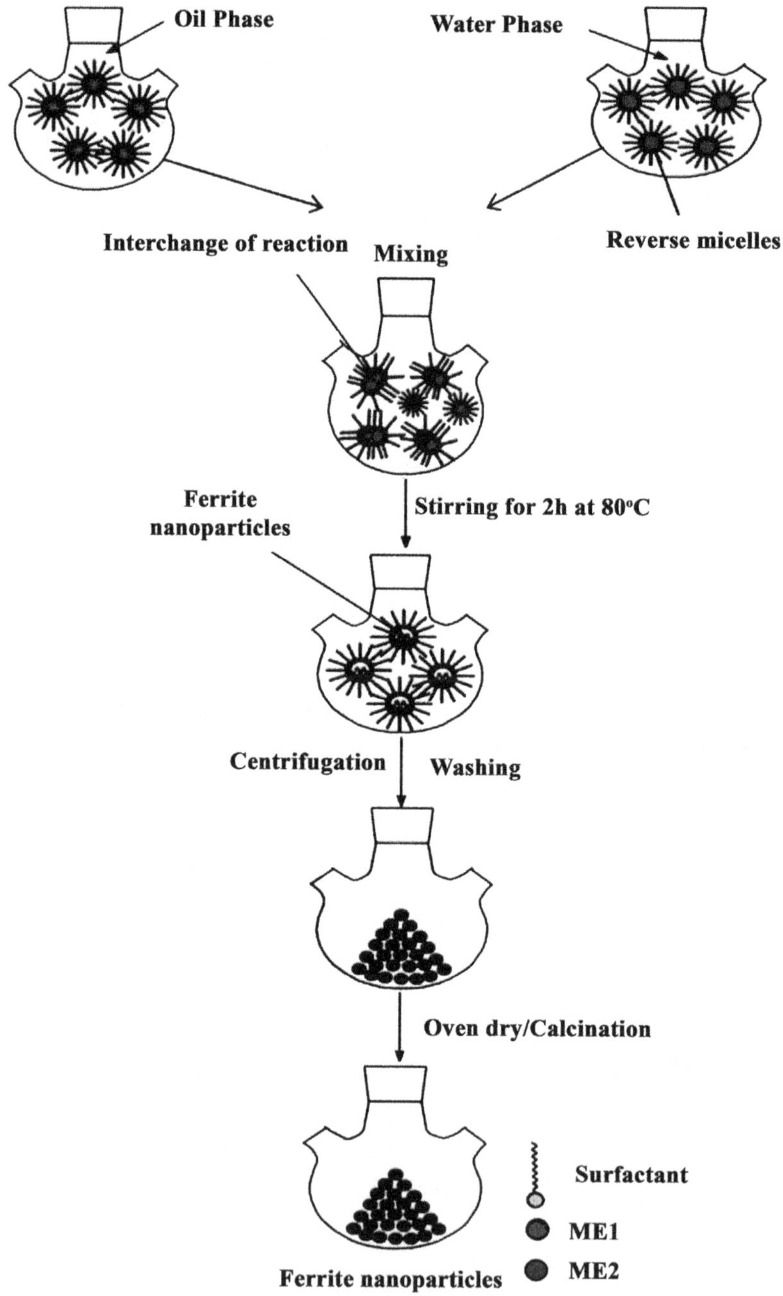

FIGURE 8.4

A schematic representation of a reverse microemulsion synthesis process.

From Kotnala, R., & Shah, J. (2015). Ferrite materials: nano to spintronics regime. In Handbook of magnetic materials *(Vol. 23, pp. 291–379). Elsevier.*

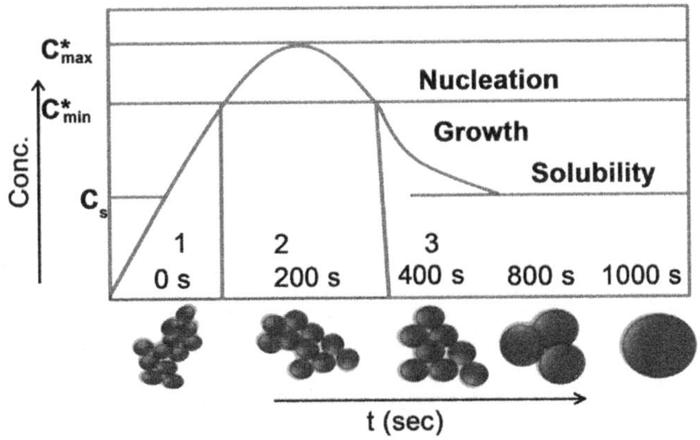

FIGURE 8.5

Nucleation and growth of nanoparticles in the La Mer model.

From Kotnala, R., & Shah, J. (2015). Ferrite materials: nano to spintronics regime. In Handbook of magnetic materials *(Vol. 23, pp. 291–379). Elsevier.*

under high-temperature and high-pressure hydrothermal processes, solvent characteristics can be considerably altered. For example, a high-temperature/high-pressure hydrothermal process can reduce the density, surface tension, and viscosity of water, whereas it increases the vapor pressure and ion product. Fig. 8.7 shows variations in the density of water, with temperature and pressure acting as the hydrothermal parameters. The ion product of water quickly increases with increasing temperature and pressure. At a temperature of 1000°C and a pressure of 25–100 kbar, the density of water is about 1.2–1.6 g/cm^3. Under these conditions, water behaves as molten salt and is converted entirely into H_3O^+ and OH^-. In addition, the viscosity of water decreases with increases in temperature. Notably, at a temperature of 500°C and pressure of 1 kbar, the water viscosity is 10% lower than under standard conditions. It should be noted that the mobility of water molecules and ions in the hydrothermal process is higher than under standard conditions (Franck, 2009). While the dielectric constant of water decreases with increases in temperature during the hydrothermal process, it increases with increasing pressure.

Fig. 8.8 shows variations in the dielectric constant of water versus temperature and pressure in the hydrothermal process. The relationship between the temperature and dielectric constant of the solvent is one of the important factors for understanding the hydration/dehydration phenomena and reactivity of solutes. In this respect, high dielectric constant values for water have been achieved at low temperatures and high pressures (Seward, 1981). Knowledge of the dependence of temperature and pressure on water characteristics is of great importance, allowing hydrothermal synthesis to be understood. The pressure under hydrothermal conditions has been found to depend on the filling factor of the sealed reactor. Fig. 8.9 shows the correlation of water pressure and temperature with the filling factor of the reactor. Generally, the filling factor ranges between 50% and 80% for pressures ranging from 0.2 to

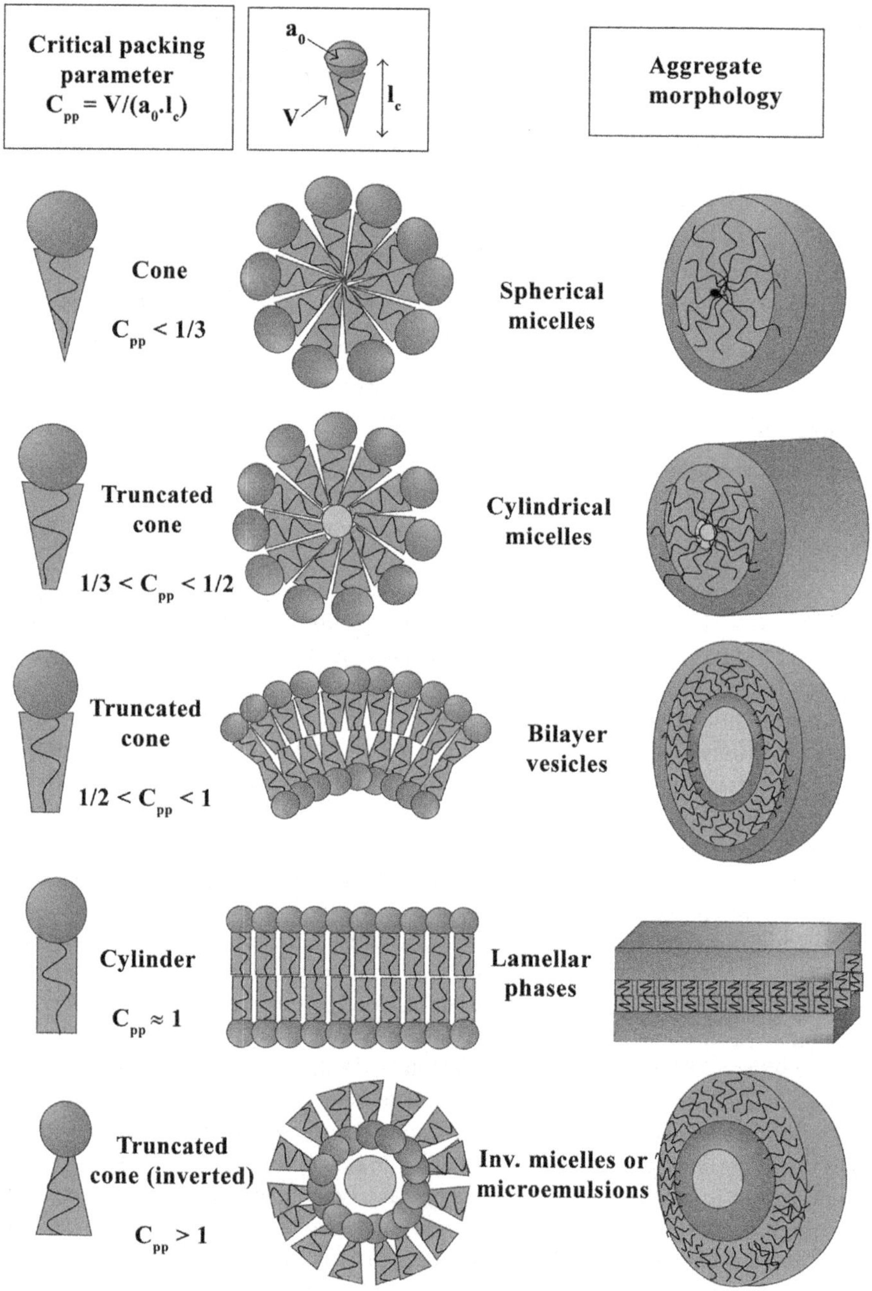

FIGURE 8.6

Packing shapes of the surfactant and structure of their aggregates.

From Choudhary, R. N. P., Lombardo, D., Kiselev, M. A., Magazù, S., & Calandra, P. (2015). Amphiphiles self-assembly: Basic concepts and future perspectives of supramolecular approaches. Advances in Condensed Matter Physics, 2015, *151683. https://doi.org/10.1155/2015/151683.*

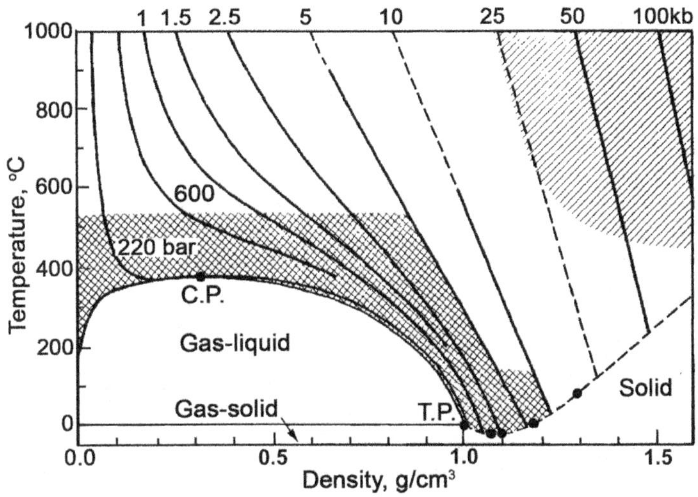

FIGURE 8.7

Variation of the density of water with temperature and pressure acting as hydrothermal parameters.

From E. U. Franck. (n.d.). Water and aqueous solutions at high pressures and temperatures. Pure and Applied Chemistry, 24*(1), 13–30. https://doi.org/10.1351/pac197024010013.*

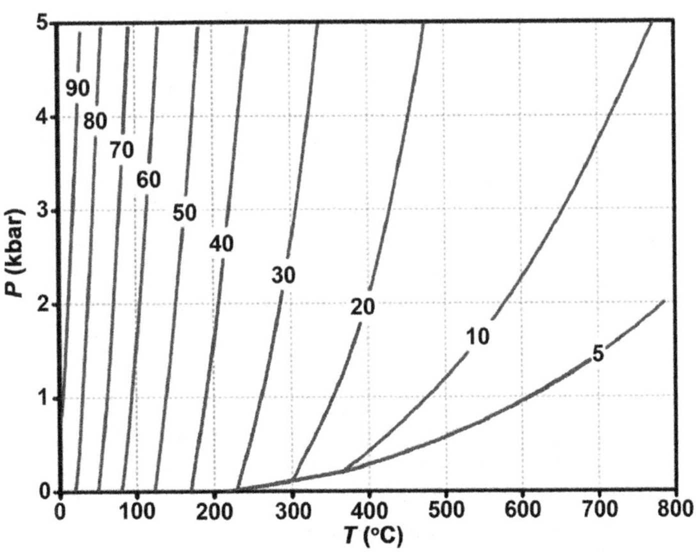

FIGURE 8.8

Variation of the dielectric constant of water versus temperature and pressure in the hydrothermal process.

From Rabenau, A. (1985). The role of hydrothermal synthesis in preparative chemistry. Angewandte Chemie International Edition in English, 24*(12), 1026–1040. https://doi.org/10.1002/anie.198510261.*

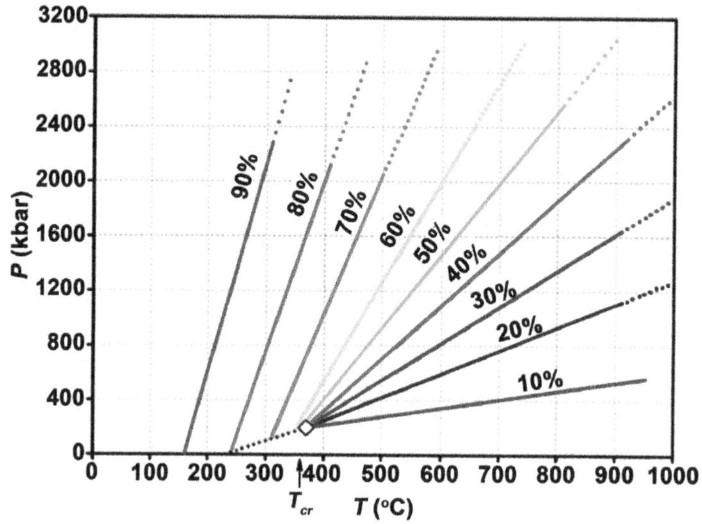

FIGURE 8.9

The correlation of water pressure and temperature with the filling factor of the sealed reactor.

From Rabenau, A. (1985). The role of hydrothermal synthesis in preparative chemistry. Angewandte Chemie International Edition in English, 24*(12), 1026–1040. https://doi.org/10.1002/anie.198510261.*

3 kbar. As previously mentioned, water is used as the reaction solution in the hydrothermal process. Conversely, the reaction solutions in the solvothermal process are composed of different organic solvents (Feng & Li, 2017; Rabenau, 1985).

To synthesize ferrite nanoparticles using a hydrothermal technique, soluble salts of divalent and trivalent transition metals are dissolved individually. The solutions are then mixed and stirred at room temperature. The resulting solution containing the metal ions is homogenized by continuing to stir while adding dropwise organic solvents such as NaOH or NH_4OH. Afterward, the solution is transferred to a reactor and placed inside an autoclave for subsequent thermal treatment in an oven. The heating temperature and time depend on the type of ferrite nanoparticles. When the thermal treatment is finished, the reactor is left to cool down to room temperature inside the oven. The resultant black powder is then separated from the solution by centrifugation, rinsed several times with deionized water and ethanol, and dried. In research work, gadolinium-substituted cobalt ferrite nanoparticles have been prepared by a hydrothermal process without subsequent annealing (Sodaee et al., 2016). $Co(NO_3)_2$, $Fe(NO_3)_3$, and $Gd(NO_3)_3$ were first converted to amorphous hydroxides by incorporating a sodium hydroxide solution, followed by their reactions to form $CoFe_{2-x}Gd_xO_4$ nanoparticles. The reactions taking place during the hydrothermal synthesis of $CoFe_{2-x}Gd_xO_4$ nanoparticles can be described as follows:

$$Co(NO_3)_2 + (2-x)Fe(NO_3)_3 + xGd(NO_3)_3 + 8NaOH \rightarrow Co(OH)_2 + (2-x)FeO(OH) + xGdO(OH) + 8NaNO_3 + 2\ H_2O \tag{8.2}$$

$$Co(OH)_2 + (2-x)Fe(OH) + xGdO(OH) \xrightarrow{120^{\circ}C,12\ h} CoFe_{2-x}Gd_xO_4 + 2H_2O \tag{8.3}$$

It is known that the size of particles and distribution of cations in the crystallographic (octahedral and tetrahedral) sites may change as a function of annealing temperature. In this regard, the crystallinity of ferrite nanoparticles has been enhanced after performing an annealing process, eliminating unwanted secondary phases. However, the random walk diffusion has caused cations to be rearranged, leading to an alternation in magnetic features.

Synthesis approaches providing a highly crystalline structure and narrow particle size distribution are preferable. In this respect, the hydrothermal technique seems to be appropriate for achieving crystalline ferrite nanoparticles without any postannealing process. Fig. 8.10 shows a schematic representation of the synthesis of a magnetic nickel ferrite-iron oxide nanocomposite ($NiFe_2O_4/Fe_2O_3$) using a hydrothermal method.

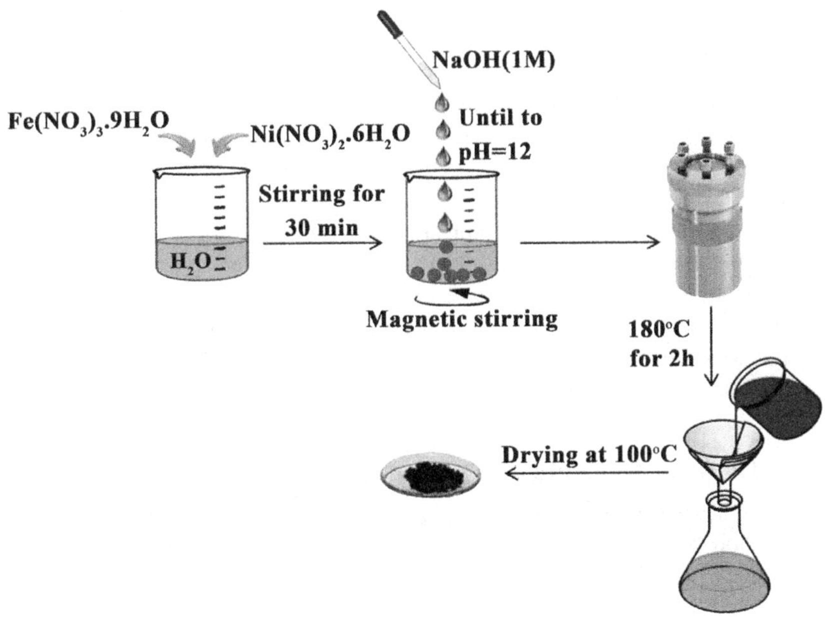

FIGURE 8.10

Schematic representation of the synthesis of magnetic nickel ferrite-iron oxide nanocomposite ($NiFe_2O_4/Fe_2O_3$) by hydrothermal method.

From Habibi, M. H., & Fakhri, F. (2017). Hydrothermal synthesis of nickel iron oxide nano-composite and application as magnetically separable photocatalyst for degradation of Solar Blue G dye. Journal of Materials Science: Materials in Electronics, 28*(19), 14091–14096. https://doi.org/10.1007/s10854-017-7261-3.*

Large-scale preparation of ferrite nanoparticles with different morphologies can also be realized using the hydrothermal technique (Wang et al., 2019). It is possible to control the size distribution of ferrite nanoparticles by choosing a suitable mixture of solvents while changing synthesis parameters such as temperature, pressure, and reaction time (Ameer et al., 2015; Nemati et al., 2014; Taniguchi et al., 2009; Tian et al., 2011). In the solvothermal synthesis technique, the size distribution, shape, and crystalline phases can be controlled as well. In this case, several parameters such as reaction temperature, reaction time, solvent, surfactant, and precursors may influence physical characteristics.

8.2.5 Mechanical milling technique

As a "top-down" route, the mechanical milling technique involves the use of a high-energy planetary ball mill for the preparation of ferrite nanoparticles. The milling process can decrease the particle size and mix them together, which in turn may result in the formation of new phases. During the milling process, the energy transferred from the ball mill to the powder can control the kinetics of the mechanical milling. This energy depends on several factors such as the type of mill, milling speed, size of the balls, the temperature and time of milling, and powder and dry or wet milling. A high-temperature milling process is needed to form crystalline phases with high atomic mobility (e.g., intermetallic materials), whereas amorphous phases can be induced at a low temperature. However, the formation of nanocrystalline phases at low temperatures is more common. Three categories are considered for powder components: (1) ductile-ductile, (2) ductile-brittle, and (3) brittle-brittle (Yadav et al., 2012).

Fig. 8.11 shows a schematic representation of the progress of the mechanical alloying method during the last decades. The main advantage of the mechanical milling technique is its capability to be scaled up to tonnage quantities in the synthesis of materials with various applications. The drawbacks of this technique are the contamination of nanoparticles during the milling, wide size distribution, and multiple morphologies (El-Eskandarany, 2015; Zhang et al., 2015).

8.2.6 Thermal decomposition technique

The thermal decomposition technique is a simple method for the preparation of ferrite nanoparticles. This technique involves the chemical decomposition of materials in the presence of surfactants at a high temperature. Accordingly, the separation of chemical bonds occurs during thermal decomposition. The preparation of magnetic nanoparticles via thermal decomposition has mostly been performed using organometallic compounds such as acetylacetonates in organic solvents (e.g., benzyl ether, ethylenediamine, and carbonyls). Some common surfactants such as oleic acid, oleylamine, polyvinylpyrrolidone, cetyltrimethylammonium bromide, and hexadecylamine have also been used in thermal decomposition technique.

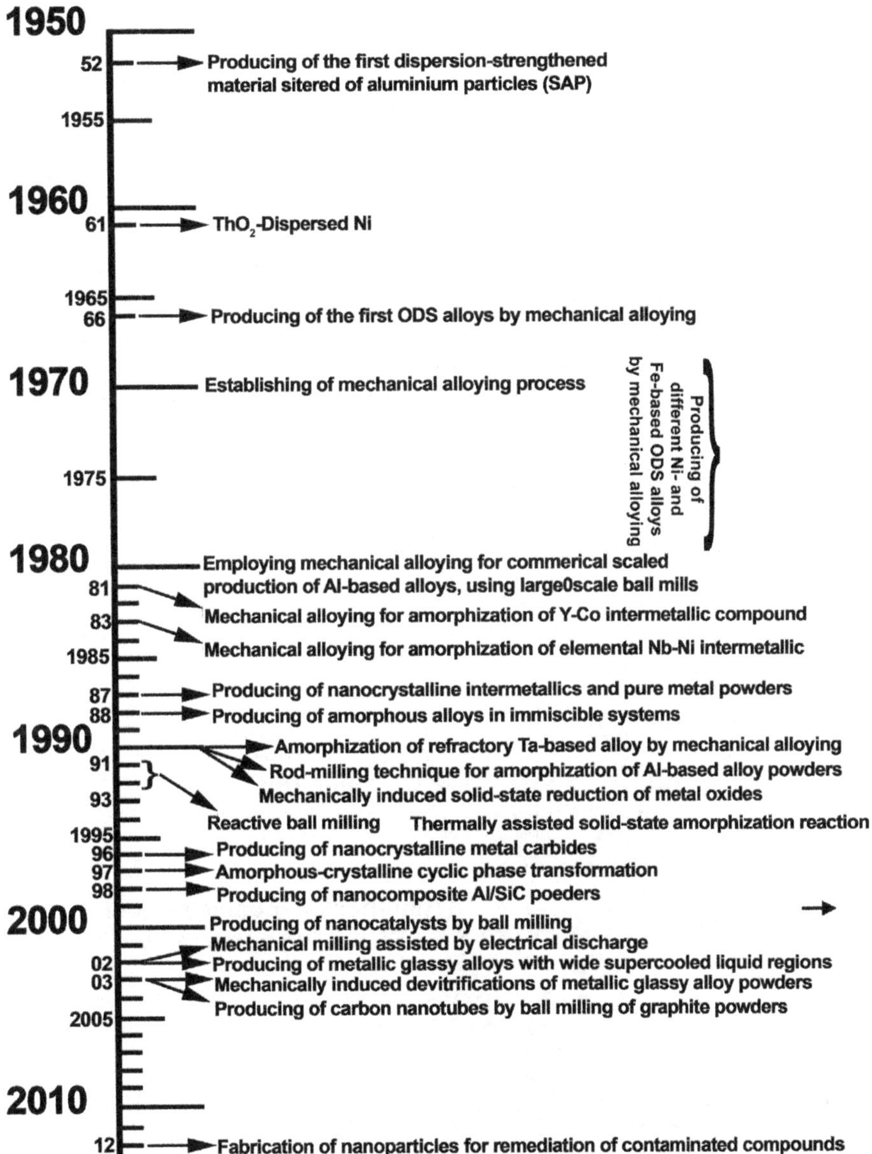

FIGURE 8.11

A schematic representation of the progress of the mechanical alloying method.

From El-Eskandarany, M. (2015). The history and necessity of mechanical alloying. Mechanical alloying (2nd ed.). Elsevier, pp. 13–47. https://doi.org/10.1016/B978-1-4557-7752-5.00002-4.

The final particle size and morphology of the magnetic nanoparticles are determined by the composition of various precursors used in the reaction (Khan et al., 2014). A schematic representation of the synthesis of MFe_2O_4 nanoparticles by the thermal decomposition technique is depicted in Fig. 8.12. The size and shape of ferrite nanoparticles can be changed by the temperature, heating rate, and concentration of precursors. The advantages of the thermal decomposition technique include high crystallinity, uniform morphology, high monodispersity, and narrow size distribution of nanoparticles. The aforementioned synthesis technique is also useful for industrial and medical applications (Amara et al., 2009; Zakiyah et al., 2015).

Monodisperse cubic and spherical maghemite (Fe_2O_3) nanoparticles have been synthesized using the thermal decomposition technique. Fig. 8.13 shows high-resolution transmission electron microscopy micrographs of monodisperse maghemite nanoparticles, indicating that the ratio of precursors and the thermal decomposition time can control their size and morphology. It has been found that

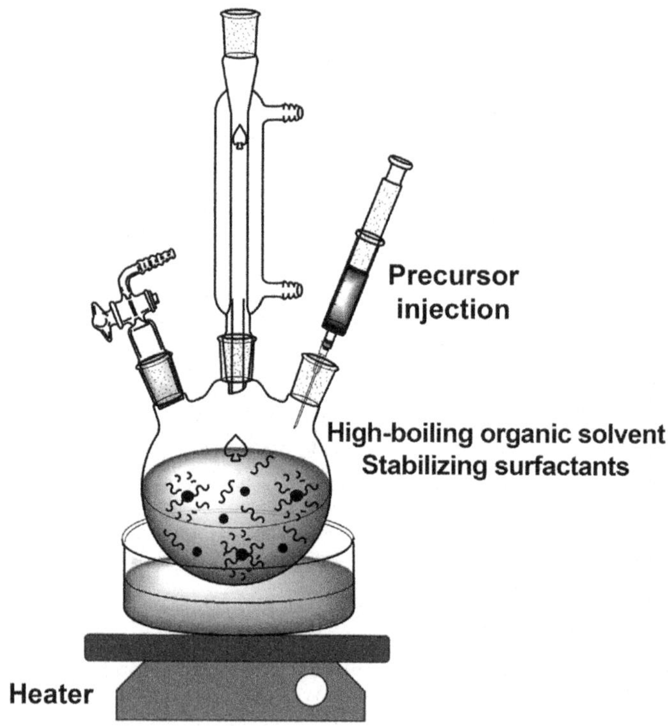

FIGURE 8.12

A schematic representation of the synthesis of MFe_2O_4 nanoparticles by a thermal decomposition technique.

From Cruz, I. F., Freire, C., Araújo, J. P., Pereira, C., Pereira, A. M., El-Gendy, A. A., Barandiarán, J. M., & Hadimani, R. L. (2018). In Micro and nano technologies *(pp. 59–116). Elsevier. https://doi.org/10.1016/B978-0-12-813904-2.00003-6.*

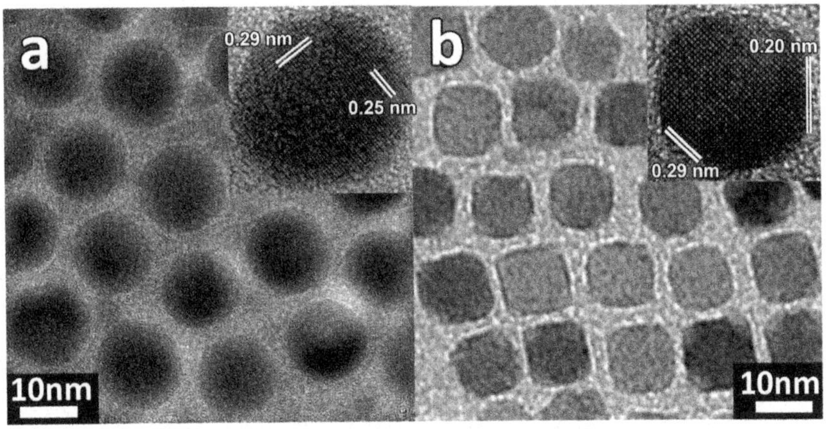

FIGURE 8.13

High-resolution transmission electron microscopy micrographs of monodisperse maghemite nanoparticles with (A) spherical and (B) cubic morphology prepared by a thermal decomposition technique.

From Salazar-Alvarez, G., Qin, J., Šepelák, V., Bergmann, I., Vasilakaki, M., Trohidou, K. N., Ardisson, J. D., Macedo, W. A. A., Mikhaylova, M., Muhammed, M., Baró, M. D., & Nogués, J. (2008). Cubic versus spherical magnetic nanoparticles: The role of surface anisotropy. Journal of American Chemical Society, 130*(40), 13234–13239. https://doi.org/10.1021/ja0768744.*

a short decomposition process (2–4 h) can form nanoparticles with spherical shapes, whereas the cubic morphology results from a longer duration of decomposition (10–12 h) (Salazar-Alvarez et al., 2008).

8.2.7 Other ferrite nanoparticle synthesis techniques

Sonochemical, microwave sintering, and electrochemical approaches are efficient techniques for synthesizing ferrite nanoparticles. These are described briefly in the following.

In the sonochemical technique, bubbles are prepared in the solvent medium during ultrasonic irradiation, thus saving the diffuse energy of ultrasound. This technique is useful for the reduction of crystal growth, uniformity of mixing, and control of size distribution. The drawback of the sonochemical technique is the poor control of the shape and dispersity (Zhu et al., 2013).

In the microwave sintering technique, energy is directly sent to materials through molecular interaction with electromagnetic radiation. The electromagnetic energy is converted to thermal energy, thereby producing heat. The reaction temperature of the microwave-assisted synthesis technique usually ranges between 100 and 200°C, whereas the reaction process occurs in a shorter time. Any vapor produced during the heating time is removed by the exhaust drain of a Teflon vessel. The advantages of this technique include cost-effectiveness, large-scale production, and quick synthesis of nanoparticles of high quality with a narrow size distribution (Dandia et al., 2015).

The electrochemical technique is similar to the coprecipitation technique, except that the ion sources result from the scarification of anode electrodes through an oxidation process. In other words, the anode electrode is oxidized during the electrochemical reaction. The advantages of the electrochemical technique include the high purity of the resultant nanoparticles and controlled particle size along with simplicity, cost-effectiveness, and environmental friendliness (Galindo et al., 2012). In fact, purity and particle size can be controlled by current density or the distance between the electrodes. In this regard, the crystallinity of nanoparticles can be improved by reducing the distance between the electrodes. On the other hand, the particle size increases with an increase in the current density or a decrease in the distance between the electrodes.

8.2.8 Comparison of ferrite synthesis techniques

Overall, it is challenging to choose the most appropriate technique for synthesizing ferrite particles. This is because the different techniques involve variable factors that can be applied to the synthesis. For example, a comparison of hydrothermal, sonochemical, and sol−gel techniques indicates that the hydrothermal technique can be proposed as the most promising one for the preparation of ferrite nanoparticles, arising from its unparalleled advantages, including the achievement of appropriate size, shape, homogeneity in composition and a high degree of crystallinity. In addition, it is feasible to synthesize nanoparticles with phase homogeneity, controlled morphology, and narrow size distribution due to the decrease in agglomeration between them when using the hydrothermal process. Similarly, the coprecipitation technique is a favorable approach to synthesizing ferrite nanoparticles, as it can provide controlled morphology and magnetic properties.

On the other hand, several researchers have suggested that the sol−gel technique is the most effective synthesis method for the preparation of ferrite nanoparticles. This is because it benefits from several advantages over other techniques, such as suitable homogeneity, high purity, and cost-effectiveness. The thermal decomposition technique is favorable for the synthesis of ferrite nanoparticles with controlled size and morphology. The microemulsion technique has proven successful in synthesizing monodisperse nanoparticles with various morphologies.

Since many parameters (e.g., the type of solvent, surfactant, concentration, pH, temperature, and stirring rate) are involved during the synthesis of nanoparticles, it is necessary to choose the most efficient method to achieve the best quality along with desired quantity. The reaction temperature in the aforementioned synthesis techniques is presented in Table 8.1. The worldwide employed synthesis techniques of magnetic nanoparticles are shown in Fig. 8.14. It is observed that 90% of magnetic nanomaterials have been synthesized by chemical methods. The use of physical methods has been limited to 8% for the preparation of magnetic nanomaterials.

Table 8.1 The reaction temperature of synthesis techniques of ferrite nanoparticles.

Synthesis techniques	Reaction temperature (°C)
Coprecipitation	25–150
Hydrothermal	100–220
Sol-gel	25–200
Thermal decomposition	150–350
Solvothermal	150–220
Sonochemical	20–50
Microemulsion	20–80
Microwave-assisted	100–210

From Kefeni, K. K., Msagati, T. A. M., & Mamba, B. B. (2017). Ferrite nanoparticles: Synthesis, characterisation and applications in electronic device. Materials Science and Engineering: B, 215, *37–55. https://doi.org/10.1016/j.mseb.2016.11.002.*

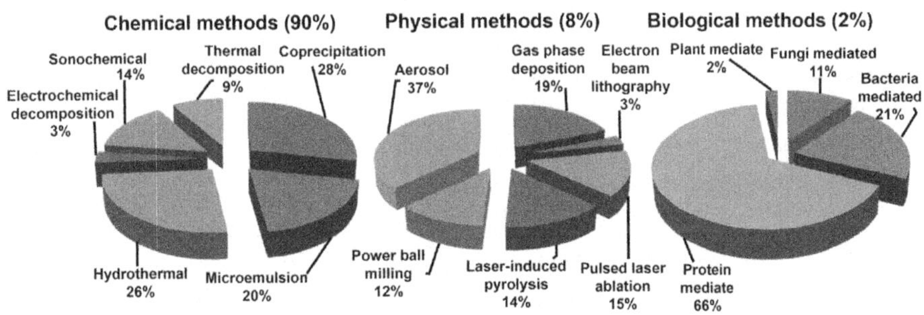

FIGURE 8.14

Synthesis techniques employed worldwide for magnetic nanoparticles divided into three main categories.

From Cruz, I. F., Freire, C., Araújo, J. P., Pereira, C., Pereira, A. M., El-Gendy, A. A., Barandiarán, J. M., & Hadimani, R. L. (2018). In Micro and nano technologies *(pp. 59–116). Elsevier. https://doi.org/10.1016/B978-0-12-813904-2.00003-6.*

Moreover, the biological methods represent the least percentage (2%). Consequently, chemical methods have mostly been used to synthesize ferrite nanoparticles. The synthesis techniques and applications of different ferrite nanoparticles, like Fe_3O_4 (Amara et al., 2009; Dandia et al., 2015; Nabiyouni et al., 2015; Nemati et al., 2014; Taniguchi et al., 2009; Tian et al., 2011; Xing et al., 2015; Zhu et al., 2013), $NiFe_2O_4$ (Galindo et al., 2012; Larumbe et al., 2012; Zhang et al., 2015), $CoFe_2O_4$ (Ameer et al., 2015; Avazpour et al., 2015; Wang et al., 2019), $CuFe_2O_4$ (Rashad et al., 2015; Zakiyah et al., 2015), and $ZnFe_2O_4$, are summarized in Table 8.2.

Table 8.2 Summary of synthesis techniques and applications of various ferrite nanoparticles.

Synthesis technique	Ferrites	T (°C)	Size (nm)	M_s (emu/g)	H_c (Oe)
Coprecipitation	Fe_3O_4	75	25–300	30.2	0
		60	<100	38.7	0.78
Hydrothermal	$ZnFe_2O_4$	180	10–50	10.2	–
	$CoFe_2O_4$	180	10	59	37.8
		120	7.3–7.9	–	–
		150	25	–	–
	Fe_3O_4	120–170	15–2000	–	–
		230	10.5	58.9	–
Sol–gel	$NiFe_2O_4$	400	11	38.5	–
	$CoFe_2O_4$	110	18.5	66	720
Thermal decomposition	Fe_3O_4	120–165	5.41–7.38	41.7	0
	$CuFe_2O_4$	500–900	24–34	–	–
Solvothermal	$CoFe_2O_4$	200	15–27	–	–
	Fe_3O_4	240	4	38.5	–
Sonochemical	Fe_3O_4	25	30–40	–	–
Microwave-assisted	Fe_3O_4	190	7	67	–
Microemulsion	$CuFe_2O_4$	600	53.7	–	–
Mechanical milling	$NiFe_2O_4$	25	–	–	–
Electrochemical	$NiFe_2O_4$	60	5–45	48	–

8.3 Bulk ferrite synthesis techniques

Ferrite powders are fired and consolidated at lower temperatures than their melting point to produce a dense body or bulk samples of ferrite particles. The formation of the bulk material by the sintering process leads to the bonding between particles due to the occurrence of diffusional mass transport. The sintering process for providing bulk can be carried out using different techniques such as spark plasma sintering (SPS), hot pressing (HP), hot isostatic pressing (HIP), and cold isostatic pressing (CIP).

8.3.1 Spark plasma sintering technique

The SPS technique is used for the synthesis of high-temperature materials with low deformability and a high-sintering temperature point. The advantage of this technique includes the use of low voltage, direct current (DC), and pulsed current–activated pressure. Bulk samples can be produced using different powder materials such as pure metals, ceramics, alloys, and composites. An electric field and external pressure are imposed simultaneously during the sintering process to enhance the densification of the powders.

It is known that the SPS process is different from the HP one. In other words, an external heating source is used in the HP technique, whereas an electric field is created during the sintering process in the SPS technique using a pulsed DC passed through the conductive die and sample. Fig. 8.15 compares the SPS process with the HP technique. Interestingly, the sintering temperature in the SPS process is a few hundred degrees lower than that in conventional sintering techniques such as HP (Cavaliere, 2019).

Nanosized powders can be sintered in a very short time without any grain growth via an SPS process, which is cumbersome to carry out using a conventional sintering technique. In addition, the short-time sintering of powders leads to the preparation of high-density samples, preventing the occurrence of coarsening and grain growth phenomena. Because of the spark discharge between the spaces of the powder particles, a local high-temperature region is created, melting their surfaces while forming products with special shapes around the contact area between the particles.

The components of the SPS technique are as follows: a furnace, pressing chamber, furnace power supply, vacuum pumping system, pressing system, water circulation system, gas inlet system, control and instrumentation, temperature measurement tools, and mold sets. The sintering process occurring inside the chamber can be carried out under different atmospheres such as vacuum and nitrogen or inert atmospheres. The important features of the chamber include resistance to the

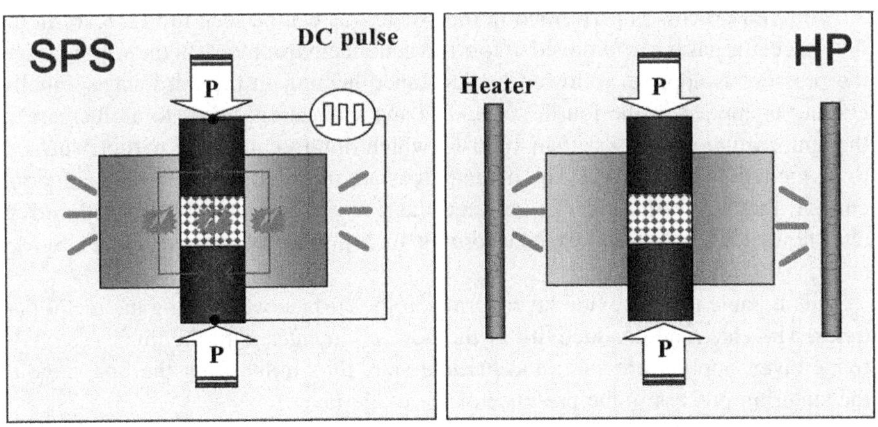

FIGURE 8.15

Comparison of the spark plasma sintering (SPS) process and hot-pressing (HP) techniques.

From Francis, L. D., Saheb, N., Iqbal, Z., Khalil, A., Hakeem, A. S., Al Aqeeli, N., Laoui, T., Al-Qutub, A., & Kirchner, R. (2012). Spark plasma sintering of metals and metal matrix nanocomposites: A review. Journal of Nanomaterials, 2012, *983470. https://doi.org/10.1155/2012/983470.*

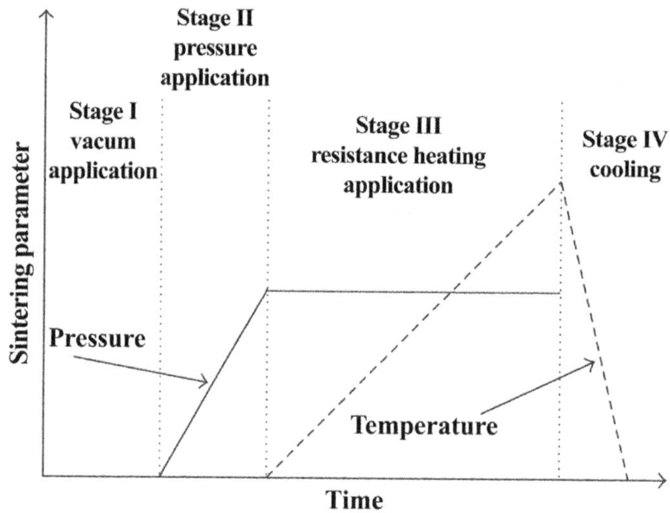

FIGURE 8.16

The performance of the spark plasma sintering process in four stages.

From Francis, L. D., Saheb, N., Iqbal, Z., Khalil, A., Hakeem, A. S., Al Aqeeli, N., Laoui, T., Al-Qutub, A., & Kirchner, R. (2012). Spark plasma sintering of metals and metal matrix nanocomposites: A review. Journal of Nanomaterials, 2012, *983470. https://doi.org/10.1155/2012/983470.*

pressure of 1×10^{-4} mbar (at least under cold conditions) and tolerating a sintering temperature as high as 2200°C.

The SPS process is performed in four stages, as can be seen in Fig. 8.16. In the first stage, the gases are removed to form a vacuum environment. In the second stage, the pressure is applied, followed by resistance heating in the third stage. Finally, cooling is applied in the fourth stage. The spark discharge leads to an increase in the temperature up to more than 1000°C, which in turn causes the particle surfaces to be melted and vaporized. The melted areas are attached together via an electron current, thereby forming necks between the powder particles. The pulsed current also flows through the powder, according to Fig. 8.17 (Cavaliere, 2019; Suárez et al., 2013).

It is notable that an oxide layer forms on the surface when using the metal particles. The electrical conductivity of the powder changes with the thickness of the oxide layer, applied pressure, and particle size, thus influencing the first stage of the sintering process in the presence of the oxide layer.

The different mechanisms involved in the SPS process are shown in Fig. 8.18. The packing of powders in the SPS process increases with increasing pressure, rearranging particles at low temperatures. Meanwhile, the applied pressure improves the contact between the particles. Theoretical investigations have shown local pressures in the range of 1000–10000 Mpa. At high temperatures, grain growth is enhanced, involving densification mechanisms such as power law creep or plastic deformation.

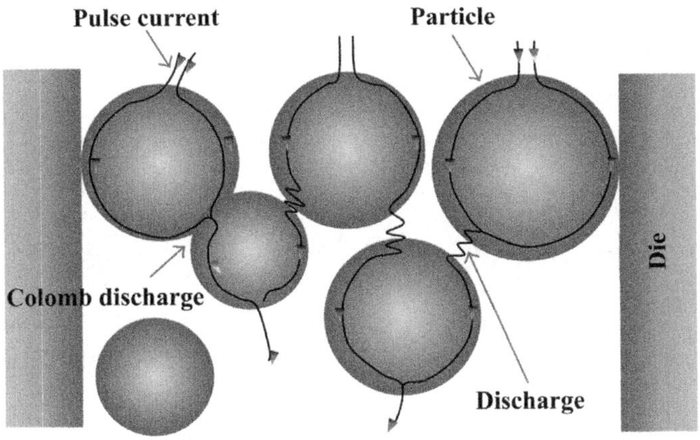

FIGURE 8.17

The flow of pulsed current through particles.

From Francis, L. D., Saheb, N., Iqbal, Z., Khalil, A., Hakeem, A. S., Al Aqeeli, N., Laoui, T., Al-Qutub, A., & Kirchner, R. (2012). Spark plasma sintering of metals and metal matrix nanocomposites: A review. Journal of Nanomaterials, 2012, *983470. https://doi.org/10.1155/2012/983470.*

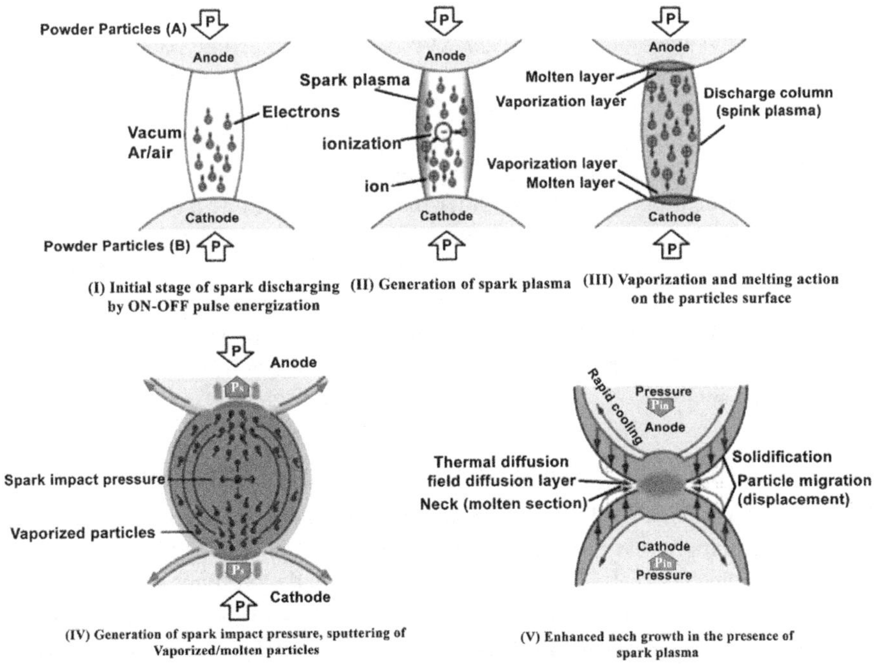

FIGURE 8.18

The various mechanisms involved in the spark plasma sintering process.

From Khoshghadam-Pireyousefan, M., Mohammadzadeh, A., Heidarzadeh, A., & Brabazon, D. (2021). Fundamentals of spark plasma sintering for metallic, ceramic, and polymer matrix composites production. *Elsevier. https://doi.org/10.1016/B978-0-12-803581-8.11907-1.*

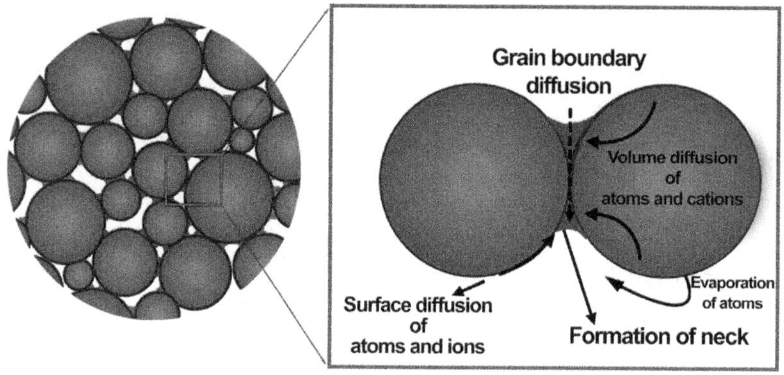

FIGURE 8.19

The mass transport mechanisms of the spark plasma sintering process.

As a result, sintering at low temperatures is preferred because of the limit on grain growth. Fig. 8.19 shows the mass transport mechanisms of the SPS process. It should be noted that the active mechanisms of neck formation during the sintering in the SPS process involve evaporation, grain boundary, surface diffusion, volume diffusion, and maybe dislocation climb.

In the initial stage of sintering, 2%–3% of particles become packed along with the neck formation. In the middle stage, the compactness of particles is 92% of the total densification, opening the greatest number of pores. The isolation of pores with spherical morphology takes place in the final stage of the sintering process. It has been evidenced that the pores are removed by the diffusion phenomenon, achieving the final densification of particles. Other mechanisms such as plastic deformation, dislocation, and diffusional creep are also involved in the elimination of the isolated pores. The generation of spark impact pressure, spark plasma, joule heating, and electric field diffusion is facilitated by an ON–OFF DC pulse. The other advantage of the SPS technique is the possibility of enhancing heating rates (generally ranging between 100°C and 600°C/min) during the sintering process (Cavaliere, 2019; Suárez et al., 2013).

8.3.2 Hot pressing and hot isostatic pressing techniques

The HP technique is used for the preparation of dense ferrites from powder samples by applying compressive stress and heat. The temperature employed in the HP technique is set to be above 1500°C, sintering materials in the presence of external pressure. The advantage of this technique includes high densities of specimens with controllable grain size distribution. In addition, it is capable of generating shapes with homogeneous density at relatively high temperatures. Brittle materials can be prepared using the HIP technique as it provides higher temperature uniformity in the heating process. However, the choice of the die material is limited in the HIP technique.

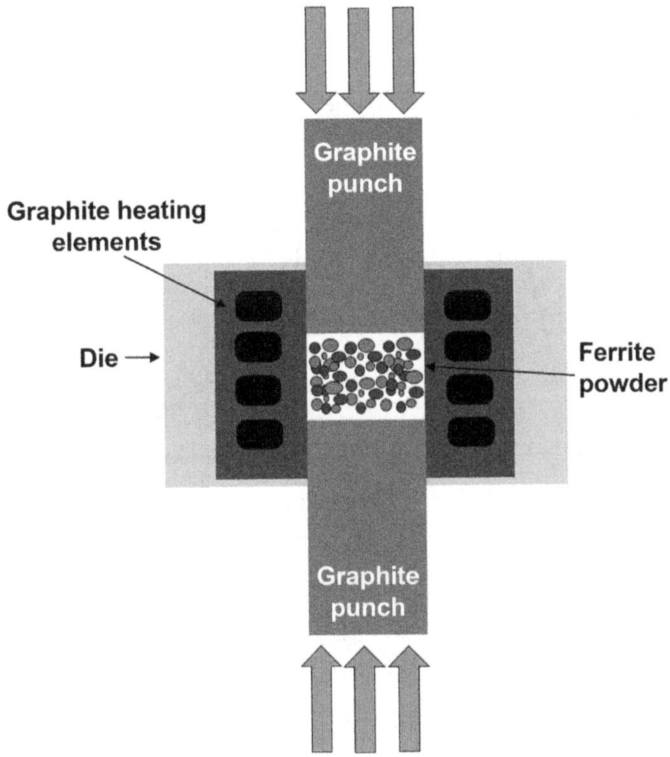

FIGURE 8.20

A schematic representation of the hot-pressing technique.

Fig. 8.20 shows a schematic representation of the HP technique, illustrating experiments carried out by graphite heating elements. The components of the HP technique consist of a simple die for mechanical loading, a punch system, and a furnace for heating the surrounding environment of the die and punch system. The punch and die are usually manufactured of graphite because of its machinability along with suitable thermal and mechanical characteristics. However, the graphite must be kept under nonoxidizing conditions such as a vacuum or an inert atmosphere. Therefore, since some ceramic oxides are reduced during the HP process, they require postannealing in oxygen to attain their appropriate stoichiometry (Francis, 2015; Loh & Sia, 1992). The maximum temperature of the heating elements is 2500°C. It has been found that the driving force for densification increases with increasing the pressure and sintering temperature. Some parameters such as temperature, pressure, and time can control the microstructure of the specimen synthesized by the HP process (Sarin, 2003). Fig. 8.21 represents the hot-pressing stages of typical substituted hexagonal ferrites (Singh et al., 2006). The effective processing

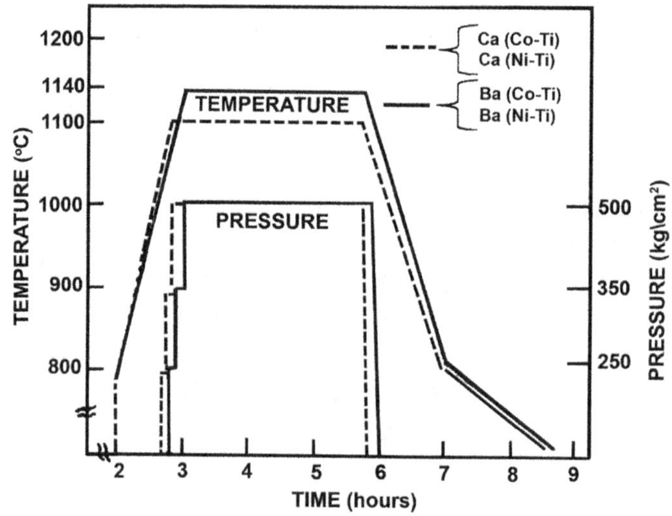

FIGURE 8.21

Definition of the hot-pressing stages for fabricating typical hexagonal ferrites.

From Singh, P., Babbar, V., Razdan, A., Srivastava, S., Agrawal, V., & Goel, T. (2006). Dielectric constant, magnetic permeability and microwave absorption studies of hot-pressed Ba-CoTi hexaferrite composites in X-band. Journal of Materials Science, 41*(21), 7190–7196. https://doi.org/10.1007/s10853-006-0921-y.*

parameters such as temperature, pressure, and time should be continuously monitored to obtain the appropriate microstructure.

It is also important to state that the densification rate of the HP process increases remarkably with an increase in pressure. The mechanism of enhancement in densification depends on the material and the temperature and pressure conditions. Diffusion and dislocation movement are involved in the HP process due to compressive stress, which in turn leads to creep deformation. The typical variation of temperature and pressure versus time for HIP is shown in Fig. 8.22. The difference between the HIP and HP techniques arises from the type of pressure applied. In other words, an isostatic pressure is applied to the material in the HIP technique, whereas a uniaxial pressure is applied to it in the HP technique, resulting in the production of simple shapes.

8.3.3 Cold isostatic pressing technique

As mentioned previously, the HIP technique is used for the preparation of high compact densities while creating complex shapes that cannot be produced by uniaxial pressing. The CIP technique is analogous to the HIP, although without the elevated temperature and the high-pressure gas applying the pressure. The two types of CIP are wet-bag and dry-bag processes. Fig. 8.23 shows schematic diagrams of the wet-bag and dry-bag CIP techniques. The former technique includes an

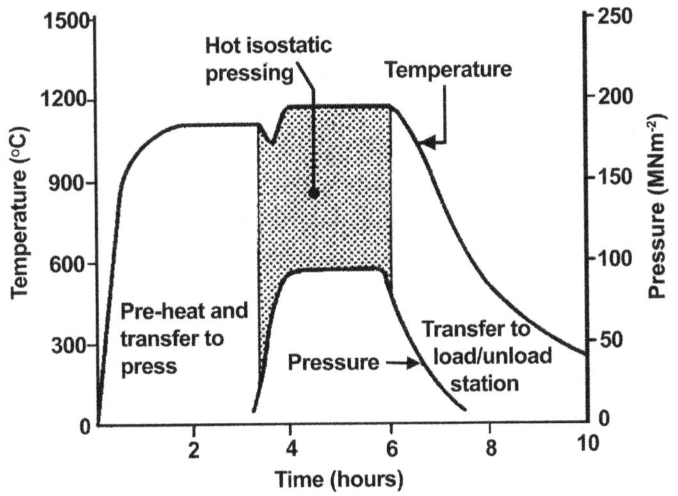

FIGURE 8.22

Typical time dependence of temperature and pressure parameter in hot isostatic pressing process.

From Loh, N. L., & Sia, K. Y. (1992). An overview of hot isostatic pressing. Journal of Materials Processing Technology, 30*(1), 45–65. https://doi.org/10.1016/0924-0136(92)90038-T.*

elastomeric mold loaded beyond its pressing pressure while immersed in the pressure vessel. The mold is ejected from the vessel after pressing and densification processes, followed by recovering and repeating the procedure.

In the latter technique, the immersion step is circumvented because of the incorporation of the mold into the pressure vessel. The mold can be filled with ferrite particles in the dry-bag CIP technique while exerting pressure. The dry-bag CIP is easier to carry out than the wet-bag CIP due to the incorporation of the mold into the pressure vessel.

The wet-bag CIP technique enables us to produce highly compact ferrite bulk arising from the omission of the die wall friction. Moreover, an open or closed hollow axisymmetric shape such as a tube can be prepared with this technique. However, it can lead to the formation of small parts with axisymmetric shapes. Due to the small wall friction in the wet-bag CIP, it is possible to achieve high densities, outperforming the dry-bag CIP in the formation of large parts. The disadvantages of the wet-bag CIP technique include reduced productivity and limited automation because of the loading and unloading of the molds.

The production of powder for the CIP process is the same as for the uniaxial pressing process. It is necessary to adjust the particle size distribution and binder content if compression is carried out by green machining (Kennard, 1991).

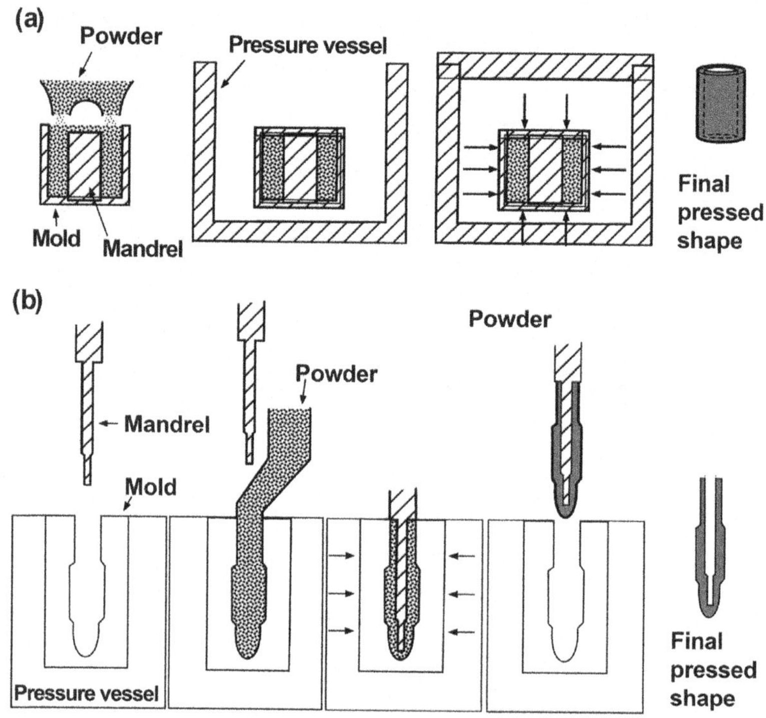

FIGURE 8.23

Schematic diagrams of (A) wet-bag and (B) dry-bag cold isostatic pressing techniques.

From Francis, L. F. (2015). Materials processing: a unified approach to processing of metals, ceramics and polymers. *Academic Press. https://doi.org/10.1016/B978-0-12-385132-1.00005-7.*

8.4 Synthesis techniques of ferrite thin films

Ferrite thin films have a variety of applications in magnetic recording media, microwave devices, and chip inductors. They could be fabricated by several techniques, including physical vapor deposition (PVD), sputtering, pulsed laser deposition, and molecular beam epitaxial growth. In the following, the key parameters and processing techniques involved in the fabrication of thin films are briefly explained.

8.4.1 Physical vapor deposition technique

The PVD technique is an atomistic coating process for the production of thin films. In the first stage of PVD, the target material is vaporized, and the atoms or molecules are formed, followed by moving the resulting vapor to the substrate through a plasma medium. Generally, the thicknesses of films deposited by the PVD process range from a few nanometers to thousands of nanometers. They are employed to fabricate graded composition deposits, freestanding structures, very thick deposits,

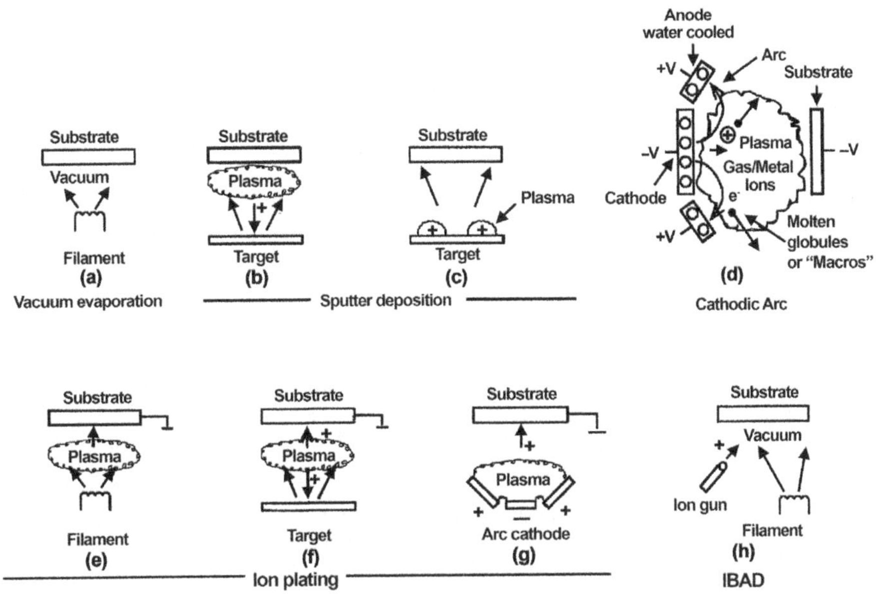

FIGURE 8.24

The main classifications of the physical vapor deposition technique: (A) vacuum deposition, (B) and (C) sputtering deposition in a plasma environment, (D) sputtering deposition in a vacuum environment, (E) ion plating in a plasma environment with a thermal vaporization source, (F) ion plating with a sputtering source, (G) ion plating with an arc evaporation source, and (H) ion beam-assisted deposition with a thermal vaporization source and ion bombardment from an ion gun.

From Mattox, D. M., & Mattox, D. M. (2010). Chapter 1 - Introduction (pp. 1–24). William Andrew Publishing. https://doi.org/10.1016/B978-0-8155-2037-5.00001-0.

and multilayer coatings with different deposition rates (ranging between 10 and 100Å per second). Moreover, using reactive PVD deposition techniques, it is possible to prepare films of elements, alloys, and compounds. The PVD technique is mainly classified into the following processes: vacuum evaporation, sputtering deposition, arc vapor deposition, and ion plating. Fig. 8.24 shows the main classifications of the PVD technique.

8.4.1.1 Vacuum evaporation

Using a thermal vaporization source (e.g., tungsten wire coil) or a high-energy e-beam heating in the vacuum evaporation process, the material is vaporized and transported to the substrate through a vacuum without any encounter with gas molecules in the space between the source and substrate. The vacuum environment can reduce the gaseous contamination in the deposition process. The gas pressure in the vacuum evaporation ranges from 10^{-5} to 10^{-9} Torr, which is influenced by the level of gaseous contamination. Additionally, the evaporation rate in vacuum evaporation is higher than that in other vaporization techniques. Optical interference coatings, electrically conducting films, wear-protective coatings, and corrosion-resistant coatings can be fabricated using vacuum evaporation.

8.4.1.2 Arc vapor deposition

In arc vapor deposition, a cathodic or anodic electrode is vaporized under high-current and low-voltage arc, thereby depositing the evaporated material on the substrate. This type of deposition is employed for the production of hard and decorative coatings. Furthermore, the ions formed by arc evaporation are effective for the ion plating process.

8.4.1.3 Ion plating

In an ion plating process, the depositing film is bombarded with atomic-sized energetic particles. These are generally inert gas, reactive gas, or film ions. Ion plating can control the characteristics of the depositing film. In this respect, several parameters such as flux, energy, and mass of the bombarding species, along with the ratio of the bombarding particles to the depositing particles, can modify the properties of the resulting films. The ion plating process can be performed in plasma and vacuum environments, preparing hard coatings of compound samples, adherent metal coatings, and high-density optical coatings. In other words, compact coatings can be produced at comparatively high gas pressures using the ion plating process (Mattox, 2010).

8.4.1.4 Sputtering deposition

In the first stage of sputtering deposition, a target is bombarded with atomic-sized energetic ions in a vacuum environment while employing an ion gun or plasma to eject the surface atoms of the target. In this way, surface particles of the target become dispersed because of their encounters with the energetic ions and are finally deposited on the substrate. In general, the distance between the source and the substrate in the sputtering deposition is lower than that in the vacuum deposition. Due to the energetic ion bombardment, surface atoms of the target are ejected with a specific rate defined by the sputtering yield (S), indicating the mean number of atoms removed from the surface per incident ions as follows:

$$S = (\text{Removed atoms})/(\text{Incident ions}) \tag{8.4}$$

Various methods such as target weight loss, target thickness reduction, sputtered material collection, and sputtered particle detection in flight can be employed to measure sputtering yield. The target weight loss method is the most common of these. It is worth noting that the variation in thickness or composition of the target can be determined using Rutherford backscattering and scanning electron microscopy techniques. Moreover, sputtering yield is influenced by several factors, such as the energy of the incident ions, target material, incident angle of the ions, and crystal orientation of the target surface.

Fig. 8.25 shows the variation in sputtering rate as a function of incident ion energy. It is found that the maximum value of the rate can be achieved in the

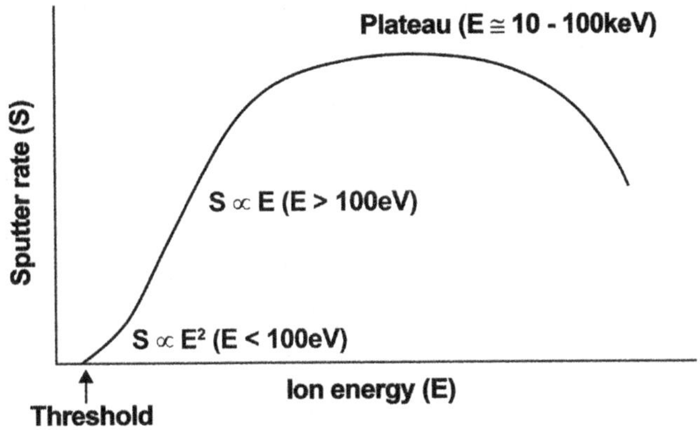

FIGURE 8.25

Variation in sputtering yield as a function of incident ion energy.

From Wasa, K., Kitabatake, M., Adachi, H., Wasa, K., Kitabatake, M., & Adachi, H. (2004). 3 - Sputtering phenomena *(pp. 71–114). William Andrew Publishing. https://doi.org/10.1016/B978-081551483-1.50004-6.*

high-energy region of the incident ions. The threshold energy of sputtering is also observed in the low-energy region, ranging between 50 and 300 eV. The threshold energy depends on the pollution of the cathode surface, the incident angle of the ions, and the crystal orientation of the cathode material. It is well known that the sputtering rate is strongly dependent on the crystal orientation of the surface of the target. Fig. 8.26 shows the variation in the sputtering rate of Ar^+ on the Cu (110), (100), and (111) planes and polycrystalline copper versus incident ion energy (Wasa et al., 2012).

Two models have been suggested for sputtering deposition: (1) Thermal evaporation model: A target is bombarded with energetic ions so that its surface is sufficiently heated for subsequent evaporation. (2) Momentum transportation model: Surface atoms are detached from the target due to the transportation of kinetic energy of the incident particles.

Several sputtering systems such as DC diode, radio frequency (RF) diode, cylindrical magnetron, planar magnetron, and ion beam have so far been used for the production of thin films. Fig. 8.27 shows the classification of the sputtering systems. It is interesting to find that the DC diode system is fundamental, meaning that the other sputtering systems can be achieved by improving the diode system. A pair of planar electrodes (cathode and anode) is used in the DC diode sputtering system. The target sample is mounted on the top surface of the cathode, and its reverse face is cooled with water. The substrate is also mounted on the anode. The sputtering chamber is filled with Ar gas. By applying a DC voltage to the electrodes, the glow discharge is started, resulting in the acceleration of the Ar ions toward the cathode. The sputtered target is consequently deposited on the substrate, forming a thin film with a uniform thickness.

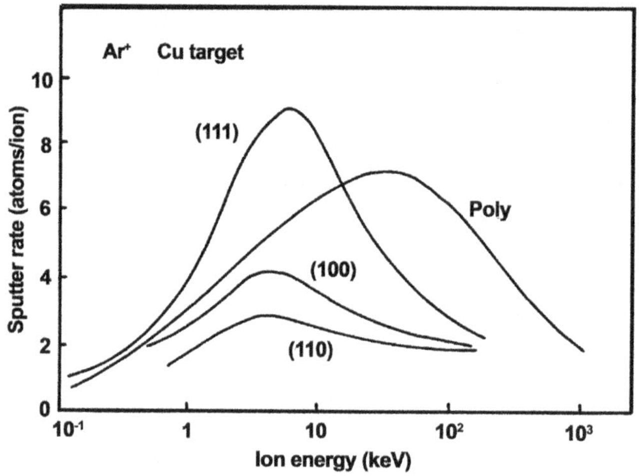

FIGURE 8.26

Variation in sputtering yield of Ar+ on the Cu (110), (100), and (111) planes versus energy of incident ions.

From Wasa, K., Kitabatake, M., Adachi, H., Wasa, K., Kitabatake, M., & Adachi, H. (2004). 3 - Sputtering phenomena *(pp. 71–114). William Andrew Publishing. https://doi.org/10.1016/B978-081551483-1.50004-6.*

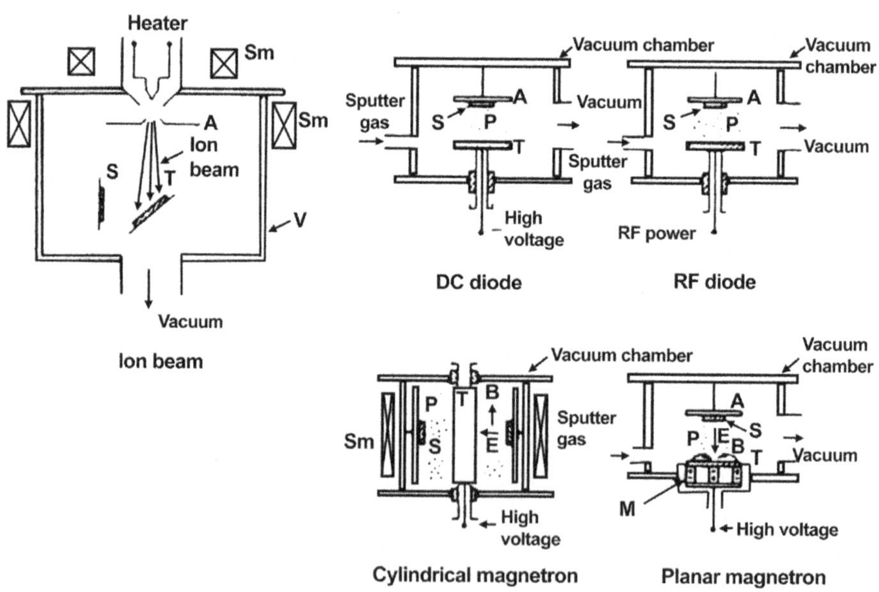

FIGURE 8.27

The classification of sputtering systems.

From Adachi, H., Wasa, K., Wasa, K., Kanno, I., & Kotera, H. (2012). 1 - Thin films and nanomaterials *(pp. 3–39). William Andrew Publishing. https://doi.org/10.1016/B978-1-4377-3483-6.00001-2.*

In the case of the RF diode sputtering system, an RF power supply is used instead of the DC voltage power supply. Moreover, an impedance-matching network is connected between the RF power supply and discharge chamber.

Basically, magnetron sputtering systems are made of coaxial cylindrical electrodes and an axial magnetic field, categorized into two types, cylindrical and planar magnetrons. These magnetron systems can be employed in various industrial applications such as microelectronics, hard coatings, automobiles, and architectural glass. It has been demonstrated that ion current density at the cathode, along with the deposition rate, increases when using a quadrupole magnetic field.

The quality of thin films depends on the chemical composition, crystal structure, crystal phase, uniformity, surface flatness, and film stress induced during their sputtering deposition process. The growth mechanism of sputtered thin films involves a nucleation process, followed by surface-diffusion growth of 3-D nuclei and the formation of a film structure with three types of morphologies (island, layer, and mixed island and layer). These are influenced by the thermodynamic parameters of the sputtering deposition and the substrate surface. Fig. 8.28 exhibits the three types of growth processes in the production of thin films (Wasa et al., 2012).

8.4.2 Pulsed laser deposition

Pulsed laser deposition is used for fabricating thin-film structured materials such as ferrite films, epitaxial interfaces, and superlattices employed in various applications, including superconducting electronic devices, optical instruments, oxide sensors, protective coatings and barriers, and polymer and organic layers (Eason, 2007). Particularly, ferrites are usually deposited using pulsed laser deposition, allowing for the control of the chemical composition. In this case, a pulsed laser with a high-power density (e.g., KrF excimer laser; $\sim 1\ J/cm^2$) is used. The target is initially heated in a vacuum by the pulsed laser to melt and vaporize the surface. The atoms removed or vaporized from the surface of the target are assembled to form thin films. The pulsed laser can also induce a plasma plume due to the production of photoemitted electrons from the target. Fig. 8.29 shows a schematic of the pulsed laser deposition process. The target used in this process can be in the form of powders, single crystals, and sintered pellets. While the design of pulsed laser deposition is simple, it does not form uniform thin films because of the presence of microsized particles removed from the target material.

The kinetic energies of ions in the ablation plume range from a few tenths to several hundred electron volts, adjusting the stress level of films by the defects created in the process. Pulsed laser deposition consists of three subprocesses, according to Fig. 8.30. It is necessary to control these subprocesses to obtain deposited layers with appropriate properties. The ablation can be realized by the interaction of the laser with the target material when the power of the beam is higher than the ablation threshold of the laser-material interaction. The concentration of particles and their density, size distribution, and chemical composition depend on the ablation

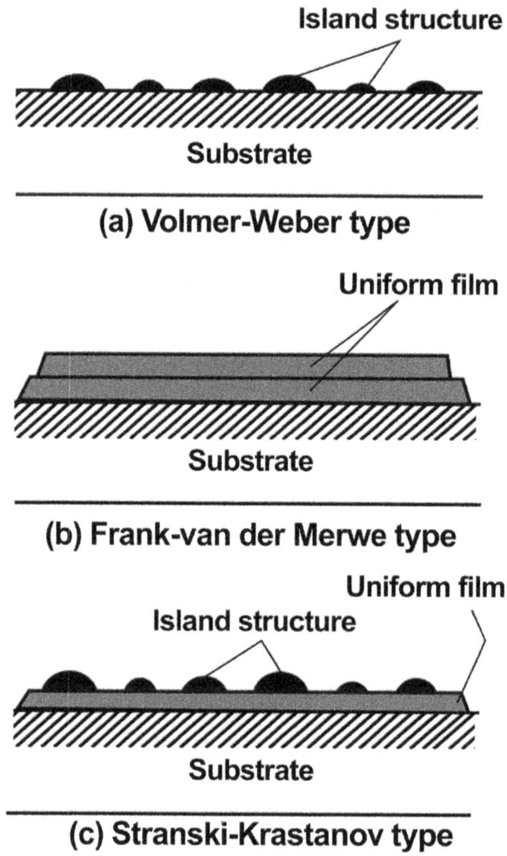

FIGURE 8.28

Three types of morphologies in the growth process of thin film.

From Adachi, H., Hata, T., Wasa, K., Wasa, K., Kanno, I., & Kotera, H. (2012). 5 - Basic process of sputtering deposition *(pp. 295–359). William Andrew Publishing. https://doi.org/10.1016/B978-1-4377-3483-6.00005-X.*

and propagation processes. Furthermore, the morphology of the resulting layers and their structural, mechanical, electric, and magnetic properties are governed by the pulsed laser deposition process (Stafe et al., 2013).

8.4.3 Molecular beam epitaxy

Molecular beam epitaxy is a thermal vaporization deposition process by which to fabricate epitaxial films of semiconductors, magnetic ferrites, insulators, and metals. The rate of vaporization of the materials is controlled by a computerized process control system. A schematic representation of the molecular beam epitaxy system is shown in Fig. 8.31, consisting of a stainless steel vacuum chamber pumped to a

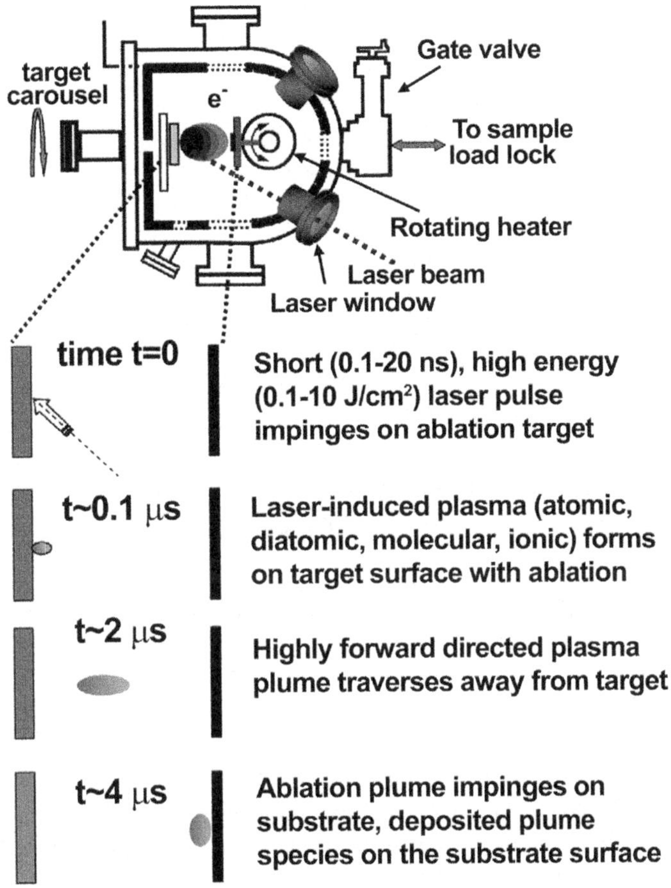

FIGURE 8.29

A schematic of the pulsed laser deposition process.

From Eason, R. (2007). Pulsed laser deposition of thin films: applications-led growth of functional materials. John Wiley & Sons. https://doi.org/10.1002/9780470052129.

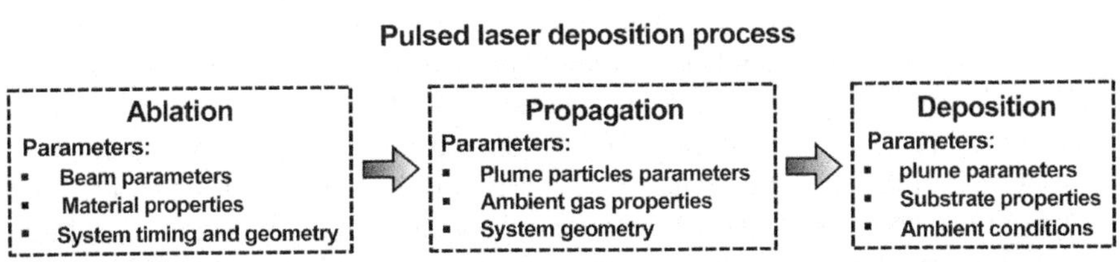

FIGURE 8.30

Three subprocesses of pulsed laser deposition.

From Stafe, M., Marcu, A., & Puscas, N. N. (2014). Material removal and deposition by pulsed laser ablation and associated phenomena. In Pulsed laser ablation of solids *(pp. 143–214). Springer. https://doi.org/10.1007/978-3-642-40978-3_5.*

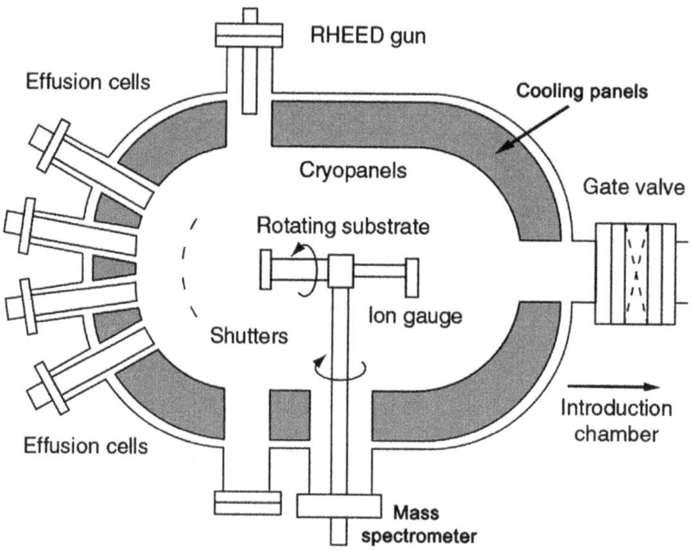

FIGURE 8.31

A schematic representation of the molecular beam epitaxy system.

From Frigeri, P., Seravalli, L., Trevisi, G., Franchi, S., Bhattacharya, P., Fornari, R., & Kamimura, H. (2011). 3.12 - Molecular beam epitaxy: An overview *(pp. 480–522). Elsevier. https://doi.org/10.1016/B978-0-44-453153-7.00099-7.*

pressure of 10^{-10} Torr. The series of effusion cells mounted on the rotating holder can produce the molecular beams, and the ion gauge modifies their intensity. The surface structure can be revealed using a reflection high-energy electron diffraction gun and a fluorescent screen. The insertion and removal of samples can be performed with the sample load-lock without allowing the system pressure to increase up to the atmospheric pressure (Orton & Foxon, 2015).

Chambers with ultra-high vacuum (UHV) pumping systems, effusion cells with high-quality components, shutter control systems with high accuracy, and methods for in-situ observation such as X-ray photoelectron spectroscopy and auger-electron spectroscopy are important factors of molecular beam epitaxy systems with high efficiency. To provide the chambers with UHV conditions, the vacuum level with pressures of the remaining gases must be lower than 10^{-7} Pa. The crucibles can be made of various materials such as pyrolytic boron nitride (pBN), fused quartz, high-melting-point metals, and graphite. High-purity pBN is employed for vaporization temperatures less than 1200°C because of its thermal and chemical stability. Depending on their chemical characteristics, metals with a high-melting point, such as tungsten, platinum, and tantalum, are employed as source materials with high vaporization temperatures. Fused quartz crucibles are often used for materials with low vaporization temperatures and for organic source materials. In fact, several parameters such as the cell temperature, configuration of the crucible, and the

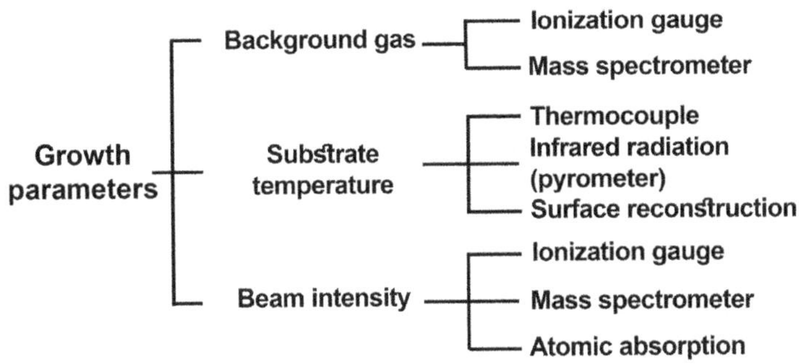

FIGURE 8.32

Parameters for performing molecular beam epitaxy growth.

From Asahi, H., & Horikoshi, Y. (2019). John Wiley & Sons.

surface area of the residual source material control the molecular beam intensity. Consequently, it is not possible to predict the intensity of the molecular beam based only on the cell temperature and crucible configuration used. In practice, the molecular beam intensity can be measured through a vacuum gauge located at the substrate position, thereby evaluating the beam equivalent pressure. It is worth noting that e-beam heating is used to vaporize refractory metals and metal oxides, requiring very high temperatures.

Fig. 8.32 shows the parameters required for performing the molecular beam epitaxy growth. Notably, background atmospheric conditions, substrate temperature, and atomic/molecular beam intensity are taken into account as the essential parameters for efficiently performing epitaxial growth. It has been found that the growth of high-purity materials and interfaces requires a very low background pressure.

A thermocouple set behind the substrate holder measures the substrate temperature. For high-energy e-beams, the scattering cross-section on the substrate surface has been observed to decrease when increasing the incident angle. Therefore, the e-beam incident angle with the substrate surface must be less than ~5degree.

8.5 Synthesis techniques for ferrite nanofibers

Ferrite nanofibers have received more consideration because of uniaxial magnetic anisotropy. The high surface-to-volume ratio of ferrite nanofibers makes them appropriate candidates for potential applications in core–shell structures, microwave absorbers, magnetoelectric components, and exchange-spring magnets. Electrospinning is frequently used to provide ferrite nanofibers. In this approach, a ferrite solution is injected while employing a high-voltage electric field to produce charge carriers with polarity similar to the electric field. Electrostatic repulsion forces the ferrite solution forward in a conelike form. By applying a high-voltage electric field, the fiber jet is formed at the cone tip and collected on the

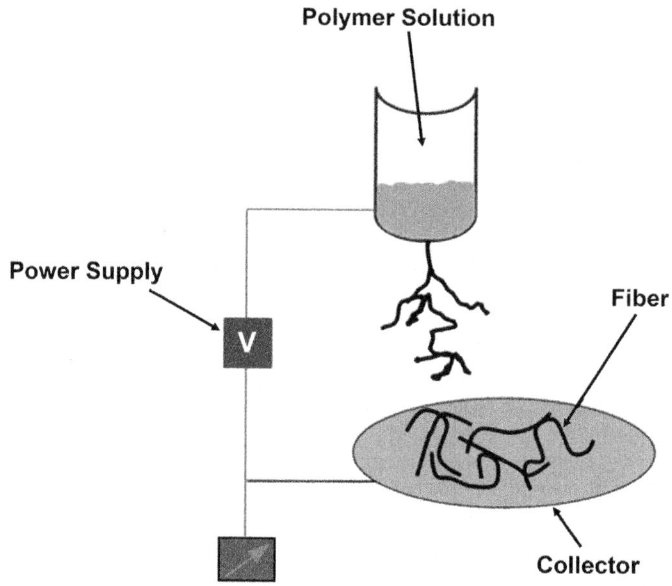

FIGURE 8.33

A schematic representation of the electrospinning process to produce nanofibers.

From Roodbar Shojaei, T., Hajalilou, A., Tabatabaei, M., Mobli, H., & Aghbashlo, M. (2019). Characterization and evaluation of nanofiber materials. In A. Barhoum, M. Bechelany, A. S. H. Makhlouf (Eds.), Handbook of Nanofibers*; (pp. 491–522). https://doi.org/10.1007/978-3-319-53655-2_15.*

oppositely charged collector, as shown schematically in Fig. 8.33. Ferrite nanofiber features are strongly dependent on the surface free energy, surface tension, concentration of solution, permeability of solvent, electrical conductivity, and viscosity (Roodbar Shojaei et al., 2019).

References

Amara, D., Felner, I., Nowik, I., & Margel, S. (2009). Synthesis and characterization of Fe and Fe_3O_4 nanoparticles by thermal decomposition of triiron dodecacarbonyl. *Colloids and Surfaces A: Physicochemical and Engineering Aspects, 339*(1), 106–110. https://doi.org/10.1016/j.colsurfa.2009.02.003

Ameer, S., Gul, I. H., & Mujahid, M. (2015). Ultra low permittivity/loss $CoFe_2O_4$ and $CoFe_2O_4$–rGO nanohybrids by novel 1-hexanol assisted solvothermal process. *Journal of Alloys and Compounds, 642*, 78–82. https://doi.org/10.1016/j.jallcom.2015.04.101

Annie Vinosha, P., Ansel Mely, L., Emima Jeronsia, J., Raja, K., Queen Sahaya Tamilarsi, D., Fernandez, A. C., Krishnan, S., & Jerome Das, S. (2016). Investigation of optical, electrical and magnetic properties of cobalt ferrite nanoparticles by naive co-precipitation technique. *Optik, 127*(20), 9917–9925. https://doi.org/10.1016/j.ijleo.2016.07.063

Avazpour, L., Zandi khajeh, M. A., Toroghinejad, M. R., & Shokrollahi, H. (2015). Synthesis of single-phase cobalt ferrite nanoparticles via a novel EDTA/EG precursor-based route and their magnetic properties. *Journal of Alloys and Compounds, 637*, 497–503. https://doi.org/10.1016/j.jallcom.2015.03.041

Buschow, K. H. J. (2015). *Handbook of magnetic materials* (Vol. 23). Elsevier Science.

Calero-Díaz del Castillo, V. L. (2005). *Synthesis and characterization of cobalt-substituted ferrite nanoparticles using reverse micelles.*

Cavaliere, P. (2019). *Spark plasma sintering of materials: Advances in processing and applications.* Springer.

Cruz, I. F., Freire, C., Araújo, J. P., Pereira, C., Pereira, A. M., El-Gendy, A. A., Barandiarán, J. M., & Hadimani, R. L. (Eds.). (2018). *Micro and nano Technologies* (pp. 59–116). Elsevier. https://doi.org/10.1016/B978-0-12-813904-2.00003-6

Dandia, A., Parewa, V., Gupta, S. L., Sharma, A., Rathore, K. S., Sharma, A., & Jain, A. (2015). Microwave-assisted Fe_3O_4 nanoparticles catalyzed synthesis of chromeno[1,6] naphthyridines in aqueous media. *Catalysis Communications, 61*, 88–91. https://doi.org/10.1016/j.catcom.2014.12.015

Eason, R. (2007). *Pulsed laser deposition of thin films: Applications-led growth of functional materials.* John Wiley & Sons. https://doi.org/10.1002/9780470052129.ch1

El-Eskandarany, M. (2015). The history and necessity of mechanical alloying. In *Mechanical alloying* (2nd ed., pp. 13–47). Elsevier. https://doi.org/10.1016/B978-1-4557-7752-5.00002-4

Feng, S.-H., & Li, G.-H. (2017). Hydrothermal and solvothermal syntheses. In *Modern inorganic synthetic chemistry* (pp. 73–104). Elsevier.

Francis, L. F. (2015). *Materials processing: A unified approach to processing of metals, ceramics and polymers.* Academic Press. https://doi.org/10.1016/B978-0-12-385132-1.00005-7

Franck, E. U. (2009). Water and aqueous solutions at high pressures and temperatures. *Pure and Applied Chemistry, 24*(1), 13–30. https://doi.org/10.1351/pac197024010013

Galindo, R., Mazario, E., Gutiérrez, S., Morales, M. P., & Herrasti, P. (2012). Electrochemical synthesis of NiFe2O4 nanoparticles: Characterization and their catalytic applications. In *International symposium on metastable, amorphous and nanostructured materials, ISMANAM-2011 (26th June to July 1st 2011), 536* pp. S241–S244). https://doi.org/10.1016/j.jallcom.2011.12.061

Kefeni, K. K., Msagati, T. A. M., & Mamba, B. B. (2017). Ferrite nanoparticles: Synthesis, characterisation and applications in electronic device. *Materials Science and Engineering: B, 215*, 37–55. https://doi.org/10.1016/j.mseb.2016.11.002

Kennard, F. (1991). Cold isostatic pressing. *ASM International, Engineered Materials Handbook, 4*, 147–152.

Khan, K., Rehman, S., Rahman, H. U., & Khan, Q. (2014). Synthesis and application of magnetic nanoparticles. In J. M. Gonzalez Estevez (Ed.), *Nanomagnetism* (pp. 135–159). One Central Press.

Kotnala, R., & Shah, J. (2015). Ferrite materials: Nano to spintronics regime. In *Handbook of magnetic materials* (Vol. 23). Elsevier.

Larumbe, S., Pérez-Landazábal, J. I., Pastor, J. M., & Gómez-Polo, C. (2012). Sol-gel $NiFe_2O_4$ nanoparticles: Effect of the silica coating. *Journal of Applied Physics, 111*(10), 103911–103918. https://doi.org/10.1063/1.4720079

Loh, N. L., & Sia, K. Y. (1992). An overview of hot isostatic pressing. *Journal of Materials Processing Technology, 30*(1), 45–65. https://doi.org/10.1016/0924-0136(92)90038-T

Mattox, D. M. (2010). *Handbook of physical vapor deposition (PVD) processing*. William Andrew.

Nabiyouni, G., Julaee, M., Ghanbari, D., Aliabadi, P. C., & Safaie, N. (2015). Room temperature synthesis and magnetic property studies of Fe_3O_4 nanoparticles prepared by a simple precipitation method. *Journal of Industrial and Engineering Chemistry, 21*, 599–603. https://doi.org/10.1016/j.jiec.2014.03.025

Nemati, A., Shadpour, S., Khalafbeygi, H., & Barkhi, M. (2014). Hydrothermal synthesis and size control of Fe_3O_4 nanoparticles in the presence of 2,2′,2″,2‴-(ethane-1,2-diylbis(azanetriyl)) tetraacetohydrazide. *Null, 44*(8), 1161–1165. https://doi.org/10.1080/15533174.2013.797443

Orton, J. W., & Foxon, T. (2015). *Molecular beam epitaxy: A short history*. Oxford University Press.

Rabenau, A. (1985). The role of hydrothermal synthesis in preparative chemistry. *Angewandte Chemie International Edition in English, 24*(12), 1026–1040. https://doi.org/10.1002/anie.198510261

Rashad, M. M., Soltan, S., Ramadan, A. A., Bekheet, M. F., & Rayan, D. A. (2015). Investigation of the structural, optical and magnetic properties of CuO/$CuFe_2O_4$ nanocomposites synthesized via simple microemulsion method. *Ceramics International, 41*(9, Part B), 12237–12245. https://doi.org/10.1016/j.ceramint.2015.06.046

Roodbar Shojaei, T., Hajalilou, A., Tabatabaei, M., Mobli, H., & Aghbashlo, M. (2019). In A. Barhoum, M. Bechelany, & A. S. H. Makhlouf (Eds.), *Characterization and evaluation of nanofiber materials* (pp. 491–522). Handbook of Nanofibers. https://doi.org/10.1007/978-3-319-53655-2_15

Salazar-Alvarez, G., Qin, J., Šepelák, V., Bergmann, I., Vasilakaki, M., Trohidou, K. N., … Nogués, J. (2008). Cubic versus spherical magnetic nanoparticles: the role of surface anisotropy. *American Chemical Society, 130*(40), 13234–13239. https://doi.org/10.1021/ja0768744

Sarin, V. (2003). *Development of hot pressing as a low cost processing technique for fuel cell fabrication*. Oak Ridge National Lab.

Seward, T. (1981). Metal complex formation in aqueous solutions at elevated temperatures and pressures. *Physics and Chemistry of the Earth, 13*, 113–132. https://doi.org/10.1016/0079-1946(81)90008-2

Singh, P., Babbar, V., Razdan, A., Srivastava, S., Agrawal, V., & Goel, T. (2006). Dielectric constant, magnetic permeability and microwave absorption studies of hot-pressed Ba-CoTi hexaferrite composites in X-band. *Journal of Materials Science, 41*(21), 7190–7196. https://doi.org/10.1007/s10853-006-0921-y

Sodaee, T., Ghasemi, A., & Razavi, R. S. (2016). Controlled growth of large-area arrays of gadolinium-substituted cobalt ferrite nanorods by hydrothermal processing without use of any template. *Ceramics International, 42*(15), 17420–17428. https://doi.org/10.1016/j.ceramint.2016.08.042

Stafe, M., Marcu, A., & Puscas, N. N. (2013). *Pulsed laser ablation of solids: Basics, theory and applications* (Vol. 53). Springer Science & Business Media.

Suárez, M., Fernández, A., Menéndez, J., Torrecillas, R., Kessel, H., Hennicke, J., … Kessel, T. (2013). Challenges and opportunities for spark plasma sintering: A key technology for a new generation of materials. *Sintering Applications, 13*, 319–342. https://doi.org/10.5772/53706

Taniguchi, T., Nakagawa, K., Watanabe, T., Matsushita, N., & Yoshimura, M. (2009). Hydrothermal growth of fatty acid stabilized iron oxide nanocrystals. *The Journal of Physical Chemistry C, 113*(3), 839–843. https://doi.org/10.1021/jp8062433

Tian, Y., Yu, B., Li, X., & Li, K. (2011). Facile solvothermal synthesis of monodisperse Fe_3O_4 nanocrystals with precise size control of one nanometre as potential MRI contrast agents. *Journal of Materials Chemistry, 21*(8), 2476–2481. https://doi.org/10.1039/C0JM02913K

Wang, Y., Cai, F., Guo, P., Lei, Y., Xi, Q., & Wang, F. (2019). Short-time hydrothermal synthesis of $CuBi_2O_4$ nanocolumn arrays for efficient visible-light photocatalysis. *Nanomaterials, 9*(9), 1257–1267. https://doi.org/10.3390/nano9091257

Wasa, K., Kanno, I., & Kotera, H. (2012). *Handbook of sputter deposition technology: Fundamentals and applications for functional thin films, nano-materials and MEMS*. William Andrew.

Xing, Y., Jin, Y.-Y., Si, J.-C., Peng, M.-L., Wang, X.-F., Chen, C., & Cui, Y.-L. (2015). Controllable synthesis and characterization of Fe_3O_4/Au composite nanoparticles. In *10th International conference on the scientific and clinical applications of magnetic carriers 10–14 June, 2014, Dresden, Germany, 380* pp. 150–156). https://doi.org/10.1016/j.jmmm.2014.09.060

Yadav, T. P., Yadav, R. M., & Singh, D. P. (2012). Mechanical milling: A top down approach for the synthesis of nanomaterials and nanocomposites. *Nanoscience and Nanotechnology, 2*(3), 22–48. https://doi.org/10.5923/j.nn.20120203.01

Zakiyah, L. B., Saion, E., Al-Hada, N. M., Gharibshahi, E., Salem, A., Soltani, N., & Gene, S. (2015). Up-scalable synthesis of size-controlled copper ferrite nanocrystals by thermal treatment method. *Materials Science in Semiconductor Processing, 40*, 564–569. https://doi.org/10.1016/j.mssp.2015.07.027

Zhang, Z., Yao, G., Zhang, X., Ma, J., & Lin, H. (2015). Synthesis and characterization of nickel ferrite nanoparticles via planetary ball milling assisted solid-state reaction. *Ceramics International, 41*(3, Part B), 4523–4530. https://doi.org/10.1016/j.ceramint.2014.11.147

Zhu, S., Guo, J., Dong, J., Cui, Z., Lu, T., Zhu, C., Zhang, D., & Ma, J. (2013). Sonochemical fabrication of Fe_3O_4 nanoparticles on reduced graphene oxide for biosensors. *Ultrasonics Sonochemistry, 20*(3), 872–880. https://doi.org/10.1016/j.ultsonch.2012.12.001

Index

Note: 'Page numbers followed by "f" indicate figures and "t" indicate tables.'

G

Made in the USA
Las Vegas, NV
17 October 2022

57489095R20361